Acknowledgements

Associated Bodies

The First World Tribology Congress is organized with the full support of:

Egyptian Society of Tribology*
Finnish Society for Tribology*
Gesellschaft für Tribologie*
Institute of Materials
Institute of Physics Tribology Group
Institution of Chemical Engineers
Italian Tribology Group of AIMETA*
Japanese Society of Tribologists*
Japan Society of Mechanical Engineers
Leeds-Lyon Symposium
Lithuanian Scientific Society Tribologija*
NORDTRIB
Polish Tribological Society*
Romanian Tribology Association*
Royal Society of Chemistry
Russian Association of Tribology Engineers*
Russian National Tribology Committee*
Scientific Society of Mechanical Engineers
Sectie Tribologie BvMK*
Slovak Society of Tribology*
Slovenian Society of Tribology*
Société Tribologique de France*
Society of Bulgarian Tribologists*
Society of Chemical Industry
Society of Tribologists and Lubrication Engineers*
Society of Tribologists of Belarus*
South African Institute of Tribology*
Tribology Division of the American Society of Mechanical Engineers*
Tribology Institution of the Chinese Mechanical Engineering Society*
Tribology Society of India*
Tribology Society of Nigeria*
Ukrainian National Tribology Society*
Wear of Materials
Yugoslav Tribology Society*

*Member of the International Tribology Council

Related Titles

Title	Author/Editor	ISBN
World Tribology Congress		1 86058 109 9
An Introductory Guide to Industrial Tribology	J D Summers-Smith	0 85298 896 6
A Tribology Casebook	J D Summers-Smith	1 86058 041 6
History of Tribology	D Dowson	1 86058 070 X
Lubrication and Lubricant Selection	A R Lansdown	1 86058 029 7
Lubricants in Operation	U J Moller & U Boor	0 85298 830 3
Lubrication of Gearing	W J Bartz	0 85298 831 1
Analysis of Rolling Element Bearings	Wan Changsen	0 85298 745 5

For the full range of titles published by MEP contact:

Sales Department
Mechanical Engineering Publications Limited
Northgate Avenue
Bury St Edmunds
Suffolk
IP32 6BW
UK

Tel: 01284 724384
Fax: 01284 718692

Contents

Foreword

Tribology is the science and technology of interacting surfaces in relative motion, and includes the study of friction, wear and lubrication. The thirty-eight papers in this volume were presented as plenary or invited papers at the World Tribology Congress in London in September 1997. The topics discussed at this historic event, the first international conference endorsed by almost every tribology society world-wide, covered a remarkably broad spectrum. The subjects of the papers published here reflect this breadth, and provide an excellent overview of current activity and interest in the interdisciplinary subject of tribology.

The first five papers, presented in plenary sessions, discuss issues of major importance in modern tribology. Dowson's review of the history of our subject in the 20th century reminds us of the enormous debt we owe to those who have laid its foundations, while Winer addresses the question of lubrication by very thin oil films under immense pressures. In the next paper, Kato discusses the wear of metals, still the materials most widely used in tribological contacts, and Neale then focuses our attention on the need to relate future tribological research to real industrial problems. Georges presents a survey of the current understanding of the molecular and atomic origins of friction and boundary lubrication.

The invited papers which follow are no less wide-ranging in their coverage. The central topic of lubrication is comprehensively reviewed, with contributions on thin elasto-hydrodynamic and boundary films (Jacobson and Spikes), lubrication in the automobile engine (Coy), solid lubricants (Gardos) and environmental aspects (Bartz). Discussion of the design of thick fluid film bearings (Tanaka) is complemented by a paper on rolling element bearings (Ioannides). Applications of tribology to the diagnosis of failures in process plant (Summers-Smith), and the roles of tribology in machine maintenance (Kimura) and hot metal working (Beynon) are covered, as well as the special problems associated with tribology in the space environment (Fleischauer).

Methods of modelling surface contact (Myshkin) and of analysing the shapes and textures of surfaces, wear debris and abrasive particles (Stachowiak) are central to a deeper understanding of friction and the formulation of models for wear, while wear itself is discussed in the context of sliding (Biswas), abrasion (Zum Gahr), and the tribochemical interactions between various types of wear and oxidation/corrosion (Stott and Ball). The recent development of wear maps of several kinds is reviewed (Lim). Tribochemical effects are also important in the behaviour of ceramics (Fischer), which are the main candidate materials for the development of oil-free automobile engines (Woydt). Another important class of materials, polymers, is covered in four papers, which include a broad overview (Zhang) as well as more detailed discussion of surface deformation (Briscoe) and the behaviour of specific tribologically important polymers: PTFE-containing composites for bearing applications (Uchiyama) and ultra-high molecular weight polyethylene for orthopaedic joint prostheses (Wang). Surface engineering, the modification or coating of surfaces to develop enhanced tribological behaviour, is reviewed in two papers (Holmberg and Bell), and the particular problems faced in certain wear tests for surface engineered components are discussed (Hutchings). More generally, the design and selection of tribological test methods are also reviewed by two contributors (Blau and Alliston-Greiner).

The increasing interest in nanotribology, the study of contact and relative motion on a sub-micrometre scale, is reflected in many of the above contributions, but is also specifically discussed in three papers (Pethica, Bhushan and Spencer). Technology transfer is vital if engineers and technologists are to benefit from tribological knowledge, and the question of how best to educate tribologists is therefore addressed (Ludema).

The papers published here originate from no fewer than 15 different countries in five different continents: a truly international collection. It gives me great pleasure, on behalf of the Programme Committee of the World Tribology Congress, to introduce this volume. I hope that its contents will be of lasting value not only as a comprehensive review of the achievements of tribology as the 20th century comes to a close, but also as inspiration for present and future colleagues seeking new directions in tribology.

Ian Hutchings
Cambridge, UK
September 1997

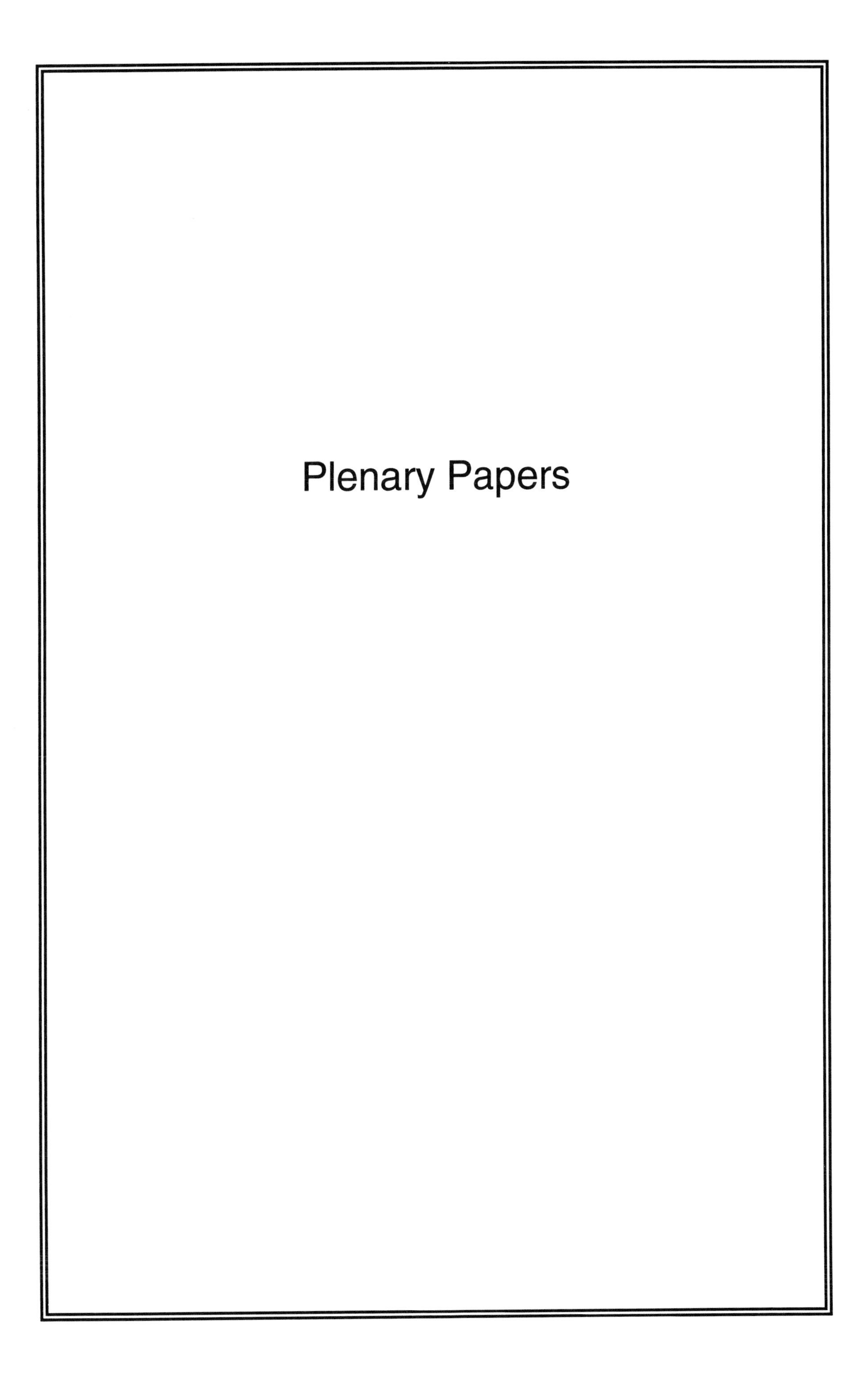

Plenary Papers

Progress in tribology: a historical perspective

D DOWSON
Department of Mechanical Engineering, University of Leeds, UK

1 INTRODUCTION

It is particularly fitting that the first World Tribology Congress is being held in London in 1997. The year marks the 150th. Anniversary of the founding of the Institution of Mechanical Engineers in Birmingham on January 27th 1847. The first President of the Institution was George Stephenson and the declared objective of the Institution was ....'*To enable Mechanics and Engineers engaged in the different Manufactories, Railways and other Establishments in the Kingdom to meet and correspond, and by a mutual exchange of ideas respecting improvements in the various branches of Science, to increase their knowledge and give an impulse to Inventions likely to be useful to the world'*. The American Society of Mechanical Engineers was founded in 1880, and its Research Committee on Lubrication was established in 1915. The present year also marks the centenary of the Japan Society of Mechanical Engineers.

It is interesting to reflect on the enormous developments in tribology in recent times. A number of engineers and tribologists now approaching retirement attained their academic qualifications and full membership of one or other of the worlds major engineering and tribological Societies or Institutions after the introduction of the word '*Tribology*' in 1966 (**1**). They have enjoyed a sense of unity of background and purpose in this field which was unknown to many generations of scientists and engineers before them. It is therefore interesting to reflect on some of the technological and scientific achievements which marked the progress towards the current unity of the subject and to review some of the exciting concepts now emerging.

2 HISTORICAL PERSPECTIVE OF TRIBOLOGY

An account of the long history of tribology has been presented elsewhere and the early developments will not be reviewed again in detail (**2**). A brief résumé of major developments prior to the 20th century is shown in Table 1. The very early studies were concerned with friction, since this was perceived to be the major problem to be overcome in transporting heavy objects on sledges. The classical studies in this field prior to the eighteenth century were those by Leonardo da Vinci (c1470), Guillaume Amontons (1699) and Charles Augustin Coulomb (1785) (see (**2**)).

The major imperatives driving studies of tribology in the late nineteenth century were the problems being experienced on the rapidly developing railways and the production of mineral oil. These industrial developments promoted the golden decade of tribological engineering science in the 1880*s*. Heinrich Hertz analysed the contact between elastic bodies and Petrov, Tower and Reynolds established the basis of fluid film lubrication (see (**2**)). By the end of the nineteenth century major Bearing Companies were being formed and the Oil Industry was established in the USA, Russia and the East.

3 THE TWENTIETH CENTURY

The major developments in tribology during the present century are becoming clearer as the millennium approaches. Scientific advances have been matched by impressive engineering applications throughout the century. It can justifiably be stated that all major developments of mechanical systems during the century were made possible by the underpinning science and technology of tribology.

3.1 The first quarter

The most spectacular development early in the 20th. century was the introduction of tilting-pad bearings, which greatly enhanced machine reliability and reduced operating costs enormously in the hydraulic machinery and marine engineering fields. Kingsbury devised and built the first tilting-pad bearing in 1898 after reading Osborne Reynolds' classical paper. Michell (**3**) solved the Reynolds' equation for inclined slider bearings during the period 1902-1904, conceived the idea of the pivoting pad

	DATE	LOCATION / NAME	ACHIEVEMENT
	c.3500	Mesopotamia	Wheel; Wood, Stone, Bone, Shell bearings
	c.2000	Egypt	Lubricant used on Sledges; Tallow on Chariot wheels
	c.1000	Mesopotamia; Egypt	Wooden Gears
	c.400	Greece	Iron Bearings in Olive Mill
	c.300	China	Tapered bronze Bearing on Vehicles
B.C	c.200	China; Rome	Bronze and Iron Bearings
------	0	Celts, Denmark	Wooden rollers in Bronze Hubs in Wagons
A.D.	c.500	Greece; Rome; Middle East	Vegetable oils & Animal Fats used as Lubricants throughout the Medieval Period
	c.1400	Denmark	Stones used to reduce wear of ploughs; axles.
	c.1470	Leonardo da Vinci	Studies of friction; wear; bearings
	1685	Robert Hooke; England	Nature of Rolling friction;
	1687	Isaac Newton	Hypothesis on Viscous Flow
	1699	Guillaume Amontons	Laws of Friction
	1710	De Mondran, France	Roller-disc bearings designed for Carriages
	1734	Jacob Rowe, England	Roller-disc bearings used on Carriages
	1734	Desaguliers, England	'cohesion'; friction 'coefficients' measured.
	1737	Bélidor,France	Analysis of friction of rough surfaces (μ=0.35)
	1750	Euler, Germany	μ=tan α
	1769	Count Carburi, Russia	Large granite block on linear ball bearing
	1770	Philadelphia, USA	Roller Bearing on Weather Vane,
	1772	Varlo, England	Cast Iron Ball Bearings for Road Carriages
	1780	Sprowston, England	Large cast iron Ball Bearing on Windmill
	1785	Charles Augustin Coulomb, France	Memoir on Friction-static and dynamic
	1804	John Leslie, England	Friction (deformation loss) and lubricants
	1822	Whitcher et al. England	Cage for Roller Bearing
	1829	George Rennie, England	Extensive Studies of Friction
	1835	Arthur James Morin, France	Rolling and Sliding Friction
	1840s	Hompesch 1841; Little 1849; England	Improved extraction of Mineral Oils
	1849	Von Pauli, Nuremberg ,Germany	Journal Bearing friction and materials
	1850	James Young, Scotland	Mineral Oil Extraction
	1853	W. Bridges Adams, England	Axle Box Lubrication
	1854	G. A. Hirn, France	Air and water as lubricants
	1854	John Ramsbottom, England	Proposals for Piston Rings
	1858	John Penn, England	Lignum vitae in marine thrust bearings
	1859	Edwin Drake, USA	Struck oil at Titusville
	1862	A. L. Thirlon, England	First Patent for Ball Bearing in Velocipede
	1863	Clark, Rockefeller, Andrews , USA	Refinery Company formed in Cleveland, Ohio
	1870	Cleveland, USA	Standard Oil Company formed
	1875	Osborne Reynolds, England	Study of Rolling Friction
	1876	V.I. Ragosine, Russia	Mineral oil produced
	1881	Heinrich Hertz, Germany	Contact between Elastic Bodies
	1883	Beauchamp Tower, England	Recognition of lubricating film in Axle-Box Bearings
	1883	Nikolai Pavlovitch Petrov, Russia	Recognition of lubricating film in Axle-Box Bearings
	1884	Osborne Reynolds, Canada	British Association Lecture, in Montreal
	1884–85	Germany; USA; England	Engler; Saybolt; Redwood viscometers
	1885	Robert H. Thurston, USA	Book on Lost Work in Machinery and Millwork
	1886	Osborne Reynolds, England	Theory of Fluid Film Lubrication
	1890	The Netherlands	Royal Dutch Company formed
	1897	Albert Kingsbury, USA	Air Lubricated Journal Bearing
	1897	John Goodman, England	Friction in Plain and Ball Bearings
	1897	England	Shell Transport and Trading Company formed
	1898	Albert Kingsbury & John Brown, USA	Built and operated tilting-pad thrust bearing
	1898	USA	Timken Roller Bearing Axle Company formed
	1898	England	Hoffmann Manufacturing Company founded
	1899	England	Glacier Metal Company founded

Table 1. A brief chronology of tribology prior to the 20th century

bearing independently of Kingsbury and patented the idea in 1905 (**4**). Kingsbury was finally awarded a patent in 1910 (**5**). The economic impact was impressive, since marine applications alone saved the United Kingdom some £500,000 in 1918. This is equivalent to at least £30m per year in the early 1990s, based upon changes in the retail price index over the period.

Journal bearings automatically presented a convergent-divergent clearance space under load, but special attention had to be given to cavitation and the associated boundary conditions. Confidence in the design, manufacture and operation of journal bearings nevertheless grew during the first quarter of the century. Specialist plain bearing companies were formed, while in Germany, Professor Stribeck investigated journal bearing friction and distinguished between boundary, mixed and fluid-film lubrication regimes.

Major ball and roller bearing companies were established in Europe and the USA to underpin the development of rail and road vehicles. The technology was supported by impressive engineering science based upon the studies of Hertz, Stribeck and Goodman (see (**2**)).

As confidence grew in the development of reliable bearings, attention was turned to gears and piston rings. Both proved to be more difficult propositions, but the early studies provided foundations for third quarter progress in these fields.

The work of Sir William Bate Hardy (see (**6**)) provided the foundations for the subject of boundary lubrication.

3.2 The second quarter

Many of the outstanding developments in the second quarter of the century were promoted by the needs of the expanding motor car and truck business. While babbitt materials dominated the plain bearing field throughout this period, the gradual move to higher mean loadings on engine bearings resulted in the development of materials like copper-lead alloys to enhance strength while preserving the excellent bearing characteristics of babbitt. Fabric, rubber and graphite bearings were also developed for special applications in the late 193*0s*.

Solutions to the Reynolds equation for infinitely wide bearings were gradually supplemented by approximate solutions for bearings of finite width. The problem was a serious one, as Swift (7) observed in 1937... '*At the present time all theories of journal lubrication suffer because they neglect side leakage.'*.... Attempts to produce approximate solutions to the full Reynolds equation for bearings of finite width received a tremendous boost in the 1940s with the application of Sir Richard Southwell's 'relaxation' method to bearing problems by Christopherson (**8**). Cameron and Wood (**9**) applied the method to journal bearings using data computed at the National Physical laboratory. These early applications of the relaxation method were undertaken by hand, or at the best with the aid of mechanical calculators, and were thus exceedingly slow and tedious.

Foundations for the analysis of dynamically loaded bearings were also established during this period by Swift (**10**), Orloff (**11**), Stone and Underwood (**12**), Burwell (**13**) and Shaw and Macks (**14**). An interesting, if limited, *'equivalent speed concept'* emerged with the recognition that hydrodynamic pressures in journal bearings were determined by the term $[2\omega-(\Omega_b - \Omega_s)]$, where ω represents the angular velocity of precession of the shaft about the bearing centre and Ω_b, Ω_s the bearing and shaft rotational speeds. When this term was used to represent the effective speed of the journal, estimates of bearing load carrying potential could be obtained directly from steady state solutions.

Rolling element bearings developed substantially in the second quarter of the 20th century, supported by studies of the accuracy, surface finish and life of such bearings by Goodman (**15**), Stribeck (**16**) and Palmgren (**17**). Palmgren proposed that the life L and load W were related as follows;-

$$W^3 L = constant \quad (1)$$

Gas and externally pressurised (hydrostatic) bearings also found increasing application.

Abbott and Firestone (**18**) of the University of Michigan introduced profilometry in 1933. The Talysurf remains a valuable tool in research and product monitoring, although new optical, contactless equipment is now rapidly being adopted.

The understanding of sliding friction during this period was dominated by the concept of adhesion. Desaguliers (see **2**) had introduced the notion that cohesion was important as early as 1734 and the subject was further advanced by Prandtl (**19**), Deryagin (**20**), Holm (**21**) and Bowden and Tabor (**22**). Deryagin proposed a binomial expression for the coefficient of friction (μ) in the form;

$$\mu = \frac{F}{a(p + p_0)} \quad (2)$$

– where F was the force of friction, a represents the real area of contact and p, p_0 the stresses introduced by external forces and molecular adhesion respectively.

Holm concluded that the force of friction was attributable to the sum of the shearing strengths of the asperity contacts, while Bowden and Tabor derived an appealingly simple relationship for the coefficient of friction;-

$$\mu = \frac{F}{P} = \frac{as}{aH} = \frac{s}{H} \quad (3)$$

– where, a represented the real area of contact and s,H the shear stress and hardness of the asperity junctions.

The early views of Reynolds (**23**) and Heathcote (**24**) on micro-slip governed thinking on the mechanism of rolling friction.

Swift (**25**) and Stieber (**26**) used stability and flow continuity concepts to determine suitable cavitation boundary conditions in the analysis of convergent-divergent bearings subjected to cavitation. Reynolds had commented on the same problem at an earlier stage and the well known Reynolds or Swift-Stieber cavitation boundary condition was thus established as;-

$$p=\frac{\partial p}{\partial x}=0 \qquad (4)$$

Taylor's (**27**) studies of vortex formation in the annulus between concentric cylinders, which was to form the foundation of stability considerations in large, high speed electricity generating sets in later years, was completed in 1923.

The development of the motor car in this period brought about a close relationship between the petroleum companies and the engine builders. The use of additives to modify lubricant properties had been established well before the dawn of the 20th century, but the increasing demands on the widely used mineral oils brought about a period of rapid progress in lubricant technology. Synthetic polymers, polymethacrylates and polyisobutylene were all introduced in the 1930s as VI improvers. The potential of synthetic lubricants was also recognised in the 1930s, although they were not fully exploited until the second half of the century. The increasing demands placed on engine lubricants were also reflected in transmissions. Mean pressures in gears moved up from about 2 MPa in 1900 to 10 MPa in 1930. The piston assembly presented increasing problems in the second quarter of the century. There was a continuing debate on the very mechanism of their lubrication, with contradictory evidence regarding boundary and fluid-film lubrication. However, it was increasingly recognised that friction between rings and cylinder liners accounted for a significant proportion of the mechanical power losses in reciprocating engines.

It was work initiated in the first half of the twentieth century that led eventually to the emergence of the subject of elastohydrodynamic lubrication. The early history has been outlined elsewhere (**28**), but there was one outstanding development just before the mid-century which should be noted. In 1949 Grubin (**29**) reported an approximate analytical solution of remarkable simplicity and effectiveness. He derived an equation which predicted film thicknesses in gears which were at least an order of magnitude greater than those revealed by hydrodynamic analysis alone. The need to take account of both elastic deformation and pressure-viscosity effects was thus established in majestic style immediately before the third quarter of the century. It appears that this major contribution to our understanding of fluid-film lubrication was undertaken by Ertel, although the publication was attributed to his senior, Grubin. The story behind the authorship of this notable work has been told by Cameron (**30,31**).

3.3 The third quarter

In 1966 the 'Jost' Report (**1**) was published in the United Kingdom. The full impact of this important event is discussed in Section 5, but it is important to record it as a third quarter development in the twentieth century. While considerable attention was focused upon the economic significance of tribology in this period, spectacular developments were taking place in both industrial applications and the enhanced understanding of basic phenomena in tribology.

3.3.1 Surfaces and contact

The nature of surfaces and surface contact attracted much attention in this period, not only because such studies were relevant to tribology, but also because of their importance to heat transfer and the flow of electricity between solids. A particularly useful concept to emerge was that of the *Plasticity Index* (ψ) (Greenwood and Williamson (**32**)). One form of the plasticity index (ψ^*) can be written as;-

$$\psi^* = \left[\frac{E}{H(1-v^2)}\right]\left[\frac{\sigma}{\beta}\right] \qquad (5)$$

– where (E, H and ν) represent the elastic modulus, hardness and Poisson's ratio and (σ, β) the standard deviation and correlation distance. The following guidelines indicate the likelihood of plastic flow.

($\psi^* \geq 1$) - **Plastic flow.**

($1 \geq \psi^* \geq 0.6$)- **Elastic/ plastic flow.**

($\psi^* \leq 0.6$) - **Plastic flow unlikely.**

3.3.2 Wear

Archard's analysis (**33**) of adhesive wear yielded a classical equation for the volume of the softer material V, of hardness H, removed in a sliding distance X under a steady load P;-

$$V = k_1\frac{PX}{3H} = k_2\frac{PX}{H} \qquad (6)$$

– where the k s represent dimensionless wear coefficients.

Analytical studies of abrasive wear revealed a similar expression and for a conical asperity it can readily be shown that ($k_2 = 2\tan\theta/\pi$). Evans and Lancaster (**34**) pointed out that a more useful form of the wear equation was;-

$$V = k\,P\,X \qquad (7)$$

– where k is a dimensional wear factor usually recorded in units of mm^3/Nm. Both equations have been used extensively throughout the latter half of the twentieth century. These simple representations of adhesive wear represent early approaches to wear modelling; a subject which was to grow in importance as the century progressed. It was soon recognised that the use of wear factors or coefficients in design, although useful, carried a major restriction. The factors range over many orders of magnitude for different combinations of materials, operating conditions and environments, such that the approach can never enjoy the precision associated with analytical approaches to fluid film lubrication problems. Furthermore, attempts to ascertain the wear factors from laboratory tests on standard equipment, such as pin-on-disc or pin-on-plate machines, often result in values which are representative of the laboratory circumstances, but not the full scale machine operating conditions. It is only when the machine operating conditions are adequately simulated that the laboratory data assumes a measure of reliability and usefulness.

3.3.3 Elastohydrodynamic lubrication

The analysis of fluid film lubrication in conforming machine elements such as plain bearings was effectively tackled in the early half of the twentieth century. Similar approaches to the analysis of highly stressed, lubricated machine elements such as gears and rolling element bearings in the same period were largely unsuccessful. The second half of the century witnessed the revelation of the remarkable mechanism of elastohydrodynamic lubrication, now known to be so important in the protection of critical machine components.

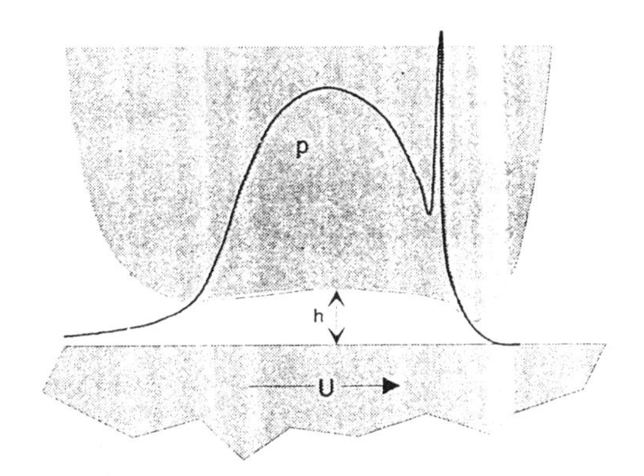

Fig. 1 Representative film shape and pressure distribution in an elastohydrodynamic line contact.

In the third quarter of the century impressive experimental work involving capacitance and optical interferometry techniques confirmed the remarkable features of EHL line contacts predicted by investigators in the USSR; namely, the near constant film thickness throughout much of the Hertzian zone and the outlet restriction in the film thickness preceded by a very high and narrow pressure peak as depicted in Fig. 1. Numerical solutions to the Reynolds and elasticity equations obtained in the late 1950s and 1960s by Dowson and Higginson (**35**) enabled the following much used minimum film thickness equation for line contacts to be derived;-

$$H = 1.6\frac{G^{0.6}U^{0.7}}{W_L^{0.13}} \qquad (8)$$

When numerical methods for the solution of the elastohydrodynamic lubrication line contact problem were being developed forty years ago, only manual computational aids were available. Mechanical or electrical calculating machines were used, but the associated handle-turning made progress exceedingly slow, such that the first solution by Dowson and Higginson took eighteen months to produce. Comparable solutions today take but a few minutes on a digital computer.

The essential features of elastohydrodynamic conjunctions for line contacts were thus well established experimentally and theoretically in the third quarter of the twentieth century. Towards the end of the period, Barwell (**36**) wrote, *'...the elucidation of the mechanism of elastohydrodynamic lubrication may be regarded as the major event in the development of lubrication science since Reynolds' own paper'.*

3.3.4 Materials

There was a major advance in the use of non-metallic bearing materials during this era. The new range of thermoplastic, thermosetting and elastomeric materials made a big impact upon the bearing field. This was particularly true of the aircraft and process industries, where lubrication by conventional means was impossible or undesirable. The use of a PV factor, where P represents the mean bearing pressure, or load for a given bearing, and V the sliding speed, became established as a material selection parameter. Valuable accounts of the development of 'dry' bearings were presented by Lancaster (**37**) and Pratt (**38**).

Thin-walled engine bearings, in which the long established white metal gave way to higher strength copper-lead, provided with an overlay of less corrodible materials such as white metal, silver and indium, became widely used.

New technologies such as nuclear power and aerospace provided an impetus for the wider use of solid lubricants. The advantages of lamellar solids such as molybdenum disulphide and graphite as lubricants for protecting sliding surfaces subjected to heavy loads, extremes of temperature and very low speeds had been known for many years. The new range of polymers greatly extended the range of solid lubricants. It was also recognised that oxide films on metals could be very effective solid lubricants and that some boundary lubricants were themselves akin to solid lubricants in their mode of operation (Braithwaite (**39**)).

3.3.5 Condition monitoring

Publication of the Jost Report (**1**) in 1966 did much to illuminate the role of tribology in determining the reliability and efficiency of machines and mechanical systems and the economic implications of failure.

The well established procedures of regular, or preventive maintenance, gave way to condition monitoring and predictive maintenance in a number of critical areas. Major applications included the large process industries, where unscheduled shutdowns were economically crippling; the aeronautical and other transport systems where safety and reliability were essential and machinery such as nuclear power plants operating in hazardous environments. Parameters frequently monitored included temperature and vibration, while periodic oil sampling was found to be particularly beneficial. The technique of Ferrography established itself during this period.

3.3.6 Bio-tribology

The Institution of Mechanical Engineers did much to promote interest in this subject when its Lubrication and Wear Group joined with the British Orthopaedic Association to present a Symposium on the subject of "Lubrication and Wear in Living and Artificial Human Joints" in 1967. A number of orthopaedic surgeons, physicians, engineers and physicists who, in future years, were to be intimately linked with progress in joint replacements and studies of joint lubrication, contributed to the Symposium and the Proceedings 181, Part 3J have developed an internationally respected position in the archives.

Rheumatism and arthritis afflict a large number of people, particularly the elderly, and in certain circumstances the surgeon recommends replacement of the worn or diseased joint by a prosthesis. Engineers, materials scientists and tribologists have all played a role in developing implants. This procedure established itself in the third quarter of the century, primarily through the work of the late Professor Sir John Charnley (**40**). His prosthesis consisted of a stainless steel femoral component and a polymeric acetabular cup. Initially PTFE was used for the cup, since this presented the lowest known friction under dry conditions, with a 7/8 in (22.225 mm) diameter femoral head to minimise the frictional torque on both the cup and the femoral stem, but wear was excessive and in 1962 an ultra-high molecular weight polyethylene acetabular cup (Fig. 2) was introduced. This low-friction arthroplasty (LFA) is still regarded as the *'gold standard'* for total hip replacements.

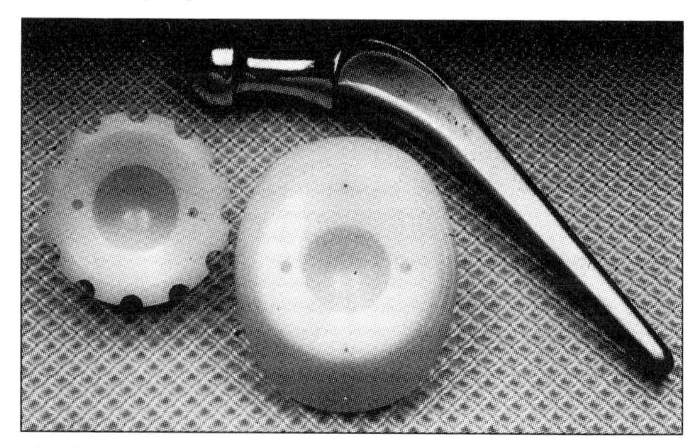

Fig. 2 Charnley prosthesis.

The term bio-tribology was introduced in 1973 (**41**) to embrace "*..all aspects of tribology related to biological systems*". In addition to synovial joints and their replacement, topics such as the wear of human dental tissues; the lubrication by plasma of red blood cells in narrow capillaries; the wear of replacement heart valves; the wear of screws and plates in and on bone in fracture repair; the tribology of skin and the friction of hair all fall within the field of bio-tribology. The field is gathering further momentum as the end of the twentieth century approaches.

4 THE FINAL QUARTER

I have discussed elsewhere (**2**) the difficulty of assessing the long term standing of recent developments in a subject. This is mainly because the impact of new discoveries in science and developments in technology often take many years to achieve their full significance. There is no doubt, however, that we are witnessing tremendous advances in the understanding of tribological phenomena and in the application of tribological knowledge to new and existing forms of machinery and systems as the twentieth century draws to a close. An important indicator of this is the impressive improvement in reliability of machinery, such as the internal combustion engine. We have come to expect levels of efficiency and reliability in engineering and domestic equipment which were unheard of but a quarter of a century ago.

I select just a few topics to convey this excitement, while recognising that the selection is a personal appraisal and that fellow tribologists will wish to add to the list. On both the fundamental and applications fronts, the final quarter of the century is dominated by nano-tribology and a number of the outlines presented below will illustrate this.

4.1 Friction

Recent revelations have greatly enhanced our understanding of friction. These have been achieved through the application of surface science concepts and techniques on an atomic scale. On the experimental front, two forms of instrument have contributed enormously to this progress; the surface force apparatus (SFA) and the atomic force microscope (AFM). These instruments enable measurements to be made on single asperity contacts.

4.1.1 The surface force apparatus (SFA) and atomic force microscope (AFM)

In the SFA two very smooth solids such as cleaved mica are pressed together (Homola et al (**42**)) to enable atomic scale friction measurements to be made. On the nano-scale, the force of friction is directly related to the real area of contact, but the concept that friction and wear were linked to adhesive junctions and the plucking out of material from one of the solids was not sustained. Friction did not correlate well with the cohesive strength of the solids and it was noted that wear-free friction could be measured. It was the irreversibility associated with the process of bringing atoms together and then separating them, rather than the force of cohesion itself, that correlated with friction. In 1929 Tomlinson (**43**) had proposed a link between friction and interacting atoms which was developed more fully by Sokoloff (**44**) in 1978. McClelland and Glosli (**45**) then developed a simple model of friction based upon vibrations of atomic lattices in which the work done in overcoming friction was dissipated through vibrations (sound) and eventually as heat.

In the AFM a very fine probe, with a tip radius in the range 10-100 nano-metre, traverses over a surface and the inter-atomic forces between the probe and the test surface can be measured to determine force components with pico-Newton (10^{-12}N) accuracy. At the atomic scale, the force of friction is no longer proportional to load, since friction depends upon the true area of contact, which in turn is determined on the atomic scale not only by the applied normal force, but also by adhesion. Even when the external applied force is zero, the contacting solids will flatten under the action of 'adhesive' forces. The physics has been outlined by Israelachvili (**46**), while the contact mechanics has been analysed by Johnson et al (**47**) and Johnson (**48**).

4.1.2 Molecular dynamics

These impressive measurements of friction on an atomic scale have been matched by equally exciting progress in the application of molecular dynamics.

In this approach Newton's equations of motion are solved for a system of particles governed by specified inter-atomic interactions. While the equations are

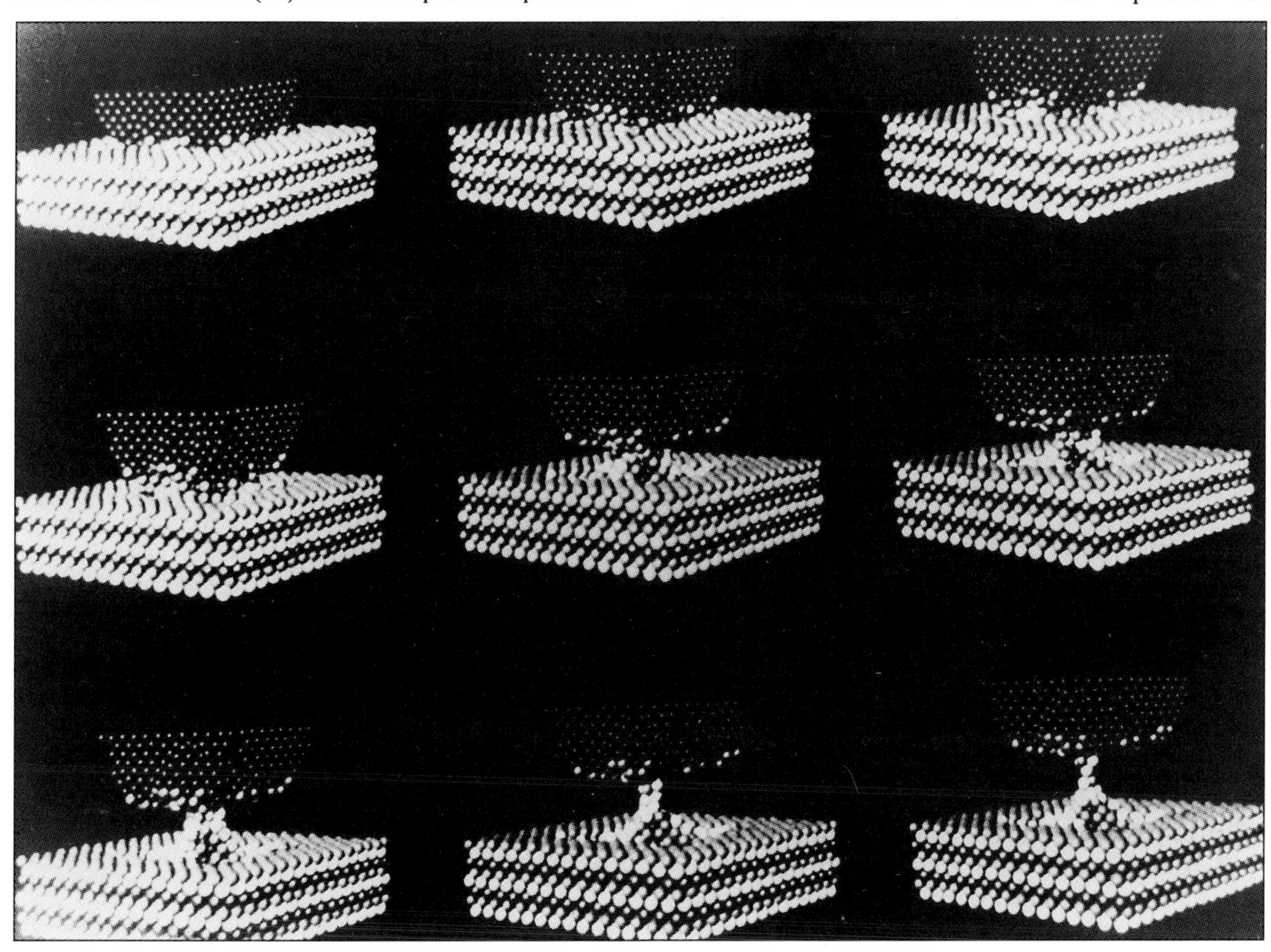

Fig. 3 Molecular dynamics simulation of retraction of a Ni tip from a Au (001) substrate (reproduced by kind permission of Kluwer Academic Publishers from reference **50**).

relatively simple, the computing power required for realistic numbers of particles is enormous. The use of supercomputers has nevertheless yielded valuable insights into the phenomena of adhesion, friction, wear and even lubrication at the atomic scale. Landman and his co-workers have demonstrated the enormous potential of this computational approach (**49,50**). A wonderful simulation of the indentation of a gold (001) substrate by a nickel indenter is shown in Fig. 3. The migration of atoms from one surface to the other is clearly shown.

Thompson and his co-workers (**51,52**) have extended molecular dynamics studies to lubricated surfaces. The work clearly demonstrates that the behaviour of thin films of molecular proportions, typically less than 40Å, cannot be related to the bulk properties of the lubricant.

A full appreciation of these molecular studies of friction has yet to emerge, making it certain that it will be an area of activity well into the twenty first century.

4.2 Elastohydrodynamic (EHL) and very thin film (VTF) lubrication

Fascinating and immensely important developments in both the fundamental and applications aspects of lubrication have been reported in the last quarter of the twentieth century.

4.2.1 Elastohydrodynamic lubrication (EHL)

In the late 1970s and early 1980s the availability of powerful computers enabled numerical solutions to be obtained for nominal point contacts. This extended EHL analysis to a wide range of machine elements. The horse shoe film shape illustrated in Fig. 4 was revealed in beautiful interferometry experiments on ball-on-disc machines and the theoretical and experimental film shapes and thicknesses appeared to be in close accord.

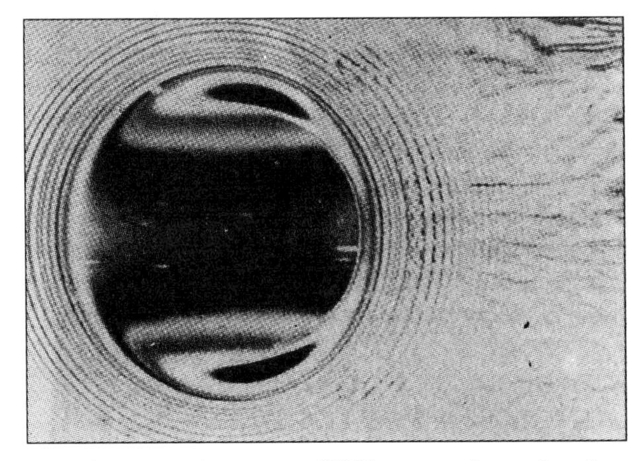

Fig. 4 Interferogram of EHL contact (reproduced by kind permission of Dr L.D. Wederven).

The 1950s and 1960s were associated primarily with studies of nominal line contact situations, while the 1970s, 1980s and 1990s saw the extension of EHL analysis to nominal point contacts representative of ball bearings and certain continuously variable transmissions (CVT's).

However, the Newtonian model for lubricant rheology greatly exaggerated the shear stress at high sliding speeds. An alternative Ree-Eyring rheological model was proposed by Evans and Johnson (**53**) while Bair et al (**54**) pointed out that the shear stress not only tended towards a limiting value at high shear rates, but that the lubricant could solidify to behave like a plastic solid.

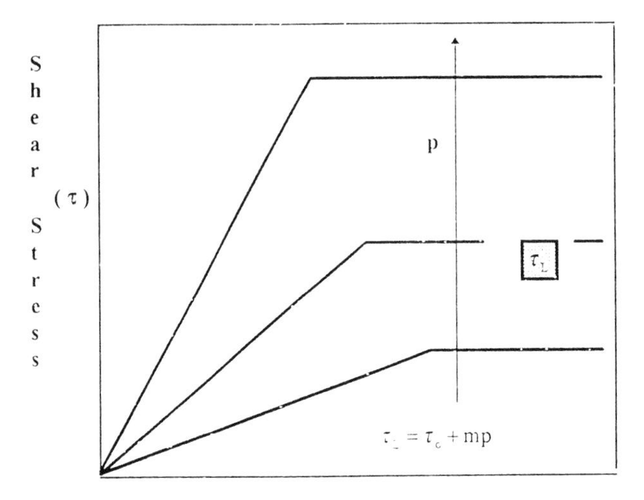

Fig. 5 Newtonian fluid-plastic solid rheological model of lubricant rheology.

Shearing of the lubricant also causes the film temperature to rise and this further reduces the shear stress attained at a given shear rate, as shown in Fig. 6.

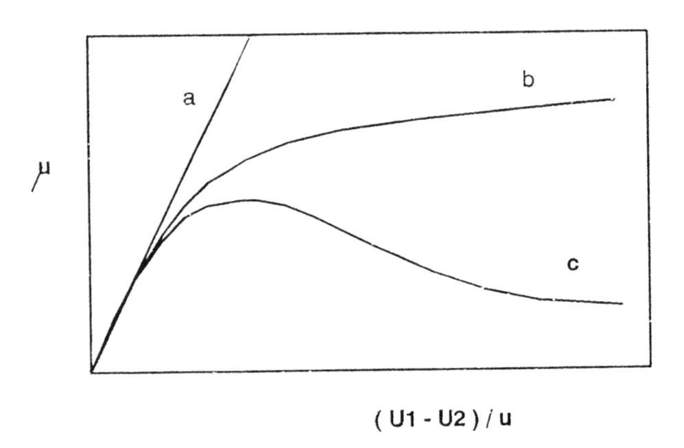

Fig. 6 Representative traction coefficients for EHL conjunctions:
(a) Newtonian-isothermal.
(b) Non-newtonian-isothermal.
(c) Non-newtonian-thermal.

While it is now possible to take account of both non-Newtonian and thermal effects with reasonable accuracy in friction and traction calculations, further work is progressing on the complex problem of lubricant behaviour under these severe conditions.

It is at first sight amazing that film thickness calculations based upon a Newtonian model are in such close accord with experimental observations. The reason for this is that film thickness is determined in the inlet to the conjunction, where the pressures and temperatures are relatively modest. For calculations of power loss, or traction, it is essential, however, to take account of non-Newtonian and thermal effects upon lubricant behaviour.

4.2.2 The thinning film

Early EHL analysis was related to components like spur gears and roller bearings, where the minimum film thicknesses are about 0.2 to 2 μm. In recent years machine operating experience and the refinement of optical interferometry equipment in the laboratory have clearly indicated that, under many circumstances, much thinner EHL films than originally envisaged can provide effective fluid-film lubrication. Indeed, it is not uncommon in the present era to measure EHL film thicknesses in nano rather than micro-metres. This impressive reduction in perceived film thickness throughout the twentieth century has been described (**55**) as '*the thinning film*'. An illustration of representative film thicknesses throughout the 20th century is shown in Fig. 7.

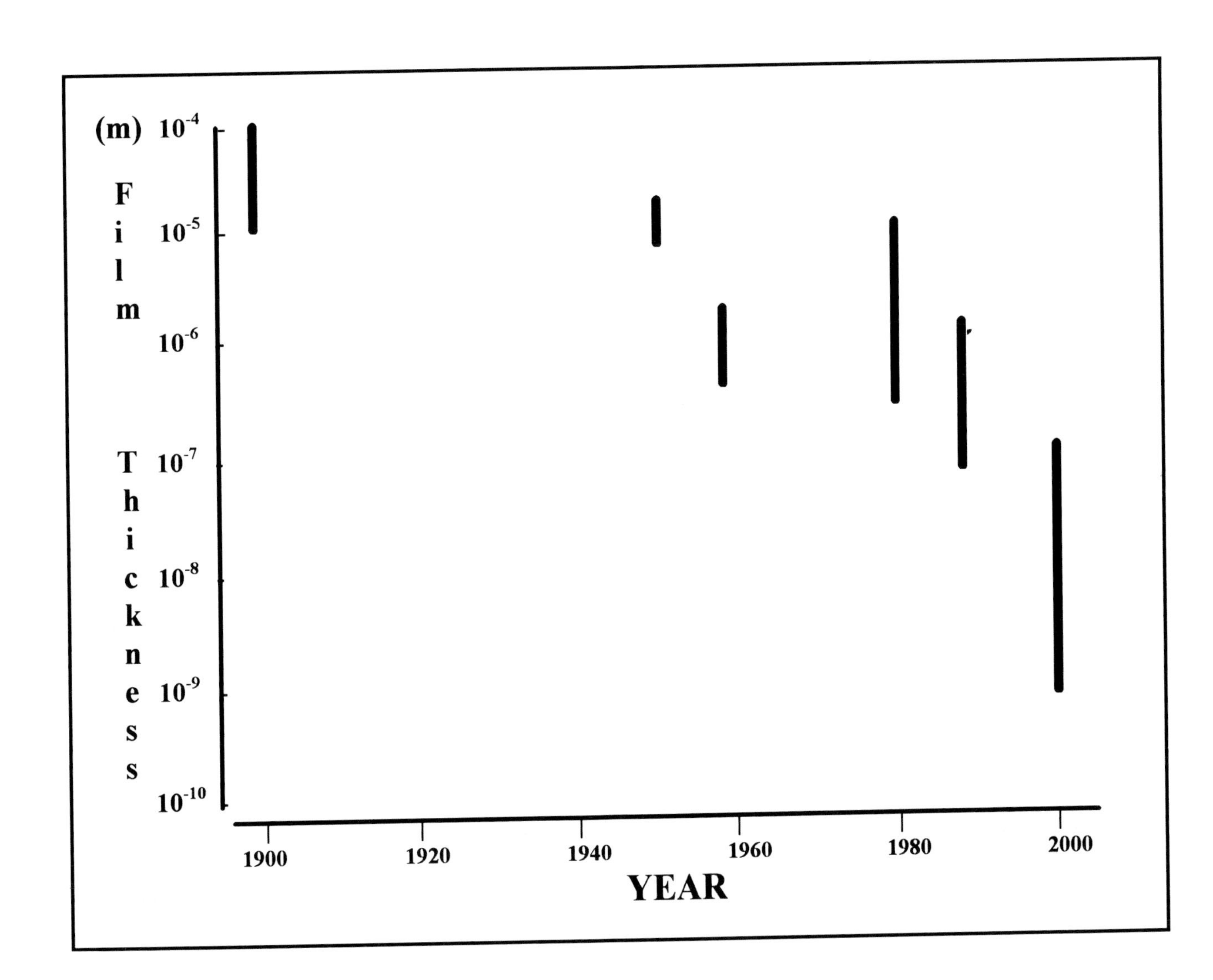

Fig. 7 The thinning film – representative film thicknesses throughout the 20th. Century.

4.2.3 Micro-EHL or asperity lubrication

An interesting feature of very thin film elasto-hydrodynamic lubrication is that the effective film thickness is often a good deal smaller than the initial, undeformed composite surface roughness of the interacting solids. One explanation of this is that, under some circumstances, the individual asperities act as local generators of hydrodynamic pressures, which in turn partially flatten elastically the very asperities causing the perturbations to the smooth surface pressure distribution.

This effect, known as micro-elastohydrodynamic (MEHL) or asperity lubrication, was deemed to be necessary to account for the fluid-film lubrication of soft-EHL conjunctions, such as those encountered in synovial joints (**56**) and elastomeric seals (**57**). The mechanism of MEHL is illustrated for a synovial joint in Fig. 8.

Furthermore, it is now recognised that MEHL might well contribute to the effective fluid-film lubrication of certain 'hard' EHL contacts (**58**), thus supporting the experimental observations of thinner and thinner EHL films as the century progressed. Initially line contacts were assessed, but in recent times the much greater problem of point contacts has been tackled (**59,60**).

The early theoretical demonstrations of MEHL were deliberately designed to exhibit the potential of the process by considering the most beneficial configuration, in which the roughness was in evidence on the stationary surface alone. The reality is more complex (**61,62**) and under some circumstances the elastic flattening of asperities is much less complete than originally proposed, yet the basic concept is now well established.

4.2.4 Very thin film (VTF) lubrication

Since EHL films persist down to film thicknesses much smaller than originally envisaged, a question arises about

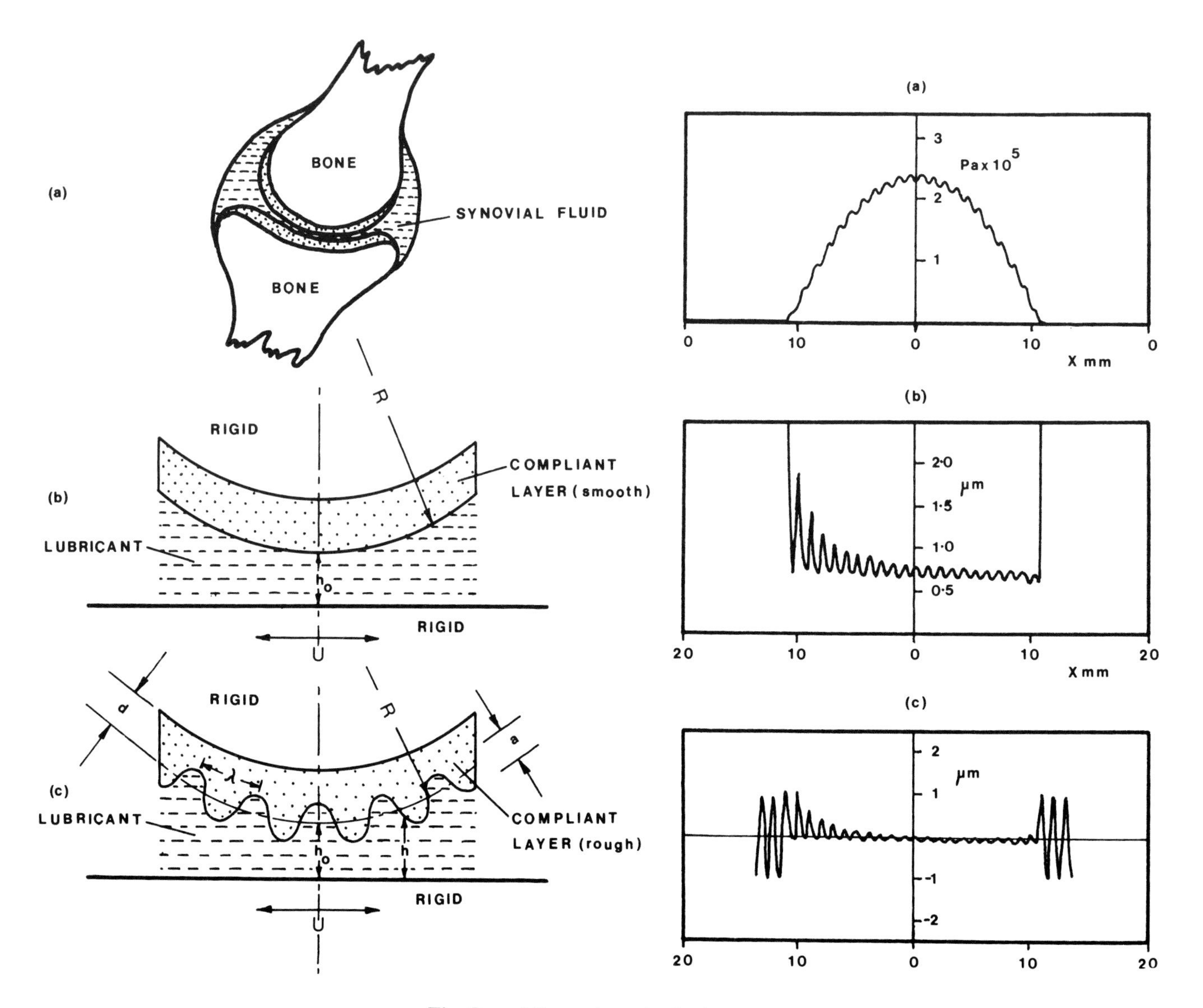

Fig. 8 Micro-elasto-hydrodynamic action in a synovial joint.

the transition to boundary lubrication. Spikes (**63**) has described a valuable 'spacer layer' technique used with optical interferometry to measure lubricating film thicknesses of nano-metre proportions. Experiments were performed to show that non-polar polyalphaolefins exhibited ever decreasing film thicknesses as the rolling speed reduced (see Fig. 9; (**66**)), much as predicted by

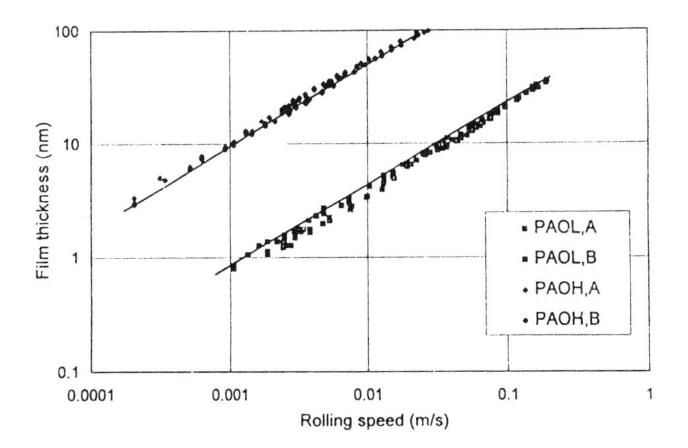

Fig. 9 Experimental film thicknesses as a function of rolling speed for polyalphaolefins (A and B represent repeat tests) (reproduced by kindpermission of Elsevier Science S.A., reference (**66**)).

EHL analysis, but that the presence of polar fluids in solution yielded an almost speed-independent separation as indicated in Fig. 10.

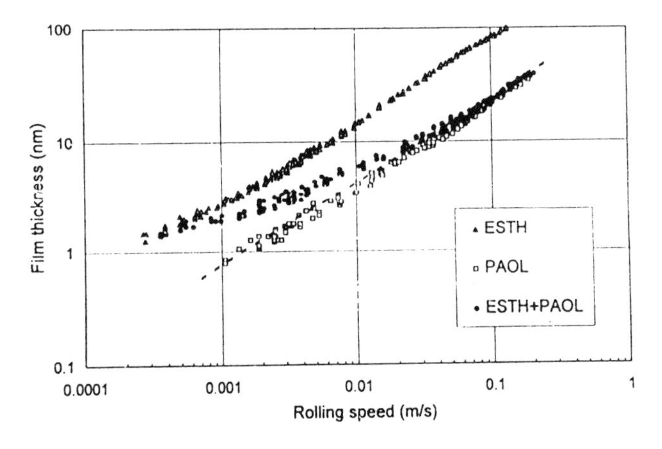

Fig. 10 Experimental film thicknesses as a function of rolling speed for polyalphaolefin with and without 10% of high viscosity ester (reproduced by kind permission of Elsevier Science, S.A., reference (**66**)).

Moore (**64**) also recorded similar behaviour. Guangteng et al (**65**) confirmed that some polymers dissolved in base fluids could form boundary lubricating films up to 30 nm thick. They also demonstrated theoretically and experimentally (**66**) that polymer additives could form surface layers exhibiting enhanced viscosity on the solid surfaces.

The concept of immobile layers of 'lubricant' on the surfaces of the bounding solids has also been discussed by Professor Georges (**67**) and his colleagues, while Bair (**68**) has recorded shear planes in solidified lubricant. Much remains to be done on shear and slip in EHL and VTF lubrication.

Experiments carried out recently under sliding conditions by Kaneta et al (**69**) have shown that a dimple can form in the normally 'flat' central region of the film under some conditions. The authors have suggested that this is attributable to the difference in elastic modulus of the two interacting solids and the neglect of surface pressure components in the plane of the conjunction. The observations certainly call for explanation and it is clear that a full appreciation of the role of rheology, thermal effects, solidification and slip in certain EHL contacts has yet to emerge.

Recent studies of lubricant behaviour are concentrated at the very interface between boundary and elastohydrodynamic lubrication. This not only demonstrates the extensive reach of elastohydrodynamic action, it more closely defines the region of boundary lubrication.

4.2.5 Lubricant technology – additives; synthetics

Mineral oils still represent the major base fluids in lubricants today. However, since the mid-century, lubricant chemists have recorded spectacular improvements through the use of additives and synthetic lubricants. The requirement to operate machinery at elevated temperatures, high loads and high shear rates has called for the introduction of additives in ever greater proportions in many fully formulated lubricants. Oxidation and degeneration of mineral oils under these severe conditions can be resisted by additives, while the viscosity index can also be improved. The development of the jet-engine promoted an interest in synthetic lubricants, which can now be found not only in aircraft, but also in cars and trucks and industrial machinery.

4.3 Computers and magnetic storage systems

Computers and magnetic storage devices have had a truly spectacular impact upon society in the final decades of the 20th century. Development of the required hardware has been accompanied by equally spectacular progress in the tribology of information storage and retrieval systems.

In modern computer tape and disc drives a load carrying air film is formed between the magnetic recording medium and the stationary or rotating magnetic head under steady state conditions. Physical contact occurs only on starting and stopping, leading to the advice to leave computers running for long periods of time rather than subjecting them to many starts and stops. Pressures for ever increasing recording densities demand that both the surface roughness and the flying height should be as small as possible. The current technology has already achieved impressive tribological performance, with surface roughnesses in the range of 1 to 5 nm and mean running clearances of 50 to 250 nm. In video tape recorders a delicate balance between flying height and light rubbing has to be achieved to provide a necessary self cleaning action on the heads. Mizoh (**70**) has estimated that this amounts to the removal of about one atomic layer for each 100 m of tape transit.

4.4 Wear

The exciting final quarter developments which we are now witnessing in the fields of friction and lubrication noted above can readily obscure the equal and perhaps even greater progress made in combating wear and understanding wear phenomena in the latter decades of the twentieth century.

4.4.1 Progress in combating wear

Everyday experience reminds us of the real progress made in combating wear. The reciprocating engine which powers our cars now enjoys greatly enhanced resistance to wear. At one stage, re-bored engines were not unknown; new piston rings had to be fitted; the notoriously severe conditions in valve trains led to rapid deterioration of cams, followers, valves and valve guides; bearings had to be replaced and the lubricant level and quality had to be monitored regularly. We now expect our engines to deliver at least 200,000 km of near trouble free service. These improvements arise from advances in lubricant technology, design, manufacture and the introduction of improved materials and surface treatments. Similar stories can be told of the advances in shoe and tyre technology, where the extended life of both footwear and tyres was so dramatic that it created severe problems through reduced demand for replacement items. New forms of machinery, such as nuclear power plants; space vehicles, computers and magnetic recording systems, robots and micro-machinery such as the tiny gear shown in Fig. 11, (See (**71**)) all rely upon tribology for their successful operation.

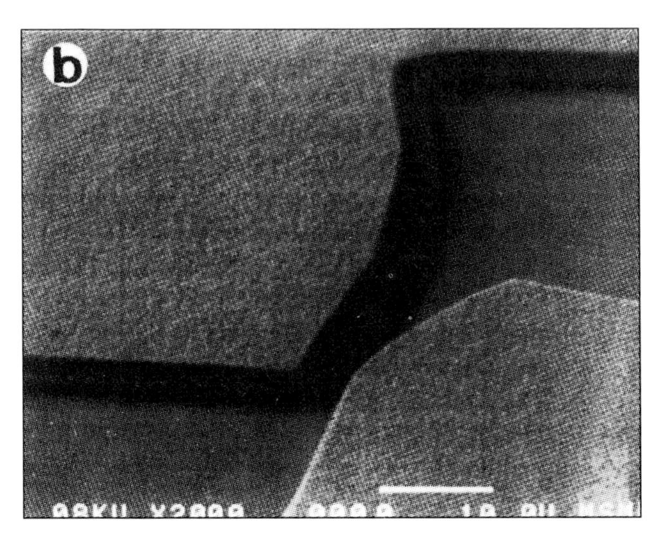

Fig. 11 Scanning electron micrographs of microgears (diameter of smaller gear ≈ 100μm) (reproduced by kind permission of Elsevier Science, B.V., reference (**71**)).

Total hip replacement is said to represent the greatest achievement in orthopaedic surgery this century. In the late 1950's PTFE was used for the acetabular cups, but the metallic femoral heads completely penetrated the cups in about three years. While femoral heads no longer penetrate fully through the polyethylene (UHMWPE) cups in current use and wear is no longer the direct cause of failure of prostheses, the effective life of implants is limited by other factors such as loosening. In the past few years tiny polyethylene wear particles, mainly in the sub-micron range, have been implicated in the loosening process. The thousands of wear particles generated at each step taken by the patient can result in the development of osteolysis, which disrupts the integrity of the implant-cement-bone interfaces. Harder, more scratch resistant metallic alloys and ceramics such as alumina and zirconia have been introduced as materials for femoral heads which yield ever lower polyethylene wear rates. Attempts have also been made to introduce materials with overall elastic properties closer to those of the host skeletal structure. While methyl methacrylate cement is still extensively used to anchor the implant in the bone, various approaches to cementless fixation have also been adopted in recent years. Another approach has been to 'customise' the femoral stem to enable it to fit more snugly within the bone with a minimum cement mantle thickness.

Considerable progress is being reported with improvements to current designs and materials, but exciting alternatives such as hard-on-hard (metal-on-metal or ceramic-on-ceramic) or fluid film (cushion) bearings are also being explored. It appears that tribologists still have much to contribute to the development of the next generation of total replacement joints. While these examples accord with everyday experience, the wider achievements in combating wear in both the established and the newer industries is widely known, but rarely acknowledged.

4.4.2 Surface engineering

Developments of surface treatments and modifications have been a noteworthy feature of the war on wear. Recognition that the bulk material in a machine could be selected on the basis of its strength and the requirements of structural integrity, while the surfaces of the interacting solids could be selected or modified by a variety of techniques to enhance resistance to wear, has underpinned developments in this field. Heat treatment through case hardening or diffusion are long established techniques used to increase surface hardness and resistance to wear. Coating techniques such as electroplating; welding; flame-spraying; ion plating and plasma coating by chemical (CVD) or physical (PVD) deposition, have attracted much attention and achieved great success in recent years. Diamond like carbon (DLC) coatings produced by chemical vapour deposition have attracted much attention since the early 1980*s*. The very high hardness of such films endows them with exceptional wear resistance, leading to applications in such devices as rock cutting drills, but they often have a relatively high roughness compared to the underlying material and hence high friction. Recent advances have, however, enabled the surface finish of the films and the resulting friction to be reduced substantially.

4.4.3 Fundamental aspects of wear

There has been marked progress in the understanding of wear since the 195*0s*. Papers by Archard and Hirst (**72**), Hirst and Lancaster (**73**), Kerridge and Lancaster (**74**) and Welsh (**75**) emphasised the work on dry friction and wear in the 1950's and 196*0s*. A popular over-arching approach was simply to classify wear as 'severe' or 'mild'. The former generally led to roughening of the surfaces and failure, with the production of relatively large wear particles some 10-100μm, or even 1mm, in length. In mild wear the roughness generally improved with running, with debris being much smaller in the range 0.01-1μm and often flake-like. Burwell and Strang's (**76**) review of wear mechanisms, in which they noted a significant increase in wear rate when the nominal contact stress exceeded about one third of the hardness of the softer material, was a useful contribution to knowledge. An interesting and wide ranging review of the understanding of wear in the 1960's was presented by Archard (**77**). The general classifications of wear as abrasive; adhesive; fatigue; erosive and corrosive were well established, but much of the work supporting these views was undertaken under dry rubbing conditions. Ceramics have assumed an important position in the spectrum of tribological materials. Their inertness, hardness and excellent surface finish make them particularly attractive in both monolithic and coating forms. If the third quarter of the century was the era of polymers in tribology, the final quarter must be associated with ceramics.

As interest moved away from the 'severe' wear end of the scale, a major step forward was recorded by Suh (**78**), when he introduced his concept of delamination wear. Quinn et al (**79**) also attacked the problem of chemical or oxidative wear. A valuable appraisal of these developments between the 1950s and 1980 was presented by Childs (**80**).

In general, there has been an increasing recognition of the importance of strain accumulation in the surfaces and fatigue in the spectrum of wear mechanisms in recent years, both for metals and polymers. Kragelski (**81**) proposed that wear was a result of fatigue, while Suh's delamination theory was linked to a limiting amount of accumulated plastic strain. A useful review of the development of work on the role of fatigue in wear was presented by Kimura (**82**). New insight into the problem emerged with the presentation by Challen and Oxley (**83**) and Black *et al.*, (**84**) of their wave model of friction and wear. They recognised some of the limitations in the widely held view of adhesive wear, such as the severity of the assumed fracture process and the problem of explaining subsequent particle detachment from the harder surface and proposed that friction was attributable to the force required to push a wave of plastically deformed softer material ahead of the hard asperities. A plane strain slip-line field was adopted to illustrate the mechanism, as depicted in Fig. 12. For small asperity angles, friction occurred without wear, until a fatigue limit was reached, while at larger angles, a cutting action and chip formation ensued.

This model provided the basis for a low cycle fatigue model of wear, in which particle formation was linked to the number of wave passes required to produce a fatigue failure in the softer material.

Kapoor and Johnson (**85**) considered the work done in plastic deformation as hard materials slide over softer ones. They drew attention to the shakedown concept in which initial plastic deformation can be followed by a totally elastic response. Kapoor et al (**86**) developed a model for the production of laminar wear debris in which plastic ratchetting was brought about by repeated pummelling of the softer material by the asperities on the harder counterface, as illustrated in Fig. 13. Torrance (**87**) assessed experimentally the low cycle fatigue and ratchetting concepts and concluded that the fracture process leading to the production of wear particles was best described by low cycle fatigue.

While the above developments have greatly advanced our understanding of wear, it remains true that the machine designer still experiences the greatest problem with predictions of wear rates. The latter range over many

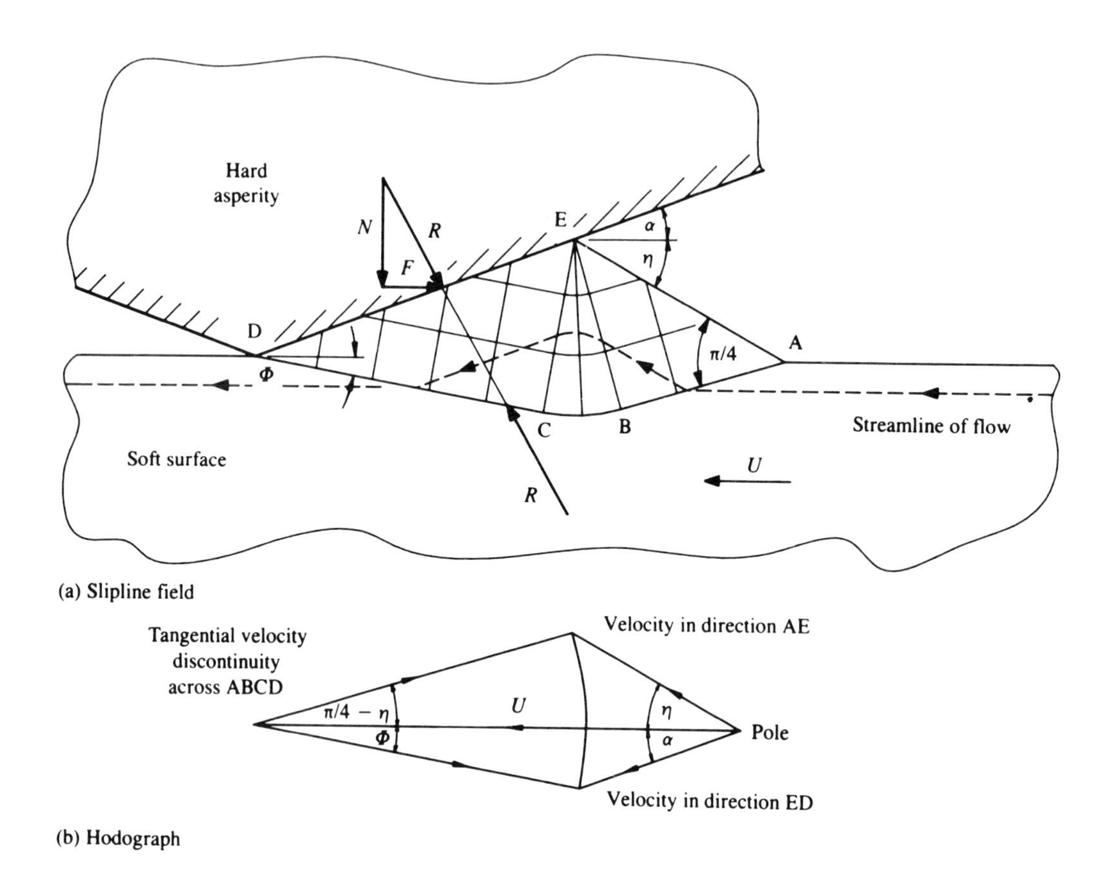

Fig. 12 Challen and Oxley's wave model of friction and wear (reproduced by kind permission, Institution of Mechanical Engineers, reference (**84**)).

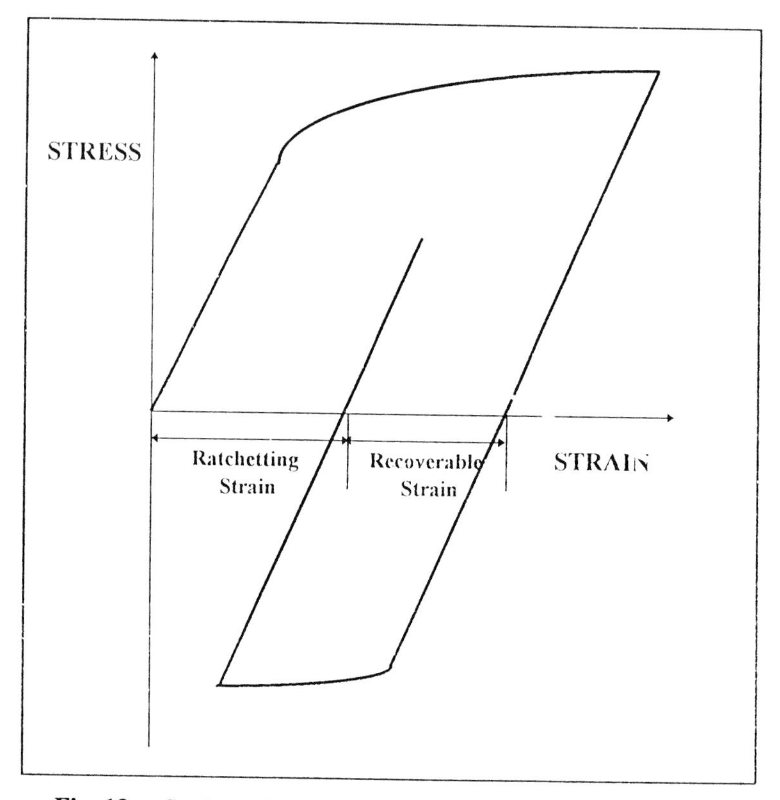

Fig. 13 Strain cycle showing ratchetting and recoverable strain.

orders of magnitude and are influenced by materials, design, environment, complex surface physical and chemical actions and the 'contact' operating conditions. A full appreciation of all these factors calls for inputs from several disciplines. It is now widely recognised that an improved approach to wear modelling is called for to provide the designer with adequate support in future years. Meng and Ludema (**88**) and Ludema (**89**) have reviewed existing wear models and called for a new approach to the development of friction and wear equations that can be used by designers of machinery.

Laboratory wear testing machines are essential to provide opportunities for studies of the basic mechanisms of wear. If the conditions of operation of machinery are well understood, it is also possible to simulate the conditions in the laboratory and to compare the relative performance of different combinations of materials or different designs of machine components. However, the road to laboratory simulation is fraught with difficulty, primarily because the operating conditions in the field are inadequately understood, or because the complexity required for adequate laboratory simulation makes the economics difficult to justify. Tests on simple laboratory wear testing machines can lead to erroneous indications of full scale plant performance if simulation of the field conditions is inadequate. It is often more difficult to design and build adequate laboratory simulators than it is to manufacture the field equipment under study! It is unfortunate that more use is not made of the opportunities to monitor wear in full scale machinery and to examine the consequences of wear in failed components. Perhaps we shall see a more rational approach to field and laboratory studies of wear in the 21st century.

4.4.4 The third body concept

Most studies of wear have concentrated upon the mechanisms responsible for the generation of wear particles, while neglecting the influence of the particles themselves upon the process. Maurice Godet (**90**) insisted that we should consider more completely the processes between the bulk (first) bodies sliding or rolling together. If wear particles exist between the interacting solids, they must transmit some of the load and modify the simple model of load transmission between well defined solids. They might also modify the surface characteristics of the solids themselves. The material between the first bodies, whether it be wear debris, oxide, coating, boundary or fluid film must 'flow' and shear to accommodate velocity differences between the counterfaces. Play and Godet (**91**) illustrated many of the third body concepts in 1977 by means of chalk debris observed through a transparent slider. The concept developed substantially over the years, as outlined by Ludema (**92**) and Berthier (**93**).

An appreciation of the value of the third body concept in guiding current thought in tribology emerges from the very extent to which the phrase '*third body wear*' is used; it has certainly become common place in many discussions and written works.

5 TRIBOLOGISTS, TRIBOLOGY SOCIETIES, LITERATURE AND MEETINGS

The timing of the *First World Tribology Congress* is significant to us all, but particularly so for the Institution of Mechanical Engineers. The first major professional meeting on Tribology organised by The Institution of Mechanical Engineers was held in London in October 1937 in the Central Hall, Westminster, where we meet today. The title was *A General Discussion on Lubricants and Lubrication* and the two volumes of 136 papers presented to 600 delegates are now classical and prized items. Some thirty one United Kingdom and twenty two Overseas Institutions, Societies and Associations co-operated in the event and it is interesting to note that many are represented here again at the First. World Congress.

Further major meetings were held in London in 1957, as a joint Conference between the American Society of Mechanical Engineers and the Institution of Mechanical Engineers and in 1967 where the emphasis was on fundamentals and application to design.

It was in the 1960s that the word tribology was introduced in the Report of a Working Group established by Lord Bowden, the Government Minister, of State for Education and Science, on *Lubrication (Tribology) Education and Research – A Report on the Present Position and Industry's Need.* Now affectionately known as the Jost Report (**1**), the publication marked a watershed for work in the field now known as Tribology. Programmes of education in the form of short, undergraduate and postgraduate courses developed throughout the world; companies sought and appointed engineers and scientists with specific knowledge and understanding of tribology; professional tribology Groups and Societies were established throughout the world; Meetings proliferated and literature, in the form of books, papers and journals, grew at an impressive rate. Contact between tribology and kindred societies was fostered through the *International Tribology Council (ITC),* formed in 1973. There are now some 33 Member Societies represented on the ITC with an additional 5 Corresponding Members in other nations. The ITC was formed at the end of the *First European Tribology Congress* held in London in September 1973. A series of *Eurotrib* Conferences to be held every four years was inaugurated at this stage, with successive meetings being held in different European nations. In 1997 the Conference has returned to London, with its status raised to commemorate a number of anniversaries under the title *First World Tribology Congress.* Reviews of the recent history of the fabric of Tribology Societies have been written by Dowson (**2**) and Wilson (**94**).

In 1969 a *Tribology Trust* was established and operated by the Institution of Mechanical Engineers in London, to recognise outstanding achievement in tribology and to

YEAR	NAME	COUNTRY	SUBJECTS
1972	D. Tabor	Cambridge, UK	Friction and Lubrication of Solids
1973	H. Blok	Delft, Netherlands	Research, Application & Teaching
1974	M. D. Hersey	Rhode Island, USA.	Teacher, mentor and researcher.
1975	I.V. Kragelski	Moscow, USSR	Friction and Wear.
1976	R.L. Johnson	Ohio, USA	Aerospace applications
1977	F.T. Barwell	Swansea, UK	Friction, Lubrication and Wear & Application to Rail Transport.
1978	D.D. Fuller	New York, USA	Research & Application of Fluid Film Lubrication
1979	D. Dowson	Leeds, UK	Elastohydrodynamic Lubrication & Bio-Engineering
1980	M.E. Merchant	Cincinnati, USA	Metal Cutting & Manufacturing Systems
1981	N. Soda	Tokyo, Japan	Friction, Lubrication & Wear
1982	G.V. Vinogradov	Moscow, USSR	Friction & Wear
1983	A. Cameron	London, UK	Lubrication research, Education & Promotion
1984	H. Peeken	Aachen, FRG.	Scientist, Engineer, Teacher
1985	K. Johnson	Cambridge, UK	Contact Mechanics
1986	W.O. Winer	Georgia, USA	Surface Temperature & Rheology
1987	F. Hirano	Tokyo, Japan	Science of Tribology & Application to Science
1988	M. Godet	Lyon, France	Wear debris & Third Body Concept
1989	G. Fleischer	Magdeburg, GDR.	Energy approach to Friction & Wear
1990	T. Sakurai	Tokyo, Japan	Tribochemistry & Tribomechanics
1991	A.V. Chichinadze	USSR	Brakes, clutches & transmissions
1992	H.S. Cheng	Chicago, USA	Fluid & Mixed film lubrication
1993	K. Ludema	Michigan, USA	Wear
1994	J.M. Georges	Lyon, France	Surface forces; molecular or nano-metric scale
1995	S. J. Pytco	Cracow, Poland	Metal Industries
1996	V.N. Constantinescu	Bucharest, Roumania	Fluid film (gas) Lubrication

Table 2. Tribology Gold Medalists

further interest in and study of the subject. The highest, international, annual award is the Gold Medal and the list of recipients and their subject specialisms is shown in Table 2. The list illustrates both the international nature of the award and the spread of disciplines covered.

6 CONCLUSIONS

The pace of activity in the field of tribology has been impressive and truly international during the latter third of the 20th century. Progress has been made in all the major disciplines, while our technological age has come to rely upon sound tribological skills. Challenges in new fields such as space, nuclear, nano-technology and bio-engineering have been accommodated, while further development of more reliable, efficient machinery has been achieved as a result of progress in tribology. Many of the scientific studies and technological advances would probably have emerged anyway, but it is interesting to speculate on the impact of the Jost report published in 1996. Historians will probably say that the most important feature of developments in tribology in the second half of the twentieth century was that the technology and science moved forward with firm links to the society it supports. One wonders if we would have been meeting today in such strength if the messages of the 1960's had not been heard by governments, academia and industry throughout the world.

7 REFERENCES

1 **Department of Education and Science** *Lubrication (Tribology) Education and Research. Report on the Present Position and Industry's Needs,* HMSO, 1966, London.

2 **Dowson, D**. *History of Tribology*, Longman Group Limited, London, 1979, 677.

3 **Michell, A.G.M.** The lubrication of plane surfaces, *Z. Math. Phy,* 1905 (a),**52**, 123-137.

4 **Michell, A.G.M.** *Improvements in Thrust and Like Bearings,* 1905, (b), British Patent **No.875.**

5 **Kingsbury, A.,** *Thrust Bearings,* 1910, US Patent No. **947242.**

6 **Hardy, W.B.** *Collected Scientific Papers of Sir William Bate Hardy,* Ed. by Sir Erik K. Rideal, published under the auspices of the Colloid Committee of the Faraday Society, 1936, Cambridge U.P.

7 **Swift, H.W.** Theory and experiment applied to journal bearing design, *Instn. Mech. Engrs., Proceedings of the General Discussion on Lubrication and Lubricants*, 1937, **1**, 309-316.

8 **Christopherson, D.G.** A New Mathematical Method for the Solution of Film Lubrication Problems, *Proc. Instn. Mech. Engrs.,* 1941, **146**, 126-135.

9 **Cameron, A.** and **Wood, W.L.,** The Full Journal Bearing, *Proc. Instn. Mech. Engrs.* 1949, **161**, 59-72.

10 **Swift, H.W.,** Fluctuating loads in sleeve bearings, *J.Instn. Civ. Engrs.,* 1937(b), **5**, 161-195.

11 **Orfoff, P.I.**, Coefficient of Friction, Oil Flow, and Heat Balance in a Complete Cylindrical Bearing, *Aeronaut. Engng. (Moscow)*, 1935, **9th year**, **No.1**, 1935, 25-56 (*Trans.S. Reiss*, NACA Tech. Mem., **1165**, Washington, October 1947).

12 **Stone, J.M.** and **Underwood, A.F.**, Load Carrying Capacity of Journal Bearings, *Soc. Auto. Engrs.*, Q. Trans., 1947, **1**, 56-70.

13 **Burwell, J.T.**, The Calculated Performance of Dynamically Loaded Sleeve Bearings, *J. Appl. Mech.*,1947 **A69**, 231-245: 1949, **Pt.II**, 71, 358-360, 1951, **Pt.III**, 393-404.

14 **Shaw, M.C.** and **Macks, F.**, *Analysis and Lubrication of Bearings*, 1949, McGraw Hill, New York.

15 **Goodman, J.**, Roller and Ball Bearings, *Proc. Inst. Civ. Engrs.*, 1912, **clxxxix**, Session 1911-1912, **Pt.III**, 4-48.

16 **Stribeck, R.**, Die Wesentlichen Eigenshaften der Gleit und Rollenlager *Z.Ver. dt. Ing.*, 1902, **46**, **38**, 1341-1348, 1432-1438, **39**, 1463-1470.

17 **Palmgren, A.**, *Ball and Roller Bearing Engineering*, 1945, SKF, Philadelphia, USA.

18 **Åbbott, E.J.** and **Firestone, F.A.**, Specifying Surface Quality, *Mech. Enging.*, 1933, **55**, 569-572.

19 **Prandtl, L.**, Ein Gedankenmodell zur kinetischen Theorie der festen Körper Z. angew. *Math. Mech* 1928., **8**, 85.

20 **Deryagin, B.V.**, *Zh. Fiz. Khim.*, 1934, **5**, **9**.

21 **Holm, R.**, *The Friction Force Over the Real Area of Contact*, Wiss.Veröff. Siemens-Werk, 1938,.**4**, 38-42.

22 **Bowden, F.P.** and **Tabor, D.**, The Mechanism of Metallic Friction, *Nature*, London, 1942, **150**, 197.

23 **Reynolds, O.**, On Rolling Friction, *Phil. Trans. Roy. Soc.*, 1875, **166**, **Pt.I**, 155.

24 **Heathcote, H.L.**, The Ball Bearing: in the Making, under Test, and on Service, *Proc.Inst. Automot. Engrs.*, 1921, **15**, 569-702.

25 **Swift, H.W.**, The Stability of Lubricating Films in Journal Bearings, *Proc. Inst. Civ. Engrs.*, 1932, **233**, 267-288.

26 **Stieber, W.**, *Das Schwimmlager*, 1933, Krayn, Berlin.

27 **Taylor, G.I.**, Stability of a Viscous Liquid Contained Between Two Rotating Cylinders, *Phil. Trans.*, 1923, **A223**, 289-343.

28 **Dowson, D.** and **Higginson, G.R.** *Elasto-Hydrodynamic Lubrication-The Fundamentals of Gear and Roller Lubrication*, 1996, Pergamon Press, Oxford.

29 **Grubin, A.N.** and **Vinogradova, I.E.**, *Investigation of the Contact of Machine Components*,1949, Kh. F. Ketova (Ed.), Central Scientific Research Institute for Technology and Mechanical Engineering (Moscow), Book No. 30, (DSIR Translation No.337).

30 **Cameron, A.**, Righting a 40-Year Old-Wrong. E.M. Ertel. – the True Author of 'Grubin's ehl Solution, 1985, *Tribology International*, **18 (2)**, 82.

31 **Cameron, A.**, Written Discussion on Session VII, Proceedings of the 23rd 1997Leeds-Lyon Symposium on Tribology, 'Elastohydrodynamics-'96, to be published.

32 **Greenwood, J.A.** and **Williamson, B.P.**, Contact of Nominally Flat Surfaces, 1966, *Proc. Roy. Soc. Lond.*, **A295**, 300-319.

33 **Archard, J.F.** Theory of Mechanical Wear, *Research*, 1952, Butterworths Publications Ltd., C1-C3.

34 **Evans, D.C.** and **Lancaster, J.K.**, The Wear of Polymers, in Treatise on Materials Science Technology, *Wear*, 1979, **13**, 85-139.

35 **Dowson, D.** and **Higginson, G.R.** New Roller Bearing Lubrication Formula, *Engineering, London*, 1961, **192**, 158.

36 **Barwell, F.T.**, The Founder of Modern Tribology, in *Osborne Reynolds and Engineering Science Today*,1970 Eds., McDowell, D.M. and Jackson, J.D., Manchester University Press, 240-263.

37 **Lancaster, J.K.** Solid Lubrication, in *Principles of Lubrication*,1966, Cameron, A., Longmans, 469-497.

38 **Pratt, G.C.**, Plastic Based Bearings, in *Lubrication and Lubricants*, 1967, Braithwaite, E.R., Ed., Elsevier, Amsterdam, 377-426.

39 **Braithwaite, E.R.**, Ed., *Lubrication and Lubricants*, 1967, Elsevier Publishing Company, 568.

40 **Charnley, J.** *Low Friction Arthroplasty of the Hip*, 1979, Springer -Verlag, 376.

41 **Dowson, D.** and **Wright,V.**, Bio-Tribology, The Rheology of Lubrication, 1973, Institute of Petroleum, 81-88.

42 **Homola, A.M.**, **Israelachvili, J.N.**, **McGuiggan, P.M.**, and **Hellgeth, J.W.**, Fundamental Studies in Tribology, *Wear*, 1990, **136**, 65-84.

43 **Tomlinson, G.A.**, A Molecular Theory of Friction, Phil. Mag., 1929. **7**, 905-939.

44 **Sokoloff, J.B.**, Theory of Atomic Level Sliding Friction Between Ideal Crystal Interfaces, *Wear*, 1991.

45 **McClelland, G.M.** and **Glosli, J.N.** Friction at the Atomic Scale, in *Fundamentals of Friction: Macroscopic and Microscopic Processes* ,1992, Ed. Singer, I.L.and Pollock, H.M., Kluwer Academic Publishers, NATO ASI Series E, Vol. **220**, 405-425.

46 **Israelachvili, J.**, Adhesion, Friction and Lubrication of Molecularly Smooth Surfaces, *in Fundamentals of Friction: Macroscopic and Microscopic Processes*, 1992, Ed. Singer, I.L.and Pollock, H.M., Kluwer Academic Publishers, NATO ASI Series E, Vol. **220**, 351-385.

47 **Johnson, K.L.**, **Kendall, K.**, and **Roberts, A.D.**, Surface Energy and the Contact of Elastic Solids, *Proc. R. Soc. Lond.*, 1971, **A 324**, 301-313.

48 **Johnson, K.L.**, Adhesion and friction Between a Smooth Elastic Spherical Asperity and a Plane Surface, *Proc. R. Soc. Lond.*, 1997,**A 453**, 163-179.

49 **Landman, U.**, **Luedtke, W.D.**,and **Ringer, E.M.**, Atomistic Mechanisms of Adhesive Contact Formation and Interfacial Processes, *Wear*, 1992, 150.

50 **Landman, U.**, **Luedtke, W.D.**, and **Ringer, E.M.**, Molecular Dynamics Simulations of Adhesive Contact Formation and Friction, in *Fundamentals of Friction: Macroscopic and Microscopic Processes*, 1992, Ed. Ringer, I.L. and Pollock, H.M., Kluwer Academic Publishers, NATO ASI Series E, Vol. **220**, 463-508.

51 **Thompson, P.A.** and **Robbins, M.O.** To Slip or Not to Slip?, *Physics World*, 1990, 35-38.

52 **Thompson, P.A. Robbins, M.O.** and **Grest, G.S.**, Simulations of Lubricant Behaviour at the Interface with Bearing Solids, *Proceedings of the 19th. Leeds-Lyon Symposium on Tribology*, 1993, Elsevier Science Publishers, 347-360.

53 **Evans, C.R.** and **Johnson, K.L.**, The Rheological Properties of Elastohydrodynamic Lubricants, *Proc. Instn. Mech. Engrs., Part C*, 1986, **200 (C5)**, 303-312.

54 **Bair,S.** and **Winer, W.O.**, The High-Pressure, High Shear Stress Rheology of Liquid Lubricants, *ASME Journal of Tribology*, 1992, **114**, 1-13.

55 **Dowson, D.** Developments in Lubrication-The Thinning Film, *J. Phys. D.: Applied Phys.*, 1992, **25,** 334-339.

56 **Dowson, D.** and **Jin, Z-M,** Micro-elastohydrodynamic Lubrication of Synovial Joints, *Eng. Med.* 1986, **15, 2,** 63-65.

57 **Miyazaki, E., Ogata, M., Mori, A.,** and **Shimotsuma, Y.,** EHL Contact of an Undulated Elastomer Surface on a Smooth Rigid Surface, *Japanese Society of Tribologists, Proceedings II International Conference*, 1995, 1039-1042.

58 **Evans, H.P.** and **Snidle, R.W.,** A Model for Elastohydrodynamic Film Failure in Contacts Between Rough Surfaces Having Transverse Finish, *ASME Journal of Tribology,* 1996, **118,** 847-857.

59 **Seabra, J.** and **Berthe, D.** Influence of Surface Waviness on Elastohydrodynamic Lubrication of Line Contacts, *ASLE Transactions,* 1987, **30, 4,** 486-492.

60 **Kweh, C.C., Patching, M.J., Evans, H.P.,** and **Snidle, R.W.,** Simulation of Elastohydrodynamic Contacts Between Rough Surfaces, *Trans. ASME,* 1992, **114,** 412-419.

61 **Morales- Espejel, G.E.,Greenwood, J.,** and **Melgar, J.L.,** Kinematics of Roughness in EHL, *Proceedings of the 22nd. Leeds-Lyon Symposium on Tribology,* 1996, 501-513.

62 **Ehret, P., Dowson, D.,** and **Taylor, C.M.,** Waviness Orientation in EHL Point Contact, *Proceedings of the 22nd. Leeds-Lyon Symposium on Tribology,* 1996, , 235-244.

63 **Spikes, H.A.,** The Behaviour of Lubricants in Contacts: Current Understanding and Future Possibilities, *Proc. Instn. of Mech. Engrs., Part J, Journal of Tribology,* 1994, 3-15.

64 **Moore, A.J., Cooper, D.,** and **Robinson, T.M.,** Rheological Properties of Engine Crankcase and Gear Oil Components in Elastohydrodynamic Oil Films, *SAE Fuels and Lubricants Meeting, Rheotribology of Automotive Lubricants and Fluids,* 1994, (SP-1055), Technical Paper, 941997, 11-23.

65 **Guangteng, G.** and **Spikes, H.A.,** Behaviour of Lubricants in the Mixed Elastohydrodynamic Regime, *Proceedings of the 21st. Leeds-Lyon Symposium on Tribology,* 1994, Lubricants and Lubrication, Ed. Dowson, D., Taylor, C., Childs, T.H.C., and Dalmaz, G., Elsevier Science Publishers, 479-485.

66 **Guangteng, G.** and **Spikes, H.A.,** Fractionation of Liquid Lubricants at Solid Surfaces, WEAR, 1996, **200,** 336-345.

67 **Georges, J-M., Millot, S., Loubet, J-L., Tonck, A., and Mazuyer, D,** Surface Roughness and Squeezed Films at Molecular Level, *Proceedings of the 19th.Leeds-Lyon Symposium on Tribology,* 1995, Thin Films in Tribology, Ed. Dowson, D., Taylor, C., Childs, T.H.C., Godet, M., and Dalmaz, G., Elsevier Science Publishers, 443-452.

68 **Bair, S., Qureshi, F.,** and **Winer, W.O.,** Observations of Shear Localisation in Liquid Lubricants Under Pressure, *Trans. ASME., Journal of Tribology,* **115,** 507-514.

69 **Kaneta, M., Nishikawa, H., Kameishi, K., Sakai, T.** and **Ohno, N.,** Effects of Elastic Moduli of Contact Surfaces in Elastohydrodynamic Lubrication, *Trans. ASME, Journal of Tribology,* 1992, **114,** *75-80.*

70 **Mizoh,Y.,** Wear of Tribo-Elements of Video Tape Recorders, *Wear,* 1996, **200,** 252-264.

71 **Komvopoulos, K.,** Surface Engineering and Microtribology for Microelectromechanical Systems, Wear, 1996, **200,** 305-327.

72 **Archard, J.F.** and **Hirst, W.,** The Wear of Metals Under Unlubricated Conditions, *Proc. Roy.Soc.,* 1956, **A236,** 397-410.

73 **Hirst, W.** and **Lancaster, J.K.,** The Influence of Speed on Metallic Wear, *Proc. Roy. Soc.,* 1960, **A259,** 228-241.

74 **Kerridge, M.** and **Lancaster, J.K.,** The Stages in a Process of Severe Metallic Wear, *Proc. Roy. Soc.,* 1956, **A236,** 250-264.

75 **Welsh, N.C.,** The Dry Wear of Tool Steel. I. The General Pattern of Behaviour; II. Interpretation of Special Features, *Phil. Trans. Roy. Soc.,* 1965, **257,** 31-70.

76 **Burwell, J.T.** and **Strang, C.D.,** On the Empirical Law of Adhesive Wear, *Journal of Applied Physics,* 1952, **23, 1,** 18-28.

77 **Archard, J.F.** Wear, *NASA Symposium on Interdisciplinary Approach to Friction and Wear,*1968, Ed. P.M. Ku., 267-304.

78 **Suh, N.P.,** The Delamination Theory of Wear, *Wear,* 1973, **25,** 111-124.

79 **Quinn, T.F.J., Sullivan, J.L.,** and **Rowson, D.M.,** Developments in the Oxidational Theory of Mild Wear, *Tribology International,* 1980, **13, (4),** 153-158.

80 **Childs, T.H.C.,** The Sliding Wear Mechanisms of Metals, Mainly Steels, *Tribology International,* 1980, 285-293.

81 **Kragelskii, I.V.,** *Friction and Wear,* 1965, Trans Ronson, L. and Lancaster, J.K., Butterworths, London, 346.

82 **Kimura, Y.,** The Role of Fatigue in Sliding Wear, *Proceedings ASM Materials Science Seminar, 'Fundamentals of Friction and Wear of Materials',* 1980, 1-52.

83 **Challen, J.M.** and **Oxley, P.L.B.,** An Explanation of the Different Regimes of Friction and Wear Using Asperity Deformation Models, *Wear,* 1979, **53,** 229-243.

84 **Black, A.J., Kopalinsky, E.M.,** and **Oxley, P.L.B.,** Asperity Deformation Models for Explaining the Mechanisms Involved in Metallic Sliding Friction and Wear-a Review, *Proc. Inst. Mech. Engrs., Part C, Journal of Mechanical Engineering Science,*1993, **207,** 335-353.

85 **Kapoor, A.** and **Johnson,** Plastic Ratchetting as a Mechanism of Metallic Wear, Proc. Roy. Soc., 1994, **A445,** 367-381.

86 **Kapoor, A., Johnson, K.L.,** and **Williams, J.,** A Model for the Mild Ratchetting Wear of metals, *Wear,* 1996, **200,** 38-44.

87 **Torrance, A.A.,** The Influence of Surface Deformation on Mechanical Wear, *Wear,* 1996, **200,** 45-54.

88 **Meng, H.C.** and **Ludema, K.C.,** Wear Models and Predictive Equations; Their Form and Content, Wear, 1995, **181-183,** 443-457.

89 **Ludema, K.C.,** Mechanism-Based Modeling of Friction and Wear, *Wear,* 1996, **200,** 1-7.

90 **Godet, M.,** The Third Body Approach. A Mechanical View of Wear, *Wear,* 1984, **100,** 437-452.

91 **Play, D.** and **Godet, M.,** Coexistence of Different Wear Mechanisms in a Simple Contact, *Wear,* 1977, **42,** 197-198.

92 **Ludema, K.C.,** Third Bodies: Perspectives on Modeling in Lubricated Contacts, etc.: Following on the Concepts of Dr. Maurice Godet, *Proceedings of the 21st. Leeds-Lyon Symposium on Tribology,* 1996, 3-19.

93 **Berthier, Y.,** Maurice Godet's Third Body, *Proceedings of the 21st. Leeds-Lyon Symposium on Tribology,* 1996,21-30.

94 **Wilson, W.,** Ruby Jubilee of Tribology Group, *Industrial Lubrication and Tribology,* 1996, 5-9.

High pressure rheology of lubricants and limitations of the Reynolds equation

S BAIR and **W O WINER**
Georgia Institute of Technology, USA
M KHONSARI
Southern Illinois University,USA

SYNOPSIS

A review of high pressure rheology leads to the conclusion that the results from rheometers may be used to generate empirical rate equations which are useful in modelling EHD traction. However, an analytical treatment of piezo-viscous liquids reveals that the Reynolds equation adequately captures the mechanics of the piezo-viscous liquid only when the shear stress is much less than the reciprocal of the pressure viscosity coefficient. Otherwise the cross film pressure gradient cannot be neglected and secondary flows result.

1 INTRODUCTION

1.1 Constitutive behaviour

In the elastohydrodynamic (EHD) regime of lubrication, the rheology of the liquid lubricant is key to the generation of a protective film and the transfer of shear across concentrated contacts in many machine elements. It is now understood that small scale EHD plays a major role in the boundary lubrication of rough surfaces. Many high-pressure metal working operations share the pressure and kinematics of elastohydrodynamics. A complete solution of the EHD problem, however, requires a thorough knowledge of the lubricant constitutive behaviour and the attending properties as functions of temperature and pressure. The Newtonian assumption alone is often inadequate, and the assumptions upon which the Reynolds equation is based must be re-examined.

Early investigations of high-pressure lubricant rheology addressed the small strain response. For small strains, viscous heating which plagues large strain measurements may be ignored. These measurements were both optical and acoustical, e.g. references (**1**) and (**2**) respectively, and provided linear viscoelastic properties. These techniques are experimentally convenient in requiring only optical or electrical paths through the high-pressure chamber. The linear viscoelastic response is, however, unable to account for any but the small slip traction in EHD.

Attention then turned to large strain investigations which could reveal non-linear behaviour and successfully predict EHD traction (**3**). The progress in this area has recently been significant in that a mechanism of apparent non-linearity (at least for low molecular weight liquids) has been shown to be a mechanically induced shear localization or shear bands (**4**). The constitutive behaviour in the material surrounding the shear bands is apparently linear viscoelastic.

Fundamental to the generation of a film in concentrated contacts is the piezoviscous property of liquid lubricants – whereby even simple low molecular weight liquids at ambient temperature attain very great viscosities under pressure. Shear rheological investigations of non-Newtonian response of lubricants (**5, 6**) have resorted to low temperatures to achieve the same level of viscosity. It is clear now that high-pressure is essential to an accurate simulation. Liquids become considerably ‘stronger’ under pressure. Peculiar to these atmospheric pressure studies was the observation of fracture. Eastwood and Harrison (**5**) observed the liquid in shear and reported cracking.

Early investigations of non-Newtonian lubricant response often studied polymer solutions. The blend of mineral base oil and polymeric viscosity index improver is representative of multi-grade motor oils. The capillary rheometer was useful for this purpose as it is very simply pressurized and although the shear stress cannot be large for this instrument, non-Newtonian flow occurs at relatively low stress for polymer solutions. The high-pressure capillary viscometer was developed to a high level by Jakobsen and Winer (**7**) who reported measurements to shear rates of 10^7 s^{-1} pressures to 600 MPa and time of shear as short as 4 μs. They reported Newtonian flow for liquid lubricant base stocks to a shear stress of 5 MPa.

1.2 Reynolds equation

The analysis of the pressure generated by lubricant films is almost exclusively performed with a form of the classical Reynolds equation. This differential equation derives from the inertialess form of the Navier-Stokes equation combined with the continuity equation with the assumption that the flow channel is small in one co-ordinate direction.

It has been generalized to incorporate variable viscosity as well as variable density and has been remarkably accurate in predicting the film thickness in nonconformal contact problems. However, there is a fundamental limitation to the Reynolds equation for problems in which viscosity varies with pressure and in particular, in the elastohydrodynamic regime. The cross-film pressure gradient cannot be neglected and secondary flows result.

Renardy (**8**) recognized that the Navier-Stokes equations for an incompressible, piezoviscous fluid may suffer from non-existence and non-uniqueness problems when the principal tensorial strain rates are not less than $(2\mu\alpha)^{-1}$. Here, μ is viscosity and $\alpha = d \ln\mu/dp$ is the local pressure viscosity coefficient. The Navier-Stokes equations can undergo a change of type – a process which has been used to characterize shear localization (**9**).

2 NON-LINEAR SHEAR RESPONSE

The search for the relevant constitutive equations which relate the lubricant stress to the flow kinematics in a concentrated contact has occupied the interest of tribologists working in EHD for at least thirty years. The goal has been to construct experiments to verify the rheological models and to provide the necessary property relations so that the complete elastohydrodynamic solution may be obtained. Various non-Newtonian models (see (**3**) for a review) have been advanced which provide accurate solutions for the traction over some operating range when the required rheological properties are obtained from the same traction data. In 1972, Dyson (**10**) warned that EHD traction research was 'enclosed within a tight circle' of fitting parameters to observations without consideration for measurements made outside of EHD. Hopefully, we have broken from the circle.

We may now generate rheological flow curves under conditions of pressure, temperature, and rate of shear which, although still rather restricted, are sufficiently within the realm of EHD to make accurate traction predictions and compare with traction experiments. We have previously interpreted these flow curves as lubricant constitutive behaviour. However, in light of the recent observation (**4**) of mechanical shear bands operating within the lubricant film concurrent with non-linear shear response, this interpretation must be accompanied by a caveat: rheological flow curves which are generated in plane shear yield an empirical rate equation which is useful in modelling Couette dominated lubrication problems. A rigorous analysis of the EHD problem would require a constitutive equation and a slip criterion such as Mohr-Coulomb (**11**).

2.1 Viscous heating in Couette viscometry

Presently, the most useful rheometer configuration for investigation of high-pressure, high-stress response of the liquid state is that of rotating concentric cylinders. When a liquid is sheared, the viscous work done raises the temperature of the liquid. While the study of this phenomenon is of itself interesting to lubrication, it is to be avoided in a rheological measurement because constitutive behaviour excludes processes which result from temperature variation. In previous work (**12**) concerning Couette devices the authors have emphasized the importance of a low Brinkman No. (through primarily a small shearing gap) to mitigate the effect of the temperature difference within the liquid film and fast instrument response to control the temperature of the surfaces of the solid boundaries to the film. An alternate experimental approach is to perform a measurement in so short a time that the temperature profile in the film has not had time to develop but at a late enough time that the velocity profile is fully formed. Winter (**13**) showed that this latter approach requires that the Prandtl No. be large.

An earlier analysis of the combined effect of instrument response time and cylinder heating (**12**) modelled the cylinders as a lumped heat capacity. It was found that to minimize errors due to this combination the measured stress history should show no thermal softening over a period of time equal to the instrument response time. Later, the evolution of the radial temperature profile within the cylinders was considered (**14**). A detailed example of a numerical thermal simulation of one of the high-pressure Couette viscometers used in this laboratory follows.

Only conduction in the radial direction is considered. An appropriate form of the energy equation for the metal cylinders is:

$$\frac{\partial^2 T}{\partial r^2} + \frac{1}{r}\frac{\partial T}{\partial r} = \frac{\rho_m C_m}{k_m}\frac{\partial T}{\partial t} \tag{1}$$

where r is the radial co-ordinate, t is time and ρ, C, and k are the density, specific heat capacity and thermal conductivity, respectively. The subscript, m, refers to the metal from which the cylinders are fabricated. Within a pressure vessel the outer surface of the outer cylinder (or cup) is in contact with low conductivity liquid. Therefore at the outermost surface the boundary condition is set at $\partial T/\partial r = 0$, namely it is an adiabatic surface. Symmetry at the axis of rotation dictates a similar boundary condition, $\partial T/\partial r = 0$ at $r = 0$.

The shearing gap is very thin compared with the working radius so that curvature of the liquid film can be ignored. A preliminary numerical solution for the liquid film with isothermal boundaries and dissipation showed that for a 1 μm film, steady state temperature distribution was achieved after 10 μs of shearing. Therefore, the storage term in the energy equation for the liquid film is dropped. Including the dissipation term, $\tau\dot{\gamma}$, we have for the liquid:

$$k\frac{\partial^2 T}{\partial r^2} + \tau\dot{\gamma} = 0 \tag{2}$$

The geometry considered is shown in Fig. 1. The liquid being sheared is in the circular gap between the outer and inner cylinders. Only the outer cylinder rotates with a velocity which yields a specified average of the rate of shear, $\dot{\gamma}$, The temperature and the radial heat flux are made continuous at the two metal/liquid interfaces. Equations (1) and (2) in their respective regions are solved numerically with the above boundary conditions and with initial uniform temperature (arbitrarily set to zero for Fig. 1) to obtain temperature distribution and shear stress versus time.

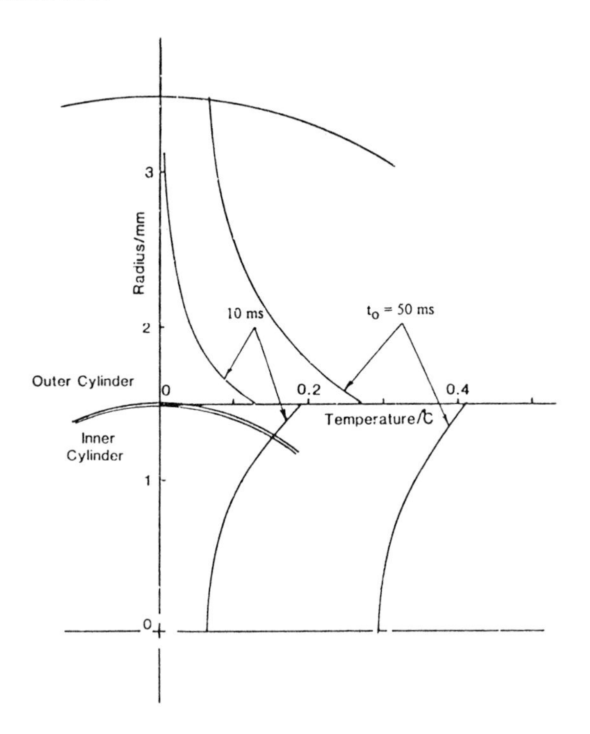

Fig. 1 Radial temperature distribution in concentric cylinders after velocity ramp lasting for time, t_o result for $t = t_o$.

The result of a sample computation is shown in Fig. 1 for two velocity histories. The velocity of the outer cylinder was increased linearly with time until an apparent or average rate of shear of 10^4 s^{-1} was reached at t_o equal to 10 and 50 ms. The liquid rheology was Newtonian with $\mu = 10^3$ Pa·s and β = 0.1°C^{-1}. The gap was 1.0 μm. The liquid and metal thermal conductivities were k = 0.15 and k_m = 225 W/m°C and for the metal ρ_m C_m = 3 x 10^6 J/m^3°C. The temperature profiles within the metal cylinders are shown in Fig. 1 for $t = t_o$. The liquid film temperature reached a maximum very near the surface of the inner cylinder. The reduction in shear stress from isothermal was less than 2% for t_o = 10 ms and less than 4% for t_o = 50 ms. The rotational velocity in this example is only about 1 revolution per second and can easily be achieved in 10 ms by a stepper motor. When a very large driving torque is required a 50 ms linear velocity ramp may be necessary.

In early high-pressure Couette devices the critical response time was often that of the torque transducer which provides the stress measurement. This transducer is immersed in a high-pressure liquid and for high viscosity the transducer responds as a first order instrument with a time constant, t_c, which is proportional to liquid viscosity. Very fast measurements may be obtained by accelerating both cylinders and then arresting the rotation of one cylinder with the torque transducer which can be immersed in a low viscosity liquid. Now, the velocity response is intimately connected with the transducer response. The undamped natural frequency for the transducer is 2.5 kHz. Theoretically, a 5 percent settling time (**15**) of 0.2 ms can be achieved for transducer response by optimizing the damping liquid viscosity. This is difficult to achieve in practice as the viscosity of the damping liquid changes with the test temperature and pressure. In any event, we have not observed an improvement in measurements typical of those reported here when done faster than 50 ms. The temperature increases shown in Fig. 1 present no problem in interpretation of results.

Also, by not considering axial conduction in the cylinders the analysis overestimates the metal temperature rise. Comparing measured stress histories for a Newtonian liquid with the computed stress history at t = 1 s we found the actual loss of shear stress compared to isothermal shear was only 64% of the predicted loss.

The temperature distribution of Fig. 1 was incorporated into a thermal-elastic numerical analysis of each cylinder using both plane-strain and plane-stress idealizations. Thermal expansion results in no substantial change in the operating film thickness at the time of the measurement.

2.2 Some results for pressures to 300 MPa

A high-pressure, high shear stress Couette viscometer was developed (**14**) in this laboratory to operate at pressures to 300 MPa. The steady shear response of various liquid lubricants (**14**) and a grease (**16**) have been investigated. For the grease, an empirical stress equation was found of the form:

$$\tau^q = \tau_y^q + (g\mu\dot{\gamma})^q, \tau \geq \tau_y \tag{3}$$

where g is a dimensionless viscosity enhancement factor approximately equal to 2, q is a dimensionless exponent and μ is the base oil viscosity. For q = 0.5, equation (3) becomes Casson's equation for oil/particulate mixtures.

Flow curves are presented for a high-traction fluid, Santotrac 50 and a mineral oil, LVI 260 in Figs. 2 and 3, respectively. Although the pressures and temperatures are comparable the shear response is quite different. The cycloaliphatic traction fluid remains Newtonian to a shear stress of at least 10 MPa whereas the mineral oil responds non-linearly above 3 MPa. Note also that for the mineral oil, the stress becomes essentially independent of rate at

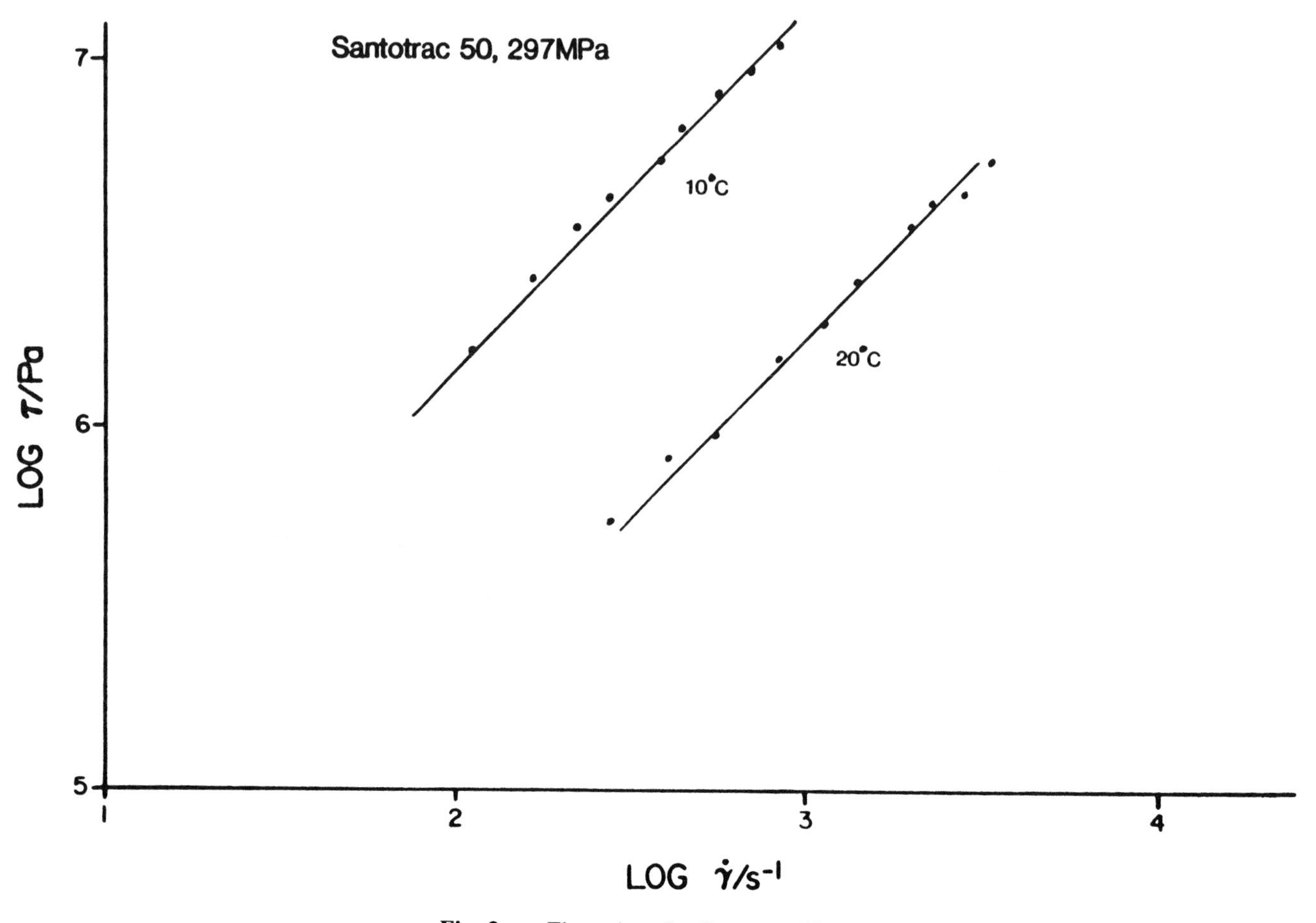

Fig. 2 Flow chart for Sanotrac 50, p = 297 MPa.

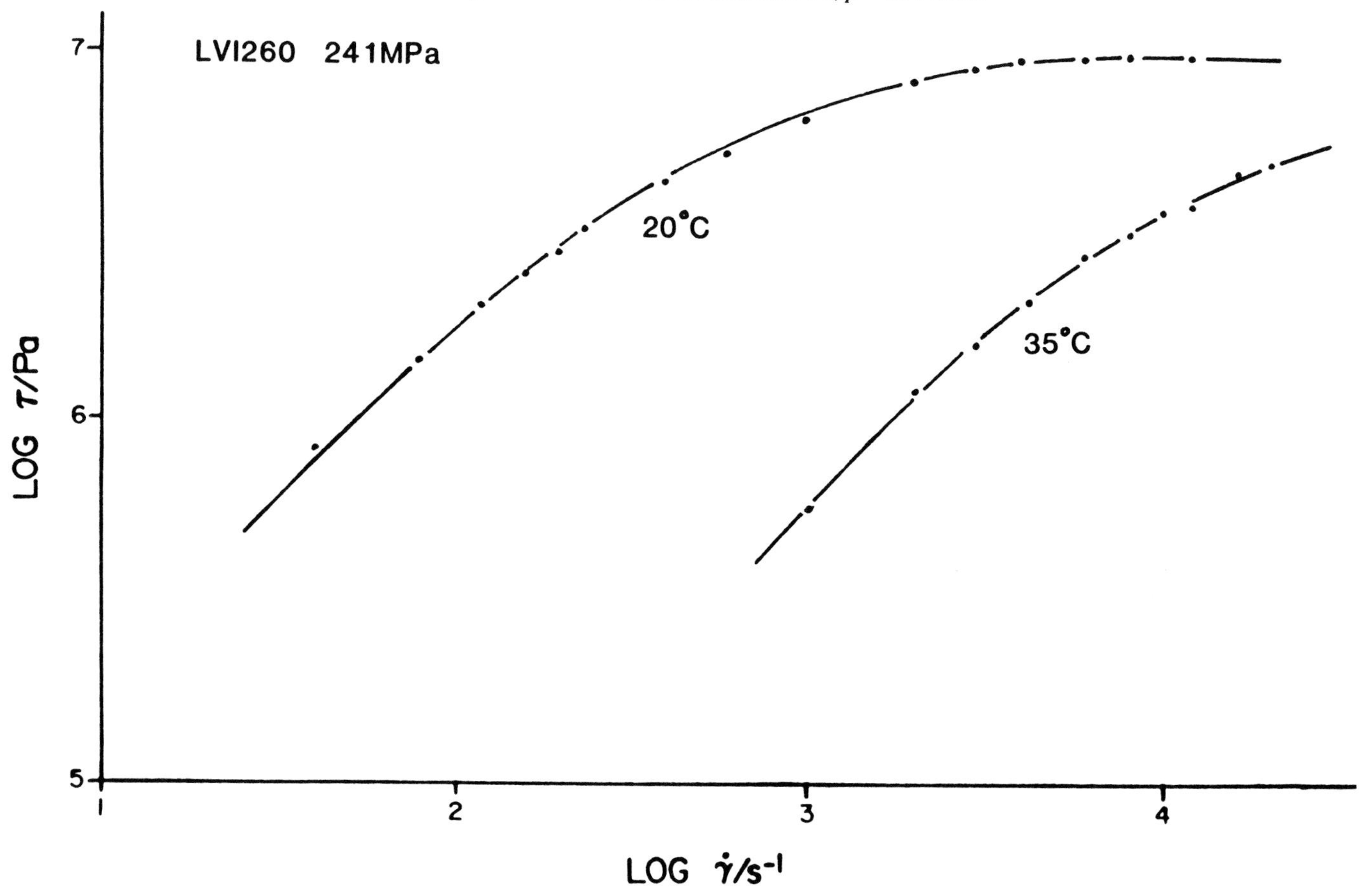

Fig. 3 Flow chart for mineral oil, LVI 260, p = 241 MPa.

about τ = 10 MPa and that the temperature dependence diminishes as the rate dependence diminishes. Because of their extended Newtonian response we have begun to use the cycloaliphatic traction fluids as viscosity standards for shear stress greater than 1 MPa in calibrating cylinder gaps.

2.3 A new high-stress Couette viscometer for 600 MPa pressure

An increase in the pressure capability of previous rotating concentric cylinder viscometers is necessary to investigate

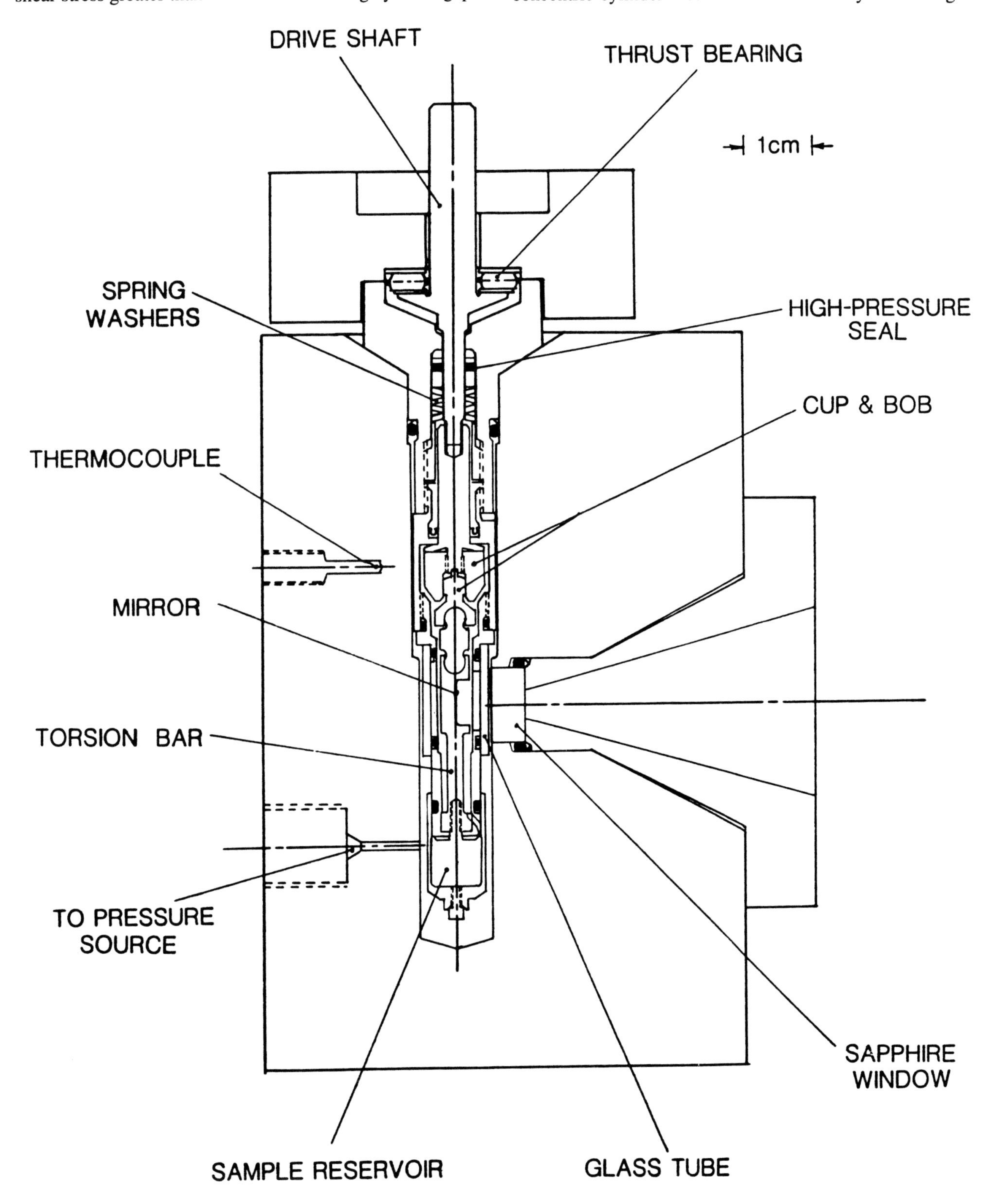

Fig. 4 High-pressure high shear stress viscometer for 600 MPa pressure.

the shear response of typical lubricants well above the glass transition temperature. Translating concentric cylinder rheometers are capable of pressures above 1 GPa; however the rate of shear is so low that the viscosity must be made very large in order to achieve an interesting magnitude of shear stress. Hence experiments are often conducted near the glass transition of the lubricant.

We recently reported (**14**) the development of a Couette viscometer for pressure to 300 MPa and the results of the previous section were obtained with that device. The pressure capability of the Couette technique is doubled with the device shown in section view in Fig. 4. The outer cylinder (cup) is driven by an external stepper motor by means of a drive shaft. The thrust bearing which prevents expulsion of the shaft from the vessel has been moved to outside the vessel so that the bearing does not run in pressurized liquid. The high-pressure seal is now a simple spring-loaded packing. There is no need for high-pressure electrical feed-throughs since the torque measurement is transmitted out of the vessel optically. The inner cylinder (bob) is restrained from rotation by a torsion bar with a mirrored surface within a glass tube. See Fig. 4.

The principle of the torque measurement is depicted in Fig. 5. The twist of the torsion bar results in a deflection of a laser beam which is detected by an optical position sensor. A glass tube (Fig. 4) provides a circular interface between the liquid which surrounds the mirror and the pressurizing fluid so that changes in refractive index of the sample have no effect on the measurement.

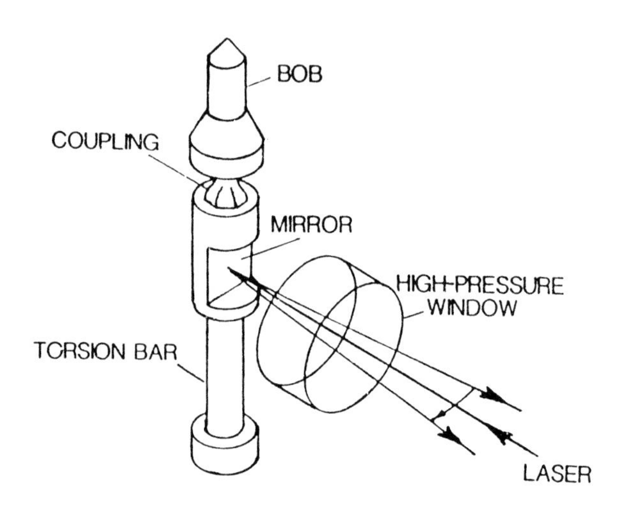

Fig. 5 Torque transducer operating principle for rheometer shown in Fig. 4.

2.4 Results for pressures to 600 MPa

A flow chart is presented for the polyalphaolefin, SHF 1001, in Fig. 6. The response is apparently Newtonian to about 2 MPa shear stress, although the effective viscosity below 2 MPa is slightly reduced compared to falling body measurements.

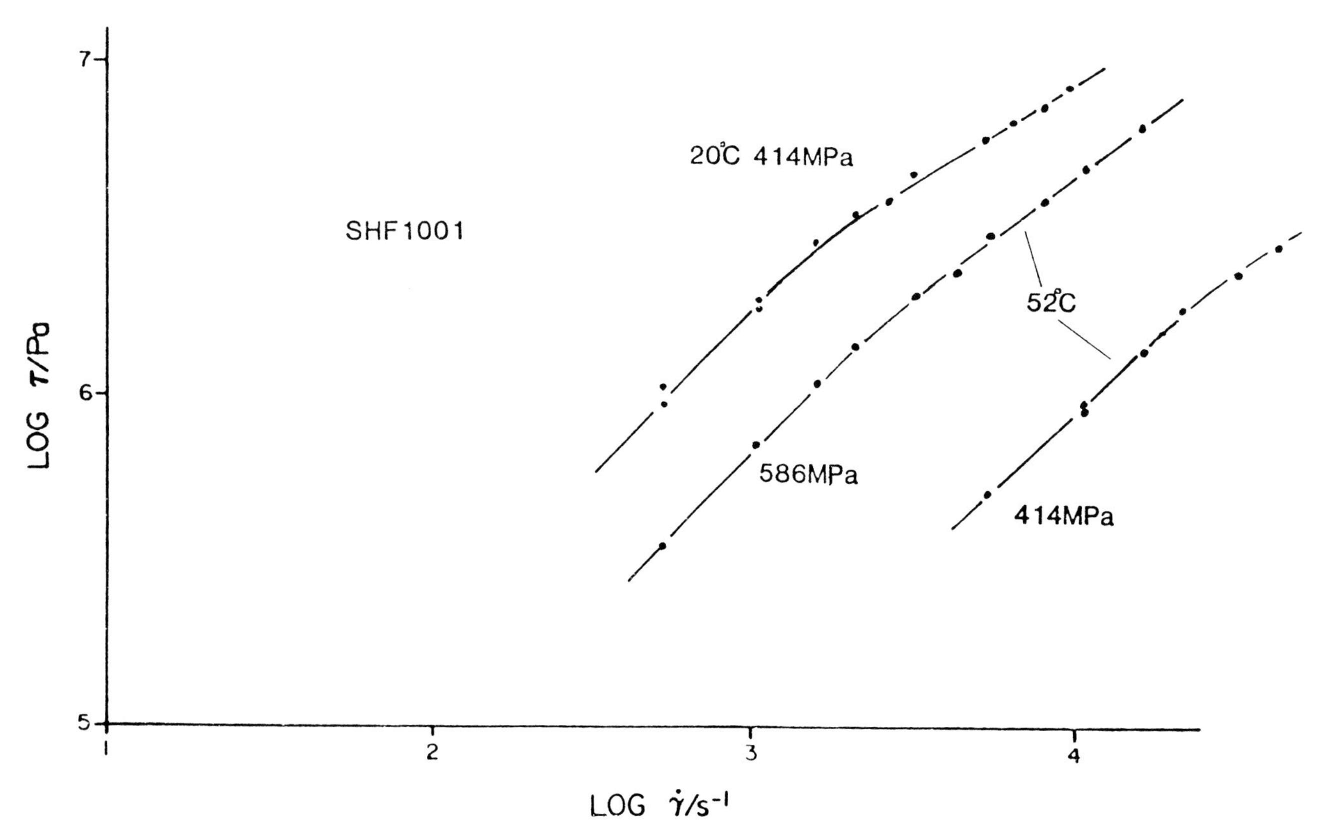

Fig. 6 Flow chart for a Polyalphaolefin.

The mineral oil, HVI 650, has been the subject of numerous traction experiments. The shear response of HVI 650 is depicted in Figs. 7a and b. The curves drawn through the data points for 23°C and 44°C represent the empirical Carreau-Yasuda equation:

$$\tau = \mu\dot{\gamma}\left[1+\left(\frac{\mu\dot{\gamma}}{\tau_L}\right)^a\right]^{(m-1)/a} \tag{4}$$

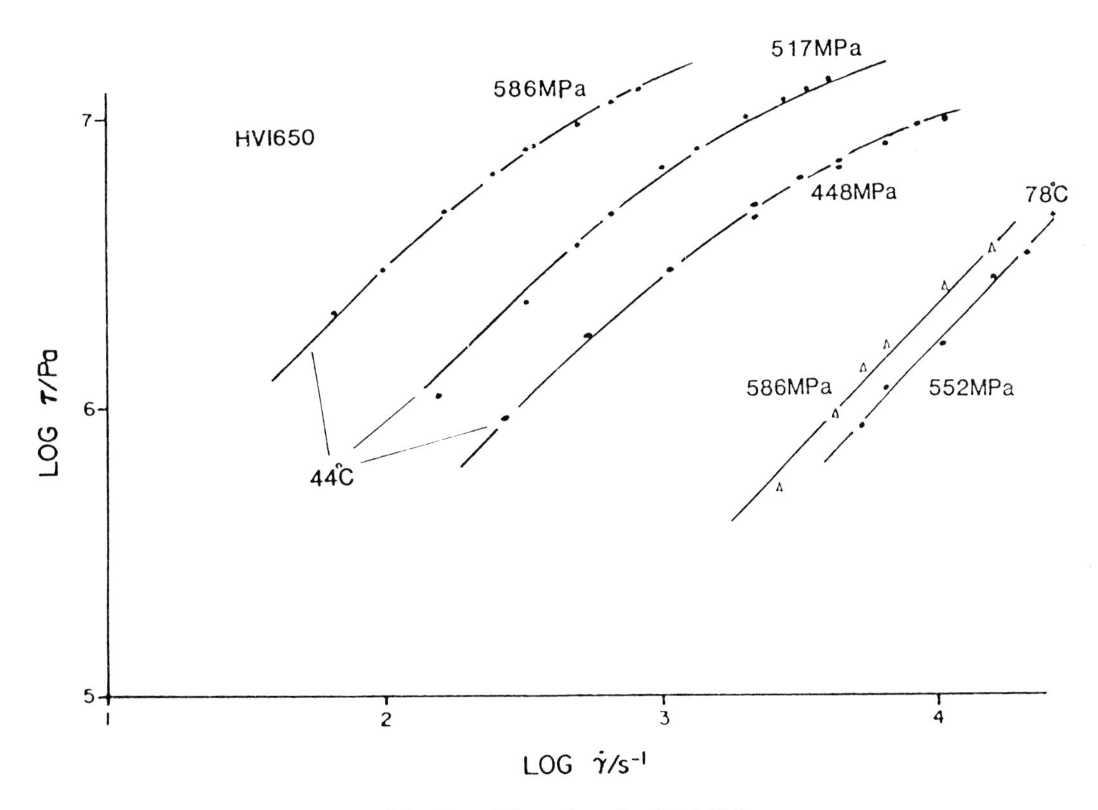

Fig. 7a Flow chart for HVI 650.

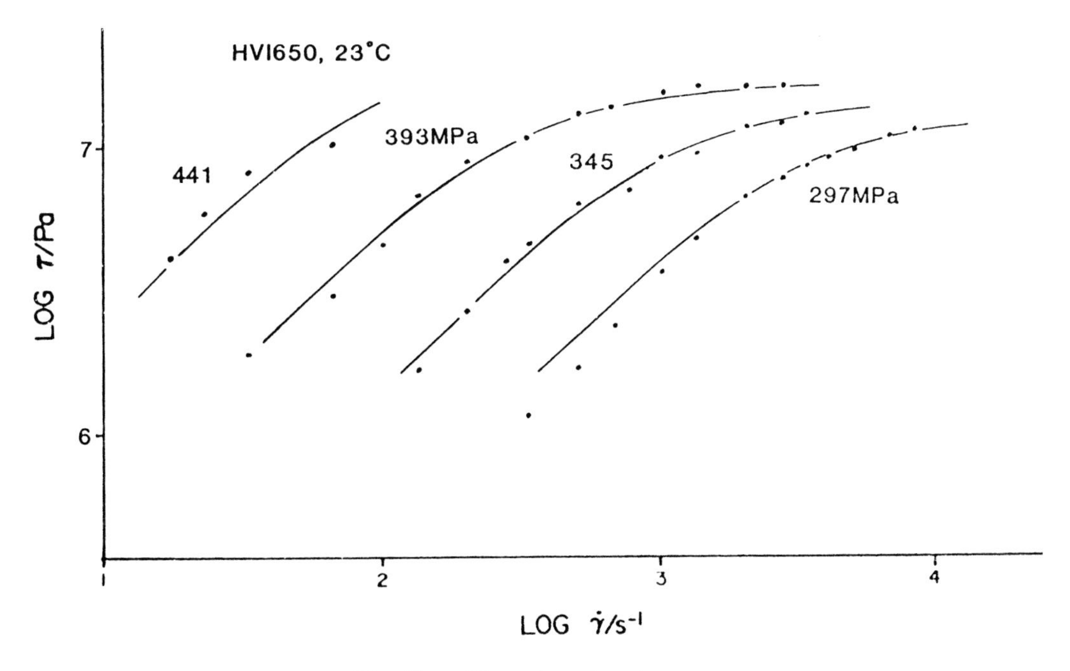

Fig. 7b Flow chart for HVI 650.

Here, a is a dimensionless parameter which controls the breadth of the transition from Newtonian to a non-Newtonian flow regime with rate sensitivity coefficient of m. For $m = 0$, τ_L is a limiting shear stress. In Figs. 7a and b, we used $a = 1.5$ at 23°C and $a = 2$ at 44°C. For the limiting stress, the form proposed by Bezot, *et al.* (**17**) was adopted:

$$\tau_L = c_0 + c_1 p + c_2 p^2 \tag{5}$$

Using the results of Fig. 7b and Ref. (**18**) and (**19**), we obtained $c_1 = 0.034$ and $c_2 = 2.1 \times 10^{-5}$ MPa with c_0 arbitrarily set to zero. Note that for HVI 650 at 44°C (Fig. 7a) the curves are also a good approximation of the Sinh Law,

$$\tau = \tau_0 \sinh^{-1}\left(\mu \dot{\gamma} / \tau_0\right) \tag{6}$$

of equation (6) together with exponential pressure-viscosity the traction gradient can be set equal to τ_0 (**18**):

$$\frac{\partial \overline{\tau}}{\partial \ell n \dot{\gamma}} = \tau_0 \tag{7}$$

Values of τ_0 obtained from traction tests using (7) are typically less than 6 MPa (**18, 20**) for cycloaliphatic traction fluids under conditions of Fig. 8. Clearly, from Fig. 8, the Newtonian limit exceeds 6 MPa and is close to 25 MPa. Therefore, any interpretation of the traction gradient as a Newtonian limit through a Sinh Law model with τ_0 independent of pressure must be suspect. However, if the pressure of a rheological measurement is chosen carefully it should be possible to obtain agreement between the Newtonian limit for that particular pressure and the

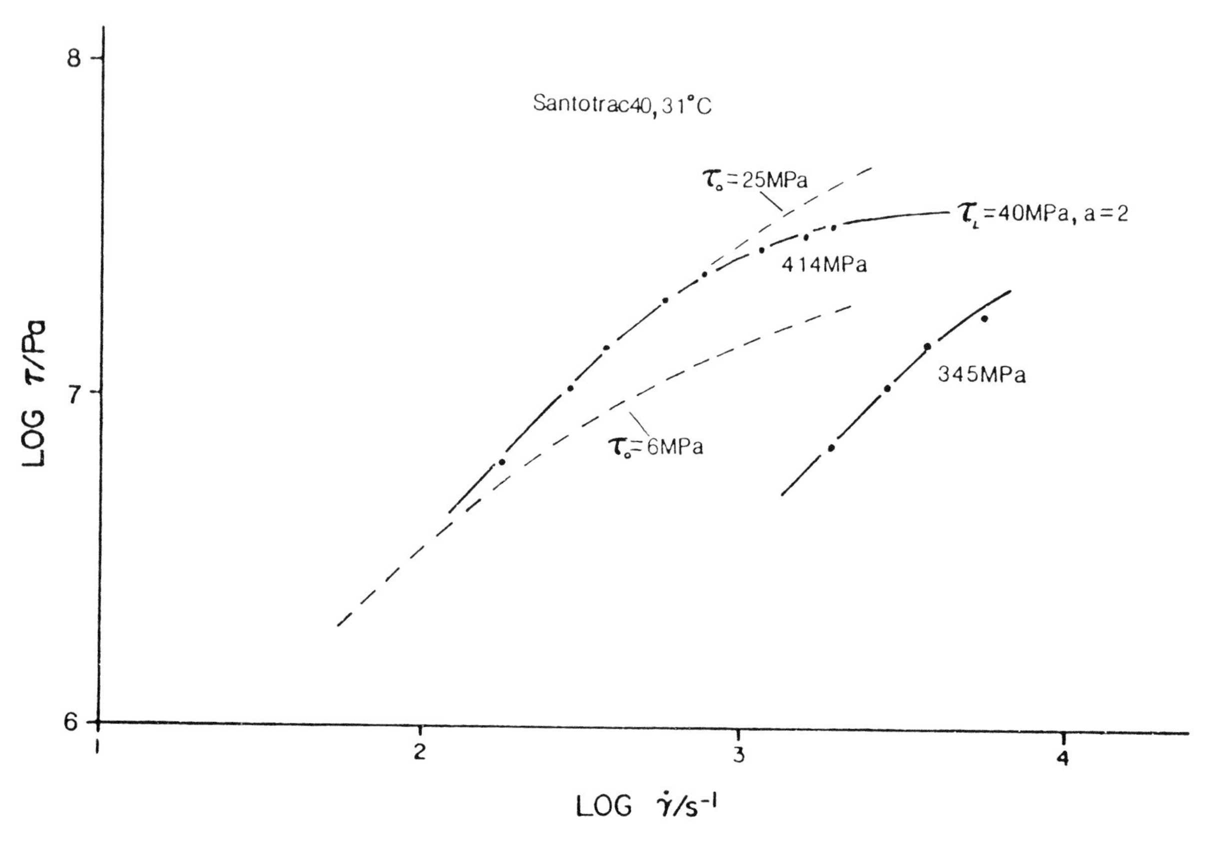

Fig. 8 Flow chart for Sanotrac 40.

where τ_0 is a stress which represents the limit of Newtonian response and is 3.3, 5.2 and 6.5 MPa for pressures of 448, 517 and 586 MPa respectively. That is, $\tau_0 \approx \tau_L/4$ at 44°C.

The shear response of the traction fluid, Santotrac 40, is shown in Fig. 8. For the pressure of 414 MPa, the solid curve is equation (4) and the two broken curves are equation (6) for τ_0 of 6 and 25 MPa respectively. When elastohydrodynamic traction curves are interpreted in terms traction gradient since the Newtonian limit is pressure dependent (e.g. HVI 650 at 448 MPa and 44°C).

We have investigated the grease from Ref (**16**) and its base oil in the current rheometer to explore departures from Newtonian behaviour at high shear stress. In Fig. 9, for room temperature, the mineral base oil, 600P, is non-Newtonian above $\tau = 2$ MPa at 310 MPa pressure. When the pressure is increased to 517 MPa this oil is Newtonian to τ = 3 MPa. The CA7000, which is a soap-thickened grease of 600P, is also shown in Fig. 9. It would appear from these limited data that the shear stress for the grease (at high stress) may be obtained by multiplying the base oil result by the previously (**16**) obtained viscosity enhancement factor, g, which is about 2. Clearly, equation (3) is not applicable here since it becomes approximately Newtonian at high shear stress. Other techniques have been developed to probe the response of liquids to high pressure and stress.

reasonable predictions of concentrated contact traction. Evans (**22**) generated isothermal, line-contact traction curves for three of the most widely investigated liquid lubricants: 5P4E, Santotrac 50 and HVI650. These data were described as isothermal since the disc temperature was adjusted to provide a constant estimated average film temperature. These traction data are presented for HVI650 in Fig. 10 for inlet temperatures of 40 and 60°C and average pressures, $\bar{p}$ of 0.47 and 0.63 GPa. The sliding velocity is ΔU and inlet temperature rise is 8°C.

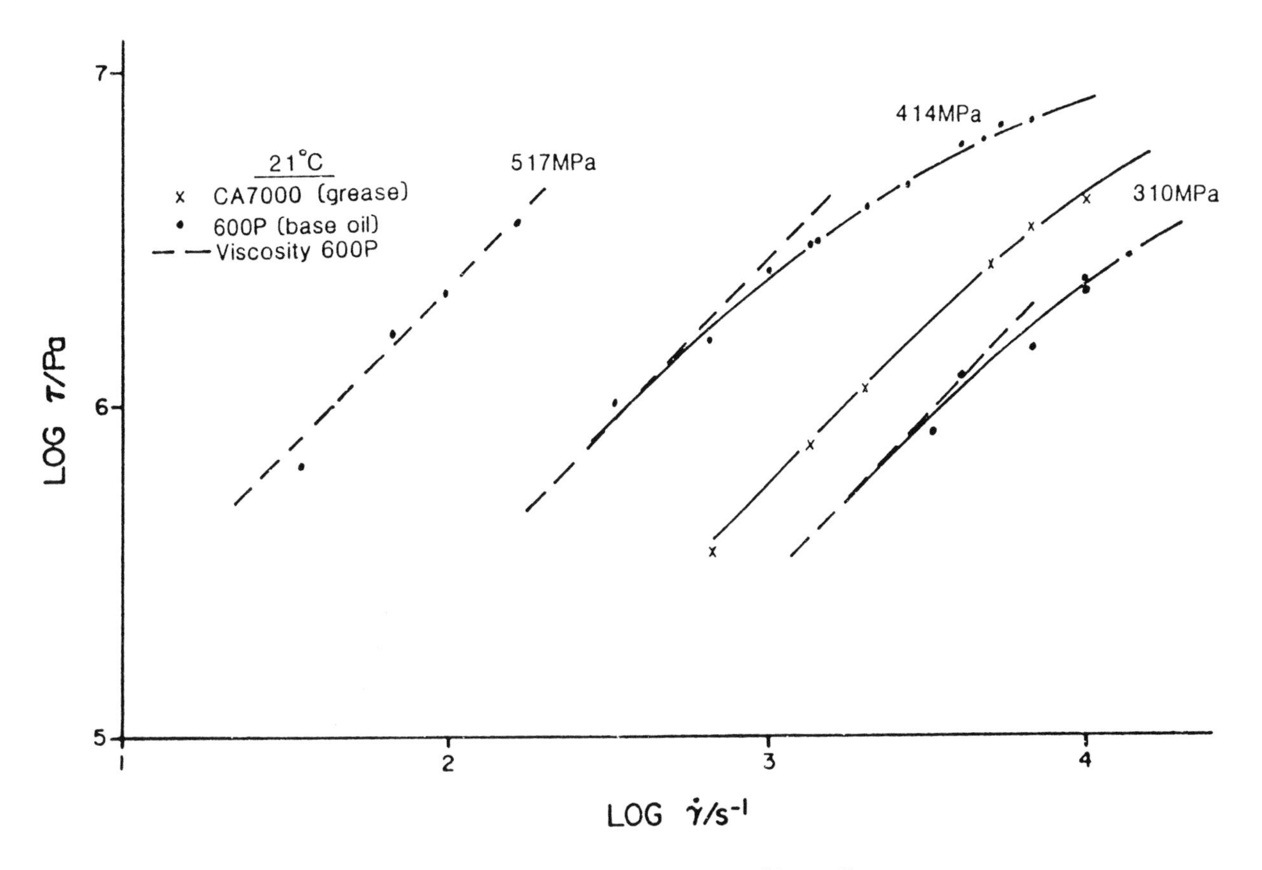

Fig. 9 Flow chart for grease and base oil.

Other laboratories have observed rate independence at elevated pressure. For example, the High-Pressure Impact Viscometer of Wong, *et al.* (**21**) entraps a quantity of liquid between a ball and a plate. Interferometry is used to determine the local flow rate of liquid leaking from the entrapment and local surface distortion which yield local pressure from elasticity theory. A Rabinowitsch Correction for slit flow gives the true shear rate at the surface. The authors concluded that the effective viscosity was found in earlier works to be a function of time because the shear rate varied with time while the shear stress remained at the limiting value. Limiting stress type behaviour was observed over 4 decades of shear rate with $m \approx 0.01$ for LVI 260.

3 EHD TRACTION CALCULATION

We should expect our property measurements to yield

For the traction calculations which are the curves in Fig. 10, we assumed the Hertzian pressure distribution and integrated the Carreau-Yasuda equation (4) across the contact area to obtain the average shear stress, $\bar{\tau}$. The dimensionless parameter, a, was specified by $a = 63/(T+19^{\circ}\mathrm{C})$ which yields the values obtained in the previous section without approaching the meaningless condition of $a = 0$ at ordinary temperatures. Viscosity was obtained from a Free Volume Model (**23**) and limiting stress was obtained from equation (5). Results were insensitive to the selection of the rate sensitivity, m, from 0 to 0.03. Although the predicted curve at 40°C rolls over more quickly than the measured traction, the general agreement is good. Considering the great differences between the two techniques and the assumptions involved (e.g. dry Hertzian pressure distribution) it might be unreasonable to expect to do better.

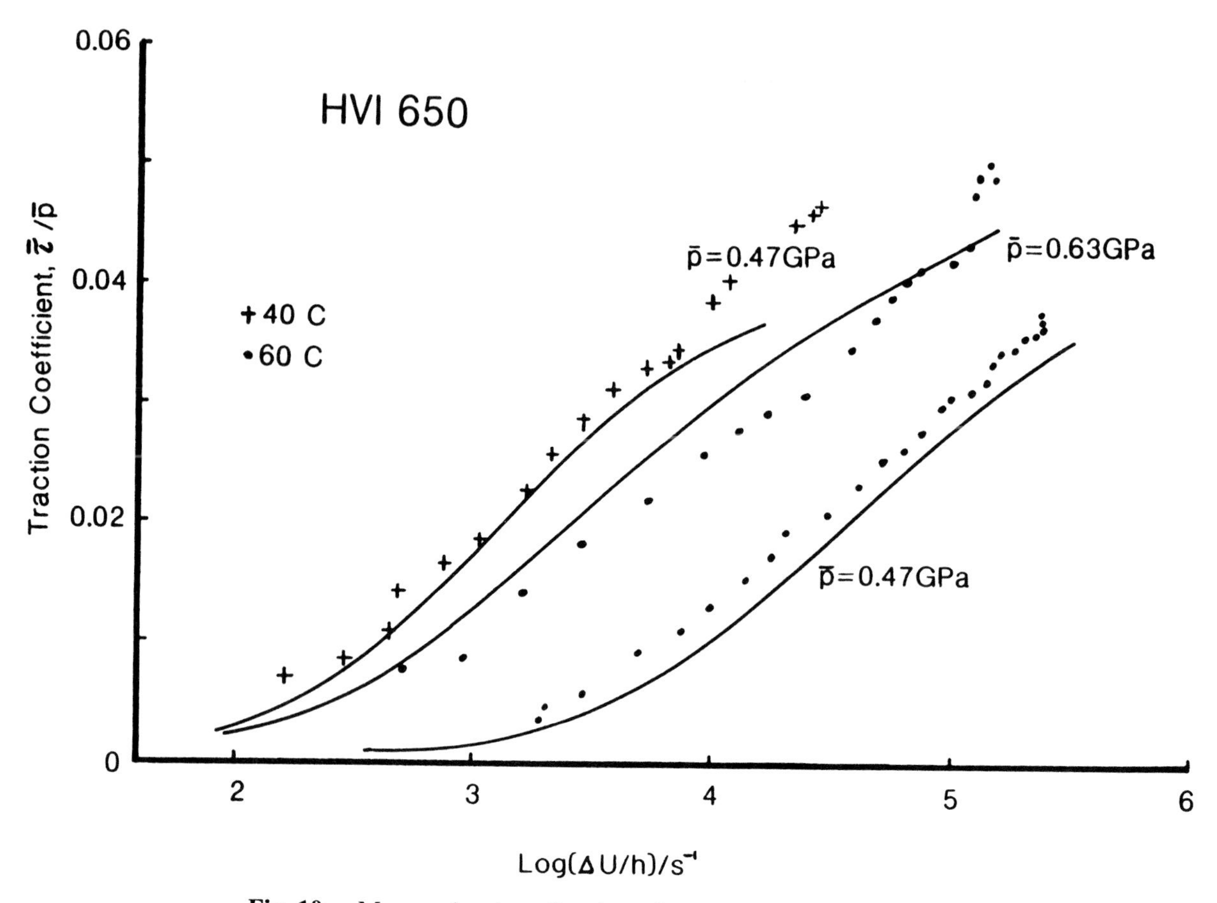

Fig. 10 Measured and predicted traction curves for HVI 650.

4 MECHANICAL SHEAR BANDS IN NON-VISCOMETRIC FLOW

It has long been suspected that the rate independent behaviour observed and discussed above was related to slip; either at the boundaries, or internal in the material. Visualization of shear bands has to date been accomplished between parallel plates (**4**). The interpretation of the flow field and stress state leading up to localization is the least ambiguous for plane Couette shear. This is not however the general case for lubrication flows. To investigate the generation of shear bands in wedge flow we fabricated a new stationary shaft for the High-Pressure Flow Visualization Cell (**4**). The stationary surface (the lower surface of Fig. 11) was ground at an angle of 5° with the direction of motion of the moving (upper of Fig. 11) surface. Observation of birefringence during shear through crossed polarizers clearly showed a shear stress gradient across the film as expected. Shear bands appeared as shown in Fig. 11 for 5P4E. These bands are similar to those observed between parallel plates. However, the Mohr-Coulomb analysis can not easily be applied because of the non-uniform stress field in the film. In solid mechanics isothermal shear bands are often associated with a change in type of the governing partial differential equations. Singularities in the governing equations accompany this change of type. As shown below, singularities do occur in the Navier-Stokes equation with realistic lubricant properties.

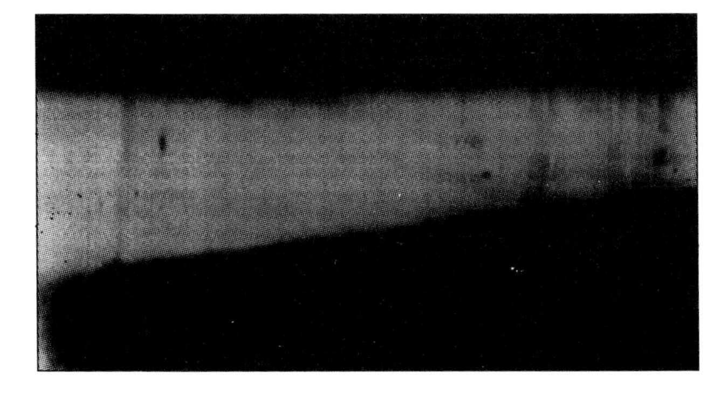

Fig. 11 Mechanical shear bands between converging plates with sliding. Lower tilted surface is stationary.

5 A FUNDAMENTAL LIMITATION OF THE REYNOLDS EQUATION FOR PIEZOVISCOUS LIQUIDS

We have just seen that shear bands will develop during non-viscometric shear of a lubricant at high pressure and high-shear stress. These visible bands effectively represent slip planes and are not predicted by current numerical analyses of lubricant behaviour.

The analysis of the pressure generated by lubricant films is almost exclusively performed with a form of the classical Reynolds equation. This differential equation derives from the inertialess form of the Navier-Stokes equation combined with the continuity equation with the assumption that the flow channel is small in one co-ordinate direction. It has been generalized to incorporate variable viscosity as well as variable density and has been remarkably accurate in predicting the film thickness in non-conformal contact problems. However, there is a fundamental limitation to the Reynolds equation for problems in which viscosity varies with pressure and in particular the regime of elastohydrodynamics. The cross-film pressure gradient cannot be neglected, and secondary flows result even for flow between parallel plates.

Renardy recognized (**8**) that the Navier-Stokes equations for an incompressible, piezoviscous fluid may suffer from non-existence and nonuniqueness problems when the principal tensorial strain rates are not less than $(2\mu\alpha)^{-1}$. Here, μ is viscosity and $\alpha = \mathrm{d}\ \ln \mu/\mathrm{d}p$ is the local pressure viscosity coefficient where p is the pressure. The Navier-Stokes equations can undergo a change of type – a process which has been used to characterize shear localization (**9**).

Recently, Bair and Khonsari (**24**) reported the occurrence of singular pressure gradients in two-dimensional, inhomogeneous flows of incompressible piezoviscous liquids. In the case of flow between parallel plates, these singularities were shown to take place when the shear stress τ approaches $1/\alpha$. This criterion is equivalent to that found by Renardy. Note that the one-dimensional Reynolds equation does not yield this singularity. In fact it predicts a trivial solution for the pressure if pressure boundary conditions are fixed at a constant value. We will show that the pressure dependence of viscosity leads to very large cross film pressure gradients and secondary flows.

5.1 Piezoviscous Reynolds equation

Consider the derivation of a generalized Reynolds equation by Dowson (**25**). The co-ordinate system is defined so that z is in the direction of the film thickness h. The x and y dimensions of the bearing are of the order of L which is several orders of magnitude greater than h. Following Dowson, the first stage of the order of magnitude analysis of the Navier-Stokes equations for a Newtonian liquid yields:

$$\frac{\partial p}{\partial x} = \frac{\partial}{\partial z}\left(\mu \frac{\partial u}{\partial z}\right) \tag{8}$$

and a similar equation for $\partial p/\partial y$.

For the cross-film direction, the appropriate equation is:

$$\frac{\partial p}{\partial z} = ... + \frac{\partial}{\partial x}\left(\mu \frac{\partial u}{\partial z}\right) \tag{9}$$

where we will assume that the terms represented by ... are of lesser magnitude than the term retained. In arriving at these equations it is apparently assumed that in general a function such as $\partial\varphi/\partial\xi$ is of the order of φ/ξ. Then if z is of the order of h_o, comparing only the right hand sides of (8) and (9) gives the result that $(\partial p/\partial z)/(\partial p/\partial x)$ is of the order of h_o/L. This result is usually used to justify the omission of $\partial p/\partial z$, but clearly the left-hand side gives the opposite result.

To illustrate the significance of the cross flow pressure gradient, we carry out the differentiation of the products in the right-hand sides of (8) and (9) as shown below:

$$\frac{\partial p}{\partial x} = \alpha\mu \frac{\partial p}{\partial x}\frac{\partial u}{\partial z} + \mu \frac{\partial^2 u}{\partial z^2} \tag{8a}$$

$$\frac{\partial p}{\partial z} = ... + \alpha\mu \frac{\partial p}{\partial x}\frac{\partial u}{\partial z} + \mu \frac{\partial^2 u}{\partial x \partial z} \tag{9a}$$

Begin with an inspection of (9a). We neglect the term involving cross derivatives since $\partial u/\partial x$ is expected to be small; however, our argument is not compromised if this term is included. Applying a similar order-of-magnitude rule as before shows that the ratio $(\partial p/\partial z)/(\partial p/\partial x)$ is of the order of $\alpha\mu U/h_o$. A conservative value of α is 10^{-8} Pa^{-1}. Taking the velocity in the x-direction, U, to be 1 m/s and h_o to be 10^{-7} m, we find that with a viscosity evaluated at low pressures, say $\mu = 10^{-2}$ Pa.s, the order of magnitude of $(\partial p/\partial z)/(\partial p/\partial x)$ is 10^{-3}. This is typical of hydrodynamic lubrication or EHL inlet zones and this order of magnitude analysis can be used to neglect the pressure gradient in the cross film direction. For pressures relevant to Hertzian zone, $\mu = 10^4$ Pa.s is not unusual and the order of magnitude of $(\partial p/\partial z)/(\partial p/\partial x)$ becomes 10^3. Clearly, therefore, at high pressures $\partial p/\partial z$ is not insignificant compared to $\partial p/\partial x$.

In deriving the classical Reynolds equation based on the reduced form of the Navier-Stokes equation, it is typical to define a dimensionless viscosity as $\bar{\mu} = \mu/\mu_o$ where $\bar{\mu}$ is assumed to be of the order of one. If μ_o is taken to be ambient viscosity, then $\bar{\mu}$ can be very large at high pressures. Even when the cross film gradient is significant, it must have an affect on the pressure profile calculation for us to question the use of Reynolds equation for elastohydrodynamic applications. Notice that in equation

(8a) the term $\partial p/\partial z$ appears explicitly and that it is coupled to $\partial p/\partial x$. This coupling can result in a singularity in the pressure gradient which the Reynolds equation is incapable of predicting.

5.2 Lubricant properties

Before developing some properties of the isothermal piezoviscous Navier-Stokes equations for the Hertzian zone, it is instructive to examine the viscosity and compressibility of liquid lubricants at very high pressure. Free volume models are useful for lubrication analyses because they provide a link between volume and viscosity. In the following example, we make use of the isothermal free volume model of Cook, *et al.* (**27**).

The Dolittle equation relates viscosity to volume, V, and occupied volume, V_{occ}:

$$\ln\mu = \frac{BV_{occ}}{V - V_{occ}} - \frac{BV_{occ}}{V_o - V_{occ}} + \ln\mu_o \tag{10}$$

The Tait equation provides an expression of the volume variation with pressure:

$$\frac{V}{V_o} = 1 - \frac{1}{K_o' + 1} \ln\left(1 + p\frac{(1 + K_o')}{K_o}\right) \tag{11}$$

At zero pressure the volume is V_o, the bulk modulus is K_o and the pressure rate of change of the bulk modulus is K'_o. The secant compressibility is then:

$$\Phi_s = -\frac{1}{V_o}\frac{dV}{dp} = \frac{1}{K_o + (1 + K_o')p} \tag{12}$$

It can be shown empirically that V_{occ} is nearly independent of pressure. Differentiating (10) yields:

$$\alpha = \frac{\beta \dfrac{V_{occ}}{V_o}}{\left(\dfrac{V}{V_o} - \dfrac{V_{occ}}{V_o}\right)^2}\Phi_s \tag{13}$$

Parameters typical of a mineral oil at a temperature of 75°C, are $B = 3$, $K_o = 1.5$ GPa, $K'_o = 10$, and $V_{occ}/V_o = 0.70$.

A plot of Φ_s and α as a function of pressure is shown in Fig. 12. First note the remarkable reduction of compressibility for increasing pressures. Although the existence of high pressure is often a justification for a compressible solution, if we restrict our attention to pressures greater than 1 GPa, an incompressible analysis is acceptable. Note also that following a slight decrease at

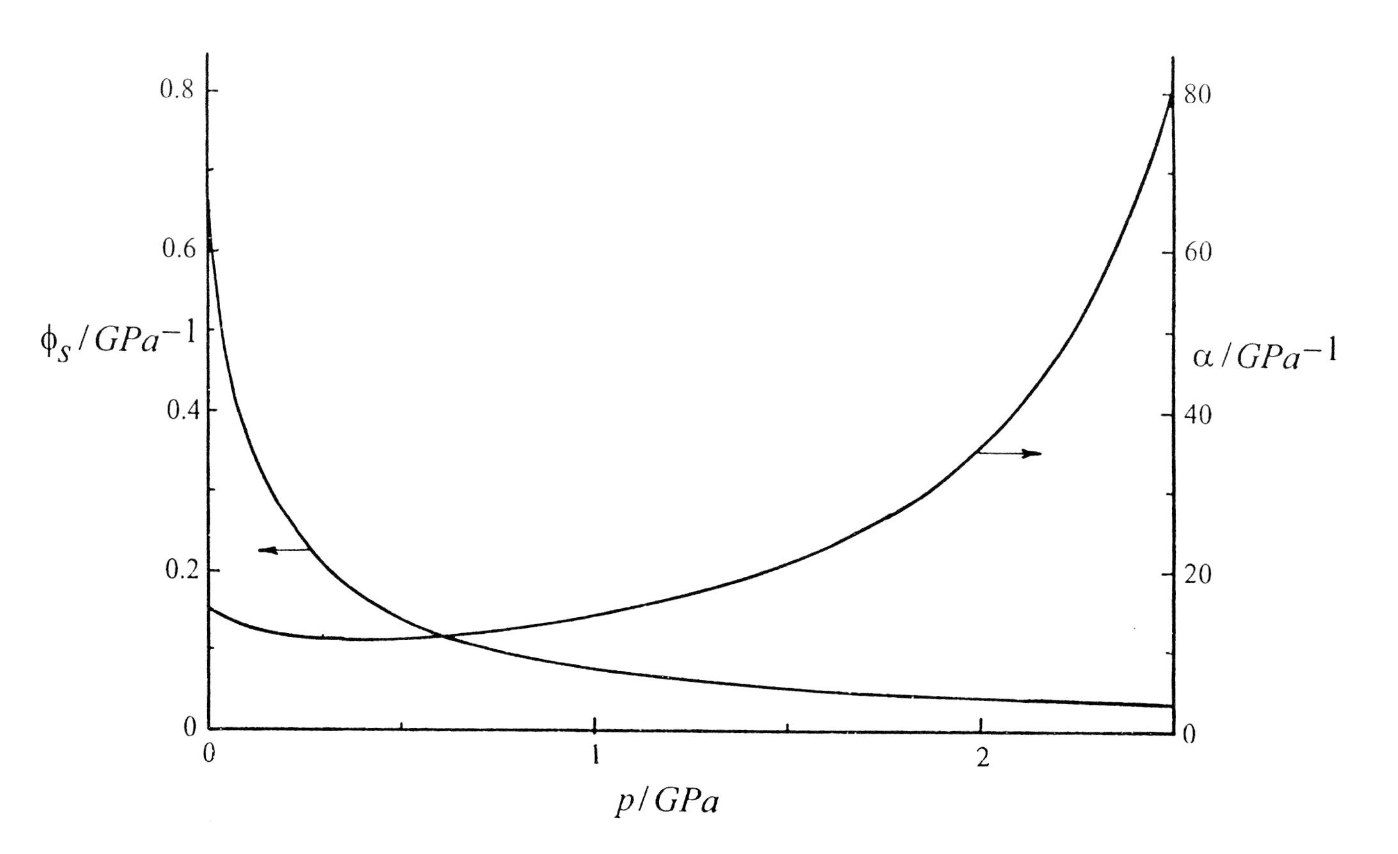

Fig. 12 The variation of compressibility and pressure-viscosity coefficient with pressure.

pyramid scratch test of ceramics. According to Evans and Marshall (**72**), the wear volume ΔV at one scratched groove is given by:

$$\Delta V = \eta \frac{W^{9/8}}{K_c^{1/2} H^{5/8}} \left(\frac{E}{H}\right)^{4/5} L \quad (16)$$

where K_c is the fracture toughness, L the sliding distance and η a material independent constant which should be determined by calibration on a material with well-characterized fracture properties. Eq. (16) can be arranged as follows:

$$\frac{\Delta V}{L} = \eta \left(\frac{E}{H}\right)^{4/5} \left(\frac{H}{K}\right)^{1/2} \left(\frac{W}{H}\right)^{1/8} \quad (17)$$

from which the Archard wear coefficient K is given by:

$$K = \eta \left(\frac{E}{H}\right)^{4/5} \left(\frac{H}{K_c}\right)^{1/2} \left(\frac{W}{H}\right)^{1/8} \quad (18)$$

It is clear from Eq. (18) that the task outstanding is to have a model for predicting η theoretically.

The exponents of E, H, K_c and W change, depending on the crack model, but they can be precisely determined. In the case of erosion of ceramics caused by the impact of hard angular particles, the wear mode of brittle fracture can be applied and the exponents of K_c and H are (-1.3) and (-0.25), respectively, for the quasi-static model, and (-1.3) and (0.1), respectively, for the dynamic wave propagation model, as shown by Hutchings (**73**). The difficulties remaining are always the same in these models of brittle fracture wear. Namely, it is to decide a so-called material-independent constant such as η in Eq. (17). This difficult problem is similar to that of deciding β in plastic grooving, and further exact analyses can be expected.

In addition to understanding the roles of some important parameters and material properties in each wear mode, mapping of all possible wear modes in one figure is a useful method for obtaining an overall view and for predicting the possible transition between two modes of wear. The idea of this mapping of wear modes was proposed in succession by Kato and Hokkiriga (**38**), Lim and Ashby (**60**) and Childs (**74**) for metals and, more recently, for ceramics by Hsu, *et al.* (**75**, **76**), and for polymers by Briscoe, *et al.* (**77**).

Some new types of theoretical maps have been introduced by Diao, *et al.* (**78**), as shown in Fig. 15, for sliding contact on a hard coating where the positions of local yield initiation are mapped in relation to the elastic moduli and yield strengths of the contacting materials, the friction coefficient, the load and the contact size. Such mapping activities will undoubtedly expand in the future and practically useful wear maps of various parameters will be developed.

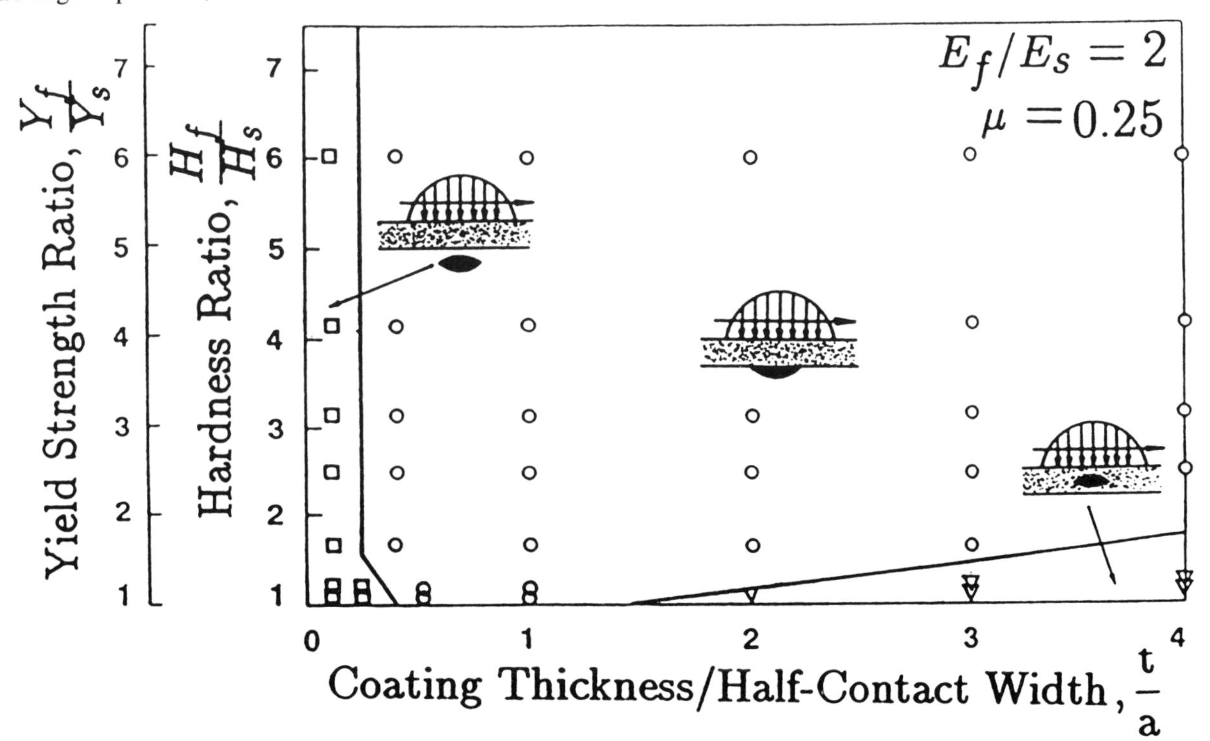

Fig. 15 Local yield map showing the yield initiation site in coating and substrate under the conditionof elastic contact of a cylinder on a flat coating surface.(t: coating thickness; a: half contact width; Y: yield strength; H: hardness, E: Young's Modulus; m: friction coefficient. Subscript f, s: film, substrate).

10 CONCLUSION

It is very clear that we are achieving a better understanding of the wear mechanisms of metals as a result of concerted efforts of tribologists world wide over the past forty years since the establishment of the four empirical wear laws.

Significant progress can be seen in the experimental and theoretical analyses of mechanical wear mechanisms. Theoretical wear coefficients have been introduced which can explain observed results to a certain extent.

Tribo-chemistry is being established and one can foresee the analyses of tribochemical wear mechanisms being developed into the 21st century. Materials science in conjunction with mechanical wear will become more positively associated with tribochemical wear. The foundations for carrying out such activities are being established alongside studies of mechanical wear.

The control of wear will become a significant requirement in industry, not only for saving materials and energy but also for developing innovative technologies and new products. Wear control will become a positive key technology, especially in some sectors of advanced technology.

11 ACKNOWLEDGMENTS

The author would like to express his most science appreciation to Dr W. H. Roberts and Prof. D. Dowson for their helpful comments and suggestions on this paper, and to Prof. K. L. Johnson and Prof. K.-H. Zum Gahr for their theoretical discussions.

The author would like to express his sincere appreciation to Profs. M. Koyanagi and K. Shyoji of Tohoku University, Prof. S. Kaneta of Kyushu Technical Institute, Mr K. Ando of Nippon Steel Co., Mr Y. Kagimoto of Mitsubishi Heavy Industry, Dr. M. Mizumoto of Hitachi Co., Mr Y. Nagasawa of Toyota Central Research Laboratory and Dr. H. Osaki of Sony Co., for their helpful discussions on their expert area.

The author's special thanks go to Mr K. Adachi of Tohoku University for his help in the preparation of material for this paper.

12 REFERENCES

1 *Proceedings of 1st International Conference on Chemical-Mechanical Polish* (CMP) for VLSI/ULSI Multi-level Interconnection, (CMP-MIC), 1996, Santa Clara, USA.

2 **Matsumoto, T., Kudoh, Y., Tahara, M., Yu, K.-H., Miyakawa, N., Itani, H., Ichikizaki, T., Fujiwara, A.,. Tsukamoto, H.,** and **Koyanagi.,** Three-dimentional integration technology based on a wafer bonding technique using micro-bumps, *Int. Conf. Solid State Devices and Materials*, 1995, Osaka, Japan.

3 *Proceedings of 2nd International Conference on Machining of Materials (MAM),* 1996, Achen, Germany.

4 **Umehara, N.** and **Kato, K.** A study of magnetic fluid grinding – 1st report : the effect of the floating pad on removal rate of Si_3N_4 balls, *Transactions of the Japan Soc. Mech. Eng.,* 1988, **54**, 503, 1599–1604.

5 **Marinescu, I. D.,** Laser-assisted grinding of ceramics, *Proc. 2nd Int. Conf. Mach. Adv. Mats.*, Achen, 1996, 297–303.

6 **Bhushan, B.,** *Tribology and mechanics of magnetic storage devices*, Second Ed., Springer, 1996.

7 *Proceedings of the ASME/STLE Tribology Conf. on Tribology of Contact/Near-Contact Recording for Ultra-High Density Magnetic Storage,* 1996 San Francisco, USA.

8 **Talke, F., Tsoi, H.,** and **Bogy, D.B.,** Critical review characterization of diamond-like carbon films and their application as overcoats on thin film media for magnetic recording, *J. Vac. Sci. Technology*, 1987, **A5**, 3287–3312.

9 **Khurushudov, A., Kato K.,** and **Sawada, D.,** Micro-tribological characterization of carbon nitride coatings, *Proc. Int. Tribo. Conference,* Yokohama, 1995, 1931–1936.

10 **Mizoh, Y.,. Yohda, H., Kita, H.,** and **Kotera, H.,** Simulation of head wear and reproduced envelope by the finite element method, *Electron Commun. Japan,* 1992, **75,** 91–101.

11 **Mizoh, Y.,** Wear of tribo-elements of video tape recorders, *Wear,* 1996, **200**, 252–264.

12 **Iwasaki, S.** and **Nakamura**, Y., An analysis of the magnetization made for high density magnetic recording, *IEEE Trans. Magn.*, 1997, 1272–1277.

13 **Osaki, H., Fukushi, K.,** and **K. Ozawa**, Wear mechanisms of metal-evaporated magnetic tapes in helical scan videotape recorders, *IEEE Trans. Magn.*, 1990, **26**, 3180–3185.

14 **Osaki, H.,** Tribology of videotapes, *Wear*, 1996, **200**, 244–251.

15 **Adachi, K., Kato K.,** and **Sasatani, Y.,** The micro-mechanism of friction drive with ultrasonic wave, *Proc. Int. Tribology Conference-AUSTRIB '94*, 1994, Australia, 235–239.

16 **Kaneko, T., Masuda H.,** and **Takezawa, K.,** *Kawasaki Steel Giho,* 1996, **28**, 108–113.

17 **Kojima, K.** and **Kamiya, S.,** Ceramics for automobiles, *Ceramics*, 1993, **28**, 1007–1021.

18 **Ishibashi, S., Yamashita K.,** and **Yonei, A.,** Application of ceramics for sliding machine parts, *Journal of JST*, 1991, **36**, 144–157.

19 **Bayer, R. G.,** *Mechanical wear prediction and prevention,* Marcel Dekker, Inc., New York, 1994, 280–291.

20 **Heiricke, G.,** *Tribochemistry,* Karl Hanser Verlag Munchen Wien, 1984.

21 **Lundberg G.** and **Palmgren, A.,** Dynamic Capacity of Rolling Bearings, *Ingeniorsveten-skapsakademiens*, 1947, Nr. 96.

22 **Lundberg G.** and **Palmgren, A.,** Dynamic Capacity of Roller Bearings, *Ingenicorsrenten-skapsakademiens*, 1952 **Nr. 210.**

23 **Weibull, W.,** A statistical theory of the strength of materials, IVA Handlinger, Royal Swedish Academy of Engineering Sciences, Proceedings 1930, **Nr. 151.**

24 **Holm, R.,** *Electric Contacts*, Almqvist and Wiksells, Stockholm, 1946.

25 **Burwell J.T.** and Strang, C. D., On the empirical law of adhesive wear, *J. Appl. Phys.* 1952, **23**, 18–28.

26 **Archard, J.F.,** Contact and rubbing of flat surfaces, *J. Appl. Phys.,* 1953, 24, 981–988.

27 **Kruschov, M.M.**, Resistance of metals to wear by abrasion, as related to hardness, *Proc. Conf. Lubrication and Wear, Inst. Mech. Eng.*, 1957, 655–659.
28 **Uhlig, H.H.**, Mechanism of fretting corrosion, *J. Appl. Mech.*, 1954, **21**, 401.
29 **Kayaba T.** and **Kato, K.**, The adhesive transfer of the slip-tongue and the wedge, *ASLE Trans.* 1981, **24**, 164–174.
30 **Vingsbo O.** and **Hogmark, S.**, *Wear of steels, Fundamentals of Friction and Wear of Materials*, D. A. Rigney Ed., ASM, 1980, 373–408.
31 **Chen L. H.** and **Rigney, D. A.**, Transfer during unlubricated sliding wear of selected metal systems, *Wear,* 1985, **105** 47-61.
32 **Cocks, M.**, Interaction of sliding metal, *J. Appl. Phys.*, 1962, **33**, 2152.
33 **Chiou Y. C.** and **Kato, K.**, Wear mode of micro-cutting in dry sliding friction between steel pairs (Part1), *T. JSLE,* 1987, **32**, 41–48.
34 **Zum Gahr, K.H.** *Microstructure and wear of materials*, Tribology series Elsevier, 1987, 132–148.
35 **Kayaba, T., Kato K.**, and **Hokkirigawa, K.**, Theoretical analysis of the plastic yielding of a hard asperity sliding on a soft flat surface, *Wear,* 1983, **87**, 151–161.
36 **Hokkirigawa K.**, and **Kato, K.**, An experimental and theoretical investigation of ploughing, cutting and wedge formation during abrasive wear, *Tribology International*, 1988, **21**, 151–57.
37 **Challen J. M.**, and **Oxley, P.L.B.**, An explanation of the different regimes of friction and wear using asperity deformation models, *Wear,* 1979, **53**, 229–243.
38 **Kato K.**, and **Hokkirigawa, K.**, Abrasive wear diagram, *Proc. Eurotrib'85,* 1985, Elsevier, Amsterdam, Vol.4–5
39 **Hokkirigawa K.**, and **Kato, K.**, Theoretical estimation of abrasive wear resistance based on microscopic wear mechanism, *Proc.Wear of Materials,* ASME, 1989, 1–8.
40 **Challen, J.M., Oxley P.L.B., and Hockenhull, B.S.**, Prediction of Archard's wear coefficient for metallic sliding friction assuming a low cycle fatigue wear mechanism, *Wear,* 1986, **111**, 275–288.
41 **Johnson, K. L.**, Contact mechanics and the wear of metals, *Wear,* 1995, **190**, 162–170.
42 **Kapoor A.**, and **Johnson, K. L.**, Plastic ratchetting as a mechanism of metallic wear, *Proc. Roy. Soc.* 1994, **A445**, 367–381.
43 **Black, A. J., Kopalinsky F. M.**, and **Oxley, P. L. B.**, Sliding metallic wear test with in-process wear measurement: a new approach to collecting and applying wear data, *Wear*, 1996, **200**, 30–37.
44 **Kapoor, A., Williams, J. A.**, and **Johnson, K. L.**, The steady state sliding of rough surfaces, *Wear,* 1995, **175**, 81–92.
45 **Akagaki T.**, and **Kato, K.**, Plastic flow process of surface layer in flow wear under boundary lubricated conditions, *Wear,* 1987, **117**, 179–186.
46 **Kapoor, A., Johnson K. L.**, and **Williams, J. A.**, A model for the mild ratchetting wear of metals, *Wear*, 1996, **200**, 38–44.
47 **Quinn, T. F. J.**, The effect of hot-spot temperatures on the unlubricated wear of steel, *ASLE Trans.*, 1967, **10**, 158–168.
48 **Fischer T. E.**, and **Tomizawa, H.** Interaction of tribo-chemistry and micro-fracture in the friction and wear of silicon nitride, *Proc. Int. Conf. Wear of Materials,* ASME, Vancouver, Canada, 1985, 22–32.
49 **Akazawa, M., Kato K.**, and **Umeya, K.**, Wear properties of silicon nitride in rolling contact, *Proc. JSLE Int. Tribo. Conf.*, Tokyo, Japan,1985, 191–196.
50 **Gee, M. G.**, The formation of aluminum hydroxide in the sliding wear of alumina, *Wear*, 1992, **153**, 201–227.
51 **Gates, R. S., Hsu S. M.**, and **Klaus, E. E.**, Tribochemical mechanism of alumina with water, *Tribology Trans. STLE,* 1989, **32**, 357–363.
52 **Quinn, T. F. J.**, *Proc. Int. Conf. on Tribology-Friction, Wear and Lubrication,* Inst. Mech. Engrs. Conf. Series 1987–5, 253–259.
53 **Krause H. and Scholter, J.** Wear of titanium and titanium alloys under conditions of rolling stress, *J. Lub. Tech.*, ASME, 1978, **100**, 199–207.
54 **Bisson, E. E.,. Johnson R. L.**, and **Swirkert. M. A.**, Friction, wear and surface damage of metals as affected by solid surface films; A review of N. A. C. A. Research, *Proc. Conf. Lub. and Wear,* 1957, Inst. Mech. Engrs. Paper No. 31, 384–391.
55 **Uetz, H., Khosrawi M. A.**, and **Fohl, J.**, Mechanism of reaction layer formation in boundary lubrication, Wear, 1984, **100**, 301–313.
56 **Perez-Unzueta A. T.**, and **Beynon, J. H.**, Effects of surrounding atmosphere on wear of sintered alumina, *Wear,* 1991, 146, 179–196.
57 **Uetz, H.** Einfluβ der luftfeuchtigkeit auf den gleitverschleiβ metallischer werkstoffe, *Werkstoffe und Korrosion*, 1968, **19**, 665–676.
58 **Mishina, H.**, Atmospheric characteristics in friction and wear of materials, *Wear*, 1992, **152**, 99–110.
59 **Iwabuchi, A.**, Fretting wear of inconel 625 at high temperature and in high vacuum, Wear, 1985, **106**, 163–175.
60 **Lim S. C.**, and **Ashby, M. F.**, Wear-mechanism maps, *Acta Metall.*, 1987, **35**, 1–24.
61 **Usui, E., Shirakashi T.**, and **Kitagawa, T.**, Analytical predication of cutting tool wear, *Wear,* 1994, **100**, 129–151.
62 **Rigney D. A.**, and **Gealser, W. A.**, The significance of near surface microstructure in the wear process, *Wear of Materials,* ASME, 1977, 41–46.
63 **Ives, L. K.**, Microstructural changes in copper due to abrasive, dry and lubricated wear, *Wear of Materials,* ASME, 1979, 246–256.
64 **Heilmann, P., Clark, W. A. T.**, and **Rigney, D. A.** Orientation determination of subsurface cells generated by sliding, *Acta Metall.*, 1983, **31**, 1293–1305.
65 **Kuhlman-Wilsdorf, D.**, What role for contact spots and dislocations in friction and wear?, *Wear,* 1996, **200**, 8–29.
66 **Nakajima K.**, and **Mizutani, Y.**, Structural change of the surface layer of low carbon steel due to abrading, *Wear*, 1969, **13**, 283–292.
67 **Garbar I. I.**, and **Skorinin, J. V.**, Metal surface layer structure formation under sliding friction, *Wear,* 1978, **51**, 327–336.
68 **Kato, K., Kayaba T.**, and **Ono, Y.**, Dislocation density and cell structure produced in the subsurface layer of aluminium during sliding wear, *Wear of Materials*, 1985, ASME, 463–470.
69 **Sasada, T.**, Formation of transfer particle by asperity rupture and aggregation, Tribology in the 80s, *NASA Conf. Pub. 2300*,. 1984, **I**, 197–218.

70 Heilmann, P., Don, J., Sun T. C., and **Rigney, D. A.,** Sliding wear and transfer, Wear, 1983, 91, 171–190.

71 Keer L. M., and **Bryant, M. D.,** A pitting model for rolling contact fatigue, Trans. ASME, *J. Lub. Tech.,* 1983, **105,** 198–206.

72 Evans A. G., and **Marshall, D. B.,** *Wear mechanisms in ceramics, Fundamentals of Friction and Wear of Materials,* ASM, D. Rigney Ed., 1980, 439–452.

73 Hutchings, I. M. *Tribology,* Edward Arnold, London, 1992, 182–186.

74 Childs, T. H. C., The mapping of metallic sliding wear, *Proc. Inst. Mech. Engr.,* 1988, *Part C,* **202**, 397.

75 Hsu, S. M., Lim, D. S., Wang Y. S,. and **Munro, R. G.,** Ceramic wear maps: alumina under various conditions, *Lub. Eng.,* 1991, **47,** 63.

76 Hsu S. M., and **Shen, M. C.,** Ceramic wear maps, *Wear,* 1996, **200,** 154–175.

77 Briscoe, B. J., Evans, P. D., Pelollo E., and **Sinha, S. K.,** Scratching maps for polymers, *Wear,* 1996, **200,**137–147.

78 Diao, D. F., Kato K., and **Hayashi, K.,** The local yield map of hard coatings under sliding contact, *Thin Films in Tribology,* D. Dowson, *et al.* Editors, Elsevier Science Publishers B. V., 1993, 419–427.

Tribology and the needs of industry

M J NEALE
Neale Consulting Engineers, Farnham, Surrey, UK

1 THE IDENTIFICATION OF TRIBOLOGY

The current needs of mankind for any product or service all depend on the availability and reliability of a wide range of machines, from textile mills and agricultural tractors to aircraft, ships and railway rolling stock. All these machines contain moving parts and as a result have components which move relative to each other. The relative movements between these components are areas of sensitivity because speed of movement is often high, together with the load. A lot of energy is therefore being transmitted across these concentrated areas and if anything goes wrong the consequences are generally severe. Tribology involves an understanding of what happens at these points of relative movement, and it is therefore of vital importance in machine design and reliability, and therefore also across the whole field of industrial activity.

This situation was recognized in the Jost study in 1966 (**1**), which also put a cost figure to the related unreliability, and thus gave the issue a political and industrial management focus. The figure for the potential savings that industry could obtain from better tribology were estimated by Jost as 1.5% of GNP for the UK. Similar studies have been carried out for West Germany (**2**), Canada (**3**), China (**4**) and the USA (**5**) and similar figures for savings in the range 1.3 to 1.6% of GNP were suggested.

A further and important contribution by the Jost study, was the recognition that many different components had similar underlying operating mechanisms and also that the technology involved, spanned a range of academic disciplines and was not therefore generally taught in a cohesive way. There was therefore considerable scope for the dissemination of better information on component design, and opportunities for relevant and useful research and teaching.

2 SUBSEQUENT DEVELOPMENTS

In the years after the Jost study up to 1997 much has been achieved, but unfortunately from an industrial point of view, the output has been of limited use. In fact in 1976 a review of progress over the first ten years showed that savings of only about 0.15% of GNP had been achieved (**6**). This was probably partly caused by the entropy of the situation, in that most of the problems were dispersed across a matrix of industrial machines. There have, however, been some major advances, such as in the understanding of the nature of the elastohydrodynamic lubrication between concentrated contacts, as in rolling bearings and gears, and in the understanding of the operation of piston rings and mechanical seals. Useful design data has also been published on the design of hydrodynamically lubricated plain bearings and rubbing bearings (**7**) and also on rotordynamics (**8**). The general subject of tribology has however moved more towards fundamental research and our library shelves are becoming filled with reports and theses on the minutae of what goes on at relatively moving surfaces. Academic members of the tribological community will probably see this a great success, particularly in terms of the number of references that become available. Unfortunately industry does not see it the same way, because the results of these detailed studies are rarely usable by themselves, for reasons which will emerge later in this paper. There are however some constructive ways forward from this position and useful guidance can be obtained by looking at some of the highlights of the development of a practical understanding of tribology, which have taken place over the last one hundred years.

3 USEFUL TECHNIQUES FROM THE PAST

At the end of the 19th century and for the first half of the 20th, the Institution of Mechanical Engineers in London had groups of their members identifying industrial problems and technological opportunities, and then taking action to achieve an appropriate outcome. This activity ceased in the 1950s possibly because the National Research Councils were becoming more established and were seen, probably in error, to be taking on this role. Also the

Institution itself then established a number of groups covering specialist subject areas and these were expected to lead the technology forward. The groups however were defined in terms of academic disciplines and from an industrial point of view progress was limited until industrial divisions were established in 1980.

A typical successful outcome of the earlier activity of the IMechE in the 1880s, was the recognition that there were problems with axle bearings on railway rolling stock, with the recommendation that some work should be done to improve the situation. Mr Beauchamp Tower was therefore given the task of learning more about how these plain oil lubricated axle bearings actually worked. Significantly, the way he did this was to take an actual railway axle bearing and run it to see what happened. He was specifically not carrying out a controlled laboratory experiment to check certain parameters in Reynolds equation. In fact, of course, what Beauchamp Tower discovered (**9**) about oil pressures in the axle bearing, caused Professor Reynolds to come up afterwards with a theoretical explanation and publish his equation (**10**). The pattern here was that a real practical problem or opportunity led the technological development and the pure research activity followed, to give a superb and simple explanation of the fundamental principles that were involved, in solving this essentially practical problem.

4 RECENT TRENDS

In more recent times investigations into Tribology have been led and conducted within the more rigorous discipline of a scientifically correct academic approach. Consequently, the investigations have been focused on narrower areas of the technology, starting in each case with a literature survey, then a theoretical study, followed by the construction of a laboratory apparatus using simplified specimens and specifically planned so that only one variable is investigated at a time. This then proves or modifies the theory. The results are then neatly written up, placed in the library and the participant leaves, appropriately doctored, to make his further contributions to the world. This procedure is all scientifically and academically correct, but from an industrial perspective it doesn't achieve very much, other than to give the participant some useful training, which might be usable in an industrial appointment.

An academic assessment of this position might be that there really isn't a problem. Anyone from industry can look up the appropriate subject reference and borrow the paper from the library. In practice it isn't like that. The engineer in industry is concerned with real components, machines and machine systems and he will not be motivated by being given a pure explanation of what might happen in the third bearing from the left, if two high spots on some perfectly defined surfaces happened to interact.

5 THE NATURE OF THE CURRENT PROBLEM

The basic principle that underlies the current problem is that if an attempt is made to produce a design method or a summary of the current technology of a tribological component, and the process of doing this is analysed, some interesting results emerge. An analysis of this process shows that even for an apparently simple component such as an oil lubricated cylindrical journal bearing, about 50-100 knowledge elements require to be combined (**11**).

An example of a practically based knowledge element might be the allowable minimum film thickness for whitemetal lined bearings related to their diameter and rotational speed. A research-based knowledge element might be an analysis of oil film extent indicating the boundaries of the full width oil film, at various eccentricities and oil feed rates. Actual examples where these knowledge elements have been identified and then combined into a design method for tribological components is shown in Fig. 1.

The design of one basic machine component is only a part of the whole machine design and an even smaller part of the design of the whole machine system. It is therefore to be expected that any individual contribution of a knowledge element, from a research programme, can be seen as infinitesimal when looked at from an industrial perspective. This is perhaps the most pessimistic view of the position and in practice with liaison between a university department and industry, the professor in charge of the research activity can develop an insight into the kind of research subjects that are likely to be particularly worthwhile in terms of their possible influence on future design and practice. There is however a problem, that the process is informal and possibly derives from contact with only one manufacturer of a particular type of machine.

6 A POSSIBLE WAY FORWARD

The points that have been discussed so far in this paper do however suggest that there may be some very positive routes forward, by which we can all learn from the past and link the academic and industrial communities to achieve some real success.

Recently in the UK there has been a government-led initiative called Technology Foresight in which experienced people are brought together to look at various industrial sectors and then suggest ways in which the technology of these sectors is likely to develop. It is, perhaps, not surprising that the people involved tend to try and look well ahead and imagine new technologies, that might revolutionize the particular sector of industry. This can be very exciting, but, perhaps, is not too realistic. There have been some major developments of this type in the past such as solid state electronics and the internal combustion engine. However on more realistic and effective timescales, over 95% of all new industrial

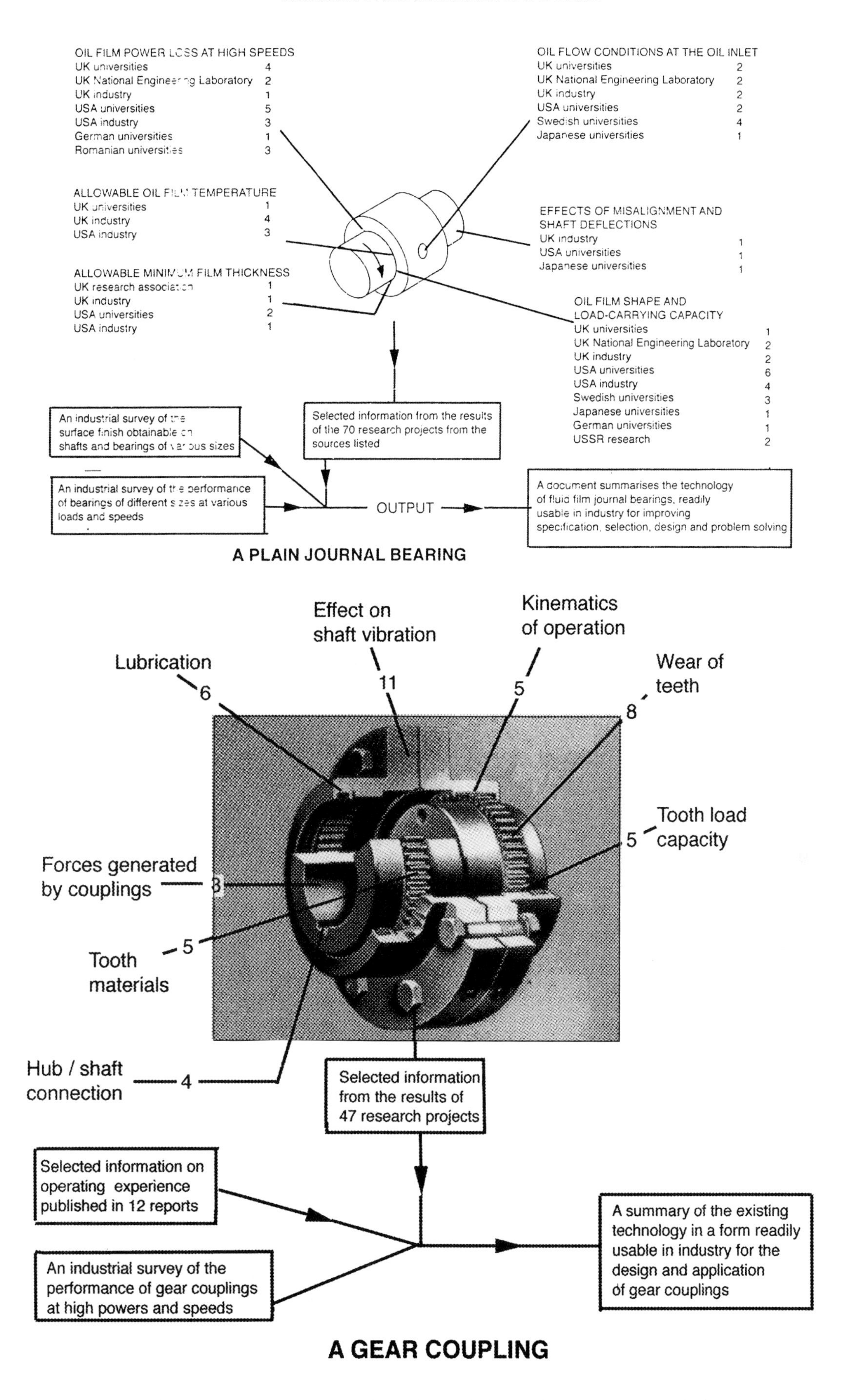

Fig. 1 The knowledge elements needed to define a component design.

developments have arisen from improvements to existing products, processes and technology.

A similar but probably more effective way forward therefore, would be to start by identifying industrial sectors, where the market position provides scope for expansion. The technology of these industries then needs to be examined in detail to identify areas where there are particular problems and/or opportunities, which could provide scope for a major improvement in market strength. This activity is close to that which was being conducted by the Institution of Mechanical Engineers in the late 19th century, and this is also where there is an opportunity for the next innovative step.

7 RESEARCH FROM ACTUAL SERVICE EXPERIENCE

The academic tradition is strongly directed by the application of scientific method to the advancement of knowledge. In this situation most experiments tend to be conducted under well-controlled and predetermined conditions.

Fig. 2 The relative performance of digger teeth of different materials after handling18,000 tonnes of gravel.

This does however restrict the subject areas that can be examined, narrows the field of the investigation, and misses many opportunities. In tribology, for example, research on wear has made limited progress and one reason is that it is very difficult to simulate the conditions that give rise to it in a laboratory. If abrasive wear from sliding contact with sharp particulate materials is to be studied, it is of little use to move specimens of components through a sample of the abrasive material because this rapidly becomes blunted. A new approach to this was pioneered by Eyre (**12**) who took an actual machine handling abrasive material in a real service environment and used it for comparative wear tests. The machine was a digger fitted with teeth and, for the experiments, teeth of various materials were fitted to it, so that their relative and simultaneous wear could be measured when digging various materials from which a sample was taken for reference and analysis. This produced some useful results as shown in Fig. 2.

An extension of this idea is the recognition that there are thousands of broadly similar machines and components of various kinds operating around us, and each is acting as an uncontrolled wear experiment, but producing results. The challenge therefore is to take these results and devise a means of analysing them from their multi-variable environment, in a way which gives insight into what is happening, and produces viable results.

An example of where this has been done is in the study of the wear of the cylinders of internal combustion engines (**13**), and some very interesting results emerge as shown in Fig. 3. It can be seen for example that the rate of wear is less on larger diameter cylinders and the practical significance of this is enhanced by the fact that a particular amount of wear is also less important as the cylinders

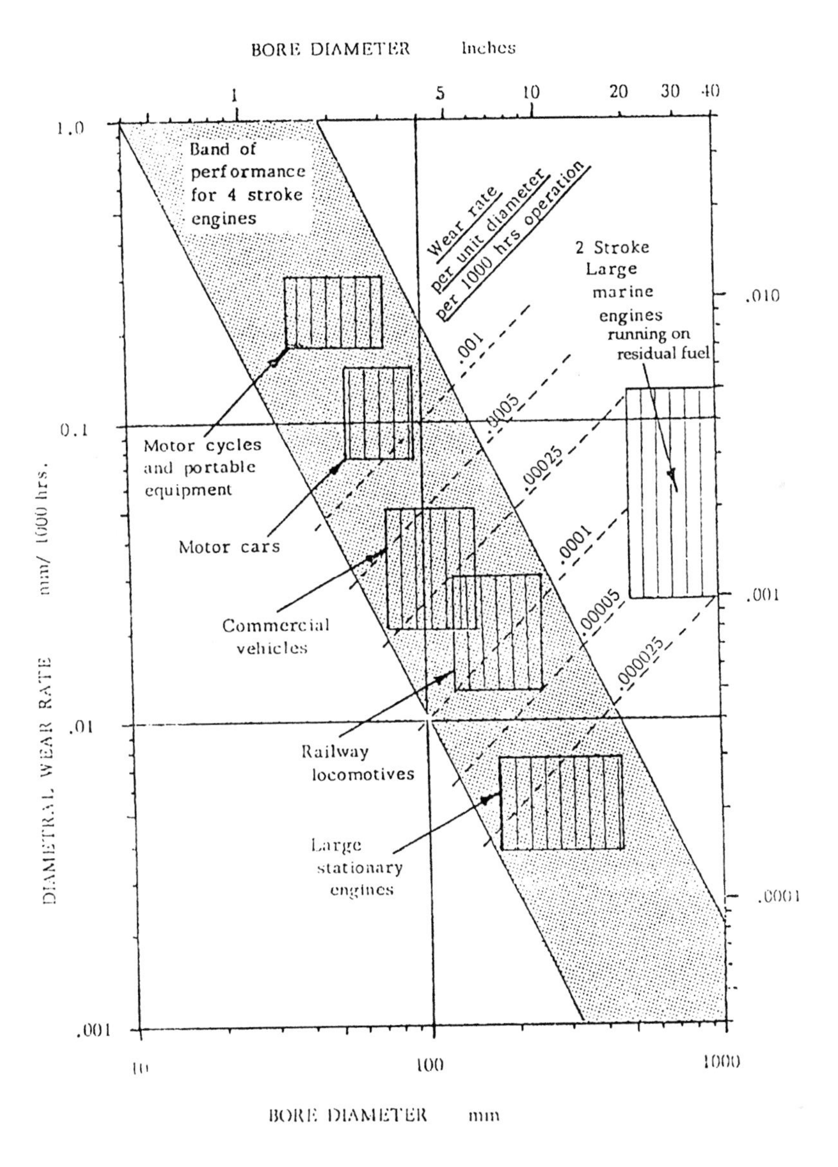

Fig. 3 Wear performance of the cylinders of internal combustion engines.

become larger. Large engines are therefore very much less affected by wear than small engines and can therefore accept operating on very inferior fuels. Fundamentally, the wear occurs primarily around top dead centre, and the number of times this area is rubbed by the piston rings depends on the rotational speed of the engine. Because of inertia forces, large engines have to rotate more slowly than small ones, and this is a major reason for their superior wear performance. There are also some patterns in the maximum cylinder gas pressures and piston ring numbers and widths, all of which affect the results.

Therefore, an important aspect of this type of investigation is that it provides a direct insight into a physical phenomenon which is relevant to a large group of machines. The results are not only of immediate use and interest to industry and professional engineers, but they can also help to identify useful research investigations, which also will be of immediate and practical use.

This wear study was a part of a much larger international investigation into the problems of piston ring scuffing, which was a major problem in the late nineteen sixties. For this study it was possible to list a number of specific fundamental research investigations that were needed to solve the problems in the industry (**14**). In the following fifteen years members of the academic community took up these ideas, with suitable modifications and additions of their own. It may be significant that from the late nineteen seventies onwards piston rings on most engines have given relatively few problems, possibly by taking advantage of the new understanding of the technology that then became available.

This technique can be very widespread in its areas of application in industry and is not confined to wear studies. Summers-Smith has carried out a number of similar investigations on machines in the process industry with problems related to lubricants and lubrication. One example is shown in Fig. 4 and relates to the failure of oil lubricated floating bush seals on centrifugal process gas compressors as used in offshore oil and gas extraction (**15**). The clearance space between these bushes and the shaft is pressurized with oil from one side, at a pressure slightly above that of the gas, which is at the other side.

The bushes do carry some radial loads arising from friction at their supports and tend therefore to operate with slight eccentricity. As in a plain journal bearing there is a risk of film cavitation downstream of the minimum film thickness and some process gas can then get admitted to the operating clearance space. This can then attack the whitemetal lining of the seal and the resulting deposit takes up the operating cl.arance and causes failure. Fig. 4 shows the improvements possible by changing the lining material of the seal and also brings out the effect of shaft peripheral speed increasing the temperature and level of corrosion. This is a major problem in gas extraction from oil fields particularly since the sulphur content of the gases tends to increase as the field becomes older. There is scope here for

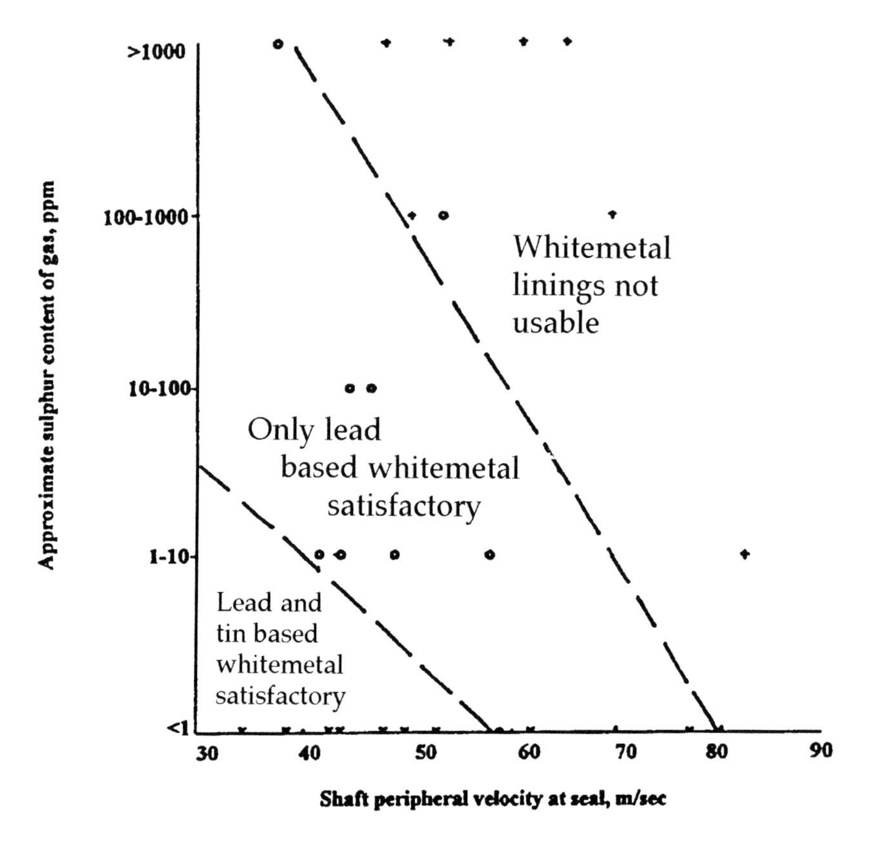

Fig. 4 The performance of whitemetal linings in bushes sealing sulphur containing gases.

research into cavitation patterns and bush bore profiles, as well as into new materials.

Similar studies have been carried out by Summers-Smith on reciprocating gas compressors. When these are compressing hydrogen (**16**) the necessary boundary lubrication is inhibited by the suppression of oxide films on the surfaces by the hydrogen. This can be eased by the addition of water to the hydrogen gas as shown in Fig. 5. With hydrocarbon gases (**17**) the gas dissolves in the oil on the cylinder walls and reduces its viscosity and Fig. 6 shows the result in terms of an effective viscosity which takes into account the effect of the dissolved gas. The important feature of these two diagrams is that they are based on experimental points from about one hundred different compressors operating in the field. They therefore provide a confirmed data starting point for any relevant more detailed research investigations into the phenomena concerned, as well as providing immediate understanding and guidance for the industry.

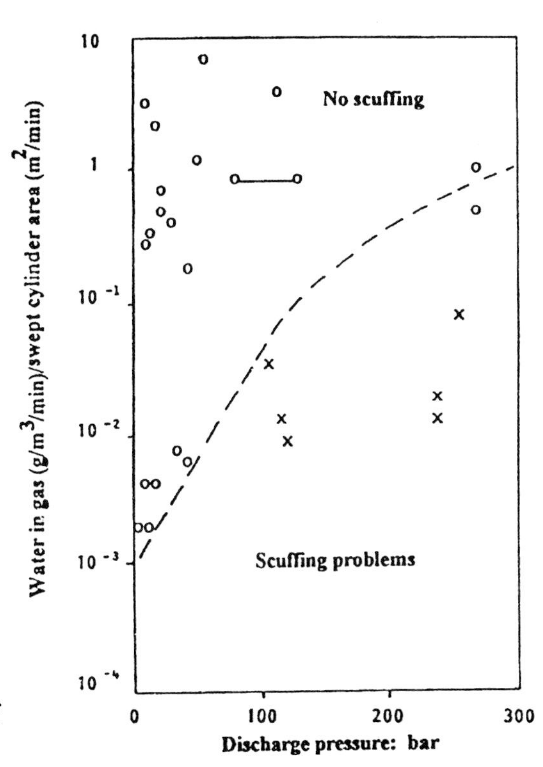

Fig. 5 The effect on the scuffing of hydrogen compressor cylinders,by the addition of water to the gas.

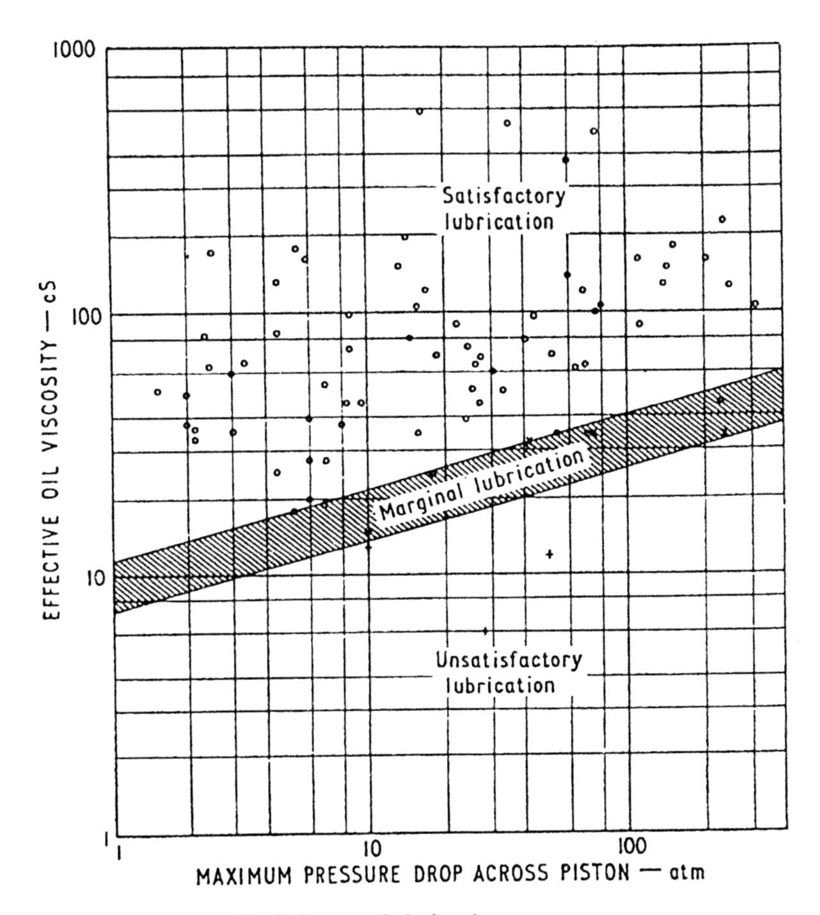

o Satisfactory lubrication.
x Scuffing during initial running.
+ Unsatisfactory lubrication.

Fig. 6 The required oil viscosity for the satisfactory lubrication of reciprocating compressors pumping gases which can dissolve in the oil.

The potential for further work in this area is very large. If a research student was to carry out such a task, with extension, for a doctorate, he would or she be very employable afterwards by the industry concerned, and the results of the investigation would be of immediate interest and applicability. The process does however contain a challenge to the academic community to devise respectable methods for multi-variable practical research investigations. This can enable useful results to be extracted in a controlled and organized way from direct operating experience in the field.

8 DESIGN METHODS AND SUMMARIES OF AREAS OF TECHNOLOGY

There is another useful practical approach, from a different angle, which can define and guide useful tribological research. This is that once a key problem or opportunity area has been identified it is worthwhile to attempt to produce a summary of the current technology or effectively a design method. When this is done as discussed earlier, it will be found that there may be of the order of 100 knowledge elements required to define the technology and enable products to be logically designed and developed. An interesting feature however is that it is found almost invariably that some of the important knowledge elements are missing and a full understanding is not available. The significant aspect however is that if these missing knowledge elements are described in a simple sentence, the wording sounds like the title of a fundamental research paper and indeed is potentially just that. This is a more structured way of defining worthwhile fundamental research projects, from which the results have every chance of receiving enthusiastic industrial recognition.

A further inference from this situation is that the standard Frascati definitions of research often used as a framework for national policies really have little validity or relevance. A scale ranging from blue skies research to near market research has no real meaning if an attempt to solve a practical industrial problem requires work worthy of the highest levels of fundamental investigation. Beauchamp Towers' work would be called near market research on railway axles, but led to the fundamental derivation of Reynolds equation and the whole basis of our understanding of hydrodynamic lubrication.

9 CONCLUSION

In summary, therefore, the industrial view of Tribology is that when first identified as a subject it cast a long needed light on key industrial problems. It then made some initial impact in providing some important understanding of the operation of machine components and some improved design methods for them. Unfortunately however in more recent years it has become more academically focused and is in great need of a redirection to enable it to make a positive contribution to the creation of real wealth for humanity from the industrial base.

It is suggested that an effective way forward should involve the following steps:-

1. Identify a sector of industry with good market potential.
2. Study the technology of the sector to identify problems and opportunities.
3. Survey, and graphically plot, the performance/problems of the machines involved.
4. Produce summaries of the technology, such as design methods, which have two uses:
   a) They are of direct use to industry and particularly small and medium sized enterprises.
   b) They can identify needed fundamental research investigations that are of direct practical significance.
5. Carry out the research that has been identified as being of significant importance.

It is hoped that these possible ways forward will be seen as exciting and interesting. They can provide a structure within which industry and the academic would come together to achieve creative results. It would be a great tribute to this conference if it were to provide the start of such a new way forward.

10 REFERENCES

1 *Lubrication (Tribology),* Education and Research: Department of Education and Science, HMSO 1966.

2 *Tribologie: Research Report T7636*, Federal Ministry for Research and Technology 1976.

3 *A strategy for tribology in Canada*: National Research Council Canada 1986.

4 *An investigation on the application of tribology in China: Report by the Tribology Institution of the Chinese Mechanical Engineering Society,* Beijing September 1986.

5 *Strategy for energy conservation through tribology*: ASME New York 1977 and 1981.

6 **Summers-Smith. D.** Ten years after Jost, the effect on industry: *Proc. IMechE,* 1976.

7 *Design Data Items:* The Engineering Sciences Data Unit London.

8 **Someya.** *Journal Bearing Databook:* Springer Verlag, 1989.

9 **Tower, B.** First report on friction experiments: *Proc. IMechE,* 1885, **39** pp58-73.

10 **Reynolds, O.** On the theory of lubrication and its application to Mr Beauchamp Tower's experiments: *Phil Trans of Roy Soc.* London, 1886, **177**, pp157-234.

11 **Neale, M.J.** The application in industry of the results of research. *Proc. IMechE,* 1977, **191,** pp333-338.

12 **Mashloosh, K.M.,** and **Eyre, T.S,**. The wear of digger teeth: *Proc. IMechE* **C343/84** pp29-34.

13 **Neale, M.J.** *IMechE Presidential Address,* 1990.

14 **Neale, M.J.** Piston ring scuffing – a broad survey of problems and practice: *Proc. IMechE,* 1970/71 **185** pp21-32.
15 **Summers-Smith, J.D.,** *Tribology Casebook,* MEP, 1997 pp195-198.
16 **Summers-Smith, J.D.,**. *Tribology Casebook,* MEP, 1997 pp68-70.
17 **Summers-Smith, J.D.,** Selection of lubricant viscosity grade for reciprocating gas compressors. *Proc. IMechE,* 1967/68, 182, pp11-19.

Some surface science aspects of tribology

J-M GEORGES
Institut Universitaire de France, Ecole Centrale de Lyon, Ecully, France

"Friction phenomena are equally interesting, for the physicist and for the engineer; their investigation belongs to a most difficult field of boundary problems of physics". *W Hardy.*

SYNOPSIS

Surface science and tribology are connected. Many of the new surface tools, recently developed, are needed for the characterisation of surfaces used, before and after tribological tests. Their uses imply a methodology established on the basis of the 'third body' analysis.

On the fundamental aspect, according the level of the contact pressure, friction and lubrication processes are described by non-condensed matter science, or condensed matter science. Some recent developments of these fields, which concern the basiç understanding of friction and lubrication, are presented.

1 INTRODUCTION

It is a great honour to be invited to give a lecture at the First World Tribology Congress. The previous four plenary lectures have dealt with the historical perspective, lubrication and rheology, wear mechanisms, industrial perspectives, that is to say they are themes familiar to engineering tribologists. Now, I will try to present some new scientific developments in the field of surface physico-chemistry in relation to tribology.

Science is a team effort. New ideas are generated by discussion and incorporated in our thought. Because of this, we sometimes cannot identify the person, who first came up with the new concept. I have had the opportunity to be surrounded by a group of talented students, numerous colleagues, and visiting scientist, with whom it is a pleasure to discuss tribology.

While the derivation of the word 'tribology' is simple, the study of friction, wear and lubrication is far from that. A tribological system is highly complex, with many variables, and involves phenomena that occur at many solid-liquid-gas interfaces. Today, the complexity of the system requires that the tribologist seeks multi-disciplinary approaches, to understand these phenomena, and to solve the tribological problems.

For instance, Fig. 1 shows in a schematic diagram, the evolution of the friction coefficient with the sliding speed, for a given lubricated system (**1**). Curve 1, represents the situation where the transition from the boundary to the hydrodynamic lubrication is very marked, as is quite often found. The question is, can we obtain a friction coefficient which does not vary with the sliding speed, as described by the curves 2 or 3, in Fig. 1? Another aspect, in modern machinery, the wear rate values, are for the researchers, very low indeed. For instance, the cast iron cam in an automotive

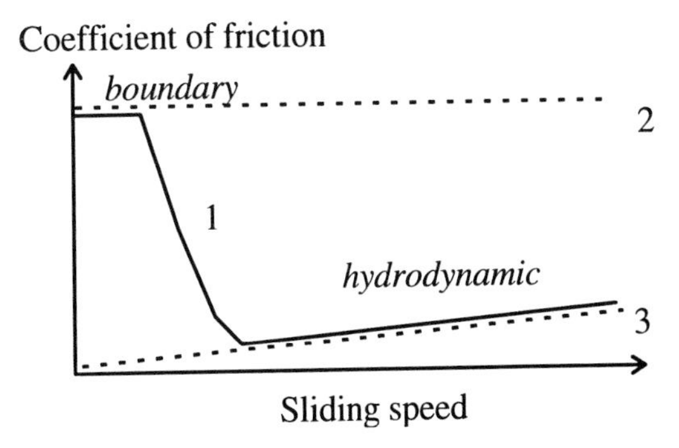

Fig. 1 Schematic diagram showing the transition from the boundary to the hydrodynamic regime.

engine loses on average 0.7 nm with a good reference oil, as its counter-face has slid a distance of 1 mm. This very low level of wear clearly shows that the wear processes are random and involve only a very small quantity of matter.

The answer to these questions requires a detailed analysis of the tribological contact, including many surface science aspects (**2**).

In this paper, the following topics are related. The second section deals with engineering approaches to examine the surface films formed in a tribological system. For this, different surface types are recognised, and surface analytical tools are used to study the nature of the surfaces. Recent development of the 'third body analysis' gives a methodological approach for engineering systems. The example of an anti-wear additive in a lubricant is taken to illustrate this approach. In the third section, fundamental parameters are briefly considered, and a list of fundamental processes recently analysed in the literature, presented. In section four, some development of nano-tribology, which allows a precise description of matter flow in the confined contact, are given. Finally, in section five, some friction laws are specified.

2 ENGINEERING APPROACH

2.1 Different types of surfaces

A solid surface is always very difficult to define. Two realms have to be considered: the realm of solid-state defect chemistry and the realm of surface physics and chemistry. Molecular films (**3**), ordered thin organic films, vary from few nanometers thick (a monolayer), to several hundred nanometers. They show considerable technological promise, and further scientific investigations may turn out to be very fruitful, not only regarding fundamental knowledge, but also for applications in tribology. Moreover, the scientific studies of molecular interactions of thin film structures lead to an understanding of their collective properties, characterized in more detail by a number of new surface science techniques (**5, 6**).

Fig. 2 represents various types of films on solid surfaces. On the first line, sub-monolayer and monolayers are important in the control, for instance, of the metallic adhesion, and therefore to the problem of seizure (4). On the second line, surfactant monolayers are considered principally for the friction with relatively low contact pressure, as found in biological tribology. On line three, relatively thick uniform films (10–100 nm) are related to different processes, and can correspond to the anti-wear films. Finally, in line four, relatively thick films (100 nm–1 µm) can also be considered, for some surface treatments.

2.2 Surface analytical tools

Macroscopic properties and performance of material surfaces are governed by surface energy, reactivity, wettability, and friction properties. These properties depend on microscopic surface structures and are recalled here after:

- chemical composition (role of trace contamination);
- nature of the surface chemical bonding, molecular functionality and forces;
- surface crystallinity, orientation of molecular groups, conformation, roughness;
- spatial distribution, heterogeneity, micro-domains.

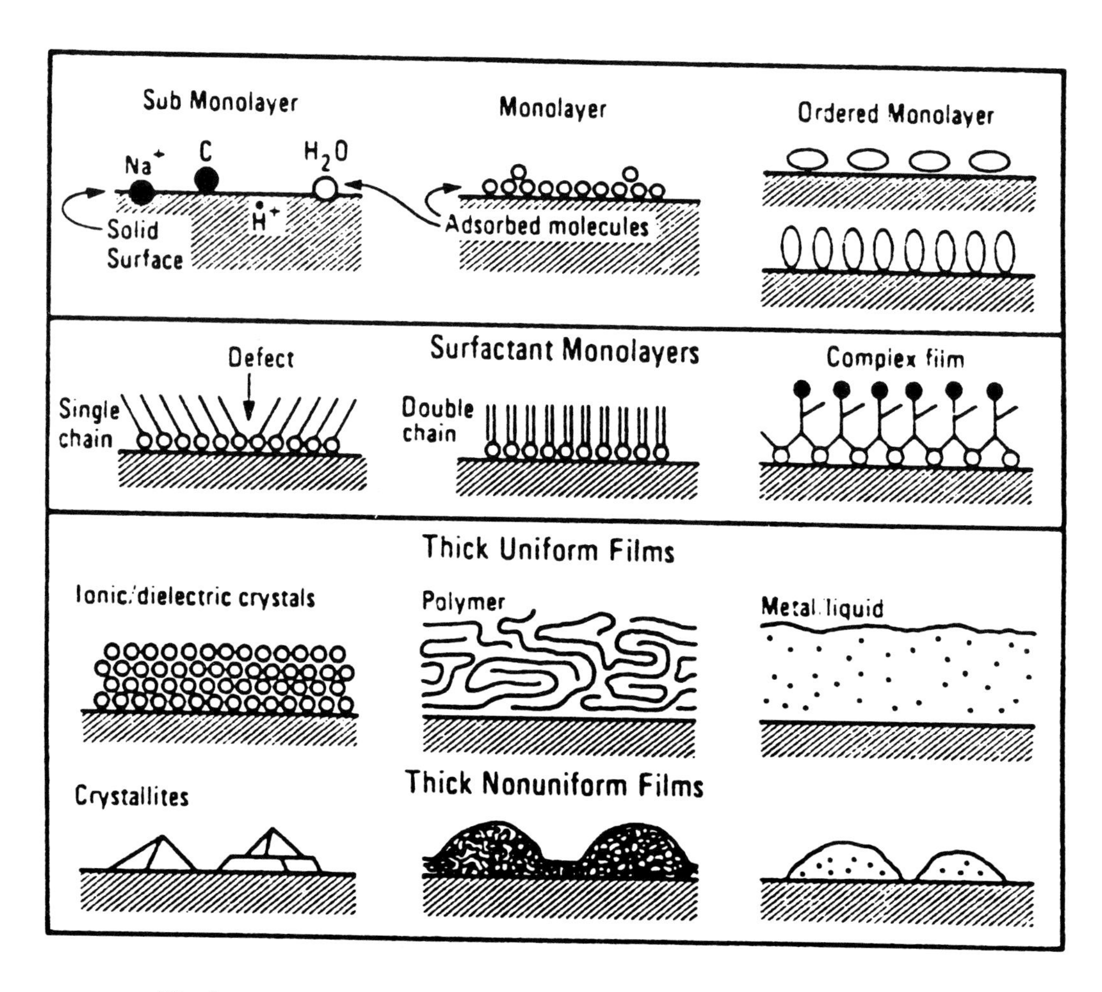

Fig. 2 Various types of films on solid surfaces shown schematically (**3**).

These are the objectives clearly assigned to surface analysis. The techniques capable of delivering such analytical information on surfaces are summarized in Table 1.

Table 1 Analytical information obtained from surface films, with the corresponding techniques. (after Tran Minh Duc and C. Donnet) (**5, 6**).

Information	Techniques
Imaging, structure, topography	Optical, electron microscopies, AFM, profilometry.
Elemental analysis	ESCA, AES, SIMS, X-ray, SIMS, LAMMA.
Chemical specification	ESCA, HREELS.
Oxidation state	TOF-SSIMS.
Molecular functionalities	RAIR, Raman.
Crystallinity	TOF-SSIMS, EXAFS, XANES.
Conformation	Grazing X ray diffraction.
Mechanical properties	SFA, STM, Nanoindentation molecular tribometer.

SEM, TEM STEM: transmission or scanning electron microscopy, XPS: x-ray photoelectron spectroscopy, AES: Auger electron spectroscopy, ESCA: electron spectroscopy for chemical analysis. SIMS: secondary-ion spectrometry, TOF, SSIMS: static secondary-ion spectrometry by time-of-flight), AFM, STM, scanning force. or atomic force microscopy, SFA: Surface force apparatus, RAIR, reflectance, absorbance infrared spectroscopy, Raman spectroscopies, HREELS: high-resolution energy electron loss spectroscopy, LAMMA: laser microprobe mass analyser, XANES X-ray absorption near edge fine structure, EXAFS: Extended X ray electron spectroscopy.

As recognized by many authors, ex situ and in situ pathologies are necessary (**7, 8**). Ex situ pathology, as the name implies, involves the analysis of surfaces before and after some tribological operations or experiments. For example, establishing the morphology of a surface prior to a wear experiment is essential in describing the characteristics and behavior of asperities. The analyses are usually conducted in separate instruments, that may require the test components to be disassembled prior to analysis. In addition to chemical analysis, detailed microstructure and mechanical property evaluation of substrates and wear particle debris are equally important. New proximal probes such as atomic force microscopy (AFM, STM, nanoindentor) are able to quantitatively measure the adhesive frictional, and mechanical properties of surfaces on the nanoscale. Tribochemistry contributes also to the complexity of tribological systems, and therefore needs in-situ pathology analysis: for instance, a controlled environment to study low friction films. An ultrahigh vacuum tribometer, equipped with *in situ* AES and XPS, has been developed (**9**).

2.3 'Third body' analysis

*2.3.1 Principles (**10–12**)*

The third body analysis establishes relations between microscopic mechanisms and the macroscopic properties such as wear, friction and load-carrying capacity of tribological systems.

From the point of view of surface degradation, the wear process under dry rubbing or boundary lubrication conditions is considered as a particle flow. It includes material detachment (by different possible mechanisms such as adhesion, abrasion, fatigue, corrosion of the first bodies). It leads to the formation of third body particles with their possible elimination from the contact as wear particles.

*2.3.2 Steps (**12**)*

The nature of material removal from the two wear tracks has to be identified. Although transfer films originate from a 'parent' material, not all films have the same composition or phase as the parent. The phases are, in fact, those predicted by equilibrium thermochemical calculations of the counterface materials plus environment components (as oxygen) (Singer (**13**), Tysoe (**14**)).

Three types of sequences have to be considered. At first, the surface films are removed from the substrate and transferred to the rider material and atmospheric gases, forming compounds with identifiable phases (**13**). Secondly, as the transfer film thickens, it is extruded from the contact area forming debris particles. This type of analysis was developed recently by Singer and Martin for MoS_2 films (**14**), by Brendlé for graphite films (**15**), and by Mischler, *et al.* for a tribocorrosion system (**16**), in which the surface degradation in the contact occurs not only by *particle* detachment (as in the case of dry rubbing), but also by *corrosion,* i.e. the transformation (by transfer of electrons) of a solid metal in metal ions dissolved in the corrosive solution.

2.3.3 Laws

The quantification of the transfer process, taking into account the mechanics and chemistry, has to be established.

The transport processes that control replenishment have to be determined, because endurance of the system is not a 'coating' wear problem, but a 'flow' problem involving third bodies. It is related to the rheological laws.

Boundary friction arises from the mechanical properties of the layers. The wear is a dynamic balance between layer formation and elimination.

Evolution of surfaces features during sliding has to take into account surface chemistry and surface roughness changes (**17, 18**).

2.4 Surface films and boundary lubrication

2.4.1 Principles

Wear protection in boundary lubrication is traditionally considered to result from the lubricants, or additives in the lubricant, reacting with the surfaces to form a load-bearing film. Shear occurs within the film, thereby minimising damage to the structural surfaces. However, it can reasonably be stated that only with zinc dialkylthiophosphate ZDTP additives has clear evidence for the creation of such films been established. More typically, surfaces that were performing satisfactorily appear to be bared. The quantity of film does not correlate with performance.

An alternative view of boundary lubrication has been developed, describing how oxides, adsorbed lubricant, additives, etc., are picked up and mixed in the convergent inlet of contacts, and then transformed into a colloidal paste by friction within the contact (**19**).

For instance, according to Coy (**20**), the following stages in the wear process with ZDTP can be recognised. First, during running-in, rough surfaces with superficial oxide have their asperities plastically deformed. Sulphide reactions give a layer which protects against adhesion. Secondly, an anti-wear film development occurs. Sulphide reactions are followed by phosphate deposition which gives an organic phosphate corrosion barrier. Thirdly, long term wear protection corresponds to a compacted inorganic layer. As described this complex development is related to the complexity of the chemistry.

Korcek, Willermet, *et al.* (**21, 22**) have, by using numerous analytical tools, comprehensively described some steps in film forming mechanisms, mainly due to a thermal, catalytic and oxidative decomposition of the molecule as shown in Fig. 3.

2.4.2 Surface film analysis

In recent years, surface analytical tools that determine structure have been combined with those measuring elemental composition and oxidation states to yield improved understanding of lubricant-derived tribofilms. The mechanical measurements of thin films have also been applied to tribofilms with thicknesses of the order of nanometers (**22, 23**). These methods, in combination with chemical and physical studies of bulk lubricant systems and assessments of lubricant behavior, made possible the description of the film-forming mechanisms (Fig. 4).

The most important reaction process has been identified to be the formation of a solid transition metal-phosphate glass material as an adherent thin film (**24, 25**). This film acts as a protective layer against wear, due to its superplastic behaviour in the contacting zone. Basically, the film also works by eliminating the abrasive contribution of iron oxides and is able to regenerate itself. In terms of chemistry, this can be explained by the ability of phosphate glasses to 'digest' oxides of transition metals, on the atomic level, within their network structure (**25**) (Fig. 5). It is reported that the films

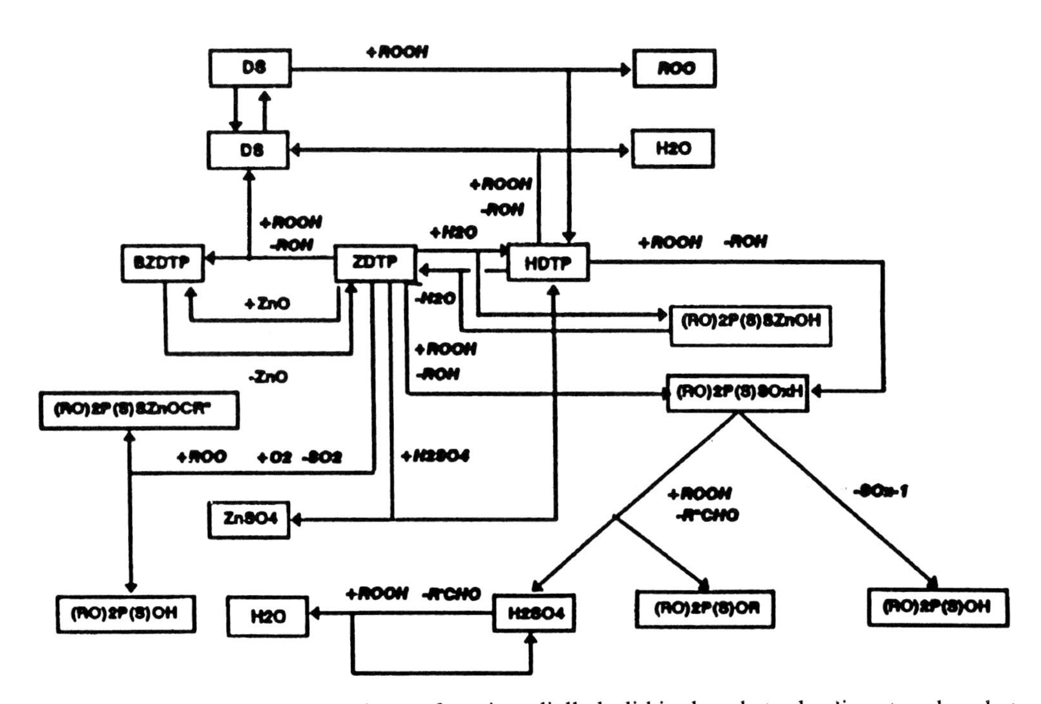

Fig. 3 Anti-oxidant reaction scheme for zinc dialkyl dithiophosphate leading to phosphate film precursors (**22**).

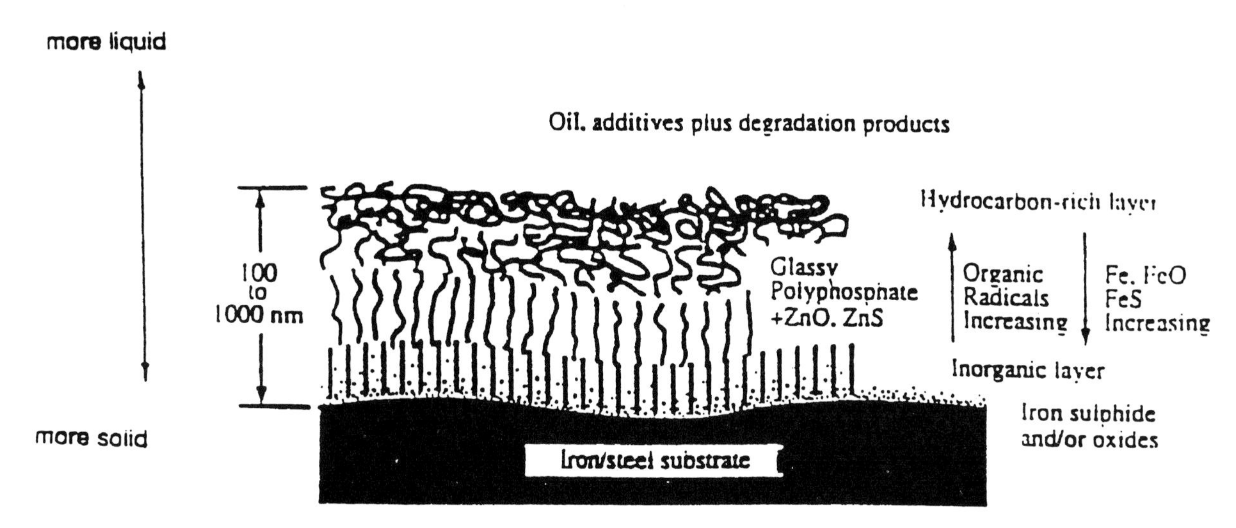

Fig. 4 Analysis with Time-of-flight Secondary Ion Mass Spectrometry, using a cold stage, of anti-wear films generated with deuterium labelled ZDTP, show positive evidence for the presence of organic material, originating from the alkyl group of the ZDTP, in the outer layers of the anti-wear film. These observations lend further support for the model proposed for a fully developed ZDTP anti-wear film (**24**).

formed from ZDDP have a dual structure with iron oxides next to the metal and organo-iron compounds on top, and are only about 10 nm thick (**26**). Whereas Cann, Johnston and Spikes (**27**), and also Belin, *et al.* (**25**) report the formation of 0.5 µm thick films of a material similar to an inorganic glass $(ZnO.P_2O_5.Fe_2O_3)_n$, which is a polyphosphate linked by bridging Fe^{3+} and Zn^{2+} ions.

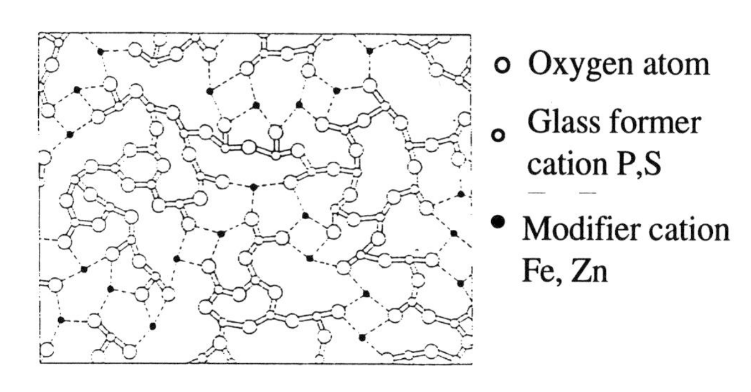

Fig. 5 Possible arrangement of atoms in an oxide glass former (**25**).

2.4.3 Mechanisms

Two basic mechanisms are proposed and probably joined.

i) As suggested (**25, 26**), the anti-wear film has EHL capability because of its surface high viscosity.
ii) Alternatively, the primary role of the friction film may be that of precursor material for the working glassy film.

Fig. 6 shows results obtained by Dr Sheasby's group (**26**), with a 4-ball test. The dependence of the wear constant with speed at four temperatures and for a given load (147 N), is studied. Optimum wear resistance at 100°C is bounded at low and high speeds by failure of partial EHL. EHL appears as a necessary condition for satisfactory boundary lubrication by surface film formation. However this work shows that EHL is not a sufficient condition, as wear can be unduly rapid within the EHL regime. Conditions for full anti-wear film formation and adhesion require a minimum temperature; i.e. a combined bulk and frictional heating, of about 125°C; i.e. 100°C with 1 mm/s, or 30°C with 100 mm/s. On the other hand the films appear to lose adherence on 52100 steel above about 250°C, (150°C with 300 mm/s). A very rapid form of wear occurred during EHL film collapse at low speeds in the presence of ZDDP. This effect is characterized by profuse amounts of carbonaceous material packed on and around essentially bare scars (Fig. 7). Apparently the ZDDP had sufficient residual surface activity to prevent the formation of an oxide glaze that would normally give some wear protection Fig. 8. The ZDDP films require a certain stability of the wearing surfaces to become established. Loss of effectiveness in the partial EHL film at low speeds was due to a thermal thinning of the oil film; this defect could be delayed or accelerated by appropriate changes in the viscosity of the oil.

These examples show that the conditions for anti-wear film formation have to be well identified. The chemical composition linked with the mechanical properties is crucial for the understanding of the wear process. Presently, these points are not well-understood and their studies are in progress.

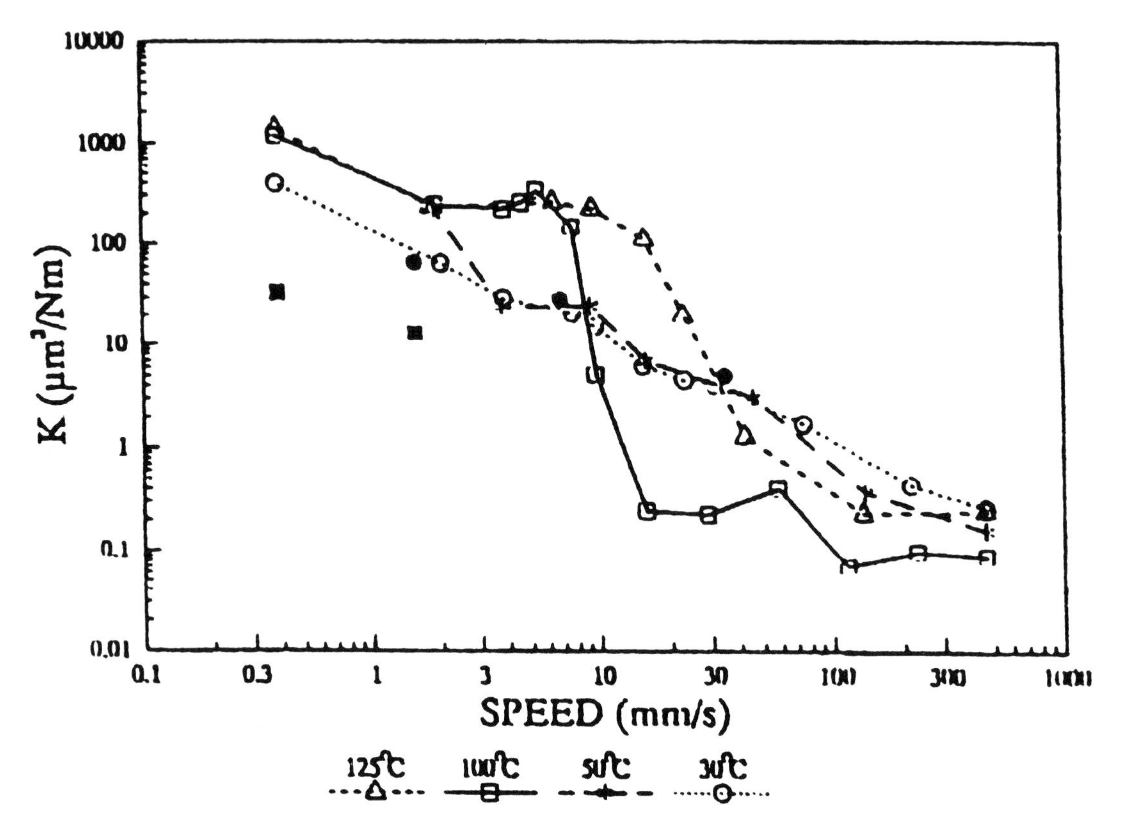

Fig. 6 Four ball test: dependence of the wear constant of 52100 steel with speed at four temperatures and for a given load 147N (**26**).

Fig. 7 Four ball test: appearance of the wear scar of 52100 steel, at 100°C, and low speed. A very rapid form of wear occurred and EHL film collapse at low speeds in the presence of ZDDP. This effect is characterized by profuse amounts of carbonaceous material packed on and around bare scars (**26**).

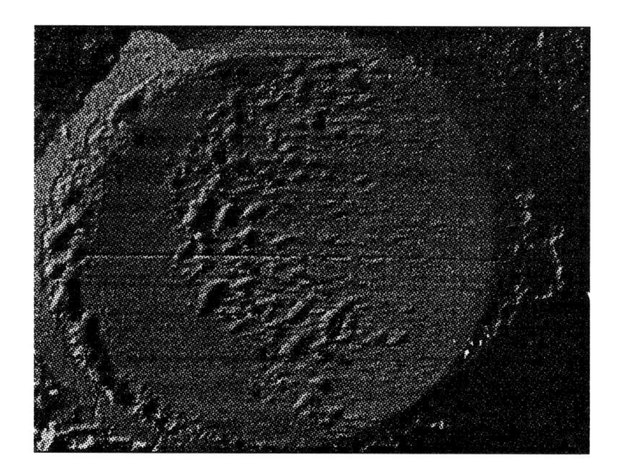

Fig. 8 Four ball test: appearance of the wear scar of 52100 steel, at 100°C, and high speed. ZDDP had sufficient surface activity to prevent the formation of a glaze, that would give some wear protection (**26**).

3 FUNDAMENTAL PARAMETERS AND PROCESSES

3.1 Energy, time, and pressure

Fundamentally, chemistry is the interaction of energy and matter. The parameters that control it are the time, the amount of energy, and the pressure, which describe a three-dimensional space.

Energy can be dissipated in breaking of molecular bonds. It can also correspond to the activation energy of all the molecules. Table 2 gives some classical values of the bond energy ε for representative interaction bonds (van der Waals, hydrogen....). But the energy of the entire molecules can be considered, and a simple measure of the intermolecular forces is provided by the effect of temperature on the viscosity η. As the temperature increases, η decreases according to an exponential law of the form:

$$\eta = \eta_o \exp(-\varepsilon / kT),$$

where k is the Boltzman constant. As presented in Table 3, the activation energy for long chains as used in lubrication remains almost constant ε = 25-30 kJ/mole. This corresponds to about the third of the vaporisation energy (**2**).

Table 2 Some energies ε per bond or per molecules (**2, 28–30**).

Energy ε	eV/bond	kJ/mole (x 96.5)
van der Waals: CH_3- CH_3, MoS_2-S_2Mo	0.08	8
hydrogen C–O–H...O = C	0.21	20-25
π-π graphite sp_2	0.4-0.8	39-77
covalent C-C	3.8-4.5	360-430
Mo-O	0.8-1.3	77-125
dimerisation H_2O	.02-0.03	1.8-2.9
benzene (solid state)	0.06	6
cyclohexane (solid state)	0.20	20
activation energy molecular chain CH_3- CH_2 -CH_2- --- (10 carbons)	(per chain) 0.10	(per chain) 10
activation energy molecular chain CH_3- CH_2 -CH_2- --- (20-30 carbons)	(per chain) 0.25-0.30	(per chain) 25-30

Table 3 The pressures in tribology.

Contact Pressure (Pa) x 10^n	Examples	Plastic flow of materials	Laws
n	interparticles		non condensed
5	rheology		matter
6	rubber friction	granular media	intermolecular forces: steric, van der Waals Laplace......
7			
8		polymer	
9	cam and tappets	metal	binding energy
10		ceramic	
11	diamond anvil		condensed matter
12	explosion		
13			

These levels of energy are expended with different time scales and pressure. Consequently, it is interesting to consider in Fig. 9, in the 3D space, where the dissipative processes present in a tribology problem are located, and to compare it with other chemical processes, such thermo-chemistry, piezochemistry, sonochemistry ... (**31**).

An important variable is the time of interaction. In tribology, it is very dependent on the size of the element considered. For thin films, it can be in the range of the inverse of mean shear rate (10^{-7} to 10^{-9} s). For an asperity, it has a much higher value.

The third important parameter is the external pressure p applied on the contact. Tribology is concerned with a large spectrum of pressure. In practical tribology, the contact pressure value between the two solids can be as low as 2×10^5 Pa, as found in hydrostatic bearings; but it can be as high as 2×10^{10} Pa , as found in the contact of two ceramic asperities. It is well known that in this latter case the plastic hardness or plastic flow of the material, which constitutes the solids in contact, controls the contact pressure (**32**) (Table 3).

Two types of descriptive laws of matter are present in the literature. In the *non-condensed matter science*, intermolecular forces are principally studied and taken into account (see for instance (**28**)). Complementary, in the *condensed matter science*, binding energy between two atoms is computed and related to basic phenomena. Quantum mechanical calculations are necessary.

Starting from a simple physical assumptions, Rose, Smith, and Ferrante (**33**) have shown that during compression, the isothermal equation of state of condensed

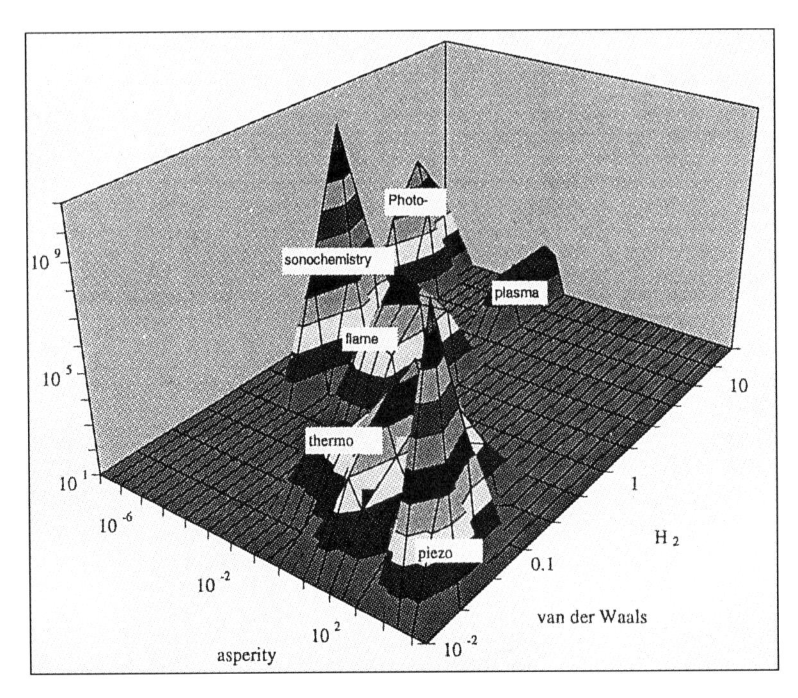

Fig. 9 Island of chemistry as a function of energy, pressure and time. Depending on the nature of the tribological problem, and on the scale of the process considered, it is beneficial to consider the other chemical disciplines which are mainly related to some volume of the space energy pressure time (adapted from 'The chemical effects of Ultrasound' by K.S. Suslick (**31**)).

materials has a universal form. This single form can describe numerous different types of materials: metals, ionic or covalent solids, and van der Waals crystals as well. They found that the adhesive binding energy W of the system versus the separation distance between the surfaces D, could be written in a form $W = \Delta E \cdot W^*(D^*)$, where ΔE is the binding energy, and the non-dimensional distance $D^* = (r - r_e)/\lambda$, where r_e is the equilibrium separation, and λ is a scaling length discussed later. $W^*(D^*)$ is the functional form for a 'universal' shape of these curves. This scaled relation is applied for adhesion of simple metals but also to a wider class of phenomena.

Fig. 10 shows, for instance, the results of scaling cohesive energies for a transition metal, bi-metallic adhesion, chemisorption, and a diatomic molecule. The scaling length λ selected for all of these cases is given by:

$$\lambda = \left[\Delta E / \left(\partial^2 W / \partial r^2 \right)_{r_e} \right]^{1/2},$$

the square root of the equilibrium binding energy divided by the second derivative of the binding energy with separation evaluated at the equilibrium position. This particular form for the length scaling λ is convenient, since the second derivative can be related to experimentally measurable properties, such as the bulk modulus for solids. Using this theory, Vinet, Ferrante, Smith, and Rose (**34**) have proposed a new isothermal equation of state and calculated the compression as a function of pressure for ionic, metallic, covalent solids and van der Waals solids such as polymers and lubricants. Their equation of state gave the pressure:

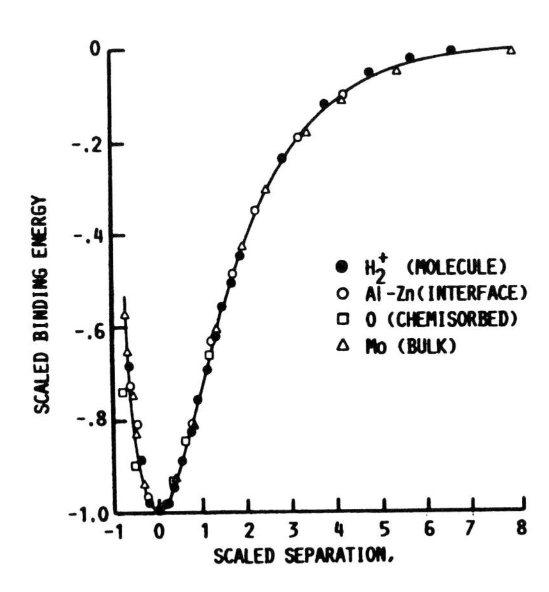

Fig. 10 The scaling cohesive energies for a diatomic molecule, bi-metallic adhesion, chemisorption, and a transition metal. Scaled binding energy $W^*(D^*)$ versus scaled separation $D^* = (r - r_e)/\lambda$. (**31**).

$$p = 3B_0 \frac{1-x}{x^2} \exp\left[\xi(1-x)\right]$$

and bulk modulus:

$$B = B_0 \frac{2+(\xi-1)x-\xi x^2}{x^2} \exp\left[\xi(1-x)\right],$$

as a function of the relative compression x, where $x = (v/v_0)^{1/3}$, and v is the volume. Here B_o is the bulk modulus at zero pressure and:

$$\xi = 1.5\left[\left(\partial B_0 / \partial p\right)_{p=0} - 1\right]$$

is a function of the pressure derivative of the bulk modulus at zero pressure.

These equations were applied with great success to describe confined lubricant films (**35**) and glassy polymeric thin films (**33**).

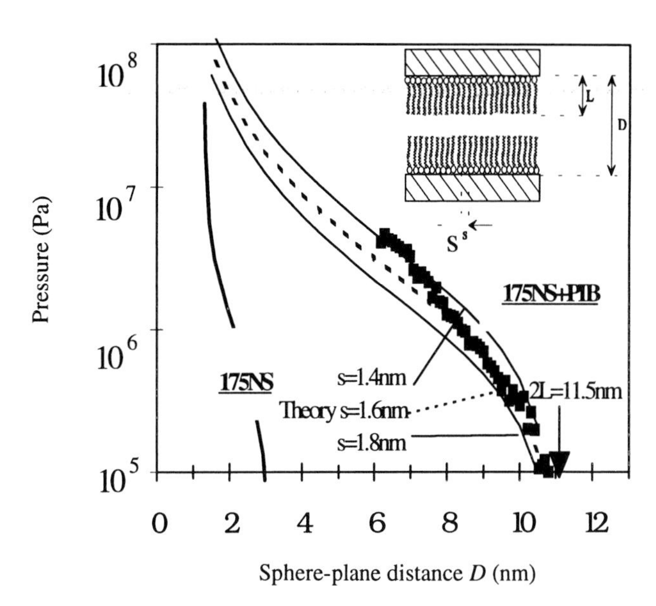

PIB 

Fig. 11 The evolution of the pressure in sphere-plane contact versus the sphere plane distance D, for a hydrocarbon solvent (base oil 175NS), and for a succinimide solution in hydrocarbon (175NS+PIB). Results are compared with the Alexander-de Gennes theory (**27**). s is the average distance between the attachment points of the PIB brush and $2L$ is twice the layer thickness. Best fit between theory and experiment is obtained for $L = 5.7$nm and $s = 1.6$nm (**61**).

3.2 Some fundamental processes studied in tribology

During the last 5 years, a great number of fundamental studies have appeared in the literature. In addition, numerous forums, workshops, and special publications discussed the relations between tribology and surface science, polymer rheology, mechanical properties and theory. Table 4 indicates some subjects studied and some references.

4 LUBRICATION AT THE ATOMIC SCALE

A better understanding of the mechanisms of adhesion and lubrication is now possible, with the recent development of new instruments such as the surface force apparatus (**28**), the molecular tribometer (**53**), and the atomic force microscope (**57**), which make it feasible to measure, at a molecular scale, the normal and tangential forces between two surfaces. Molecular simulations clarify these. Some examples follow.

4.1 Static equilibrium of an adsorbed film

It is now possible to have a clear picture of the resistance of two polymer layers adsorbed on solid surfaces and squeezed in a solid contact. The mechanical properties of surfaces coated with adsorbed polymers control not only phenomena such the tribological properties of solids, but also colloidal stability.

Polymer stabilisation is one of the means to prevent the aggregation and sedimentation of particles such as wear particles, or soots, which are essentially carbon particles. The adsorption on carbon surfaces of a polyisobutene succinimide (PIB) in dilute solution with hydrocarbon solvent (175N) at 25°C was investigated with an SFA (**61**). Fig. 11 presents a comparison of the contact pressure between two carbon surfaces in pure solvent and in PIB solution, and the resulting interactions between them are also characterised. PIB creates a 5.5 nm thick anisotropic brush layer on the carbon surface,. These double brushes resist an external pressure of 10^7 Pa. The results are compared with the Alexander – de Gennes theory which is typically a non-condensed matter theory (**65**). They permit a description of the amount of polymer molecules adsorbed on the surface.

4.2 Shear properties of hexadecane

The numerical simulations reveal nanoscale processes, which include spatial and temporal variations in the density and pressure of the lubricant, particularly in the region confined by the approaching asperities. The contact asperity can induce molecular layering transitions, which are reflected in oscillatory patterns of the friction force.

Luedtke and Landmann (**59**) have studied the atomic-scale structure, dynamics, flow and response characteristics of a thin molecular film of confined hexadecane. The normal force on the nickel tip versus the distance between the tip and the gold surface are recorded during the process of lowering

Table 4 Fundamental processes studied in tribology.

Nature of the processes	Some references
Interfacial forces and the true area of contact	
• Interfacial solid junctions formation of asperities.	**(7, 18, 41, 50)**
• Static equilibrium properties of liquid films in a contact	**(3, 36, 37)**
• attractive : van der Waals, capillary	
• repulsive : double layer, sterics, entropic, electrostatics, oscillatory	**(28, 48)**
• consequences : external pressures versus equilibrium interface pressure Π	**(53)**
• solidification freezing	**(54)**
• Surface roughness ; single versus multiple asperities	**(40)**
Dissipative processes	
• At the Solid-Solid interface	
• Atomic dissipation	**(3, 4, 37)**
• Molecular dissipation : molecular movement during the sliding process	**(39, 44, 52, 53)**
• time scale and chain length scale matching	**(28, 48)**
• Granular dissipation	**(54)**
• At the Liquid-Solid interface	
• Shear flow near the solid	**(45–47, 50)**
• Confined shear flow	**(49)**
• Dissipative regions of a solid -solid contact	
• Behaviour in inlet region	**(40)**
• Behaviour in the contact region,	**(54)**
• Cavitation in the outlet region	**(18)**
• Shear induced ordering transitions	**(18, 53)**
Thermodynamic criteria	**(12)**

the nickel tip onto the hexadecane covered Au surface. Force oscillations are associated with the layering transformations in the film. It is noticed that during the process the hexadecane film can resist pressures of 1 GPa.

When a liquid is confined in a such narrow gap in the lubricated contact between solids, new dynamic behavior emerges (**62**). The effective shear viscosity is enhanced compared to the bulk, relaxation times are prolonged, and nonlinear responses set in at lower shear rates. These effects are more prominent as the liquid film becomes thinner. They appear to be the manifestation of collective motions. The flow of liquids under extreme confinement cannot be understood simply by intuitive extrapolation of bulk properties. Fig. 13 shows the shear behavior of a few layers of hexadecane squeezed between two molecularly flat mica surfaces (**63**).

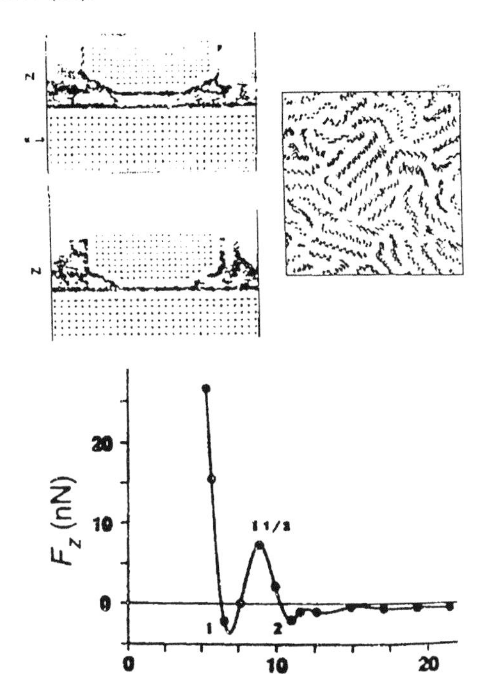

Fig. 12 **Side and top:** Atomic configurations of a nickel tip, an Au(001) surface and an n-$C_{16}H_{34}$ (hexadecane) film, starting at a tip-to-surface spacing corresponding to two layers of the confined film 1.1 nm (upper) and ending with a one layer film d = 0.65 nm (lower). The linear dimension of the system along the x axis is 7.75 nm. The coordinate axes are normalized such that $x = z = 1$ corresponds to a linear dimension of 7.75 nm. Short-time trajectories of hexadecane molecules in the layer next to the Au surface, corresponding to the one layer confined film shown at the bottom illustrating intra-layer ordered domains.
Bottom: Normal force on the nickel tip versus the distance between the tip and the gold surface (**18, 58**).

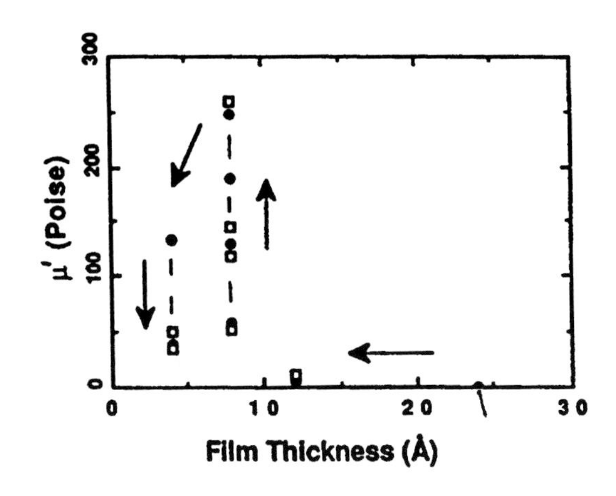

Fig. 13 Apparent dynamic shear viscosity of hexadecane plotted against film thickness. Arrows denote direction of increasing normal load (**62**).

The thin film molecular hexadecane lubricant is now confined and sheared topographically between non-uniform solid gold surfaces, which slide at a relative velocity of 10 m/s. The investigation uses molecular dynamics simulations.

While the relaxation process involves a spectrum of relaxation times associated with various inter- and intra-molecular degrees of freedom, an overall relaxation time of 250 ps is measured. The relaxation process involved structural rearrangements coupled to some molecular diffusion inside the confined zone. The time taken by the asperities to close the gap between them is 100 ps. Consequently the Deborah number is 2.5 for the overlap of asperities. The lubricant response is visco-elastic, with a large viscous component; the effect of enhanced confinement in the overlapping-asperity case is to create dynamically a transient medium which may be characterized as a highly visco-elastic, perhaps even elasto-plastic or waxy solid.

This layering effect is due to the flexibility of the linear alkane chain. Experiments with a more rigid chain, indicate that the confined layer corresponds to a randomly organised network of rigid molecules (**66**).

These simulations extend macro-continuum investigations into the nano-scale regime and provide molecular-scale insights into the fundamental mechanisms of ultra-thin film lubrication phenomena under extreme conditions.

4.3 Nano-rheology of polymer films

The ability of adsorbed polymer layers to reduce both friction and wear has long been suggested, although early work in the area of concentrated contacts was contradictory. Some researchers found evidence of viscosity lubricant enhanced on the contact surface, while others did not find it even at film thicknesses of the order of 10 nm. The main problem with these early studies is the difficulty in accurately measuring very thin films. Recently the technique of ultra-thin film interferometry (**54**) has been used to study the film thickness of polymer-containing oils and established the existence of adsorbed boundary films.

Fig. 14 shows two different rheological experiments with adsorbed layers of poly-isoprene solutions. A polyisoprene

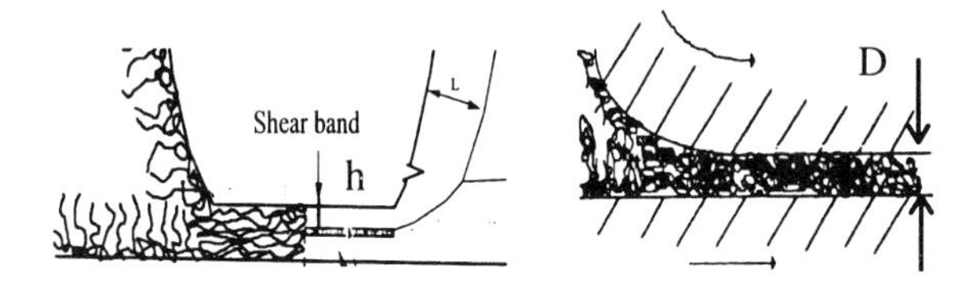

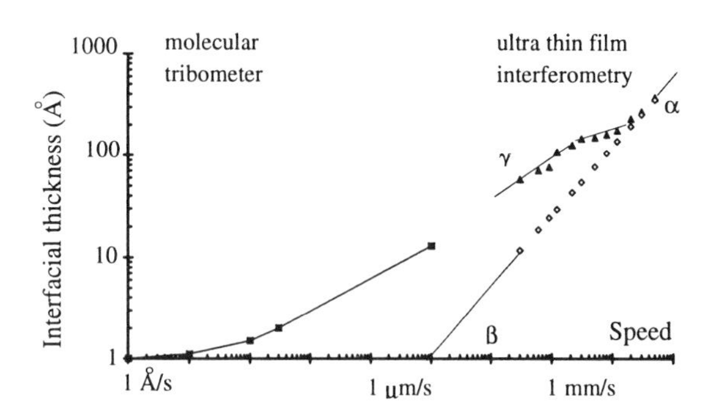

Fig. 14 Two different rheological experiments with adsorbed layers of poly-isoprene solutions:
Above: elastohydrodynamic film thickness D measurements using a thin film optical rheometer.
Below: rheology of adsorbed layers of polyisoprene between two cobalt surfaces, investigated with a molecular tribometer (**56, 64**).

layer is obtained on the metallic surface by adsorption of the polymer from a semi-dilute solution of cis 1-4 poly-isoprene in 2,4, dicyclohexyl-2-methylpentane, which is a small hydrocarbon molecule and a good solvent of the poly-isoprene at 23°C.

First, elastohydrodynamic film thickness D measurements using a thin film optical rheometer are considered. Randomised measurements of fluid film thickness were taken over a range of low rolling speeds (0.0003 m/s to 0.0500 m/s). Bulk temperature (30°C) and contact pressure (0.4 GPa), were held constant throughout the measurements. Curve αβ corresponds to that of pure solvent. The curve follows the classical EHD law:

$$D \,\alpha\, V^{0.67}$$

Curve αγ corresponds to the semi-dilute solution of polymer. For small rolling speed V, the classical EHD law is not followed. This is due to a concentration increase of polymer in the contact entry. It was shown that at low speeds, the boundary films which have an enhanced viscosity (between 10 to 50) generate EHD-type films many times that of the bulk fluid (**63, 64**).

Second, on the same graph are reported the rheology of adsorbed layers of poly-isoprene between two cobalt surfaces, investigated with a molecular tribometer. The two layers of poly-isoprene are compressed between a smooth sphere and a plane. During the compression process, the solvent molecules are repelled from the polymer network and formed a compressed polymer 'mesh' which is not connected. The mean 'mesh' size is lower than the one corresponding to the 'rubber' plateau of a polyisoprene melt. The contact pressure is 10^7 Pa, and the film thickness of the two compressed layers is in the range of D = 193 Å. A sliding experiment is conducted, and the variations of D, accurately measured. The film thickness variations follow those of the friction force and are the sum of two contributions (**56**). One is a thickness decrease due to creep of the layers themselves. Another is a very small increase of the interfacial thickness h between the two layers, which was found to be dependent on the sliding speed. This interfacial thickness h is that of the shear band. Its value is reported as a function of the sliding speed (**56**).

It is interesting to note that over 8 speed decades two major processes control the sheared zone. At low speeds, movements of molecular groups are responsible for the lift of the two compressed layers. This is the consequence of best trajectories taken by the molecular groups during the sliding as also described by Harrison's simulations (**66**) . At higher speeds, the global increase of the viscosity in the inlet zone provokes the lift.

5 FRICTION

The topic of interfacial sliding has experienced a burst of interest and activity since 1988. The stabilised friction force F_x^{st} (**13**) is considered as the result of three contributions: the ploughing of surfaces by the asperities, the adhesion, and the shear of the interfacial film. If the surface roughness is very small, the first contribution is negligible in comparison with the two others. Although many processes can simultaneously be produced in the contact zone, a mean effect is inquired and the stabilised tangential force F_x^{st} is expressed as:

$$F_x^{st} = \tau \cdot A \qquad (1)$$

where A is the real area of contact, and τ is the mean shear strength.

Measurements of friction cannot be easily compared to an atomic scale, because such measurements are averaged over large heterogeneous samples.

5.1 Atomic models

Low friction, including zero friction, can be achieved at low contact pressure, with weak interface interactions and with small atoms at the interface. The mechanical principle that explains this behavior follows from the simple, one-dimensional independent oscillator model (**43, 67, 68**). The strain energy transmitted by interfacial atoms during the first half of the cycle is returned to them during the second half of the cycle. This behaviour is also observed in the more realistic, three dimensional simulations. The third dimension itself contributes an additional friction reduction. It provides the interfacial atoms with an extra degree of freedom. For example, H terminated atom can rotate around each other in aligned collisions or 'zigzag' along potential minimal channels; these trajectories are not available to atoms described in two dimensional models (**44, 69**).

Friction is increased by many factors:

i) Strong interfacial interactions (corrugations), according to the simple oscillator model, give a finite static friction force, then stick-slip motion between atoms (**43**).
ii) Coupling of excited modes establishes multiple pathways for energy dissipation, thereby increasing friction coefficients. An example is the simulation of alkanes, where torsion modes become allowed, providing a new pathway for energy dissipation (**69**).
iii) Defects in the interfaces create atomic rugosity (**43**).
iv) Surfaces charges in insulators are responsible for breaking material and influence friction (**70, 71**).
v) Commensurate lattices increase friction forces. In principle, 'low friction trajectories' can be found along selected directions in real crystals having anisotropy interaction potentials (corrugations). These possibilities are treated quantitatively by Hirano and Shinjo (**72**) and Sokoloff (**73**). The low friction of two basal planes of the sulfur rich (0001) MoS_2 is attributed to the atomic non-commensurability of two superposed stackings. The atomic commensurability of two superposed stackings disappears in the sixfold symmetry of the hexagonal lattice (**74**).
vi) The size and shape of molecules can also influence friction behavior. Small atoms or molecules may follow low-friction trajectories whereas larger atoms or molecules may not 'fit' (**43**).
vii) The energy dissipation is maximum, when the time (and length) scales of contact match the intrinsic time and length scales of molecular interactions (**53, 75**).

5.2 Microscopic models

5.2.1 Adhesive controlled contact

The approach corresponds to that described by many authors (Tabor (**2**), Israelachvili (**75**)). It considers first, the simplest model consisting of two oriented fatty acid or soap monolayers sliding on one another (Fig. 15). The sliding interface consists of CH_3 groups, each group occupying an area a of approximately 0.20 nm^2. It assumes, secondly, that most of the interaction energy occurs between the end-groups. However, it avoids all detailed analysis on the grounds that such a treatment does not match the inadequate knowledge on the processes which actually take place during sliding. It considers, thirdly, in the simplest possible way the energy expended in dragging one end-group over the other. If the two monolayers are to be separated to infinity, two new surfaces exposing CH_3 groups would be formed. Such surfaces have free surface energies γ of the order of 25 to 30 mJ/m^2. If the interaction energy between the monolayers is dominated by CH_3 groups across the interface and if a is the area occupied by such groups, it can be written: $\varepsilon/a = 2\gamma$, where ε is the interaction energy between two CH_3 groups. In the energy expended in dragging one CH_3 group over its neighbour, it is assumed, fourthly, that only a very small fraction of ε is involved, because the end groups are not separated to infinity. Thus, it is assumed, that a small fraction β, is involved. If F_x is the tangential force required to displace the end group by a distance X to its next equilibrium position, we may equate the work done $F_x \cdot X^*$ to the energy: $F_x \cdot X^* = \beta \cdot \varepsilon$. Therefore, the effective shear stress τ_o is:

$$\tau_o = F_x^{st} / a = [\beta \cdot a \cdot (2\gamma)] / [a \cdot X^*] \tag{2}$$

Because the distance X^* is of the order of $a^{1/2}$, and on an empirical basis the fraction β is taken equal to 1/30. Therefore:

$$\tau_o = (1/30) \cdot (\varepsilon / a^{3/2}) \tag{3}$$

Numerical application of equation (3) gives τ_o = 2 MPa, in agreements with results of experiments with mica sheets coated with a calcium stearate monolayer shown in Fig. 16 (**75**). In these experiments, the adhesive pressure value Π_a = 500 MPa is much more important than the external pressure, which varies between 6 to 20 MPa. It is an 'adhesive controlled' experiment (Fig. 17). The approach presented for a stearic acid monolayer can be generalized to other systems.

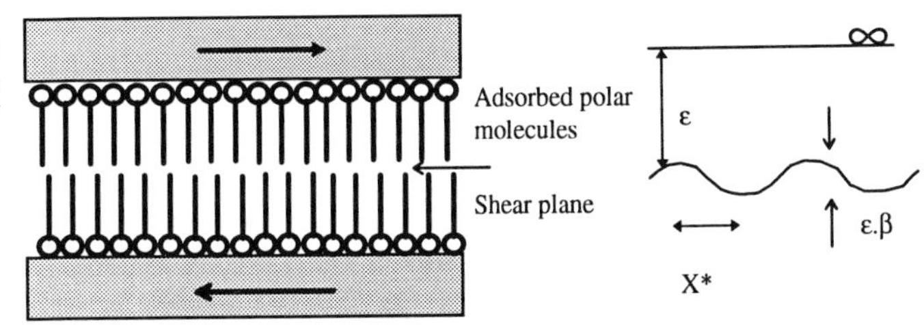

Fig. 15 Microscopic model.

5.2.2 Load controlled contact

If an external pressure p is applied to the system, the shear stress τ is:

$$\tau = \tau_o + (\delta z / X^*) \cdot p \quad \text{or} \quad \tau = \tau_o + \alpha p \tag{4}$$

where, dz corresponds to atomic bumpiness of the surface group. A similar equation was derived by Briscoe and Evans

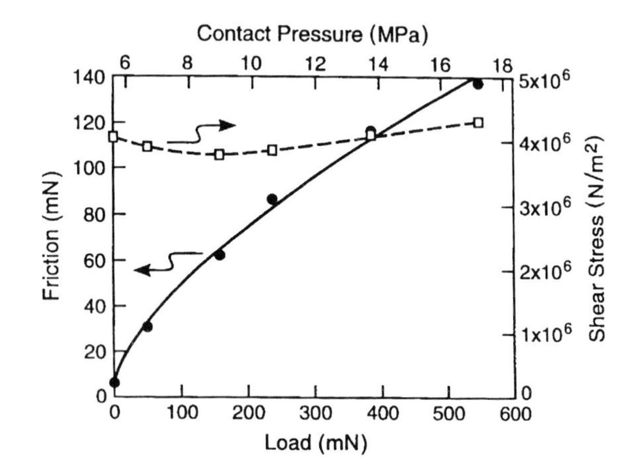

Fig. 16 Experimental results of adhesion-controlled friction. The friction and shear stress of mica surfaces coated with monolayer of calcium stearate plotted against external load Fz. Non-zero friction at $Fz = 0$ is indicative of an adhesive contribution to the friction. In this system the shear stress is independent of load and the friction depends only of the contact area (**75**).

(**76**), with a different physical significance. Therefore, the stabilised tangential force can be written as:

$$F_x^{st} = A \cdot \tau_o + \alpha F_z \quad (5)$$

Although the area A and the normal force F_z are dependant, in many cases, where the first term is negligible in comparison with the second, the friction coefficient is:

$$F_x^{st}/F_z = \alpha \quad \text{(Amontons law).}$$

As an example, Fig. 17 shows the friction coefficient measured over a range of pressure for different compounds. Experimental and theoretical results show that high friction is obtained, when the molecules interlock with each other under high pressure. Low friction is obtained with aromatic compounds such as benzene, because it is difficult for them to get close to each other, due to the repulsion of π-electrons and they have little unevenness for interlocking.

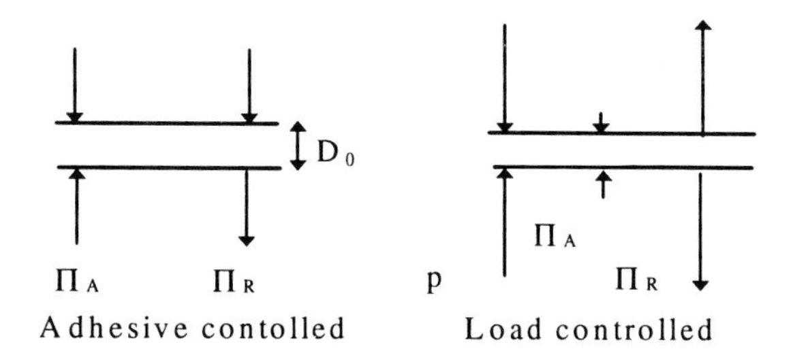

Fig. 17a Schematic 'adhesive controlled' contact where the adhesive pressure is dominant and 'load controlled' contact where adhesive pressure is negligible.

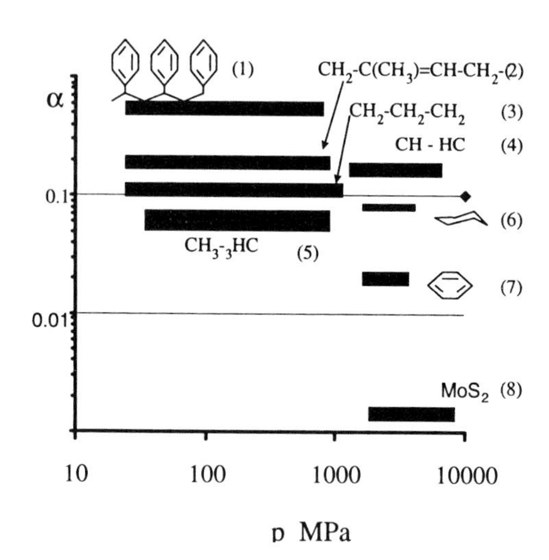

Fig. 17b Friction coefficient α versus the pressure p. 1 polystyrene (**77**), 2 polyisoprene (**56**), 3 polythene (**77**), 4 hydrogen terminated on diamond (**78**), 5 CH_3 versus CH_3 (**76**), 6 cyclohexane (**79**), 7 benzene (**79**), 8 MoS_2 (**74**).

6 ACKNOWLEDGEMENTS

The author wishes to thank Denis Mazuyer, Christophe Donnet, Sebastien Hollinger and Ton Lubrecht for their help during the preparation of the manuscript.

7 REFERENCES

1 **Cameron, A.**, *Principles of Lubrication*, J. Wiley, 1967.
2 **Tabor, D.**, in *Microscopic Aspects of Adhesion and Lubrication*, Georges, Ed. Tribology, Ser. 7, 1982, 651.
3 **Swalen, J.D., Allara, D.L., Andrade, J.D., Chandros, E.A., Garoff, S., Israelachvili, J., McCarthy, T.J., Murray, R., Peazse, R.F., Rabolt, J.F., Wynne, K.J.,** and **Yu, H.**, *Langmuir*, 1987, **3**, 932-950.
4 **Miyoshi, K. and Chung, Y.W.**, *Surface Diagnostics* in Tribology series in modern Tribology, World Scient., Singapore, 1993.
5 Tran Minh Duc, *Surf. Sci. Rev. Letters*, 1995, **2**, 833–858.
6 **Donnet, C.**, *Problem Solving Methods in Surface Analysis* Riviere, Myhra S. eds M. Dekker Inc. 1997.
7 **Colton, R.J.**, *Langmuir*, 1996, **12**, 4574-4582.
8 **Moore, D.J.**, *Leeds-Lyon symp.* Tribology series 33, Dowson, D., *et al.* ed., 1996, 21.
9 **Martin, J.M.** and **Le Mogne, Th.**, *Surf. Coatings Technol.* 1991, **49,** 427. **Donnet, C.**, *Condens. Matter News*, 1995, **4,** 9.
10 **Godet, M.**, *Wear*, 1984, **100,** 437. *Wear*, 1990, **136,** 29.
11 **Berthier, Y.**, in The Third body concept, *Leeds-Lyon symp.* 1995.
12 **Singer, I.**, *Langmuir*, 1996, **12,** 4486-4500.
13 **Singer, I.L., Le Mogne, T., Donnet, C.,** and **Martin, J.M.**, *J. Vac. Sci. Technol.*, 1996, **14,** 38.

14 **Tysoe, W.T., Surerus, K., Lara, J., Blunt, T.J.,** and **Kotvis, P.V.,** *Tribology letters,* 1995, **1,** 39.
15 **Brendlé, M., Turgis, P.,** and **Lamouri, S.,** *Tribology trans.* 1996, **39,** 157–165.
16 **Mischler, S., Debaud, S.,** and **Rosset, E.A.,** Landolt, in The Third body concept *Leeds-Lyon symp* Tribology series 31, Dowson D., *et al.* ed., 1995, 623.
17 **Shu, S.,** *Langmuir,* 1996, **12,** 4482-4485.
18 **Bhushan, B., Israelachvili, J.N.,** and **Landman, U.,** *Nature* 1995, **374,** 6.
19 **Georges J.M.,** in Fundamentals of friction: Macroscopic and Microscopic Process: Singer, I.L. and Pollock, H.M., Eds., *NATO ASI Series, Series E: Appl Science,* Vol. 220; Kluwer Academic Publishers: Dordrecht, 1992.
20 **Coy, R.C.** and **Jones, R.B.,** *ASLE trans,* 1981, **24,** 91.
21 a) **Korcek, S., Jensen, R., Johnson, M.,** and **Clausing E.,** *Proc. Int. Conf.* Yokohama, 1995, 733.
b) **Tohyama, M., Ohmori, T., Shimura, Y., Akiyama, K.,** and **Ashida, T.,** *Proc. Int. Conf.* Yokohama, 1995, 733.
22 a) **Willermet, P.A., Carter, R.O., Schmitz, P.J., Everson, M., Scholl, D.J.,** and **Weber, W.H.,** *Proc. Techn. Acad.* Esslingen 1996.
b) **Willermet, P.A., Carter, R.O., Schmitz, P.J., Zhu, P.J., Bell, J.C.,** and **Park, D.,** *Tribology Inter.,* 1995, **28,** 163–175.
23 **Bec, S.** and **Tonck, A.,** In The Third body concept *Leeds-Lyon symp.* Tribology series 31, Dowson D., *et al.* ed., 1995, 173.
24 **Bell, J.C.** and **Delargy, K.M.,** in Wear particles *Leeds-Lyon symp.* tribology series 29, Dowson D., *et al.* ed., 1992, 387.
25 **Belin, M., Martin, J.M., Mansot, J.L., Dexpert, H.,** and **Lagarde, P.,** *ASLE, Trans.* 1985, **29,** 523,.
26 **Sheasby, J.S.,** and **Caughlin, T.A.,** in The Third body concept *Leeds-Lyon symp.* Tribology series 31, Dowson D., *et al.* ed., 1995, 685. To be publ. In Tribology trans.
27 **Cann, P.M.E., Johnston, G.J.,** and **Spikes, H.A.,** *I.Mech.E. Conf. On Tribology,* paper C208/87, 1987, 543.
28 **Israelachvili, J.N.,** *Intermolecular and Surface Forces,* second edition Academic Press, 1992.
29 **Pauling L.,** *The Nature of Chemical Bond,* Cornell Univ. Press 1960.
30 **Mori, S., Hori, K.,** and **Tamai, Y.,** *J. JSLE, int.,* 1983, **4,** 73–76.
31 **Suslick.** *The chemical effects of Ultrasound,* Scientific American Inc., 1989.
32 **Bowden, F.P.** and **Tabor, D.,** *The friction and Lubrication of solids* Vol. I and II Oxford 1950, 1954.
33 **Ferrante, J.** and **Pepper, S.,** Fundamentals of tribology at atomic level in *New materials approaches to Tribology,* Pope, L., Fehrenbacher, L.L,. and Winer, W.O., MRS, 1989, 110.
34 **Vinet, P., Ferrante, J., Smith, J.R.,** and **Rose, J.H.,** *J. Phys. C* 1986, **19L** 467.
35 **Jacobson, B.O.** and **Vinet, P.,** *ASME J. of Tribology,* 1987, **109,** 709–713.
36 Workshop on *Physical and Chemical Mechanism in Tribology* Bar Harbor, Maine 1995, Paper in *Langmuir* 12, 1996.
37 **Singer, I.L.,** and **Pollock, H.M.,** Eds., Fundamentals of friction: Macroscopic and Microscopic Process, *NATO ASI Series, Series E: Appl Science,* Vol. 220; Kluwer Academic Publishers: Dordrecht, 1992.
38 **Nichols, F.,** Ed. *Materials Tribology.* MRS Bull. 1991, **16,** (O 30).
39 **Fischer, T.E.,** *Research Needs and Opportunities in Friction.* Special Publication CRTD, Vol. 28; The American Society of Mechan. Engineers, New York, 1994.
40 **Busham, B.,** *Microtribology: Proceedings of the 1st International Workshop.* October 12–13, 1992, Morioka, Japan. **Kaneko, R.,** and **Enomoto, Y.E.,** *Wear* 1993, 169.
41 **Belak, J.F.,** Ed. *Nanotribology;* MRS Bull. 18, 1993.
42 **Bhushan, B.,** Ed. *Handbook of Micro/Nato Tribology;* C. Series: Mechanics and Materials Science; CRC Press: Boca Raton, 1995.
43 **Person, B.N.J.** and **Tosatti, E.,** Eds. *Physics of Sliding Friction NATO ASI Series, Series E: Appl Science,* Vol. 311; Kluwer Academic Publishers: Dordrecht, 1995.
44 **Harrison, J.A.,** and **Brenner, D.W.,** *J. Am. Chem. Soc.,* 1994, **116,** 10399. **Harrison, J.A., White, C.T., Colton, R.J.,** and **Brenner, D.W.,** *Thin Solid Films,* 1995, **260,** 205.
45 **El Kissi, N.,** and **Piau, J.M.,** *C.R. Acad. Sci;* Paris, 1981, 314, Ii, 7.
46 **de Gennes, P.G.,** *C.R. Acad. Sci;* Paris, 1979, **288** B, 219.
47 **Brochart Wyard, F. and de Gennes, P.G.,** *C.R. Acad. Sci;* Paris, 1992, **314** II, 873.
48 **Crassous, J., Charlaix, E.,** and **Loubet, J.-L,** *Europhysics letter,* 1994, **28,** 37–42.
49 **Granick, S.,** *Science,* 1991, **253,** 1374–1379.
50 **Pethica, J.B.** and **Oliver, W.C.,** *Physi. Scr.* 1987, **T19,** 61.
51 **Carpick, R.W., Agraït, N., Ogletree, D.F.,** and **Salmeron, M.J.,** *Vac. Sci. Technol.* B, 1996, **14,** 1289.
52 **Krim, J., Solina, D.H.,** and **Chiarello, R.,** *Phys. Rev. Lett.,* 1991, **66,** 181.
53 a) **Chen, Y.L., Helm, C.A.,** and **Israelachvili, J.,** *J.Phys. Chem.,* **95,** 10736–10747.
b) **Robbins, M.O.** and **Smith, E.D.,** *Langmuir,* 1996, **12,** 4543–4547.
54 **Spikes, H.A.,** *Langmuir,* 1996, **12,** 4567–4573.
55 **Silberzan P., Léger, L., Ausséré,** and **Benattar, J.J.,** *Langmuir,* 1991, **7,** 1647–1651.
56 **Georges, J.-M., Tonck, A.,** and **Mazuyer,** *et al. J. Phys., II,* France, 1996, **6,** 57-76.
57 **Meyer, E., Overney, R., Brodbeck, D., Howald, L., Lüthi, R., Frommer, J.,** and **Güntherodt, H.J.,** *Phys. Rev. Lett.* 1992, **69,** 1777–1780.
58 **Luedtke, W.D.** and **Landman U.,** *Comput. Mat. Sci.,* 1992, **1,** 1.
59 **Luedtke, W.D.** and **Landman, U.,** in Physics of Sliding Friction, Person, B.N.J. and Tosatti E. Eds. *NATO ASI Series, Series E: Appl Science,* Vol. 311, p325; Kluwer Academic Publishers: Dordrecht, 1995.
60 **Landman, U.,** and **Luedtke, W.D.,** MRS. Bulletin 1993, **17,** 36.
61 **Georges, E., Georges, J.-M.** and **Hollinger, S.,** submitted to *Langmuir* 1997.
62 **van Alsten, J.** and **Granick, S.,** *Phys. Rev. Lett.* 1988, **61,** 2570. **Granick, S.,** *Science,* 1991, **253,** 1374.
63 **Cann, P.M.** and **Spikes, H.A.,** *STLE ASME,* 1993-TC-**1B**-1.
64 **Millot, S., Tonck, A., Georges, J.-M. Coy, R.C., Schlipper, A,** and **Williamson, B.P.,** to be published 1997.
65 **de Gennes, P.G.,** *Macromolecules,* 1981, **14,** 1639–1644. **de Gennes, P.G.,** *Ad. Colloid Interface,* 1987, **27,** 189–194.
66 **Georges, J.M., Millot, S., Loubet, J.L.,** and **Tonck, A.,** *J.Chem. Phys.,* 1993, **98,** 7345–7360.

67 **Tomlison G.A.,** *Phil. Mag. S.* 1929, **7,** 905.
68 **McClelland, G.M.,** in *Adhesion and Friction,* Grunze, M. and Kreuser, H.J. Eds. Springer series in Surface Sciences, 1985.
69 **Brenner, D.W.,** *Phys. Rev.,* 1990, **B42,** 9458. **Harrison, J.A., White, C.T., Colton, R.J.,** and **Brenner, D.W.,** *Surf. Sci.,* 1992, **271,** 57. **Harrison, J.A., White, C.T., Colton, R.J.,** and **Brenner, D.W.,** *J. Phys. Chem.* 1993, **97,** 6573.
70 **Blaise, G., le Gressus, C.,** *IEEE trans. Elec. Ins.* 1991, **27,** 427–479. **Blaise, G. le Vide,** 1995, **275,** 164–169.
71 **Fayeulle, S., Berroug, H., Hamzaoui, B., Treheux, D.,** and **le Gressus, C.,** *Wear,* 1993, 162–164, 906–912.
72 **Hirano, M.** and **Shinjo, K.,** *Phys. Rev.,* 1990, **B 41,** 11837.
73 **Sokolov, J.B.,** *Phys. Rev.* 1992, **B 42,** 1262.
74 **Martin, J.M., Donnet, C., le Mogne, T.,** and **Epicier, T.,** *Phys. Rev.* 1993, **B 48,** 10583.
75 **Israelachvili, J.N.,** In *Handbook of Micro Nano Tribology:* Bhushan, B., Ed.; CRC Series: Mechanics and Materials Science, CRC Press: Boca Raton, FL, 1995; pp 267–319.
76 **Briscoe, B.J.** and **Evans, D.C.B.,** *Proc. Roy. Soc. Lond.* 1982, **A380,** 389–407.
77 **Briscoe, B.J.,** in Fundamentals of friction: Macroscopic and Microscopic Process, Singer, I.L., Pollock, H.M., Eds *NATO ASI Series, Series E: Appl Science,* Vol. 220, Kluwer Academic Publishers, Dordrecht, 1992.
78 **Gardos M.N.** *Carbon,* 1990, **28,** 783. Final report, Vol. 1, 2, 3, WRDC-TR-90-4096, Huges Air. Corp. El Segundo, CA, 1990.
79 **Tsubouchi, T., Hata, H.,** *Tribology Int.* 1995, **28,** 335-340.

Invited Papers

Test methods in tribology

A F ALLISTON-GREINER
Plint and Partners Limited, Wokingham, UK

SYNOPSIS

What level of test method is required in order to be sure that the data generated regarding the friction and wear performance of a material is relevant to the final application? This paper examines the issues at stake and seeks to provide a framework for making a better informed choice. Crucial to this is the process of defining the contacting environment. Traditional test machine types are examined in terms of the overall energy dissipation patterns they produce and the implications of these for machine selection are examined.

1 INTRODUCTION

A great deal of effort is expended in the development, analysis and revision of test methods in the field of tribology.

Such methods are a vital part of solutions engineering because they give design and development information about the performance of materials and lubricants that cannot be obtained through numerical analysis or other modelling techniques. It is an unavoidable fact that the response of any material to sliding is dependent both on the material properties (surface and subsurface) and on the contacting environment.

The process of defining the contacting environment is something that has occupied engineers, physicists, chemists and materials scientists for many decades and still remains perhaps the greatest single challenge for this multi-disciplinary subject of tribology.

The current need to exploit new materials, lubricants and surface engineering solutions demands test machines and procedures which can produce reliable and relevant information. Established equipment and methods may or may not be of use, but how can this be determined?

1.1 The testing options

The test options can be divided into four categories and in broadest terms into price and timescales as detailed in Table 1.

1. Field trials are the real thing, the complete mechanisms running under a range of real life conditions. All other types of test ultimately have as their goal the prediction of the performance of the component in the final application. Field trials are the most expensive to run, the most time consuming and, unfortunately, are the least controllable in terms of having access to the test pieces and the performance parameters: but at least the contact environment is correct. The field trial is seldom used as a pure development tool.
2. Component or assembly testing takes a defined part of a system (e.g. an automobile transmission) and subjects it to a simulation of the real performance conditions. In the case of the transmission system this has the benefit of being stationary and one can quite easily measure powers, temperatures etc.

   The timescale of tests can be much reduced through accelerated testing (i.e. by running continuously rated at high stress conditions). While the overall performance of the system can be controlled more closely, the conditions in individual contacts are not controlled (e.g. the temperature between the gear teeth which is all important in terms of lubricant protection) nor can individual contacts be studied readily (e.g. the wear performance of the shaft seals). The contact environment is close to reality but there is still this lack of control over the conditions in the critical contact areas.
3. Bench tests are where individual contacts are investigated either by the selection of materials and contact conditions to model the real situation or as a means of providing a database on the response of the material or lubricant to different situations. These test methods are relatively cheap, providing indicative information and design information quickly and easily. The level of control over test conditions is increased, because single contacts are being studied. But the practical issue to face here is how to relate what is seen in the bench test to how the product might perform in practice. How significant is the fact that the whole contact environment is not reproduced?

   The apparent simplicity of standard test methods (e.g. those adopted by ASTM, ISO, IP, DIN, JASO etc.) belies the complexity of the contacts they are deemed to be modelling, or wear processes they are simulating. Many of the established test methods predate the theoretical understanding of the contacting conditions and wear processes found in machine elements. This fact alone can severely limit their relevance.

4. Physical tests are well controlled, cheap and repeatable tests to assess the quality or nature of a product. Their relation to final product performance is remote but they form the basis of screening out definite no-hopers and as an aid to quality control. Their importance may be judged from a consideration of the number of standards related to such tests that must be passed for approval. Examples include basic material properties such as hardness, bending strength, ductility etc., chemical properties such as elemental composition, corrosion resistance, volatility, flash point etc. These physical tests are written up as standard test methods and provide very important design and quality information on products.

   Standard test methods are very attractive because everyone carries out the same test procedure (often, but not always on the same make of machine) and there is a great deal of effort put into defining the level of 'certainty' which can be read into the data by running inter-laboratory tests on repeatability and reproducibility. But while this works admirably for the physical tests described above, all is not quite so simple for the bench tests for friction and wear because these are not intrinsic properties, but a function of the environment (which must include the test machine) as well.

1.2 The contacting environment

At a superficial level the contacting environment can be defined in terms of the contact pressure, the contact speed, some representative temperature and conditions of lubrication and atmosphere. However none of these is completely straightforward to define – either for the practical contact or in the test machine.

If the materials in contact are stationary then the apparent contact pressure is simply determined from a knowledge of the normal load, the bulk elastic properties and the geometry of the surfaces. This is relatively straight forward for bulk materials and more difficult to define for layered or coated systems.

When the surfaces are in motion the picture can change dramatically. What was a steady normal load becomes a variable influenced by the stiffness of the material and the supporting structure and driven both by the dynamics of the system and the evolution of the surface roughness and wear particles in the contact. Blau (**1**) suggested a hierarchy of interaction levels which are responsible for the evolution of the measured friction and wear. At the largest scale (most real contacts and many test machines), the friction results from the interaction of all three levels: interfacial media, the bounding solids and the machine and its environs. The friction and wear response is not separable from its environment.

Table 1 Test options cost and benefits.

Type of Test	Examples	Duration	Cost	Comments
Field trial	Production units	Months	£250,000 plus	The real thing, subject to the variation of the real world
Component	Gear box test Engine test Bearing test	Weeks	£10,000 – £50,000	Better overall control than field trial, still subject to variations of the real world.
Bench	Four ball Falex Pin on disc Reciprocating	Days	£500 – £5,000	Well controlled, repeatable, and can be related to practical performance
Physical	Tensile strength Hardness Composition	Hours	£50 – £500	Well controlled, repeatable, but relates to specific properties not performance

In practical terms this means that the mechanical environment influences the wear and friction processes in the contact. For the test machine, which are in many cases simpler and lighter systems than the practical mechanism, this means that the method of loading, the method of specimen support, the method of friction measurement and the method of driving the specimens could all have an influence on the data obtained.

The level of the sensitivity will of course depend on the materials being tested. A classic example of this is in the wear testing of alumina in ball on disc contact where Gee (**2**) found three orders of magnitude change in a linear wear measurement according to the type of loading system (and therefore vertical stiffness of specimen support) adopted. This work has been extended (**3**) with the aim of making recommendations for the reporting of contact dynamic stiffness as an additional parameter along side friction coefficients and wear rates.

Contact speed is obvious in the case where one object is moving over a fixed surface. Where both surfaces are moving then the contact speed is defined by the relative velocity. However, this cannot then be used simply as a sliding speed for a uni-directional test since this ignores how the frictional energy is dissipated in the latter case.

In some applications there are cyclic speed variations as determined by the drive mechanism but in others there may be a nominally constant speed with higher frequency variations superimposed. As with normal load these are caused both by the stiffness of the material and supporting structure, but in this case also by the friction conditions in the contact.

The ability to adjust the stiffness of the specimen support can be advantageous: for example in the study of the transitions from static to dynamic friction encountered in machine tool slideways or wet clutch systems (judder). If the specimen supports are too rigid, the stick-slip action will be suppressed and no information will be obtained. Plint and Plint (**4**) and Coates (**5**) have pioneered such methods as lubricant development tools and these have yielded a much higher degree of sensitivity to additive types and treatment levels than an older method which used an intrinsically stiffer system (**6**).

An often overlooked parameter is the contact frequency. This is an important factor to consider in contacts where there is tribochemical activity – either in the form of lubricant additive reactions with the surfaces or with the involvement of water vapour or oxygen in the formation of gels and oxides. There is often a dynamic interplay between the removal of such films during the contact and their reformation or repair during the non-contact time. If the frequency of contact is too high, there may be insufficient time for the protective films to form and thus failure will be accelerated.

The conditions of lubrication may also be difficult to define, especially when the machine components are operating with a mixture of boundary and hydrodynamic lubrication. In these circumstances it is not possible to define the proportion of the load carried by the fluid, the surface films and the surface asperities. The lubrication regime can be related to the measured friction coefficient in some manner (**7**, **8**) or to measurements of the electrical contact resistance (**9**), but neither of these can determine how the load is shared.

Lubricated mechanisms with cyclic loading and/or speed characteristics are a challenge to model in a test machine. However if the focus of the test program is to look at the wear performance of the material then it is only the critical parts of the cycle that must be modelled – the places where the fluid film lubrication breaks down or where the contact is most distressed. This will be discussed later in the paper.

For the test machines where the contact is lubricated, the atmosphere does not play a strong role in determining wear behaviour. However as soon as such traditional contacts are subjected to the conditions found in aerospace or even space environments it becomes clear that ambient air can provide some vital sources of lubrication – oxygen and water vapour. Low atmospheric water is encountered in many land-based systems and this forces the experimenter to take the atmospheric environment into account. An example of this problem is being presented at the WTC (**10**). In fretting conditions the contact is starved of oxygen and this results in a different wear mechanism from contacts that are open to the atmosphere.

1.3 The scale of the contact

When building a model test considerations of scale are also paramount. This is where it is hazardous simply to attempt to define the real life conditions (load, speed, temperature etc.) and apply them to a small test piece on a test machine. The first major concern is to do with the temperatures reached by the test specimens.

If large amounts of energy are being dissipated in small test specimens with supporting structures that do not allow the heat to escape then it is clear that the specimens will become very hot. Hence the bulk temperature may exceed what is experienced in practice and this will undoubtedly produce transitions in wear or frictional response. The temperature reached at the surface of the contact (the flash temperature) is also strongly influenced by the width of the contact. These flash temperatures are responsible for many wear and friction effects. Lim and Ashby (**11**) used these concepts to build their models for the wear transitions in steel and such models are being applied other materials.

Material structure and microstructure is another scale consideration. Where the scale of the microstructure is much less than the contact area, the material can be considered to be homogeneous. But where the phase size is of the same order as the contact width, then the surface roughness on the counterface material and the properties of the different phases will influence the wear and friction behaviour of the material. A classic example of this is in reinforced plastics where there is an interplay between a ploughing wear mode for low levels of glass filler and a smearing or patching mechanism for higher filler contents (**12**).

In abrasion this is reflected in the size of the abrasive particles relative to the matrix of the composite (**13**). The relative hardness of the abrasive to the material also has a strong influence over the type of abrasive action: whether high stress or low stress.

1.4 The criteria for a successful bench test

The purpose of tribological bench tests is to improve accessibility to the friction or wear processes and thus to facilitate a more detailed investigation than is possible with the real mechanism. A key objective in the design of the test is to change dependent variables of the real process into independent or non-interacting and hence controllable variables in the test.

Drawing conclusions from single tests is also perilous and the normal method of providing a level of certainty would be ensure that certain bench-mark or reference materials that have a known good, bad or indifferent performance in practice are ranked in the correct order of merit on the bench test.

Most would accept that, even given all the issues outlined above, the most important criterion is that the test should reproduce the wear and/or failure mechanisms of the application. This is a necessary but not sufficient criterion since it is possible to produce such wear mechanisms by a variety of routes.

2 ENERGY DISSIPATION AND THE ENERGY PULSE CONCEPT

All wear processes are driven by temperature, be they the formation of oxides on the surfaces, the transformation of microstructure, the formation or break-down of lubricant additive or other tribochemical films, the melting of the surface (the *PV* limit of the material) or thermal stress induced failure. To be more specific, wear occurs as the result of the dissipation of frictional energy in the contact and this is irresistibly accompanied by a rise in temperature.

2.1 Friction power intensity

The friction power intensity (FPI) as presented by Matveesky (**14**) is simply defined as of the amount of energy pumped into the rubbing surfaces as they pass through the contact zone. The temperature achieved in the contact and in the bulk is directly related to the FPI and the size and thermal characteristics of the materials and their supports.

Friction Power Intensity,

$$Q_F = \frac{\mu P v_s}{A} \quad W/mm^2$$

μ is friction coefficient, P is the normal load, v_s is the sliding speed and A is the apparent area of contact. Typical practical contacts such as gears have FPI's in the range 5,000 to 20,000 W/mm^2. The FPI defines only the rate of energy generation and does not take into account the timescale over which this energy can be lost to the contacting materials. This timescale clearly has implications for the amount of damage caused in the contact.

2.2 Energy pulse

The Energy Pulse (EP) as presented by Plint (**15**) is the product of the FPI and the contact transit time. The EP therefore takes into account the length of time during which the material is subjected to energy input during its transit of the contact zone. Typically the time taken for a point on the moving surface to traverse the contact zone is of the order of one millisecond. Successive transits may occur perhaps every 50 milliseconds.

Energy Pulse,

$$EP = \left[\frac{\mu P v_s t_t}{2A}\right] \quad J/mm^2$$

t_t is the transit time and the factor of 2 makes the arbitrary assumption that the energy generated in the contact zone is equally split between the two surfaces. In practice, the slower moving surface will have the higher Energy Pulse.

The EP can be seen to be analogous to the Archard Wear Law, but using the friction force rather than the applied load. This is perhaps more logical since it takes into account the actual rubbing conditions (but assumes that μ can be measured on the test machine).

Archard Wear Law,

$$\Delta V = \left[\frac{kPv_s t_t}{A}\right] \quad mm^3$$

Each Energy Pulse can be regarded as an incremental contribution to wear or surface damage in the contact. The sum of the Energy Pulses can be used as a measure of the total wear.

This fact is illustrated by the work of Bell and Colgan (**16**). Using a derivative of the Archard Wear Law they produced very accurate predictions of wear patterns between cam and finger followers in engines. This confirmed that maximum damage occurred not at the point of minimum film thickness (although the wear is also high at that point), but at the point of maximum contact dwell time (Energy Pulse in the terms of this paper).

In machine components there can be very high FPI's but because the contact durations are short, the Energy Pulse is low and hence the incremental damage is low. In simple sliding test machines there is no Energy Pulse as such: one specimen receives a continuous flow of energy and therefore the rate of damage will be correspondingly high. The calculation of the FPI is still of use in such cases. Mohrbacher et al. (**17**) for example have demonstrated that the cumulative friction energy dissipated in a fretting contact is directly proportional to the volumetric wear loss. The Energy Pulse parameter can be used to define the severity of contacts both in the application and on the test machine.

3 TRIBOLOGY TEST MACHINE GROUPS

Using these concepts, tribology test machines can be divided into three basic categories (**18**) and their limitations defined.

3.1 Group 1 machines – continuous energy pulse, thermally self-regulating

Here the point of contact is stationary with respect to one of the specimens and subject to constant speed uni-directional sliding. This Group includes pin-on-disc, block-on-ring, ring-on-ring, pin-on-plate, crossed cylinder, journal bearing and sliding 4-ball.

In Group 1 machines, the EP for the fixed point of contact specimen is continuous: it lasts for the duration of the test. Instead of brief rubbing episodes frequently repeated, the Group 1 machine subjects one specimen to continuous rubbing and the associated temperature field dominates the situation. The test configuration defines the thermal conditions in the contact. The contact temperature is self-regulating and cannot be controlled as an independent variable.

The fact that the Energy Pulse is continuous implies that these contacts are effectively accelerating the wear process by the ratio between the test duration and the contact transit time in the real contact. This factor could be many thousands. This highlights the perils of applying 'real contact conditions' to a simple sliding test.

3.2 Group 2 machines – cyclic energy pulse, thermally self-regulating

Here the point of contact moves over both contacting surfaces. This Group includes a number of component test machines, using idealised or standardised components such as gears (e.g. FZG and Ryder), cam/follower (MIRA (**19**)) and simplified machines such as two roller, Amsler, reciprocating Amsler and other slide/roll devices.

In common with Group 1, in Group 2 machines the test configuration defines the thermal conditions in the contact. The contact temperature is thus self-regulating and cannot be controlled as an independent variable. So while the machines can emulate real contacts, there is still a lack of control over this key parameter.

3.3 Group 3 machines – minimal energy pulse, independently thermally controlled

Here the point of contact is stationary with respect to one of the specimens and subject to reciprocating sliding. This Group includes pin-on-plate, ball-on-plate and piston ring-on-liner, with amplitudes that vary from fretting to long strokes.

By comparison to Group 2, the Group 3 machines (except in the case of the piston ring on liner contact near end stroke) do not set out to be the same as the real contact under investigation, but aim to simulate the intimate contact conditions in a controllable and accessible way. In this respect these machines come closest to fulfilling the requirements of an effective bench test.

In Group 3 machines the sliding velocities are deliberately maintained at low levels in order to minimise frictional heating and, in the case of lubricated tests, to promote boundary lubrication. This reduction of frictional heating means that there can be independent control of contact temperature by external control of the bulk temperature. These principles were originally promoted by Mills and Cameron (**20**).

3.4 Some implications

Because of the entrainment conditions associated with constant speed sliding, Group 1 machines usually require heavily loaded contacts in order to overcome hydrodynamic or mixed lubrication regimes and thus promote lubricant film failure. These machines are thus not good models of real lubricated contacts.

In the case of the 4-Ball Extreme Pressure test (**21**) (a Group 1 machine) one of the test procedures lasts as little as 10 seconds and the 1/2" test balls can be blued or welded together.

By contrast in the FZG Gear test (**22**) (a Group 2 machine) a load stage lasts 15 minutes but the bulk temperature of the test pieces is only just above the temperature of the oil in which they are immersed.

The difference is that the FZG specimens are 50 times heavier, have 20 times the surface area and have a very much larger contact size than the 4 balls. Heat dissipation by conduction, convection and churning are important loss mechanisms for the gears. Similar loss mechanisms are not there for the 4 balls. Therefore, although the FPIs are much greater for the FZG test, the 4-Ball is the more severe test because the Energy Pulse is continuous.

The FPIs for a range of standard Group 1 machines for lubricant evaluation are shown in Fig. 1. The 4-Ball and Timken machines have a greater FPI than the Falex or Reichert devices. This energy input is critical in determining the contact temperature and therefore the action of lubricant additives. It is not surprising therefore to find that the different tests rank specific additives in a quite different order.

Sulphur additives will perform well in a 4-Ball test because the contact conditions are extremely severe: the FPI is 40,000 W/mm^2. Sulphur additives require activation temperatures of several hundred degrees. By contrast phosphorus additives perform well on the milder Reichert test since the activation temperatures are much lower.

The problem with liquid lubrication arises at high speeds where the lubricant is either thrown from the moving specimen or forms a hydrodynamic wedge: it is difficult to control the lubrication condition.

The issue of the dynamic response of the machine and test specimen support structure is very important, not least because the simplicity of the contact encourages many to design and build their own machines. Each has a unique frequency response.

This has proved a major problem in finding agreement between pin-on-disc type machines working with ceramic materials. The international round-robin under the VAMAS program determined up to 500% variation in friction and wear data between laboratories and the two primary variables were conditions of humidity and machine dynamic response (**23**, **24**).

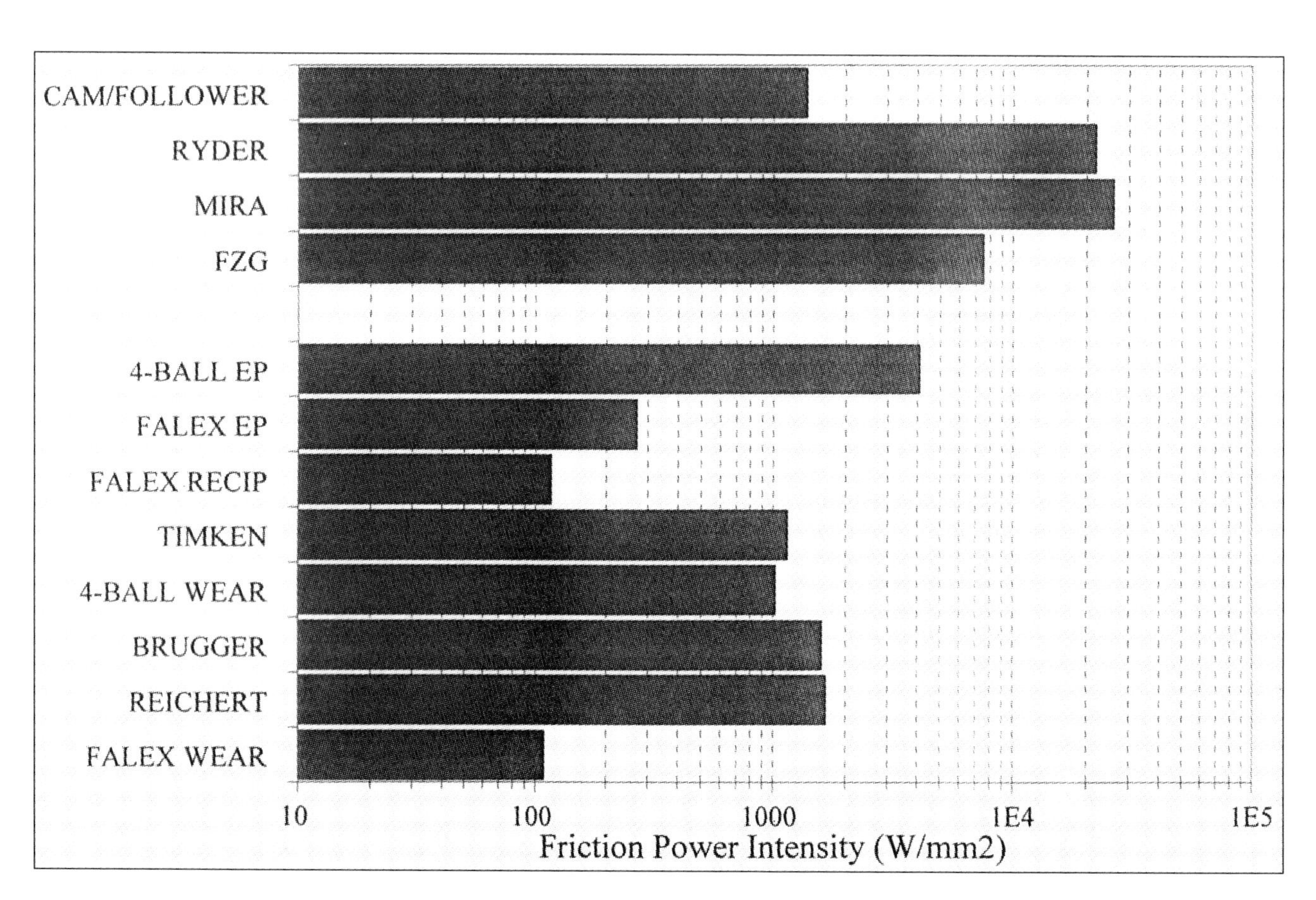

Fig. 1 Friction power intensity – under initial contact conditions.

3.5 Uses of Group 1 machines

Group 1 machines are often used for fundamental wear studies of materials including metals, plastics, composites and ceramics both coated, solid lubrication or dry. They are less successful with liquid lubrication. Their principal advantage is that it is easy to cover wide load and speed ranges and therefore obtain a broad sweep of material performance. Wear mapping and parametric studies are readily performed.

Provided it is acknowledged that such problems exist and that they are more important for some materials then for others, then it seems that progress can be made.

3.6 Uses of simplified Group 2 machines

In addition to full component tests, Group 2 includes

devices with simplified specimens which can achieve a close approximation to the motion in machine components (e.g. gears, cams, plunging joints and mechanisms). Test-piece production costs can be less than if using the full-scale components.

1. Pure Rolling Machines
   Here two specimens (often rollers) are loaded together and are rotated at the same speed. The motion is pure rolling and such machines are used to address particular problems in the lubrication of gears and drives associated with lubricant performance in the piezo-viscous region. They are also used to study pitting failure (rolling contact fatigue), caused by the cyclic stressing of the surfaces.
   These are important machines because there is no other reasonable test to assess these conditions and, with appropriate choice of conditions they can simulate practical contacts. However the thermal characteristics of the specimens influence how the heat is conducted away and this can have a profound effect on the lubrication conditions in the contact.

2. Combined Rolling/Sliding
   Here two specimens (usually rollers) are loaded together and are rotated with a locked-in speed difference between them. These are used to emulate the conditions found in highly loaded contacts and to assess the performance of materials, coatings and treatments under the tractive conditions, as well as to test the effectiveness of lubricants.

3.7 Uses of Group 3 machines

Group 3 machines have been used for fundamental and applications studies in the lubricants industry with a particular emphasis on the lubricant chemistry and lubricated wear mechanisms. These rigs have an obvious similarity to the motion experienced by many practical components and are therefore being used more and more for wear studies of coatings and surface treatments.

Similar parametric studies to the Group 1 machines can be performed with limitations on the absolute sliding speed. But this is the whole basis of this Group. Sliding speed is simply a heat generator so it is much better to control temperature by external means than to have uncontrollable, unmeasurable values.

In Group 3 machines, the Energy Pulse has been replaced by external heating and this clearly has implications for the development of wear in the contact. But with the contact operating under dry or boundary lubrication conditions, the build-up of surface damage will only take time. Both friction and wear have been found to be highly repeatable and reproducible in Group 3 machines (**25**).

These machines are also limited in their dynamic responses. To achieve good, smooth operation at high amplitude and high frequency is difficult and requires very careful balancing of the reciprocating masses.

In addition wear debris involvement in the contact is a factor that can effect the friction conditions. This is present in the case of Group 1 machines but the influence of the third body is stronger in the reciprocating contact. The particles are more readily entrained and held within the contact.

3.8 Summary of sliding wear test groups

Group 1 machines are the least representative of practical contacts, the contact temperature is out of the control of the experimenter and the continuous energy dissipation on one specimen means that the build-up of surface damage is very much faster than in practical contacts. This very fact, however, makes then useful devices for exploring the limits of the material and the transitions in wear.

Group 2 machines are representative of many contacts where there is a combination of sliding and rolling (i.e. the point of contact moves over both surfaces). However, since there is a significant energy input into the contact the temperature is again out of the control of the experimenter. These machines can be used as simulators by selecting contacting conditions to match the practical application. Their advantage can come from having smaller and sometimes cheaper test components but definitely better controlled test conditions than in a practical contact.

Group 3 machines offer the highest degree of control over the contact temperature. The reciprocating contact models some practical contacts closely, and the ability to control the thermal conditions makes them useful devices in simulating conditions of intimate contact in many mechanisms.

3.9 Rubbery materials: additional considerations

In rubber the friction force depends strongly on the real contact area between the rubber and counter surface, the interfacial shear strength and the deformation properties of the rubber (**26**). Each of these factors depends on others. The area of contact relates to the hardness and surface roughness, the applied load and the relative radius of curvature of the contacting surfaces. The interfacial shear strength depends on the polymer itself and the presence or absence of lubricants, solid or liquid.

For the sliding of rubber against a rough (abrasive) surface such as silicon carbide paper there is a gradual change in friction with sliding rate. This is understood in terms of a combination of effects including adhesion, ploughing hysteresis and tearing. The friction force F is determined by contributions from four different sources:

$$F = F_a = F_h = F_v = F_t$$

where F_a is the interfacial adhesive component, F_h is the hysteretic component arising from energy losses associated with bulk deformation, F_v is the viscous component arising from viscous dissipation in any lubricant, and F_t is due to any tearing.

To make measurements of the friction coefficient of rubber it is important to have a well-defined contact geometry. A flat-on-flat geometry tends to give poor repeatability due to problems associated with achieving and holding the surfaces flat against each other. This requires both the production of smooth, flat surfaces and also a rigid test machine. Even if the specimens can be held in the correct orientation there are still problems associated with frictional tilt during motion.

This tilt is a direct result of the flexibility or low elastic modulus of the rubber material. A metal ball on a pivoted arm loaded against a rubber flat will indent to a considerable distance. This indentation gives rise to two effects. First the line of action of the friction force shifts out of the horizontal plane. Second the large contact area gives rise to high friction force, even at quite modest normal loads. In both these respects rubber differs from harder materials such as metals and ceramics and this must be catered for in the machine design (**27**, **28**).

With rubber and other polymeric materials there is also a strong influence of sliding direction on the evolution of the wear surface. The unidirectional sliding of natural rubber for instance produces much rubber debris and what is known as pattern abrasion: the pattern consists of a series of millimetre sized ridges and valleys on the surface resulting from the tearing of the surface.

In reciprocating sliding this abrasion pattern is not evident. The change in direction of sliding suppresses the formation of the pattern abrasion and the wear rate is lower. A finer scale of roughness (microns rather then millimetres) is produced and this is termed intrinsic abrasion.

3.10 Abrasion and erosion: special considerations

Abrasion and erosion testing of materials require different types of machine since a third body is deliberately involved. Material loss is still due to the dissipation of energy, and the abrasion resistance of a material can be defined in this way (**29**, **30**). Abrasion tests either use loose grit or sand of a well defined size or abrasive paper bonded to a substrate. One of the principal problems with these tests is controlling the state of the abrasive.

If a pin-on-disc test is carried out with abrasive paper, after a few rotations the abrasive becomes blunt or the paper becomes clogged with wear particles. One way around this is to run the pin across the disc in a spiral pattern, ensuring that fresh abrasive comes into contact all the time.

Some methods use a random-walk type motion to even out the effect of clogging and produce particle motion. Reproducibility in these cases is good, but the relevance to the 'real' application may be less clear. All standard methods define the grit to be used, but wear rates are profoundly affected by the relative hardness of the grit to the material being abraded.

With hopper fed abrasives it is important to ensure a uniform supply rate. As the feed rate is increased, the greater amount of abrasive will result in particle loading and clog the contact, possibly changing the process from low to high stress abrasion.

As with the sliding wear tests the most important criterion of success is that the test method reproduces the predominant mechanism of surface damage. Attention to the contacting environment remains is also crucial in abrasion testing (**31**).

In erosion tests the flow rate and angle of impact both affect the dynamics of the erosive particle and the wear mechanism produced. The sharpness or angularity of the particle and its hardness and fracture toughness also have a large influence on the wear process (**32**).

A number of standards exist for abrasion and erosion but these are usually specific to a material or product group. This is an area where current research is helping our understanding and may lead to better laboratory test methods.

4 CONCLUSION

The selection of a test machine or test method for modelling tribological contacts in a practical application depends on having a clear picture of the contacting environment and the scale of the contact.

Given that the wear in any contact is as a result of energy dissipation, it is important that this becomes a consideration in the selection of a test machine for wear testing. The Friction Power Intensity and the Energy Pulse have been shown to be useful global parameters in defining the severity of contacts both in the application and on the test machine. Procedures can be developed on standard test machines by giving attention to the way energy dissipation drives wear or failure in the contact.

The knowledge is available for those involved in the development of lubricants, materials and surface treatments to make an informed choice of a test machine, to design a suitable test procedure and to have confidence that the data obtained is relevant to the application.

5 REFERENCES

1 **Blau, P. J.** Scale Effects in Steady-State Friction. *Tribology Transactions*, 1991, **34**, 335-342.

2 **Gee, M. G.** Effect of Machine Dynamics on the Sliding Wear of Alumina. *Wear Testing of Materials, ASTM STP 1167*, 1992, 24-44.

3 Development and Validation of Test Methods for Coatings. *European Commission FASTE Project Number SMT4-CT 95-2029*, 1995.

4 **Plint, A.G., Plint, M.A.** A New Technique for the Investigation of Stick-Slip. *Tribology International*, 1985, **18**(4), 247-249.

5 **Coates, D. A.** The Development of a Reciprocating Rig Technique to Assess the Stick-Slip Properties of Slideway Lubricants. *Presented at I. Mech. E., The Mission of Tribology Research*, December 1992.

6 CM Stick-Slip Test, Cincinatti Milacron Marketing Company, Cincinatti, OH 4509-9988, USA.

7 **de Gee, A. W. J., Begelinger, A. and Salomon, G.** Mixed Lubrication and Lubricated Wear. In *Proc. 11th Leeds-Lyon Symposium on Tribology*, Dowson, D., Taylor, C. M., Godet, M. and Berthe, D., Eds., Butterworth, 1985, 105-116.

8 **Schipper, D. J.** Transitions in the Lubrication of Concentrated Contacts. Ph. D. Thesis, Delft University of Technology, *ISBN 90-9002448-4*, 1989.

9 **Furey, M. J.** Metallic Contact and Friction Between Sliding Surfaces. *ASLE Transactions*, 1961, **4**, 1-11.

10 **Bayliss, R. W., Stirling, C. A., Alliston-Greiner, A. F. and Plint, A. G.** The Development of Testing of Polymer-Matrix Composites Running Under High-Speed Reciprocating Conditions. *I. Mech. E. World Tribology Congress*, Session TH2, 1997.

11 **Lim, S. C. and Ashby, M. F.** Wear-Mechanism Maps. *Acta metall.*, 1987, **35**(1), 1-24.

12 **Friedrich, K.** Wear Models for Multiphase Materials and Synergistic Effects in Polymeric Hybrid Composites. In *Advances in Composite Tribology*, Friedrich, K., Ed., Elsevier Science Publishers, 1993, 209-273.

13 **Hutchings, I. M.**, ABrasive and erosive wear of metal matrix composites, *Proc. 2nd European Conf. on Advanced Materials and Processes, Euromat '91*, **2**, Institute of Materials, 1992, 56-64.

14 **Matveesky, R. M.** The Critical Temperature of Oil with Point and Line Contact Machines. *ASME Transactions*, 1965, **87**, 754.

15 **Plint, M A.** The Energy Pulse: A New Criterion and its Relevance to Wear in Gear Teeth and Automotive Engine Valve Trains. *Proceedings of the XI NCIT*, January 22-25, 1995, 185-192.

16 **Bell, J. C. and Colgan, T. A.** Critical Physical Conditions in the Lubrication of Automotive Valve Train Systems. *Tribology International*, 1991, **24**(2), 77-84.

17 **Mohrbacher, H., Blanpain, B., and Celis, J.-P.** Friction and Wear Mechanisms on CVD Diamond and PVD TiN Coatings under Fretting Conditions. In *ASTM STP 1278*, Bahadur, Ed., American Society for Testing and Materials, 1996, 76-93.

18 **Plint, A. G.** Machines and Methodologies for Wear Testing Extreme Pressure and Anti-Wear Properties of Lubricants. *Proceedings of the XI NCIT*, January 22-25, 1995, 375-386. CEC L-38-T-87 Gasoline Engine Valve Train Scuffing Test.

19 **Mills, T. N. and Cameron, A.** Basic Studies on Boundary, EP and Piston-Ring Lubrication Using a Special Apparatus. *ASLE Transactions*, 1982, **25**, 117-124.

20 ASTM D4172-88. Standard Test Method for Wear Preventive Characteristics of Lubricating Fluid (Four Ball Method), *ASTM Annual Book of Standards*, Volume **5**.

21 ASTM D4998-89 Test Method for Evaluation of Wear Characteristics of Tractor Hydraulic Fluids, *ASTM Annual Book of Standards*, Volume **5**.

22 **Czichos, H., Becker, S. and Lexow, J.** MultiLaboratory Tribotesting: Results from the Versailles Advanced Materials and Standards Programme on Wear Test Methods. *Wear*, 1987, **114**, 109-130.

23 **Gee, M. G.** Results from a UK Interlaboratory Project on Dry Sliding Wear of Alumina, in *Wear Testing of Advanced Materials, ASTM STP 1167*, Divakar R and Blau, P. J., Eds., 1992, 129-150.

24 ASTM G133-95 Standard Test Method for Linearly Reciprocating Ball-on-Flat Sliding Wear. *ASTM Annual Book of Standards*, Volume **3**.

25 **Barquins, M. and Roberts, A. D.** Rubber Friction Variation with Rate and Temperature: Some New Observations. *J. Phys. D. Appl. Phys.*, 1986, **19**, 547-563.

26 **Alliston-Greiner, A. F.** Friction Test Machines for Rubbery Materials. *Tribotest Journal*, 1994, **1**, 63-75.

27 BS 903: Part A61: (92/43803DC) Determination of the Frictional Properties of Rubber.

28 **Scieszka, S. F. and Dutkiewicz, R. K.** Testing Abrasive Wear in Mineral Comminution. *International Journal of Mineral Processing*, 1991, **32**, 81-109.

29 **Clark, H. McI.** Test Methods and Appplications for Slurry Erosion: A Review. In *Tribology: Wear Test Selection for Design and Application, ASTM STP 1199*, Ruff, A. W. and Bayer R. G., Eds., American Society for Testing and Materials, 1993, 113-132.

30 **Williams, J. A.** The Laboratory Simulation of Abrasive Wear. *Tribotest Journal*, 1997, **3**, 267-306.

31 **Hutchings, I. M.** Tribology: Friction and Wear of Engineering Materials, Chapter 6, Edward Arnold, London, 1992.

Combating abrasive-corrosive wear in aggressive mining environments

A BALL
Department of Materials Engineering, University of Cape Town, Republic of South Africa

SYNOPSIS

In-situ underground tests and laboratory assessments of a large number of standard grade, proprietary and experimental steels have led to an evaluation of the contributions of abrasion and corrosion to wear in mine environments and the recognition of synergistic effects. The study has led to an understanding of the compositional and microstructure factors which provide resistance to simultaneous severe abrasive wear and aggressive corrosive conditions. The development of reliable laboratory simulation of these conditions will also be discussed. It is now possible to formulate materials for specific conditions and pre-assess their performance in the laboratory.

1 INTRODUCTION

The wearing out of mining equipment is a universal problem which in the past has to a great extent been accepted as a 'normal' consequence of the aggressive conditions. The extent of this accepted loss was particularly large in the deep gold mines of South Africa where abrasion by quartzitic rock and corrosion by acidic waters cause rapid deterioration. As a consequence of depreciating ore quality, increasing labour and safety costs, considerable effort was made to introduce mechanised mining and rock handling systems. (**1**) This investment in machinery and the need to have reliable and continuous operation led to a co-operative programme of materials directed research by the Chambers of Mines Research Organisation of South Africa. We were given the task of 'designing' a steel or a set of steels which give optimum resistance to abrasion and corrosion and at the same time satisfy the normal requirements of a construction steel. These include machinability, formability, weldability, strength and toughness at an acceptable cost penalty over mild steel.

It was quickly realised that the performance of a large number of candidate materials could not be assessed *in-situ* underground. A reliable laboratory test is required which can accurately simulate the real conditions and rank the performance of any material against a standard material. This important aspect of the research will be outlined in this paper. The results emphasised that the contribution of corrosion to the total wear loss is appreciable for plain carbon steels. More significantly, abrasive and corrosive actions are found to be synergistic. This synergism is a function of the frequency of abrasive action and the passivity of the corrosion products. This aspect of the work will also be recalled in this paper. The research has led to conclusions concerning the chemical composition, microstructural condition and micromechanical behaviour which provide optimum wear resistance in a given environment.

2 THE DEVELOPMENT OF A LABORATORY TEST

The wear resistance of a component is not controlled by material properties alone but also by the loading and environmental conditions. Information obtained from standard laboratory testing is often therefore of little value. One solution is to undertake a series of lengthy *in-situ* tests, which can be expensive, inconvenient, and time consuming; a strict continuous control is necessary to ensure reliable results.

A more acceptable and convenient approach is to design a laboratory experiment to closely simulate the real conditions under which the materials have to perform. It is essential, however, that such a test can be verified as a true simulation of the problem and must therefore initially be carried out in parallel with *in-situ* experiments. An advantage of laboratory testing is the possibility of separating the variables which control corrosive-abrasive wear. This can be achieved by careful design of laboratory testing procedures. Wear can be studied as a function of variables such as load, speed of abrasion, pH of the environment, and frequency of corrosive-abrasive actions. The case of a prototype rock conveying system (**1**) in a gold mine will be used to illustrate the procedures.

As stated, there is a need to verify a laboratory test, and, it is therefore necessary to conduct an *in-situ* experiment. However, this experiment may only require the testing of a short list of materials. Generally, the experiment will involve the placement of material samples in the item of equipment under consideration and the monitoring of mass loss as a function of running time. It is necessary to ensure that the equipment experiences typical conditions during the experiment and that the time of exposure is sufficiently long to include all variations normally influencing the wear.

A parallel laboratory test should be set up only after acquiring knowledge on the *in-situ* mechanisms of deterioration. An obvious requirement is that the surface topography and surface hardening are similar to those produced *in-situ* in the component under investigation. These can be determined by profilometry, scanning electron microscopy and microhardness traverses.

A successful laboratory test must also provide the correct ranking order of the short list of materials and, in addition, the ratio of performance of the materials should be similar to that obtained from the *in-situ* test. It is convenient to present the wear resistance of the material relative to that of mild steel, which is a general engineering material used for a wide spectrum of engineering applications. Having, in this manner, verified that the laboratory test accurately simulates the real *in-situ* situation, it can now be used to assess a range of available materials and, in addition, new materials can be designed by a metallurgist to have suitable abrasive and corrosive properties.

framework in the mining stope. Conveyors have been manufactured in the past from mild steel; it was obvious that loss of metal by abrasion was occurring but the lack of a visual corrosion product might have indicated that corrosion was not an important factor in the wear process.

The necessary *in-situ* tests were performed by placing batches of specimens in the form of tablets (75 mm × 50 mm) in the bottom of the trough of a working conveyor. The weight loss suffered by each specimen was determined after a known working time and, in each array of specimens, several mild steel tablets were included as standards.

The *in-situ* relative wear resistance RWR is expressed as a ratio:

RWR = Volume loss of mild steel/Volume loss of alloy.

Simultaneously with the *in-situ* test, a laboratory test was undertaken on the same short list of specimens. After a number of attempts at copying the abrasive actions of the rock against the conveyor, it was found that a simple pin-on-abrasive belt machine (**2**) was the most convenient and reproducible test. However, the results of dry abrasion do not correlate with those obtained *in-situ*. This is illustrated

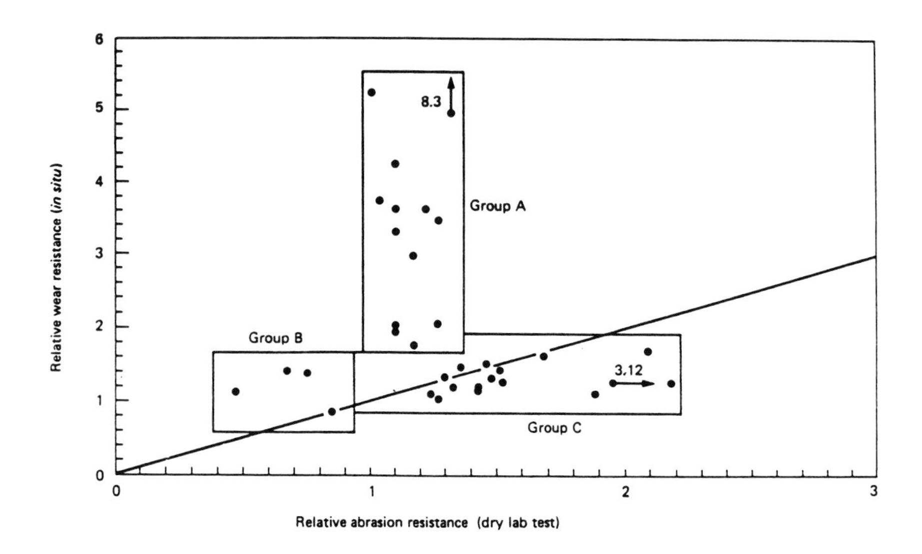

Fig. 1 Comparison of relative wear resistance (in-situ) and relative abrasionresistance (dry lab test) for number of alloys: group A – stainless steels: group B aluminium alloys: group C general purpose and proprietary wear-resistant steels.

In the case of the prototype rock conveying system, the contact stresses are generally less than the crushing stress of quartzites and, thus, the conditions have been termed low stress. It was also noted that the water which lubricated the conveyor, and which was present throughout the mining environment, contained up to 300 ppm chlorides, 500 ppm sulphates, and had a pH which varied from 4.2 to 9.8. The relative sliding velocity of the rock was less than 1 m/s. The existing equipment consists of a shallow trough, welded in convenient lengths, supported on an inclined

in Fig. 1, where the *in-situ* relative wear resistance RWR is plotted against the relative dry abrasion resistance, RWR. Clearly, the stainless steels of group A perform markedly better than would be predicted from the laboratory test. It was obvious, therefore, that a corrosive element had to be included in the laboratory test in order to provide close simulation.

After some experimentation with the variables of abrasion and corrosion in synthetic mine water, it was found that a test comprising of components of abrasion and corrosion produced a correct ranking of the short list of materials. The procedures described in previous papers (**3, 4, 5**) comprise a dry abrasion and corrosion test. For the abrasion-corrosion test the synthetic mine water contains 520 ppm dissolved solids comprising 343 ppm sulphate and 175 ppm chloride at a pH of 6.8. At intervals the specimens are taken out of the rig, washed free of the edge lacquer and any oxide product by ultrasonic agitation in a buffered mild acid solution. They are dried and weighed and then abraded a short distance on the abrasion rig before cleaning, reweighing and being put back in the corrosion rig. This abrasion-corrosion procedure is repeated for four cycles. The volume loss for each steel is plotted *versus* the time of corrosion and the distance of abrasion (Fig. 2) and the wear resistance (RWRL) is calculated relative to that of mild steel (070M20.)

The plot of relative wear resistance *(in-situ)* RWR against the relative wear resistance(laboratory), RWRL (Fig. 3), falls on or near the straight line passing through coordinates 0,0, 1,1, and 2,2; an acceptable laboratory test has therefore been devised. Furthermore, examination of the worn surfaces by surface profilometry and scanning electron microscopy confirmed that the correct topography had been produced by the test. Microhardness tests on taper sections revealed that similar work-hardening characteristics are obtained for specimens tested *in-situ* and in the laboratory. The successfully developed laboratory test has now been used to rank a wide range of materials for their suitability in the construction of shaker conveyors.

The advantage of the testing procedure as outlined above, in contrast with the predictive assessment of wear resistance based on alloy hardness is illustrated in Figs. 4

Test condition	RAR	RWRL#1	RWRL#2
Abrasive path length	4 x 3.0 m	4 x 0.25 m	4 x 1.0 m
Duration of corrosion	zero	4 x 46 hrs	4 x 22 hrs

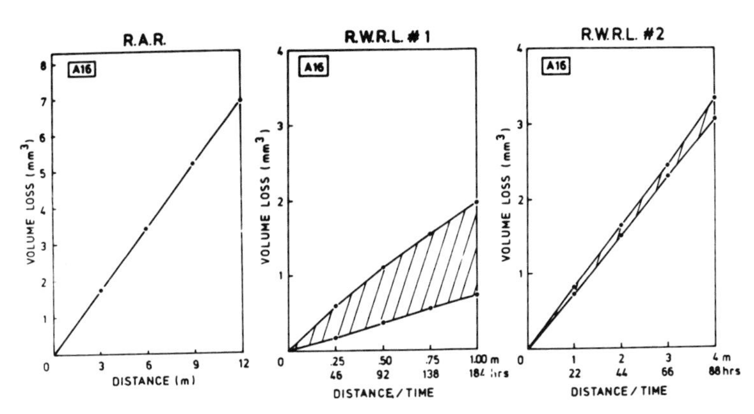

Fig. 2 The relative abrasion resistance (RAR) and relativewear resistances RWRL#1 and RWRL#2 for a martensitic steel containing eight percent chromium. The shaded portions indicate the contribution due to corrosion.

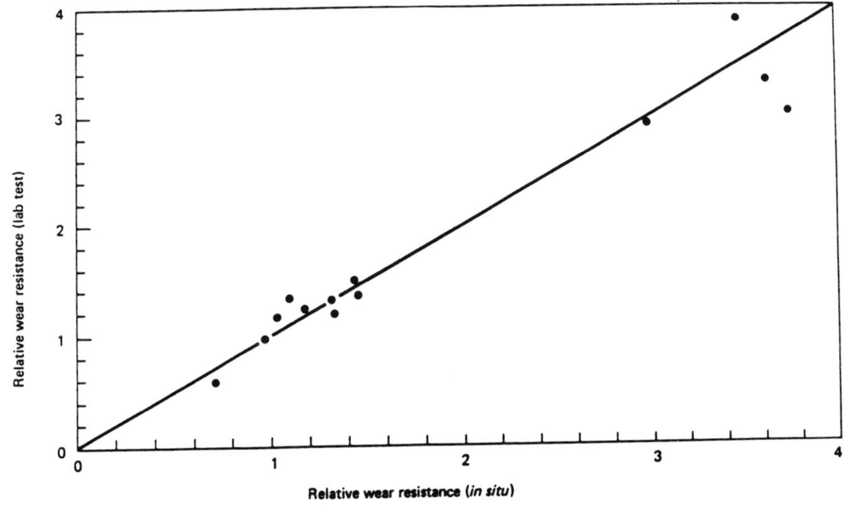

Fig. 3 The close correlation obtained between the relative wear resistance RWLR obtained in the laboratory and the relative wear resistance RWR measured *in-situ* on the shaker conveyer in the mine.

and 5. It can be seen (Fig. 4) that alloys with hardnesses ranging from 200-700 HV_{30} possess a similar relative dry abrasion resistance. Thus, hardness is no criterion for wear performance. The inclusion of a corrosive action, as expressed in the relative wear resistance, shows even less dependence on hardness. The engineer cannot, therefore, base the selection of materials solely on hardness, but must resort to procedures similar to those outlined. It should be emphasized that it is only necessary to validate a laboratory test by undertaking an *in-situ* test on a short list of materials.

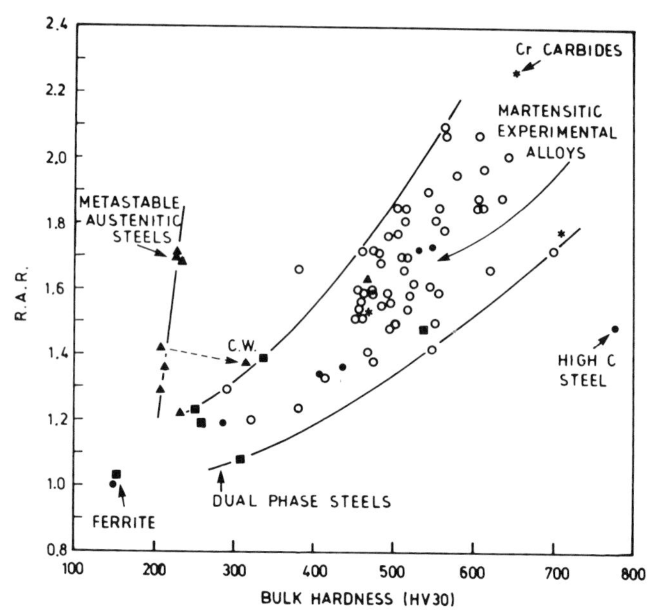

Fig. 4 The relationship between the relative dry abrasion resistance (RAR) and bulk hardness. (•) carbon steels; (o) martensitic steels; (■) dual phase steels; (*) precipitation hardened steels; (▲) metastable austenitic steels.

The laboratory test chosen in the case of the rock conveyor was considerably versatile, since the variables of load, relative velocity of sliding, corrosion periods, and frequency of corrosion-abrasion, can be systematically changed. In this way, it has been shown (**6**) that the percentage role of corrosion decreases with increasing abrasive load. An increase in the frequency of abrasive and corrosive actions may lead to greater overall wear. Thus laboratory information enables one to predict the importance of corrosion for an engineering component working under variable abrasive conditions.

3 THE SYNERGISM BETWEEN ABRASIVE SURFACE DEFORMATION AND CORROSION FOR STEELS CONTAINING VARYING AMOUNTS OF CHROMIUM

It is necessary to establish the mechanisms which constitute the wear process and attempt to quantify their

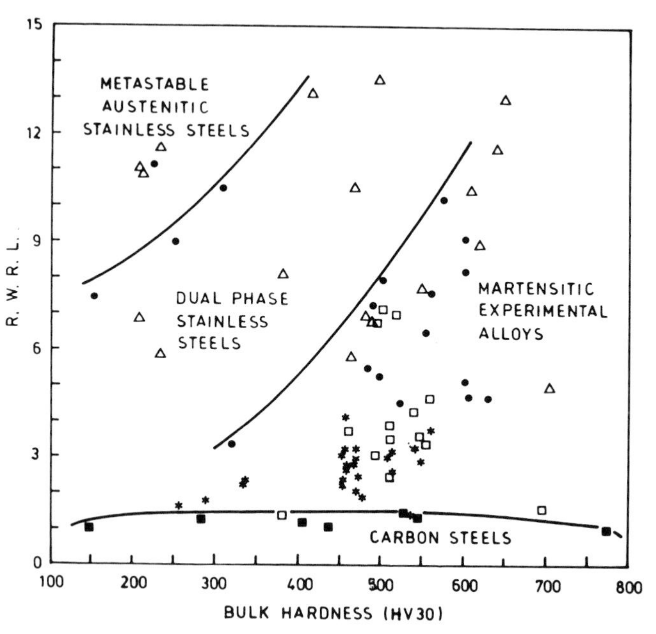

Fig. 5 The relative wear resistance obtained in the laboratory RWRL as a function of bulk hardness for the four types of steels containing increasing amounts of chromium (■) 0-5%; (*) 6-8%; (□) 8-10%;(•) 10-12%; (Δ) 12-18% Cr.

contributions. The role of corrosion of metals which are continuously or intermittently abraded is often underestimated because the corrosion product is not observed. Several authors (**6. 7, 8**) have attempted to determine the contributions of abrasion and corrosion. The combined effect is synergistic and a function of the tribological system. This section of the paper concerns the relative contributions of corrosion and abrasion for a conveyor for which laboratory simulations were devised; we address the selection of a material without resorting to expensive highly alloyed steels.

Commercial and experimental steels were tested in their as-received condition or in a specified heat treated condition. One group of steels is based on the high strength, tough, chromium containing martensitic steel with a lath structure. A second approach is based on the corrosion resistant dual phase ferritic-martensitic steel 3CR12 (**9**). It is presently used in ore-cars, walkways and chutes where abrasion and corrosion occur concurrently. A third approach was a metastable austenitic stainless steel with high work hardening and impact energy absorbing capabilities. Details of compositions of these steels are given in the paper by Barker and Ball (**10**).

Variations of the RWRL test were used to investigate the synergistic and frequency effects of abrasion and corrosion. One experiment involved changing the per cycle distance of abrasion and period or corrosion while maintaining the same abrasive distance and corrosion time overall. In another test specimens underwent the identical preparation and abrasion before being placed in the corrosion rig. At intervals over the following 60 hours, a specimen was removed and the volume lost was determined.

An experiment quantified the effect of abrasion on corrosive loss. Two specimens of the same material were run-in on the abrasion rig, annealed in a vacuum furnace at 1100°C for 30 minutes and quenched under argon gas. One specimen was re-abraded and then both specimens were cleaned and weighed. They were subsequently placed in the corrosion rig until the first visual signs of rust had formed, removed, cleaned and re-weighed.

Fig. 4 demonstrates that bulk hardness of the steels tested is not a good indication of dry abrasion resistance; the metastable austenitic steels show outstanding performance despite their low bulk hardness. The strong dependence of the corrosion-abrasion performance on chromium content is demonstrated in Fig. 5.

The dual phase steels do not perform well under dry abrasive conditions (Fig. 4). If the chromium level is kept above 11 per cent, these alloys show potential in corrosive-abrasive wear situations when the abrasion occurs under light loading conditions. The optimum properties of the metastable austenitic steels are given by the annealed, fully austenitic structure. For the martensitic steels, the optimum heat treated condition is an oil quench from 1050°C to 1100°C followed by a double temper at 200°C. This produces a fine lath martensitic microstructure with small equiaxed prior austenite grain size.

Scanning electron microscopy of the abraded surfaces revealed that material was removed by ductile microfacture after ploughing had caused extensive plastic deformation. Evidence of brittle fracture events was noted on the abraded surface of the hardest steels. The debris was in the form of swarfs of extruded material originating from the ridged material. The nature of corrosion, namely general, local and pitting, of an abraded surface was dependent on the chromium content (Fig. 6). Steels containing less than 11 per cent chromium are not fully passive in the mine water and corrode.

There is a local increase in the anodic activity and attack occurs at an abraded groove on a polished surface of mild steel; (see Fig. 7). The initial rates of corrosion of abraded surfaces of four steels with different chromium contents were compared to annealed surfaces of similar roughness (Fig. 8) The free corrosion potential of a freshly abraded surface was below that of a polished surface and changes with time in the mine water. The three types of corrosion behaviour are identified as Type I, II and III (Fig. 9). Visual observation revealed that the Type I corrosion behaviour (e.g. for mild steel) is an immediate general attack although accelerated attack of the abrasion ridges occurs during the initial stages. Type II corrosion behaviour is initially passive but within 4 hours there is locaslized breadown and corrosion then spreads from isolated regions. After 67 hours the entire surface is covered by a thick oxide layer. However, the anodic sites remain as localised crevices under precipitated oxide

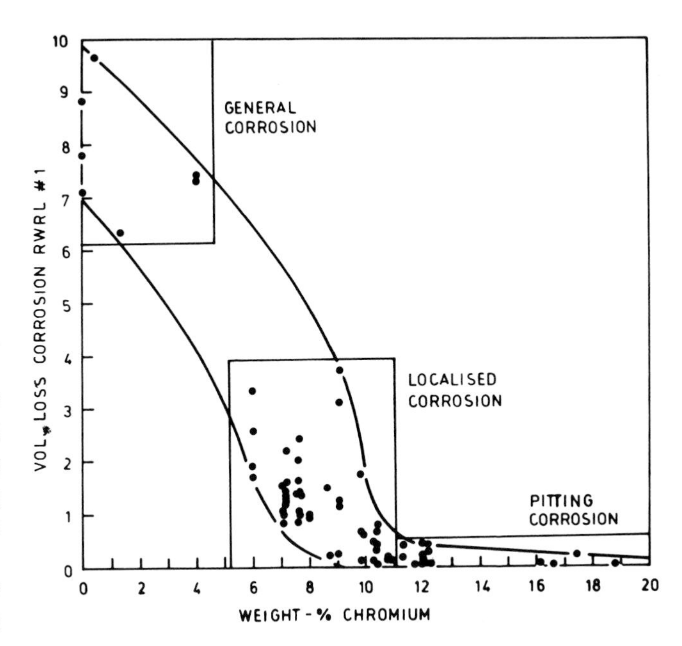

Fig. 6 The volume lost due to the corrosion of an abradedsurface as a function of chromium content. Three types of corrosion are indicated.

product. Type III corrosion behaviour was observed on the stainless steels, isolated pitting sites are active for short durations under the aerated, flowing conditions.

Fig. 7(a) Corrosion product on the abrasivewear track on mild steel after 10 min of exposure to synthetic mine water.

Since the rate of corrosion and material loss are a function of the time following abrasion, the frequency with which the abrasive action disrupts the corrosion process must affect the overall rate of wear. The consequence of abrasive action every 22 hours (high frequency) and every

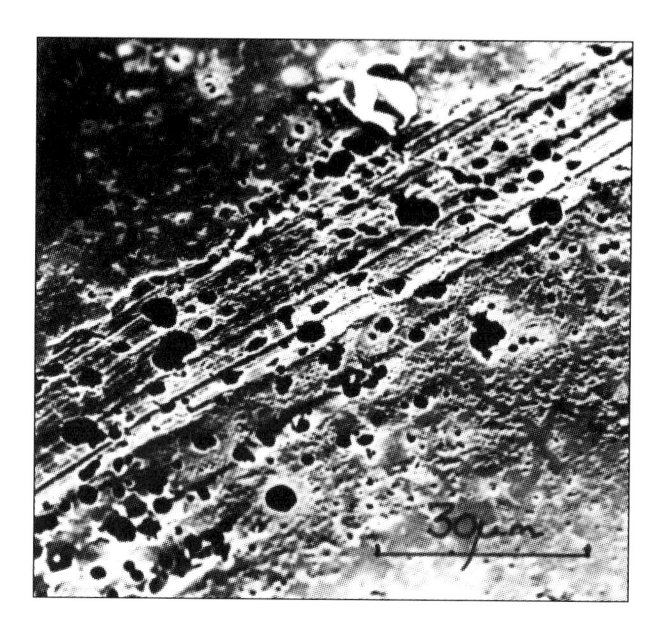

Fig. 7(b) Corrosion pitting along a wear track such as (a) after removal of the corrosion product.

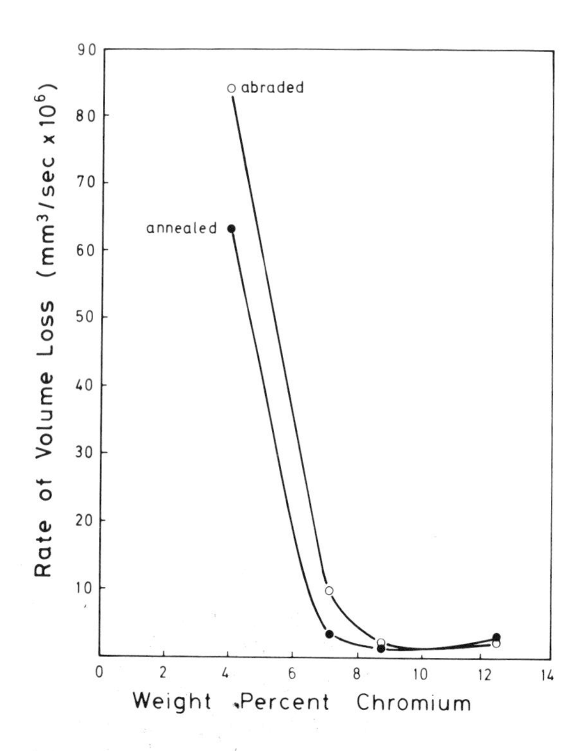

Fig. 8 The influence of abrasion on the rate of corrosion of a roughened surface for four steels of different chromium contents.

46 hours (low frequency) with increasing chromium content of steels has been established. The high frequency testing results in a larger contribution of corrosion to volume loss for low chromium steels but the reverse is true as the chromium level increases above about 7 percent.

4 DISCUSSION

The versatile laboratory testing procedure described in the first part of this paper has enabled us to research the dry abrasion, synergistic corrosion effects and the consequences of the frequency of abrasive and corrosive events for a wide range of steels with potential for use in the mining industry.

The metastable austenitic steels demonstrate the inadequacies of hardness as a guide to material selection (Figs. 4 and 5). Several authors (**2, 11–14**) have associated work hardening with abrasion resistance. It is postulated (**11**) that the deformation in the surface regions leads to micro-fracture and wear after a critical strain has been exceeded. Abrasion is a dynamic process and material is removed at this critical strain whilst material below the surface is yielding. Thus materials with high work hardening capacities can attain ultimate hardness while plastically accommodating the imposed stresses and resisting microfacture. This is confirmed by the excellent performance of metastable austenitic steels and the fact that cold worked metals do not perform as well as annealed samples.

These metastable austenitic steels transform to martensite during abrasion and retain adequate ductility. The martensitic steels with the greater work hardening capacities were found to have the better RWR. A minimum bulk hardness of 500 HV plus a work hardening capacity of 20 per cent is required of a martensitic microstructure for good hard-rock abrasive wear resistance.

The increase in internal free energy of an abraded surface is small and does not significantly change the thermodynamics of corrosion. However, an abraded surface has, in comparison to a smooth, polished and annealed surface, a greater surface area and a greater dislocation density due to work hardening. It is in a higher state of disorder with a lower average atomic co-ordination. Thus the physical and chemical activity of an abraded surface is higher and there is an increased probability of corrosive reaction. Abrasion can also influence the rate of corrosion through the heat of friction, agitation of the electrolyte and increased transport rates of reactants. As a consequence of these factors, abrasion does change the kinetics of the surface reactions. The activation energy is lower and thus according to the Arrhénius Rate Law, abrasion will increase the rate of corrosion.

Subsequent abrasion will remove the corrosion film and expose newly deformed material to the environment. The regrowth of a thin impermeable protective film occurs in tens of milliseconds for stainless steels and there is no measurable volume loss through general corrosive action. A similar protective film does not form on the low alloyed steels and corrosion proceeds even after a thick oxide layer has been established. The oxide layer on a low alloy steel is a diffusion barrier which gradually reduces the rate of corrosion. A constant rate of corrosion is reached after 20 hours for an abraded surface of mild steel grade 070M20.

Initially a stoichiometric oxide film is formed on stainless steel but selective dissolution of the iron oxide results in concentrates of the more noble elements in the passive oxide. The accelerated kinetics of growth on an abraded surface results in a more rapid formation of a passive film (**15**). Conversely the increase in roughness and surface area also increases the number of potential sites for pitting and local breakdown. Thus if the steel does not have the alloying levels to sustain passivity, the period the abraded surface can remain passive is short and the rate of material loss shows as a maximum. Polished coupons of a steel containing 8.7 percent chromium show no increase in material loss after 120 hours in the same electrolyte whereas an abraded surface shows a maximum after 15 hours.

Three types of behaviour designed Type I, II and III at an abraded surface have been identified and are summarized in (Fig. 9). Steels like 070M20 show general corrosion immediately after abrasion and have no induction period (Type I) whereas steels containing 7 to 10 percent chromium have a short induction period prior to localised action (Type II). Stainless steels passivate immediately and have an infinitely long induction period and never lose material except by pitting corrosion (Type III). Thus an increase in the chromium content will improve the corrosion resistance by increasing the induction period and decreasing the rate of volume loss from an abraded surface.

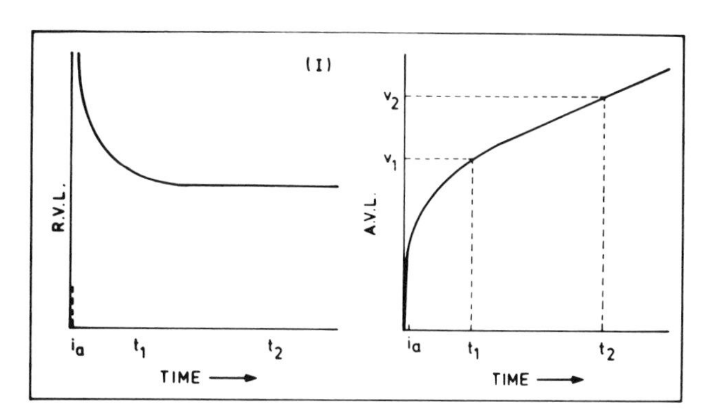

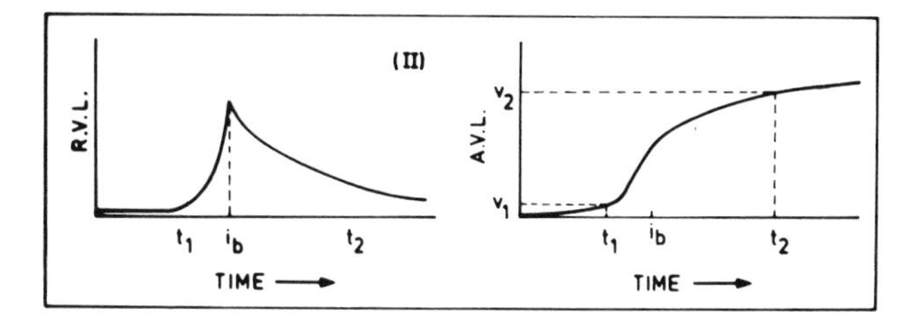

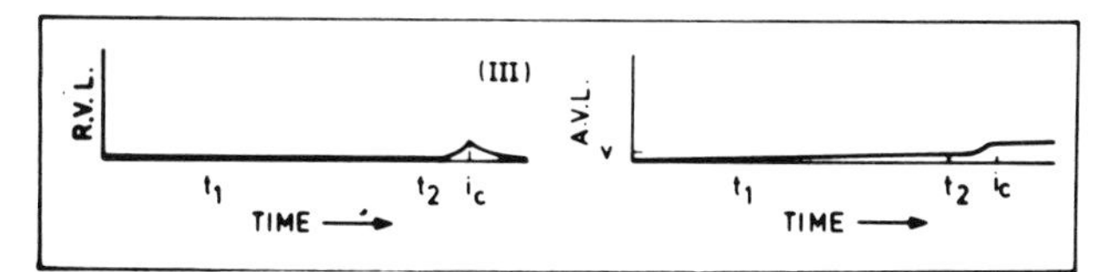

Fig. 9 Schematic representations of the of the rate of volume loss (RVL) by corrosive action and the accumulative volume loss (AVL) as a function of time for the three types of steel containing increasing chromium contents in the abraded condition.

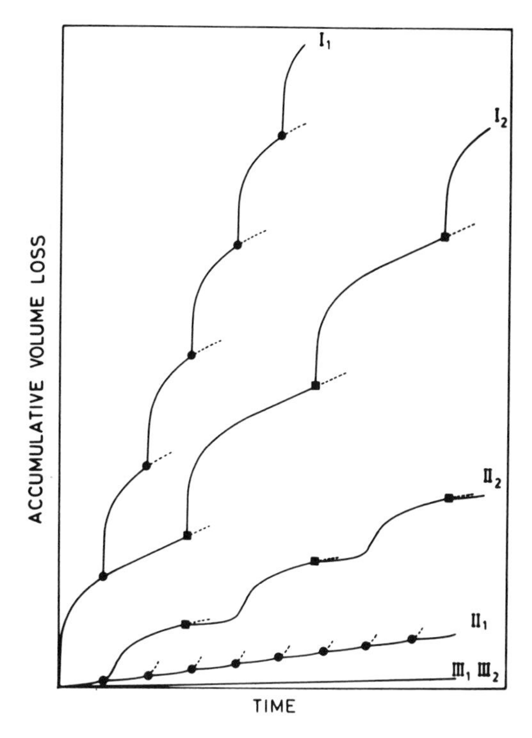

Fig 10 The accumulative volume loss (AVL) due to corrosion for high and low frequency abrasion tests for the three types of steel.

Abrasion may be intermittent, viz: a semi continuous abrasive action with short corrosive periods or occasional abrasive actions within corrosion periods. The induction period in relation to the extent of the corrosion periods is thus important when considering abrasive-corrosive wear as a cyclic phenomenon. It is evident from a consideration of Figs. 9 and 10 that high frequency corrosive-abrasive wear results in an increased contribution of corrosion for materials of Type I whereas the reverse is true for materials of Type II provided the corrosion period is less than their induction periods. Thus for a given abrasive action and equivalent dry abrasion resistance an increase in chromium contact provides an improvement in performance when high frequency or semi-continuous abrasive-corrosive action is experienced. However, inexpensive alloys which are lean in chromium can provide acceptable performances approaching those of Type II steels if the component experiences low frequency abrasive-corrosion actions. The corrosion performances of steels of Type II containing moderate amounts of chromium can approach those highly alloyed stainless steels if the frequency of the abrasive-corrosive action is high and the corrosive period is less than the induction period. These effects have been confirmed experimentally and could have important economic consequences in practice.

5 CONCLUSIONS

1. A versatile abrasion-corrosion laboratory testing procedure has been developed which simulates *in-situ* behaviour of steels in a mining environment.

2. Steels which contain at least eight per cent chromium and have martensitic, dual phase or metastable austenitic microstructures with high work hardening capacities, have good abrasion-corrosion properties.

3. Abrasive and corrosive actions are synergistic because abrasion affects the kinetics of corrosion.

4. The level of chromium determines the effect of the frequency of abrasion on corrosion. Low chromium steels perform better under low frequency conditions while medium to high chromium steels performs better under high frequency or semi-continuous abrasive-corrosive conditions.

6 ACKNOWLEDGEMENTS

The author acknowledges the considerable contributions to the work described of Professor C. Allen, Dr R. Noel and Mr B. Protheroe.

7 REFERENCES

1 **Joughin N.C.** *J.S. Afr. Inst. Min. Metall.,* 1976, **76(6),** 285–300.
2 **Allen C., Protheroe B.E., Ball A**. *Wear,* 1981 - 82, **74**, 287–305.
3 **Allen C., Protheroe B.E., Ball A**., *J.S. Afr. Inst. Min. Metall* 1981, **81,** 289.
4 **Noel R.E.J., Allen, C., Ball A**., *IMechE,* 1984 C358/84 p.23.
5 **Ball A., Ward J.J**., *Tribol. Int.,* 1985, **18,** 347.
6 **Noël R.E.J., Ball A**., *Wear,* 1983, **87,** 351–361.
7 **Postlethwaite J., Jawrylak M.W.** *Corros. NACE,* 1875, **31,** 237.
8 **El-Koussy R., El-Raghy S.M., El-Mehairy A.E.,** *Tribol. Int.,* 1981, **14,** 323.
9 **Ball A., Hoffman J.P.,** *Met. Technol.* 1981, **8,** 329.
10 **Barker K.C., Ball A.,** *Br. Corros. J.,* 1989, **24,** 222.
11 **Ball A.,** *Wear,* 1983, **91,** 201.
12 **Moore M.A., Richardson R.C.D., Attwood D.G.,** *Metall. Trans.,* 1972, **3,** 2485.
13 **Mutton P.J., Watson J.D.,** *Wear,* 1978, **48,** 385.
14 **Sundararajan G.,** *Wear 1987,* **117,** 1.
15 **Burstein G.T., Marshall P.I.,** *Corros. Sci.* 1984, **24,** 449.

Lubricants and the environment

W J BARTZ
Technische Akademie Esslingen, Ostfildern, Germany

SYNOPSIS

About 1 % of the total mineral oil consumption is used to formulate lubricants. Everywhere the production, application, and disposal of lubricants has to cover the requirements of the best possible protection of our nature and the environment in general and of living beings in particular. Often health hazards do not affect human beings directly, more often they follow indirect routes through our environment.

For all cases of direct contact between lubricants on the one side and human beings and nature on the other, their compatability has to be checked. The increasing need for environmental compatability tests has to be understood by all those working in the fields of production, application, and disposal of lubricants.

In simple terms it can be stated that health hazards and water hazards must be minimized. Using the three examples 'Coolants and Metal Working Lubricants', 'Engine Oils', and 'Fast Biodegradable Lubricants and Operational Fluids' some of the most important aspects of environmenal damage will be discussed.

1 INTRODUCTION

About 1% of the total mineral oil consumption is used to formulate lubricants.

Everywhere the production, application, and disposal of lubricants has to cover the requirements of the best possible protection of nature and the environment in general and of living beings in particular. Often health hazards do not affect human beings directly, more often they follow indirect routes through our environment.

For all cases of direct contact between lubricants the on one side and human beings and nature on the other side, their compatability must be checked. The increasing need for environmental compatability tests must be understood by all those who are working in the fields of production, application, and disposal of lubricants.

2 LUBRICANT MARKET – ENVIRONMENTAL DAMAGE

Fig. 1 illustrates the volume of the worldwide lubricant market, showing that about one third of all lubricants are consumed in Europe, America and Asia, respectively.

Between 13% (EC countries) and 32% (USA) of all used lubricants return into the environment more or less changed in properties and appearance (**1**). First of all, these

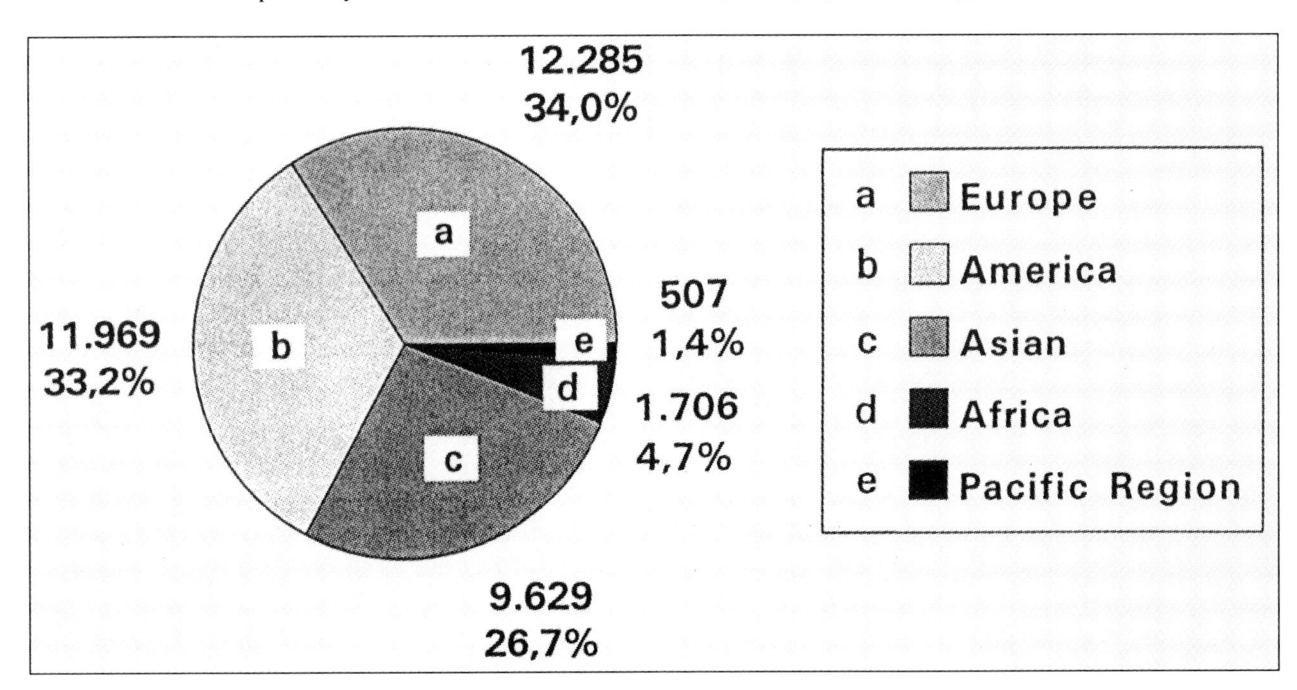

Fig. 1 World-wide comsumption of lubricants.1995 = 36.096.000t

are lubricants used in total loss lubrication of frictional contacts – about 40,000 tons annually in Germany – as well as for those lubricants used in circulation systems which are not collected and disposed. In addition, lubricants lost from leaks and the amounts remaining in filters and containers or empties have to be taken into account. Altogether, the environment, e.g. in Germany, is exposed to about 150,000 tons annually, based on the 13% share mentioned above, of the total lubricants volume which return to the environment. A calculation based on the actual lubricant consumption in Germany and the disposal rates for the different lubricant types results in about 250,000 tons annually. Including the undefined rest volume as well as the volume for lost lubrication, the total amount of lubricants in Germany returning to the environment might be of the order of magnitude of at least 300,000 tonnes per year.

Table 1 Fields of application for environmental acceptable lubricants.

- Outboard two stroke engine oils
- Chain saw and saw frame oils
- Railway wheel tread lubricants
- Mould parting compunds
- truck centralized system lubricants
- Hydraulic oils for machinery in bridge construction, deep workings and underground workings
- Hydraulic oils for forest and agricultural equipment
- Lubriants for sewage treatment plants, water weir plants and lock gate mechanisms
- Lubricants for food machinery
- Lubricants for snow-mobiles and ski run maintenance equipment
- Metal working and metal forming processes
- (Internal combustion engines and Hydraulic systems in general

Taking into account Germany's share of the worldwide lubricant consumption, as well as the fact that in parts of the world the collecting and recycling rate of used lubricants is lower than in Europe or in Germany, the total amount of lubricants returning into the environment will be of the order of magnitude of 12 million tonnes annually.

Therefore all measures must be taken in order to keep the impairment of the environment to the lowest possible level. For evaluating the allowable detrimental effect upon the environment, the benefits of lubricants, e.g. their performance or their economic properties must be considered, on one hand, and the risks caused by these lubricants, eg. their ecological properties, have to be taken into account on the other (**2, 3**).

Table 1 shows the applications of lubricants to which the aspects mentioned above apply. In this paper, the problems related to the ecological and the economic aspects of lubricant use will be discussed, taking the situation in Gemany as an example (**4**).

3 DRIVING FORCES FOR REDUCING ENVIRONMENTAL DAMAGE

Ways and means to reduce environmental damage which intend to overcome or at least to reduce the problems caused by contact of lubricants with the environment in general, are initiated or enhanced by the following driving forces:

Environmental Facts
Public Awareness
Government Directives and Regulations
Globalisation of Markets
Economic Incentives.

The environmental facts and public awareness resulting from these facts were described above. Information on government laws and regulations will be given below.

Simply, it can be stated that the behaviour regarding ecotoxicity and biodegradability governs the impairment of the environment by lubricants which are unaccounted for when returning to the environment. The preferred course of action is to reduce toxicity and to increase biodegradability.

In general terms biodegradability means the tendency of a lubricant to be ingested and metabolized by micro-organisms. Complete biodegradability indicates that the lubricant has essentially returned to nature. Partial biodegradability usually indicates one that or more component of the lubricant is not degradable.

Regarding ecotoxicity, a general rule of thumb exists according to which materials with an LD_{50} value > 1000 ppm/kg are low or non-toxic. In fact ecotoxicity represents the toxic effect of a lubricant on plants and animals (not on human health).

4 LAWS AND REGULATIONS

4.1 International activities

Environmental regulation or even legislation to regulate the use of lubricants exists in Austria, Canada, Hungary, Japan, Poland, Scandinavia, Switzerland, the USA, and the EU to mention a few countries and areas of the world.

In the USA currently, there are no laws requiring the use of environmentally compatible lubricants. But two regulations may have a significant impact on the use and disposal of conventional lubricants (**1**).

– Executive Order 12873 (EO 12873)
– Great Lakes Water Quality Initiative (GLWQI).

The executive order 12873 encourages the use of recovered materials and environmentally preferable products. Possible implications of EO 12873 may cover the:

– use of recycled oils for government contracts (military lubricants)
– use of environmentally compatible oils where possible to meet requirements.

The Great Lakes Water Quality Initiative is intended to maintain, protect and restore the unique Great Lakes resource (e.g. water quality). Within the GLWQI are proposals zinc limitations to such low values that this essentially means a ban on the use of zinc compounds in the Great Lakes Basin.

Austria is the only country with a law that bans the use of mineral oil-based lubricants in particular applications, e.g. chain saw oils.

The other countries mentioned above have at least established regulations to evaluate lubricant-caused impairment of the environment, e.g. labelling schemes.

Recently the European Community (EU) released the Dangerous Substances Directive. This establishes criteria for a product's potential hazards to the aquatic environment. This hazard potential is determined through assessment of aquatic toxicity, biodegradability, and bioaccumulation potential. Compounds which fall into the highest risk categories are required to carry a special symbol (Fig. 2). The criteria are shown in Table 2.

4.2 German activities

Germany has some of the best known guidelines for environmentally acceptable lubricants with an oriented interest in the lubricants market. But, at the moment there exist no laws only quasi-legal provisions and regulations which encourage the use of environmentally acceptable lubricants.

The production, application, and disposal of lubricants must be performed without any, or at least without detectable damage to human beings or the environment. In order to ensure that this requirement be met the German legislative code has enacted a framework of laws (**5**) which comprise:

– Chemical Substances Act
– Waste Material Act
– Water Protection Act

to mention only some of the most important laws.

4.2.1 Chemical Substances Act

According to this Act substances are classified with regard to their endangering potential. Based on this law, the marketing and application of lubricants are regulated in

Fig. 2 EC Dangerous Substance Directive – label.

Table 2 EC Dangerous Substance Directive – Labelling Criteria.

Aquatic Acute Data (LC_{50})	Ready Biodeg.	Bioaccum. Potential (log P_{ow})	Dead Tree-Fish Symbol	Risk Phrases Required
If compound solubility >1 mg/l				
>1 mg/l	NO	>3.0	YES	R50, R53
1-10	NO	>3.0	YES	R51, R53
10-100	NO*	n/a	NO	R52, R53
>100	n/a	n/a	NO	NONE
If compound solubility <1 mg/l				
n/a	NO*	n/a	NO	R53

* Unless other convincing scientific evidence to demonstrate no long-term effects from substance or its degradation products.

R50 – very toxic to aquatic organisms
R51 – toxic to aquatic organisms
R52 – harmful to aquatic organisms
R53 – may cause long-term adverse effects in aquatic environment

detail by the so-called Regulation for Dangerous Materials.

4.2.2 Waste Material Act

For the disposal of used lubricants the regulations of the law apply. According to their endangering potential, waste materials are to be classified as follows:

– Waste material
– Special waste material.

Special waste material is defined as material requiring special control. Within the Waste Material Act the following regulations have to be covered:

– Waste Allocation Regulation
– Waste and Remainder Material Monitoring Regulation
– Used Oil Regulation.

In addition, there is the so-called Technical Regulation Waste which constitutes an administrative instruction and not a law.

Regarding the allocation of rape seed oils within this framework of laws and regulations in Germany as an example there is the following discrepancy:

a) Used Oil Regulation:
Rape seed oils are used oils not to be regarded as special material.
b) Technical Regulation Waste:
Rape seed oils are special waste materials requiring special control.

4.2.3 Water Protection Act

Within this act, substances are classified according to their water endangering potential. For this purpose, they are classified according to a water hazard classification (Table 3). The water endangering potential is evaluated using the criteria shown in Table 4.

It can be recognised from the framework of laws that the evaluation of environmental impairment lubricants in Germany is very complex and not at all unequivocal.

Table 3 Water hazard classification.

WGK 0 = Generally not water endangering

WGK 1 = Slightly water endangering

WGK 2 = Water endangering

WGK 3 = Severely water endangering

Table 4 Criteria for evaluating the water endangering potential.

- ♦ Acute toxicity, especially against mammals, fishes and bacteria
- ♦ Decomposition properties
- ♦ Long term effects and physical-chemical characteristics

5 ENVIRONMENTAL ASPECTS OF LUBRICANTS

5.1 Definitions, Terms, and Agreements

In attempting to classify lubricants with regard to their impairment of the environment, many different terms have been established regarding the fact that the whole matter is not at all clear at the moment (Table 5) (**2**, **6**). It seems useful to graduate the environment-related terms as in Table 6. According to this graduation, no lubricant can be environmentally friendly, that is improve the environmental conditions. At best, the lubricant can remain neutral against the environment. Normally one has to be content with the fact that a lubricant is environmentally acceptable, that is that it affects the envirnment only to a less pronounced degree. Under this point of view, all aspects of impairment between the production of the lubricant, its application and its disposal have to be taken into account (Table 7).

Table 5 Terms regarding lubricants and environment.

- Environmentally positive
- Environmentally friendly
- Environmentally sociable
- Environmentally justified
- Environmentally careful
- Environmentally neutral
- Environmentally protective
- Environmentally conformable
- Biologically eliminatable

The consideration of environmental aspects of lubricants is focussed on:

- Health Hazards
- Water Hazards.

There is often an agreement to define lubricants as environmentally acceptable if they cover the following requirements:

- Fast biodegradable
- Non toxic against human beings
- Non toxic against fishes
- Non toxic against bacteria etc.

Increasing Impairment ↓

Friendly (improving the environment)
Neutral (unimportant, harmless)

- **Threshold of Perception**
  Sociable (low, unsuitable)
- **Start of Legal Regulations**
  Annoying (disagreeable, unpleasant, impairing)
  Irksome (troublesome, inconvenient)
- **Limit of Burdening**
  Endangering (excessive, unimputable)
  Harmful (dangerous, irreversible effects)

Table 6 Graduating environment related terms.

At production environmentally neutral - low energy consumption, no waste materials, no emissions If possible using regeneratable resources - no depletion of resources, no greenhouse effect Physiologically harmless - non toxic, not cancerous etc. Without toxic decomposition substances - non bio-accumulative potential Eco-toxicological acceptable - non water endangering, non water miscible, lower density than water After usage fast biologically degradable - no toxic or unpleasant decomposition products No disposal problems - possibility of simple recycling processes

Table 7 Aspects of environment impairment by lubricants.

5.2 Health Hazards

Health hazard aspects are characterized by the endangering of living beings by fresh and used lubricants. Improvement of the refining processes for mineral oils and the optimized selection of additives have reduced the endangering of human health to only few a remaining points. Nevertheless, the advantages and disadvantages of formulating lubricating oils and greases, using additives will be explained using coolants for metal-working processes as an example (Fig. 3).

Health hazards to human beings mainly arise where close contacts exist between the lubricant and the human being, eg. in the case of using coolants. Table 8 reveals some aspects related to the health discussion in connection with lubricants. The most important aspects of health hazards by lubricants are their endangering potential and their toxicity.

The endangering potential can be classified based on physico-chemical properties as follows:

Explosive
Oxidizing
Extremely flammable
Highly flammable
Flammable.

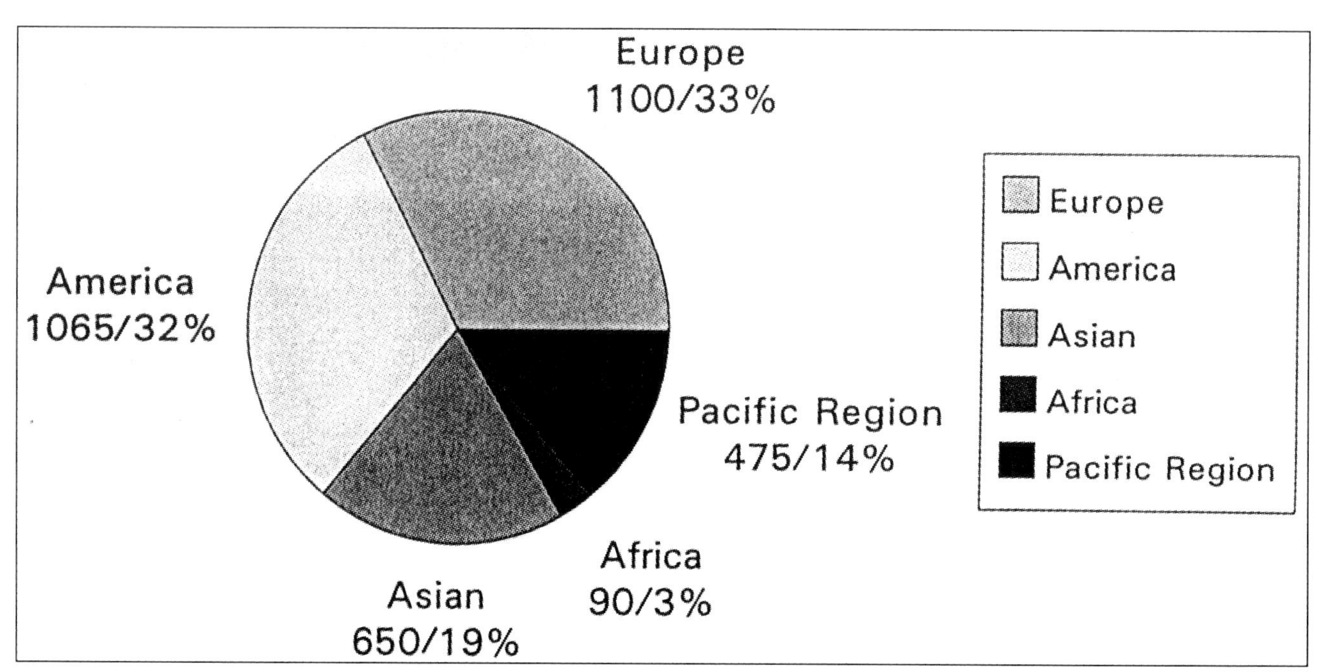

Fig. 3 Additive consumption (estimated) (ORONITE). World-wide additive market.

- Development of Nitrosamines in Coolants
- Effect of Oilfogs and -fumes
- Skin Diseases Caused by Contacts with Coolants
- Bactericide Effects in Connection with Coolants
- Cancer Generating Substances in Used Engine Oils
- Solvent Containing Products
- Heavy Metal Compounds in Additives

Table 8 Health discussion related to lubricants.

The toxicity of lubricants can be classified as follows:

Very toxic
Toxic
Harmful
Corrosive
Irritant
Carcinogenic
Mutagenic
Toxic to reproduction
Dangerous for the environment.

The description of the health hazard potential of lubricants has to be done in accordance with the EU Safety Data Sheet defined in the EU-Regulation 91/155/EWG and in the TRGS 220 (for Germany).

Regarding toxicity, the aspects listed in Table 9 have to be taken into account. The allocation to different toxicological potential is shown in Table 10.

- Acute Toxicity (LD_{50}, LC_{50}– Value)
- Irritation/Caustic Effect
- Sensitivity Effects
- Effect After Repeated and Long Duration Exposition (Subacute or Chronic Toxicity)
- Carcinogenic, Mutagenic and Reproduction Endangering Effects
- Specific Symptoms/Practical Experiences

Table 9 Safety data sheet according to TRGS 220 (toxicological effects).

LD_{50} (Oral, Rat)	< 25 mg/Kg	Very Toxic
LD_{50} (Oral, Rat)	> 25 and < 200 mg/kg	Toxic
LD_{50} (Oral, Rat)	> 200 and < 2000 mg/kg	Harmful

Table 10 Safety data sheet according to TRGS 220 (toxicological potential).

5.3 Water Hazards

Simplified, the definition environmetally acceptable lubricants means:

- Water hazard classification = 0
- Fast biodegradable.

The Water Hazard Classification is calculated using three evaluation numbers for acute mammal toxicity, acute bacterial toxicity and acute fish toxicity Table 11, Table 12. Based on this classification number, fast biodegradable lubricants can be allocated to the next lower and persistent substances to the next higher classification number. Table 13 reveals this relationship. This calculation is based on the allocation of different lubricants within the Water Hazard Classification (Table 14).

6 LUBRICANTS AND THEIR HEALTH AND WATER ENDANGERING POTENTIAL

The following two aspects will be discussed:

Fast biodegradability,
Toxicity.

6.1 Fast Biodegradability

6.1.1 General Relationship

The period needed to degrade a certain percentage of the lubricant defines its biodegradablity. Several testing methods have been developed to evaluate the biodegradability of the base fluid as well as of the additives

Acute Mammals Toxicity (LD_{50} Oral, Rat, mg/kg)
Fish Toxicity (LC_{50} or LC_0 , Fish, mg/l)
Bacteria Toxicity (EC_{50} or EC_0 , Pseudomonas putida, mg/l)
Fast Biodegradability

Table 11 Influences on the Water Hazard Classification.

LD 50 (mg/kg)	>2000	>200–2000	>25–200	<25
BWZ a	1	3	5	7
LC resp. EC (mg/l)	>10000	>100–10000	>1–100	< = 1
BWZ b and c	<1.9	2–3.9	4–5.9	> = 6

$$\frac{BWZa + BWZb + BWZc}{3} = \text{WGZ Water Hezard Number}$$

WGZ	<1.9	2–3.9	4–5.9	> = 6
WGK	0	1	2	3

Table 12 Calculation of the Water Hazard Classification.

Biodegradability OECD 301E			≥ 70%	≤ 70%
$WGZ = \frac{ST + BT + FT}{3}$	WGZ	WGK	biodegra-dable WGK	resistent WGK
ST: Toxicity against mamals	0 bis 1,9	0	0	1
BT: Toxicity against bacteria	2 bis 3,9	1	0	2
FT: Toxicity against fishes	4 bis 5,9	2	1	3
	> 6	3	2	3

	WGK	
White Oil	1	(changed 1992)
Base Oil	1	
Lubricating Oil with Addition	2	
Lubricating Oil Water Miscible	3	
Used Oil	3	

WGZ = Water Hazard Number
WGK = Water Hazard Classification

Table 13 Water Hazard Classification.

• WGK 0 – Not Water Endangering	Vegetable Olis
• WGK 1 – Low Water Endangering	Plain Lubricating Oils, Base Oils, White Oils
• WGK 2 – Water Endangering	Additivated Lubricating Oils, Engine and Industrial Oils
• WGK 3 – High Water Endangering	Additivated Water Miscible Lubricating Oils, Water Miscible Coolants

Table 14 Allocation of lubricants within the Water Hazard Classification.

of the lubricant. Widely used to evaluate the behaviour of lubricants – base oils and additives – are the tests mentioned in Table 15.

6.1.2 Base Fluids for Fast Biodegradable Lubricants

According to Table 16, three different groups of substances are available which can be used as base fluids to formulate fast biodegradable lubricants and operationl fluids. They can be allocated to the following classes (**6**):

– Watermiscible fluids
– Vegetable Oils
– Synthetic ester oils

Of the group of polyalkyleneglycols (PAG), polyethyleneglycols (PEG) are used most exclusively, because they are eco-toxicologically acceptable and fast biodegradable. The latter properties last mentioned apply only for the low molecular weight types as shown in Fig. 4.

	Duration, Days	Measured Parameter	Fast Degradable According to OECD	RAL
For Components with > 5 wt - %				
CEC Decomposition Test CEC-L-33-A-94	21	IR		>70 %
For Components with < 5 wt - %				
Zahn-Wellens-Test OECD 302 B	28	DOC	> 20 %	> 20 %

DOC - Reduction of DOC (Dissolved Organic Carbon)
IR - Measuring the reduction of organic part of test substance by IR

Table 15 Testing methods for evaluating the biodegradability.

Watermiscible Fluids

- Monoethyleneglycol
- Monopropyleneglycol
- Polyethyleneglycol (M-wt 200-1500)
- Ethylene/propylene mixed polymers

Non Watermiscible Fluids
Vegetable Oils
- Rape seed oil
- Castor oil
- Soya oil
- Peanut oil

Synthetic Ester Oils (with steric hindrance)
- Trimethylpropane
- Pentaerythrite
- Neopenthritepolyol

Table 16 Groups of substances for fast biodegradable lubricants.

The advantages of these fluids are good oxidation stability, viscosity-temperature behaviour, excellent low temperature behaviour, and good mixed film lubrication properties. The disadvantages are characterized by their nonmiscibility with mixed oils and their behaviour against seal materials.

Of the group of synthetic ester oils, diester and polyol esters are used especially to formulate fast biodegradable oils. Their advantages are characterized by their good miscibility with mineral oils, their good low temperature behaviour, and their high oxidation stability. Poor hydrolytic stability resulting in problematic corrosion

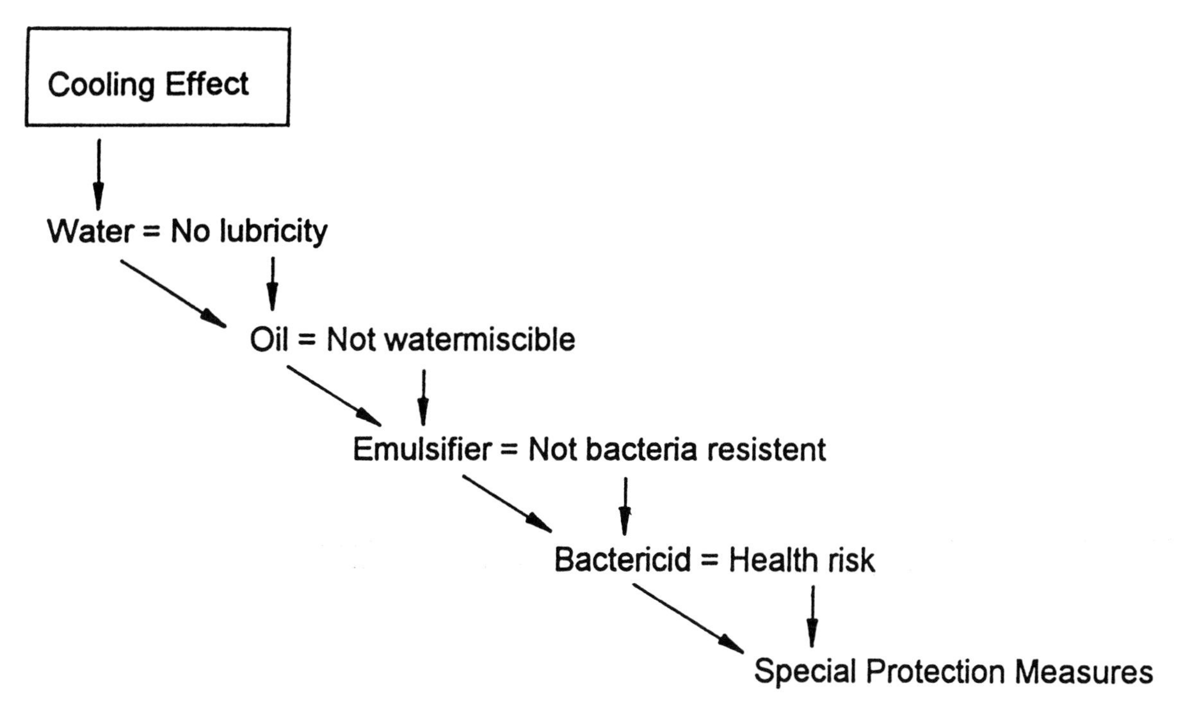

Fig. 4 Advantages and disadvantages of additives.

protection is one of the disadvantages of this group of fluids.

Vegetable or natural oils are triglycerides of natural fatty acids, e.g. palmitic acid, stearic acid, oleic acid, vegetable oils, linoleic acid etc. One of the most important natural oils is rape seed oil. Due to the rather high content of unsaturated fatty acids, natural oils tend to have a low oxidation stability. A high content of unsaturated acids is characterized by lower iodine numbers which in turn tend to result in higher solidification temperatures. This means a worse cold flow behaviour. Rape seed oils exhibit a rather good compromise between low temperature behaviour and oxidation stability (Fig. 5) (**6**).

Low temperature flow properties and oxidative

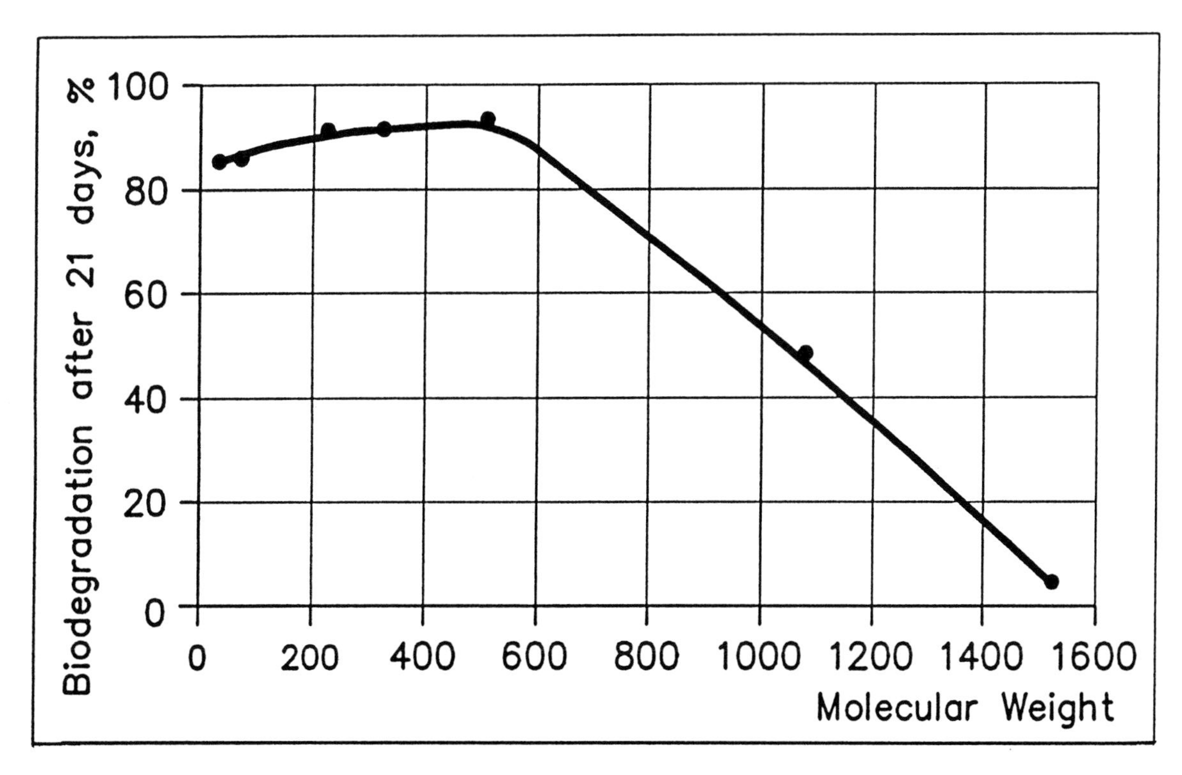

Fig. 5 Biodegradability of Polyethyleneglycols.

stability, and especially the relationship between both, are related to an oil's fatty acid profile. Oils containing high levels of mono-unsaturated fatty acids, like high oleic acid rapeseed, combine relatively high oxidative stability with good low-temperature performance.

Fig. 6 compares the biodegradabilty of different oils including those allocated to the group of fast biodegradable oils evaluated according to the CEC-L-33-A-94 Test. A relative evaluation and rating of different important properties is given in Fig. 7.

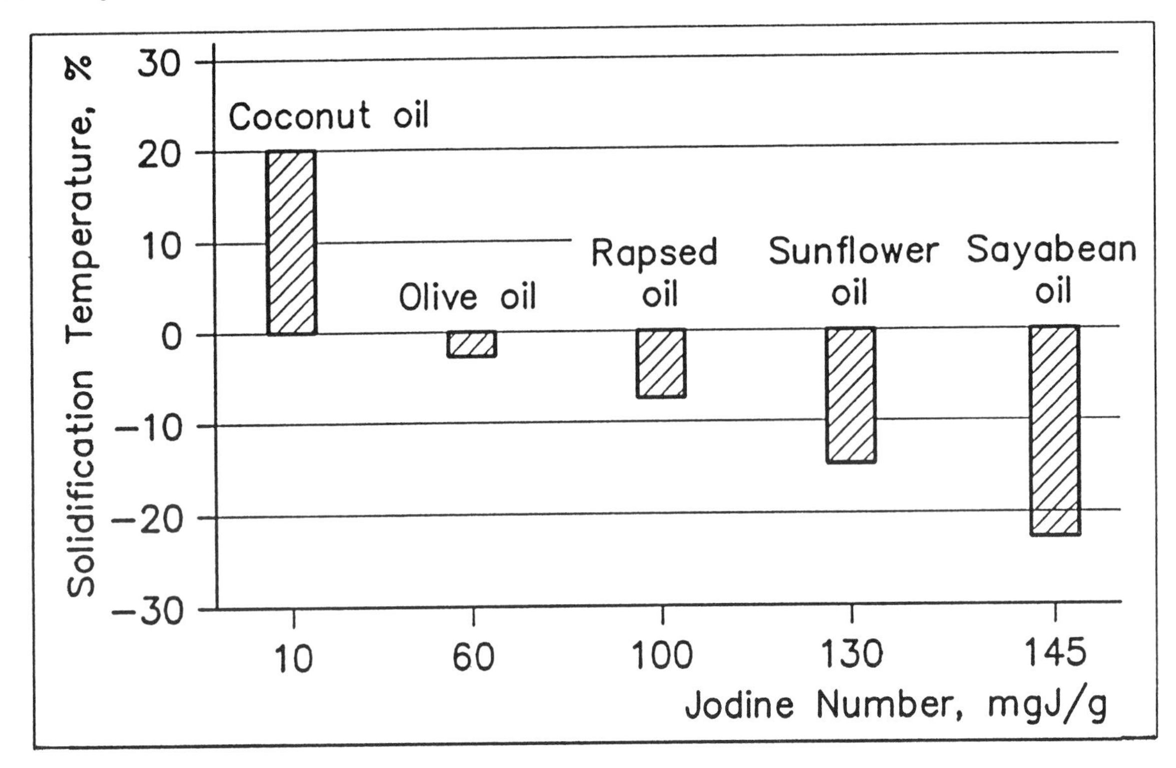

Fig. 6 Solidification temperatures of natural oils with different Jodine numbers.
Decreasing Jodine number improves oxidation stability.

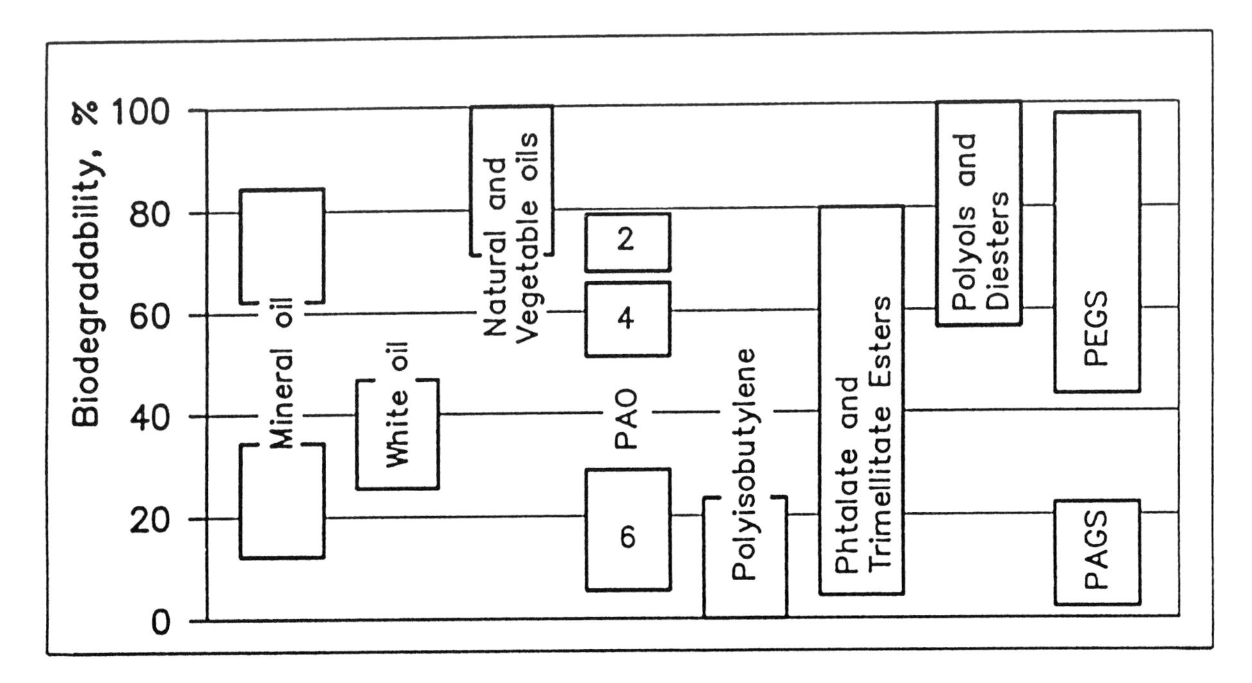

Fig. 7(a) Biodegradability of different base oils (CEC-L-33-A-94 test).

Evaluation 1 – Excellent 2 – Very Good 3 – Good 4 – Moderate 5 – Poor	Mineral Oils	Polyalphaolefines	Polyalkyleneglycols	Dicarboxylic Acid Esters	Neopentyl Polyesters	Rape Seed Oils
Viscosity Temperature Behaviour (VI)	4	2	2	2	2	2
Low Temperature Behaviour (Pourpoint)	5	1	3	1	2	3
Liquid Range	4	2	3	2	2	3
Oxidation Stability (Ageing)	4	2	3	2/3	2	5
Thermal Stability	4	4	3	3	2	4
Evaporation Loss, Volatility	4	2	3	1	1	3
Fire Resistance, Flash Temperature	5	5	4	4	4	5
Hydrolytic Stability	1	1	3	4	4	5
Corrosion Protection Properties	1	1	3	4	4	1
Seal Material Compatability	3	2	3	4	4	4
Paint and Lacquer Compatability	1	1	4	4	4	4
Miscibility with Mineral Oil	–	1	5	2	2	1
Solubility of Additives	1	2	4	2	2	3
Lubricating Properties, Load Carrying Capacity	3	3	2	2	2	1
Toxicity	3	1	3	3	3	1
Biodegradability	4	3/4	1/2	1/2	1/2	1
Price Relation against Mineral Oil	–	3–5	6–10	4–10	4–10	2–3

Fig. 7(b) Relative comparison and rating of different oils.

6.2 Toxicology of Lubricants

Regarding the toxicological potential of lubricants, base oils and additives must be considered.

6.2.1 Base Oils

The most useful general information on the toxicological behaviour of base oils can be found in the CONCAWE report 'Health Aspects of Lubricants' (**7**). The findings published in this report have been confirmed by a Shell study recently presented (**8**).

The results of the CONCAWE report can be summarized as follows:

- The acute dermic LD_{50} for rats was higher than 2000 mg/kg. This is the maximum dose which can be applied and is normally classified as harmless. For human beings no detrimental consequences are known for short-time dermal contacts.
- For solitary application to the skin of rabbits, base oils exhibited no to mild irritations. For human beings skin irritations as the result of short-time contacts have been proven, without any importance in practice.
- For solitary application into rabbit eyes, base oils exhibited no to mild irritations. For human beings no problems were found.
- Using the so-called skin sensitiveness test with guinea pigs, base oils exhibited no sign of sensitivity, for human beings too, no sensitivity was observed.
- For all oils tested the accute oral LD_{50} for rats was higher than 10 000 mg/kg. This value represents a low toxity. However, swallowing by accident low viscosity products will cause serious effects, because aspiration after vomiting can produce chemical pneumonitis.

The results of the Shell study can be summarized as follows:

- The acute toxicity of mineral oil base oils, extracts and distillates, is so low that any health danger by accute effects can practically be excluded, with the exception that low viscosity products were swallowed.
- Testing many base oils of different sources and grades of refining, the products mentioned below may exhibit carcinogenic potentials for long-time and repeated applications:

  Vacuum distillates
  Acid-refined oils from distillates
  Aromatic extracts from distillates
  Mild hydrogenated/refined base oils.

As a conclusion it can be stated that:

Conventional mineral base stocks have low toxicity by inhalation and by skin absorption. Eye irritation is not usually a problem, but skin irritation leading to dermatitis may occasionally occur from prolonged exposure to automotive lubricants, although these oils do not represent nearly as great a hazard as soluble cutting oils. Inhalation of vapour or oil mist for short periods may cause mild irritation of the mucous membranes of the upper respiratory tract.

6.2.2 Additives

The most useful information about the toxicity of additives can be found in a report released by the Additive Technical Committee (ATC) (**9**). The results can be summarized as follows (**10**):

In general, lubricant additives exhibit few, if any, physical hazard. Due to their low mammalian toxicological potential, the majority of lubricant additives are not classified as dangerous. To facilitate risk assessment and labelling, ATC has developed a classification system, naming 35 classes of additives divided into six broad groups of classification. All similar additives from different manufacturers are classified in the same way.

Of the most important additive types only zinc dialkyl dithiophosphate and some long chain calcium alkaryl sulphonates are classified as irritant. According to European legislation they must be classified "dangerous". Some, but not all sulphonates have been shown to be skin sensitizers in laboratory tests.

Some more details regarding the acute toxicity of additives are shown in Table 17.

Experience has shown that the toxicological properties of fully formulated lubricants are related to those of the base oil and additive components. There is also a close relationship between human health and environmental hazards. Measured toxicity of mixtures is generally found to be close to the arithmetic sum of component toxicities.

Due to the different fate of lubricants during their use – e.g. contamination by fuel and combustion products of engine oils – the toxicity of used lubricants may be significantly different than that of fresh oils.

Additives[a]	Typical LD50		Typical Irritancy Classification [b]	
	Oral (rat) mg/kg	Dermal (rabbit) mg/kg	eye	skin
Zinc alkyl dithiosphosphate	≥2000	>2000	irritant	irritant
Zinc alkaryl dithiophosphate	≥5000	>2000	not irritant	not irritant
Calcium long chain alkaryl sulfonate	>5000	>3000	not irritant[c]	not irritant[c]
Calcium long chain alkyl phenate	>10000	>2000	not irritant	not irritant
Calcium long chain alkyl phenate sulfide	>5000	>2000 (rat)	not irritant	not irritant
Polyolefin amide alkeneamine	>10,000	>2000	not irritant	not irritant
Polyolefin amide alkeneamine borate	>2000	>3000	not irritant	not irritant
Olefin/alkyl ester co-polymer	>10,000	>3000	not irritant	not irritant
Poly alkyl methacrylate	>15,000	>3000	not irritant	not irritant
Polyolefin	>2000	>3000	not irritant	not irritant

Notes:
(a) The additive names are based on the nomenclature system developed by the Technical Committee of Petroleum Additive Manufacturers in Europe.[4]
(b) Eye and skin irritancy potential is classified according to the EU criteria.
(c) This comment is anticipated to be updated.

Table 17 Acute toxicity of petroleum additives (source: CONCAWE report 5/87) (Acc. to Caines/Haycock).

7 ENVIRONMENTAL LABELLING

7.1 International activities

In order to characterize and label environmetally acceptable products, authorities and institutions in different countries have developed environmental labelling schemes and special signs or labels. It is the aim of these signs to draw the attention of lubricant users to environmentally acceptable lubricants.

Table 18 lists environmental labels of some countries, some of them having been developed on a trial basis. Fig. 8 shows some of these labels.

Country	Label
USA	Green Cross/Green Seal
Canada	Maple Leaf
Japan	Eco Mark (Arms embracing the globe)
Scandinavian Countries	White Swan
Austria	Symbol developed by the famous painter Hundertwasser
Germany	Blue Angel
Poland	Beetle on Leaf
European Union	Blossem-Symbol
Hungary	Green Tree

Table 18 Some environmental labels.

Fig. 8 Some environmental labels.

7.2 Blue Angel in Germany

In Germany, the Institute for Quality Assurance and Labelling (RAL) grants an environmental label called 'Blue Angel', according to regulations laid down by the 'Federal Environment Agency'. The Blue Angel is granted for a period of three years. At the end of this period, a committee has to check whether the user is still entitled to use the label. If necessary the label can be withdrawn or new requirements for prolongation, laid down. At the moment, there are regulations for granting the Blue Angel for the following products:

- Lubricants for chain saws
- Sheeting board oils
- Hydraulic oils.

In order to grant the Blue Angel environment label, the following properties are evaluated (Table 19).

- Eco-toxicological properties
- Technical properties
- Production processes and equipment.

8 SUMMARY

These statements can be summarized as follows:

- To protect the environment, lubricants which are environmentally more acceptable will gain importance.
- In some countries, e.g. Germany, several laws and regulations have been enacted to control the production, application, and disposal of lubricants. Unfortunately the requirements and definitions in this framework of laws are not unequivocal.
- The consideration of environmental aspects of lubricants is focussed on health hazards and water hazards. To define and classify the impairment of the environmental damage, numerous different terms are used, revealing the fact that the whole matter is still confusing at the moment and not at all clear.
- Water-miscible and non-water-miscible fluids are used as base fluids for environmentally acceptable fluids. The watermiscible fluids are characterized by polyethyleneglycols. Prominent non watermiscible fluids are synthetic ester fluids and natural or vegetable oils, e.g. rape seed oils as natural ester oils.

A. Eco-Toxicological Properties

Prohibited constituents

Dangerous materials according to the Dangerous Materials Regulation
Dangerous materials according to the Chemicals Act
Materials with a water hazard classification WGK > 2
Nitrate
Organically bonded chlorine

Base oil requirements

Fast biodegradable to > 70 % of constituents with over 5 % content
WGK < 1

Additive requirements

Total additive content < 5 %
Potential biodegradability of every single additive > 20 %
Toxicological behaviour
Immobility in soil

B. Technical Usability

Wear protection properties
Cold flow properties
Behaviour against other materials
Contamination tendency (gum/deposit generating)
Flash point
Approved by the
„Board of Forest Technology"

C. Production Processes and Equipment

Evaluation whether during the production process any contamination with toxic material can be excluded.
Approved by the „State Office for Nature, Environment and State Development"

Table 19 Granting regulations for the environment label 'Blue Angel'.

- A relative comparison of these base fluids reveals that they are characterized by advantages and disadvantages.
- Based on the fluids mentioned above, lubricating greases can be formulated covering the requirements for fast biodegradability.
- In order to label environmentally acceptable lubricants, in some countries labelling schemes and special signs have been developed and defined. In Germany, the 'Blue Angel' is gaining importance and acceptance.
- In addition to technical requirements, aspects of toxicology and industrial medicine will exert an important influence on the formulation of coolants. Higher disposal costs will result in the use of products with high service life and constant long term properties as well as in better maintenance procedures.
- It will become more and more difficult to find a balance between economic possibilities and ecological requirements. Of course, toxicologically and ecologically questionable products must be excluded from futher use in lubricants, if they pose any significant health risk under the conditions of application. On the other hand, must be taken into account that the technological level of lubricants will decrease, if unnecessary restrictions are built up.

9 REFERENCES

1 **Hamblin, P.**, Environmentally Compatible Lubricants: Trends, Standards and Terms. Proc. Environmental Aspects in Production and Utilization of Lubricants. *Sopron*, 1995.

2 **Randles, S.J.**, Formulation of Environmentally Acceptable Lubricants. *49th STLE Annual Meeting*, 1994, Pittsburgh,USA.

3 **Steber, J.**, Native Esters: Base Oils for High Performance and Environmentally Acceptable Hydraulic Fluids. Part B: Ecological Aspects (in German) *Proc. 8th International Colloquium 'Tribology 2000'*, 1992, TAE, Ostfildern, Germany.

4 **Bartz, W.J.**, Biologically Fast Degradable Lubricants and Operational Fluids (in German), *TAE-Course Tribologie und Schmierungstechnik*, Germany.

5 **Schmoltzi, M.**, Framework of Laws and Economic Aspects of Rape Seed Oils as Lubricants (in German). *In: Biologically Fast Degradable Lubricants and Operational Fluids.* W.J. Bartz, Ehningen, expert-verlag 1993.

6 **Lehmann, B.**, Environmentally Acceptable Synthetic Operational Fluids and Lubricating Greases (in German) *In: Biologically Fast Degradable Lubricants and Operational Fluids.* W.J. Bartz, Ehningen, expert-verlag 1993.

7 'Health Aspects of Lubricants', *CONCAWE*-Report 5/87.

8 **Pocklington, J.**, Zur Frage der Toxizität von Schmierstoffgrundölen. TAE-Seminar 21591, "Ökonomische Aspekte von Schmierstoffen", Nov. 1996, Ostfildern.

9 **Raddatz, J.H.** and **Ch. v. Eberan-Eberhart.**, Lubricant Additives and the Environment, *TAE-Seminar 21591 'Ökonomische Aspekte von Schmierstoffen'*, Nov. 1996, Ostfildern, Germany.

10 **Caines A.** and **Haycock, R.**, *Automotive Lubricants Reference Book*, 1996, MEP, UK.

Realising the potential of duplex surface engineering

T BELL
School of Metallurgy and Materials, The University of Birmingham, Birmingham, UK

SYNOPSIS

We are now close to the new millennium and on the threshold of an era of rapid change. Limitations to the further advance of manufacturing industry in the 21st century are most likely to be surface-related. Many mechanical systems will operate under ever more severe application conditions, such as intensive loads, high speeds and harsh environments, in order to achieve high productivity, high power efficiency and low energy consumption. Consequently, many challenging complex design situations have emerged where the combination of several properties (such as wear resistance, load bearing capacity, and fatigue performance) is required. These new challenges can be met only through realising the potential of duplex surface engineering. Indeed, there are thought to be great technical and economic benefits available through the application of duplex surface engineering technologies in many new market sectors. The present paper is a synthesis of several strands of recent surface engineering research at the University of Birmingham, including the duplex ceramic coating-nitrided steel system and the duplex DLC coating-oxygen diffusion treated titanium system. The prediction of the performance of duplex systems based on advanced contact mechanics modelling is also discussed.

1 INTRODUCTION

Notwithstanding the fact that the formal definition of the discipline of surface engineering was given in the European Journal of Engineering Education only a decade ago (**1**), surface engineering has become an established technology and has had a very significant technological, economical and environmental impact on modern science and technology through a reduction in capital investment, increased profitability, design changes and technical innovation (**2**). Indeed, it is now widely recognised that surface engineering is an important technology that will underpin virtually all industrial sectors, and that surface engineering is to Value Management as Gearing to Financial Management (**3**).

The past 10 years have seen many advances in the field of surface engineering: optimisation of traditional processes (e.g., electro- and electroless plating, weld surfacing, thermal spraying and thermochemical treatments); commercialisation of the modern techniques (e.g., CVD, PVD, plasma thermo-chemical processes, plasma spraying and ion implantation); developments of innovative hybrid technologies (e.g., plasma immersion ion implantation, and plasma source ion implantation); emergence of new coating materials (e.g., diamond and diamond-like coatings). Despite the fact that great achievements have been made in the domain of existing first generation surface engineering technologies, real designed surfaces with economically viable technically enhanced performance were rarely produced until duplex surface engineering, also referred to as second generation surface engineering, emerged. For instance, thin coatings such as PVD TiN can provide a surface with dramatically improved tribological properties in terms of low friction and high resistance to wear, but catastrophically premature failure will occur if the substrate plastically deforms under a high applied load; on the other hand, deep hardened layers produced by such surface modification techniques as energy beam surface alloying can sustain high contact stresses but still exhibit higher friction and wear rates when compared with most ceramic coatings. It is the combination of such surface engineering technologies that constitutes duplex surface engineering.

2 PRINCIPLES OF DUPLEX SURFACE ENGINEERING

2.1 Definition and classification

Duplex surface engineering, as the name implies, involves the sequential application of two (or more) established surface technologies to produce a surface composite with combined properties which are unobtainable through any individual surface technology. According to the interactions between the two individual processes and their relative contributions to the combined effects of the composite layer, duplex surface engineering may fall into two general groups: in the first group (Type I), two individual processes complement each other and the combined effects result from both processes; in the second group (Type II), one process supplements and reinforces the other, thus serving as pre- or post-treatment, and the resultant properties are mainly related to one process. PVD treatment of pre-nitrided steel (**4**) is a typical example of the first group while electron beam surface melting of a sprayed overlay (**5**) is a typical example of the second group. Some examples of duplex surface engineering technologies are listed in Table 1.

Table 1 Typical duplex surface engineering technologies.

TYPE I Complementary Technology	TYPE II Supplementary Technology
PVD coating of pre-nitrided steel	Energy beam melting of overlay coatings
Plasma nitriding of energy beam alloyed Ti	Nitriding of pre-laser-alloyed steels
PVD coating of Ni/Cu diffusion treated Al	Nitriding of pre-laser-hardened steels
Nitriding of pre-carburised steels	Sprayed MoS_2 on electroless Ni coatings
PVD coating of electroplated deposits	SiC intermediate layer for DLC coatings

2.2 Metallurgical aspects

Although the possible combinations of surface technologies are virtually unlimited and the list of duplex surface technologies could be endless, to date only a limited number of duplex treatments have been developed, and few of them have yet found real applications (**6, 7**). It should be pointed out that duplex treatments are not simply mixing two surface treatments which may individually produce desirable properties. This is because a duplex treated component is typical of a multi-layer system and the resultant performance of a duplex system depends more on the combined effects from the two individual processes, rather than the expected effect provided by individual processes, i.e. synergy of the processes usually occurs. For instance, inappropriate combination and/or incorrect control will lead to worse rather than improved combination effects (**4**). Accordingly, it is essential to predict correctly the metallurgical reactions so that the effects resulting from the first process will not be deteriorated by the second process.

2.3 Mechanical aspects

Furthermore, with such numerous possibilities comes the problem of designing an optimum duplex system for a specific application. This can be overcome only through developing mathematical models to simulate surface engineering processes and to predict the service behaviour of the resultant duplex surface engineered systems. The design of duplex surface engineering systems based on experience or empirical formulae achieved by *ad hoc* or trial-and-error approaches will severely reduce the maximum uptake of the benefits which duplex surface engineering offers.

Based on FEM analysis (**8**), it has been found that in most coating systems plastic deformation initiates in the substrate near the coating-substrate interface when subject to relatively high intensity loading, and plastic deformation does not initiate in the coating until a large plastic zone has been developed in the substrate. The load bearing capacities of coating-substrate systems thus increase with substrate properties. Clearly, deep case hardening can significantly enhance the load bearing capacity of a coating-substrate system.

FEM analysis of the sliding contact of a layered duplex coating system has also demonstrated that (**9**), according to the local shear strain or stress criterion, which may be most suitable for compression-dominated contact, an interface crack or adhesive failure may initiate in the layered media if the plastic shear strain at the interface is beyond a critical value. The magnitudes of shear stress and strain along interfaces increase significantly with the friction coefficient. Thus low friction coatings such as nitrides and oxides used as the top coating layer for duplex systems can not only increase wear resistance but can also diminish interfacial shear stress and strain, and thus the tendency for debonding of top coatings. In this respect, DLC or diamond coating, could be more effective since they possess the lowest friction against most engineering surfaces. All these principles will be further highlighted by way of examples in the following sections.

3 THE DUPLEX CERAMIC COATING-NITRIDED STEEL SYSTEMS

During the past decade, various duplex systems have been explored to meet the ever-increasing requirements in surface and subsurface properties. Of these duplex systems, PVD treatment of pre-nitrided steel is the most widely researched and well documented (**4, 10, 11**). A typical duplex process involves combined plasma nitriding and PVD ceramic coating treatment of steels. Plasma nitriding produces a relatively thick (~500 μm) and hard (900-1000 HV) subsurface, and at the same time a thin iron nitride compound layer is formed at the outmost surface. The thickness of the compound layer is a function of the active nitrogen capacity of the plasma and the processing temperature and time. It has been shown (**4**) that the nature and thickness of this iron nitride phase can have a profound effect on the quality of the titanium nitride product deposited and on the bonding strength to the nitrided subsurface. In one duplex system after plasma intruding, the nitrided surface was coated with titanium nitride about 3 μm thick by various PVD processes. Depending on the nitriding and the coating process conditions, as well as surface modification prior to coating, a variety of coating-nitrided surface combinations were produced (**4**), including (a) ceramic coating/dense compound layer, (b) ceramic coating/diffusion layer (c) ceramic coating/decomposed compound layer (black layer).

As a result of the combined effects of the two processes engineering components exhibit low friction and wear, a characteristic of ceramic coatings, and a high load bearing capacity and high fatigue strength, characteristics of the nitrided subsurface. In addition improved coating-subsurface adhesion strength can be also achieved provided these two processes are properly combined and carefully controlled. On the other hand, incorrect process control will lead to the formation of a soft 'black' layer below the coating by thermochemical decomposition of the outer part of the previously present iron nitride layer, which severely deteriorates the load bearing capacity of the composite through reduced bonding strength between the outermost titanium nitride and the nitrided subsurface. The problem of the formation of undesirable 'black' layer has been addressed by grinding off the compound layer prior to coating or avoiding the formation of compound layer during the nitriding process ('bright nitriding'). Reducing the PVD process temperature to < 450°C is effective in avoiding the undesirable decomposition of iron nitride layer.

A comparison of the wear volume (weight loss) from various specimens, duplex treated or individually treated, after a ball-on-wheel sliding test using a load of 20 N is given in Figure 1. As expected the TiN/untreated system exhibited very poor wear resistance due to severe subsurface plastic deformation. The as-nitrided sample was worn via an oxidative wear mechanism. Among all the specimens tested, the duplex treated specimens showed by far the lowest wear rate. Clearly, the duplex TiN coating/nitrided steel system is more effective in terms of improving wear resistance than both individual plasma nitriding and individual PVD TiN coating.

Experimental work (**12**) has also demonstrated that the combined plasma nitriding and PVD TiN coating treatment can significantly improve the corrosion resistance of the duplex system. As can be seen in Figure 2, the corrosion current density of duplex treated En40B steel is several orders of magnitude lower than either plasma nitrided or TiN coated En40B steel within the practical potential range (< 1V vs SCE). The dense compound layers formed during plasma nitriding play an important role in determining the overall corrosion performance of the resultant duplex system.

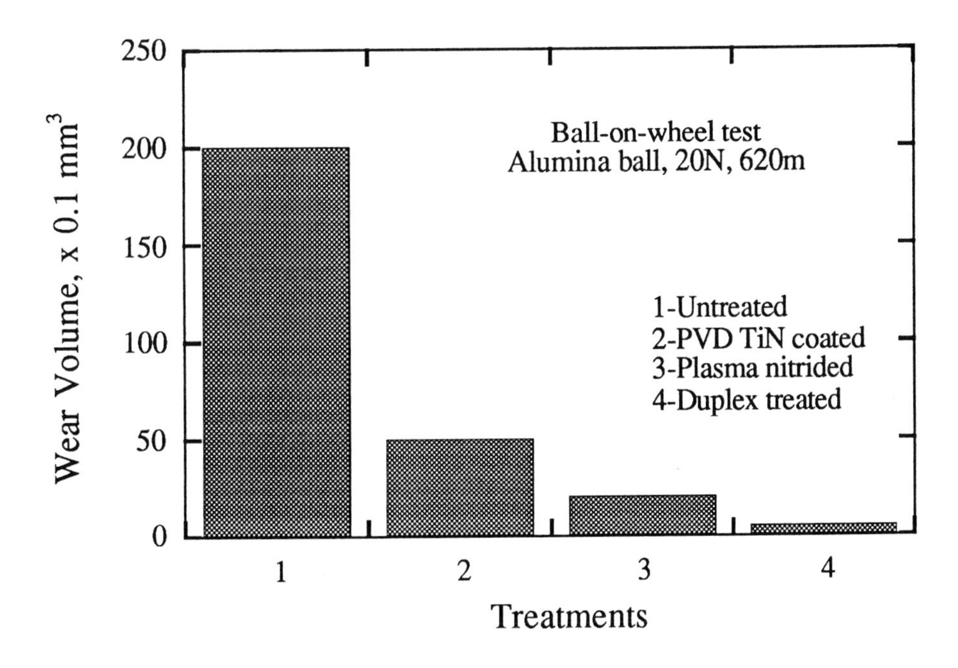

Fig. 1 Ball-on-wheel wear test results for untreated, PVD TiN coated, plasma nitrided and duplex treated En40B steel.

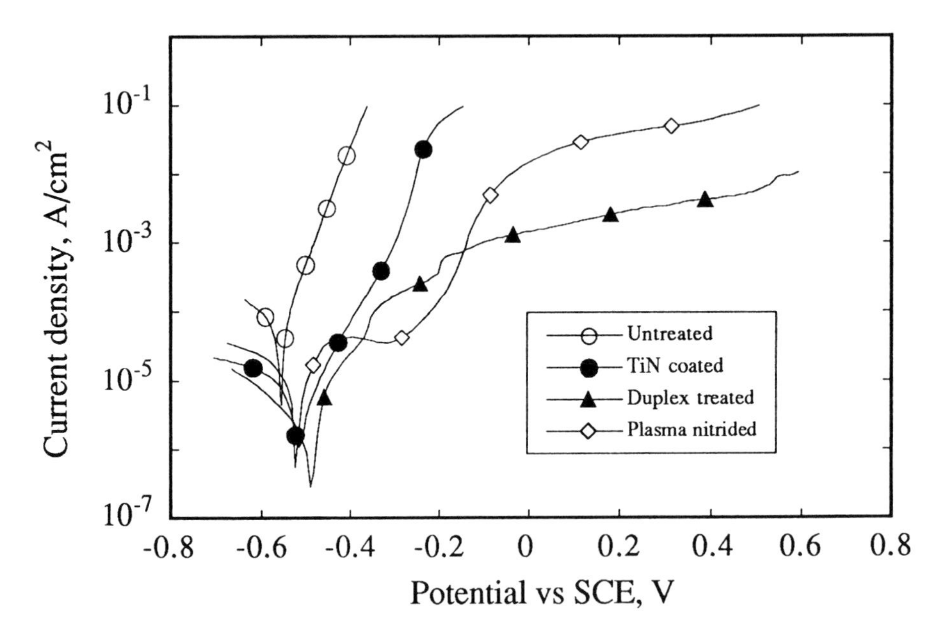

Fig. 2 Typical anodic polarisation curves of untreated, PVD TiN coated, plasma nitrided and duplex treated En40B steel.

High performance gears in modern machines are required to operate at ever-increasing torque and speed for power transmission which can not be achieved by individual carburising or nitriding treatments. Unlubricated, self-matched gear test results have further demonstrated the great potential of such duplex treatments for high performance gear applications. As can be seen from Table 2, duplex treated En40B gears not only show significantly improved wear resistance but also exhibit greatly enhanced load bearing capacity.

Table 2 Gear test results.

Test Piece	Roughness (R_a, μm)		Weight loss	Load carrying capacity (Nm)	
	before test	after test	(mg)	predicted	tested
En40B	1.187	8.213	550	10	12.5
TiN/En40B	2.315	7.754	250	33	38
Plasma	1.548	1.189	50	65	72
Duplex	1.938	1.903	10	80	86

Clearly, the ideal PVD ceramic coating-nitrided steel duplex system can thus be designed with a view to achieving combined improvements in tribological behaviour, corrosion resistance and fatigue strength based on the above discussion, as follows: (1) hardening and tempering of low alloy steel to obtain a combination of good core properties and nitriding response; (2) plasma nitriding to produce a dense compound layer which is essential for achieving excellent corrosion resistance, and also to produce a deep hardened case which serves as strong support for the hard ceramic coating, and to form an intensive near surface compressive stress, which assists in conferring excellent fatigue strength (**13**); (3) removing the superficial part of the iron nitride formed in the plasma nitriding process by micro-blasting or polishing to secure high adhesion strength between the iron nitride and the ceramic coating, and to eliminate a possible negative effect on the corrosion resistance associated with interfacial porosity; and (4) PVD-nitride coating at temperatures below 450°C to produce a very hard, wear and corrosion-resistant ceramic coating without appreciably impairing the beneficial effects resulting from the plasma nitriding treatment. Figure 3 schematically shows such a design system.

In addition to the discontinuous duplex process mentioned above, continuous processes have recently been

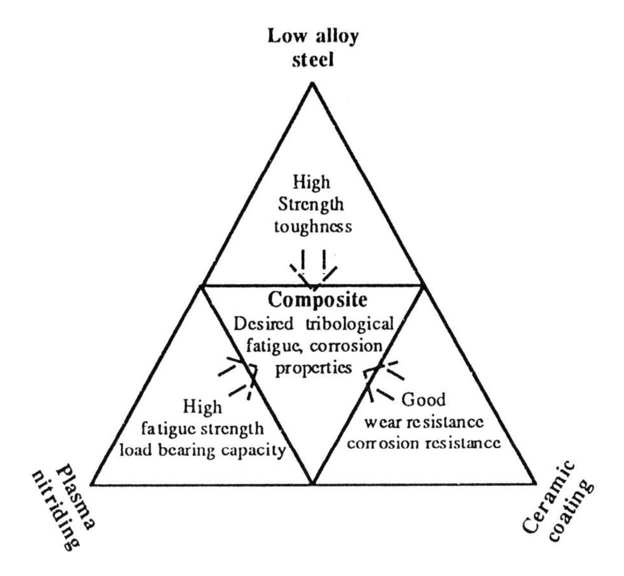

Fig. 3 Towards combined improvements in tribological, fatigue and corrosion properties.

investigated to produce such ceramic coating/nitrided steel systems (**14**). A typical continuous process involves plasma nitriding followed by ceramic coating (PVD or PCVD) in the same equipment without interruption (**15**). Although the integrated duplex processes have the advantages of possibly better process control, simple logistics and better delivery time, they have the economic disadvantage of using an expensive PVD coating unit for a nitriding treatment which can be up to 60 hours.

In order to fully realise the maximum uptake of the benefits available from the duplex ceramic coating/nitrided steel system, it is essential to select optimum nitriding parameters. In view of this, a mathematical model which simulates the plasma nitriding process has been successfully developed. The model can simulate both the plasma and gaseous nitriding processes to predict the following features: (1) total nitrogen content profile; (2)

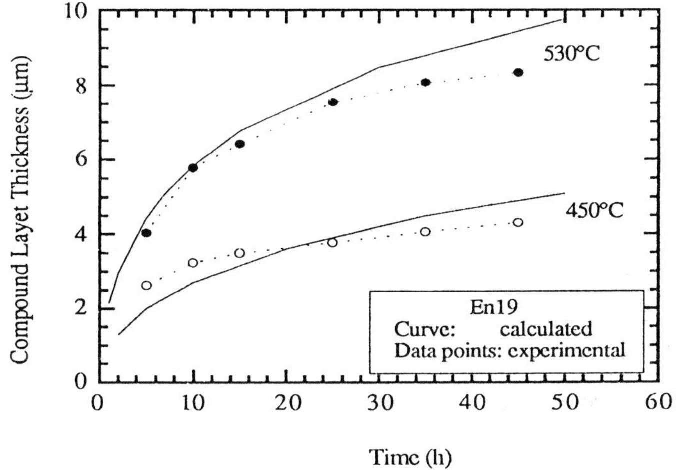

Fig. 4 Comparison of calculated and measured compound layer thickness for En19 steel.

nitrogen in solution profile; (3) nitrogen in the form of nitride precipitates profile; (4) amount of alloy element in solution; (5) amount of nitride precipitates; (6) incubation time for γ' phase initiation and (7) γ'-layer thickness. The model can also be extended further to simulate multi-stage nitriding processes and in principle, to predict the hardness profile. The model can also predict the growth kinetics of the compound layer, taking into account the effect of sputtering. For example, Figure 4 compares the calculated and experimentally measured Fe_4N layer thickness for En19 steel nitrided at 450°C and 530°C for various times. Clearly, good agreement has been achieved between the calculated values and the experimentally measured values (**16**).

4 THE DUPLEX DLC COATING-OXYGEN DIFFUSION TREATED TITANIUM SYSTEM

Titanium alloys are attractive owing to their outstanding combination of properties in terms of a high strength to weight ratio, exceptional resistance to corrosion, and excellent biocompatibility. These attractive properties make titanium alloys suitable for aerospace, chemical and recently, biomedical applications. But current commercial titanium alloys are characterised, especially in sliding situations, by poor tribological properties as well as low load bearing capacity. Consequently, the use of titanium alloys traditionally has been restricted to non-tribological applications.

However, with the end of the 'cold war', defence spending has been reduced, resulting in cutbacks in the aerospace industry which used to consume over half of the production of titanium alloys. At the same time, there is ever increasing interest in the applications of titanium alloys in such sectors as automotive, medical, power generation and general engineering, in which tribological behaviour is often a major concern. Therefore, how to overcome the tribological limitations of titanium alloys is a timely task, which presents a major challenge to the surface engineers and researchers from both a theoretical and practical viewpoint.

Over the past 10 years significant progress has been made to overcome these inherent tribological problems of titanium alloys by means of surface engineering techniques (Figure 5) including PVD and plasma nitriding, as well as electron beam surface alloying techniques (**17–20**). Thin coatings such as TiN generated by plasma nitriding or by PVD deposition can provide a titanium alloy surface with greatly improved tribological properties in terms of low friction and high resistance to wear. However, premature failure will occur if the substrate deforms plastically under a high load; on the other hand, deep hardened layers produced by electron beam surface alloying can withstand a high contact stress, but do not endure sliding contact. Consequently, these techniques are technologically only suitable for components used under moderate loads or with a low or moderate sliding ratio. In this respect, technological advances have been made in the development of advanced surface engineering techniques for titanium, such as combining deep hardening with low friction high wear resistant thin coatings, which can provide titanium surfaces with good tribological behaviour to meet different requirements arising from diverse application conditions, especially pure sliding or high contact loads or a combination of both.

To meet the above design requirement, a novel duplex system combining an oxygen diffusion treatment (OD) with a diamond-like coating (DLC), has been developed (**20**). The microhardness profile measured in the OD layer shows a high hardness (> 700HV0.05) plateau in the first 80 μm and the total hardened layer is about 300 μm (Figure 6). A diamond-like coating is then deposited on the OD treated Ti6Al4V, employing a rf-reactive Teer 450/4 unbalanced

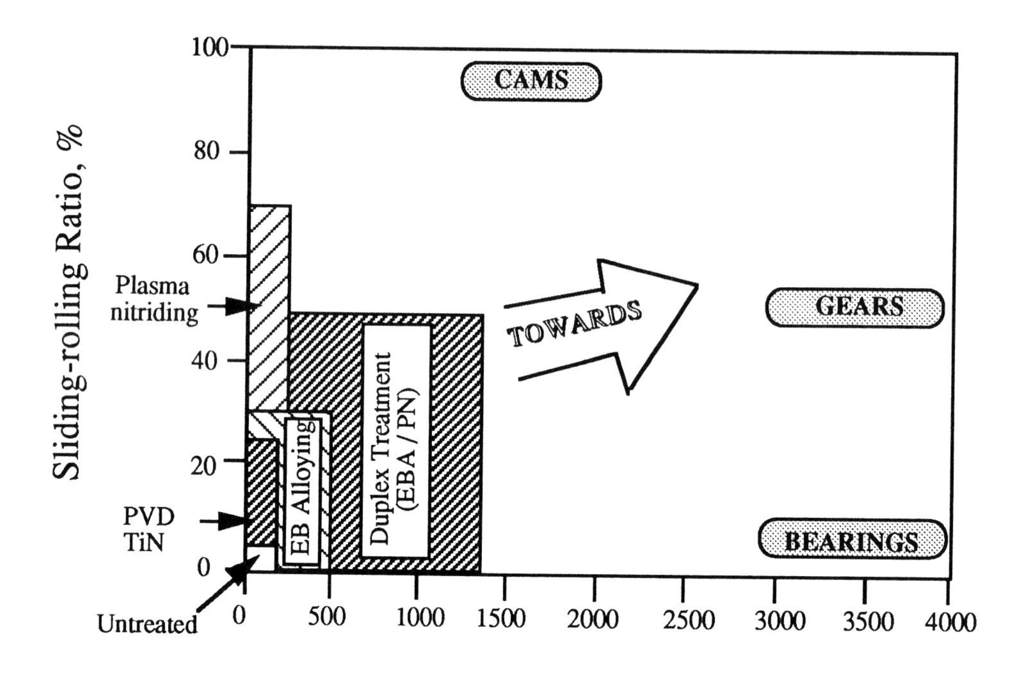

Fig. 5 The current status of the load bearing capacities of surface engineered titanium alloys.

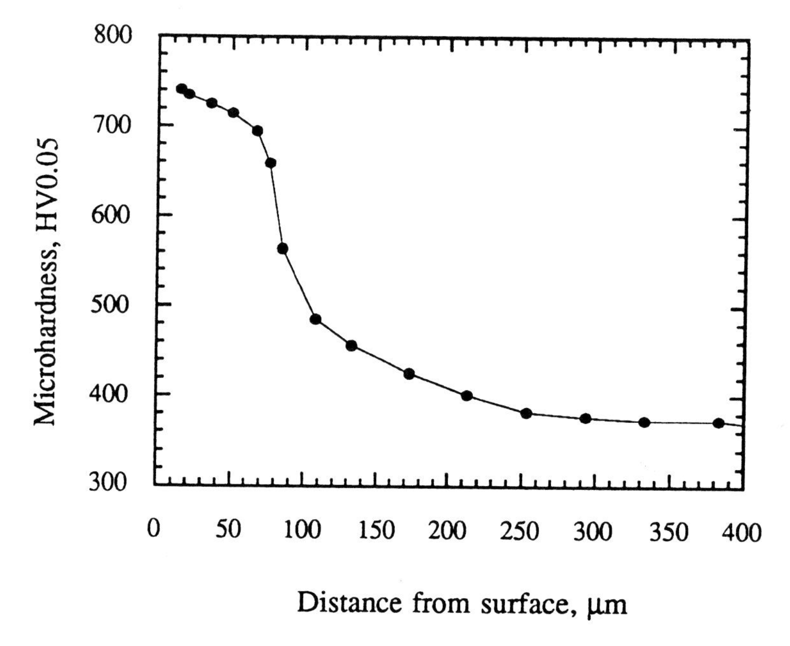

Fig. 6 Microhardness profile for the OD treated Ti-6Al-4V specimen.

magnetron sputtering system. A graded intermediate layer, Ti/TiN/TiCN/TiC, between the OD treated zone and the DLC coating was used to ensure good adhesion of the DLC coating at low temperature. Both the DLC coating and the graded intermediate layer are very dense, and good bonding between the surface coating and the OD zone has been achieved, as is evidenced in Figure 7.

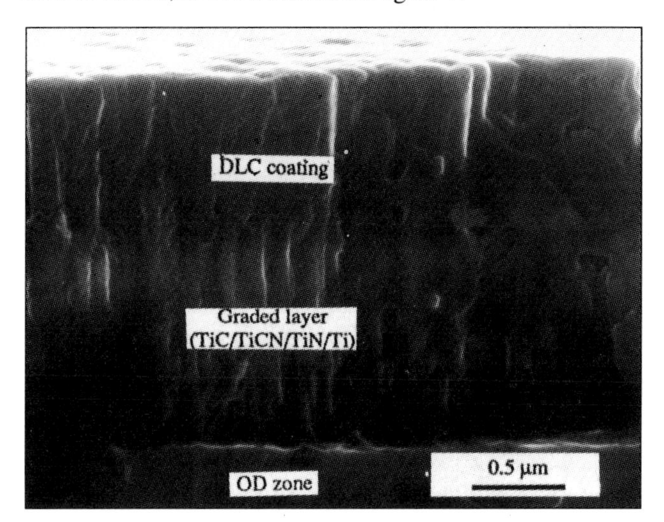

Fig. 7 SEM micrograph showing overall cross-section structure of duplex treated Ti-6Al-4V specimen.

The load bearing capacities of the duplex treated, as well as DLC coated, Ti6Al4V specimens have been evaluated, both statically and dynamically, employing a friction monitored scratch tester, a ball-on-disc tribometer and an indentation tester.

The static load bearing capacity of a duplex treated Ti6Al4V alloy has also been evaluated using an indentation method. Ring cracks around the indentation imprints were observed when the applied load was increased to a critical value (the so-called critical load) but no radial cracks were noticed for all the specimens tested. It was also found that ring cracks initiated at a load of 50 gf (0.49 N) for the DLC coated specimen, while the corresponding critical load for the duplex treated specimen was about 200 gf (1.96 N), indicating a 4-fold improvement in load bearing capacity. Clearly, the duplex treatment can significantly improve the static load bearing capacity of Ti6Al4V alloy.

Scratch testing can be used to evaluate the load bearing capacity of coatings under both normal and tangential forces. Therefore the dynamic load bearing capacity has been assessed by scratch testing and verified by SEM examination of the scratch. Figures 8 (a) and (b) show respectively the friction response of a duplex treated specimen, and a DLC coated specimen under varying loads. It can be seen from Figure 8 (a) that the DLC coated specimen failed at a load of about 8 N, a clear indication of its low load bearing capacity, which probably resulted from plastic deformation in the substrate. In contrast, the friction force increases at first linearly with the applied normal load, implying that the friction coefficient was constant. On increasing the applied load further to about 45 N, a sudden increase in friction force occurred, as is clearly indicated by the first derivative of the friction force-load curve, reflecting the collapse of the DLC coating (Figure 8 (b)). These findings were further verified by SEM examination of resultant scratches. For the duplex treated specimen when the applied load was increased to about 30 N, typical tensile cracking occurred, but there were neither spalling nor chipping failure; when the load was increased further to about 40 N sideways lateral flaking failure occurred; and finally forward lateral flaking failure happened at a critical load of about 45 N. When scratched, the DLC coating buckled at a load as low as 5 N and peeled off entirely.

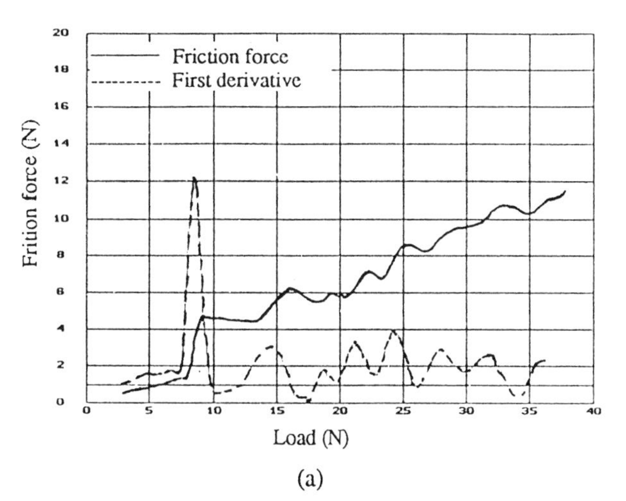

(a)

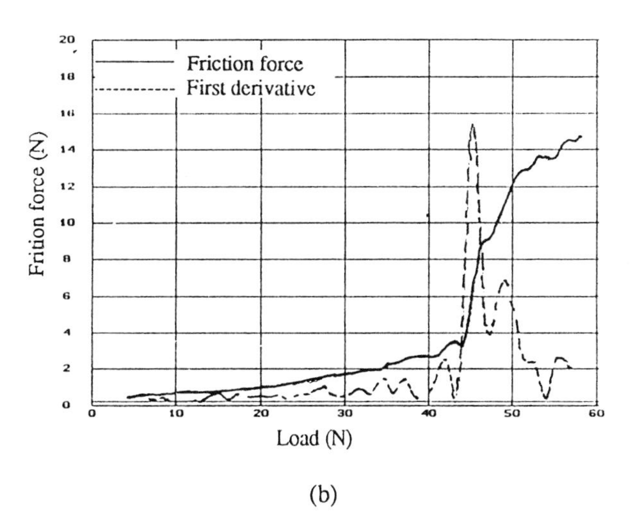

(b)

Fig. 8 Variation of friction forces with applied loads during scratch tests on (a) DLC alone coated and (b) duplex treated Ti6Al4V specimens.

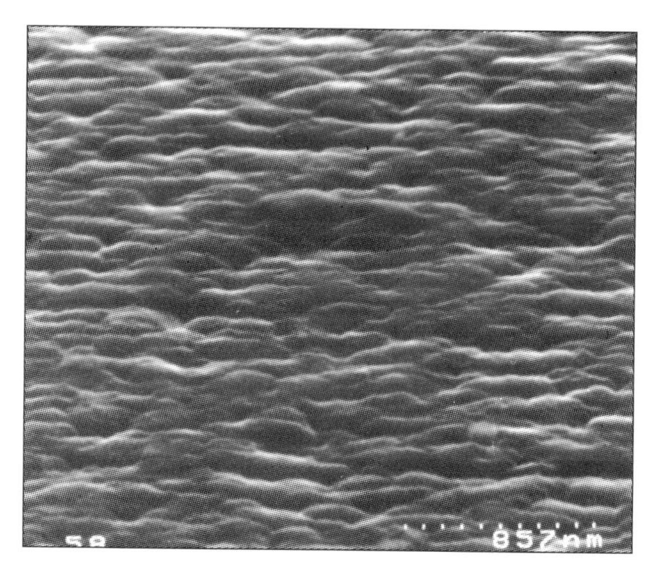

Fig. 9 SEM micrograph showing worn track on duplex treated Ti6Al4V specimen under a load of 120 N.

Dynamic load bearing capacity has also been evaluated under conditions similar to that found in a bearing. The load bearing capacities for a DLC coated Ti6Al4V specimen, a duplex treated Ti6Al4V specimen, and a DLC coated hardened En 19 steel specimen were 50 N, 130 N and 160N, respectively. Figure 9 shows a wear track on a duplex treated Ti6Al4V specimen tested under a load of 120 N. Although the surface was worn slightly, no cracks were observed. On the other hand, there were many cracks in the wear track of a DLC coated Ti6Al4V specimen after testing, as is illustrated in Figure 10, which shows severe tensile cracking.

The load bearing capacities of duplex treated materials are summarised in Table 3, showing an improvement by a factor of 4, 5.6 and 2.6 as measured by indentation, scratch and ball-on-disc methods respectively, when compared with that of DLC coated material alone.

In the present work, it has been shown that the novel duplex system, combining the oxygen diffusion treatment with DLC coating, possesses excellent load bearing capacity, irrespective of the evaluation method. In short, this novel duplex system has effectively extended the window of the design service conditions for titanium components, in terms of both high sliding ratio and/or

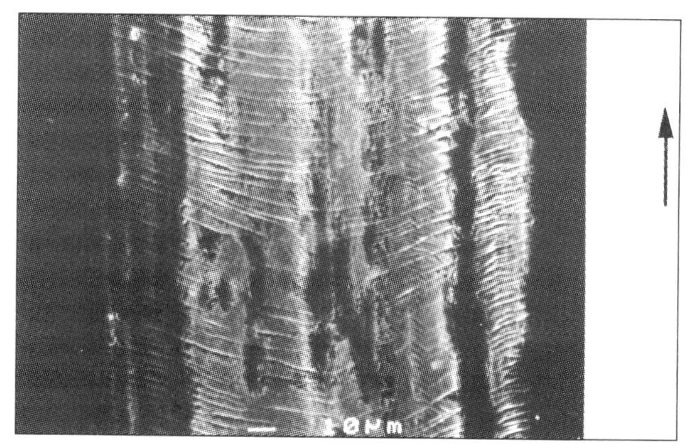

Fig. 10 SEM micrograph showing worn track on DLC alone coated Ti6Al4V specimen under a load of 50 N.

higher loads (Figure 11), and it is thus seen as an important step towards titanium designer surfaces.

This outstanding performance can be attributed to the optimised design of the duplex system (Figure 12). DLC coating is a unique material which has both a high hardness and a extremely low friction coefficient against most engineering materials. In the layered coating system, friction has significant effects on the stress distribution: (a) the maximum Von-Mises stress not only increases in value but also moves towards the surface when the friction coefficient increases; (b) the maximum shear stress at the interface also increases with increasing friction coefficient. Thus it follows that the low friction of DLC could effectively reduce both the interface stress and the stresses near the surface. However, DLC is usually characterised by high levels of residual stress and poor adhesion when coated directly onto most engineering materials. These problems have been successfully addressed by adapting a graded intermediate layer between the DLC and the OD treated sublayer. The compositionally graded intermediate layer, Ti/TiN/TiCN/TiC, eliminates interfacial cracking by

Table 3 Load bearing capacity of duplex treated Ti6Al4V.

Test Method	Load Bearing Capacity		Increase
	DLC alone Coated	Duplex Treated	%
Indentation	0.49 N	1.96 N	400
Scratch	8 N	5.6 N	560
Ball-on disc	50 N	130 N	260

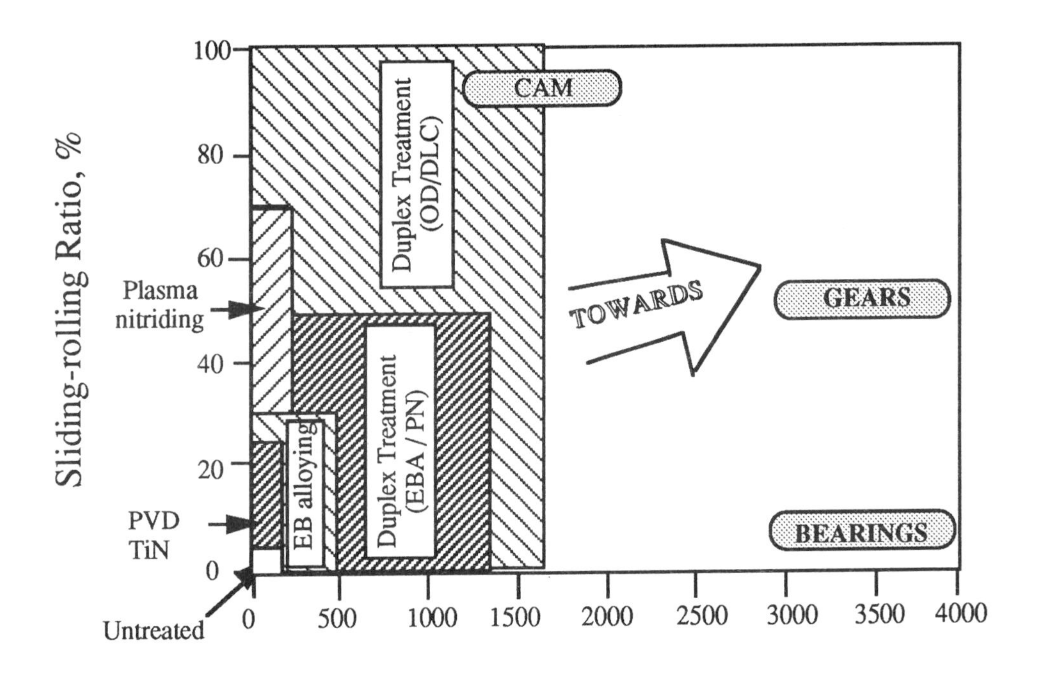

Fig. 11 The effect of duplex surface engineering in extending the areas of application of Ti6Al4V.

homogenising the stress distribution. Experimental results, however, have revealed that the modified DLC coating still showed a relatively low load bearing capacity compared with the duplex system. Clearly, a significant improvement in load bearing capacity can be achieved only when the DLC coating is deposited on a deep hardened sublayer (about 300 μm) produced by the OD treatment, thus mitigating against plastic deformation. Therefore, it is clear that the duplex treatment is essential to achieve a high load bearing capacity in this titanium based system.

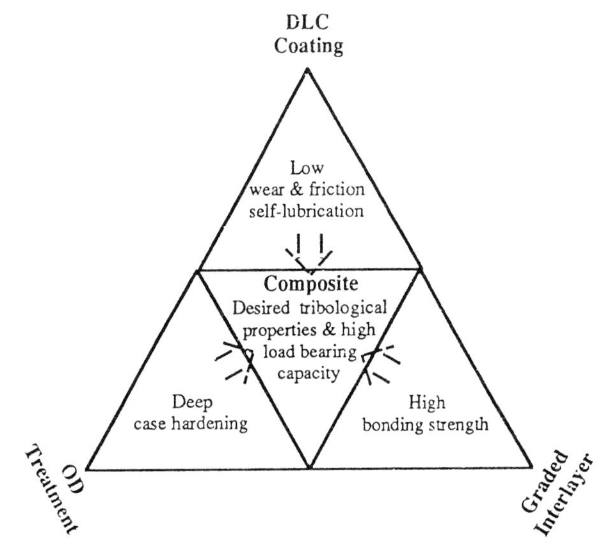

Fig. 12 Towards titanium designer surfaces with desired combinations of properties through a novel duplex surface engineering system.

5 PREDICTION OF THE PERFORMANCE OF DUPLEX SYSTEMS

Designing with duplex surface engineering requires a knowledge of the stress distribution in layered surfaces, which is affected by the elastic and plastic properties of the layers and substrate, the friction coefficient of the surface, as well as the residual stress distribution. The conventional approach to this problem is based on classical Hertz theory of elastic contact, which is restricted to frictionless smooth surfaces of homogeneous properties and perfectly elastic solids. However, when layered surfaces such as nitrided surfaces and nitrided/TiN coated surfaces are involved in sliding or rolling-sliding contact motion, the situation is much more complicated. Accordingly, attempts have been made to develop mathematical models to simulate the contact behaviour of layered surfaces, taking into account (a) different properties in the surface layers and substrate; (b) friction coefficient; (c) surface roughness parameters.

Such a model, based on recent theories of multi-layered surface contact, takes into account multi-layered structures, real surface roughness and friction effects. It is on the basis of the following major assumptions that research was undertaken, but some of these assumptions may be relaxed in the future: (1) The structure of the contact bodies is considered to be composed of one or more elastically homogeneous surface layers rigidly bound to each other. (2) The contact is assumed to give rise to a state of plain strain and the strains are assumed small, leading to the usual assumptions of linear elastic theory. (3) The contact is considered to be dry, i.e. no specific lubricant is present.

For the co-ordinate system and the 'n' layered structure shown in Figure 13, the boundary conditions are as follows:

1. At the surface the normal and shear stresses must equal the applied pressures.
2. At every interface in the system, the normal and shear stresses and the normal and transverse deflections on each side of the interface must be equal.
3. The stresses fall to zero at a large distance from the load.

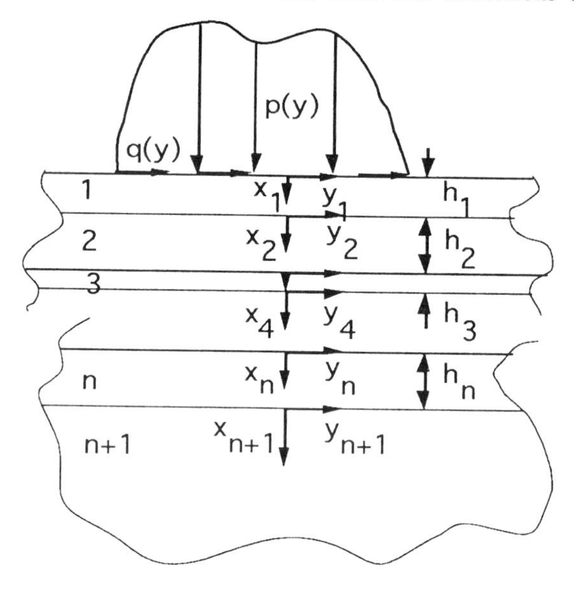

Fig. 13 Co-ordinates and notations for multi-layered elastic solid.

During loaded contact motion, the contact surfaces are subject to material loss due to the applied forces and their relative movement. Virtually any engineering component can fail as a result of the applied stress exceeding the strength of the material. Initial surface failure is associated with such phenomena as wear due to the shear force experienced at the contact when one surface slides, or rolls with sliding, over another. On the other hand, subsurface initial failure is the result of the high stresses developed beneath the surface exceeding the fatigue strength or yield strength of the material, leading to fatigue fracture. In the present research, distortion-energy theory (Figure 14) has been used and the failure criterion obtained: $\sigma_{vms} > k\sigma_y$, where k is fatigue ratio, σ_{vms} is the maximum Von Mises stress and σ_y is the yield strength of the material (**22**).

It may be noted that the Lundberg and Palmgren (**23, 24**) orthogonal shear stress criterion has not been applied here. This is because the orthogonal shear stress criterion is only applicable to smooth and frictionless contact and the subsurface stress field becomes very complex when the roughness and friction are considered (**25**).

To evaluate the load bearing capacity of engineering surface contacts, the strength and stress profiles need to be established. The required strength profiles can be found

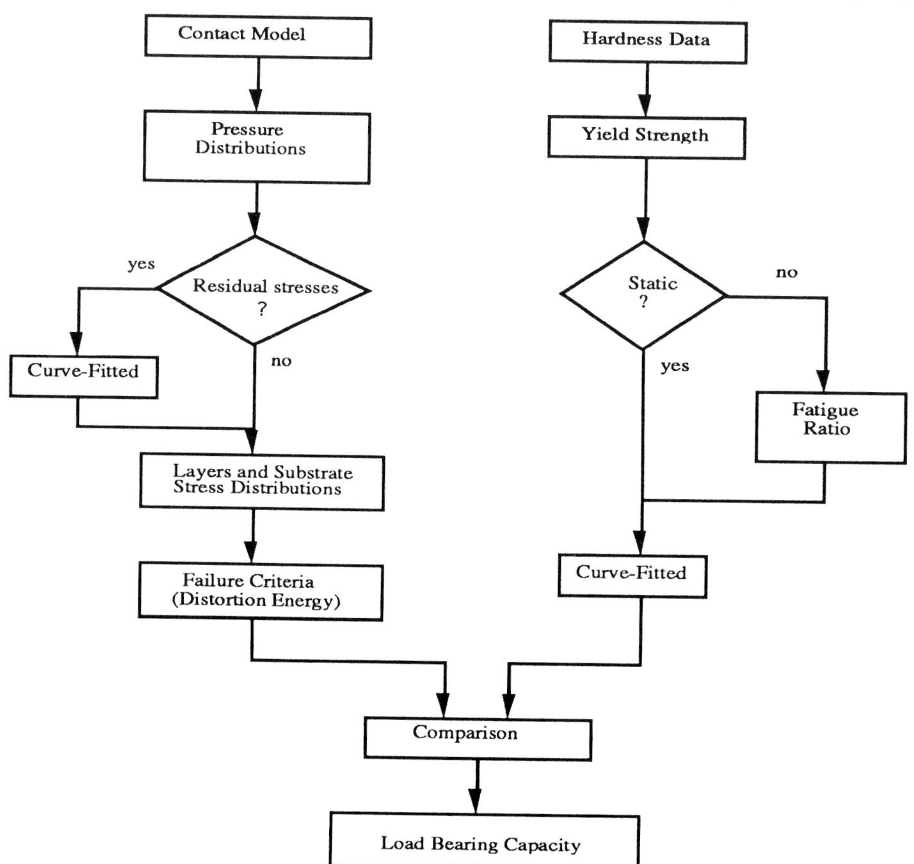

Fig. 14 Flow diagram for load bearing capacity prediction.

from the work of Bell and Sun (**26**), whereby the stress profile distribution in the layers, and the subsurface stresses for real surface roughness with friction can be obtained (**27, 28**). It may be noted that the presence of residual stress can also be accounted for in the present model.

5.1 Applications of a mathematical model for duplex surfaces

The model was first applied to the Amsler test results of Bell and Sun (**29**). In their experimental work, the hardness/depth profiles were measured for En40B after plasma nitriding at 500°C for 2.5, 10 and 35 hours. Corresponding samples were then contact fatigue tested on an Amsler machine in mixed rolling/sliding under a load of 0.2 N/μm and with a measured coefficient of friction of 0.6. After 60,000 cycles the sample nitrided for 2.5 hours showed cracking of the nitrided case due to substrate plastic deformation, whereas no cracking was observed in the 10 and 35 hour samples. On the basis of a fatigue ratio of 2, the model predicts failure of the 2.5 hour nitrided specimen upon the application of a load of 0.1725 N/μm, which is in reasonable agreement with the experimentally observed failure at 0.2 N/μm.

The model has been used also to predict the subsurface crack failure of spur gears (**30**) in En40B material. The tests included untreated, plasma nitrided and duplex treated (TiN coated on top of a plasma nitrided substrate) gears. All the gears were made to the following specification: Pinion = 15 teeth; wheel = 15 teeth; module = 4 mm; face width = 10 mm and pressure angle = 20 degrees. All the tests lasted 30,000 cycles and the teeth surface roughness change and weight loss were examined before and after tests. The results are shown in Table 2. From the table it can be seen that the experimentally tested performance improvement by duplex treatment and the predicted values from the model are in good agreement. The failure modes, as expected, were spalling around the pitch point due to low friction and pitting far away from the pitch point, due to a high sliding effect.

Finally, the model has been tested in relation to the friction and wear behaviour of TiN, TiN/MoS_2 and TiN/TiCN/DLC coatings on tool steel (**31**). Figure 15 shows the tool steel subsurface stress distributions under three different coating conditions. First, consider the difference in stress distribution between the DLC/TiCN/TiN/tool steel system and the TiN/tool steel system. The stress in the TiN/tool steel system is about three times higher than that of the DLC/TiCN/TiN/tool steel system. This reduction in stress may be due to two reasons. One is the low friction of the DLC/TiCN/TiN/tool steel system (μ = 0.12) compared to μ = 0.4 for the TiN/tool steel system. The other is the low modulus of elasticity of the DLC which will increase the contact width and reduce the normal and tangential contact pressures. Further improvements were achieved for the MoS_2 /TiN/tool steel system, as in shown in Figure 15. Those improvements are again due to the low friction (μ = 0.05) of the MoS_2 and its low modulus of elasticity. Scratch tests have shown a similar trend regarding stress distribution (**31**). In addition, the model has also been used to investigate the interfacial fatigue strength of hard coatings (**32**). Recently, the model has been used for predicting the industrial performance of titanium alloy components, and initial results look promising.

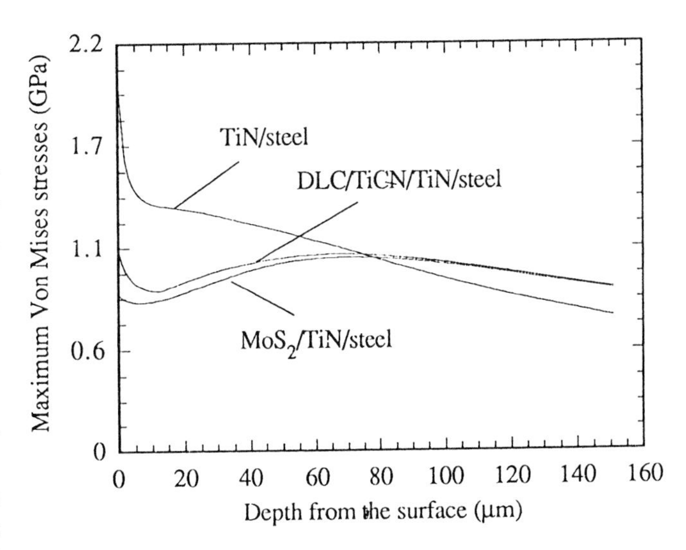

Fig. 15 Subsurface stress distributions for TiN coated, DLC coated and duplex treated (MoS_2/TiN) tool steel.

6 CONCLUSIONS

1. Under specific tribological conditions the TiN PVD treatment of En40B steel can improve its wear resistance by a factor of 2.5, while plasma nitriding improves it by a factor of 11. However, duplex surface engineering involving a combination of the plasma nitriding and PVD technologies improves the wear resistance by a factor of 55. The duplex system also shows much better corrosion performance than either the TiN coated or nitrided material.

2. While untreated titanium and its alloys can hardly sustain 0.1 GPa at less than 5% sliding-rolling ratio, the novel titanium duplex system combining an oxygen diffusion treatment with a DLC coating can withstand 1.7 GPa under 100% sliding conditions. Clearly, the development of the novel duplex system is a big step towards designing dynamically loaded titanium engineering components, such as bearings and gears, in the near future.

3. FEM contact mechanics modelling can successfully simulate the frictional elastic contact behaviour of real duplex surface engineered systems. The experimentally

tested load bearing capacity and the predicted values from the model are in good agreement. Based on the contact mechanics model, it is possible to design an optimum duplex system for a specific application.

4. The last two decades of the present millennium have seen the emergence of many new surface engineering technologies and many incremental improvements in traditional techniques for surface treatment, which have stimulated the emergence of the new multi disciplinary subject of surface engineering. The integration of the concepts of materials engineering, contact mechanics and tribology, with the basic sciences of chemistry and physics, has demonstrated significant potential for the enhanced performance of engineering components for use in all sectors of the machine building industries, and other industries such as bio-engineering. This potential, technically, economically, and environmentally, will be increasingly realised in the early decades of the next millennium through the further development and widespread industrial acceptance of duplex surface engineering in the component design process.

7 ACKNOWLEDGEMENTS

The author wish to acknowledge the Commission of the European Union for its support of the programme 'Finite element modelling of the plasma nitriding process and the resultant load bearing capacity of low alloy steels (BRITE/EURAM Project BE 4242)', and DTI/EPSRC for its support of the LINK project 'Advanced surface engineering of titanium alloy components (AdSurfEngTi)'. I would also like to extend my many thanks to all members of the Wolfson Institute for Surface Engineering within the School of Metallurgy and Materials, particularly Drs A. Bloyce, Y. Sun, H. Dong, K. Mao and P. Buchhagen, for their valuable contributions.

8 REFERENCES

1 **Bell, T.,** Surface Engineering: a rapidly developing discipline, *European Journal of Engineering Education*, 1987, **12,** 27–32.

2 **Bell, T.,** Wealth creation through surface engineering, *Surface Engineering News,* LINK Surface Engineering Programme Secretariat, 1994.

3 **Bell, T.,** Realising the potential of surface engineering in the 21st century, *Executive Engineer*, 1997, **5,** 22–23.

4 **Sun Y.** and **Bell, T.,** Combined plasma nitriding and PVD treatments, *Transactions of the Institute of Metal Finishing*, 1992, **70,** 38–44.

5 **Tucker T.R.,** and **Ayers, J.D.,** Consolidation of metal coatings by electron beam melting, in C. W. White and P. S. Peer (eds): *Laser and Electron Beam Processing of Materials, Academic Press,* New York, 760–765.

6 **Dingremont, N., Dergmann E.,** and **Collignon, P.,** Application of duplex coatings for metal injection moulding', *Proceedings of 9th international congress on Heat Treatment and Surface Engineering*, PYC EDITION 5, IVRY-SUR-SEINE CEDEX, France, pp 131–137.

7 **Staines, A.M.,** Trends in plasma-assisted surface engineering processes, *Heat Treatment of Metals,* 1991, No 4, 85–92.

8 **Sun, Y., Bloyce. A.,** and **Bell, T.,** Finite element analysis of plastic deformation of various TiN coating/substrate systems under normal contact with a rigid sphere, *Thin Solid Films*, 1995, **271,** 122–131.

9 **Tian. H.,** and **Saka, N.,** Finite element analysis of an elastic-plastic two-layer half space: sliding contact, *Wear*, 1991, **148** 261–285.

10 **Staines. A.M.,** and **Bell, T.,** Multiple surface treatments for low-alloy steels, paper presented at the conference of 'Engineering the Surface' (14–16 May, 1986, London), The Institute of Metals.

11 **Spies, H.-J., Larisch, B.,** and **Hoeck, K.,** Optimisation of TiN hard coatings on prenitrided low alloy steels, Surface Engineering, 1995, **11,** 319–323.

12 **Dong, H., Sun, Y.,** and **T. Bell,** Enhanced corrosion resistance of duplex coatings, *Surface and Coatings Technology*, in press.

13 **Bell, T.** and **Loh, N.L.,** The fatigue characteristics of plasma nitrided three Pct Cr-Mo steel, *Journal of Heat Treatment*, 1982, **2,** 232–237.

14 **Dingremont, N., Bergmann, E., Collingnon. P.,** and **Michel, H.,** Optimisation of duplex coatings built from nitriding and ion plating with continuous and discontinuous operation for construction and hot working steels, *Surface and Coatings Technology*, 1995, **72,** 163–168.

15 **Huchel, U., Bramers, S., Crummenauer, J., Dressler S.,** and **Kinkel, S.,** Single cycle, combined layers with plasma assistance, *Surface and Coatings Technology*, 1995, **76–77,** 211–217.

16 **Bell, T., Sun, Y., Mao K.,** and **Buchhagen, P.,** Mathematical modelling of the plasma nitriding process and the resultant load bearing capacity, *Advanced Materials and Processes*, 1996, No 4, 40Y–40BB.

17 **Morton, P.H.** and **Bell, T.,** Surface engineering of titanium, *Memories et Etudes Scientifiques, Revue de Metallurgie*, No.**10**, 1989, 639–646.

18 **Bell, T., Morton P.H.,** and **Bloyce, A.,** Towards the design of dynamically loaded titanium engineering components, *Materials Science and Engineering*, 1994, **A184**, 73–86.

19 **Bloyce, A., Dong H.,** and **Bell, T.,** Comparative evaluation of treatments for wear protection of Ti-6Al-4V, In P.A. Blenkinshop, W.J. Evans and H. M. Flower (eds): *Titanium '95: Science and Technology*, The Institute of Materials, London, 1996, 1975–1982.

20 **Dong, H.,** *Development of novel surface engineering technologies for Ti6Al4V*, PhD thesis, 1997, The University of Birmingham, UK.

21 **Mao, K., Sun Y.,** and **Bell, T.,** A numerical model for the dry sliding contact of layered elastic bodies with rough surfaces, *Tribology Transaction of STLE*, 1996, **39,** 416–424.

22 **Ioannides, E.,** and **Harris, T. A.,** A New Fatigue Life Model for Rolling Bearings, *Journal of Tribology, ASME*, 1985, **107** 367–378.

23 **Lundberg, G.** and **Palmgren, A.,** Dynamic Capacity of Rolling Bearing, *Acta Polytechnic Mechanical Engineering Series*, 1947,**1**, Stockholm, Sweden, 7–14.
24 **Lundberg G.** and **Palmgren, A.,** Dynamic Capacity of Rolling Bearing, *Acta Polytechnic Mechanical Engineering Series*, 1952, **2,** Stockholm, Sweden, 5–32.
25 **Bailey D. M.** and **Sayles, R. S.,** Effect of Roughness and Sliding Friction on Contact Stresses, *Journal of Tribology, ASME*, 1991, **113,** 729–738.
26 **Bell T.** and **Sun Y.,** Load Bearing Capacity of Plasma Nitrided Steel under Rolling-sliding Contact, *Surface Engineering*, 1990, **6,** 133–139.
27 **Mao, K., Sun Y.,** and **Bell, T.,** *Modelling of Layered Engineering Surface Contacts,* Contact Mechanics 2, Computational Mechanics Publications, Southampton, Boston, 1995, 73–80.
28 **Mao, K., Bell T.,** and **Sun, Y.,** Effect of Sliding Friction on Contact Stresses for Multi-layered Elastic Bodies with Rough Surfaces, *ASME/STLE Tribology Conference,* San Francisco, 1996.
29 **Bell, T.** and **Sun, Y**. Load Bearing Capacity of Plasma Nitrided Steel under Rolling-sliding Contact, *Surface Engineering,*1990, **6**, 133–139.
30 **Mao, K., Bell T.,** and **Sun, Y.,** An Application of the Contact Stress Model to the High Performance of Duplex Treated Gears, *ASME/STLE Tribology Conference*, San Francisco, 1996.
31 **Ma, K. J., Bloyce, A., Mao, K., Bell T.,** and **Teer, D.,** Friction and Wear Behaviour of TiN/Au, TiN/MoS_2 and TiN/TiCN/DLC Coatings, *International Conference of Metallurgical Coatings and Thin Films*, San Diego, 1996.
32 **He, J. W., Hendrix, B. C., Hu, N. S., Xu, K. W., Bell, T., Sun, Y.,** and **Mao, K.,** Interfacial Fatigue Limit as Measure of Cyclic Bonding Strength of Hard Coatings, *Surface Engineering*, 1996, **12,** 49–54.

Tribology of hot metal forming

J H BEYNON
Department of Mechanical Engineering, University of Sheffield, UK

SYNOPSIS

Many aspects of the tribology of hot metal forming are poorly understood. This leads to difficulties when modelling the forming operations to develop process improvements. It also handicaps the development of new tool materials, since the operating conditions with which they will have to cope are not fully described. Several key issues in friction, heat transfer, lubrication, wear and fatigue are discussed. This is an area of tribology with many opportunities for research. It is concluded that progress will be made through combination of techniques, including laboratory simulation, computer modelling and industrial trials.

1 INTRODUCTION

The tribology of hot metal forming involves a wealth of unresolved problems for the research community. There is considerable and growing industrial interest in these problems which will require great ingenuity to solve. It is an intellectually challenging area that deserves more attention.

This paper reviews outstanding issues in friction, lubrication and wear arising from the hot working of metals. The examples cited are taken primarily from the rolling industry, but most issues are common to other forming operations such as forging. Apologies are offered in advance to those authors who are not represented here despite their contribution to this area of tribology. This paper is not intended to be a comprehensive review, rather a personal view of key issues which is illustrated by selective reference to the literature.

2 FRICTION AND HEAT TRANSFER

Friction is beneficial to most hot metal forming operations. In rolling it allows the metal to be drawn into the gap between the rolls. In open-die forging it prevents the metal escaping from between the tools and in closed-die forging it provides the back pressure in the flash to ensure filling of the die cavity. In the extrusion of steel, which is well lubricated by using glass, friction serves little useful purpose, but does not contribute excessively to the operating forces. The extrusion of aluminium alloys is often conducted without lubrication and friction can nearly double the extrusion pressure.

The interface between tool and workpiece in hot metal working is crucial to both friction and heat transfer. Since the intimacy of the contact between the cold tool and the hot workpiece is critical to both the traction and the rate of heat transfer, the friction and thermal behaviours are inter-related and must be considered together.

The hot metal invariably starts with an oxide on its surface. Aluminium has a thin adherent oxide film on its surface when at high temperature in air. With steel, the thickness and composition of oxide scale depend on temperature, time at temperature, steel composition and availability of oxygen. Below 570°C FeO does not form, and the scale comprises a layer of Fe_3O_4 next to the metal with a layer of Fe_2O_3 on top. Above 570°C, FeO is present, forming the innermost layer next to the metal. Thus the oxide richest in oxygen is at the outer surface where the availability of oxygen is greatest. The rate of scale formation is complicated by the composition of the underlying steel (**1**). For instance, at low temperatures (below 500°C) increasing carbon content increases the rate of oxidation, but lowers the rate at higher temperatures (above about 700°C). Other alloying elements change the rate of oxidation and also affect the metal-oxide bond. For example, elements alloyed in the steel that are more noble than iron, such as nickel and copper, encourage a strong bond which makes the scale difficult to remove between reheating and hot forming (**1**). In practical descaling, a thin adherent oxide layer usually remains after the bulk of the scale has been removed. Even if the whole layer is removed, growth of fresh oxide begins immediately after descaling. Thus an oxide layer should be assumed to be always present during hot forming.

The plastic straining imposed during hot metal forming causes deformation or fracture of the oxide scale. If such fracture occurs when the hot metal is in contact with the cold tool, clean metal to metal contact can occur. This will produce a sharp rise in the friction, and a dramatic rise in the heat transfer coefficient compared with contact through an oxide barrier. What is not known is under what conditions such fracture occurs.

Making measurements on industrial plant is fraught with difficulties. Since the hot stock is often very large, moves large distances and, in the case of rolling, accelerates to high speeds, implanting transducers into such material is very difficult. The rolls in a hot rolling mill are regularly removed from service for regrinding of the surface; thus access is easier. Embedding thermocouples into the roll surface and sub-surface would be relatively easy to achieve, but it would be expensive and the author knows of no industrial rolling operation where this has been done.

Embedding transducers within a roll can be used in the form of so-called pressure pins to measure normal load, or as a cantilever with its axis normal to the roll surface to measure shear forces (**2,3,4**). Unfortunately the gaps which are needed between the pin and roll are easily penetrated by the hot, soft metal, hampering accurate measurement. (Even when used in cold metal forming operations, local hydrodynamic action at the pins strongly affects the friction measurements (**5**).)

Industrial hot forming is difficult to simulate in the laboratory for several reasons. The temperatures involved are readily achieved, but the speed of the operation can often be too high to simulate under controlled laboratory conditions. Even in the industrial hot rolling of flat metal strip and plate, contact times between roll and stock vary from 100 ms down to 0.5 ms, a very short time to make detailed measurements. The arc of contact in such rolling can also be short, ranging between 10 and 150 mm. The surface states of both stock and roll are difficult to replicate. The hot stock will oxidise on its surface, often in a complicated manner as described earlier. The roll will pick up oxide from the stock, the amount depending on the roll metallurgy as well as operating conditions. The roll metal will also oxidise itself due to the high temperature excursions which it suffers when in contact with the workpiece. Again the amount of oxidation is affected by the roll composition (**6**).

Perhaps not surprisingly given the complexity of the roll stock interface, measured values of heat transfer coefficient vary enormously: 18-38 kW/m^2K (**7**), 23-81 kW/m^2K (**8**), 2-20 kW/m^2K (**9**), 5-50 kW/m^2K (**10**) and 200 kW/m^2K (**11**). One reason for the variability is that heat transfer will be sensitive to the real area of contact across the interface. Pressure sensitivity of the heat transfer coefficient has been reported with local values in the arc of contact varying from 10 to 260 kW/m^2K (**12**). For the hot rolling of aluminium alloys, the stock is so much softer and is likely to produce intimate roll-stock conformity early in the roll contact with little change thereafter (**13**). The reported range of measured values is similar to those for steel: 19-22 kW/m^2K (**14**), 10-50 kW/m^2K (**15**), 15-20 kW/m^2K (**16**), 15 kW/m^2K (**17**) and 100-210 kW/m^2K (**13**).

With such hot metal surfaces, there is usually plenty of water used to keep the rolls cool. In rolling this water is drawn into the roll-stock interface. Although not producing hydrodynamic lubrication, the water does interfere with more intimate metal to metal contact in the roll gap, and may well form steam pockets in the asperity valleys. However, this is currently speculation since direct observations have not been made.

Thus the details of events occurring in the roll-stock interface are very poorly understood. This has led those people responsible for modelling hot metal forming operations to resort to correspondingly simple descriptions of friction and heat transfer. Single, coulombic coefficients of friction are often used over a wide range of conditions as reported by Montmitonnet and Buessler (**18**). Likewise heat transfer is normally described by a single coefficient throughout. This is despite the continually changing nature of the interface.

What is needed is a computer-based model of the interface supported by whatever sparse information is available. The model can be used to demonstrate the likely sensitivity of friction and heat transfer to the different variables. This will allow for better definition of further experiments, which is particularly worthwhile given the cost of working under these conditions. Such a model will also allow the constitutive behaviour of the interface to be defined, which will in turn form the next generation of friction and heat transfer formulae.

Detailed modelling of the tool-workpiece interface means working at the asperity scale. Slip line field solutions of asperity deformation (**19,20**) are limited in the phenomena which they can encompass. Temperature and strain rate sensitive flow stresses, oxide scale fracturing during deformation and non-plane strain stress states are just some of the conditions with which slip line field theory cannot cope. Upper bound plasticity analyses (**20,21,22**) provide useful insight, particularly for parametric sensitivity. More recently the finite element method has been applied to the deformation of asperities for cold forming (**23**) and the hot compression of a single asperity (**24**).

The finite element method is capable of incorporating a wide range of phenomena in each model, as exemplified by recent attempts to model friction and heat transfer during the relatively high speed rolling of thin (about 1 mm) flat steel strip in the last stand of a seven-stand tandem mill (**25**). Here the deformation of a group of asperities under complicated mechanical and thermal loading conditions, including a pressurised fluid in the asperity valleys, is used to model friction and heat transfer. The micro-scale modelling is linked iteratively with the macro-scale modelling of the whole roll gap. While such multi-level finite modelling is very intensive in its use of computers, and would not be used on a regular basis, it is a valuable means of gaining insight by dealing with the interactions of the various events.

3 LUBRICATION

As mentioned in the previous section, the need to keep the tools cool in hot metal forming means that there is plenty of cool liquid introduced around the piece being worked.

In steel rolling this is usually just water, since the vaporisation of oils can be difficult to control to keep a clean environment. However, such control can be achieved and more rolling mills, notably in Japan, are using lubricants to reduce friction and to improve consistency of the rolling operation. The amount of oil used is 30 - 40 g per tonne of steel rolled per rolling stand (**26**). It has particular benefit in the rolling of non-flat sections where targeted application can alleviate wear where the local conditions are particularly aggressive.

Unlike hot steel rolling, the hot rolling of aluminium alloys requires lubrication to prevent excessive transfer of aluminium to the roll (**27**). Inadequate lubrication allows the aluminium surface to come into contact with the steel roll to which it readily adheres, creating so-called "pick-up" (**29**). This results in a poor surface finish on the rolled product, both by uneven removal of metal and the subsequent tearing of the surfaces by particles previously deposited on the roll surface. Also, these particles on the roll can detach and become rolled into the stock surface. However, as mentioned above, some friction is required to allow the rolls to "bite" the stock and draw it into the roll gap. This requirement limits the amount of lubricant which can be used.

In practice, therefore, the hot rolling of aluminium is lubricated at a compromise level. Hot rolling lubricants usually comprise 5-10% lubricant emulsified in demineralised water (**27**). Since pick-up is difficult to control, nylon bristle brushes are often used on the work rolls to remove excess transferred aluminium. Some oxidised aluminium inevitably remains on the work roll surface, but a roll which has no such roll coating has poor biting characteristics.

Thus in the hot tandem rolling of aluminium alloys, lubricants based on the oil/water emulsion reduce friction. Series AA3000 and AA1000 alloys can be modelled satisfactorily using an interface shear factor, ranging from about 0.6 to 1 at the first finishing stand F1 (of three) to around 0.2 at F3. This variation appears to be due partly to speed, which rises about five-fold through the tandem mill, with the fastest exit speeds being near 500 m/min, but 100 m/min being common. The higher speed probably draws more lubricant into the roll bite, aiding lubrication. The lubrication comes from the oil (5-6%) in water emulsion which contains additives such as oleic acid, which forms a complex soap on contact with aluminium fines from the surface of the stock. Boundary lubrication becomes operative below about 350-400°C, the higher temperature corresponding to higher rolling speed due to greater ingress of lubricant into the roll bite. Such temperatures are found during finish rolling in the tandem, resulting in a significant contribution to the drop of friction factor from stand 1 to 3.

This appearance of boundary lubrication according to an as yet unquantified combination of temperature and speed could explain an anomalous observation made by Alcan when rolling aluminium - 4.5 wt. % magnesium alloy (series AA5182) as used for the ends of beverage cans (**29**). Some hot rolling mills are equipped to cool the strip just before it enters the finishing tandem, the main aim being to control the coiling temperature. On such a mill, AA5182 alloy was regularly rolled with and without this cooling, other conditions being very similar. This represented a reasonably well controlled industrial experiment. For the same speed and reduction, the first finishing stand, F1, was entered at a lower mean strip temperature (typically 50K lower) yet the rolling load was some 20% lower than for the equivalent rolling without the prior cooling. This is despite the hot strength being higher at the lower rolling temperature. Stands F2 and F3 showed no difference between the two cases.

One possible explanation is that at the lower temperature of rolling with the prior cooling switched on, boundary lubrication becomes operative in stand F1, reducing the friction and thus reducing the rolling load. Without the prior cooling the strip is hotter and boundary lubrication may well not operate until stand F2, by which time the mean strip temperature will have fallen by about 50K.

Although inevitably somewhat ill-defined, being a report of industrial experience rather than a controlled experiment, this case does demonstrate that lubrication in hot rolling poses many interesting problems and is ripe for further attention by tribology researchers. An interesting aside: hot mills for rolling aluminium use large volumes of emulsion, of the order of thousands of gallons per day. The treatment and safe disposal of spent emulsion presents its own problems (**30**).

4 WEAR AND FATIGUE

There is a wealth of literature on the wear of tools used in metalworking operations, Schey's classic text (**31**) cites most of the papers up to the early 1980s. For metal rolling, the handbook published by the Iron and Steel Society in the USA in 1990 (**32**) provides a valuable source on the wide range of characteristic damage found on the surface of rolls. Recent years have seen considerable developments in roll metallurgy (**33**) including much current interest in the use of high speed steel rolls. These rolls are expensive to manufacture but offer the potential of considerably enhanced life.

The repeated dramatic cycles of heating and cooling imposed in work rolls and forging tools by contact with hot metal stock inevitably leads to damage by thermal fatigue. "Firecracks" of themselves are not necessarily a problem, but can lead to the propagation of larger cracks and the spalling of metal from the tool surface (**34**). Resistance to firecracking can be enhanced by increasing the toughness of the tool material (often difficult when trying to increase the hardness of a tool) and by reducing the size that individual firecracks will grow to by putting crack arresting features into the microstructure (**6**).

The other common type of damage to the tool surface is abrasive wear, primarily by cooled particles of oxide. Harder tool materials have been developed which reduce this type of wear enormously. However, as steel producers increasingly form metals at lower temperatures for improved metallurgical properties, there is evidence that the different oxide mix may increase tool wear (**35**).

Perhaps surprisingly, the mechanisms of wear and the interaction with tool metallurgy are only sketchily understood. It is not possible to use a description of the operating conditions which a particular roll will encounter and then design a roll ideally suited to the task. The procedure for introducing new roll metallurgies takes the inverse course of extensive trawling of promising developments in the microstructure of the roll metal to see where the, hopefully, enhanced performance can be best exploited.

There are two fronts to enhance understanding of the wear of metalworking tools. First, computer-based modelling of the tool surface and the thermal and mechanical loading which it suffers. The earlier section on friction and heat transfer illustrated this approach. A complementary tack is to conduct carefully controlled experiments. This is difficult to perform with industrial production plant, particularly capital intensive equipment such as a rolling mill. Not only is experimental design secondary for production requirements, the scope for measurement in most industrial equipment is limited and "specimens" are expensive to manufacture.

Much more attractive is laboratory simulation where test conditions are well controlled and reproducible monitoring events is straightforward, specimens are cheaper to manufacture, and results are quicker to obtain than on industrial plant. As will be discussed below, laboratory simulation is not a perfect replica of industrial conditions, so it remains a partial insight, complementary to computer-based modelling and to the final arbiter, industrial trials.

The principles of the laboratory simulation of tribological processes in metalworking operations are thoroughly described in Schey's book (**31**). Attempting to mimic the mechanical circumstances at the tool-workpiece interface inevitably requires compromise unless the whole forming operation is to be copied. The usual set-up for simulating metal rolling is the use of two discs loaded and rolled together with their axes of rotation parallel (**35–39**). Such two-disc machines can provide good simulation of many features, notably the thermal fatigue cycle times. However, unlike in metal rolling, the surface shears in a two-disc machine are undirectional. Also, if the two discs are to run for an extended period, the plastic deformation of the hot disc must be restricted. This means that the surface expansion which occurs during large plastic straining will be severely restricted. Since such surface expansion will lead to fracture of the oxide scale and exposure for clean metal-metal contact, the simulation may well miss key events which occur in actual forming operations.

Nevertheless, laboratory simulation is an essential means of accumulating evidence for what happens at the tool-workpiece interface. While no single approach will provide all the necessary answers, the contribution of computer and laboratory simulation, allied to industrial trials, has to be the basis for further investigations.

5 CONCLUSIONS

Many unsolved problems remain in the tribology of hot metal forming. Considerable uncertainty prevails in the description of friction and heat transfer across the tool-workpiece interface, exemplified by the enormous variation in measured coefficients of heat transfer. The complexity of the events occurring at the interface, together with the hostility of one surface being very hot, makes lubrication problematic. An example of a puzzling issue in the lubrication of the hot rolling of aluminium has been cited. The damage mechanisms at the interface reflect the complexity of events and remain poorly described.

Thus there is a wealth of issues for the interested researcher to tackle. It is suggested that insight will come from the accumulation of evidence from different sources: laboratory tests and simulation, computer modelling and industrial trails. The end-point will be a constitutive description of the tool-workpiece interface which encompasses friction, heat transfer, lubrication and damage mechanisms. Once such descriptions begin to appear, tool design can be based on detailed knowledge of the operating conditions which the tool will experience, and the tool material be optimised to suit those conditions. Such a procedure will reverse the current trial and error approach.

6 ACKNOWLEDGEMENTS

The helpful discussion with Dr Keith Waterson of Alcan International Ltd. is gratefully acknowledged, as are the helpful comments of Prof. Mike Sellars of the University of Sheffield.

7 REFERENCES

1 **Chang, Y.-N.** and **Wei, F.-I.** *J. Mater.Sci.*, 1989, **24**, 14-22.
2 **Van Rooyen, G.T.** and **Backofen, W.A.** J. *Iron Steel Inst.*, **186**, 1957, 235-244.
3 **Yoneyama, T.** and **Hatamura, Y.** *JSME Int.*, **32**, 1989, 113-117.
4 **Jeswiet, J.** and **Nyahumwa, C.** *J. Mat. Proc. Tech.*, 1993, **39**, 251-268.
5 **Tabary, P.E., Sutcliffe, M.P.F., Porral, F.** and **Deneuville, P.** *Trans. ASME, J. Tribology*, 1996, **118**, 629-636.

6 **Harper, P.** in *Rolls for the Metalworking Industries* edited by R B Corbett, published by Iron and Steel Society Inc., 1990, 87-93.

7 **Stevens, P.G., Ivens, K.P.** and **Harper, P.** *J. Iron Steel Inst.*, **209**, 1971, 1-11.

8 **Murata, K., Morise, H., Mitsutsuka, M., Haito, H., Kumatsu, T.** and **Shida, S.** *Trans. ISIJ*, 1984, **24**, 9, B309.

9 **Malinowski, Z., Lenard, J.G.** and **Davies, M.E.** *J. Mat. Proc. Tech.*, **41**, 1994, 125-142.

10 **Pietrzyk, M.** and **Lenard, J.G.** *J. Mat. Shaping Tech.*, 7, 1989, 117-126.

11 **Sellars, C.M.** *Mat. Sci. Tech.*, **1**, 1985, 325-332.

12 **Chen, W.C., Samarasekera, I.V.** and **Hawbolt, E.B.** *33rd MWSP Conference Proceedings,* Vol. XXIX, 1992, 349-357.

13 **Hlady, C.O., Samarasekera, I.V., Hawbolt, E.B.** and **Brimacombe, J.K.** Heat transfer in the hot rolling of aluminium alloys in *Proceedings of the 32nd Annual Conference of Metallurgists*, Quebec City, Canada, 1993.

14 **Pietrzyk, M.** and **Lenard, J.G.** in *Computational plasticity: models, software and applications,* Proc. Int. Conf., edited by D R J Owen, E Hinton and E Oñate, 1989, 947-958.

15 **Chen, B.K., Thomson, P.F.** and **Choi, S.K.** *J. Mat. Proc. Tech.*, **30**, 1992, 115-130.

16 **Semiatin, S.L., Collings, E.W., Wood, V.E.** and **Altan, T.** J. *Eng. Ind.*, **109**, 1987, 59-67.

17 **Timothy, S.P., Yiu, H.L., Fine, J.M.** and **Ricks, R.A.** *Mat. Sci. Tech.*, 7, 1991, 255-261.

18 **Montmitonnet, P.** and **Buessler, P.** *ISIJ Int.*, **31**, 1991, 525-538.

19 **Kopalinsky, E.M.** and **Oxley, P.L.B.** *Wear* **190**, 1995, 145-154.

20 **Sutcliffe, M.P.F.** *Int. J. Mech. Sci.* **30**, 1988, 847-868.

21 **Wilson, W.R.D.** and **Sheu, S.** *Int. J. Mech. Sci.*, **30**, 1988, 475-489.

22 **Wilson, W.R.D.** and **Chang, D-F.** *Trans. ASME J. Tribology*, **118**, 1996, 83-89.

23 **Korzekwa, D.A., Dawson, P.R.** and **Wilson, W.R.D.** *Int. J. Mech. Sci.*, **34**, 1992, 521-539.

24 **Lin, G-J., Kikuchi, N.** and **Takahashi, S.** in *Contact problems and surface interaction in manufacturing and tribological system.* ASME, 1993, 105-113.

25 **Fletcher, J.D.** and **Beynon, J.H.** in *Modelling of metal rolling processes, Proc. 2nd Int. Conf.*, edited by J H Beynon et al., publ. Inst. Materials, 1996, 202-212.

26 **Fivash, J.** *Lubrication (Steel),* lecture notes from "The Technology of Flat Rolling" course at the University of Sheffield, September 1996.

27 **Poole, R.L.** in *Rolls for the Metalworking Industries* edited by R B Corbett, published by Iron and Steel Society Inc., 1990, 307-316.

28 **Ball, J., Treverton, J.A.** and **Thornton, M.C.** *Lubrication Engineering,* **50**, 1994, 89-93.

29 **Waterson, K.**, Alcan International, Banbury, UK, Private Communication, 13 August 1996.

30 **Budd, M.K.** *The surface quality of rolled aluminium and the role of lubrication in achieving it,* lecture notes from "The Technology of Flat Rolling" course at the University of Sheffield, September, 1996.

31 **Schey, J.A.** *Tribology in Metalworking: Friction, Lubrication and Wear,* ASM, 1983.

32 *Rolls for the Metalworking Industries* edited by R B Corbett, publ. Iron and Steel Society, USA, 1990.

33 *Rolls 2000* 1996, Birmingham, UK, Proc. Int. Conf., publ. by Inst. Materials, 1996.

34 **Tait, W.H.** in *Rolls for the Metalworking Industries* edited by R B Corbett, publ. by Iron and Steel Society Inc., 1990, 135-149.

35 **Lundberg, S-E.** and **Gustafsson, T.** *J. Mat. Proc. Tech.*, **42**, 1994, 239-291.

36 **Magnee, A., Gaspard, C.** and **Gabriel, M.** *CRM Metallurgical Reports,* no. 57, 1980, 25-39.

37 **Noguchi, H., Hiraoka, H., Watanabe, Y.** and **Sayama, Y.** *Trans. ISIJ*, **28**, 1988, 478-484.

38 **Spuzic, S., Strafford, K.N., Subramanian, C.** and **Savage, G.** *Wear,* **176**, 1994, 261-271.

39 **Goodchild, J.** and **Beynon, J.H.** Laboratory simulation of work roll wear in hot mills *First World Tribology Congress,* 8-12 September 1997, London.

Micro/nanotribology using atomic force/friction force microscopy: state of the art

B BHUSHAN
Computer Microtribology and Contamination Laboratory, Department of Mechanical Engineering, The Ohio State University, Columbus, Ohio, USA

SYNOPSIS

Atomic force microscopy/friction force microscopy (AFM/FFM) techniques are increasingly used for tribological studies of engineering surfaces at scales, ranging from atomic and molecular to microscales. These techniques have been used to study surface roughness, adhesion, friction, scratching/wear, indentation, detection of material transfer, and boundary lubrication and for nanofabrication/nanomachining purposes. Micro/nanotribological studies of materials of scientific and engineering interests, have been conducted. Commonly measured roughness parameters are found to be scale dependent, requiring the need of scale-independent fractal parameters to characterize surface roughness. Measurement of atomic-scale friction of a freshly-cleaved highly-oriented pyrolytic graphite exhibited the same periodicity as that of corresponding topography. However, the peaks in friction and those in corresponding topography were displaced relative to each other. Variations in atomic-scale friction and the observed displacement has been explained by the variations in interatomic forces in the normal and lateral directions. Local variation in microscale friction is found to correspond to the local slope suggesting that a ratchet mechanism is responsible for this variation. Directionality in the friction is observed on both micro- and macro scales which results from the surface preparation and anisotropy in surface roughness.

Microscale friction is generally found to be smaller than the macrofriction as there is less ploughing contribution in microscale measurements. Microscale friction is load dependent and friction values increase with an increase in the normal load approaching to the macrofriction at contact stresses higher than the hardness of the softer material. Wear rate for single-crystal silicon is negligible below 20 μN and is much higher and remains approximately constant at higher loads. Elastic deformation at low loads is responsible for negligible wear. Mechanism of material removal on microscale is studied. At the loads used in the study, material is removed by the ploughing mode in a brittle manner without much plastic deformation. Most of the wear debris is loose. Evolution of the wear has also been studied using AFM. Wear is found to be initiated at nano scratches. AFM has been modified to obtain load-displacement curves and for measurement of nanoindentation hardness and Young's modulus of elasticity, with depth of indentation as low as 1 nm. Hardness of ceramics on nano scales is found to be higher than that on micro scale. Ceramics exhibit significant plasticity and creep on nanoscale. Scratching and indentation on nanoscales are the powerful ways to screen for adhesion and resistance to deformation of ultrathin films. Detection of material transfer on a nanoscale is possible with AFM. Boundary lubrication studies and measurement of lubricant-film thickness with a lateral resolution on a nanoscale have been conducted using AFM. Self-assembled monolayers and chemically-bonded lubricant films with a mobile fraction are superior in wear resistance.

Friction and wear on micro-and nanoscales at low loads have been found to be generally smaller compared to that at macroscales. Therefore, micro/nanotribological studies may help define the regimes for ultra-low friction and near zero wear.

1 INTRODUCTION

The recent emergence and proliferation of proximal probes, in particular tip-based microscopies the scanning tunneling microscopy and atomic force microscopy (**1–3**) has allowed the study of surface topography, adhesion, friction, wear, lubrication and measurement of mechanical properties all on a micro- to nanometer scale, and to image lubricant molecules and availability of supercomputers to conduct atomic-scale simulations has led to development of a new field referred to as Microtribology or Nanotribology (**4–6**). This field concerns experimental and theoretical investigations of processes ranging from atomic and molecular scales to microscale, occurring during adhesion, friction, wear, and thin-film lubrication at sliding surfaces. At most solid-solid interfaces of technological relevance, contact occurs at numerous asperities; a sharp atomic force microscope tip sliding on a surface simulates just one such contact.

The micro/nanotribological studies are needed to develop fundamental understanding of interfacial phenomena on a small scale and to study interfacial phenomena in micro- and nano structures used in magnetic storage systems, microelectromechanical systems (MEMS) and other industrial applications (**6–10**). Friction and wear of lightly-loaded micro/nano components are highly dependent on the surface interactions (few atomic layers). These structures are generally lubricated with molecularly-thin films. Micro- and nanotribological studies are also valuable in fundamental understanding of interfacial phenomena in macrostructures to provide a bridge between science and engineering (**8, 11–13**).

Atomic force microscopy (AFM) can be used for measurement of all engineering surfaces which may be either electrically conducting or insulating. AFM has become a popular surface profiler for topographic measurements on micro to nanoscale. AFM has been modified to measure both normal and friction forces and this instrument is generally called "Friction Force Microscope" (FFM) or "Lateral Force Microscope" (LFM). By using a standard or a sharp diamond tip mounted on a stiff cantilever beam, researchers have used AFM for scratching, wear, and measurements of elastic/plastic mechanical properties (such as load-displacement curves, indentation hardness and modulus of elasticity). Boundary lubrication studies of molecularly-thick lubricant films can be conducted using FFMs.

Surface roughness, adhesion, friction, wear and lubrication at the interface between two solids with and without liquid films have been studied using the AFM and FFM. AFM and its modifications have also been used for nanomechanical characterization. Status of current understanding of micro/nanotribology of engineering interfaces follows.

2 EXPERIMENTAL TECHNIQUES

2.1 AFM/FFM

Atomic Force Microscope relies on a scanning technique to produce very high resolution, 3-D images of sample surfaces. AFM measures ultrasmall forces (less than 1 nN) present between the AFM tip surface and a sample surface. These small forces are measured by measuring the motion of a very flexible cantilever beam having an ultrasmall mass. The deflection can be measured to with ±0.02 nm, so for typical cantilever force constant of 10 N/m, a force as low as 0.2 nN can be detected. AFM is capable of investigating surfaces of both conductors and insulators on an atomic scale. In the operation of high resolution AFM, the sample is generally scanned, however, AFMs are now available where the tip is scanned and the sample is stationary (**4**). To obtain atomic resolution with AFM, the spring constant of the cantilever should be weaker than the equivalent spring between atoms. A cantilever beam with a spring constant of about 1 Nm or lower is desirable. Tips have to be as sharp as possible. Tips with a radius ranging from 10 to 100 nm are commonly available.

A commercial AFM/FFM commonly used to conduct studies of friction, scratching, wear, indentation, and lubrication from micro- to atomic scales and nanofabrication/nanomachining is shown in Fig. 1 (**4**). Simultaneous measurements of surface roughness and friction force can be made with this instrument. In the AFM/FFM, the sample is mounted on a PZT tube scanner which consists of separate electrodes to precisely scan the sample in the X-Y plane in a raster pattern and to move the sample in the vertical (Z) direction. A sharp tip at the end of a flexible cantilever is brought in contact with the sample. Normal and frictional forces being applied at the tip-sample interface are measured using a laser beam deflection technique. A laser beam from a diode laser is directed by a prism onto the back of a cantilever near its free end, tilted downward at about 10 deg with respect to a horizontal plane. The reflected beam from the vertex of the cantilever is directed through a mirror onto a quad photodetector (split photodetector with four quadrants). The differential signal from the top and bottom photodiodes provides the AFM signal which is a sensitive measure of the cantilever vertical deflection. Topographic features of the sample cause the tip to deflect in the vertical direction as the sample is scanned under the tip. This tip deflection will change the direction of the reflected laser beam, changing the intensity difference between the top and bottom photodetector (AFM signal). In the AFM operating mode or the 'height mode,' for topographic imaging or for any other operation in which the applied normal force is to be kept a constant, a feedback circuit is used to modulate the voltage applied to the PZT scanner to adjust the height of the PZT, so that the cantilever vertical deflection (given by the intensity difference between the top and bottom detector) will

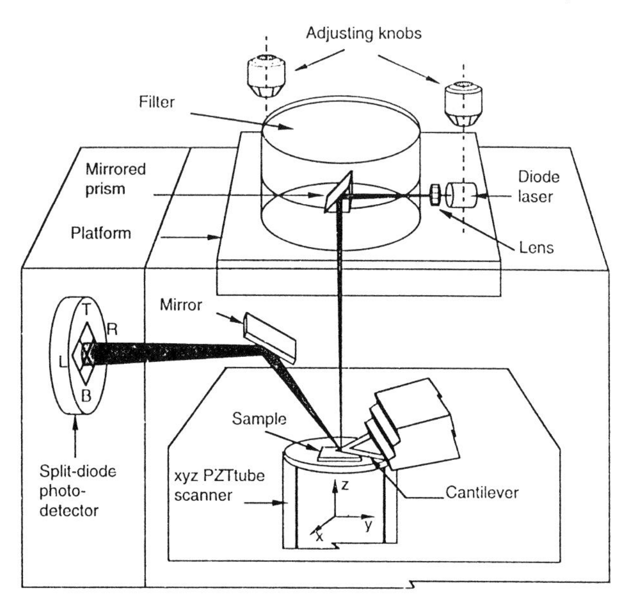

Fig. 1 Schematic of a commercial atomic force microscope/friction force microscope (AFM/FFM) using laser-beam deflection method.

remain almost constant during scanning. The PZT height variation is thus a direct measure of surface roughness of sample.

For measurement of friction force being applied at the tip surface during sliding, the other two (left and right) quadrants of the photodetector (arranged horizontally) are used. In the "friction mode", the sample is scanned back and forth in a direction orthogonal to the long axis of the cantilever beam. Friction force between the sample and the tip will produce a twisting of the cantilever. As a result, the laser beam will be reflected out of the plane defined by the incident beam and the beam reflected vertically from an untwisted cantilever. This produces an intensity difference of the laser beam received in the left and right quadrants of the photodetector. The intensity difference between the left and right detectors (FFM signal) is directly related to the degree of twisting and hence to the magnitude of friction force. One problem associated with this method is that any misalignment between the laser beam and the photodetector axis would introduce error in the measurement. However, by following the procedures developed by Ruan and Bhushan (**14**), the average FFM signal for the sample scanned in two opposite directions is subtracted from the friction profiles of each of the two scans to eliminate the misalignment effect. This method provides 3-D maps of friction force. By following the friction force calibration procedures developed by Ruan and Bhushan (**14**) and Bhushan (**4**), voltages corresponding to friction forces can be converted to force units.

Topographic measurements are typically made using a sharp tip on cantilever beam with normal stiffness of about 0.58 N/m at a normal load of about 10 nN and friction measurements are carried out in the load range of 10–150 nN. The tip is scanned in such a way that its trajectory on the sample forms a triangular pattern. Scanning speeds in the fast and slow scan directions depends on the scan area and scan frequency. A maximum scan size of 125 μm x 125 μm and scan rate of 122 Hz typically can be used. Higher scan rates are used for small scan lengths.

For nano-scale boundary lubrication studies, the samples are typically scanned over an area of 1 μm x 1 μm at a normal force of about 300 nN, in a direction orthogonal to the long axis of the cantilever beam. The samples are generally scanned with a scan rate of 1 Hz and the scanning speed of 2 μm/s. Coefficient of friction is monitored during scanning for a desired number of cycles. After scanning test, a larger area of 2 μm × 2 μm is scanned at a normal force of 40 nN to observe for any wear scar.

For micro-scale scratching, micro-scale wear and nano-scale indentation hardness measurements, sharp single-crystal natural diamond tip mounted on a stainless steel cantilever beam with normal stiffness of about 25 N/m is used at relatively higher loads (1 μN – 150 μN). For scratching and wear studies, the sample is generally scanned in a direction orthogonal to the long axis of the cantilever beam (typically at a rate of 0.5 Hz) so that friction can be measured during scratching and wear. The tip is mounted on the beam such that one of its edge is orthogonal to the long axis of the beam; therefore, wear during scanning along the beam axis is higher (about 2 × to 3 ×) than that during scanning orthogonal to the beam axis. For wear studies, typically an area of 2 μm × 2 μm is scanned at various normal loads (ranging from 1 to 100 μN) for selected number of cycles.

For nanoindentation hardness measurements the scan size is set to zero and then normal load is applied to make the indents. During this procedure the diamond tip is continuously pressed against the sample surface for about two seconds at various indentation loads. Sample surface is scanned before and after the scratching, wear or indentation to obtain the initial and the final surface topography, at a low normal load of about 0.3 μN using the same diamond tip. An area larger than the indentation region is scanned to observe the indentation marks. Nanohardness is calculated by dividing the indentation load by the projected residual area of the indents.

For measurements of surface roughness, friction force, nanoscale scratching and wear, a microfabricated square-pyramidal Si_3N_4 or silicon tip with a tip radius ranging from 10 to 50 nm (Fig. 2a) is generally used at loads ranging from 10 to 150 nN. For measurements of microscale scratching and wear and for nanoindentation hardness measurements and nanofabrication, a three-sided pyramidal single-crystal natural-diamond tip with a tip radius of about 100 nm (Fig. 2b) is generally used at relatively high loads ranging from 10 to 150 μN.

2.2 Nano/Picoindenter

As described earlier, conventional AFMs have been used for indentation studies on nanometer-scale depths. In these studies the hardness value is based on the projected residual area after imaging the indent. Direct imaging of the indent allows one to quantify piling up of ductile material around the indenter. However, it becomes difficult to identify the boundary of the indentation mark with great accuracy. This makes the direct measurement of contact area somewhat inaccurate. A technique with the dual capability of depth-sensing as well as in-situ imaging which is most appropriate in nanomechanical property studies, has been developed (**15–18**). This indentation system is used to make load-displacement measurement and subsequently carry out in-situ imaging of the indent. A schematic of the nano/picoindenter system used is shown in Fig. 3. The indentation system consists of a three-plate transducer with electrostatic actuation hardware used for direct application of normal load and a capacitive sensor used for measurement of vertical displacement. The AFM head is replaced with this transducer assembly while the specimen is mounted on the PZT scanner which remains stationary during indentation experiments. The transducer

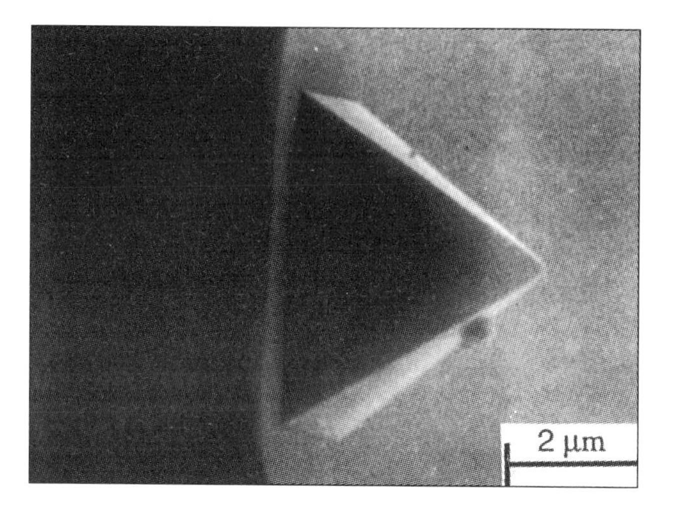

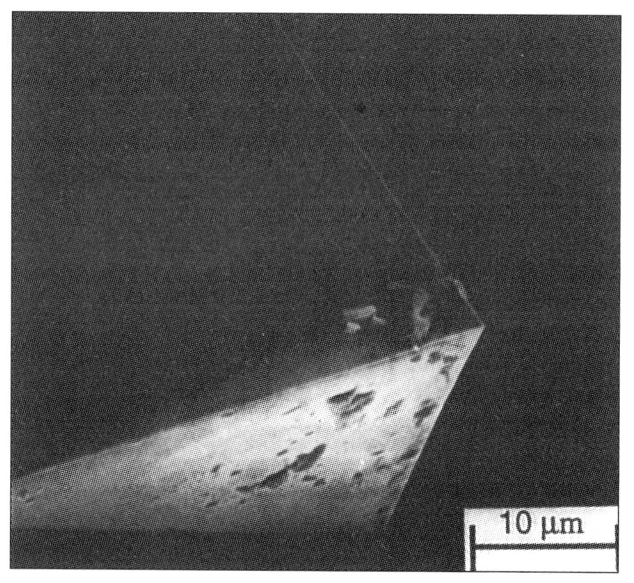

Fig. 2 SEM micrographs of a PECVD Si_3N_4 cantilever beam with tip and a stainless steel cantilever beam with diamond tip.

consists of a three (Be-Cu) plate capacitive structure which provides high sensitivity, large dynamic range and a linear output signal with respect to load or displacement. The tip is mounted on the center plate. The upper and lower plates serve as drive electrodes. Load is applied by applying appropriate voltage to the drive electrodes thereby generating an electrostatic force between the center plate and the drive electrodes. Vertical displacement of the tip (indentation depth) is measured by measuring the displacement of the center plate relative to the two outer electrodes using capacitance technique. The load resolution is 100 nN or better and the displacement resolution is 0.1 nm. At present, a load range of 1 µN to 10 µN can be employed. Loading rates can be varied by changing the load/unload period. The AFM functions as the platform providing an in-situ image of the indent with a lateral resolution of 1 nm and a vertical resolution of 0.2 nm. The load-displacement data can be acquired and displayed on the display monitor. Hardness value can be obtained from the load-displacement data as well as from direct measurement of the projected residual area of the indent after imaging. The Young's modulus of elasticity is obtained from the slope of the unloading curve.

A three-sided Berkovich indenter with tip radius of about 0.1 µm has been generally used for the measurements (**4**). Sharper diamond tips with included angle of 60-90 and tip radii of 30-60 nm are sometimes employed for shallower indentation (on the order of 1 nm). In order to obtain an accurate relation between the indentation depth and the projected contact area, tip shape calibration needs to be done (**4**). Also for surfaces with rms roughness on the order of indentation depth, the original (unindented) profile is subtracted from the indented profile (**19**).

In a typical indentation experiment the tip is lowered close to the sample (ideally <l00µm). Scan size and scan rate are selected. The tip is engaged to the sample surface by a stepper motor with a set point of 1 nA (about 1 mN). A desired image area is captured prior to indentation. The feedback is set to zero to disable the scanner, the scan size is set to zero so that the indenter will be positioned at the center of the image. An appropriate set point for pre load condition is selected. The indentation rate can be varied by changing the load/unload period.

3 SURFACE ROUGHNESS

Solid surfaces, irrespective of the method of formation, contain surface irregularities or deviations from the prescribed geometrical form. When two nominally flat surfaces are placed in contact, surface roughness causes contact to occur at discrete contact points. Deformation occurs in these points, and may be either elastic or plastic, depending on the nominal stress, surface roughness and material properties. The sum of the areas of all the contact points constitutes the real area that would be in contact, and for most materials at normal loads, this will be only a small fraction of the area of contact if the surfaces were perfectly smooth. In general, real area of contact must be minimized to minimize adhesion, friction and wear (**8,12**).

Characterizing surface roughness is therefore important for predicting and understanding the tribological properties of solids in contact. The AFM has been used to measure surface roughness on length scales from nanometers to micrometers. Roughness plots of a glass-ceramic disk measured using an AFM (lateral resolution ~15 nm), noncontact optical profiler (lateral resolution 1 µm) and stylus profiler (lateral resolution of ~0.2 µm) are shown in Fig. 4(a). Figure 4(b) compares the profiles of the disk obtained with different instruments at a common scale. The figures show that roughness is found at scales ranging from millimeter to nanometer scales. Measured roughness profile is dependent on the lateral and normal resolutions of the measuring instrument (**20–23**). Instruments with different lateral resolutions measure features with different

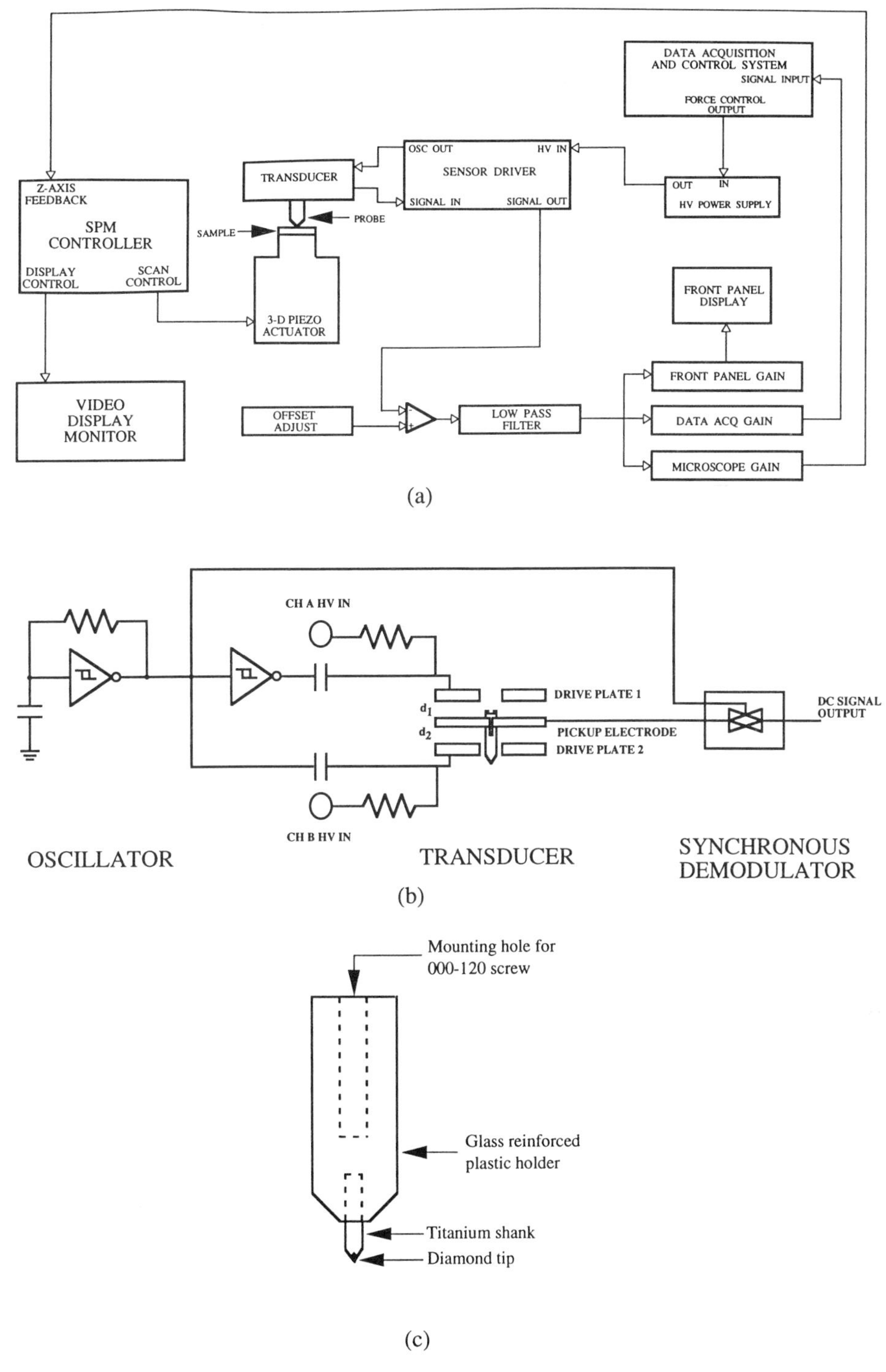

Fig. 3 Schematics of three-plate transducer with electrostatic actuation hardware and capacitance sensor (**15**).

scale lengths. It can be concluded that a surface is composed of a large number of length scales of roughness that are superimposed on each other.

Surface roughness is most commonly characterized by the standard deviation of surface heights which is the square roots of the arithmetic average of squares of the vertical deviation of a surface profile from its mean plane. Due to the multiscale nature of surfaces, it is found that the variances of surface height and its derivatives and other roughness parameters depend strongly on the resolution of the roughness measuring instrument or any other form of filter, hence not unique for a surface (**22–24**) see, for example, Fig. 5. Therefore, a rough surface should be characterized in a way such that the structural information of roughness at all scales is retained. It is necessary to quantify the multiscale nature of surface roughness.

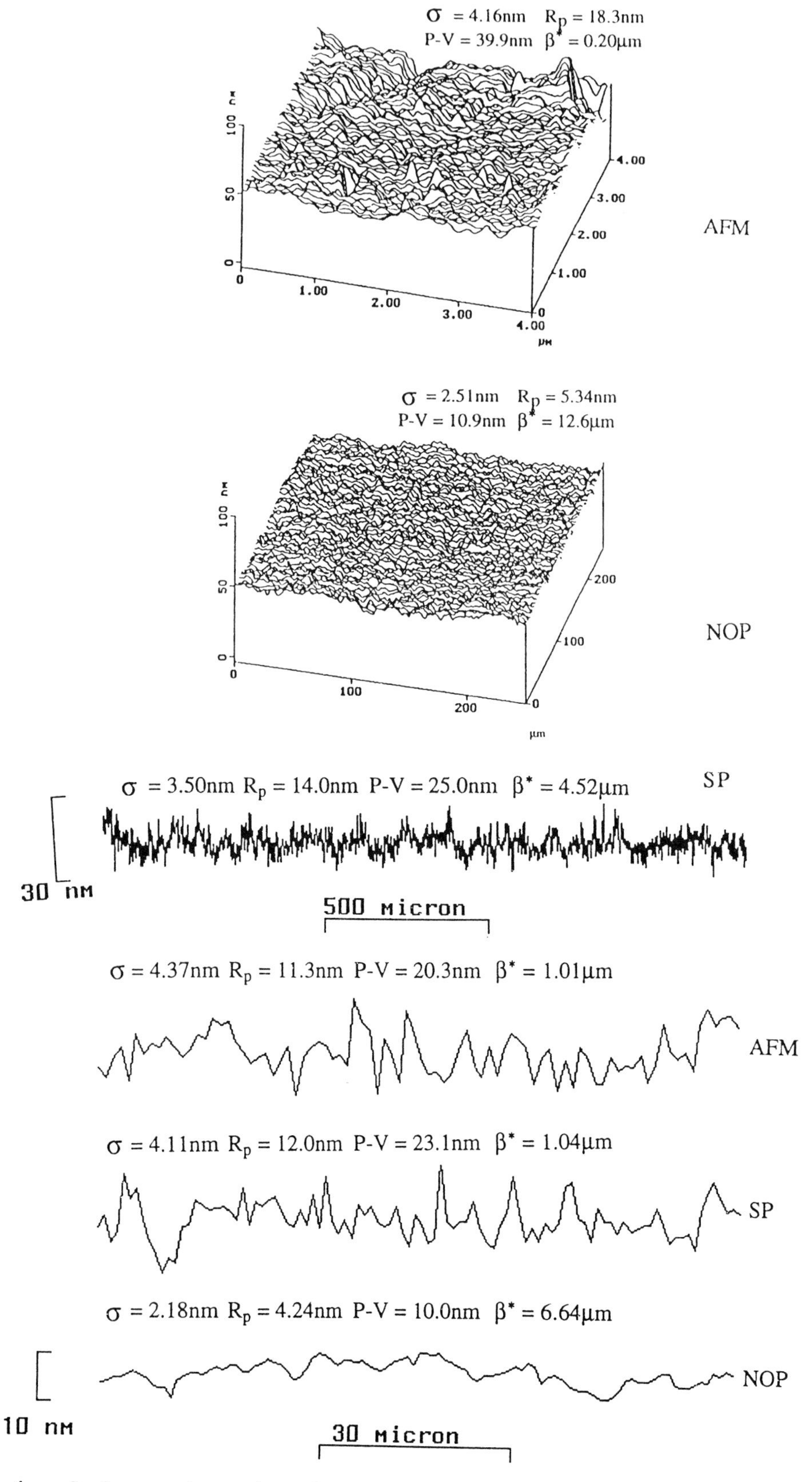

Fig. 4 Surface roughness plots of a glass-ceramic disk (a) measured using an atomic force microscope (lateral resolution ~15 nm), noncontact optical profiler (NOP) (lateral resolution ~1 μm) and stylus profiler (SP) with a stylus tip of 0.2-μm radius (lateral resolution ~0.2 μm) and (b) Measured using an AFM (~150 nm), SP (~0.2 μm), and NOP (~1 μm) and plotted on a common scale (**23**).

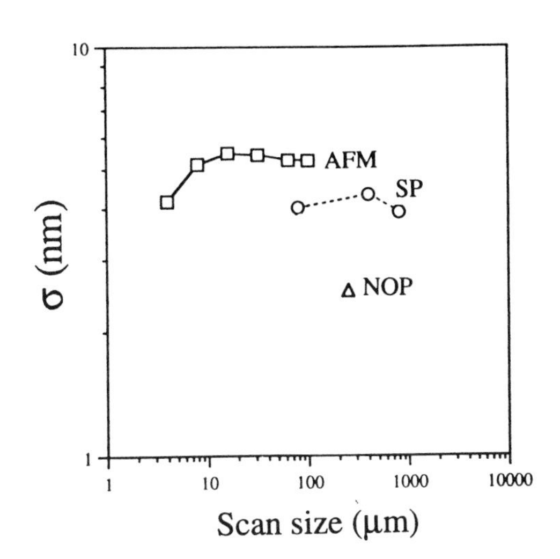

Fig. 5 Scale dependence of standard deviation of surface heights for a glass-ceramic disk, measured using atomic force microscope (AFM), stylus profiler (SP), and noncontact optical profiler (NOP).

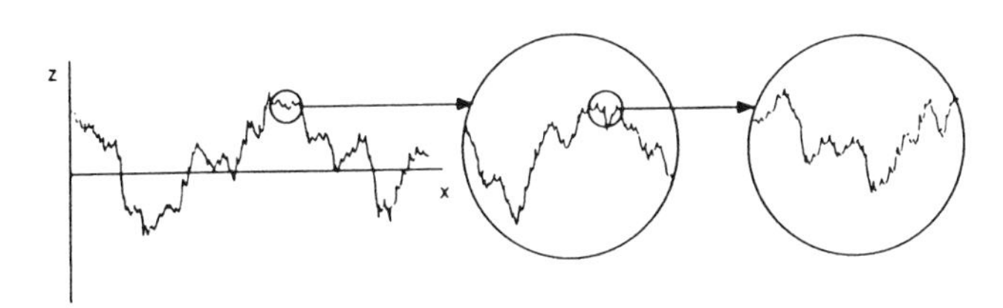

Fig. 6 Qualitative description of statistical self-affinity for a surface profile.

A unique property of rough surfaces is that if a surface is repeatedly magnified, increasing details of roughness are observed right down to nanoscale. In addition, the roughness at all magnifications appear quite similar in structure, as qualitatively shown in Fig. 6. That statistical self-affinity is due to similarity in appearance of a profile under different magnifications. Such a behavior can be characterized by fractal analysis (**22, 25**). The main conclusion from these studies are that a fractal characterization of surface roughness is *scale independent* and provides information of the roughness structure at all length scales that exhibit the fractal behavior.

Structure function and power spectrum of a self-affine fractal surface follow a power law and can be written as (Ganti and Bhushan model):

$$S(\tau) = C\,\eta^{(2D-3)}\,\tau^{(4-2D)} \quad (1)$$

$$P(\omega) = \frac{c_1 \eta^{(2D-3)}}{\omega^{(5-2D)}} \quad (2a)$$

and:

$$\frac{c_1 \Gamma(5-2D)\sin[\pi(2-D)]C}{2\pi} \quad (2b)$$

The fractal analysis allows the characterization of surface roughness by two parameters D and C which are instrument-independent and unique for each surface. D (ranging from 1 to 2 for surface profile) primarily relates to relative power of the frequency contents and C to the amplitude of all frequencies. η is the lateral resolution of the measuring instrument, τ, is the size of the increment (distance), and ω is the frequency of the roughness. Note that if $S(\tau)$ or $P(\omega)$ are plotted as a function of ω or τ respectively, on a log-log plot, then the power law behavior would result into a straight line. The slope of line is related to D and the location of the spectrum along the power axis is related to C.

Figure 7 presents the structure function of a thin-film rigid disk measured using AFM, non-contact optical profiler, and stylus profiler. Horizontal shift in the structure functions from one scan to another, arises from the change in the lateral resolution. D and C values for variouss can lengths are listed in Table 1. We note that fractal dimension of the various scans is fairly constant (1.26 to 1.33), however C increases/decreases monotonically with σ for the AFM data. The error in estimation of η is believed to be responsible for variation in C. These data show that the disk surface follows a fractal structure for three decades of length scales.

Majumdar and Bhushan (**26**) and Bhushan and Majumdar (**27**) developed a new fractal theory of contact between two rough surfaces. This model has been used to predict whether contacts experience elastic or plastic deformation and to predict the statistical distribution of contact points. For a review of contact models, see Bhushan (**10, 28**).

4 ADHESION AND FRICTION

To study friction mechanisms on an atomic scale, a well characterized freshly-cleaved surface of highly oriented pyrolytic graphite (HOPG) has been studied by Mate, et al. (**29**) and Ruan and Bhushan (**30**). The atomic-scale friction force of HOPG exhibited the same periodicity same as that of corresponding topography (Fig. 8a), but the peaks in friction and those in topography were displaced relative to each other, (Fig. 8b). A Fourier expansion of the interatomic potential was used to calculate the conservative interatomic forces between atoms of the FFM tip and those of the graphite surface. Maxima in the interatomic forces in the normal and lateral directions do not occur at the same location, which explains the observed shift between the peaks in the lateral force and

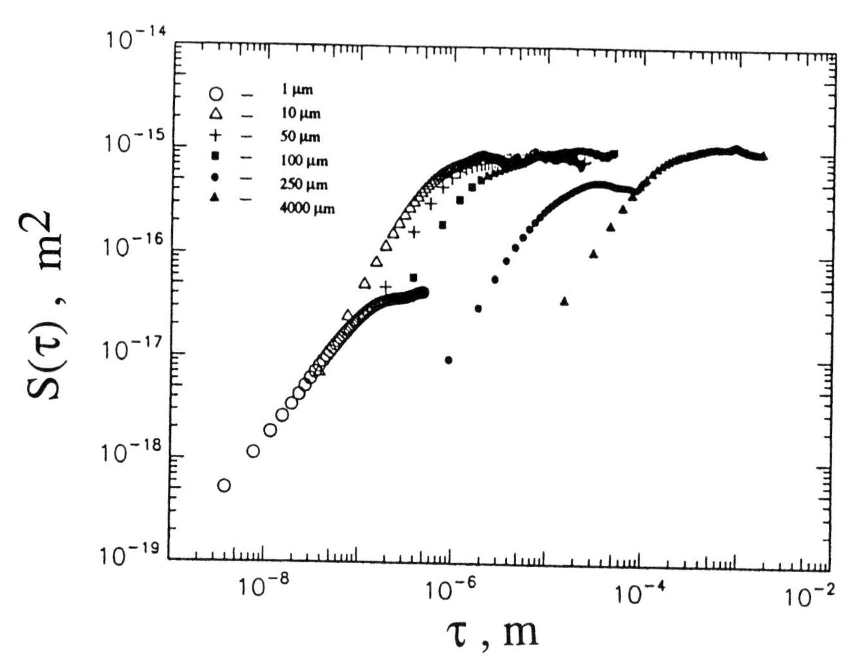

Fig. 7 Structure functions for the roughness data measured using AFM, SP and NOP, at various scan lengths for a thin-film rigid disk (**22**)

Table 1 Surface roughness parameters for a polished thin-film rigid disk (**22**).

Scan size (μm x μm)	σ (nm)	*D*	*C* (nm)
1 (AFM)	0.7	1.33	9.8×10^{-4}
10 (AFM)	2.1	1.31	7.6×10^{-3}
50 (AFM)	4.8	1.26	1.7×10^{-2}
100 (AFM)	5.6	1.30	1.4×10^{-2}
250 (NOP)	2.4	1.32	2.7×10^{-4}
4000 (NOP)	3.7	1.29	7.9×10^{-5}

*AFM – Atomic force microscope.
†NOP – Noncontact optical profiler.

those in the corresponding topography. Furthermore, the observed local variations in friction force were explained by variation in the intrinsic lateral force between the sample and the FFM tip (**30**) and these variations may not necessarily occur as a result of atomic-scale stick-slip process (**29**), but can be due to variation in the intrinsic lateral force between the sample and the FFM tip.

Friction forces of HOPG have also been studied. Local variations in the microscale friction of cleaved graphite are observed, which arise from structural changes occuring during the cleaving process (**14**). The cleaved HOPG surface is largely atomically smooth but exhibits line-shaped regions in which the coefficient of friction is more than order of magnitude larger. Transmission electron microscopy indicates that the line-shaped regions consist of graphite planes of different orientation, as well as of amorphous carbon. Differences in friction have also been observed for multi-phase ceramic materials (**31**) and for organic mono- and multi-layer films (**32**), which again seems to be the result of structural variations in the surfaces. These measurements suggest that the FFM can be used for structural mapping of the surfaces. FFM measurements can be used to map chemical variations, as indicated by the use of the FFM with a modified probe tip to map the spatial arrangement of chemical functional groups in mixedorganic monolayer films (**33**). Here, sample regions that had stronger interactions with the functionalized probe tip exhibited larger friction.

Local variations in the microscale friction of scratched surfaces can be significant, and seen to depend on the local

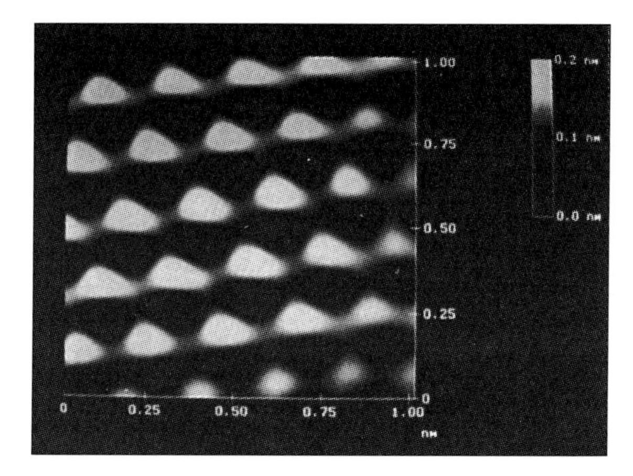

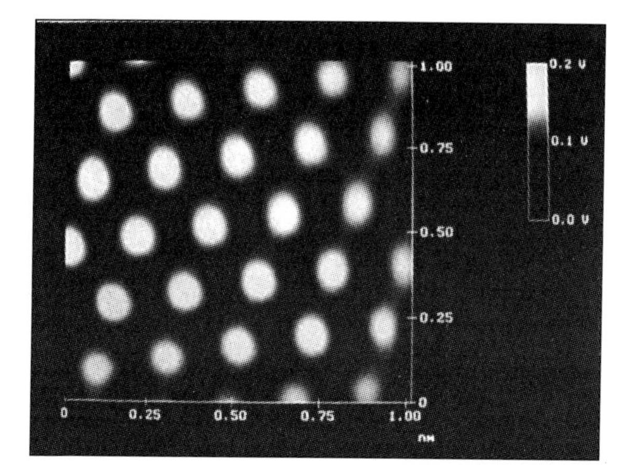

Fig. 8(a) Gray-scale plots of surface topography (left) and friction profiles (right) of a 1 nm x 1 nm area of freshly cleaved HOPG, showing the atomic-scale variation of topography and friction,

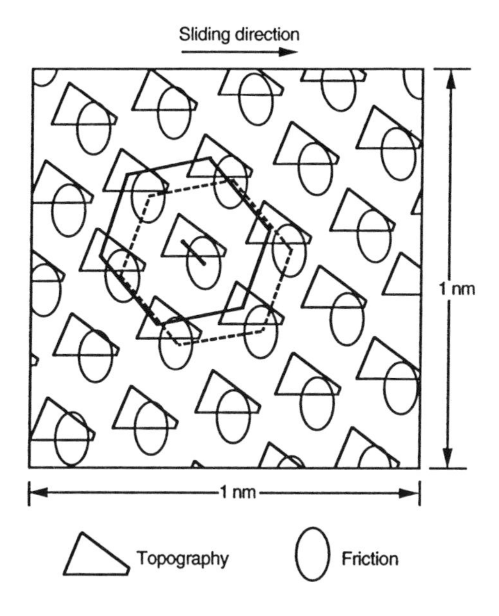

Fig 8 (b) Diagram of superimposed topography and friction profiles from (a); the symbols correspond to maxima. Note the spatial shift between the two profiles (**30**).

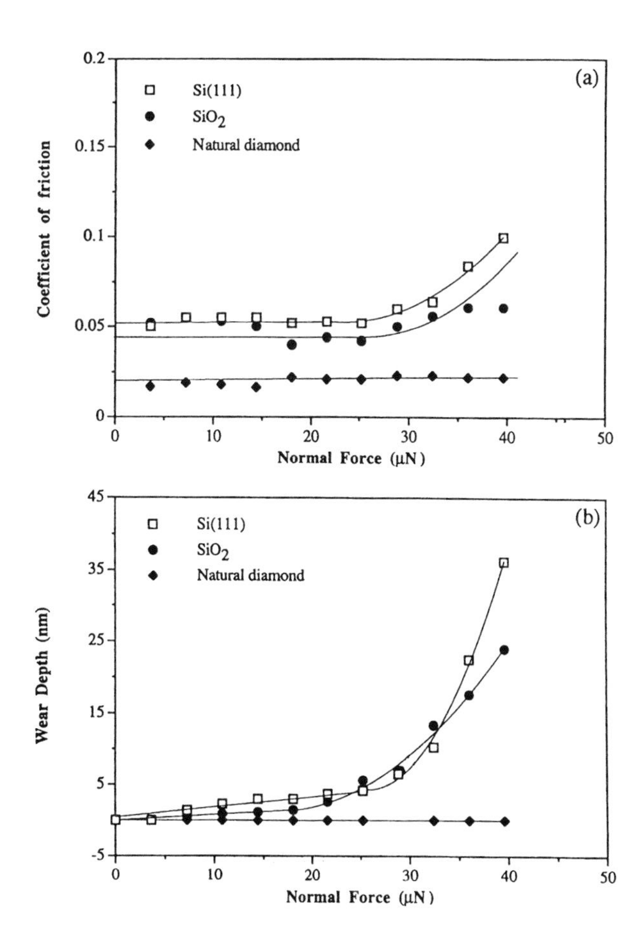

Fig. 9 (a) Coefficient of friction as a function of normal force and (b) corresponding wear depth as a function of normal force for silicon, SiO_2 coating and natural diamond. Inflections in the curves for silicon and SiO_2 correspond to the hardnesses of these materials (**34**).

surface slope rather than the surface height distribution (**4, 19, 24**). Directionality in friction is sometimes observed on the macroscale; on the microscale this is the norm (**4, 19**). This is because most engineering surfaces have asymmetric surface asperities so that the interaction of the FFM tip with the surface is dependent on the direction of the tip motion. Moreover, during surface finishing processes material can be transferred preferentially onto one side of the asperities, which also causes asymmetry and directional dependence. Reduction in local variations and in directionality of frictional properties therefore requires careful optimization of surface roughness distributions and of surface-finishing processes.

Table 2 Surface roughness and micro- and macro-scale coefficients of friction of various samples.

Material	R.M.S. roughness, nm	Micro-scale coefficient of friction versus Si_3N_4 tip[1]	Macro-scale coefficient of friction versus alumina ball[2]
Si (111)	0.11	0.03	0.18
C^+ -implanted Si	0.33	0.02	0.18

1 Tip radius of about 50 nm in the load range of 10-150 nN (2.5 – 6.1 GPa), a scanning speed of 5 µm/s and scan area of 1 µm x 1 µm.
2 Ball radius of 3 mm at a normal load of 0.1 N (0.3 GPa) and average sliding speed of 0.8 mm/s.

Table 2 shows the coefficient of friction measured for two surfaces on micro- and macroscales. The coefficient of friction is defined as the ratio of friction force to the normal load. The values on the microscale are much lower than those on the macroscale. When measured for the small contact areas and very low loads used in microscale studies, indentation hardness and modulus of elasticity are higher than at the macroscale. This reduces the degree of wear. In addition, the small apparent areas of contact reduce the number of particles trapped at the interface, and thus minimize the 'ploughing' contribution to the friction force.

At higher loads (with contact stresses exceeding the hardness of the softer material), however, the coefficient of friction for micro-scale measurements increases towards values comparable with those obtained from macroscale measurements, and surface damage also increases, Fig. 9 (**34**). Thus Amontons' law of friction, which states that the coefficient of friction is independent of apparent contact area and normal load, does not hold for microscale measurements. These findings suggest microcomponents sliding under lightly loaded conditions should experience very low friction and near-zero wear.

5 SCRATCHING, WEAR AND INDENTATION

The AFM can be used to investigate how surface materials can be moved or removed on micro- to nanoscales, for example in scratching and wear (**4**) (where these things are undesirable), and nanomachining/nanofabrication (where they are desirable). The AFM can also be used for measurements of mechanical properties on micro- to nanoscales. Figure 10 shows microscratches made on

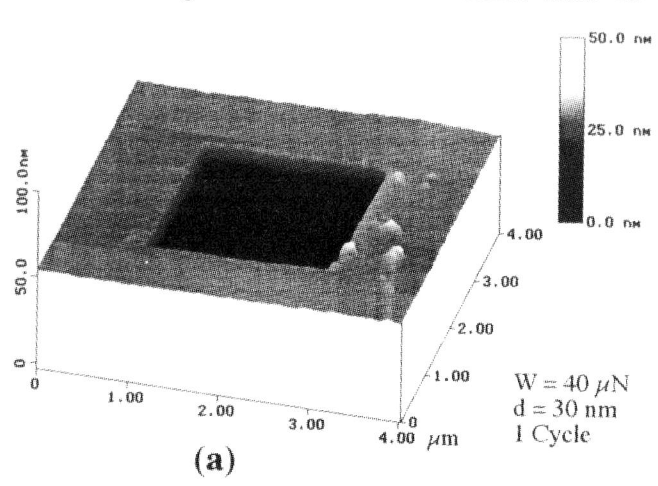

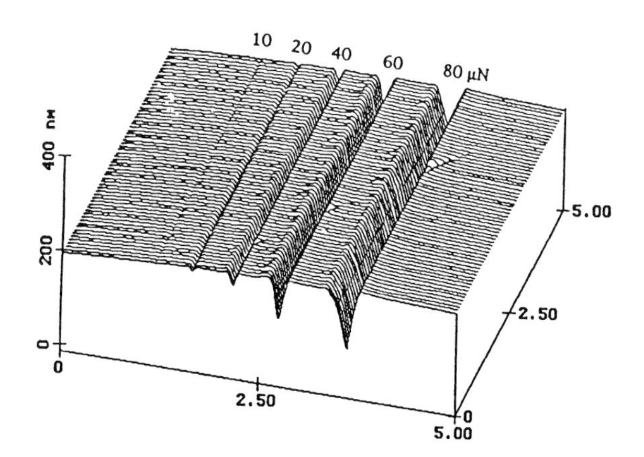

Fig. 10 Surface plots of Si(111) scratched are various loads. Note that x and y axes are in µm and z axis is in nm (**36**).

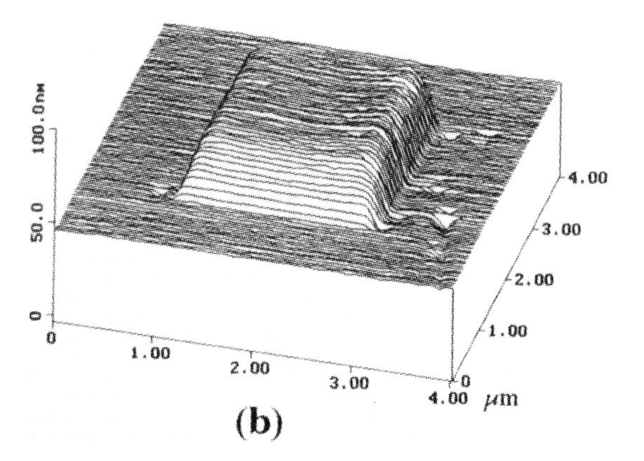

Fig. 11 (a) Typical gray scale and (b) inverted AFM images of wear mark created using a diamond tip at a normal force of 40 µN and one scan cycle on Si(111) surface (**36**).

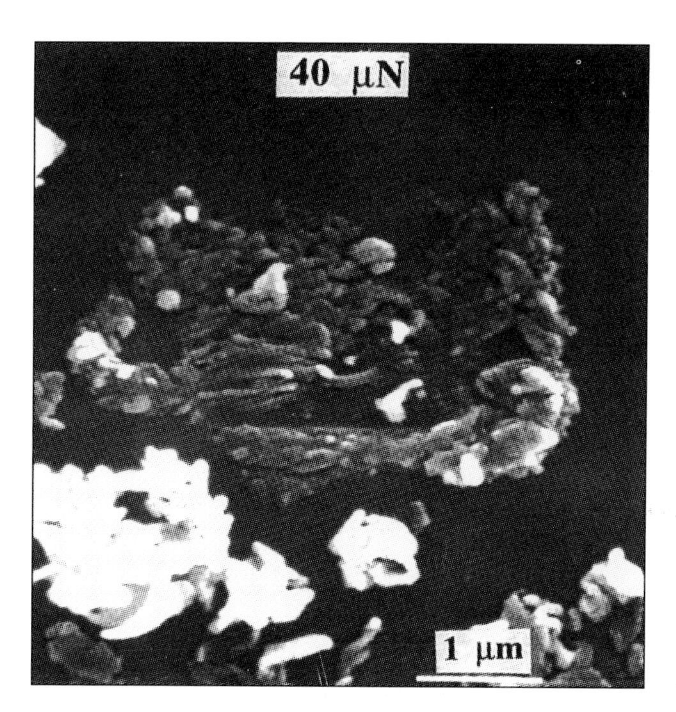

Fig. 12 Secondary electron image of wear mark and debris particles for Si produced at a normal force of 40 μN and one scan cycle (**36**).

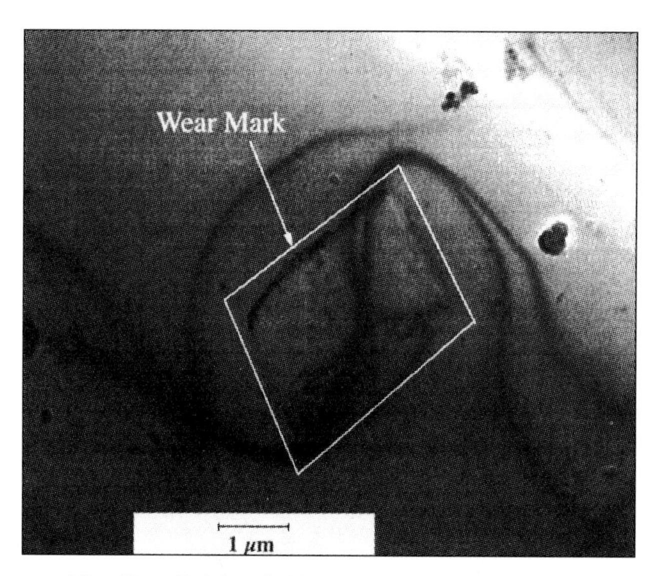

Fig. 13 Bright field TEM micrograph showing wear mark and bend contour around and inside wear mark in Si produced at a normal force of 40 μN and one scan cycle (**36**).

Si(111) at various loads after 10 cycles (**35**). As expected, the depth of scratch increases with load. Such microscratching measurements can be used to study failure mechanisms on the microscale and to evaluate the mechanical integrity (scratch resistance) of ultra-thin films at low loads.

By scanning the sample in two dimensions with the AFM, wear scars are generated on the surface. Typical wear mark generated at a normal load of 40 μN for one scan cycle and imaged using AFM at 300 nN load, is shown in Fig. 11(a). Inverted line plot of wear mark is shown in the Fig. 11(b) showing the uniform material removal at the bottom of the wear mark. Next we examine the mechanism of material removal at light loads on microscale in AFM wear experiments (**36**). Figure 12 shows a secondary electron image of wear mark and associated wear particles. The specimen used for the SEM was not scanned after initial wear, to retain wear debris in the wear region. Wear debris is clearly observed. AFM image of the wear mark shows small debris at the edges, swiped during AFM scanning. Thus the debris is 'loose' (not sticky) and can be removed during the AFM scanning. SEM micrographs show both cutting type and ribbon-like debris. TEM studies were performed to understand the material removal process. TEM micrograph of the worn region in Fig. 13 shows evidence of bend contours passing though the wear mark. The bend contours around and inside the wear mark suggests that there are some residual stresses around and inside the wear mark region. There is no dislocation activity or cracks observed inside the wear track. The dislocation activity and/or cracking probably occurs at the subsurface. Thus it is believed that the material in the experiment described here is removed by the ploughing mode in a brittle manner without much plastic deformation (dislocation acitivity).

Figure 14 shows the effect of normal load on the wear rate. We note that the wear rate is very small below 20 μN of normal force. A normal force of 20 μN corresponds to contact stresses comparable to the hardness of the silicon. Primarily, elastic deformation at loads below 20 μN is responsible for low wear.

The evolution of wear of a diamond-like carbon coating on a polished aluminum substrate is shown in Fig. 15 which illustrates how the micro-wear profile for a load of 20 μN develops as a function of the number of scanning cycles (**19**). Wear is not uniform, but is initiated at the nanoscratches indicating that surface defects (with high surface energy) act as initiation sites. Thus, scratch-free surfaces will be relatively resistant to wear.

Mechanical properties, such as hardness and Young's modulus of elasticity can be determined on micro- to picoscales using the AFM (**4, 37**) and a new nano/pico indentation system used in conjunction with an AFM (**15–18, 38**). Indentability on the scale of sub-nanometers can be studied by monitoring the slope of cantilever deflection as a function of sample traveling distance after the tip is engaged and the sample is pushed against the tip. For a rigid sample, cantilever deflection equals the sample traveling distance; but the former quantity is smaller if the tip indents the sample. Figure 16 shows the load-displacement curves at different peak loads for Si(100). Load-displacement data at residual depths as low as about 1 nm can be obtained. Loading/unloading curves are not

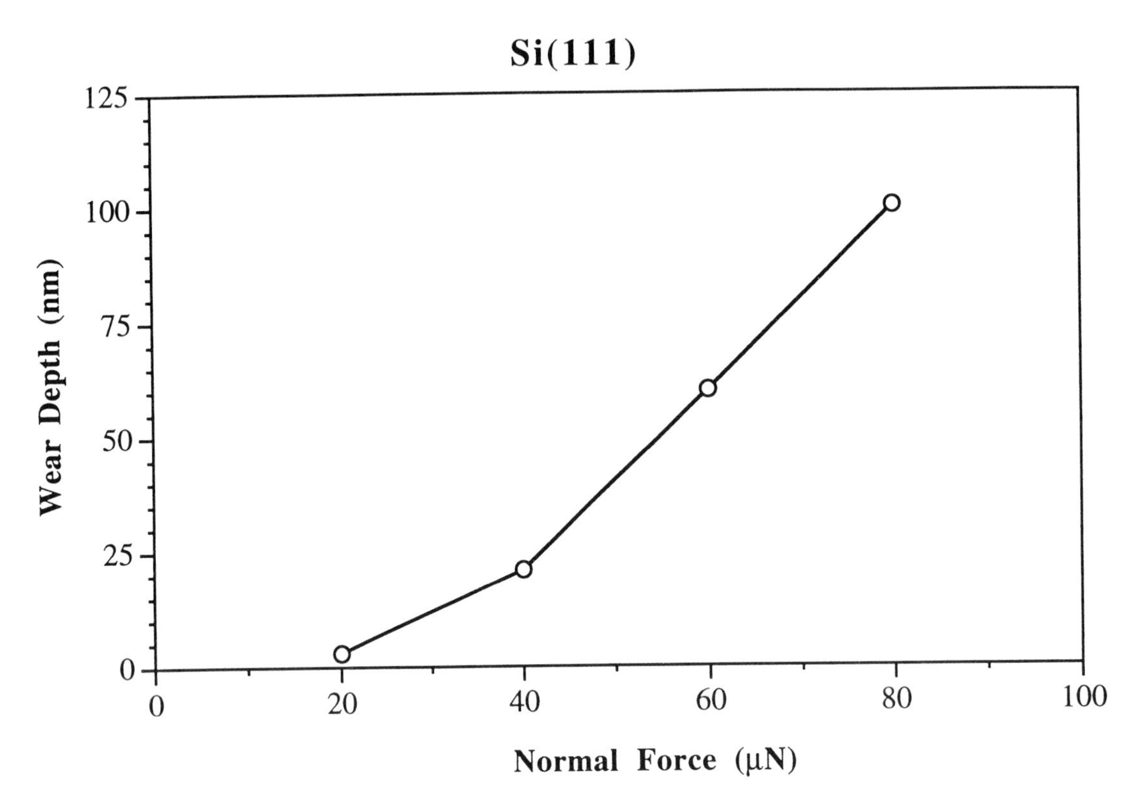

Fig. 14 Wear depth as a function of normal force for Si after one cycle (**36**).

smooth, but exhibit sharp discontinuities particularly at high loads (shown by arrows in the figure). Any discontinuities in the loading part of the curve probably results from slip. The sharp discontinuities in the unloading part of the curves are believed to be due to formation of lateral cracks which form at the base of median crack which results in the surface of the specimen being thrust upward.

The indentation hardness of surface films with an indentation depth of as small as about 1 nm has been measured for Si(111) (**15, 37**). Triangular indentations are observed for shallow penetration depths. The hardness of silicon on nanoscale is found to be higher than on microscale, Fig. 17. This decrease in hardness with an increase in indentation depth can be rationalized on the basis that as the volume of deformed material increases, there is a higher probability of encountering material defects. Bhushan and Koinkar (**37**) have used AFM measurements to show that ion implantation of silicon surfaces increases their hardness and thus their wear resistance. Formation of surface alloy films with improved mechanical properties by ion implantation is of growing technological importance as a means of improving the mechanical properties of materials. Hardness of 20-nm thick diamond like carbon films have been measured by Kulkarni and Bhushan (**18**).

The Young's modulus of elasticity is calculated from the slope of the indentation curve during unloading (**4, 15–18**). Maivald, et al. (**39**) used an AFM in "force modulation mode" to measure surface elasticities. AFM tip is scanned over the modulated sample surface with the feedback loop keeping the average force constant. For the same applied force, a soft area deforms more, and thus causes less cantilever deflection, than a hard area. The ratio of modulation amplitude to the local tip deflection is then used to create a force modulation image. The force modulation mode makes it easier to identify soft areas on hard substrates.

The nano/picoindentation system has been used to study the creep and strain-rate effects of ceramics. Bhushan, et al. (**15**) and Kulkarni, et al. (**17, 18**) have reported that ceramics exhibit significant plasticity and creep on nanoscale.

Detection of transfer of material on a nanoscale is possible with the AFM. Indentation of C_{60}-rich fullerene films with an AFM tip has been shown (**40**) to result in the transfer of fullerene molecules to the AFM tip, as indicated by discontinuities in the cantilever deflection as a function of sample traveling distance in subsequent indentation studies.

6 BOUNDARY LUBRICATION

The classical approach to lubrication uses freely supported multimolecular layers of liquid lubricants (**8, 11**). The liquid lubricants are chemically bonded to improve their wear resistance (**4, 18**). To study depletion of boundary layers, the micro-scale friction measurements were made

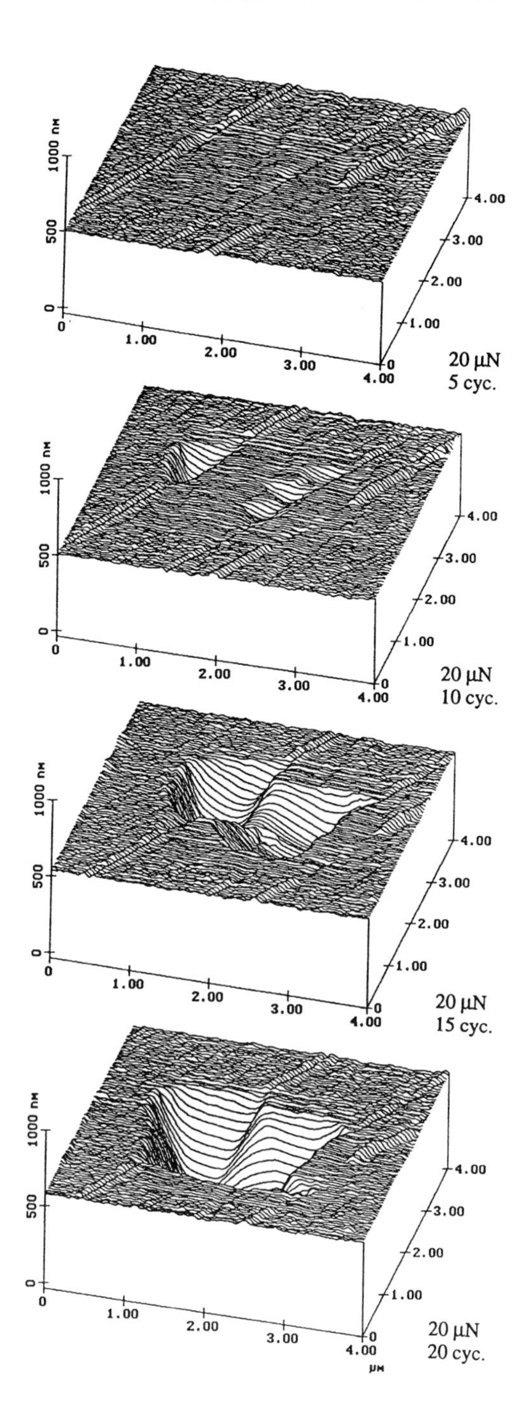

Fig. 15 Surface plots of diamond-like carbon-coated thin-film disk showing the worn region; the normal load and number of test cycles are indicated (**37**).

as a function of number of cycles of virgin Si(l00) surface and silicon surface lubricated with Z-15 and Z-Dol PFPE lubricants, Fig. 18 (**41,42**), Z-Dol is PFPE lubricant with hydroxyl end groups. Its lubricant film was thermally bonded at 150°C for 30 minutes (BUW- bonded, unwashed) and, in some cases, unbonded fraction was washed off with a solvent to provide a chemically bonded layer of the lubricant (BW) film. In Fig. 18(a), the unlubricated silicon sample shows a slight increase in friction force followed by a drop to a lower steady state value after 20 cycles. Depletion of native oxide and possible roughening of the silicon sample are believed to be responsible for the decrease in friction force after 20 cycles. The initial friction force for Z-15 lubricated sample is lower than that of unlubricated silicon and increase gradually to friction force value comparable to that of the silicon after 20 cycles. This suggests the depletion of the Z-15 lubricant in the wear track. In the case of the Z-Dol coated silicon sample, the friction force starts out to be very low and remains low during the 100 cycles test. It suggests that Z-Dol does not get displaced/depleted as readily as Z-15. The nanowear results for BW and BUW/Z-Dol samples with different thicknesses are shown in Fig. 18(b). The BW with thickness of 2.3 nm exhibits an initial decrease in the friction force in the first few cycles and then remains steady for more than 100 cycles. The decrease in friction force possible arises from the alignment of any free liquid lubricant present over the bonded lubricant layer. BUW with a thickness of 4.0 nm exhibits the similar behavior to BW (2.3 nm). Lubricated BW and BUW samples with thinner films exhibit a higher value of coefficient of friction. Among the BW and BUW samples, BUW samples show the lower friction because of extra unbonded fraction of the lubricant.

Effect of the operating environment on coefficient of friction of unlubricated and lubricated samples is shown in Fig. 19. Silicon (100) samples were lubricated with 2.9 nm thick Z-15 and 2.3 nm thick Z-Dol bonded and washed (BW) lubricants. Coefficient of friction in dry environment is lower than that at high humidity environment. We believe that in the humid environment, the condensed water from the humid environment competes with the liquid film present on the sample surface and interaction of the liquid film (water for the unlubricated sample and polymer lubricant for the lubricates sample) to the substrate is weakened and boundary layer of the liquid forms puddles. This dewetting results in poorer lubrication performance resulting in high friction. Since Z-Dol is a bonded lubricant with superior frictional properties, dewetting effect in humid environment for Z-Dol is more pronounced than Z-15.

Effect of scanning speed on the coefficient of friction of unlubricated and lubricated samples is shown in Fig. 20. Coefficient of friction for unlubricated silicon sample and lubricated sample with Z-15 decreases with the logarithm increase in the scanning velocity in ambient environment. These samples are insensitive to scanning velocity in dry environments. Samples lubricated with Z-Dol do not show any effect of scanning velocity on the friction. Alignment of liquid molecules (shear thinning) is believed to be responsible for the drop in friction with an increase in scanning velocity for samples with mobile film and exposed to ambient environment.

For lubrication of microdevices, a more effective approach involves the deposition of organized, dense

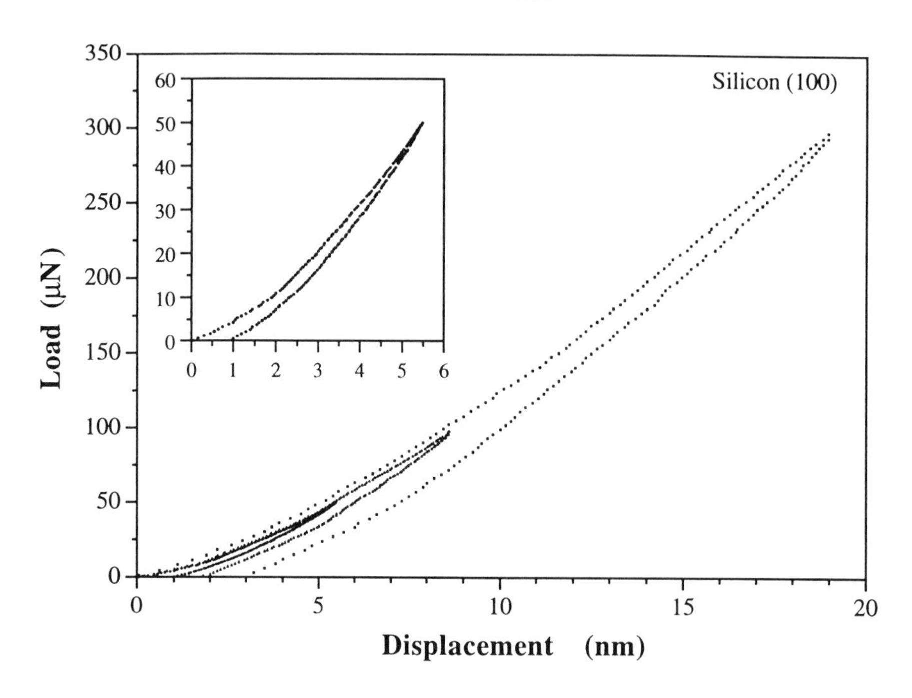

Fig. 16 Load-displacement curves at various peak loads for Si(100) (**15**).

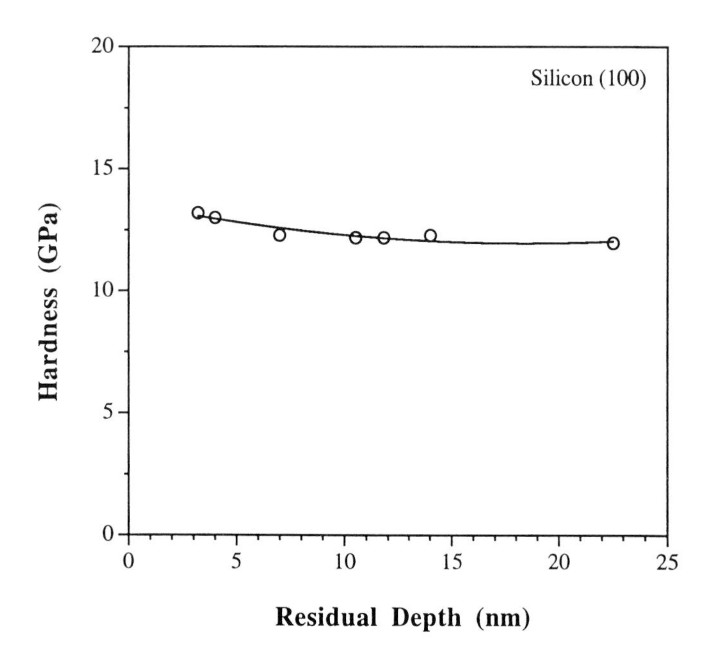

Fig. 17 Indentation hardness as a function of residual indentation depth for Si (100) (**15**).

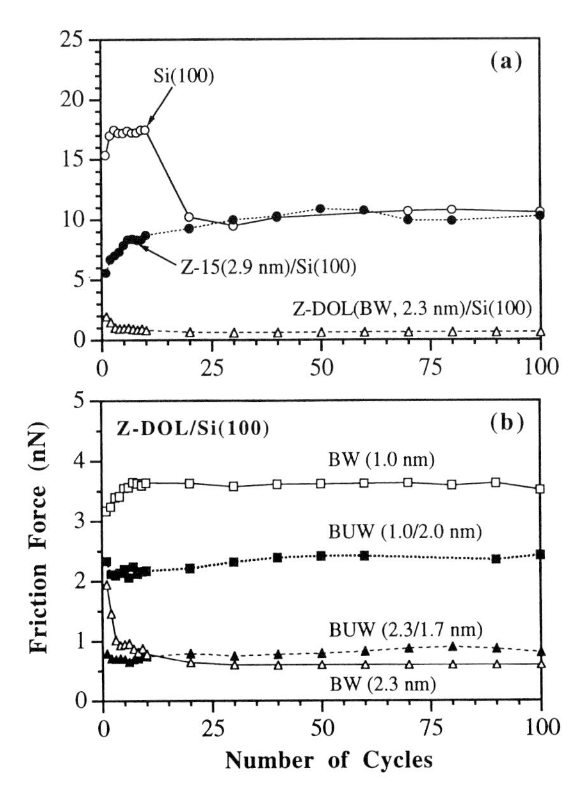

Fig. 18 Friction force as a function of number of cycles using silicon nitride tip at a normal force of 300 nN for (a) unlubricated silicon, Z-15 and bonded washed (BW) Z-Dol and (b) bonded Z-Dol before washing (BUW) and after washing (BW) (**42**).

molecular layers of long-chain molecules on the surface contact. Such monolayers and thin films are commonly produced by Langmuir-Blodgett (L-B) deposition and by chemical grafting of molecules into self-assembled monolayers (SAMs). Based on the measurements, SAMs of octodecyl (C_{18}) compounds based on aminosilanes on an oxidized silicon exhibited lower coefficient of friction of 0.018 and greater durability than LB films of zinc

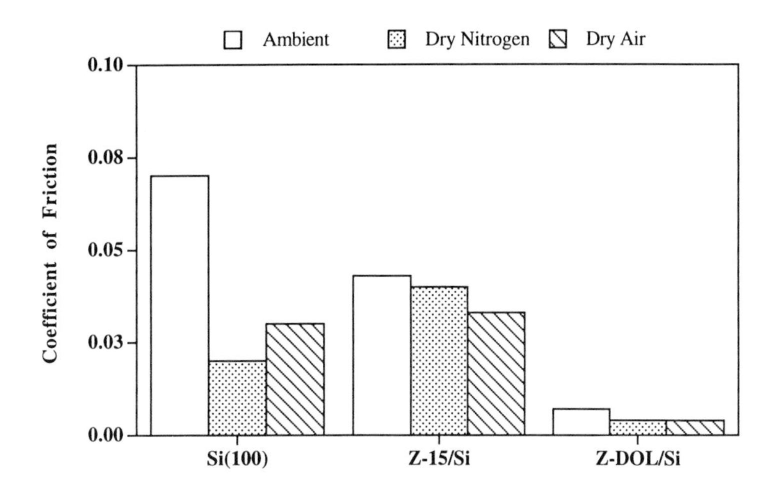

Fig. 19 Coefficient of friction for unlubricated and lubricated Si(100) samples in ambient (~50% RH), dry nitrogen (~5% RH) and dry air (~5% RH) (Koinkar and Bhushan, 1996b).

arachidate adsorbed on a gold surface coated with octadecylthiol (ODT) (coefficient of friction of 0.03) (Fig. 21) (**43**). LB films are bonded to the substrate by weak van der Waals attraction, whereas, SAMs are chemically bound via covalent bonds. Because of the choice of chain length and terminal linking group that SAMs offer, they hold great promise for boundary lubrication of microdevices.

Measurement of ultra-thin lubricant films with nanometer lateral resolution can be made with the AFM (**4**). The lubricant thickness is obtained by measuring the force on the tip as it approaches, contacts and pushes through the liquid film and ultimately contacts the substrate. The distance between the sharp 'snap-in' (owing to the formation of a liquid of meniscus between the film and the tip) at the liquid surface and the hard repulsion at the substrate surface is a measure of the liquid film thickness. This technique is now used routinely in the information-storage industry for thickness measurements (with nanoscale spatial resolution) of lubricant films, a few nanometers thick, in rigid magnetic disks.

Lubricant film thickness can also be measured using

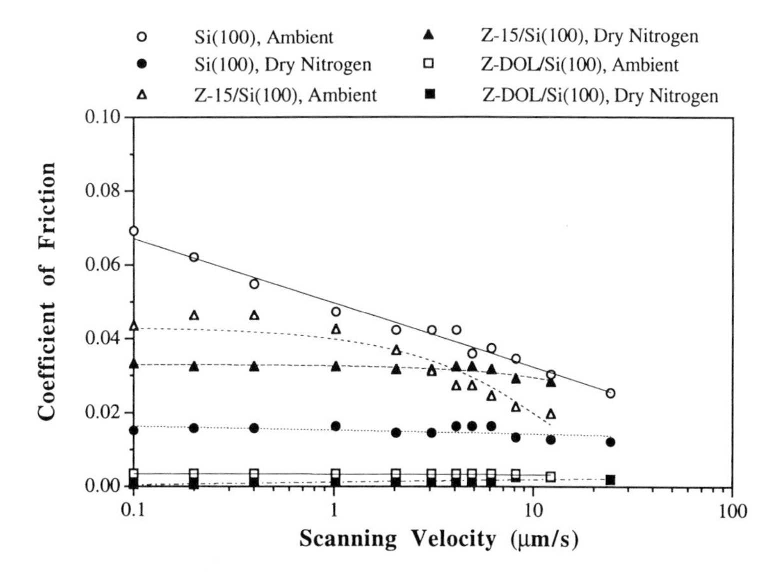

Fig. 20 Coefficient of friction as a function of scanning velocity for unlubricated and lubricated Si (100) samples in ambient (~50% RH), dry nitrogen (~5% RH) and dry air (~5% RH) (**41**).

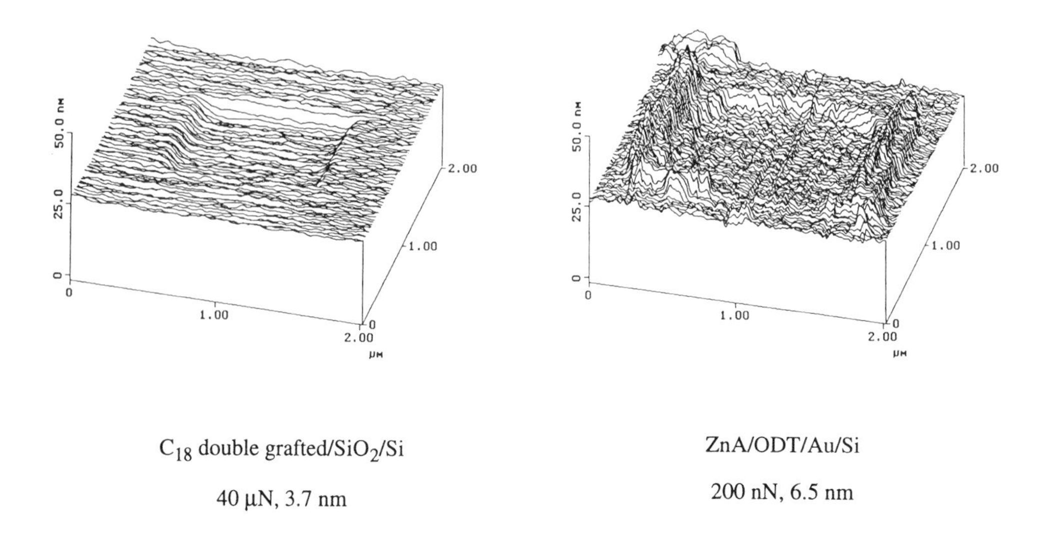

Fig. 21 Surface plots showing the worn region after one scan cycle for self-assembled monolayers of octodecyl silanol (C_{18}) (left) and zinc arachi date (ZnA) (right). Normal force and wear depths are indicated. Note that wear of ZnA occurs at only 200 nN as compared to 40 mN for C_{18} film (**43**).

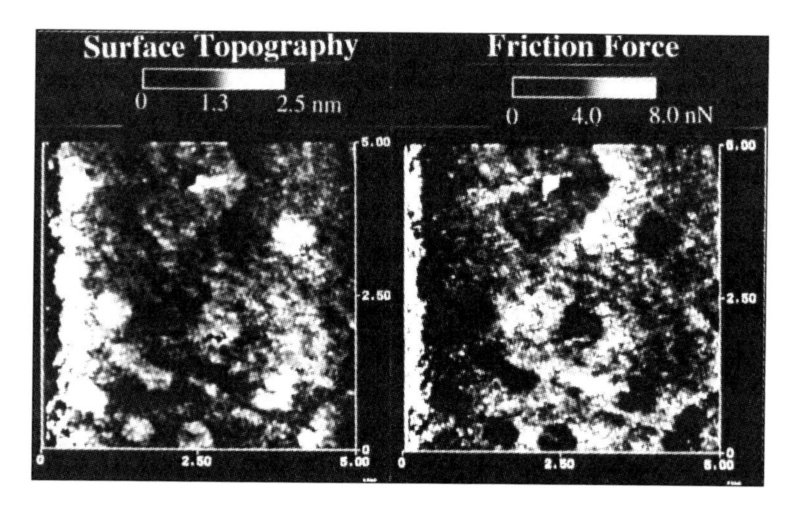

Fig. 22 Gray scale plots of the surface topography and friction force obtained simultaneously for unbonded perfluoropolyether lubricant film on silicon (**42**).

friction force microscopy (**42**). Fig. 22 shows gray scale plots of the surface topography and friction force obtained simultaneously for unbonded Dennum type perfluoropolyether lubricant film on silicon. The friction force plot shows well distinguished low and high friction regions roughly corresponding to high and low regions in surface topography (thick and thin lubricant regions). A uniformly lubricated sample does not show such a variation in friction. Friction force imaging can thus be used to measure the lubricant uniformity on the sample surface which cannot be identified by surface topography.

7 CONCLUSIONS AND FUTURE OUTLOOK

AFM/FFM has been developed as a versatile instrument to conduct a variety of micro/nanotribological studies. These studies are crucial to develop fundamental understanding on micro- to nanoscales. These studies are of significant interest in magnetic storage and MEMS industries. We expect micro/nanotribology to grow rapidly and dominate the tribology research arena.

8 ACKNOWLEDGEMENTS

Financial support for this research was provided by the Office of Naval Research, Department of the Navy (Contract No. N00014-96-1-10292). The information herein does not necessarily reflect the position or policy of the government, and no official endorsement should be inferred.

9 REFERENCES

1 **Binnig, G., Rohrer, H., Gerber, Ch.**, and **Weibel, E.** Surface Studies by Scanning Tunnelling Microscopy, Phys. Rev. Lett. (1982), **49,** 57-61.

2 **Binnig, G., Quate, C.F.** and **Gerber, Ch.**, Atomic Force Microscopy, *Phys. Rev. Lett.* (1986), **56**, 930–933.

3 **Binnig, G., Gerber, Ch., Stoll, E., Albrecht, T.R.**, and **Quate, C.F.**, Atomic Resolution with Atomic Force Microscope, *Europhys. Lett.* (1987), **3**, 1281–1286.

4 **Bhushan, B.**, *Handbook of Micro/Nanotribology,* (1995), CRC Press, Boca Raton, Fl.

5 **Bhushan, B., Israelachvili, J.N.**, and **Landman, U.**, Nanotribology: Friction, Wear and Lubrication at the Atomic Scale, *Nature* (1995a), **374**, 607–616.

6 **Bhushan, B.** Micro/Nanotribology and its Applications, *NATO ASI Series E: (1997), Applied Sciences* – Vol. 330, Kluwer Academic Pub., Dordrecht, Netherlands.

7 **Bhushan, B.** Micro/Nanotribology and its Application to Magnetic Storage Devices and MEMS, *Tribol. International* (1995), **28**, 85–95.

8 **Bhushan, B.**, *Tribology and Mechanics of Magnetic Storage Devices,* (1996), second edition, Springer–Verlag, New York.

9 **Bhushan, B.** Nanotribology and Nanomechanics of MEMS Devices, *Proc. MEMS (1996), '96,* I EEE, NY, 91–98.

10 **Bhushan, B.** Contact Mechanics of Rough Surfaces in Tribology: Multiple Asperity Contact, (1997), *Tribology Letters* (in press)

11 **Bowden, F.P.** and **Tabor, D**, The *Friction and Lubrication of Solids,* (1950 & 1964), Parts I & 11, Clarendon, Oxford, UK.

12 **Bhushan, B**. and **Gupta, B.K.** *Handbook of Tribology: Materials, Coatings and Surface Treatments,* (1991), McGraw-Hill, New York.

13 **Bhushan, B.** *Mechanics and Reliability of Flexible Magnetic Media,* (1992) Springer-Verlag, New York.

14 **Ruan, J.** and **Bhushan, B.** Frictional Behavior of Highly Oriented Pyrolytic Graphite, *J. Appl. Phys.* (1994), **76,** 8117–8120.

15 **Bhushan, B., Kulkarni, A.V., Bonin, W.** and **Wyrobek, J.T.,** Nano/Picoindentation Measurements Using a Capacitive Transducer System in Atomic Force Microscopy, *Philos. Mag.* (1996), **74**, 1117–1128.

16 **Kulkarni, A.V.** and **Bhushan, B.**, Nanoscale Mechanical Property Measurements Using Modified Atomic Force Microscopy, *Thin Solid Films,* (1996),

17 **Kulkarni, A.V.** and **Bhushan, B.** Nano/picoindentation Measurements on Single-Crystal Aluminum Using Modified Atomic Force Microscopy, *Materials: Letters* (1996), **29**, 221–227.

18 **Kulkarni, A.V.** and **Bhushan, B.** Nano/picoindentation Measurement of Amorphous Carbon Coatings Using a Capacitive Transducer System in Atomic Force Microscopy, *J. Mat. Res.* (1997), (submitted for publication).

19 **Bhushan, B., Koinkar, V.N**. and **Ruan, J.** Microtribology of Magnetic Media, *Proc. Inst. Mech. Engrs., Part J: J. Eng. Tribol.* (1994), **75**, 5741–5746.

20 **Bhushan, B.** and **Blackman, G.S.**, Atomic Force Microscopy of Magnetic Rigid Disks and Sliders and its Applications to Tribology, *ASME Journal of Tribology* (1991), **113**, 452–458.

21 **Oden, P.I., Majumdar, A., Bhushan, B., Padmanabhan, A.** and **Graham, J.J.,** AFM Imaging, Roughness Analysis and Contact Mechanics of Magnetic Tape and Head Surfaces, *ASME Journal of Tribology* (1992), **114**, 666–674.

22 **Ganti, S.** and **Bhushan, B.,** Generalized Fractal Analysis and its Applications to Engineering Surfaces, *Wear* (1995), **180**, 17–34.

23 **Poon, C.Y.** and **Bhushan, B**., Comparison of Surface Roughness Measurements by Stylus Profiler, AFM and Non-contact Optical Profiler, *Wear* (1995), **190**, 76–88.

24 **Koinkar, V.N.** and **Bhushan, B.,** Effect of Scan Size and Surface Roughness on Microscale Friction Measurements, *J. Appl. Phys.* (1997),(in press).

25 **Majumdar, A.** and **Bhushan, B.,** Role of Fractal geometry in Roughness Characterization and Contact Mechanics of Surfaces, *ASME Journal of Tribology* (1990), 112, 205–216.

26 **Majumdar, A.** and **Bhushan, B.,** Fractal Model of Elastic-Plastic Contact Between Rough Surfaces, *ASME Journal of Tribology* (1991), **113**, 1–11.

27 **Bhushan, B. and Majumdar, A.,** Elastic-Plastic Contact Model for Bifractal Surfaces, *Wea* (1992), **153**, 53–64.

28 **Bhushan, B.,** Contact Mechanics of Rough Surfaces in Tribology: Single Asperity Contact, *Appl. Mech. Rev.* (1996), **49**, 275–298.

29 **Mate, C.M., McClelland, G.M., Erlandsson, R.,** and **Chiang, S.**, Atomic-Scale Friction of a Tungsten Tip on a Graphite Surface, *Phys. Rev. Lett.* (1987), **59,** 1942–1945.

30 **Ruan, J.** and **Bhushan, B.,** Atomic-Scale and Microscale Friction of Graphite and Diamond Using Friction Force Microscopy, J. *Appl. Phys.* (1994), **76,** 5022–5035.

31 **Koinkar, V.N.** and **Bhushan, B.**, Microtribological Studies of Al_2O_3, Al_2O_3-TiC, Polycrystalline and Single-Crystal Mn-Zn Ferrite and SiC Head Slider Materials, Wear,'(1996) **202** (in press).

32 **Meyer, E.,** *et al.* Friction Force Microscopy of Mixed Langmuir-Blodgett Films, *Thin Solid Films,* (1992), **220,** 132–137.

33 **Frisbie, C.D., Rozsnyai, L.F., Noy, A., Wrighton, M.S.,** and **Lieber, C.M.,** Functional Group Imaging by Chemical Force Microscopy, *Science* (1994), **265,** 2071–2074.

34 **Bhushan, B.** and **Kulkarni, A.V.,** Effect of Normal Load on Microscale Friction Measurements, *Thin Solid Films,* (1996), **278,** 49–56.

35 **Bhushan, B.,** and **Koinkar, V.N.,** Tribological Studies of Silicon for Magnetic Recording Applications, *J. Appl. Phys.* (1994), **75,** 5741–5746.

36 **Koinkar, V.N.** and **Bhushan, B.**, Scanning and Transmission Electron Microscopies of Single-Crystal Silicon Microworn/machined Using Atomic Force Microscopy, *J. Mat. Res.* (1997),(submitted for publication).

37 **Bhushan, B.** and **Koinkar, V.N.,** Nanoindentation Hardness Measurements Using Atomic Force Microscopy, *Appl. Phys. Lett,* (1994),. **64,** 1653–1655.

38 **Koinkar, V.N.** and **Bhushan, B.,** Microtribological Properties of Hard Amorphous Carbon Protective Coatings for Thin Film Magnetic Disks and Heads, *Proc. Instn Mech. Engrs. Part J:J. Eng. Trib.* (1997), (submitted for publication).

39 **Maivald, P., Butt, H.J., Gould, S.A.C., Prater, C.B., Drake, B., Gurley, J.A., Elings, V.B.,** and **Hansma, P.K.,** Using Force Modulation to Image Surface Elasticities with the Atomic Force Microscope, *Nanotechnology* (1991), **2,** 103–106.

40 **Ruan, J.** and **Bhushan, B.,** Nanoindentation Studies of Fullerene Films Using Atomic Force Microscopy, *J. Mat. Res,* (1993). **8,** 3019–3022.

41 **Koinkar, V.N.** and **Bhushan, B.,** Microtribological Studies of Unlubricated and Lubricated Surfaces Using Atomic Force/Friction Force Microscopy, *J. Vac. Sci.* (1996), Technol. **A14,** 2378–2391.

42 **Koinkar, V.N.** and **Bhushan, B.,** Micro/Nanoscale Studies of Boundary Layers of Liquid Lubricants for Magnetic Disks, *J. Appl. Phys.* (1996), **79,** 8071–8075.

43 **Bhushan, B., Kulkarni, A.V., Koinkar, V.N., Boehm, M., Odoni, L., Martelet, C.** and **Belin, M.,** Microtribological Characterization of Self-Assembled and Langmuir-Blodgett Monolayers by Atomic and Friction Force Microscopy, *Langmuir* (1995), **11,** 3189–3198.

Sliding wear of materials

S K BISWAS
Department of Mechanical Engineering, Indian Institute of Science, Bangalore, India

SYNOPSIS

We review here our understanding of the sliding wear phenomenon: some generalities have emerged in the last 50 years of research, these can now be taken as established principles and be used for practical design and maintenance. Other issues related for example to nano-wear, the role of microstructure on wear or mechanism of crack nucleation require renewed efforts, for greater predictivity in wear. The review is based on published literature with examples principally drawn from our work on sliding wear of metals and ceramics.

1 PREAMBLE

To minimise wear it is crucial to understand why the wear occurs. From such an understanding can emerge predictive models which can help in designing a component and selecting the material to minimise the loss of material by wear.

Wear is caused by mechanical, electrical or chemical interactions. Due to mechanical interaction cracks may nucleate by; breakage of molecular level bonds in polymers, shear banding in metals, grain boundary failure in ceramics or interfacial failures in composites and multiphase materials. These cracks may propagate to detach a wear debris out of the parent material. This is possible when the interaction is violent as in erosion and impact wear or relatively non-violent when a traction is generated due to the relative motion of two solid bodies already in contact and transmitting normal force. Of the latter further sub-divisions are possible according to the dynamics of contact; partial slip occurs in rolling and fretting while sliding involves gross slip, fretting has reciprocating motion while rolling and sliding may be unidirectional or reciprocating. One may make further divisions based on the morphology of contact; adhesion and abrasion. We will refrain from the latter on the ground that most slid surface show telltale marks of both, sliding therefore involves the relative motion of two bodies, such that these motions occur in planes which are parallel and separated by the roughness of bodies. In the present paper we only consider the case where directions of motion are collinear but opposite.

The early history of our understanding of sliding wear is mainly about the wear of metallic systems. With growing practical experience, industrial awareness and basic research some general principles, of importance not only to metallic wear but to other materials as well, started to emerge. We would in this article try to elucidate these principles in their generalities but with the help of specific examples drawn principally from our own work on metals and to a limited extent from our work on ceramics.

2 SOME GENERALITIES

Wear is a highly complex phenomenon. The character of a surface is determined by the bulk chemical and physical structure as well as environmental chemistry. The topography is determined by the method of manufacture. The short range interactions between surfaces are dominated by physical forces while tractive interaction gives rise to stresses and strains which bring about damage which is ultimately responsible for wear. Wear is thus a phenomenon which straddles many disciplines of study. The complexity of the phenomenon inhibitted any comprehensive modelling for a long time. The mechanical engineers who faced the immediate problem of wear control in machinery had however to respond to the situation by formulating first order models based on purely mechanical concepts.

It was recognised in the early nineteen fifties that when metals wear out 1) asperities are involved in this interaction and 2) large strains are incurred at the asperity level or below the asperity root which, in case of metals at least take the material well past the yield point. The first models of adhesive and abrasive wear thus invoked asperity contact and considered the soft asperity or substrate to undergo plastic flow.

3 FUNDAMENTAL BASIS OF WEAR: ADHESION, ABRASION

Adhesive wear occurs as junctions form and grow (**1**), when asperities from the two surfaces contact, transmit tangential motion and fracture. The transferred material on detachment emerges as wear debris. Assuming the contact between asperities to be circular Archard formulated that the wear volume: V is given by:

$$V = K\frac{W}{H}S \qquad (1)$$

where K is the probability of a contact producing a wear particle, H-hardness of the softer, material, W is the normal load, and S is the sliding distance.

Abrasive wear may be caused by a hard asperity or a third body grit cutting and removing the softer material. It can be shown (**2**) (also see (**3**)) that a single abrasive interaction also leads to:

$$V = K\frac{W}{H}S \tag{2}$$

where $K = \dfrac{2\tan\alpha}{3\pi}$, a geometric factor related to average strain, and α is the asperity attack angle.

For adhesive interaction the friction coefficient due to the above physical model is *S/P*. where S is the yield shear stress and P is the flow stress. For abrasive interaction the coefficient of friction is given by:

$$\mu = \frac{\tan\alpha}{\pi} + \frac{S}{P} \tag{3}$$

4 EFFECT OF LOAD, SPEED AND HARDNESS ON WEAR

The efficacy of these models lies in experimental data satisfying the relationships; wear is proportional to hardness inversely, load directly, sliding distance directly and strain directly.

Lancaster (**4**) found the wear to be inversely proportional to hardness only when the load is high and the sliding speed low. Further, when the experimental variables such as load or speed change, at some critical values of these variables the mechanisms of wear change, bringing about a drastic change in wear rate. Thus the wear rate over a wide range of load may be non-linear, while that between two transitions may indeed be linear – (see Fig. 1). Rigney (**5**) has discussed the possible sources of deviation from the wear rate-hardness relationship predicted by eqn (1). Fig. 2 shows such a deviation in the mild wear regime. One of the reasons for this deviation is the development of in situ protective layer of transferred and mechanically mixed fine material. Irrespective of the matrix hardness, the hardness of these layers for different Al-Si alloys were found to be roughly the same. During sliding wear different materials develop different hardness gradients; positive (work

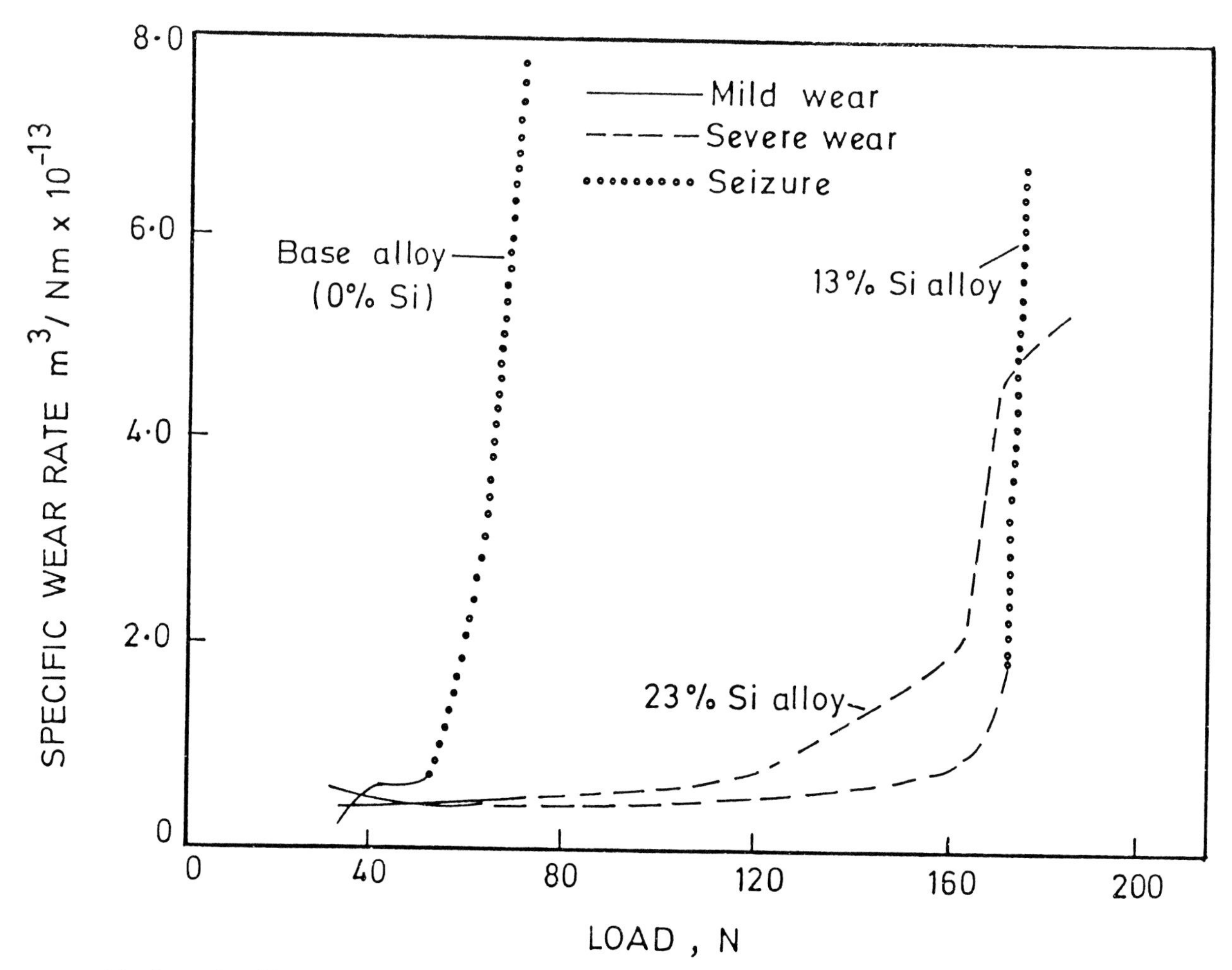

Fig. 1 Specific wear rate of Al-Si alloys demarcating three modes of wear – mild, severe and seizure.

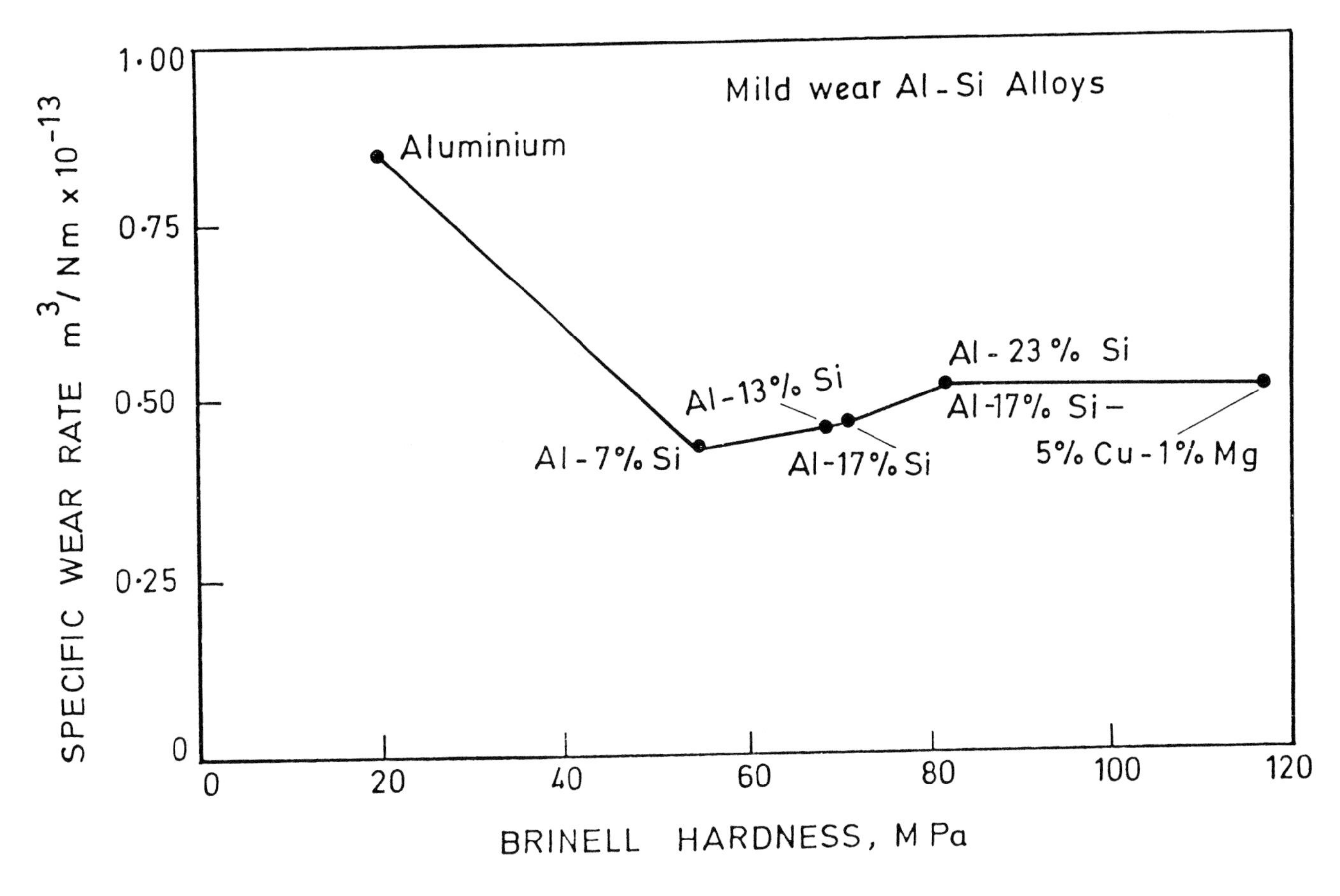

Fig. 2 Mild wear of Al-Si alloys slid against EN24 steel vs hardness.

hardening) and negative (work softening). The actual gradients can affect the wear rate greatly. The wear rate as will be shown later is also related to the proneness of a material to undergo shear instabilities. Fig. 3 shows the wear rate of Ti and Cu rubbed against steel. Titanium in spite of being 2 times as hard is much less wear resistant than copper. It will be argued that this is because of the more acute strain rate response of titanium than copper. Such strain response gives rise on traction, to adiabatic shear bands. For a given substrate hardness, the other parameters which may affect wear are relative hardness of the substrate and the counterface.

The wear rate in abrasion is also profoundly affected by the geometry of the grit. Suh, Sin and Saka (**6**) have demonstrated experimentally that the wear coefficient due to eqn (2) is a gross overestimate for a range of grit geometry. Finally, contrary to expectation from the above models the debris from the sliding wear are generally laminar and chip-shaped.

5 WEAR TRANSITIONS AND MECHANISMS

Fig. 1 shows that wear occurs in stages, each stage may be broadly related to one mechanism, the mechanism changing from stage to stage. Two such principal stages are; mild wear regime where the wear rate is low, (K in eqn (1) $\approx 10^{-6}$ to 10^{-4}) and the severe wear regime where the wear rate is high (K in eqn (1) $\approx 10^{-4}$ to 10^{-2}) (see Table 1 (**1**)) It was realised that the mild wear is mild because the wear event, resulting in tiny flake-like debris is confined to a thin and hard oxidised film on the surface. The severe wear is severe because big lumps and laminates, the underside of which lay originally in the bulk subsurface, emerge with each wear event giving rise to a large rate of removal. The mild wear was thus largely investigated in its relation to oxidation kinetics while the severe wear research probed the stresses, flow and fracture in the subsurface.

It was also realised fairly early on that the transition from one mechanism to the other depend critically on the load and speed combination, the absolute value of such a combination being related to the physical properties such as thermal diffusivity and thermal conductivity and mechanical properties such as hardness. While the exact mechanism for this dependency is still under debate and discussion Lim and Ashby (**7**) proposed a wear map for steel which differentiated mild oxidational, delamination, melt seizure and severe oxidational wear in a normalised force-velocity space. While empirical data could be assigned to different wear compartments situated in definite pockets of the map, heat transfer calculation over idealised rough contact domain, given the frictional work, yields flash and bulk temperature rise estimates. Such estimates

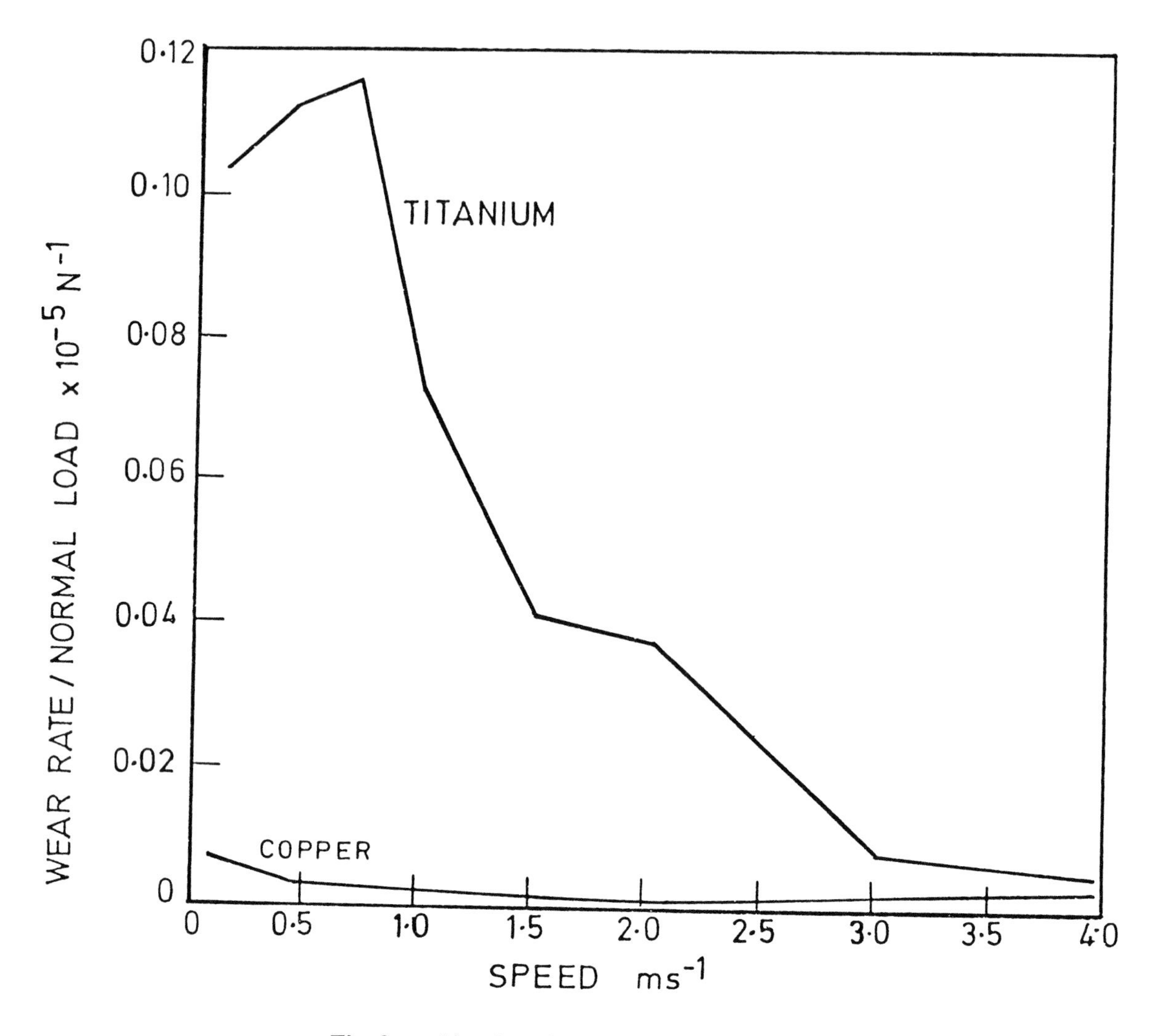

Fig. 3 Abrasion of Ti and Cu against Al_2O_3 disc.

provide the rationale for delineating oxidative and melt wear as well as seizure. A more stress analysis-based approach, yielding critical subsurface plastic strain at inclusions and void interfaces provides the basis for marking out delamination/severe wear. More fundamental problems related to complex material responses to strain, strain rate and temperature, which may have profound bearing on the criteria used for the demarcation of the different wear regimes, however remain and need careful investigations in the future.

In this essay we will look at some of these issues in the perspective of our general state of knowledge of sliding wear. I intend to discuss wear mechanisms as a function of load and speed. In this the stages, mild wear ($K \approx 10^{-6}$-10^{-4}), severe wear (10^{-4}-10^{-2}) and seizure (gross material removal) are well characterised, symptomatically and delineated. Quinn (**8**) discusses another stage of severe wear which precedes mild wear. It is not a very well understood stage except that at a notional level it is associated with asperity removal leading to increase in contact area, elastic contact and mild wear. There has in the very recent period a great spurt in the investigation of ultra low load tribology where a further stages of no wear followed by this severe wear has been observed. At loads less than 1mN there is thus a great scope for further investigation and observation. As we will see that this stage is today associated with many important industrial application and thus deserve a definite mention in this presentation. I thus consider four stages of wear:

1. Nano-scale Wear
2. Mild Wear
3. Severe Wear
4. Seizure

5.1 Nano-scale wear

One of the main reasons for the recent spurt of interest (**9–18**) in nano-tribology is the problems associated with the performance and life of magnetic data storage systems. The tribology of interaction between different components; the ceramic head, the particulate or the metal evaporated smooth magnetic tapes and the polymeric substrates is being investigated in different ambient and contact environments, under ultra low load conditions. The other major motivation for this area of research has come from efforts to develop efficient micromechanical devices; microsensors, micromotors, microactuators, micropumps and microgears. In this context the tribology of silicon and carbon films is being actively. investigated A wealth of data and observation have already emerged, most of it not yet well understood. What is however clear is that the tribophysics of these ultra low load phenomena contain unique elements and need special attention.

The recent emergence of proximal probes such as Surface Force Apparatus, Atomic Force and Friction Force microscopes (**19**), all capable of nanometric and nano Newton levels of resolutions has made it possible to investigate ultra low load tribological phenomenon with confidence and repeatability. Two very interesting results to have emerged from such studies are.

1. At very low loads the wear is zero and the friction is low (**20-22**). Increasing the load, the wear becomes more conventional with ploughing and cutting and the friction obeys the Amonton's law again. There are many speculations as to the genesis of this phenomenon. There is a school which attributes this to atomic level asperity tip deformation and fracture while a second school asserts that this happens because the contact is elastic due to the high hardness associated with small deformation volumes. A more realistic explanation may lie in the fact that at these loads there is capillary condensation from the ambient, the contact is therefore a liquid junction and shearing of these junctions may be done easily and without loss of material (Fig. 4) (**17**). Some recent calculations (**23**) show that for a 150 nm tip the experimentally measured contact stiffness at these loads lie in the liquid junction regime and not in the adhesive contact regime. This view is supported by the fact that tribology under these conditions is extremely sensitive to ambient humidity, even when there is a layer of lubricant present at the interface. Further support for the view comes from molecular dynamic simulation of tip/surface sliding interaction which shows that in the presence of a lubricant layer at the interface the tip undergoes zero wear beyond a running-in phase (**24**).

   A somewhat contrary experimental observation however comes from a study of steel balls sliding on flat test piece in the 0.8 to 2350 μN load range. The work shows that under these conditions the friction force is high at low normal loads and the tip surface adhesion force affects the friction force the same way as the normal load.

2. At low loads where the displacements and perturbations may be an order less than the main asperity dimension, it is the subasperities which cut and plough and cause nanowear (**21**). This is interesting as it points to the direct impact of the fractal nature of surfaces on wear.

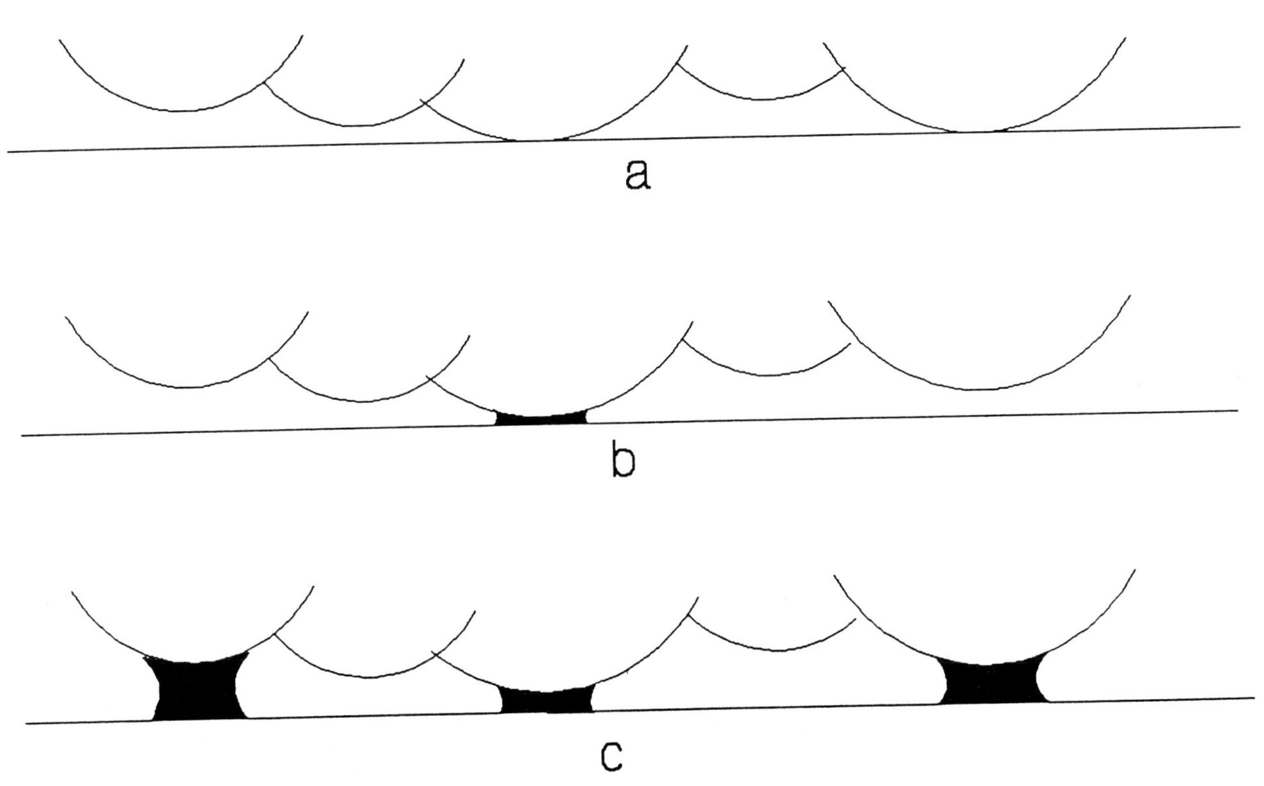

Fig. 4 Capillary condensation at nano-contact of rough surfaces. (a) dry contact, (b) low humidity, (c) high humidity.

5.2 Mild wear

Unlubricated mild wear co-efficients generally fall in the 10^{-6} to 10^{-4} range, (see Table 1). The debris in this region are equiaxed and small. The established opinion is that in this regime metals wear by corrosion. When the wear takes place in the ambient, the metal reacts readily with the environmental oxygen. Mild wear is thus traditionally held to be synonymous with oxidational wear. What we will argue here is that the above order of wear co-efficients and debris morphology are also possible with the in-situ generation of any protective layer be that a product of a chemical or a mechanical interaction.

Table 1

Sliding wear; metals, ceramics	Wear coefficient, K
Lubricated : full fluid film	$< 10^{-13}$
Lubricated : solid lubricant	$\sim 10^{-6}$
Unlubricated : mild wear	10^{-6} - 10^{-4}
Unlubricated : severe wear	10^{-4} - 10^{-2}

5.2.1 Oxidational wear

After the very early (time or distance-wise) traverse, where the asperity tips are readily removed (**8**) the contact becomes more conformal, and a large contact area consisting of thermally expanded smooth plateaux comes into existence. Frictional heating of asperity tips (flash temperature) give rise to oxidation of the plateaux surfaces. These plateaux continue to transmit traction, in the process get abraded and crack, generating very small size debris. When the thickness reaches a limit the oxidised platelets spall and are removed as debris. This exposes a nascent substrate which again starts to oxidise. Given this, if the metal can oxidise quickly the wear rate is low. Traditionally the oxidational wear rate is thus related to activation energy of oxidation. In more recent times the rate has been found to correlate strongly with the ability of a material to chemiadsorb atmospheric oxygen. Accordingly transition metals and lanthanides as well as intermetallic components such as of copper and zinc show very low rates of oxidational wear. What has to be kept in mind is that given the ability of the metal to repair the film quickly the wear rate is also influenced by the mechanical

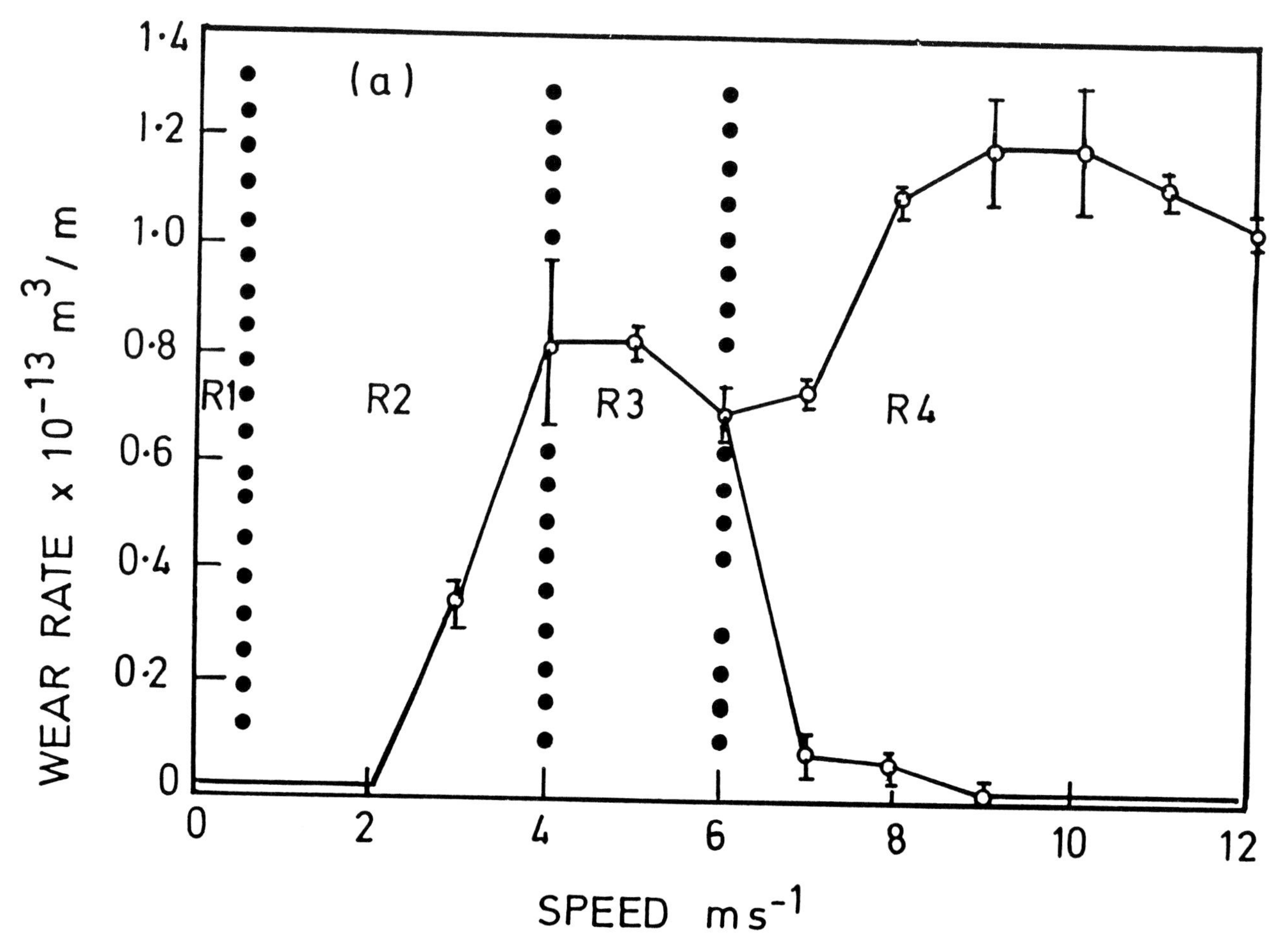

Fig. 5 Wear rate as a function of speed at 15.5 MPA.(a) Al_2O_3 against steel, (b) Si_3N_4 against steel.

properties of the oxide film and the substrate. If the substrate is weak as for example for grey iron, (**25**), fast oxidation protects the iron from being worn by adhesion and cutting while for a hard and tough iron like vermicular iron, oxidation enhances wear under normal atmospheric conditions.

The unit volume of oxides being different from that of the substrate, they carry substantial deleterious residual stresses, which under tractions lead to interfacial failure (blistering) or cracking in the oxide itself. The role of frictional heat on oxidational wear is of interest. For steel Lim and Ashby (**7**) show that a flash temperature of about 400°C, independent of normal load, is needed to trigger a mild oxidational regime. Given this the raised temperature affect the mechanical property of the oxide film, the interfacial adhesion as well as the activation energy. The process is complex and continues to draw the attention of tribologists.

The chemical nature of mild sliding wear is not confined to metals. Fig. 5 shows the wear of Al_2O_3 (**26**) and Si_3N_4 (**27**) slid against steel discs in a pin-on-disc machine. The regimes R1 and R2 are low temperature (bulk temperature = 120°C, flash temperature = 900°C) regions where the wear rate is low. Oxidation here occurs at the disc counterface producing γ-Fe_2O_3. The debris from this layer fills out the valleys on the ceramic surface, gets compacted under traction and protects the ceramic from wear. For Si_3N_4 in the RB regime, the SiO_2nH_2O film generated in situ (which for example remains hard in the RA regime) softens (flash temperature = 700°C) and provides easy shear in the film and correspondingly low friction and wear (also see (**28**) for tribochemical-reaction related transitions in SiC sliding).

5.2.2 Mechanical mild wear

A mechanism somewhat more complex than the above and a combination of chemical and mechanical events was suggested by Razavizadeh and Eyer (**29**, **30**) for aluminium silicon alloys sliding on a steel counterface. The oxidised aluminium in this case fractures and gets compacted into the alloy valleys to form a smooth layer. This layer

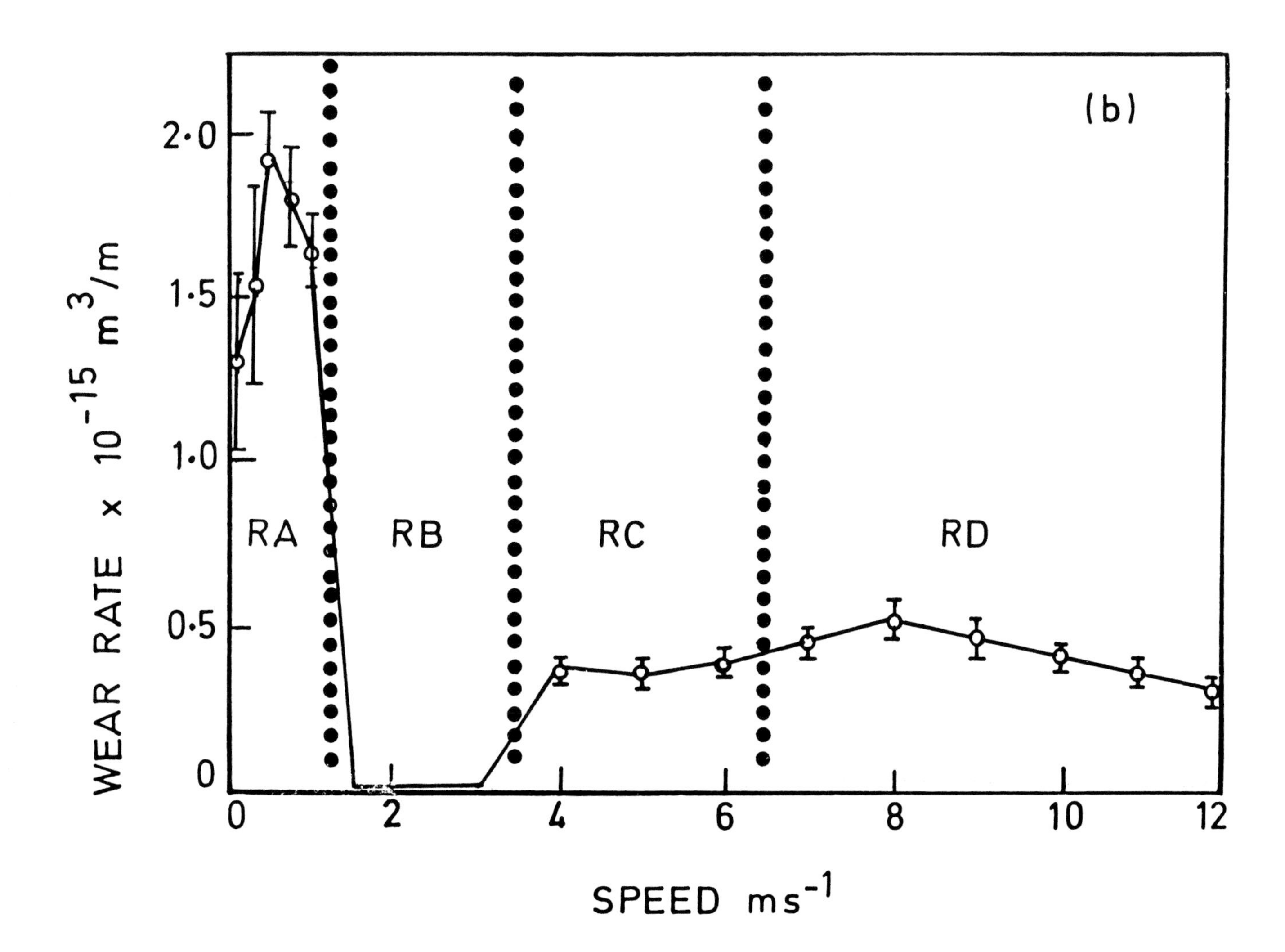

delaminates under traction to produce plate like debris. We found the protective layer to be rich in iron. The layer is a compacted mechanical mix of fine particles of iron abraded from the counterface and the aluminium alloy debris (**31-36**). X-ray (**32, 36, 37**) and TEM (**38**) studies supported the observation that these particles have been removed from the metals and the layer contains very little oxide. Such fine grain protective layer consisting of a mixture of the counterface and slid material have also been found by Rigney (**39**) in his work on sliding wear of OFHC copper. The layer forms by transfer and back transfer of the counterface and alloy debris and becomes mixed in a process similar to the early stages of mechanical alloying (**5**). Its ability to be protective (prolonging mild wear) however depends on its hardness, integrity and its attachment to the substrate. Rigney (**5**) points out that if the hardness of this layer is more than that of the substrate it can press into the substrate to form a strong attachment, and not otherwise. We (**32**) have found for example that the addition of Mg to Al-7% Si alloy and subsequent heat treatment increases the ratio of the hardness of the layer to the substrate hardness from 3 to 5, as the rate of mild wear ($K \approx 10^{-6}$-10^{-7}) is almost halved. Even under mild wear conditions we found eutectic silicon near the surface fragmented and equiaxed, the cracks nucleating at particles/matrix interfaces and propagating through the protective layer to the surface. Cracks may also nucleate in the compacted layer itself, leading to the delamination of this layer. Clearly under mild wear conditions the removal and generation of this layer are in equilibrium. Integrity of this layer, as mentioned above, and the ability of the substrate to resist nucleation of cracks are additional factors which control mild wear rate. Both these factors to a great extent can be controlled by alloy addition and heat treatment. We have found Cu and Mg addition to Al-Si alloy and subsequent heat treatment, (**32**) spherodises silicon and renders the matrix tough. The protective layer is also made continuous and smooth by such alloy addition. The abrasive resistance of the alloy increases by the alloy addition, the layer is difficult to destabilise due to the limiting of plastic flow in the subsurface and the spheroidal silicon particles discourage crack nucleation. The overall effect is to reduce the mild wear rate and postpone transition to the severe wear regime. Similar observations were made and similar conclusions drawn by Harun *et al* (**40**) in their study of the influence of Ce, Zn and Zr additions on the sliding wear of Al-13% Si alloy.

5.3 Severe wear

The wear coefficient for severe wear in sliding fall in the range 10^{-4} to 10^{-2}. The debris are generally chunky or laminate shaped particles, compositionally a large proportion of which is the substrate material. At a critical value of load or speed large plastic strain occurs in the subsurface destabilising the protective layer and ushering in direct contact between the counterface and surface. This results in a rise in friction, large stresses extending to considerable depths and temperature rise. Cracks now nucleate and propagate at substantial depths, and when they intersect thick debris are generated.

5.3.1 Elastic analysis and ceramic wear by fracture

Because the traction is directly transmitted to the substrate tribologists have always found severe wear, as opposed to other types of wear, more tractable quantitatively. Assuming Hertzian contact and Amonton's law to hold in a sliding contact, the steady state elastic stresses may be written out (**41**). The stresses at the rear of the contact are tensile, the maximum shear stress is located deep in the subsurface and somewhat to the rear of the contact, and there are large compressive hydrostatic stresses in the near surface region.

The elastic analysis made it possible to achieve some early success in the modelling of wear of brittle materials such as ceramics (**42**). Lateral and radial cracks were found to lead to spalling when hard asperities pressed on to the ceramic surface. Ceramics are however prone to grain boundary failures which are stress-driven but thermally activated processes (**43**) and to severe corrosive wear. For example when an alumina pin slides against a steel disc at speeds more than 4 m/s (Pressure = 15.5 MPA) the wear is primarily due to the easy removal of tribochemical layer consisting of Fe_3O_4, Fe Al_2O_4 and $FeAlO_3$ (**26**). Plastic deformation is also a known cause of ceramic wear under severe sliding conditions. Hsu has recently (**44**) reported a comprehensive study of ceramic wear, in which he comes back to the conclusion that it is fundamentally the contact stresses which predominantly determines how material is removed in the sliding of ceramics. At low contact stresses plastic deformation-induced microfracture controls the removal of asperities. At higher stresses partial cone, lateral and radial cracks propagate and yield debris when they intersect.

5.3.2 Elastic plastic analysis – delamination, shakedown

For metals it was important to extend the elasticity analysis to include yield and plastic flow. Two approaches developed. Suh used the Merwin-Johnson analysis (**45**) to determine the conditions under which the accumulated plastic strain at the particle/matrix-interface of a two-phase material reaches criticality and causes separation and crack nucleation. This leads to delamination wear. While there is enough experimental evidence for laminate shaped debris the model has never been fully accepted as it has not been made clear how a soft dislocation-free subsurface region (as assumed by Suh) can accumulate large plastic strain to initiate void nucleation. It is also not clear how a crack may propagate parallel to the sliding direction to give rise to the laminate debris. Suh clearly does not address the issue of how laminate-shaped debris are also generated in the severe wear of single phase materials.

The other interesting development in elastic plastic analysis of wear is due to Kapoor and Johnson (**46**). They show that plasticity may be easily obtained in sliding contact but the state reverts back to elasticity due to `shakedown'. If the nominal contact pressure exceeds the shakedown limit there is further accumulation of plastic strain and severe wear, the shakedown limit being dependent on the surface roughness, hardness and coefficient of friction. In the severe wear regime there is an irreversible component of plastic shear strain, which in unidirectional sliding gives rise to rachetting failure generating fine plate-like debris. Alternatively for low attack angle asperity/grit there is ploughing and for sharp asperities there is wear by cutting (**47**). If the sliding is reciprocating an additional reversible plastic strain give rise to low cycle fatigue wear. Our work with aluminium silicon alloy has shown (**48**) that in reciprocating sliding wear can be anything between 2 to 3 times more though the friction is somewhat less than that in unidirectional sliding. The effect of alloy addition and speed on wear, however, is the same as that observed in unidirectional sliding.

5.3.3 *Deformation damage in sliding wear and the role of microstructure*

Moore and Douthwaite (**49**) (experimentally) estimated the subsurface strain incurred due to single point abrasion. They showed that the energy absorbed in deforming the grooved material is approximately 50% of the total work done on the system. Zum Gahr (**51**) by modelling subsurface deformation in wear could relate wear to fracture strain of a material. The model accorded with experimental results of Suh, Saka and Sin (**6**) who showed that less than 50% of the interaction events produced wear. Challen and Oxley (**51, 52**) constructed a slip line field model of single point abrasion in plane strain. The analysis yielded a criterion which was able to distinguish plastic flow form cutting given the asperity geometry and skin friction. According to this model the process is insensitive to material properties. The skin friction may however be sensitive to temperature, strain and strain rate. Later work by Oxley and group (**53**) incorporates strain hardening. More refined models using similar approach has been given by Komvopoulus (**54**), Petryk (**55**) Torrance (**56**) , Koplanisky (**57**) and Williams (**58**). The latter based on an upperbound method also affords an estimate of average strain and strain rate.

Some of our (**59, 60**) work using single point abrasion shows the coefficient of friction of Pb, Al, Cu and Ti to decrease in that order although their skin friction when rubbed by a SiC wedge remains roughly the same. By the Oxley criterion Pb is strictly a candidate for the ploughing or the rubbing model whereas Ti is a candidate for the cutting model. Increasing the temperature of the abraded Ti to 400°C takes the coefficient of friction well into the ploughing or the rubbing regime. Profilometry showed that at room temperature Pb is definitely cut whereas no groove is made on Ti, suggesting the mode of interaction purely rubbing in the latter case. This is true even at temperatures up to 400°C. Further, lubrication makes little difference to the coefficient of friction of Ti whereas the coefficient of friction is lowered significantly by lubrication in the case of aluminium

The role of the initial microstructure and that of the specific changes in the microstructure generated in response to actual conditions of strain, strain rate and temperature in the subsurface on wear, have become issues which distinguish the wear resistance of different materials. In steels, the work of Wang and Lei (**61**) has shown the wear resistance to increase in the following microstructural order, spherodised carbide, martensite, bainite and lamellar pearlite. Vardavoulias et al (**62**) have discussed the role of in situ traction induced martensitic phase transformation on the wear of austenitic stainless steel and its composites. For aluminium alloys the role of microstructure, especially that of silicon morphology, on wear has been studied extensively by Eyre (**63**) and others (**32, 40**).

The question remains as to why different microstructures respond to traction in different ways.

5.3.4 *Plasticity-induced microstructural instabilities in wear of single phase metals*

Interesting investigations, theoretical and experimental, have established that:

1. compressive and shear stresses dominate (**64, 65**) in the deep subsurface of a worn track,
2. complex microstructure (**66–71**) and consequent cracks and voids (**72**) nucleate under this state of stress,
3. given this state of stress and the microstructural weaknesses it is possible for a crack to propagate parallel to the sliding direction and,
4. there is a surface layer of mechanically mixed composite (worn and counterface materials) of ultrafine grains (**25, 35, 38, 39**),

There appears to be enough evidence now to suggest that the fracture process which ultimately leads to the wear debris has its origin related to the dynamic evolution of microstructure in the subsurface in response to the conditions which prevail in a general sliding wear situation. Dislocation cell walls, micro-bands of dislocations evolving into shear bands, recrystallised grains, and micro-twins are such products of these evolution which can act as the source of cracks and shear instabilities.

The question remains as to how such evolutionary processes are related to the parametric conditions which prevail in the sliding wear situation.

In the 1980s an important development took place in the field of material processing by large scale deformation, (**76-79**). These works demonstrate, in compression tests, the development of intrinsic instabilities such as adiabatic shear banding, twinning, wedge cracking and dynamic recrystallisation. An implicit premise of this work is that given a temperature-strain rate regime the microstructural response of a metal to the strain rate is unique and the response varies with initial microstructure of the metal.

5.3.5 Strain rate response – wear and friction

Soderberg *et al.* (1985) performed a pendulum scratch experiment on a variety of aluminium alloys the microstructures of which were deliberately varied by heat treatment. For precipitation hardened alloy they observed adiabatic shear banding in the chip. The hardness of these bands was low. The authors suggested dynamic recrystallisation as the operating mechanism in these bands. It is of importance to note that there is little evidence of these bands in chips gouged out from other microstructures. Adiabatic shear bands are readily seen in some steels, impacted by particles or projectiles (**80**). If they occur in the surface and subsurface white layers they lead to spalling. They also occur in the subsurface deformed layer. The role of the latter in wear is not clear at this stage.

5.3.6 Why does titanium wear out faster than low hardness copper?

It was to investigate the relation of microstructural response to strain rate, that we (**59, 81**) conducted a series of scratch and abrasion tests on four metals; lead, aluminium, copper and titanium. The metals were initially characterised in compression. The flow stress data recorded in regimes of 10^{-5} to 10^{2}/s strain rate and 25°C and 400°C temperatures yielded processing and instability maps. Friction data were recorded in scratch tests done with silicon carbide wedges. The abrasion tests were done by sliding the metal against rough alumina discs. Friction and wear data were recorded in situ.

Fig. 6 shows the schematic representation of the strain rate response of titanium and copper. For titanium adiabatic shear bands are generated at room temperature and high strain rates. Tendency for adiabatic shear banding decreases with increasing temperature and strain rate response is more acute at half the strain rate. It decreases with increasing and decreasing strain rates. It should be noted that the strain rate response of titanium at room temperature is much sharper than that of copper. The strain rate response of aluminium and lead is minimal at room temperature as the deformation is homogenuous.

The bow wave of titanium in scratching was found to consist of shear bands while the subsurface deformation is limited to twinning. In contrast the bow wave of copper was plastically deformed and plastic deformation extended to large subsurface depths. Cracks parallel to the sliding direction propagate deep in the subsurface of copper.

Fig. 3 shows the wear of titanium and copper in multiple point abrasion by alumina discs. Without going into the finer details of the wear characteristics it should be noted that at low speeds the wear of titanium is substantially higher than that of copper. This is most interesting if one considers that the hardness of titanium is significantly greater than that of copper.

The above results indicate a legitimate role of strain rate response in the wear of metals. In sliding wear strain rates decreases rapidly near the surface, with depth. The fact that instabilities such as adiabatic shear banding and inhomogeneous deformation occur at high strain rate in the case of titanium and at low strain rate in the case of copper would assign such instabilities to the surface region in titanium samples and deep subsurface in case of copper. Such near-surface instabilities promotes easy shear. The resulting coefficient of friction may be low and the wear high. When the strain rate response of a material is weak as for example in the case of copper, the deformation would tend to be more homogeneously plastic. The results of lubricated and high temperature scratch experiments support this thesis (**59, 60**).

A detailed understanding of the role of strain rate response on wear is lacking at this stage. Preliminary results however suggest that if there are large-scale instabilities near the surface and the stresses required to drive cracks originating from these instabilities are available the wear of the material is likely to proceed at a high rate. If on the other hand the instabilities are occurring in the depth of the subsurface, the stresses there may not be high enough to drive cracks prodigiously. In such a case the wear may be expected to be low. The high wear rate of titanium in comparison to the low wear rate of copper at low velocities where the interface temperatures can be expected also to be low, would tend to lend credence to this argument. Increasing the velocity would increase the interface temperature and take titanium away from the effect of decreasing the wear rate. With increasing velocity the wear rate of titanium would approach that characteristic of copper. This trend is seen in Fig. 3.

5.4 Seizure

The final stage of sliding wear is seizure. The two surfaces adhere to each other so strongly that the driving motor stalls. This is a somewhat unusual situation as in most practical situations there is lubrication present in the interface. When the lubrication fails there is solid-solid contact at the asperity level. This is scuffing. Of the limited amount of work undertaken in this area of gross contact, most has been done looking into the parameters which affect scuffing and to evolve criteria for scuffing.

The failure of lubricant is associated with temperature. When the adsorbed molecules have sufficient energy to escape against the pressures which develop in the narrow gaps, solid-solid contact occurs in the desorbed areas (**82**). This raises the temperatures further encouraging more molecules to escape. Both plasticity and thermoelastic instabilities are held responsible for further progress of scuffing (**83**). The ability of the lubricant molecule as well as the solid surfaces to form chemiadsorbed bonds which can withstand thermal activation, in the final analysis, determines whether there will be any scuffing or not. At a more engineering level the relative magnitude of the asperity height and thickness of the liquid layer is clearly another parameter which influences scuffing. The scuffing process is therefore dependent on the surface roughness (**84**) and the mechanical properties of the material couple

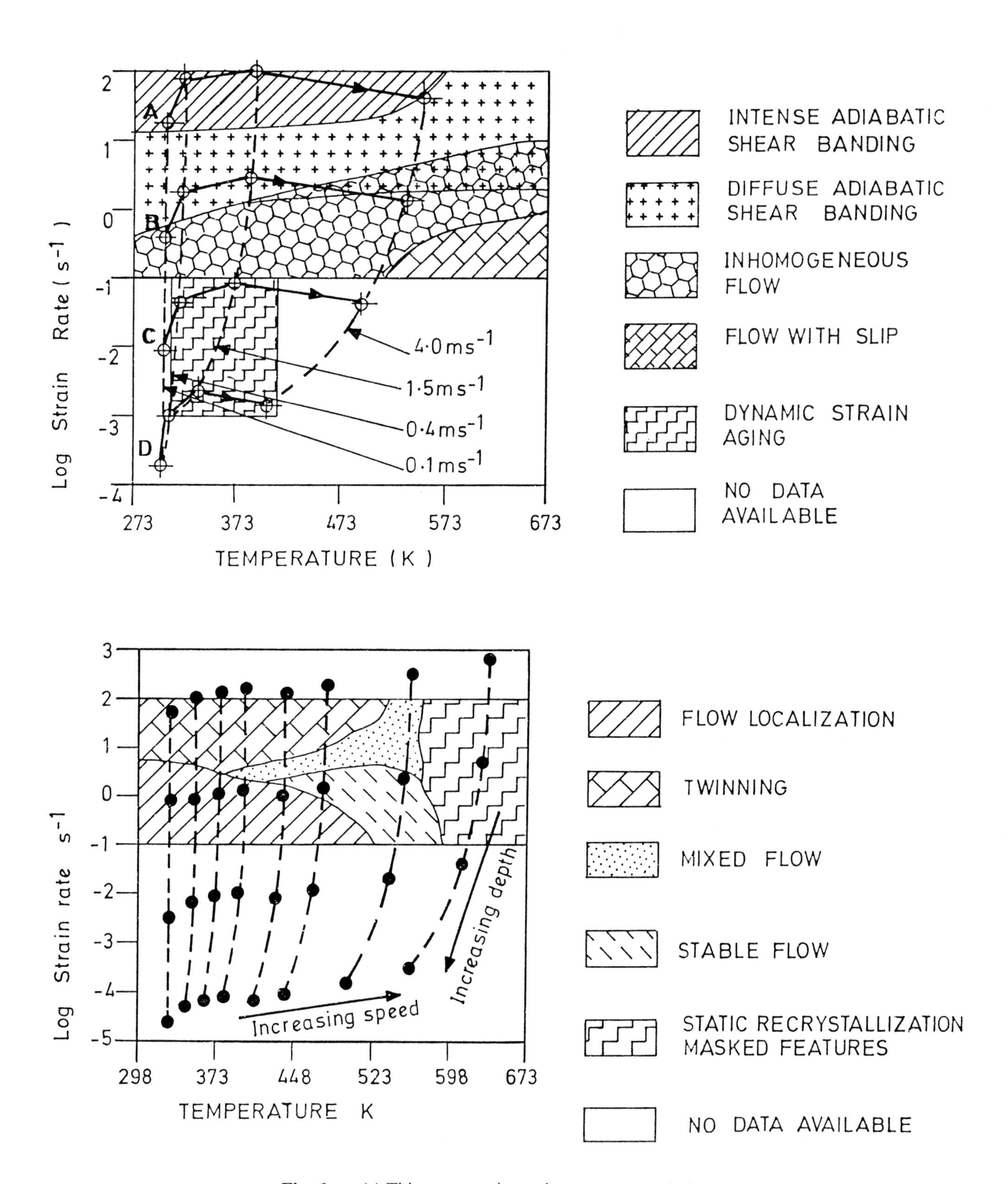

Fig. 6 (a) Ti in compression, microstructure evolution,
(b) Cu in compression, microstructure evolution.

and the asperity geometry the latter influences the propensity of a surface to undergo plastic flow.

Two main scuffing criteria have emerged. There is a plasticity index for scuffing which is proportional to the Youngs modulus and average asperity slope directly and hardness inversely. There is some contrary experimental evidence (**85**) on the hardness proportionality which observes that the harder the substrate more the difficult it is to disturb the thermally insulating oxide layer. The temperature at the asperity tip is therefore high, initiating early scuffing. This is contrary to the plasticity index criterion which would have the softer materials scuffing early. The issue is not really resolved. Following a detailed experimental investigation Park and Ludema (**83**) conclude that neither the asperity slope nor the plasticity index are a good scuffing criteria. The second criterion which is also the older one (**86**) is a more general one which states that for each system consisting of the oil and the material couple there is a critical temperature at which the scuffing occurs. This has been interpreted by different workers in different ways, some looking into the thermal properties of the lubricants and characteristics of adsorption (**82**) while the others have considered the softening of asperities to facilitate plastic flow.

Some of our own work on aluminium-silicon alloy (**48, 75**) is concerned with seizure as the final state of wear where a substantial portion of the surface of the soft member of the interacting couple gets sheared, generating wear slabs which may be about 100μm thick. The rate of wear is an order higher than more severe wear, the machine vibrates violently and if the loads are sufficiently high the driving motor stalls.

5.4.1 Mechanism of seizure

Figures 7, and 8 show that:

1. at low sliding speeds addition of silicon increases the seizure load in unidirectional as well as reciprocating modes,
2. the seizure load decreases with sliding speed,
3. in unidirectional sliding, 23% Si alloy is highly seizure resistant. In reciprocating sliding at high speeds the seizure load for the hypereutectic alloy is low compared to the eutectic alloy,
4. addition of alloying elements enhances the seizure resistance of the alloys.

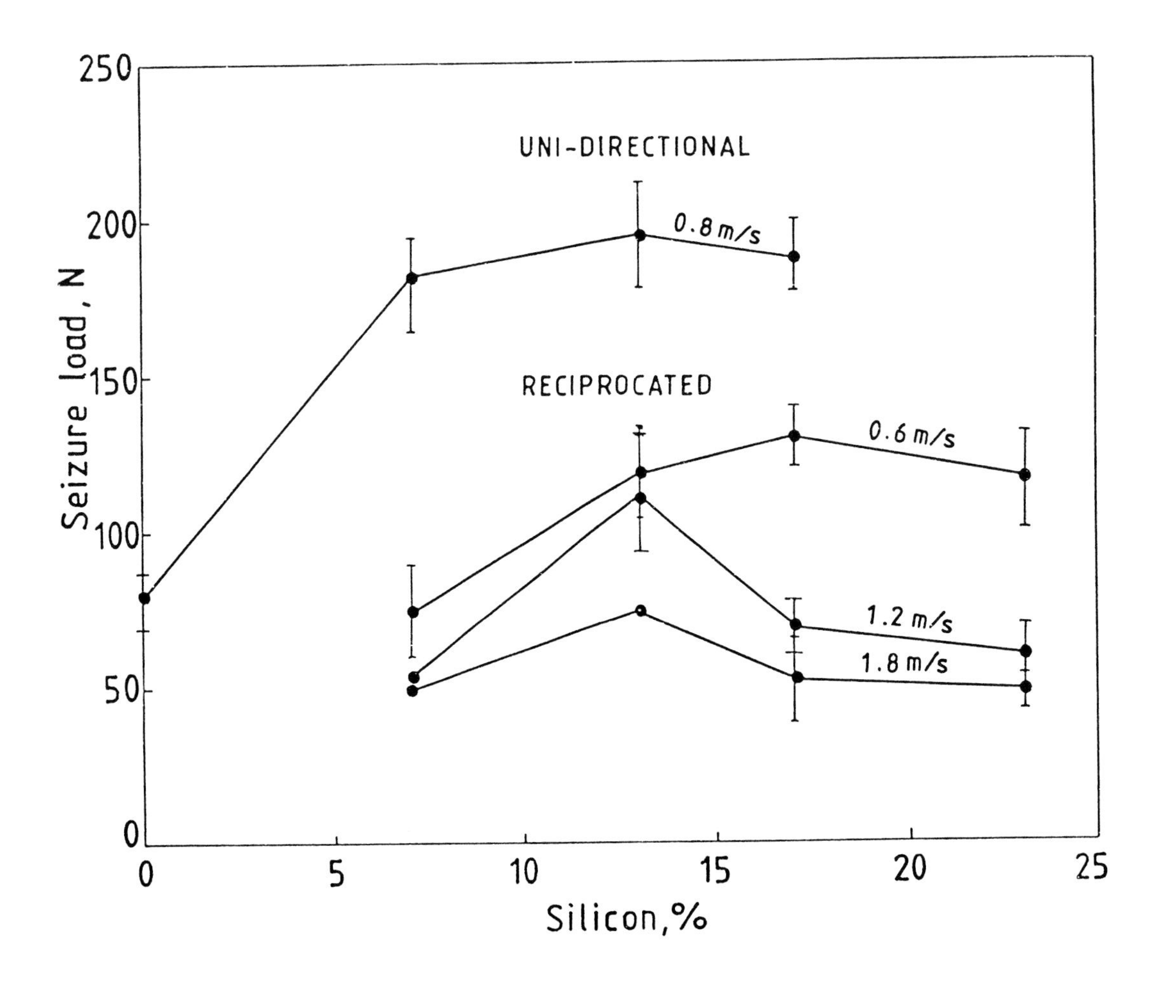

Fig. 7 The load (seizure load) at which seizure occurred for binary alloys.

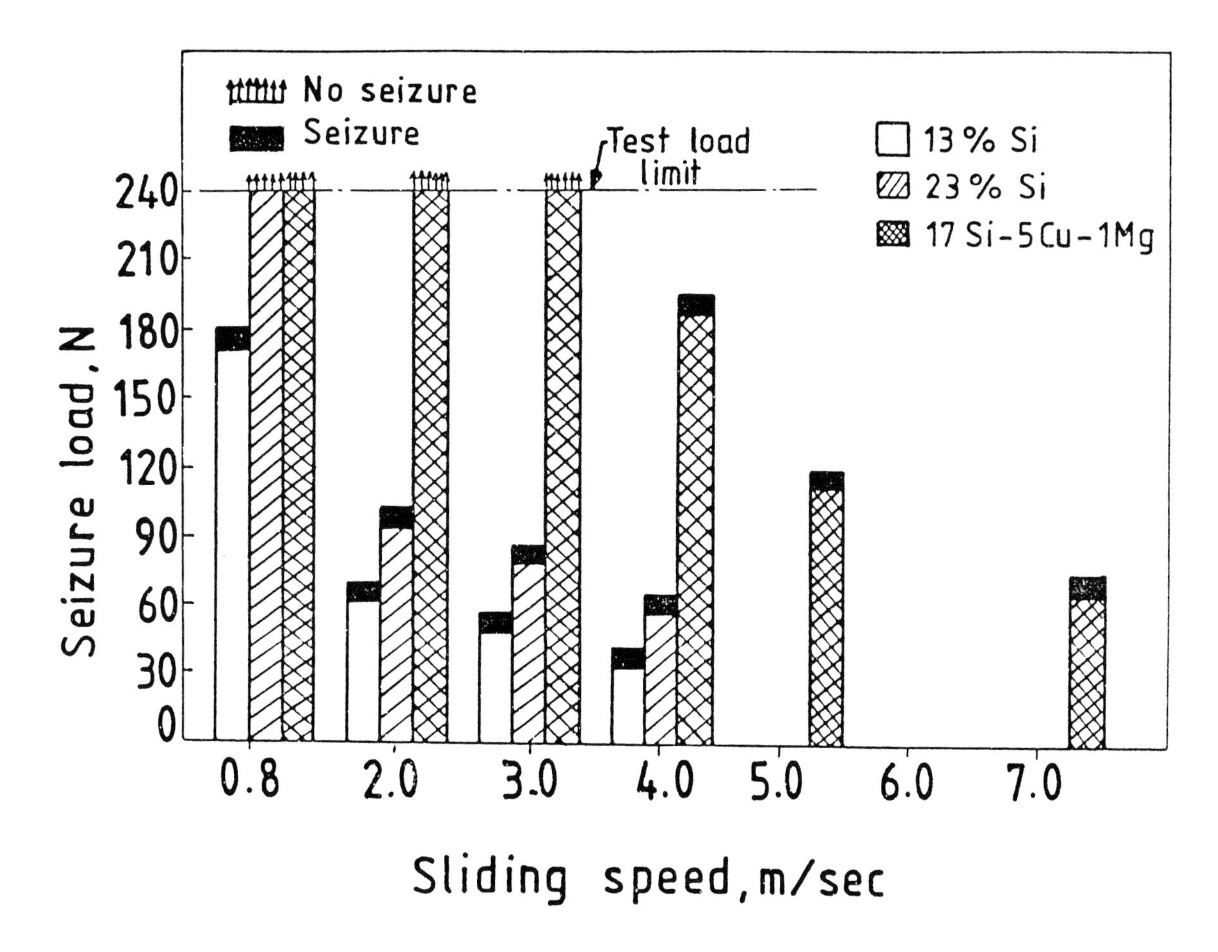

Fig. 8 Load at seizure of test alloy as a function of sliding speed.

5.4.2 Postulate

To account for the above results it is postulated that there exists a temperature for a material at which gross plastic flow criterion of shear stress/hardness (0.2-0.166) is satisfied at a subsurface depth. When this happens the material seizes and gross shear takes place at this depth in a plane parallel to the sliding plane.

5.4.3 Analysis

If s is the adhesive component of the shear stress experienced in an isothermal scribing it is possible to deconvolute this from the measured tangential force. Figure 9 shows that when s/p, where p is the hardness at a certain temperature, is plotted against temperature the seizure criterion is always satisfied for an alloy with no silicon. The temperature at which the criterion is satisfied for other alloys appears to increase with silicon content. When copper and magnesium are added to the binary alloys the criterion is not satisfied even up to 400°C.

This argument is now extended to the sliding wear tests and an s/p ratio is computed. If s is assumed to remain constant with temperature, Fig. 10 shows that the seizure criterion for 0% Si alloy is satisfied at the lowest temperature. The seizure temperature increases with the silicon content, so much so that the 23% Si alloy does not seize even at 400°C.

This model has the following implications which appear to agree with the observation on seizure made earlier. Increasing the silicon content pushes the seizure temperature up. This means that the addition of silicon requires an increase in load or speed or both to yield the higher subsurface temperature required to bring the new material to the point of seizure. The argument may be extended to the addition of alloying elements.

Figure 7 shows that the seizure load decreases as the operating mode is changed from unidirectional to reciprocating and decreases further when the reciprocating speed is increased. When alloying additions are made the friction coefficient decreases (Fig. 11) from that of the base alloy but the strength increases (**75**) considerably. The addition of alloying elements would thus reduce the s/p ratio.

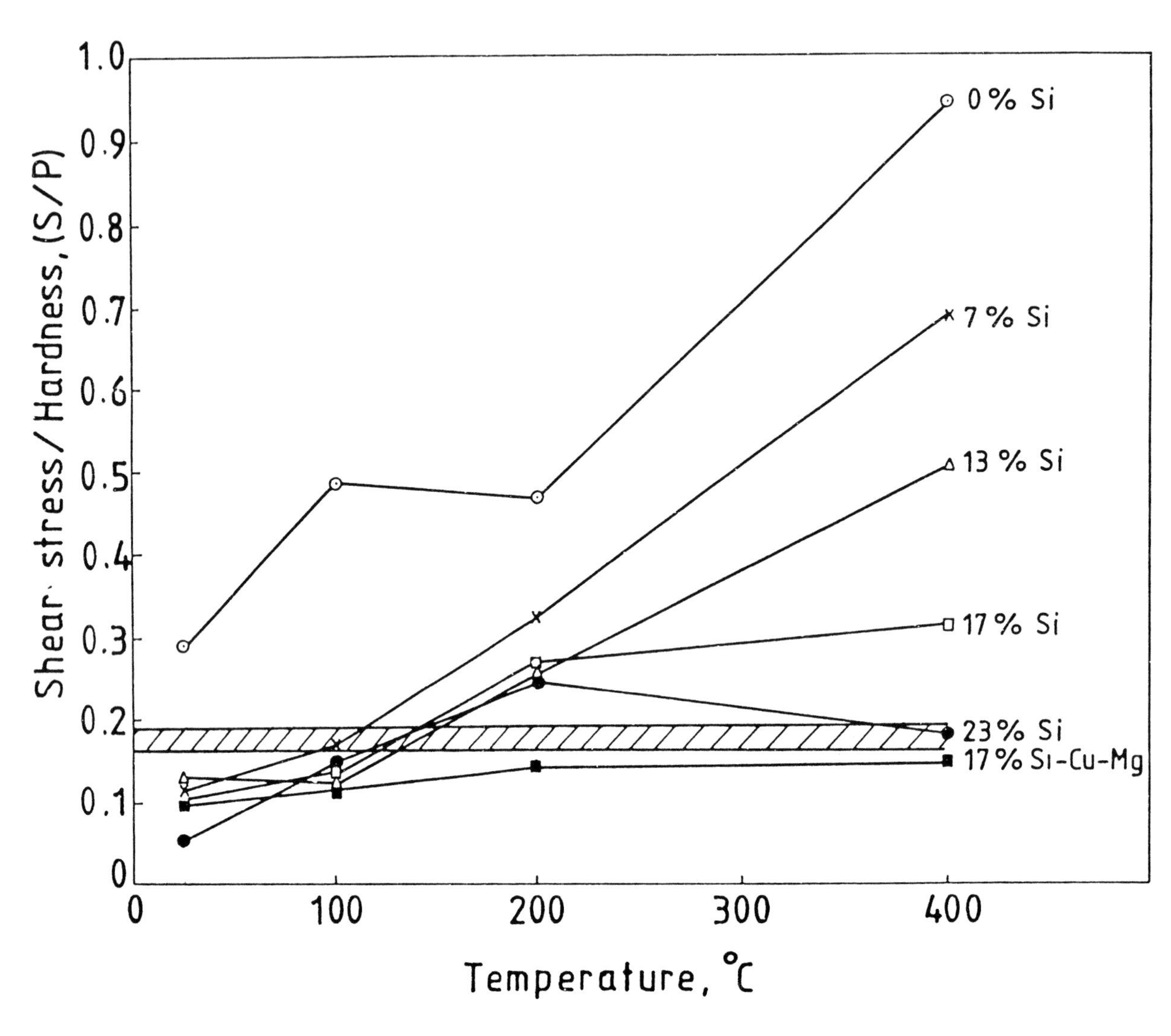

Fig. 9 Shear stress-hardness (*s*-*p*) ratio of test alloys as a function of temperature as obtained using scribing test data, sliding speed, 3×10^{-5} ms^{-1}.

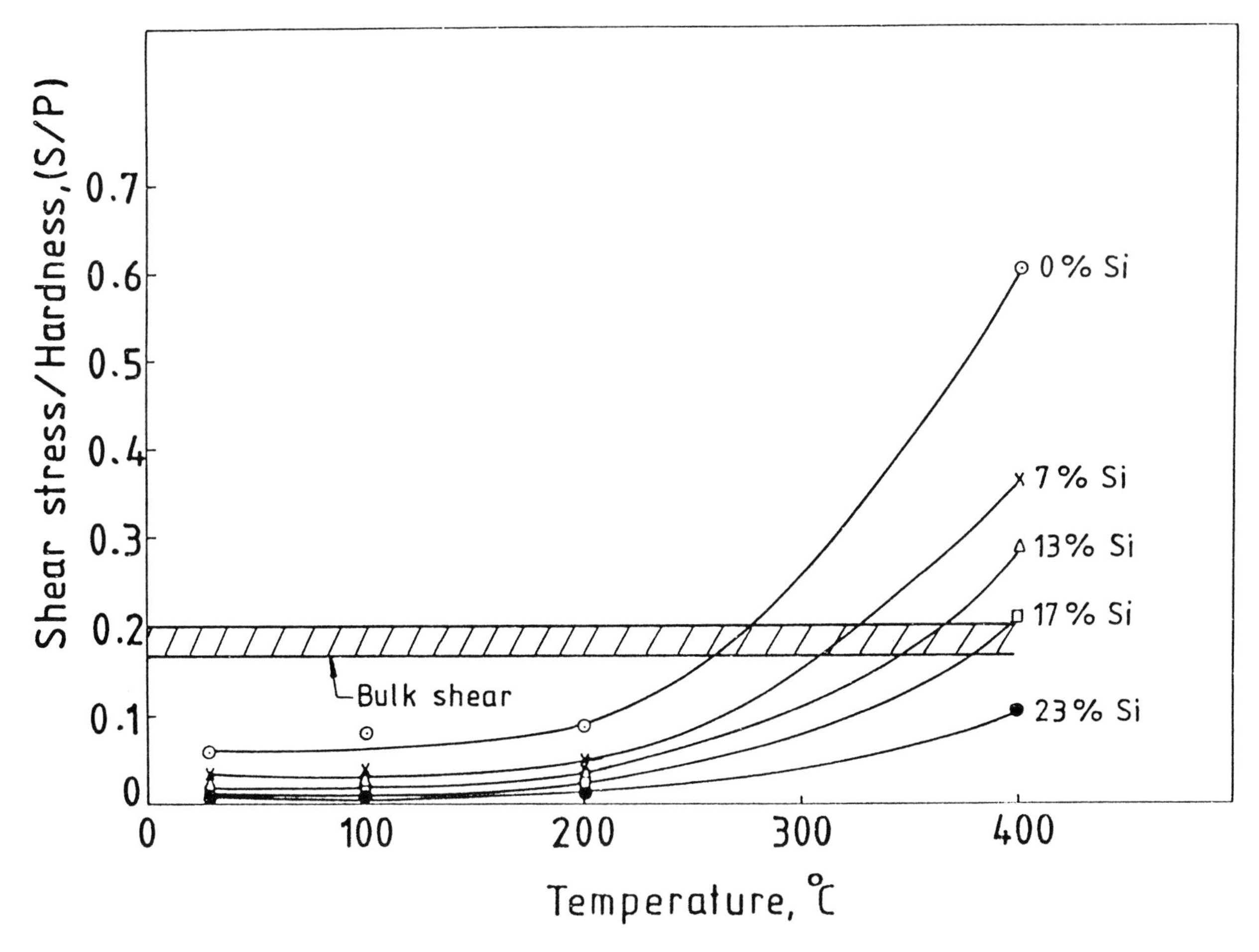

Fig. 10 Shear stress-hardness (s-p) ratio of different test alloys as a function of temperature in unidirectional sliding, sliding speed, 0.8 ms^{-1}.

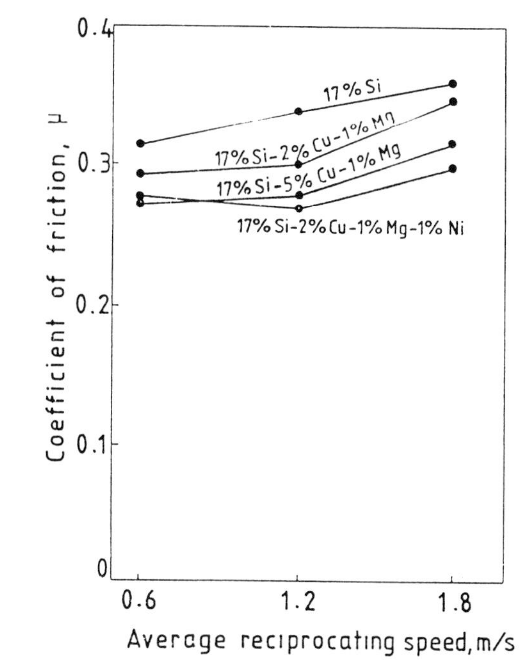

Fig. 11 Coefficient of friction of Al-17Si alloys in the mild wear regime of reciprocating sliding.

9 REFERENCES

1 **Bowden, F.P.** and **Tabor, D.**, *The friction and lubrication of solids Part I,* 1950,Oxford Press, Cambridge
2 **Rabinowitz, E.**, *Friction and wear of materials*, 1965, John Wiley, NewYork.
3 **Goddard, J.** and **Wilman, H.**, *Wear,* 1965, **5**, 115.
4 **Lancaster, J.K.**, Wear, 1990, **141,** 159.
5 **Rigney, D.A.**, *Wear,* 1994, 175, 63.
6 **Suh, N.P, Sin, H.L.**, and **Saka, N.**, *Fundamentals of Tribology,* 1981, MIT Press, 493.
7 **Lim, S.C.** and **Ashby, M.F.**, *Acta Metall,* 1987, **35** (1), 1.
8 **Quinn, T.F.J.**, *Fundamentals of Tribology*, 1981, MIT Press, 477.
9 **Kawakubo, Y.** and **Yahisa, Y.**, *J. of Tribology*, 1995, **117,** 297.
10 **Koka, R.**, *Trib. Trans*, 1995, **38** (2), 410.
11 **Shueh, S.B.**, 1995, *Trib Trans*, **38** (4), 863.
12 **Ando, Y.**, **Ishikawa, Y.**, and **Kitahara, T.**, *J. of Tribology,* 1995, **117,** 569.
13 **Koka, R.**, Trib. Trans., 1995, **38**(2), 417.
14 **Bhushan, B.** and **Joinkar, V.N.**, *Trib. Trans.,* 1995, **38**(1), 119.
15 **Gerber, C.T**, **Bhushan, B.**, and **Gitis, N.V.**, *J. of Tribology,* 1996, **118**, 12.
16 **Yang, M.** and **Talke, F.E.**, *Trib Trans.,* 1996, **39**(3), 615.
17 **Yang, M.** and **Talke, F.E.**, *Trib Trans.*, 1996, **39**(3), 691.
18 **Patton, S.T.** and **Bhushan, B.**, *J. of Trib.,* 1996, **118,** 21.
19 **Bhushan, B.**, **Israelachvili**, and **Landman, U.**, *Nature,* 1995, 374, 607.
20 **Wei, B.** and **Kpmvopoulos, J.**, *J. of Trib.* 1995, **117,** 594.
21 **Xu, J.** and **Kato, K.**, *Trib. Trans.,* 1995, **39**(3), 621.
22 **Zhaoguo, J**, **Lu, C.J**, **Bogy, O.B.** and **Miyamoto, T.**, *J. of Tribology,* 1995, **117,** 328.
23 **Biswas, S.K**, **Jarvis, S.P.** and **Pethica, J.B.**, *App. Phy. Letter,* Communicated.
24 **Pethica, J.B.** and **Sutton, A.P.**, H.J. Guntherodtetal (eds), *Forces in Scanning Probe Methods,* 1995, Khuwer Academic Pub., 353.
25 **Liu, Y.C.**, **Schissler, J.M.**, and **Mathia, T.G.**, *Trib. Int.,* 1995, **28**(7), 433.
26 **Ravikiran, A**, **Nagarajan, V.S**, **Biswas, S.K.**, and **Pramila Bai, B.N**, 1995, *J.Am.Ceram. Soc,* **78**(2), 356.
27 **Ravikiran, A.**, 1994, *PhD Thesis*, Indian Institute of Science, Bangalore, India.
28 **Dong, X**, **Jahanmir, S.**, and **Ives, L.K.**, Trib. Int, 1995, **28**(8), 559.
29 **Razavizadeh, K.** and **Eyre, T.S.**, *Wear,* 1982, **79,** 325.
30 **Razavizadeh, K.** and **Eyre, T.S.**, *Wear,* 1983, **87**, 261.
31 **Pramila Bai, B.N.** and **Biswas, S.K.**, *J.Mater.Sci.*, 1984, **19,** 3588.
32 **Pramila Bai, B.N.** and **Biswas, S.K.**, *ASLE Trans.,* 1986, **29,** 116.
33 **Pramila Bai, B.N.** and **Biswas, S.K.**, *Lubric. Engng,* 1987, **43,** 57.
34 **Pramila Bai, B.N.** and **Biswas, S.K.**, *Wear,* 1987,, **120**, 61.
35 **Pramila Bai, B.N.** and **Biswas, S.K.**, *Acta Metall.*, 1991, **39,** 833.
36 **Somi Reddy, A**, **Pramila Bai, B.N**, **Murthy, K.S.S.**, and **Biswas, S.K.**, *Wear,* 1994, **171,** 115.
37 **Yen, B.K.** and **Ishihara, T.**, *Wear,* 1996, **198**, 169.
38 **Antoniou, R.** and **Borland, B.W.**, *Mat.Sc.Eng.,* 1987, **93,** 57.
39 **Rigney, D.A.**, *Ann. Mater. Sci,* 1988, **18**, 141.
40 **Harun, M**, **Talib, I.A.**, and **Daud, A.R.**, *Wear,* 1996, **194,** 54.
41 **Johnson, K.L.**, *Contact Mechanics*, 1985, Cambridge University Press, Cambridge.
42 **Evans, A.G.** and **Marshall, D.B.**, *Fundamentals of friction and wear of materials,* ed by D.A. Rigney, ASM, Metals Park, Ohio, 1981, 439.
43 **Brezhak, J.**, **Breval, E.**, and **Macmillan, N.H.**, J. Mater. Sci, 1985, 4657.
44 **Wang, Y.** and **Hsu, S.M.**, *Wear,* 1996, **195**, 112.
45 **Suh, N.P.**, *Tribophysics,* 1986, Prentice Hall.
46 **Kapoor, A**, **Williams, J.A.**, and **Johnson, K.L.**, *Wear,* 1994, **175,** 81.
47 **Xie, Y.** and **Williams, J.A.**, Wear, 1996, **196,** 21.
48 **Somi Reddy, A**, **Murthy, K.S.S.**, and **Biswas, S.K.**, *Proc. I. Mech. E., J.Eng. Tribol,* 1995, **29,** 287.
49 **Moore, M.A.** and **Douth Waite, R.M.**, *Metallurgical Transactions*, 1976, **A7**, 1833.
50 **Zum Gahr, K.H.**, *Microstructure and wear of materials, Tribology Series Vol.10,* 1987, Elsevier, Amsterdam.
51 **Challen, J.M.** and **Oxley, P.L.B**, *Wear,* 1979, **53**, 229.
52 **Challen, J.M. Mclean, L.J.** and **Oxley, P.L.B.**, *Proc. R. Soc. London,* **A394**, 161.
53 **Challen, J.M.**, **Koplanisky, E.M.**, and **Oxley, P.L.B.**, *1987, Proc. I Mech E.*, **C 156**, 958.
54 **Komvopoulos, K.**, **Saka, N.**, and **Suh, N.P.**, 1986, *J. of Tribology,* **108**, 301.
55 **Petryk, H.**, Proc. Inst. Of Mech Eng. 1987, *Int. Conf. on Tribology - Friction, Lubrication and Wear, Fifty years on –* **II**, 987.
56 **Torrance, A.A.** and **Buckley, T.R.**, *Wear,* 1996, **196,** 35.
57 **Koplanisky, E.M.** and **Oxley, P.L.B.**, *J. of Tribology*, 1995, **117**, 315.
58 **Williams, T.A.** and **Xie, Y.**, *J. of Phys. D., Applied Physics*, 1992, **25**(1A), 158.
59 **Kailas, S.V.** and **Biswas, S.K.**, *Wear,* 1993, **162**, 110.
60 **Kailas, S.V.** and **Biswas, S.K.**, *Wear,* 1995, **181**, 648.
61 **Wang, Y.** and **Lei, T.**, *Wear,* 1996, **194,** 44.
62 **Vardavoulias, M.**, **Jeandin, M.**, **Velasco, F.**, and **Torralba, J.M.**, *Trib. Int,* 1996, **29(6)**, 499.
63 **Davis, F.A.** and **Eyre, T.S**, *Trib. Int.,* 1994, **27(3)**, 171.
64 **Oliver, A.V.**, **Spikes, H.A.**, **Bower, A.F** and **Johnson, K.L.**, *Wear,* 1986, **107**, 151.
65 **Ho, H.W**, **Noyan, C.**, **Cohan, J.B.**, **Khanna, V.D.**, and **Eliezer, Z.**, *Wear,* 1983, **84,** 183.
66 **Divakar, R.**, 1987, *PhD Thesis*, Ohio State University, 148.
67 **Kuo, S.M.**, 1987, *PhD Thesis,* Ohio State University, 183.
68 **Ives, L.K.**, *Proc. Int.conf. Wear of Materials,* 1979, ASME, New York, 246.
69 **Don, J.**, 1987, **PhD Thesis**, Ohio State University, 248.
70 **Dupont, F.** and **Finnnie, I.**, *Wear,* 1990, **140,** 93.
71 **Dautzenberg, J.H.**, *Wear,* 1980, **60,** 401.
72 **Alpas, A.T.**, **Hu, H.**, and **Zhang, J.**, ,*Wear,* 1993, **162**, 188.
73 **Rosenfield, A.R.**, *Wear,* 1987, **116,** 319.
74 **Rosenfield, A.R.**, *Wear,* 1981, **72,** 97.
75 **Somi Reddy, A.**, **Murthy, K.S.S.**, and **Biswas, S.K.**, *Wear,* 1995, **181**, 658.
76 **Semiatin, L.**, *Formability and workability of metals; plastic instability and flow localisation* ASM, 1984, Metals Park, Ohio.
77 **Rishi Raj**, *Met. Trans.,* 1981, **12A**, 1089.

78 **Gegel, H.L., Malas, J.C., Doraivelu, S.M.,** and **Shande, V.A.,** *Metals Hand Book*, 1988, **14,** 417.
79 **Prasad, Y.V.R.K.,** *Indian J. of Technology*, 1990, **28**, 435.
80 **Zhang, B., Liu, Y., Wang, Y., Tang, X.,** and **Wang, X.**, *Wear,* 1996, **198,** 287.
81 **Kailas, S.V.** and **Biswas, S.K.,** *J. of Tribology*, 1997, **119**, 1, 31.
82 **Lee, C.** and **Chen, H.,** , *Trib. Trans.,* 1995, **38(3),** 738.
83 **Park, K.B.** and **Ludema, K.C.,** *Wear,* 1994, **175**, 123.
84 **Horng, J.H., Lin, J.F.,** and **Li, K.Y.,** *J. of Tribology*, 1996, **118,** 669.
85 **Rapoport, L.,** *J. of Tribology,* 1996, **118,** 527.
86 **Dyson, A.,** *Tribology International*, 1975, **8**, 77, 117.

Design and validation of laboratory-scale simulations for selecting tribomaterials and surface treatments

P J BLAU
Metals and Ceramics Division, Oak Ridge National Laboratory, USA

SYNOPSIS

Engineering approaches to solving tribology problems commonly involve friction, lubrication, or wear testing, either in the field or in a laboratory setting. Since wear and friction are properties of the materials in the larger context of the tribosystem, the selection of appropriate laboratory tribotesting procedures becomes critically important. Laboratory simulations must exhibit certain key characteristics of the application in order for the test results to be relevant, but they may not have to mimic all operating conditions. The current paper illustrates a step-by-step method to develop laboratory-scale friction and wear simulations based on a tribosystem analysis. Quantitative or qualitative metrics are established and used to validate the effectiveness of the tribosimulation. Sometimes standardized test methods can be used, but frequently a new type of test method or procedure must be developed. There are four factors to be addressed in designing effective simulations: (1) contact macrogeometry and the characteristics of relative motion, (2) pressure – velocity relationships, (3) thermal and chemical environment (including type of lubrication), and (4) the role of third-bodies. In addition, there are two typical choices of testing philosophy: (1) the worst-case scenario and, (2) the nominal-operations scenario. Examples of the development and use of simulative friction and wear tests are used to illustrate major points.

1 INTRODUCTION

Laboratory-scale friction and wear tests often play an important role in tribomaterials and lubricant selection. Since wear and friction are characteristics of the materials in the larger context of the tribosystem, the selection of the proper testing methodology is vitally important when testing is used to screen materials for specific applications. Laboratory-scale simulations must therefore exhibit key characteristics of the intended application in order to be useful. It is possible, and even probable, that an application which must meet several performance specifications will require more than one type of screening test. Thus, a 'balance of properties', which could include not only tribological performance but other factors, such as fatigue and corrosion resistance, may be required in order to complete the material evaluation.

History fails to record when engineers first began using laboratory simulations to solve friction and wear problems, although it is probable that laboratory-scale friction experiments predated those of wear (**1**). It is reasonable to assume, however, that the first experiments in tribomaterials selection were performed under field conditions, often with disastrous results, both in terms of the machine's response and even perhaps, the fate of the test engineer. Later, three factors – the ability to explore alternatives in private before drawing conclusions, the avoidance of the risks associated with full-scale tests, and the convenience of making measurements on a controllable, model system – began to make laboratory-scale testing appealing. As engineers withdrew into their testing laboratories they faced the challenge of connecting their results convincingly to the behavior of actual machines in the field. This central issue in material and lubricant screening still exists today.

Those charged with selecting materials or lubricants for tribological applications undoubtedly would prefer simply to use handbooks, simple formulae, tables of data, or computer databases. While a number of comprehensive tribology handbooks have been published (**2, 3, 4, 5,**) these cannot address every possible friction or wear situation, and experience shows that it is prudent not to extrapolate friction and wear data much beyond the conditions under which they were obtained. Despite the recent growth and sophistication of computer programmes for design and materials selection, physical properties data, like friction coefficients and wear rates, must still be found largely by experiment.

A principal question in tribosimulation is: with what level of exactitude must the laboratory test method duplicate the operating conditions of the machine in order to produce useful information for selecting materials and/or lubricants? Sometimes, useful design information can be obtained from relatively simple testing geometries, like the pin-on-disk or block-on-ring. Sometimes, however, such simple systems inadequately simulate the tribological behavior of the component under its normal conditions of use.

Godet, *et al.* (**6**) once observed: '(A) simulation can be compared to a black box: entry conditions are fed to it and results are transposed to the application with a minimum amount of interpretation. The difficulty resides in carefully designing the black box.' The design and validation of laboratory-scale simulations for selecting tribomaterials are therefore issues of major consequence in mechanical systems engineering. This paper discusses methods to characterize tribosystems and to develop and validate simulative tests.

2 FACTORS IN THE DEVELOPMENT OF TRIBOSIMULATIONS

The five basic steps in developing an effective tribosimulation are: (1) reaching a clear understanding of the tribosystem, (2) developing metrics by which the problem can be quantified and its solution tested, (3) selecting the necessary conditions for the simulation, (4) selecting or developing a testing procedure to reproduce those necessary conditions, and (5) validating the simulation.

2.1 Understanding the nature of the tribosystem

The first step in the design of a tribosimulation involves identifying the primary causes for concern about the tribocomponent's behaviour. Concern may have arisen during the design review of a new, yet untested machine, or by a documented series of functional problems in the field. Perhaps the problem is merely a perceived one based on surface appearance and not with a measured loss of function. Understanding the origins of concern over the problem can better focus the initial approach to its solution.

Several formal approaches have been developed to define the attributes of tribosystems (**7, 8**). These offer varying degrees of detail and are somewhat different in structure. In the present discussion, we shall consider sixteen elements of tribosystem analysis. These are listed in Table 1 and can be subdivided into five sets. Elements 1 and 2 concern the macrogeometry of the contacting surfaces. Elements 3–5 affect the manner in which forces are transmitted in the vicinity of the tribocontacts. Elements 6–9 concern the interfacial environment. Temperature includes both frictional heating effects and the temperature of the surroundings. Elements 10 and 11 are associated with the contacting materials. Element 10 includes the composition, microstructure, processing conditions, and surface finishing of the tribomaterials. Element 11 has implications for such aspects as running-in, the operable lubrication regime, and the cost of fabricating the components. Elements 12–16 better define the problem in terms of how the failure is determined and what kinds of prior observations have been made. They also address the issue of whether additional requirements on the materials, beyond tribological concerns, must be met.

Despite their separate listing in Table 1, the elements of tribosystem analysis are not independent of one another. For example, the contact geometry, size, and applied load are needed to ascertain the contact pressure. The lubrication regime and its stability are determined by several of the elements. Thus, interactions of the relevant tribosystem elements must be considered.

Table 1 The elements of tribosystem analysis

1	Contact geometry
2	Apparent area of contact
3	Type of motion: steady, reciprocating, intermittent, etc.
4	Sliding speed: magnitude, constancy, etc.
5	Loading on the contact (stress or force)
6	Temperature of the contact region
7	Atmospheric environment: relative humidity, gas composition, pressure
8	Lubrication regime, composition, and condition of lubricants
9	Third-body environment: sizes, shapes, mechanical properties of particles
10	Presently-used materials/surface treatments on the contact surfaces
11	Surface finish of contacting surfaces
12	How failure is determined
13	Other property or operating requirements
14	Type(s) of surface damage/wear observed
15	Frictional requirements: steady-state, stability, stick-slip, static
16	Failure history and related observations

Information for tribosystem analysis is gathered from a variety of sources: operating records, specification drawings, verbal reports, and interviews. It is not uncommon, however, to discover that no one really knows the precise operating conditions of the subject component(s). Depending on the type of component, even such things as contact pressure or normal force on the surface may not be accurately known. Perhaps there is little or no information on exactly how the component was used or how it was maintained. The latter is commonplace for builders of processing equipment who are faced with customer complaints about the wear of their equipment, but are unable to extract exact information from those customers on the relevant conditions of use. Certainly, it is desirable but not usually possible, to instrument and monitor operating machines or to develop detailed computer simulations. Desirable as it might be to begin with all the information contained in Table 1, one is much more likely to be faced with the challenge of developing a tribosimulation for an incompletely-defined tribosystem.

One should next identify which types of surface phenomena are of primary importance. For example, wear or surface damage of several types can be present at different areas of the same part, and some of these may not be critical to part performance. The author has had occasion to examine the wear of metal-coated extrusion machine screws and to design an appropriate tribosimulation. Such screws turn slowly (e.g., tens of revolutions per minute), forcing hot, molten plastic through the bore of the extrusion machine barrel and through an orifice at the outlet end. On a single screw, it is common to see a variety of wear types at different locations: polishing abrasion, scoring, erosion, and scuffing as well as pitting due to corrosive chemicals in the plastic. Possible causes of some of the wear damage include abrasion by mineral fillers in the plastic or by wear debris particles produced on other components within the machine which worked their way downstream (**9**). In the present case, it was necessary to select the most performance-reducing form of wear on which to base the design of a meaningful laboratory-scale simulation, and to design the simulator for maximum flexibility in order to test a variety of potentially harmful conditions.

More than one type of screening test will probably be needed to assess a material's resistance to several wear modes. For example, Czichos, *et al.* (**10**) demonstrated that four surface treatments for steel ranked oppositely in their resistance to sliding and abrasive wear. Thus, confining screening tests to only one of the operative wear types would be inadvisable and potentially disastrous.

2.2 The development of metrics

The next step is to develop one or more specific measures of friction, wear, or surface damage (i.e., 'metrics'). These will be used for judging the success of the simulation. Metrics are such quantities as wear volume, friction coefficient, and the change in diametral bearing clearance per unit of machine operating time. Sometimes it is not possible to obtain quantitative metrics. Instead, the success of the simulation could be judged by examining field components and comparing their surface features to those produced in laboratory tests. In that case, the metric is a visual inspection rather than a dimensional measurement. Other, more indirect measures include comparing the type of wear debris produced or, less commonly, comparing the subsurface microstructures which are induced by the surface contact history (**11**). Fischer (**12**) has stated that it is essential to clearly identify the acting wear processes in order to select materials for an application. In sliding contact situations, it is helpful to examine both members of the contacting pair to confirm the identification of the wear modes.

Examples of both quantitative and qualitative metrics are given in Table 2. Quantitative metrics are preferable to qualitative metrics because the degree of material suitability can be more precisely specified; however, the ability to define quantitative metrics is related to the wear mode and to the uniformity of contact. For example, galling is a severe and crippling form of surface damage, yet it is difficult to measure quantitatively (**13**). Engineers have developed standard test methods (**14**) and developed a 'critical load for galling' criterion. This is an interesting metric because it combines quantitative measurements (critical load) and qualitative observations (whether or not galling has occurred at a given load). Similarly, scuffing is of great concern to manufacturers of internal combustion engines, yet it is difficult to measure since it tends to be localized and non-uniform. A standardized set of photographs or photomicrographs is an alternative for qualitatively assessing or ranking the severity of wear.

The units which are selected to measure wear or surface damage should be meaningful in the context of the problem at hand. For example, it might be possible to obtain a wear rate for mechanical face seal materials in units of mm^3/N-m; however, that parameter may be less meaningful to the manufacturer or user of the component than the leak rate in liters per hour. Many units for wear have been used (**15**) and each has its proper place.

Studies of the wear of railroad rails provide an example of the selection of the appropriate metrics. Data on the wear of railroad rails have been collected for many years, indicating that curved track sections experience the most severe wear. Clayton (**16**) concluded that fundamental models for wear were not useful for solving the problem, and that laboratory testing was needed. The three conditions to be fulfilled before the laboratory data would be used to predict rail wear behavior were: (1) the wear mechanisms of the service components had to be identified as well as possible, (2) the laboratory test had to generate surface damage representative of the field components, and (3) the relative performance of the materials had to agree in laboratory and field results. Clayton stated: 'Laboratory testing with different test machines over the last fifty years shows that even small changes in the operating environment can alter the wear mechanism, often with large adjustments in wear rate.' Data obtained from an Amsler machine (roller-on-roller), designed in the late 1920s, agreed best with full-scale test track results (see Table 3). Three different wear regimes were identified: I. was a combination of oxidative and metallic flake formation, II was mainly metallic wear, and III involved a break-in period in which large debris fragments became embedded and caused abrasion of the opposing surfaces. Clayton's approach clearly exemplifies the process of identifying the primary cause(s) of concern, characterizing the material response, and developing useful metrics.

Table 2 Examples of metrics for use in tribosimulation.

Application	Quantitative Metrics	Qualitative Metrics
wear of a hard coating on a contact surface	wear rate of the coating; critical pressure or time for wear-through; percent of original coating remaining after specific exposure time; change in friction force; time until the acoustic emission increases	presence of wear-through; transfer of material to the counterface; type of microfractures in the coating; evidence for delamination
sliding wear of a lubricated journal bearing	wear rate; change in clearance; temperature rise in the bearing; concentration of debris in the lubricant	presence or absence of 'mild' or 'severe' wear; appearance of wear debris in the lubricant; noise or vibration
rolling element bearing	number of cycles to failure; critical load to failure (step test); maximum DN number	excessive noise or vibration; presence of pits or spalls; comparison to photographs of various types of surface damage severity
solid lubricant coating on a sliding surface	average (and variation) of friction force versus time; remaining below a critical value of friction coefficient; time to loss of function; maximum temperature for acceptable life or friction coefficient	appearance of the coated surface; tendency to transfer to the counterface; manner of loss during wear
polymer bushing material	wear rate; friction coefficient; PV-limit; maximum operating temperature; surface roughness after exposure	noise; surface appearance and presence or absence of deformation

Table 3 Comparison of laboratory wear rankings with test track results for four steels* (Laboratory tests: 1220 MPa contact pressure, 35% slide/roll ratio, no lubricant)

Steel	Relative Wear Resistance (Laboratory)	Relative Wear Resistance (Track)
CrMo I	2.02	2.13
CrMo II	1.78	2.06
MnSiCrV	1.31	1.85
standard carbon steel (reference)	1.00	1.00

*After P. Clayton (1995).

The foregoing example described a comprehensive study involving extensive data collection over a period of years. Such a data-rich situation is probably more the exception than the rule. In fact, many wear problems are identified with a relatively small amount of anecdotal or incomplete information. Some of the information might be first-hand, through direct observations of wear or friction behaviour, but in other cases, awareness of a tribological problem may come from second-hand sources such as a customer complaint or the report of a wear failure by people who are not trained in tribology or failure analysis. Therefore, a method like tribosystem analysis is needed to better define the problem before metrics can be established.

2.3 Selection of the test methodology

The fact that friction and wear are tribosystem properties, rather than exclusively materials properties, makes the selection of testing methodology important. When friction researchers test different material combinations on the same tribometer, and using that data alone, develop models to predict frictional behavior, they implicitly ignore that point. Figure 1 shows that the friction coefficients for AISI 52100 steel sliding on Ni_3Al alloy varied by a factor of nearly eight on three different testing machines, and that even within the same testing machine, the friction coefficient was load-dependent (**17**). These differences were due to the changing relative role of ambient surface films, third bodies, and the ploughing component of friction. Varying the test machine stiffness characteristics while keeping the materials pairing constant has been used by mechanical engineers to evaluate such things as the effects of system stiffness and vibrations on friction and wear (**18**).

DIN standard 50 322 (**19**), describes various levels of simulation, ranging from field trials to simple laboratory tests. Likewise, the test methods used for the simulation can be systematically characterized. For example, ASTM has produced a standard guide for the presentation of sliding wear data (**20**). Seven categories of entries comprise that guide. They are test identification, test type, test conditions, material definitions, specimen identifications, test results, and documentation. Despite its detail, it does not address all the issues of tribosystem analysis.

As the test method approaches the real operating conditions more closely, running the tests usually becomes more complicated and expensive, and the results may become harder to interpret. However, even when one uses simple tests, there may be difficulties in selecting the appropriate one. For example, Budinski (**21**) compared four different methods for measuring the sliding friction of polyester films against stainless steel in order to obtain information for their manufacture and handling. The first method was a sled test in which a small flat steel specimen was pulled on the film. The second was a pin-on-disk test using the film as the disk material. The third was a capstan test in which the films was slid over a steel cylinder

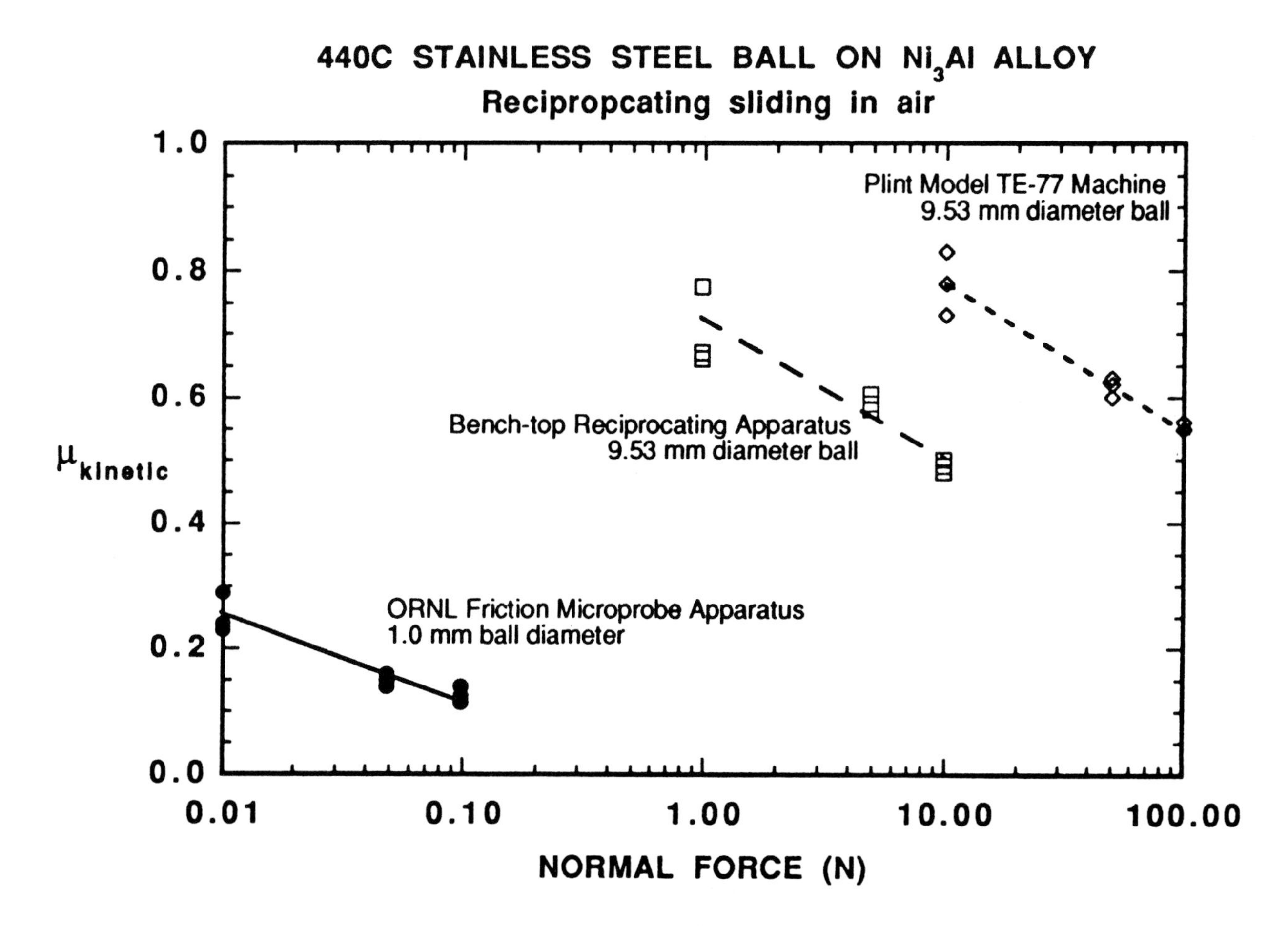

Fig. 1 The influence of the tribosystem is apparent when measuring the friction coefficient of the same two materials in three different testing machines.

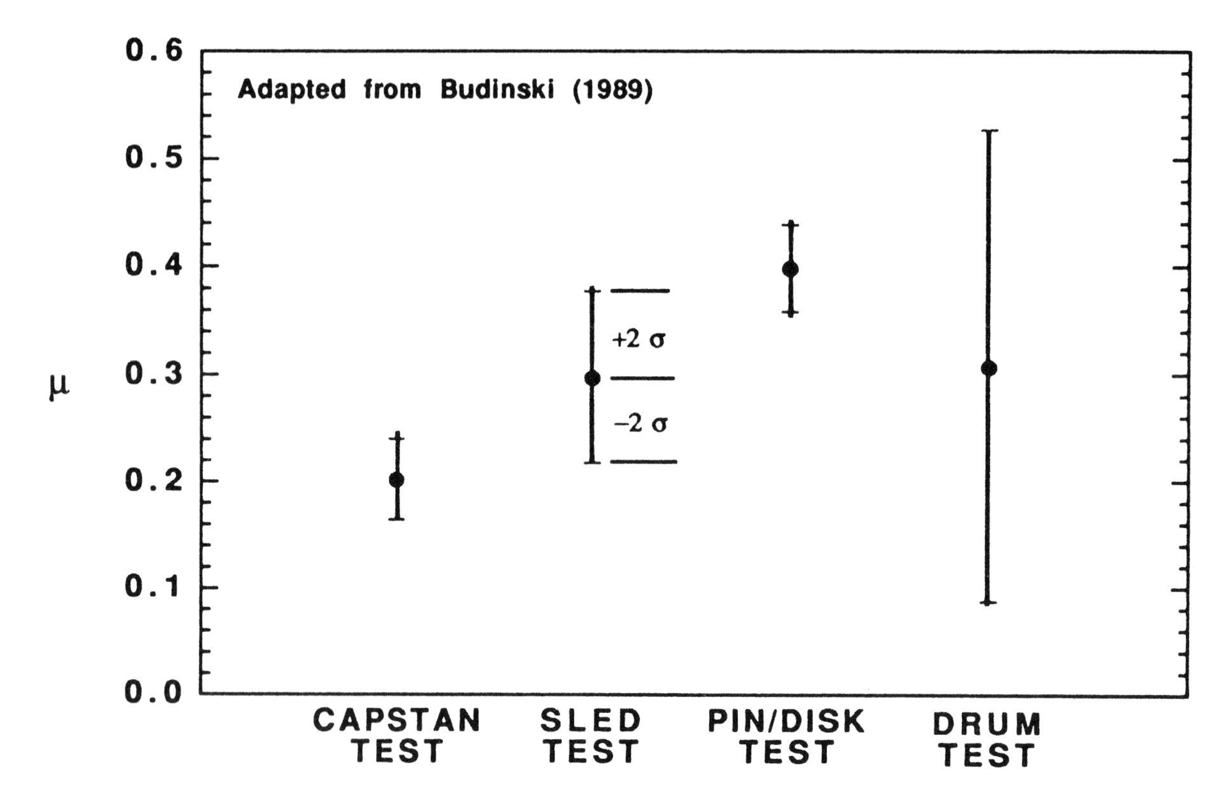

Fig. 2 The range and average values for the sliding friction coefficient of polyester films against stainless steel varied significantly among four testing machines.

(capstan). The fourth was a flat-on-drum test in which a specimen of the film was pressed tangentially against a rotating stainless steel drum. Friction force traces were recorded for each testing system and exhibited differing degrees of fluctuations and average friction coefficients. Fig. 2 summarizes Budinski's results, showing the average and ± 2 standard deviations from the average value for each test method. The pin-on-disk test gave the least variability in results, but it also damaged the films during the test, making it unsuitable. The drum-type test also tended to damage the films. Budinski found that the various tests may each be useful under certain circumstances of contact pressure and sliding speed, but no one test was completely satisfactory in all respects.

A first-order simulation is one used primarily for preliminary material or lubricant screening. It should contain only the most basic elements of simulation, yet provide a simple and straightforward means to obtain quantitative, meaningful data. Sometimes, a first-order simulation is sufficient to solve the problem. Other times, more exhaustive trials of the leading materials or lubricants in closer-to-actual conditions are needed. The following four factors should be taken into account when designing a successful first-order laboratory simulation:

1. Contact macrogeometry and the characteristics of relative motion.
2. The pressure-velocity relationships
3. Thermal and chemical factors (including type of lubrication)
4. The participation of third-bodies

Inattention to any of these factors could seriously reduce the effectiveness of the simulation. Factors that go beyond those of first-order simulations include matching the contact vibrational characteristics, contact stiffness, heat flow paths, and the thermal-mechanical interactions with the surroundings that occur under field conditions.

In sliding contacts, test geometry selection should consider whether the contact is conformal, such as a brake shoe on a drum, or non-conformal, such as a sphere on a flat surface. Contact conformity affects the distribution of lubricant in the interface, heat flow, and the movements of third-bodies. The speed constancy and reversibility of direction of motion has an effect on at least three related tribosystem elements: the regime of lubrication in the contact, the texturing of the surface and subsurface of the material, and the influence of third-bodies. Contact geometry and motion characteristics also affect heat retention and dissipation in the vicinity of the contact. It should not be surprising, therefore, to observe differences between unidirectional and reciprocating sliding friction and wear even for the same material couple and macrogeometry (e.g., see Table 4). In other forms of wear, such as erosive wear, relative motion considerations involve such things as the angle of impingement, the

Table 4 Effects of sliding motion on the friction and wear silicon nitride (Type NBD 100 silicon nitride ball and polished flat, 0.925 mm diameter, 10 N normal force, average speed 0.2 m/s, average data for 2 tests per condition).

	Unidirectional	Reciprocating
Steady-state kinetic friction coefficient	0.72	0.87
Ball wear volume, mm^3 x 10^{-2}	1.07	5.04
Flat specimen wear volume, mm^3x 10^{-2}	not measurable	2.71

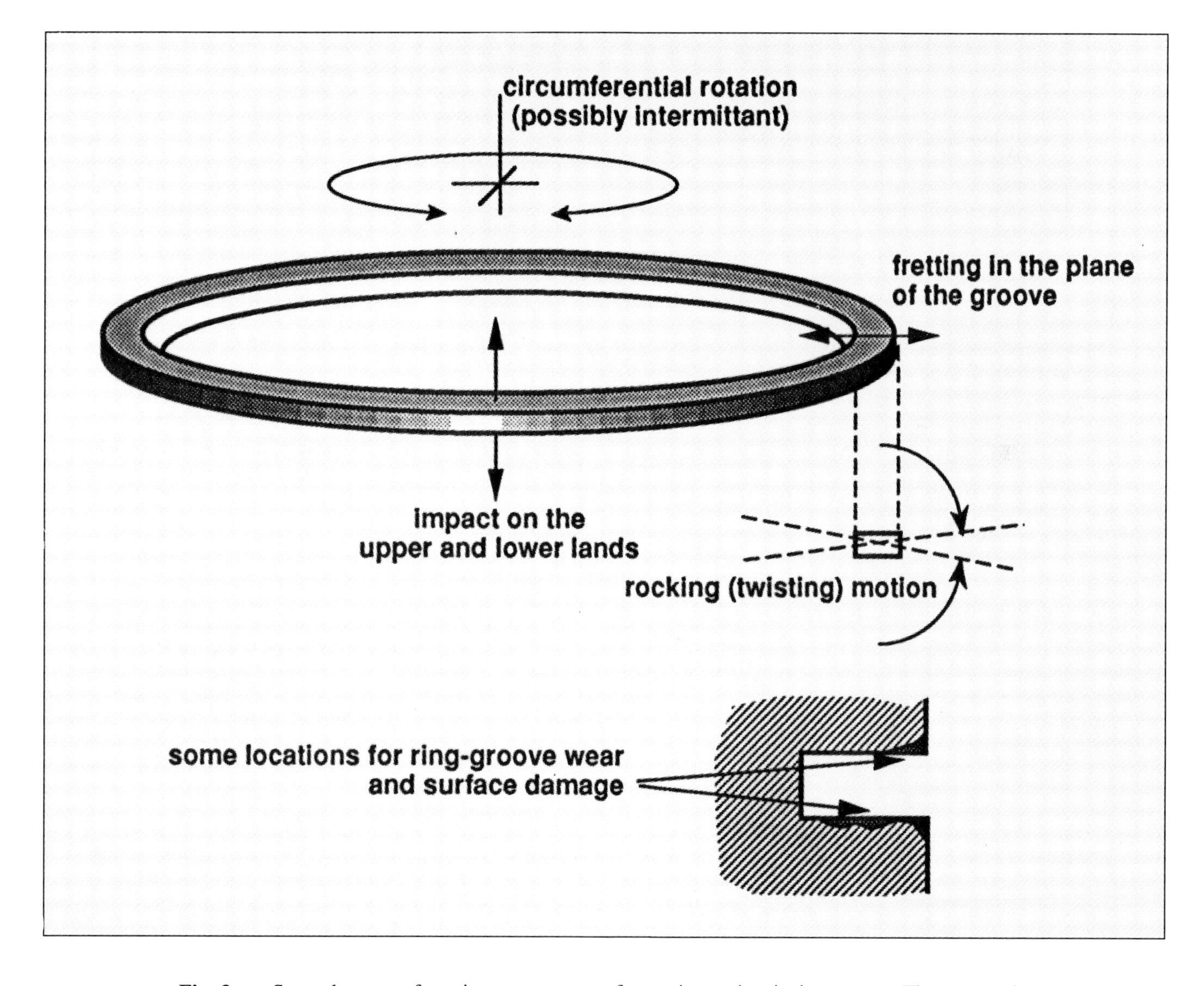

Fig. 3 Several types of motions are present for a piston ring in its groove. The appropriate motion needed for the simulation is determined by examining piston cross-sections and the contact surfaces of used rings and ring grooves.

particle shape and rebound characteristics, particle flux screening, and the particle velocity distribution.

Simulators have been commonly used for friction and wear tests of internal combustion engine parts (**22-24**). Piston ring-groove wear is one example in which the type of motion is complex and affects the localized wear processes. Wear of the uppermost piston ring groove in internal combustion engines can result in excessive crevice volumes which trap and expel unburned fuel, thereby increasing engine hydrocarbon emissions. The motions of the piston ring in its groove are complex (**25, 26**). There is radial fretting in-and-out of the groove, impact on the top and bottom groove surfaces, tilting (rocking), and circumferential ring rotation (Fig. 3). The question becomes one of deciding which motions or combinations thereof are responsible for the most insidious forms of ring groove wear, and how these motions can best be incorporated into a tribosimulation. Bialo, *et al.* (**27**) used low-amplitude sliding contact in kerosene at room temperature in their studies. Wacker (**28**), on the other hand took a more comprehensive approach, trying to simulate as many aspects of the ring motion as possible. Different engines behave differently in terms of the prevalence of such features as ring-groove bell-mouthing or lip-recession, etc., so that the type of tribosimulation should be tailored to replicating the wear mode which occurs within the specific engine of interest. Whether it is possible to reduce the complexity of the motion and yet still produce the proper form of wear must be determined in preliminary, 'scoping' trials with the proposed test apparatus.

Contact pressure and sliding velocity are two of the most critical aspects of simulation design. They are treated together because their affects on wear are often interrelated. If a given pressure p is applied to a contact which is undergoing relative motion with velocity v, then the energy input rate to the surface, and the temperature rise of the contact, are proportional to pv. Energy inputs and outputs to tribosystems were discussed systematically by Czichos (**29**). Frictional energy can produce heat, sound, or fracture with the subsequent creation of new surfaces (debris). Thus, it makes sense to consider matching the contact pressure, the relative velocity, and consequently the pv product, of the application. In fact, the pv product is often specified when selecting polymeric bearing materials (**30**), and engineering design diagrams which plot the locus of the pv limit across various pressure and velocity ranges have been developed for material selection.

The temperature and surrounding environment of the tribological contact are sometimes important elements in achieving successful simulations. For example, whether the temperature of the contacting surfaces is above or below the boiling point of water can affect the adsorption of moisture, a key factor in both friction and wear of many materials, including metals and ceramics. In addition to its effects on water and other adsorbates, temperature can affect the shear strength and tribochemistry of the materials at the contact surface.

Wear and corrosion have been shown to have synergistic effects. In fact, ASTM test methods have been developed specifically to address this point (**31**). Grobler and Mostert (**32**), for example, conducted abrasive wear tests with corrosive solutions to help screen and develop new alloys for the mining industry.

The type and characteristics of third-bodies can have an important influence on the success of a simulation. For example, the use of carefully-characterized used engine oil has been shown to produce a better simulation of the wear of candidate ring and liner materials than the use of fresh oil. In fact, many practical lubricated wear problems are the result of the contamination of oils and other fluids by third bodies. For example, studies of the abrasive particle contamination of engine oils of varying viscosities showed that the abrasive wear rates of piston rings depended on the size of the abrasive, reaching a maximum at 20 μm within the particle size range of 5–65 μm (**33**).

The characteristics of third bodies are obviously important when developing abrasive and erosive wear simulations. Not only size, but the shapes of particles can affect abrasive wear rate, as demonstrated by Swanson (**34**) in three-body abrasion tests of steels with both sharp and rounded silica grains. Mechanical properties of particles are another aspect of third-body effects. A recent paper by Shipway and Hutchings (**35**) clearly established the importance of the relative hardness of the erodent, with respect to the target material, on erosion rate.

2.4 Selecting or developing a testing system

After analyzing the tribosystem, developing a set of metrics, and isolating the testing conditions, the testing machine is selected. Voitic (**36**) has developed one approach for matching the application characteristics to those of the testing machine. The Tribological Aspect Number (TAN Code) consists of four elements, each of which has several choices: (1) velocity characteristics (4 choices), (2) contact area characteristics (8 choices), (3) contact pressure characteristics (3 choices), and (4) entry angle characteristics (10 choices). While these help establish the mechanical characteristics, they do not explicitly address the thermal and chemical aspects of the simulation.

Ideally, an existing testing machine or established method can be used, but the specific details of the problem at hand may require modifications to an existing method or that a new machine or method be developed. Perhaps, an existing system can be modified to encompass all key elements of the simulation.

Under the auspices of the Versailles Project on Advanced Materials and Standards, a compilation of international friction and wear testing standards was prepared (**37**). It contains a listing and index to hundreds of friction, lubrication, and wear test methods from seven participating countries. ASTM alone had 115 standards, practices, and guidelines listed. The advantages of using published standards wherever possible are that the repeatability and reproducibility of the methodology has generally been verified by extensive testing, and data are available with which to compare your results. A few of the published standards were developed on the basis of extensive international interlaboratory testing programs, providing a degree of confidence in the precision and accuracy of the data which would be nearly impossible to obtain at a single laboratory (**38**). Some ASTM published standards involve simple geometries, such as the ASTM G-77 block-on-ring test or the ASTM G-99 pin-on-disk test. Others are designed for specific applications such as ASTM D-821 'Methods for Evaluating Degree of Abrasion, Erosion, or a Combination of Both in Road Service Tests of Traffic Paint.'

U.S. Military Specifications also contain standard procedures for wear testing with very specific metrics. For example, MIL-R-46762, as described by Wood and Taylor (**39**), specifies wear criteria for road tires as follows: 'The average height of the rubber chevrons or pads at the end of the road test shall not be less than 50 percent of their original height. Not more than 30 percent of the number of shoes or pads shall lose more than 3 square inches of rubber bearing area, and the original total bearing area of the rubber chevrons or pads shall be reduced by not more than 7 percent.' Standard test methods can also be modified to more closely approach the actual component conditions. For example, Perez, *et al.* (**40**) used a modified four-ball wear test method to obtain component-correlated data on the wear of steels in various hydraulic fluids for pumps. If used frequently and successfully, such modified procedures become *de facto* standards themselves.

If no standard test method or suitable testing machine can be found, one confronts the task of designing and constructing a new wear or friction simulator. The disadvantages of this approach are (1) there are no pre-existing data with which to directly compare results, and (2) any new testing machine needs to be characterized to determine the peculiarities of its behavior and the repeatability of its results.

The importance of understanding the operating characteristics of a custom-designed testing machine, even one as simple as a pin-on-disk tribometer, are illustrated in Fig. 4. In that case, a 9.525 mm diameter silicon nitride ball was slid on a TiN hard-coated tungsten carbide disk under a normal force of 5.0 N and a sliding speed of 0.5 m/s in room temperature air. The friction force trace in Fig. 4 (a) was obtained using a ball holder whose design did not fix its mounting shaft tightly enough, resulting in a certain amount of wobble and a continually-rising trend in friction. When replaced with a pin holder of alternate design, the friction trace for the same two specimen materials exhibited a different shape and reached steady-state with a smaller amplitude of variation in the friction force (Fig. 4(b)). The mechanical design of wear testing machines goes beyond the scope of this paper; however, it is advisable to allow sufficient flexibility in the design so that a range of operating conditions, containing within them those required by the current work, is provided.

Just as small changes in the design of testing machines can produce significantly different test results, so can the specimen mounting procedure. In line contact tests, such as the block-on-ring test, small specimen alignment errors can significantly affect the repeatability of tests, the amount of wear, and the frictional running-in characteristics. The importance of assuring that the specimen is properly aligned is illustrated for aluminum bronze-on-steel block-on-ring tests in Fig. 5, adapted from Blau and Whitenton (**41**). As the tilt of the block was intentionally increased, the running-in period shortens and the total wear recorded increases. For a block tilt of only 0.84 degrees, the total wear of the block specimen was 2.6 times higher than when it was properly aligned. Note that the steady-state wear rates (slopes) of all tests, after running-in, were quite similar indicating that once running-in had occurred, wear progressed approximately linearly.

There are two popular approaches for selecting the operating conditions: (1) the worst-case (or 'factor of safety') scenario and, (2) the nominal operations scenario. In the former approach, one tries to determine the most severe conditions under which the subject component must operate and conduct tests under conditions which are equal to or greater than those. An even more conservative approach is to apply an engineering factor of safety which increases the severity of conditions still further. If the materials pass such a test repeatedly, they are expected to perform without problems in the actual, albeit less severe environment of the application.

One drawback of the worst-case approach is that the materials may tend to wear in a way that is uncharacteristic of the application, making it impossible to compare worn features to the type of surface damage observed on normally-operating components. The same consideration is true for so-called 'accelerated' testing strategies which apply extreme operating conditions. A more straightforward approach is to operate the simulation at the middle or at the upper margin of the range of normal operating conditions (i.e., the 'nominal operations scenario'). In this way, the metrics which were established based on the actual component characteristics are more straightforward to compare.

Accelerated tests, as noted above, must be used with due caution, but they can be quite informative if designed and applied properly. Forrest, et al. (**42**) used the established relationship between the grit size of an abrasive and the wear rate to estimate the wear of magnetic recording head materials in a video tape system. By using larger grit sizes, wear was accelerated by up to a factor of four, leading to faster screening of candidate materials. The effects of abrasive particle shape changes could likewise be studied.

As a final point, there are statistical tools available to

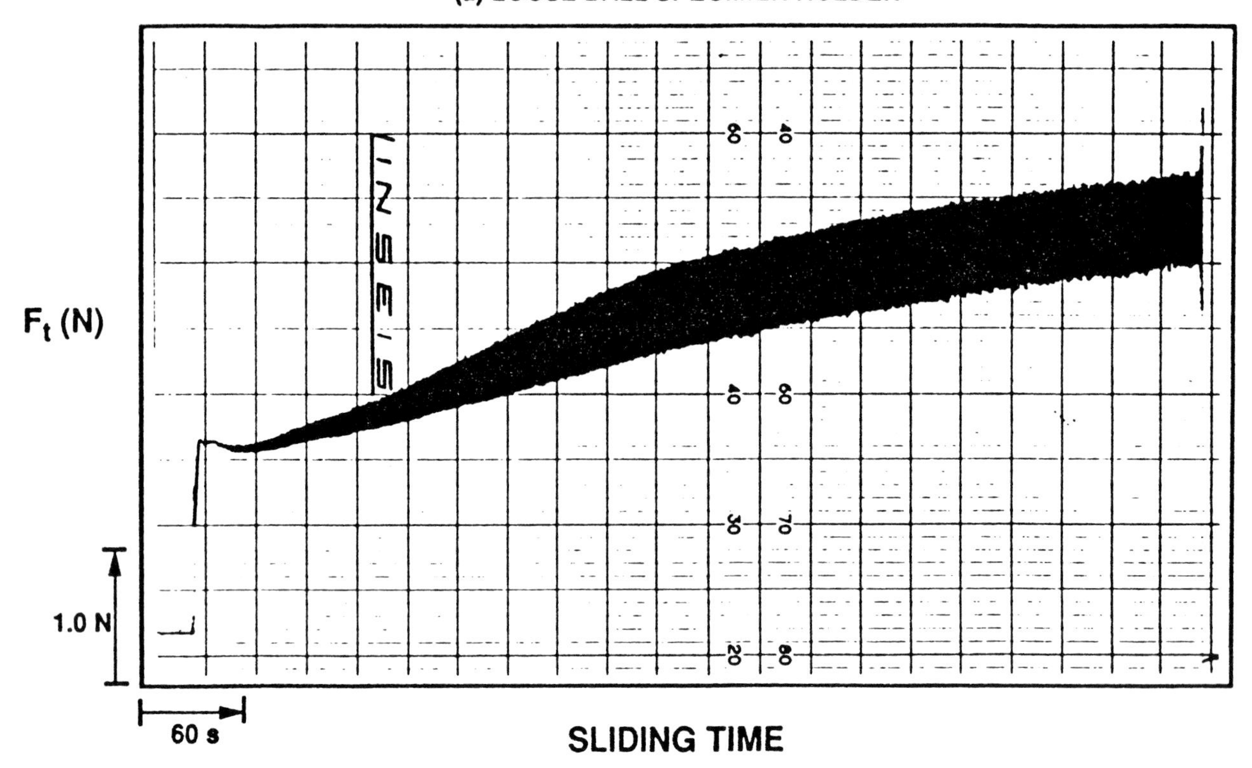

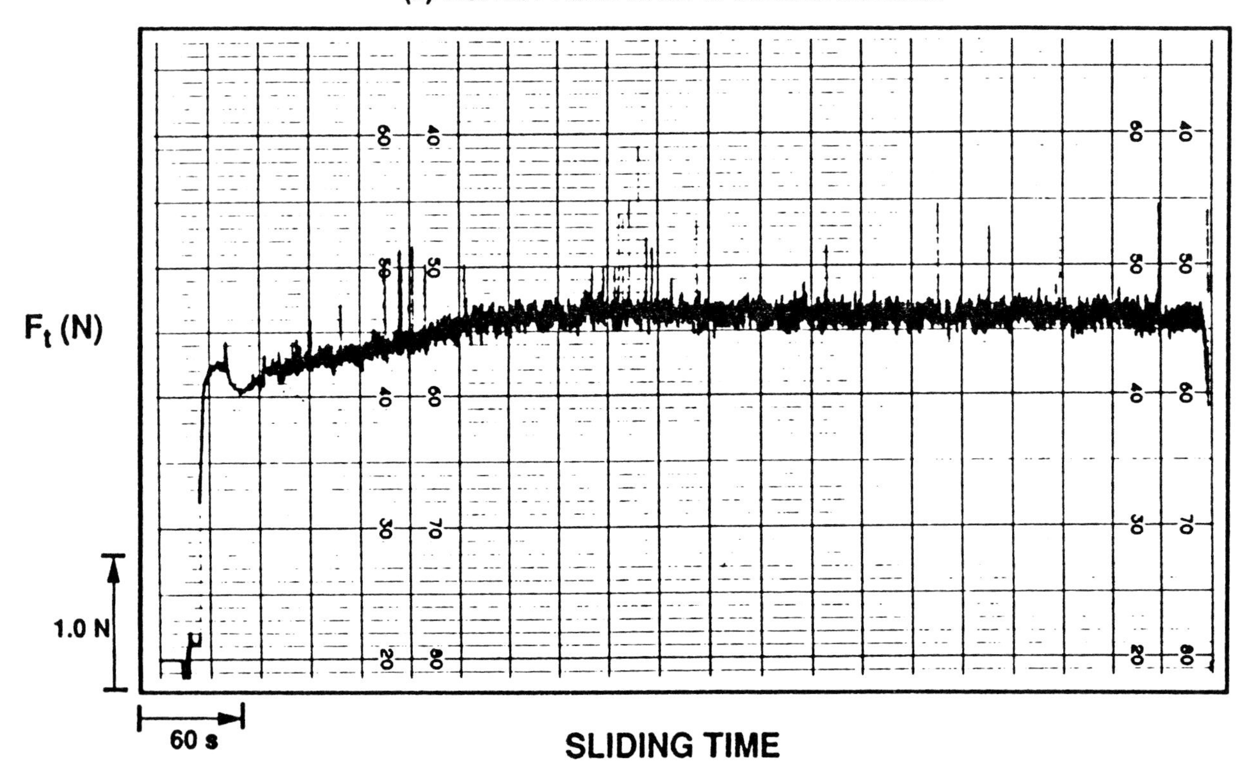

Fig. 4 Friction versus time behavior for silicon nitride sliding on a TiN hard-coated tungsten carbide disk under a normal force of 5.0 N and a sliding speed of 0.5 m/s in room temperature air. (a) effects of slight rocking of the ball specimen holder, (b) same experiment with a rigidly-held ball specimen.

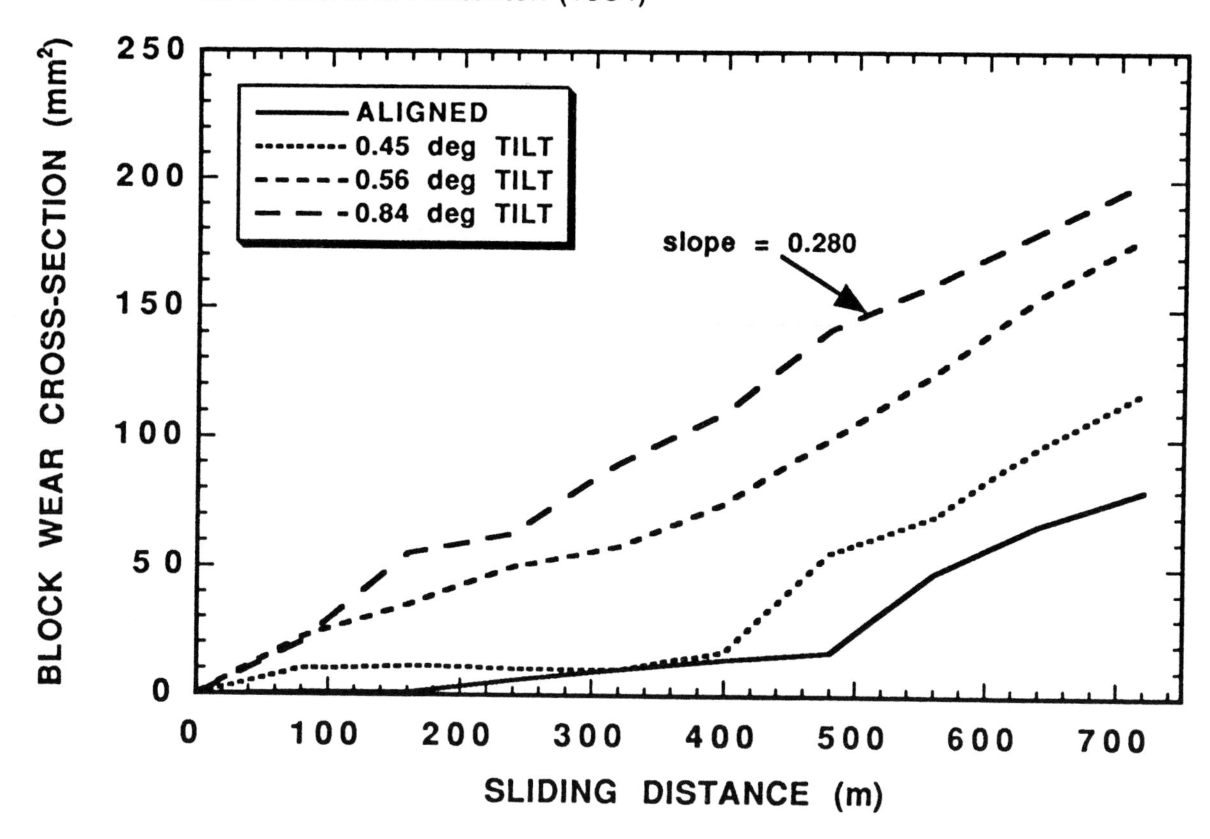

Fig. 5 Suppression of the wear-in period and the resultant increases in cumulative wear amount resulting from intentional misalignment of the block specimen in block-on-ring tests.

help design sets of simulative experiments in which specific combinations of operating variables can be investigated. Such tools can identify which variables and their combinations seem to have the most influence on material performance. In particular, the methodology developed by Taguchi (**43**) in the 1980s is receiving widespread attention. This type of approach has been used by Mok and Gorman (**44**) to investigate the effects of five factors, at two levels, on the wear of rubber materials for mud pumps. The dedicated simulator they constructed was based on an oscillating design and allowed for the introduction of flowing mud at the desired temperature. Comprehensive analysis of the data was coupled with observations of wear surface features not only to identify significant wear-influencing variables, like load and sand concentration, but also to better understand the material responses which accompanied these phenomena.

2.5 Validating the simulation

If the metrics have been well defined and the simulation developed with them in mind, the process of validating a simulation becomes a relatively straightforward one; namely, comparing, item by item, the results of the laboratory test with the attributes of the actual components. To illustrate this approach, consider the hypothetical example given in Table 5. A plain bearing might experience a certain amount of wear in 100 hours of operation, and a block-on-ring test might be selected as a laboratory simulation. Both quantitative and qualitative metrics are listed in the table. The quantitative metrics for wear do not agree well in this example, but most of the observed features on the surfaces of parts removed from service have been reproduced in the simulation. Mixed results like these may not allow one to clearly establish the success of the simulation unless one or two key metrics in the list are seen to be more important than the others, and are used as the primary measure of simulation success.

Work by Zhou, et al. (**45**) provides another example of validating a simulation by the observation of surface features. A single-wire, bench-scale laboratory fretting test was developed to simulate the contact conditions a wire in a many-strand cable would experience. This test produced the same. form of surface features as observed on actual cables: similar fretting zone size, similar delamination

Table 5 Validation table for a hypothetical tribosimulation

Metric	Field Components	Test Specimens
Steady state kinetic friction coefficient	estimated from torque readings; $\mu \sim 0.2$	Ave. $\mu = 0.31 \pm 0.1$
Wear loss at the mid-line of the shaft (after 100 hours of exposure)	0.180 mm (shaft diameter reduction)	0.064 mm (change in diameter of rotating cylinder specimen)
Presence of transferred patches of material	yes	yes
Surface delaminations	yes	minor indications but not exactly like the actual case
Scuffing at one end of the part	yes	not observed
Heat indications (burn marks)	yes	yes
Metallic flakes in the lubricant	yes	yes

features, similar crack nucleation modes, and number of cycles to nucleate cracks.

The relative ranking method is a more convincing validation than confirmation by the observation of characteristic surface features alone. In this approach, one determines to what extent a series of materials ranks in performance in the same order in field and laboratory trials. The earlier example of rail wear illustrates that approach (Table 3). The numerical values, such as wear rates or threshold conditions for the onset of surface damage, associated with the rankings may differ in magnitude between field trials and laboratory tests, but at least the order of merit of various material combinations or lubricants is similar. Of course, it is not possible to use the relative ranking method when only one materials couple has been used in practice.

In summary, the following attributes, listed in descending order or desirability, can be used to validate a simulation:

1) Numerical wear rates, characteristics of surface damage, and friction coefficients produced in the simulation correspond directly with those observed in the application.
2) The relative ranking and the magnitude of the differences between several candidate materials is the same in the test and the application.
3) The relative ranking (but not necessarily the magnitude of the difference) of several candidate materials is the same in the test and the application.
4) The wear type, wear particle characteristics, or surface damage features have nearly-identical characteristics in the simulation and in the application.

3 SIMULATION HIERARCHIES AND SUPPORTING TESTS

The value of laboratory-scale tribosimulations depends on their ability to provide necessary data for supporting decisions on the selection of materials, surface treatments, and lubricants. It may be cost-effective to screen a large number of candidate materials using a simple, first-order simulation and then conduct more realistic tests on a smaller number of leading candidates under conditions which more accurately approach the field conditions. As noted earlier, German DIN standard 50 322 lists six levels of testing which vary in simplicity and correspondence to actual component conditions.

Sometimes, conducting simulative wear and friction tests alone is not sufficient to address the problem at hand. For example, Santner and Meier zu Köcker (**46**) published a recent article on the subject of tribosimulation in which three examples were provided to illustrate their points. One in particular involved graphite-filled PTFE slider bearings for a tumble drier. There was a question regarding whether the supplier had changed anything when bearings began to perform poorly. Simple pin-on-disk tests were run using pins of the two materials. Results showed differences between the behavior of the two suspect lots of polymer composites both in terms of friction coefficients and in terms of the wear surfaces of the disk specimens (steel cut from the tumbler drums). The composite performing well in service had fine wear marks and transfer of the PTFE while the poorly-performing lot showed abrasion marks on the disks similar to those observed on the drum surfaces. Chemical analysis, X-ray diffraction, differential thermal analysis, and water absorption tests were needed in addition to the tribotests to better identify the source of the problem. It was found that there was no clear difference in chemical compositions of 'good' and 'bad' materials, and that major evidence for poor performance seemed to be a difference in the adhesion between the graphite filler and the matrix. While there was no compositional basis for screening out bad materials, a simple friction test could be used by the tumble drier manufacturer prior to installing the bearing materials. Using a combination of test methods may therefore pave the way to an answer if evidence from any single test method is not in itself sufficient.

4 SUMMARY

The solution of friction and wear problems by simulative testing involves five steps: (1) developing a clear understanding of the tribosystem involved, (2) developing metrics by which the problem can be quantified, (3) selecting the necessary conditions for the simulation, (4) selecting or developing a testing system and procedure to provide those necessary conditions, and (5) validating the simulation based on the metrics established earlier. Tribosystem analysis is the first step in defining the problem. It is important, at minimum, to simulate the macrogeometry of contact, the type of motion, the contact pressure and velocity, the thermal-chemical environment (temperature, lubricant, and gaseous environment), and the possible influence of third-bodies, but it should not be necessary to simulate every aspect of the application. Validation of the tribosimulation can be based on any of several criteria: friction coefficient values or wear rates, relative rankings of specific material combinations, the observation of key surface features, or the wear debris characteristics. It may be necessary to make adjustments in the testing procedure to satisfy the established metrics. The worst-case scenario and the normal operating condition scenario are options for developing friction and wear simulations. More than one type of screening test may be required to evaluate materials or surface treatments depending on service requirements and the required balance of properties. Other types of materials analysis, in addition to tribotesting alone, may be needed to obtain a satisfactory solution for some types of friction and wear problems.

5 ACKNOWLEDGMENTS

The author would like to acknowledge the support of the U.S. Department of Energy, Assistant Secretary for Energy Efficiency and Renewable Energy, Office of Transportation Technologies, High Temperature Materials Laboratory User Program, under contract DE-AC05-96OR22464 with Lockheed Martin Energy Research Corp.

Research sponsored by the U.S. Department of Energy, Assistant Secretary for Energy Efficiency and Renewable Energy, Office of Transportation Technologies, as part of the High Temperature Materials Laboratory User Program, under contract DE-AC05-96OR22464 with Lockheed Martin Energy Research Corp.

6 REFERENCES

1 **Dowson, D.,** *History of Tribology,* Longman, 1979.

2 *ASM Handbook, Friction, Lubrication and Wear Technology, Vol. 18,* ASM International, Materials Park, Ohio, 1992.

3 **Booser, R.,** ed., *Handbook of Lubrication: Theory and Practice of Tribology, Vol. I* (1983), II (1984), III, CRC Press, Boca Raton, Florida.

4 **Peterson, M.B.** and **Winer, W.O.,** ed., *Wear Control Handbook,* Amer. Soc. of Mech. Engr., New York, 1980.

5 **Bhushan, B.** and **Gupta, B.K.,** *Handbook of Tribology,* McGraw-Hill Pub., New York, 1991.

6 **Godet, M., Berthe, D., Dalmey, G., Flamand, L., Floquet, A., Gadallah, N.** and **Play, D,** *Tribo-Testing,* in Tribological Technology, Vol. II, ed. P. B. Senholzi, Martinus Nijhoff Pub., The Hague, 1982, pp. 525–609.

7 **Czichos, H.,** *Basic Tribological Parameters,* in *ASM Handbook, Friction, Lubrication and Wear Technology, Vol. 18,* ASM International, Materials Park, Ohio, 1992, pp. 473–479.

8 DIN standard 50320, *Wear, Terms; System Analysis of Wear Processes; Classification of the Field of Wear,* Deutsches Institut fur Normung e.v., Berlin, Germany, 1979.

9 **Blau, P.J.,** *Tribocommunication,* in *Friction and Wear Transitions of Materials,* Noyes Pub., Park Ridge, NJ, pp. 418–430.

10 **Czichos, H.,** *Tribology: A Systems Approach,* Elsevier Pub., Lausanne, 1978, pp. 322–325.

11 **Xu, L., Clough, S., Howard, P.,** and **StJohn, D.** Laboratory assessment of the effect of white layers on wear resistance of digger teeth, *Wear,* Vol. **181–183**, 1995, pp. 112–117.

12 **Fischer, A.,** Well-founded selection of materials for improved wear resistance, *Wear,* Vol. **194,** pp. 238–245, 1996.

13 **Peterson, M.B., Bhansali, K.J., Whitenton, E.P.** and **Ives, L.K.,** *Galling Wear of Metals, Proc. Wear of Materials,* ASME, New York, 1985, pp. 293–301.

14 ASTM G-98-91, *Standard Test Method for Galling Resistance of Materials,* ASTM Ann. Book of Standards, Vol. 03.02, pp. 382–384.

15 **Blau, P.J.,** The Units of Wear – Revisited, *Lubrication Engineering,* Vol. **45** (10) pp. 609–614.

16 **Clayton, P.** Predicting the wear of rails on curves from laboratory data, *Wear,* Vol. **181–183**, 1995, pp. 11–19.

17 **Blau, P.J.,** *Scale Effects in Friction: An experimental study, in Fundamentals of Friction,* ed. I. L. Singer and H. M. Pollock, Kluwer Pub., Dordrecht, 1992, 523–534.

18 **Aronov, V., D'Souza, A.F., Kalpakjian,** and **Shareef, I.,** Experimental investigation of the effect of system regidity on wear and friction-induced vibrations, J. of Lubr. Tech., Vol. **105**, pp. 206–211.

19 DIN std 50 322,....Deutsches Institut fur Normung e.v., Berlin, Germany,

20 ASTM G-118-93, *Standard Guide for Recommended Data Format of Sliding Wear Test Data Suitable for Databases,* ASTM Book of Standards, Vol. **03.02**, West Conshohocken, Pennsylvania, 1996, pp. 489–492.

21 **Budinski, K.G.,** Friction of plastic films, *Proc. International Conf. on Wear of Materials,* ASME, New York, 1989, pp. 459–468.

22 **Blau, P.,** A Retrospective Survey of the Use of Laboratory Tests to Simulate Internal Combustion Engine Materials Tribology Problems, in *Tribology: Wear Test Selection for Design and Application,* ed. Ruff, A. W. and Bayer, R. G., ASTM Spec. Tech. Pub. 1199, Pennsylvania, 1993, pp. 133–148.

23 **Patterson, D.J., Hill, S.H.,** and **Tung, S.C,** Bench Wear

Testing of Engine Power Cylinder Components, *Lubr. Engrg.*, 1993, pp. 89–95.

24 **Malatesta, M.J., Barber, G.C., Larson, J.M.,** and **Narsimhan, S.L.,** Development of a Laboratory Bench Test to Simulate Seat Wear of Engine Poppet Valves, *Tribology Trans.*, Vol. **36** (4), 1993, pp. 627–632.

25 **Ma, M-T., Smith, E.H.,** and **Sherrington, I.,** A three-dimensional analysis of piston ring lubrication – Part 1: modeling, *J. of Engineering Tribology*, Vol. **209**, 1995 (J1), pp. 1–14.

26 **Ma, M-T., Smith, E.H.,** and **Sherrington, I.,** A three-dimensional analysis of piston ring lubrication – Part 2: sensitivity analysis, *J. of Engineering Tribology*, Vol. **209**, 1995 (J1), pp. 15–28.

27 **Bialo, D., Duszczyk, J., de Gee, A.W.J., van Heijningen, G. J.J.,** and **Korevaar, B.M.,** Friction and Wear Behaviour of Cast and Sintered Al-Si Alloys Under Conditions of Oscillating Contact, *Wear*, Vol. **141**, 1991, pp. 291–309.

28 **Wacker, G.,** The Use of a Testing Machine for Simulating Piston Ring Groove Wear, in *Metallurgical Aspects of Wear*, ed. Hornbogen, E. and Zum-Galır, K. H., Deutsche Gesell. fur Metalkunde, 1981, pp. 247–267.

29 **Czichos, H.,** Ref. 10, pp. 33–42.

30 *Dixon Bearing Manual,* Furon Advanced Polymers Division, Bristol, Rhode Island, 1993, 23 pp.

31 ASTM G-119-93, *Standard Guide for Determining Synergism Between Wear and Corrosion,* Annual Book of Standards, Vol. **03.02**, ASTM, West Conshohocken, Pennsylvania, 1996, pp. 493–498.

32 **Grobler, P.E.** and **Mostert, R.J.,** Experience in the Laboratory and Commercial Development of Abrasion-Corrosion Resistant Steels for the Mining Industry, Proc. Intern. Conf. on Wear of Materials, ASME, New York, 1989, pp. 289–295.

33 ASM Metals Handbook, *Service Characteristics of Carbon and Alloy Steels: Wear Resistance*, ASM International, Materials Park, Ohio, 1985, p. 465.

34 **Swanson, P.A.** and **Klann, R.W.,** Abrasive Wear Studies Using the Wet Sand and Dry Sand Rubber Wheel Tests, *Proc. Intern. Conf. on Wear of Materials*, ASME, New York, pp. 379–381, 1981.

35 **Shipway, P.H.** and **Hutchings, I.,** The role of particle properties in the erosion of brittle materials, *Wear*, Vol. **193**, pp. 105–113, 1996.

36 **Voitik, R.,** The Tribological Aspect Number, in *Wear Test Test Selection for Design and Application*, ed. A. W. Ruff and R. G. Bayer, ASTM, Spec. Tech. Pub. 1199, Philadelphia, 1993, pp.

37 **Blau, P.J.,** ed., A Compilation of International Standards for Friction and Wear Testing of Materials, Versailles Agreement on Advanced Materials and Standards (VAMAS), Technical Working Area 1, report No. 14, National Institute of Standards and Technology, Gaithersburg, MD, 1993, **49**, pp.

38 Versailles Project on Advanced Materials and Standards, Technical Working Area 1, Wear Test Methods, conducted a series of pin-on-disk tests on steel and ceramics during the 1980s. These resulted in several new ASTM and DIN standards.

39 **Wood, F.W.** and **Taylor, J.W.,** A Review of DOD Wear and Friction Tests for Rubber Products, in *Wear and Friction of Elastomers,* ed. Denton, R. and Keshavan, M. K., ASTM Special Tech. Pub. 1145, Amer. Soc. for Testing and Materials, Philadelphia, 1992, pp. 12–29.

40 **Perez, J.M., Hausen, R.C.,** and **Klaus, E.E.,** Comparative Evaluation of Several Hydraulic Fluids in Operational Equipment, A Full-Scale Pump Stand Test and the Four-Ball Wear Tester. Part II. Phosphate Esters, Glycols, and Mineral Oils, *Lubrication Engr.*, Vol. **46** (4), pp. 249–255, 1990.

41 **Blau, P.J.,** and **Whitenton, E.P.,** Effect of Flat-on-Ring Alignment on the Sliding Friction Break-in Curves for Aluminum Bronze on 52100 steel, *Wear*, Vol. **94**, 1984, p. 201.

42 **Forrest, D., Matsuotka, K., Tse, M.-K,** and **Rabinowicz, E.,** Accelerated wear testing using the grit size effect, *Wear*, Vol. **162–164**, 1993, pp. 126–131.

43 **Taguchi, G.,** System of experimental design, Krans, New York, 1987.

44 **Mok, S.H.** and **Gorman, D.G.,** Using Taguchi experimental design to investigate operating variables that significantly affect wear in mud pumps, *J. of Engineering Trib.*, Vol. **209**, 1995, pp. 29–40.

45 **Zhou, Z.R., Goudreau, S., Fiset, M.,** and **Cardou, A.,** Single wire fretting tests for electrical conductor bending fatigue evaluation, *Wear*, Vol. **181–183**, 1995, pp. 537–543.

46 **Santner, E.** and **Meier zu Kocker, G.**, Utility and limitations of tribosimulation for quality control and material preselection, *Wear*, Vol. **181–183**, 1995, pp. 350–359.

Isolated contact stress deformations of polymers; the basis for interpreting polymer tribology

B J BRISCOE
Department of Chemical Engineering, Imperial College, London, UK

SYNOPSIS

The paper reviews the use of what are essentially extensions of the scratch hardness method as a means of probing the surface mechanical properties of polymers for applications in tribology studies. There is a short history and a summary of more recent work. The various damage regimes such as ironing, ploughing, brittle cracking and so on are introduced and the role of the important contact mechanical variables described. A number of 'scratch' or damage maps are provided to illustrate the types of material which may be derived and portrayed using this approach.

1 INTRODUCTION

Ever since it became generally appreciated that the contact between solids comprised asperities touching each other in discrete areas there has been a desire to understand the consequences of these specific interactions. The initial approaches sought to simulate the asperity engagements using indentation and scratching experiments on a relatively large scale compared with that which might exist in a typical sliding process. The advent of nano-hardness and nano-scratching probes has provided the opportunity to at least reduce the scale of the deformation but perhaps not realistically model other facets of the dynamic asperity interactions in practical tribological contacts.

In the context of organic polymers the first extensive systematic application of this idea was probably practised by Schallamach and his colleagues in the search for better understandings of the wear processes in rubbers (**1, 7**). He drew metallic indentors of various geometries over the surfaces of typical rubbers; he called them 'isolated stress' intensifiers. With needles he produced a characteristic tearing process which he was able to interpret by speculating upon the stress fields at fracture and the relaxation of the deformed contact zone upon the strain release provided by the tearing process (**2**). When blunt indentors were used the traction cracking commonly seen for brittle or ceramic substrates was observed.

Schallamach's basic premise was that the inspection and study of these types of damage modes would provide the basis for interpreting the abrasion of elastomers and, in particular, the formation of the abrasion patterns frequently observed for rubber automobile tyres. Subsequent work on elastomers sought to simulate the effects of multiple asperity engagements and correlate the abrasion rate to the fatigue life of the elastomer. The experiment most frequently cited is one where a two-dimensional asperity, actually a razor blade, moves over a rubber roller (**8, 9**). The experiments simulated the abrasion patterns commonly seen in practice and also indicated that radical reactions occurred with free radical emissions in some cases (**10**).

This example, drawn from elastomer tribology, has been repeated in other areas of polymer tribology (**11**). The present author, in conjunction with Evans and Lancaster, attempted to relate scratch hardness data to the abrasive wears of γ-damaged PTFE and poly(methyl methacrylate), PMMA (**12–14**). Similar correlations were sought for dense phase flow erosion for the same system. More recently, the present author and others have attempted to interrelate liquid jet erosion rates with interfacial toughness values derived from indentation hardness for PMMA composites (**15, 16**).

These types of studies with thermoplastics have also been extended in order to attempt to elucidate the influence of the environment, say lubricants, upon the abrasive or erosive properties (**17, 18**). Here, the focus has actually been somewhat different and goes back to the ideas embodied in the 'Adhesion Model' of friction as expounded by Bowden and Tabor (**19**). Basically indentation, or indeed scratch hardness, methods are employed to probe the surface mechanical property gradients which may be incorporated into a frictional model to rationalise the effects of frictional heating and environmental surface plasticisation.

These types of studies illustrate the value of using 'isolated stress' probes to mimic and characterise the deformations which occur in the asperity regions of polymers. Essentially, the experiments provide descriptions of the damage modes, the work dissipated, and a range of mechanical properties including estimates of the plastic yield stress and the elastic modulus. The present review examines what is currently known or established regarding the influences of the important variables and addresses the significance in the context of polymer tribology.

2 INDENTING AND SCRATCHING POLYMERS

2.1 Introduction

It is now fairly well documented that, providing that the modes of deformation are similar, the mechanical response in indentation is similar to that in scratching (**20**). Thus, if economy of effort is a factor, then it would not be normal practice to use both indentation and scratching methods. Each has merits in comparison with the other and Table 1 provides a summary of these. The basic difference is that scratching is a good means of obtaining sliding deformation modes but does not provide such unequivocal values of material properties such as those which might be required in contact area calculations; for example the time dependent Young's Modulus is rarely abstracted from a scratching experiment. Also, clearly the scratching method more directly corresponds to the deformation produced at asperities in practice. For the present review it is useful to cite, where necessary, indentation studies but to focus upon scratching.

Table 1 A comparison of the relative merits of indentation and scratch hardness methods

Characteristic	Indentation	Scratching
Elastic response	Satisfactory	Limited
Plastic response	Satisfactory	Often satisfactory
Time-dependent modes	Somtimes limited	Flexible but limited
Brittle/ductile	Obscruded	Good
Damage modes	Limited to extremes	Subjective but effective
Lubrication/surface gradients	Potentially problematical	Sensitive

The first step is to introduce the significant experimental variables and also to indicate how they influence the response of the system. The main external variables are as follows, assuming rigid conical indentations, Figure 1:

(i) the cone angle,
(ii) the load and the depth of penetration,
(iii) the state of interfacial lubrication,
(iv) the sliding velocity.

The influence of each variable will be reviewed.

2.2.1 The contact strain; the asperity slope angle

The interrelationship between imposed contact strain and indentor geometry has been quite well established for normal indentation; the mean contact strain is proportional to the contact slope; see Figure 1(i) (**20, 21**). For metals the strain is about 0.2 tan θ for conical indentors (**22**); Figure 1. The situation for organic polymers is less clear but a similar relationship appears to apply (**23**). The major, and unsubstantiated, assumption is that a similar relationship may be adopted for the corresponding case of sliding. As will become apparent, this assumption appears to be a reasonable one at a first level of examination.

The variation of the contact strain will, for the case of organic polymers at least, produce two main effects. First, the elastic/plastic response (excluding other effects such as time and temperature) will often be very strain dependent; specifically here the plastic or viscous response will be a marked function of the imposed strain. Second, for some polymers the level of the imposed strain will demark a ductile from a brittle response.

These various effects are shown in Figure 2 in the form of a 'Deformation Map' for a poly (carbonate) (PC) (**24**). The map, which is a relatively simple one, shows the various regimes of deformation produced by conical indentations at different normal loads; the ambient temperature, sliding velocity and the state of lubrication are fixed. The regimes of deformation may be assessed from SEM and laser reflectivity studies, although the identification of the precise deformation modes involves a subjective appraisal of these data. Table 2 provides a summary of the main pictorial characteristics of each regime.

Several points are worth noting in considering Figure 2, not least the fact that all the polymers that have been investigated show comparable behaviour in a qualitative but not quantitative sense. Some semicrystallline polymers exhibit other and additional types of response but the behaviour is broadly similar (**25**). A subsequent section will deal with these influences of other variables such as sliding velocity and ambient temperature but here the behaviour is not markedly altered in a qualitative way. It will also be inferred from the form of Figure 1 that the cone angle and the normal load or penetration depth are the key external variables and this is generally the case.

Thus, referring again to Figure 2 and taking an arbitrary fixed load of about 1.8 N, we may consider the change in the nature of the surface damage as the cone angle is reduced; a summary is given in Figure 3 along with the corresponding friction data. Initially, for small asperity slopes (large included cone angles) there is no measurable permanent deformation. Further increases induce asperity ironing and then ductile ploughing processes where permanently deformed material is moved to the sides of the moving indentor to produce pronounced grooves. The final stages of the progressive increase in the attack angle induces a variety of surface legions and tears which are accompanied by ductile and brittle machining processes.

2.2.2 Depth of penetration; normal load

A careful study of indentation hardness, using conical

Table 2 The influence of cone angle α, upon the types of damage produces in scratching. This type of response is typical for polymers but the scale given corresponds to Figs. 2 and 3. For clarity of presentation the pictorial view does not show the variation in attack angle α, (Figure 1).

Response (Pictorial)	Generic	α
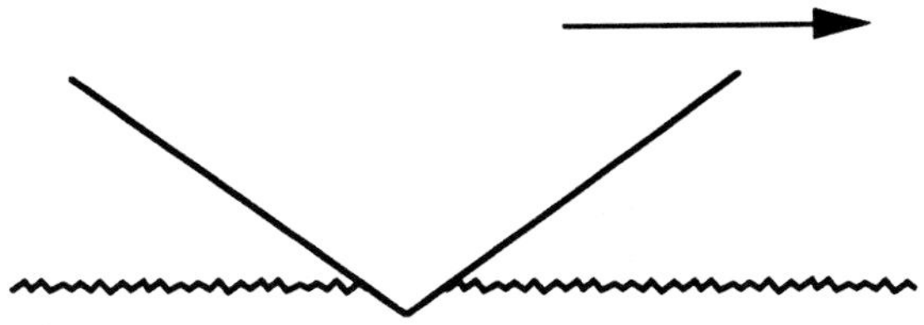	Elastic	180°–150°
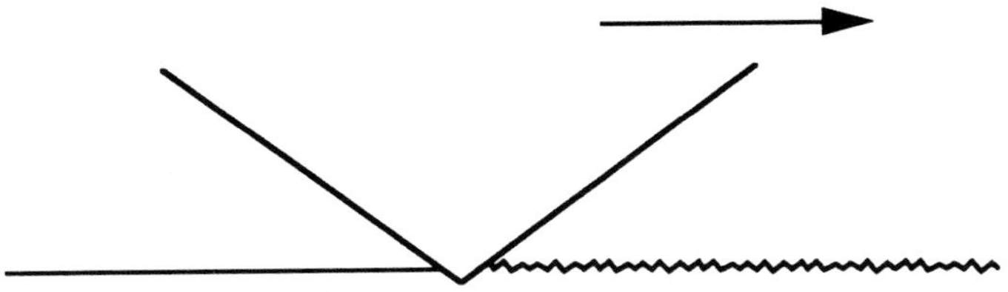	Ironing	150°–120°
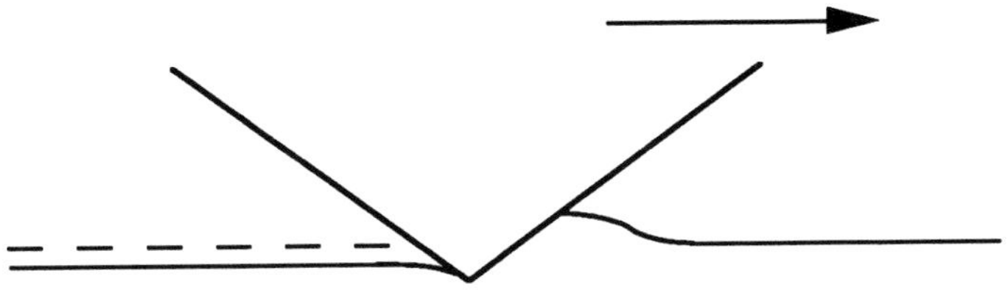	Ductile Ploughing	120°–90°
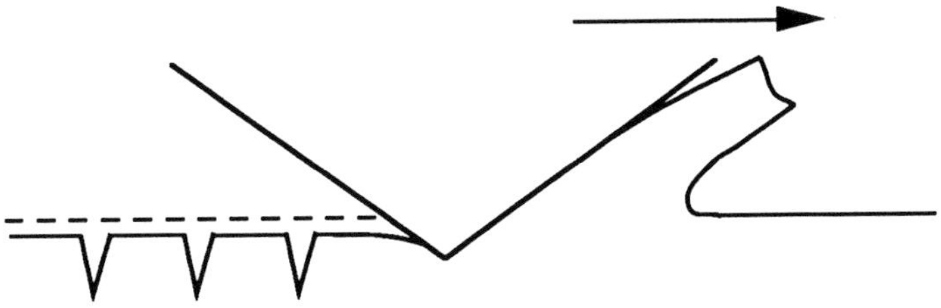	Ductile Machining + Cracking	90°–60°
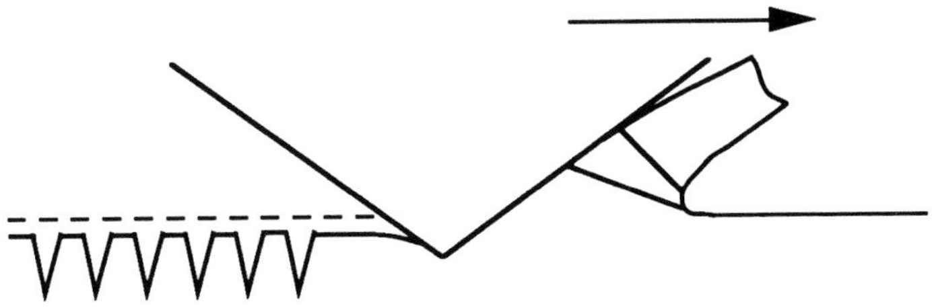	↕	60°–30°
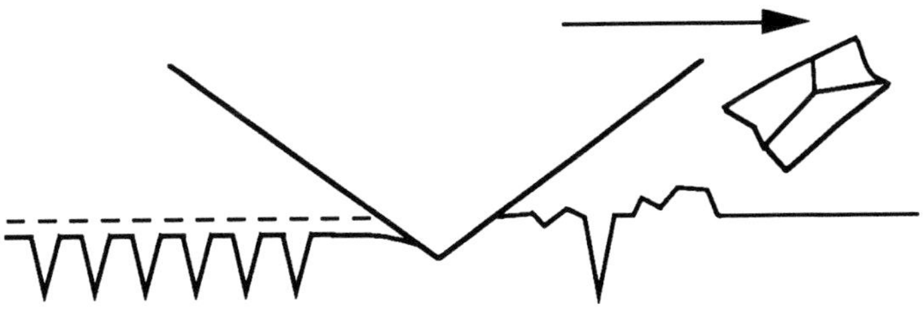	Brittle Machining	30°–0°

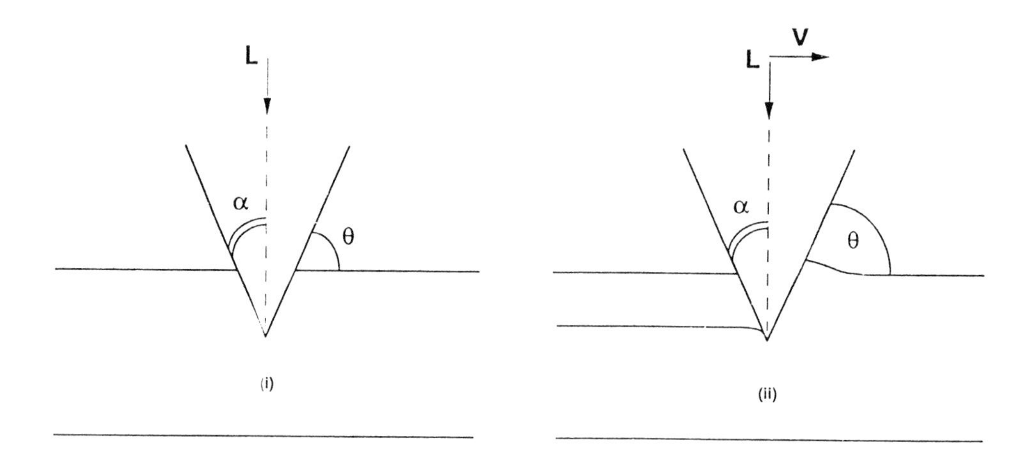

Fig. 1 (i) Pictorial view of an indentation hardness deformation where a ridged cone, semi angle α, indents a soft solid under a load L.
(ii) as for (i) but now sliding with a velocity V.

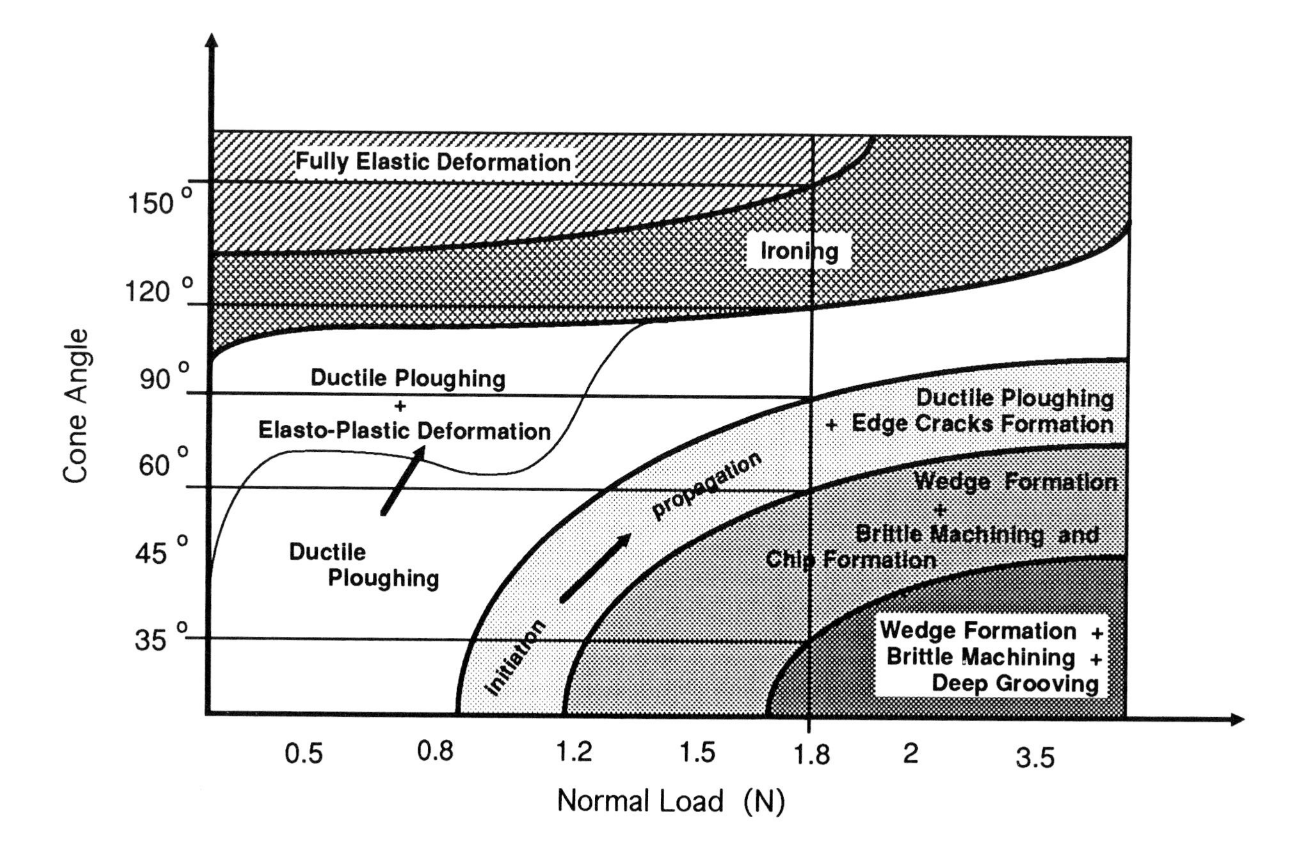

Fig. 2 Deformation map for a poly (carbonate) resin. The diagram shows results from scratching performed at room temperature for a range of cone angles and normal loads and at a scratching velocity of 0.0026 mm/s.

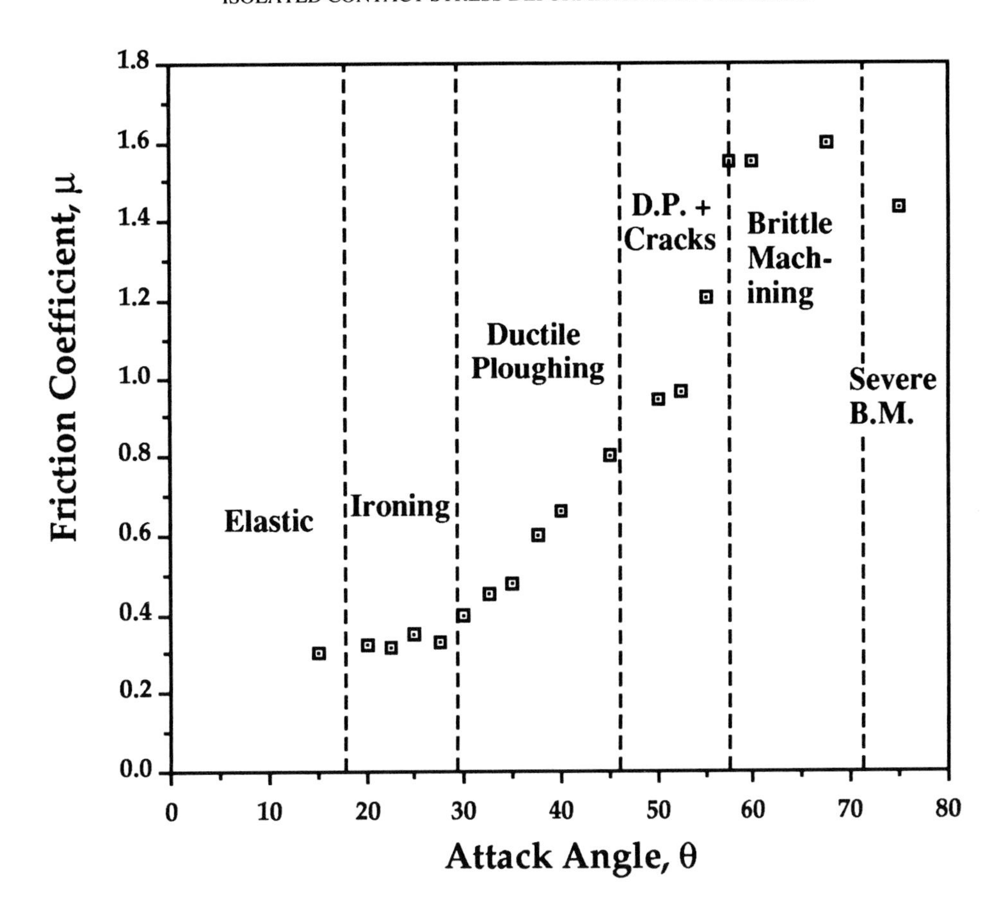

Fig. 3 Friction coefficient, μ, as a function of attack angle, θ,; see Fig. 2, the normal load is 1.8 N. After $\theta \geq 30°$, the major irreversible deformation modes begin. The decline in μ after $\theta = 60°$ is associated with the development of internal shear angles in the machinery modes. B. M., brittle machining, D.P. ductile ploughing.

indentors, as a function of penetration depth with usually provide invariant values of the plastic yield stress and the Young's Modulus; that is providing that the tip defect on the cone (the natural radius at the apex) is small and there are no gradients of mechanical properties produced by, say, plasticisation or transcrystalline growths (**26**). The case for scratching is rather different as may be noted by reference to Figure 2. Thus, whilst the contact strain is nominally fixed, since θ is constant, the damage varies greatly with depth; Lamy (**27**) provides nice data for this variation using the pendulum scratching device where the action of the pendulum naturally provides a range of depths and also the corresponding depth dependent damage modes. The reason for the pronounced penetration or load dependence is not entirely clear but it is nevertheless of major significance. The simplest argument is to suppose that the volume of material deformed is not directly proportional to the depth of penetration; the volume increases less rapidly than the increase in the depth of penetration. As a consequence, the energy density, or strain, density clearly increases with the penetration depth. We may recall, see for example Figure 3, that the friction coefficient for these cases of conical indentors is almost independent of the material and contact conditions; $2/\pi \tan\theta$ (**28**).

2.2.3 *Effects of lubrication*

The general influence of interfacial lubrication is to suppress traction-induced failures such as brittle cracking (**20**). Perhaps of more interest is the generally observed fact that lubrication reduces indentation hardness whilst it increases the scratch hardness. Thus, for virgin poly(methylmethacrylate) PMMA, the normal and scratch hardness values are comparable (**20**). For a lubricated contact, the former is about a half of the latter. There are also special influences of environments on polymer surfaces which are seen in a pronounced way in scratching but less so in indentation; this is the case for surface plasticisation (**29**).

2.2.4 *Deformation times; sliding velocity and loading times*

Perhaps of all the major variables which may influence the deformation modes, time is the most significant. In indentation, major variations in elastic and plastic behaviour are apparent. For the scratching configurations the observed responses are invariably complicated by what appears to be adiabatic frictional loading effects even at the most modest sliding velocities; say 1 mms^{-1} or less. Thus, an increase in the velocity will usually provide an enhanced ductility, or at least suppress the brittle responses, rather than promote brittle ruptures (**25**). As a consequence, increasing the scratching velocity often has the same effect as increasing the ambient temperature.

3 IMPLICATIONS IN POLYMER TRIBOLOGY

One of the central tenets of tribology is the two-term non-interacting model of friction which may be extended to include damage and wear processes. The point deformation experiments provide the following sources of information:

- The critical angle for the onset of gross deformation which may correspond to the transition from the adhesion to the ploughing modes of friction;
- The characteristic damage processes which provide an indication of the appropriate material characteristic for wear modelling;
- Some measure of surface mechanical properties which may be of value for contact area calculation;
- An indication of the presence of transformed surface layers.

Thus these methods provide much of the key information which is needed to interpret and rationalise the tribological behaviour of polymers. The major challenge is to incorporate these data into appropriate models for tribological behaviour.

4 ACKNOWLEDGEMENTS

The author is grateful to Enrico Pelillo for the provision of the diagrams and tables included in this paper and also for useful discussions on the scratching of polymers.

5 REFERENCES

1 **Schallamach, A.,** *J.Polymer Sci.,* **9,** 385–396, 1952.
2 **Schallamach, A.,** *Wear,* **1,** 384, 1958.
3 **Schallamach, A.,** *Wear,* **6,** 375–382, 1963.
4 **Schallamach, A.,** in *The Chemistry and Physics of Rubber-like Substances,* L. Bateman (ed.), Ch. 13, McLaren & Sons, London, 1965.
5 **Schallamach, A.,** *Rub.Chem.Tech.,* **41,** 209-223, 1968.
6 **Schallamach, A.,** *Wear,* **17,** 301, 1971.
7 **Lancaster, J.K..** Friction and Wear, in *Polymer Science. A Materials Science Handbook,* A.D.Jenkins (ed.), Ch. 14, North-Holland Pub., 1972.
8 **Champ, D.H., Southern, E.,** and **Thomas, A.G.,** in *Advances in Polymer Friction and Wear,* L.H. Lee, Editor, Vol. **5A,** p.113, Plenum Press, New York, 1974.
9 **Brodskii, G.I.** and **Reznikovskii, M.M..** in *Abrasion of Rubbers,* D.I. James, Editor, M.E. Jolley, Translator, MacLaren & Sons, London, 1967.
10 **Gent, A.N.** and **Pulford, C.T.R..** *Wear,* **49,** 135 (19780.
11 **Adams, M.J., Biswas S.K.,** and **Briscoe B.J..** (Eds), *Proc. UK/India Royal Society/Unilever Fora on Solid-Solid Interactions,* Imperial College Press, 1996.
12 **Briscoe, B.J., Evans, P.D.,** and **Lancaster, J.K.,** *J.Phys.D:Appl. Phys.,* **20,** 346–353, 1987.
13 **Briscoe, B.J., Evans, P.D.,** and **Lancaster, J.K.,** *Proc. 12th Leeds-Lyon Tribology Symp.,* 1985, Elsevier, Amsterdam, 1985.
14 **Briscoe, B.J., Evans, P.D.,** and **Lancaster, J.K.,** *Proc. ASME Conf Wear of Materials 87,* p.607.
15 **Briscoe, B.J., Pickles, M.J., Julian K.S.,** and **Adams, M.J.,** *Wear,* **181-183,** 759-765, 1995.
16 **Briscoe, B.J., Pickles, M.J., Julian K.S.,** and **Adams, M.J.,** to appear in *Wear of Materials* 97.
17 **Briscoe, B.J., Pelillo, E., Ragazzi, F.,** and **Sinha, S.K.,** in press *Polymer,* 1996.
18 **Briscoe, B.J.** and **Stuart, B.H.,** *Proc. Leeds/Lyon Meeting 1995 Third Body Phenomena,* Elsevier, p.69, 1996.
19 **Bowden F.P.** and **Tabor, D.,** *The Friction and Lubrication of Solids,* Clarendon Press, Oxford, Part 1, 1950; Part 2, 1964.
20 **Briscoe, B.J., Evans, P.D., Biswas, S.K.,** and **Sinha, S.K.,** *Tribology International,* **29,** 2, 31-49, 1996.
21 **Williams, J.A.** *Tribology International,* **29,** No.8, 675-694, 1996.
22 **Johnson, K.L.,** *Contact Mechanics,* Cambridge University Press, Cambridge, UK, 1985.
23 **Briscoe B.J.** and **Sebastian, S.K.,** *Proc Roy.Soc.Lond A,* **452,** 439-457, 1996.
24 **Briscoe, B.J., Pelillo E.,** and **Sinha, S.K.,** *Poly.Eng.Sci.,* **36,** 24, 1996.
25 **Briscoe, B.J., Evans, P.D., Pelillo E.,** and **Sinha, S.K.,** *Wear,* **200,** 137-147, 1996.
26 **Balta Calleja, F.J.,** *Trends Polym. Sci,* **2,** 419, 1994.
27 **Lamy, B.,** *Tribology International,* 17, 1, 35-38, 1989.
28 **Tabor, D.,** *Hardness of Metals,* Clarendon Press, Oxford, 1951.
29 **Briscoe B.J.** and **Stuart, B.H.,** *Polymer International,* **38,** 95-99, 1995.

Practical applications of lubrication models in engines

R C COY
Shell Research and Technology Centre, Thornton, Chester, UK

SYNOPSIS

As automobile fuel efficiency has grown in importance over the past few years so the questions concerning friction and wear (durability) in engines have grown in significance. This paper describes practical ways in which lubrication models have been developed and applied to key sub-systems (valve trains, piston assemblies and bearings) in engines to gain insight into the optimisation of overall performance, that is the balance between reducing friction whilst maintaining acceptable durability. The effects of lubricant rheology, hydrodynamic, elastohydrodynamic and boundary conditions on friction and wear will be explored under both steady state and transient conditions and comparisons made with experimental results.

1 INTRODUCTION

The automotive industry has made great strides over the past 20 years to improve the fuel efficiency of their vehicles. This has been achieved by many detailed design improvements (**1, 2**) to the engine, drive train, tyres, aerodynamics etc. From an engine viewpoint, and the lubricant in particular, this has required an optimisation so that both friction and wear are minimised. Ever more detailed and demanding specification tests for fuel economy (**3, 4**) and engine durability (**5**) are being developed to determine whether the lubricant meets the requirements laid down by legislators. These tests are expensive and frequently imprecise thus any insight that can be gained through modelling engine performance will aid the lubricant formulator (and the engine designer) and thus help to contain the cost of lubricant (and engine) development. To gain insight into the key parameters determining performance, lubrication models of increasing complexity have been developed that provide predictive capability (**6–16**). A wide parameter space has been explored to determine which parameters are crucial to performance. In this paper a range of rheological models for multigrade and Newtonian oils are described. For Newtonian oils at high shear stresses and shear rates Non Equilibrium Molecular Dynamics simulations (NEMD) have been used to develop a 'molecular' constitutive equation to calculate friction under elastohydrodynamic (EHD) conditions. For multigrade oils a White-Metzner (**11**) model has been used to explore viscoelastic effects in dynamically loaded journal bearings. Transition models (from {elasto}- hydrodynamic to boundary lubrication) have been developed for both friction and wear and these have been applied to engine sub-systems.

The content of this paper is an overview of the work carried out at Shell Research and Technology Centre, Thornton over the past few years and is drawn from a wide range of publications.

2 LUBRICANT RHEOLOGY

The effects of temperature and pressure on lubricant viscosity can be described by many functions (**17**). In this paper Vogel's equation has been used to describe the variation of viscosity with temperature:

$$\eta_o = \kappa \cdot exp\left(\frac{\theta_1}{\theta_2 + T}\right) \tag{1}$$

where η_o is the low shear rate viscosity (mPa.s), T (°C) is the oil temperature, and κ (mPa.s), θ_1 (°C), and θ_2 (°C) are constants for a given lubricant.

The effect of pressure on viscosity is treated simply using Barus's equation:

$$\eta = \eta_{atm} \exp\,(\alpha p) \tag{2}$$

or for more detailed calculations of EHD friction via a modified Roelands law:

$$\eta_0(p) = \eta_{atm}\, exp\left\{\frac{\alpha p_0}{z}\left[\left(1 + \frac{p}{po}\right)^z - 1\right]\right\} \tag{3}$$

The value of η_{atm}, the viscosity at atmospheric pressure, is available experimentally. α is the pressure coefficient of viscosity and z is a parameter that modifies the increase in viscosity at high pressures. These parameters are obtained by fitting experimental data and, if treated independently of each other, can give excellent fits for a wide range of fluids. p_o is a universal constant.

The effect of shear rate on the viscosity of multigrade oils can be described by a shear thinning equation of the following form:

$$\eta(\gamma) = \eta_\infty + \frac{\eta_o - \eta_\infty}{1 + \left(\frac{\gamma}{\gamma_c}\right)^n} \quad (4)$$

were the low shear viscosity at a particular temperature is η_o, and the limiting high shear viscosity is η_∞, γ is the shear rate and γ_c is the shear rate at which the viscosity is mid-way between η_o and η_∞. The reciprocal of this, λ, is a characteristic relaxation time of the polymer under these conditions.

$$\lambda = \frac{1}{\gamma_c} \quad (5)$$

n has been taken to be one in this case.

The Cross Equation (**18**) has the correct features to describe the flow curve $\eta(\gamma)$. Figure 1 shows flow curves for two multigrade oils that are both 10W50 grades. However, it is clear that the two oils which are nominally the same (according to the SAE J300 classification) can have quite different viscometric behaviour. These curves demonstrate that for multigrade oils the effect of shear rate as well as temperature and pressure need to be incorporated into lubrication models. Indeed, under the conditions where the multigrade lubricant is non-Newtonian, viscoelasticity may also be important.

For the solution of dynamically loaded journal bearings, such as main and big end bearings, the lubrication approximation is generally invoked. The effectiveness of the lubrication approximation has been supported by experimental evidence in a wide range of lubrication studies (**19–22**) but if elastic effects (**23–26**) are to be incorporated (for instance when the relaxation time of a viscoelastic {multigrade} lubricant is sufficiently high, for example due to pressure thickening) then the full set of coupled equations governing the flow of the lubricant taking proper account of the moving parts of the geometry must be analysed (**11, 27, 28**).

To model the performance of transiently loaded journal bearing/lubricant systems requires a knowledge of the rheology of multigrade lubricants combined with a method of estimating the bearing kinematics and dynamics. Only with the advent of modern high speed computational power combined with efficient and accurate numerical methods can the calculations be performed. For journal bearing load capacity and friction, the lubrication analysis was performed using the following governing equations:

The dependence of viscosity on γ and p is given by combining equations (2) and (4):

$$\eta(\gamma, p) = \left\{\eta_\infty + (\eta_0 - \eta_\infty)/\left[1 + (\lambda\gamma)^n\right]\right\} . exp\,\alpha . p. \quad (6)$$

were η_∞, η_o, λ and n are obtained by best fit experimental data.

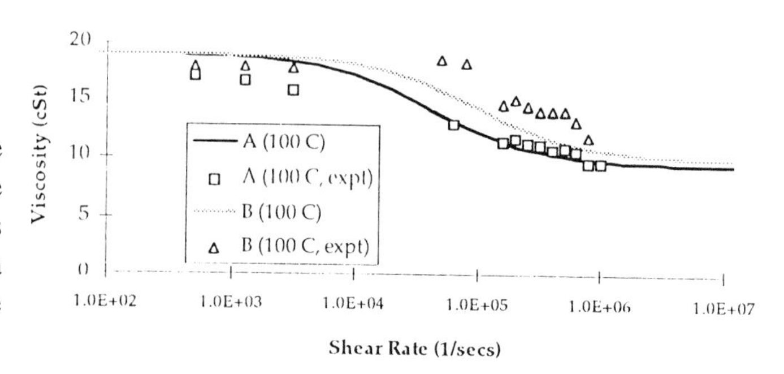

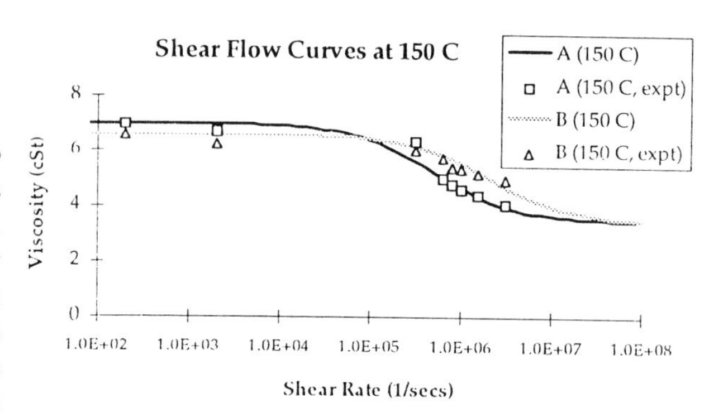

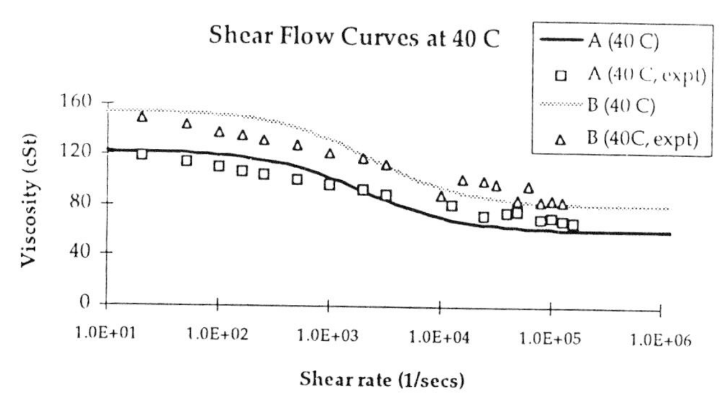

Fig. 1 Viscosity shear rate relationships for two 10W50 multigrade oils.

The influence of cavitation is modelled by modifying Eqn. (2) for pressures below a negative pressure threshold p^-. At pressures below p^- it is assumed that the lubricant viscosity quickly but smoothly decreases to an asymptotic value close to that of air. The influence of lubricant viscoelasticity is quantified by the introduction of a characteristic relaxation time (λ) for the lubricant. It is known empirically that, for lubricating oils, under viscometric conditions:

$$N_1 = 2\,k_t\,(\eta\gamma)^q \quad (7)$$

where N_1 is the First Normal Stress Difference and k_t and q are constants, ($q \approx 2$) for multigrade oils. The White–Metzner model under viscometric conditions yields:

$$N_1 = 2\,\eta\,\lambda\,\gamma^2 \quad (8)$$

Combining Eqns (7) and (8), when q = 2, gives:

$$\lambda = k_t\,\eta. \tag{9}$$

The above equation (9) has been adopted as the simplest model for relaxation time.

Using the above, both 2 and 3 D treatments of the lubricant flow can be calculated together with estimations of the heat flows in and out of the journal bearing to give the effect of lubricant rheology on load bearing capacity and hydrodynamic friction.

Experimental measurements (**29, 30**) have shown that at high pressures and shear rates 'Newtonian' base oils become non-Newtonian and also have flow curves. The flat part of the flow curves in Figure 1, at high shear rates ($>10^6\,s^{-1}$) represents the viscosity of the base oil (eg for the multigrade becomes η_o for the base oil) which at higher pressures and shear rates itself also shear thins. This non-Newtonian behaviour of the base oil is particularly important under severe EHD conditions as it determines the friction. NEMD simulations of simple 'Newtonian' fluids (**31–34**) have also shown shear thinning behaviour which can be described by a Cross equation. The equation for a simple base oil can be written in the same form as equation (4):

$$\eta(\gamma,\eta_0,\eta_\infty,n,\lambda) = \eta_\infty + \frac{\eta_0(p)-\eta_\infty}{1+(\lambda\gamma)^n} \tag{10}$$

which has four parameters: the asymptotic low and high shear-rate values of viscosity (denoted $\eta_o(p)$ and η_∞) and n and λ, which govern the slope and position (in terms of shear-rate) of the shear-thinning region in which the bulk of the transition between the two asymptotes takes place. The parameters n and η_∞ are material constants for the given class of molecular structures, independent of the state-point and the molecular weight. This can be physically justified because at high shear-rates the molecules are aligned along the direction of flow so that the shear stresses are virtually independent of the length of the molecule. η_∞ is fitted from the simulation data. The shear-rate at which shear-thinning takes place is governed by the characteristic relaxation time of the molecule, marking the point at which aligning effects of the hydrodynamic flow are sufficient to overcome the thermodynamic relaxation of the molecule. Accordingly, we have estimated the parameter λ by the Rouse relaxation time (**35**):

$$\lambda = \frac{1}{\gamma_c} = \frac{K\eta_0(p)}{T\rho(p)} \tag{11}$$

K is a structure dependent constant, T is the temperature and $\rho(p)$ is the density given, according to Dowson and Higginson (**36**), by

$$\rho(p) = \rho_{atm}\left(1+\frac{\mu p}{1+\nu p}\right) \tag{12}$$

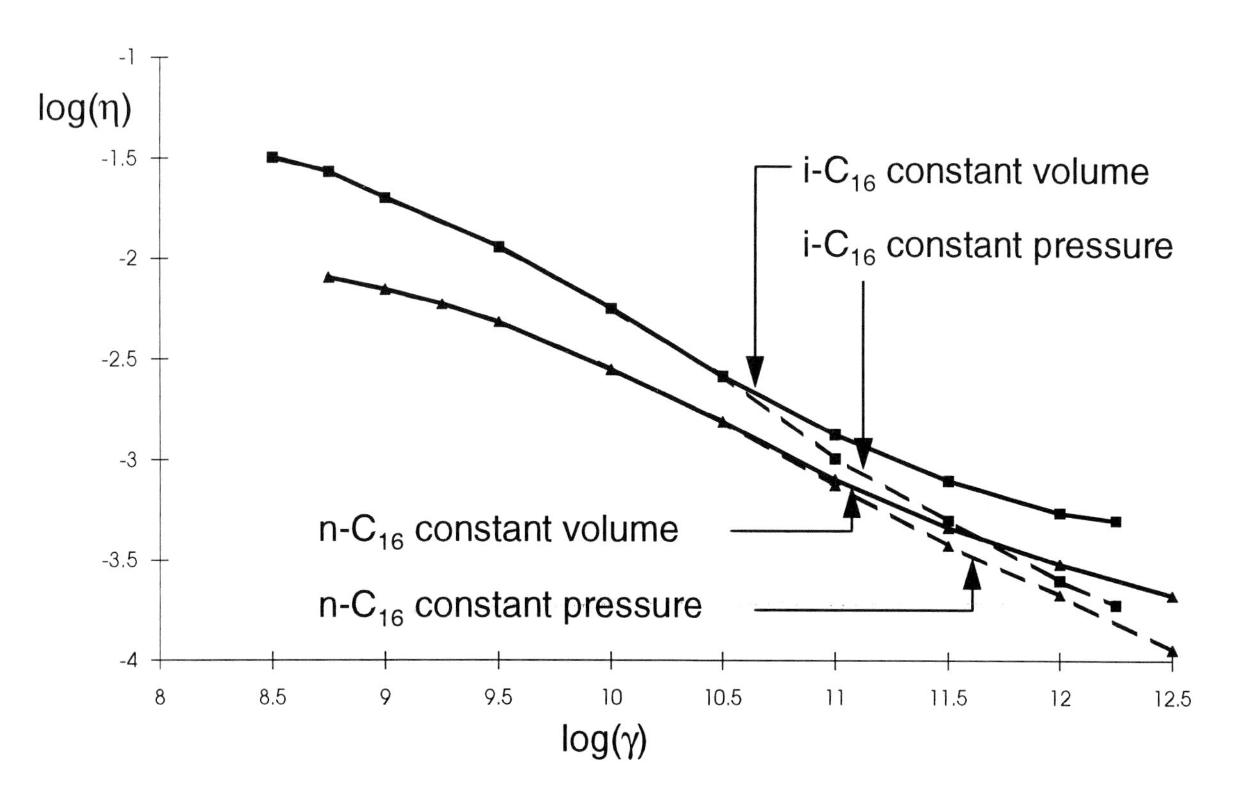

Fig. 2 NEMD simulation – effect of shear reate and constant density/pressure on velocity.

μ and v are parameters obtained by fitting experimental data and ρ_{atm} is the density at atmospheric pressure.

Combining Equations (10–12), we obtain viscosity as the function

$$\eta = \eta(\gamma, \eta_{atm}, \eta_{\infty}, n, K, \rho_{atm}, T, p, \alpha, z, \mu, v) \quad (13)$$

This constitutive equation for the fluid has been used as input for EHD calculations of friction. Figure 2 (**34**) illustrates the simulated flow curves for two simple hydrocarbons, n-Hexadecane, (n-C_{16}) and 2,2,4,4,6,8,8-Heptamethyl-nonane (i-C_{16}).

3 HYDRODYNAMIC/BOUNDARY FRICTION AND WEAR TRANSITION MODELS

Under full hydrodynamic or EHD lubrication conditions wear is minimal and to a first approximation can be considered to be zero. As the moving surfaces become closer the effect of surface roughness becomes apparent and there are asperity interactions. Figure 3 illustrates the way in which surfaces come together (**37**). The λ ratio, which is defined as the ratio of oil film thickness, *h*, divided by the combined surface roughness, *S*, gives an indication of surface interactions. For $\lambda > 5$ then full fluid film lubrication occurs and the wear rate can be taken as zero. In figure 4 the wear rate for h > hc is zero and λ can be considered to be > 5. When $h_c > h > h_o$, the λ ratio can be considered to be between 1 and 5 and there is mixed lubrication. The wear coefficient is taken to be a linear function of film thickness in this region. For $h<h_o$ then boundary lubrication prevails and a steady wear coefficient k_0 is used. The problem is determining what the roughness is. The initial roughness has been widely used to calculate the λ ratios but as can be seen from Figure 3 once surfaces are run-in they may be correlated (**38, 39**) and hence the effective roughness may be rather small and full film lubrication may exist for λ ratios nominally <1. Also the elastically deformed roughness is substantially lower than the undeformed roughness and micro-EHD analysis can be used to explore these effects. Thus the wear transition model is qualitative rather than quantitative and these simple transition models hide a great deal of complexity which has not yet been fully analysed.

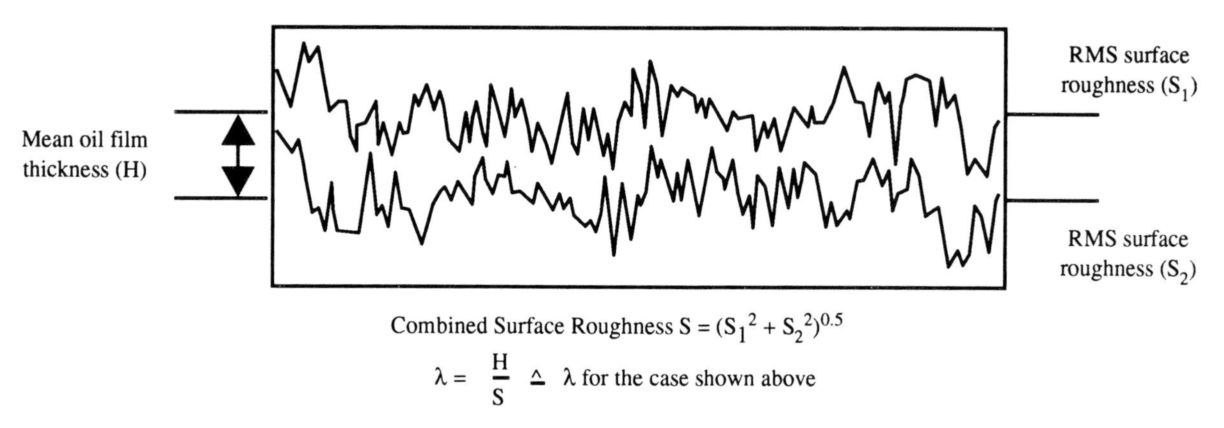

Fig. 3 Schematic illustration of an elastohydodynamic contact.

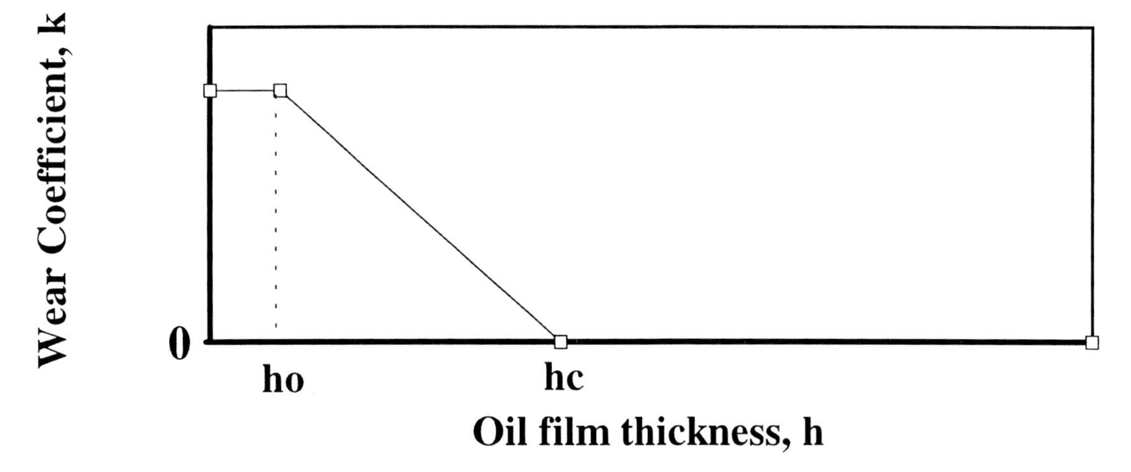

Fig. 4 Wear transaction model.

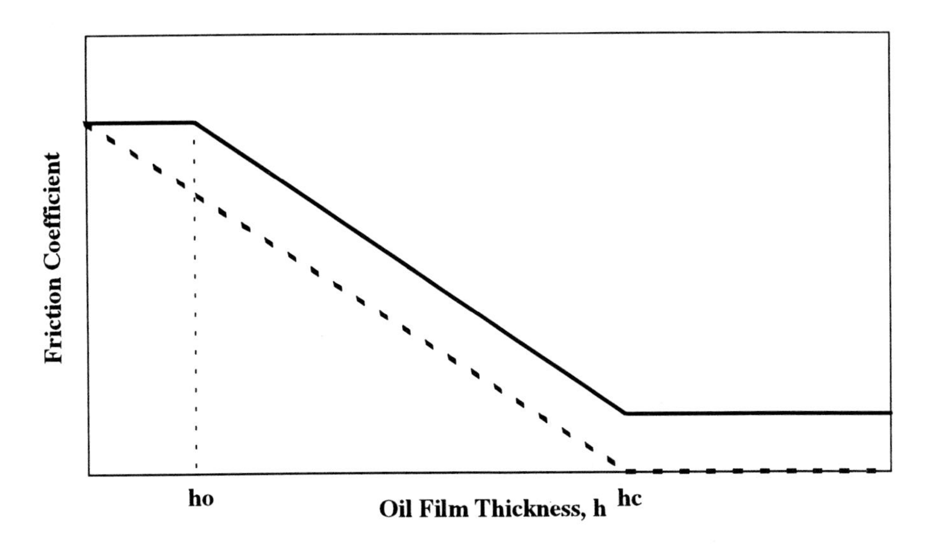

Fig 5 Friction transition model.

The transition friction model is illustrated in Figure 5. The boundary coefficient of friction μ is for many systems of the order of 0.1 whereas the hydrodynamic friction can be of the order of 0.0002 and hence if there is appreciable boundary lubrication hydrodynamic friction may be neglected. Under EHD conditions the friction can be higher and in the case of traction fluids comparable to boundary friction hence is taken as finite for $h>h_c$. These two cases are represented in Figure 5 by dotted and solid lines respectively.

4 VALVE TRAIN WEAR AND FRICTION MODELLING

Wear of valve train components is one of the most critical factors limiting the life of automotive engines and in the performance evaluation of lubricants. Finger/rocker follower systems are particularly wear-prone and the Sequence VE and PSA TU3M tests are designed to assess cam and follower wear. The kinematics of these systems are rather complex (**10, 38, 40**) and lead to wide changes in loading, sliding speed and hence oil film thickness. Figure 6 shows typical velocities across the active cycle of a valve train system. Using static EHD solvers (Static EHD solvers neglect squeeze lubrication effects, which are most important at small oil film thicknesses. Inclusion of squeeze effects in dynamic solvers gives more accurate solutions in these regions) the Oil Film Thickness (OFT) can be computed, see Figure 7. Examination of worn cams and followers from several engines indicated that in most cases steady, non-catastrophic wear processes prevail. Based on these observations, a mathematical model was developed (**10, 41**) based on a simple archard wear law (**42**) in which the wear coefficient k_0, obtained from subsidiary well controlled experiments, is made a linear function of OFT through the mixed lubrication regime, see Figure 4. This model is capable of predicting the wear profiles of both cams and followers.

$$W = \int_{t}^{t+\Delta t} k(h).F.|V_s|dt \qquad (14)$$

where W is the wear volume in m^3, $k(h)$ the wear coefficient in $m^3N^{-1}m^{-1}$, F the applied force in N and Vs the sliding speed in ms^{-1}. Comparison between theory and experiment, Figure 8, shows that the model includes the essential wear features and that changes in conditions such as loss of base circle clearance or 'jack-up' of hydraulic lifters, Figure 9, can be accounted for in failed components. Experimental measurement (**43**) of OFT between cam and follower in the TU3M showed major discrepancies with static EHD theory particularly around positions of zero entrainment velocity indicating that transient effects are important, see Figure 7. Currently the incorporation of dynamic EHD into the wear and friction models is progressing. Figure 10 shows the dynamic OFT as a function of cam angle which compares more closely with experiment.

In a similar way the friction in valve trains can be calculated using a transition model. For a given temperature, the oil viscosity (at the high shear stresses and shear rates in the EHD contact the viscosity of multigrade oils can, to a first approximation, be derived from the base oil viscosity (**44**)) is calculated and used to estimate the OFT. As soon as the OFT drops below hc (typically of the order of 0.15 to 0.2 μm dependent on the combined roughness) boundary lubrication is assumed to occur. The appropriate friction coefficient, f, is assumed to vary from a boundary value at h=0 (typically ~ 0.08 for friction modified oils to ~ 0.12 for conventional formulations) to a value of zero at $h=h_c$, see dotted curve in Figure 5. Above this value of OFT there is considered to be no contribution from hydrodynamic friction. This approximation has been found to be reasonable for the relatively low engine speeds that occur under fuel economy test conditions (**45**). For high speed operation and for low temperature cold start

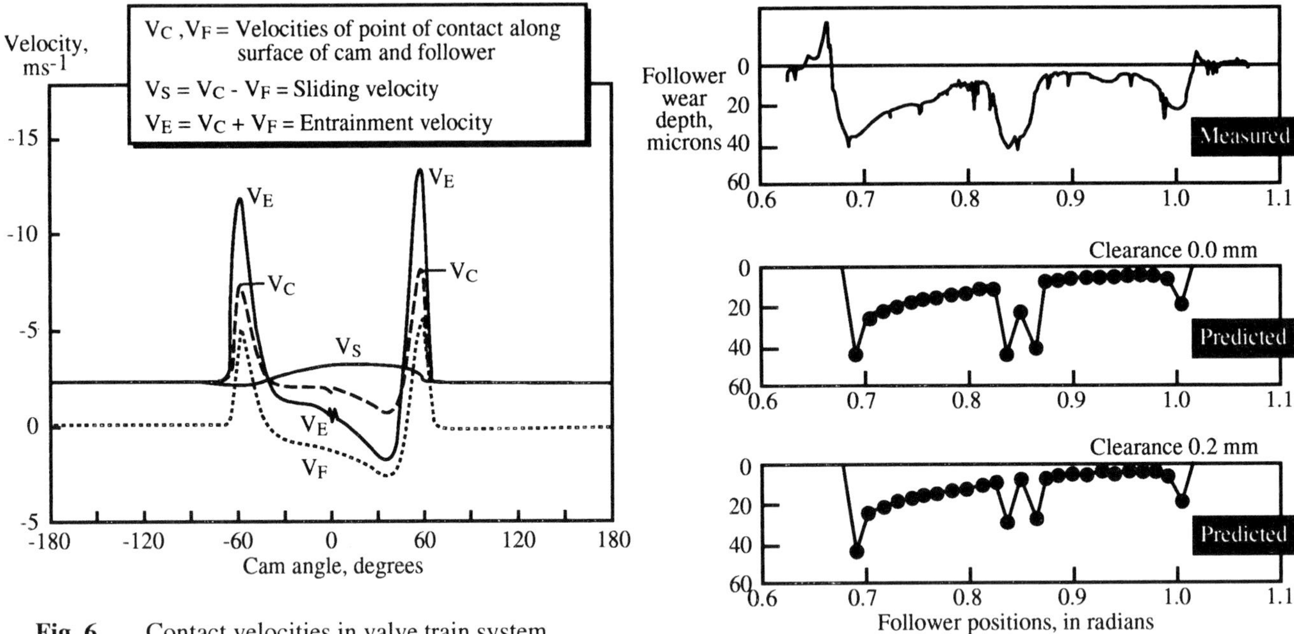

Fig. 6 Contact velocities in valve train system.

Fig. 8 Measured and predicted rocker pad wear.

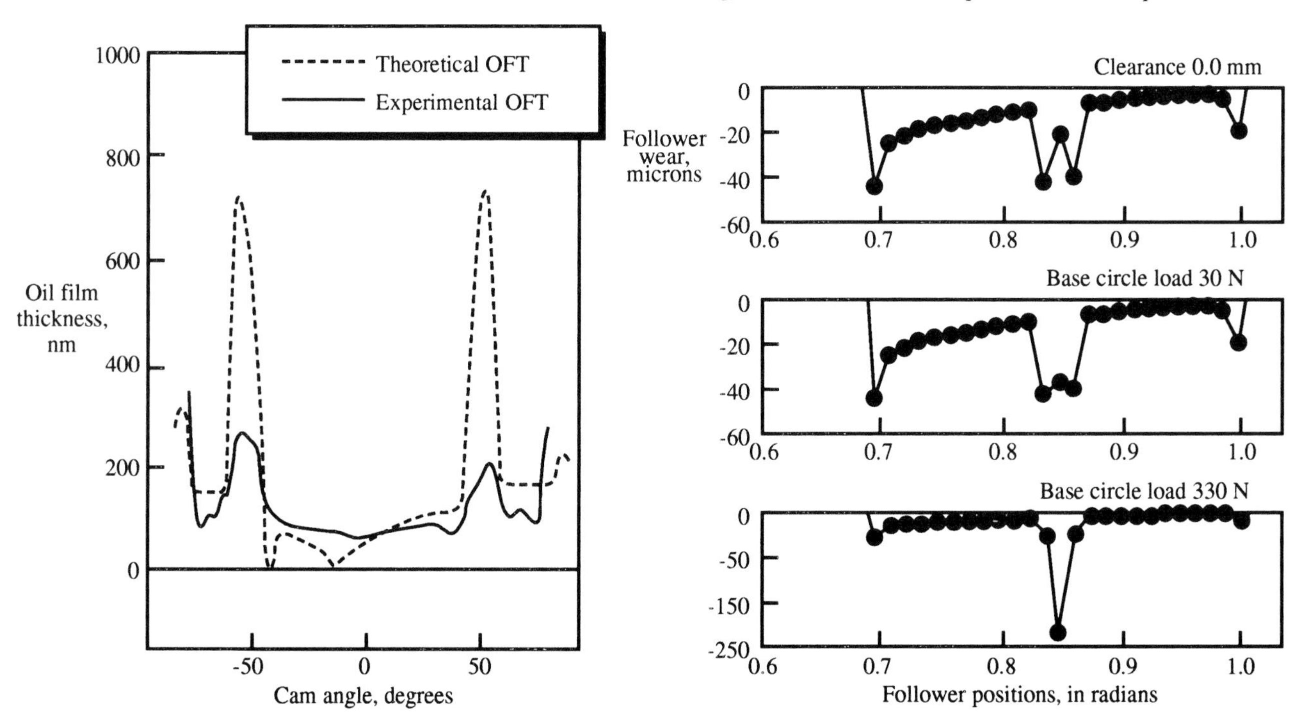

Fig. 7 Valve train oil film thickness.

Fig. 9 Effect of base circle loading.

conditions this assumption is not valid. Figure 11 shows a plot of measured versus modelled torque for a Mercedes Benz M111 valve train.

Piston ring-pack lubrication models need to take into account both lubricant viscosity/temperature and viscosity/shear rate variations but the viscosity/pressure effect can generally be ignored (**46, 47**). In addition, lubricant starvation of the upper piston rings, due to restriction of oil supply by the lower rings must be included. Figure 12 illustrates the effect of these variables on modelled OFT compared to experimentally measured OFT using a Laser-induced Fluorescence technique. Using the transition wear model it is possible to produce liner wear profiles for top ring contact, see Figure 13. The wear

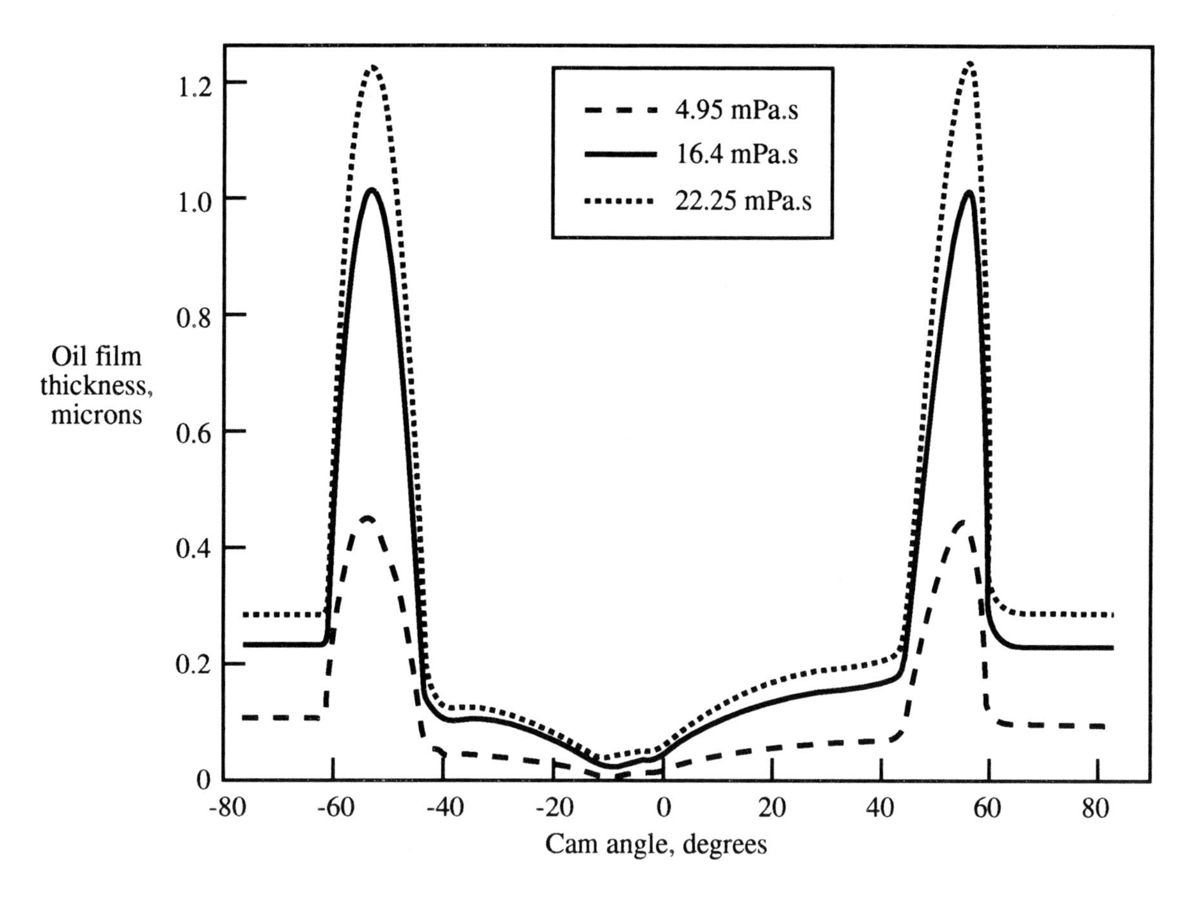

Fig 10 Peugeor TU3 valve train, 2000 r/min. Dynamic analysis.

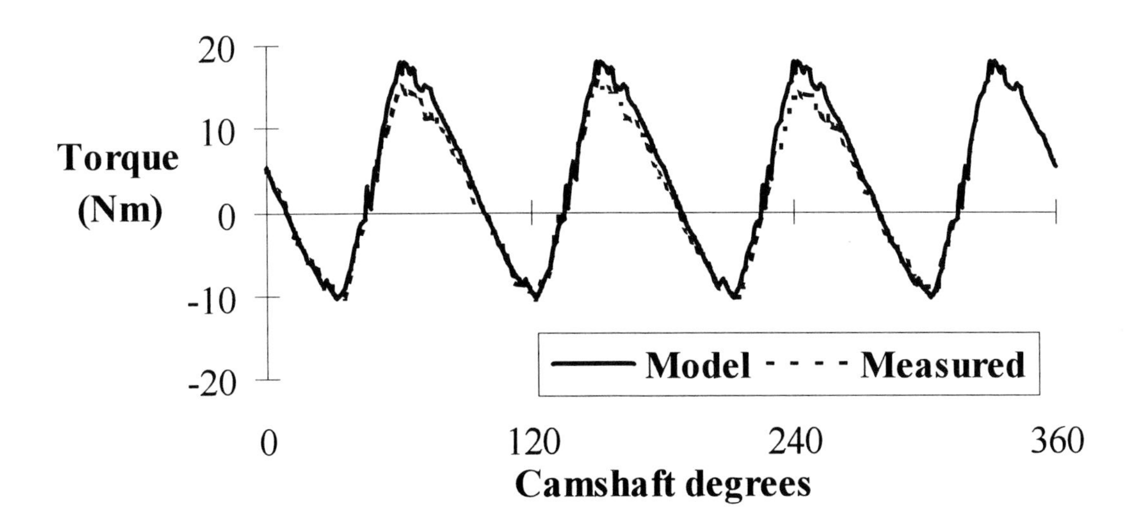

Fig 11 Comparism of measures and modelled torque in Mercedes Benz M111.

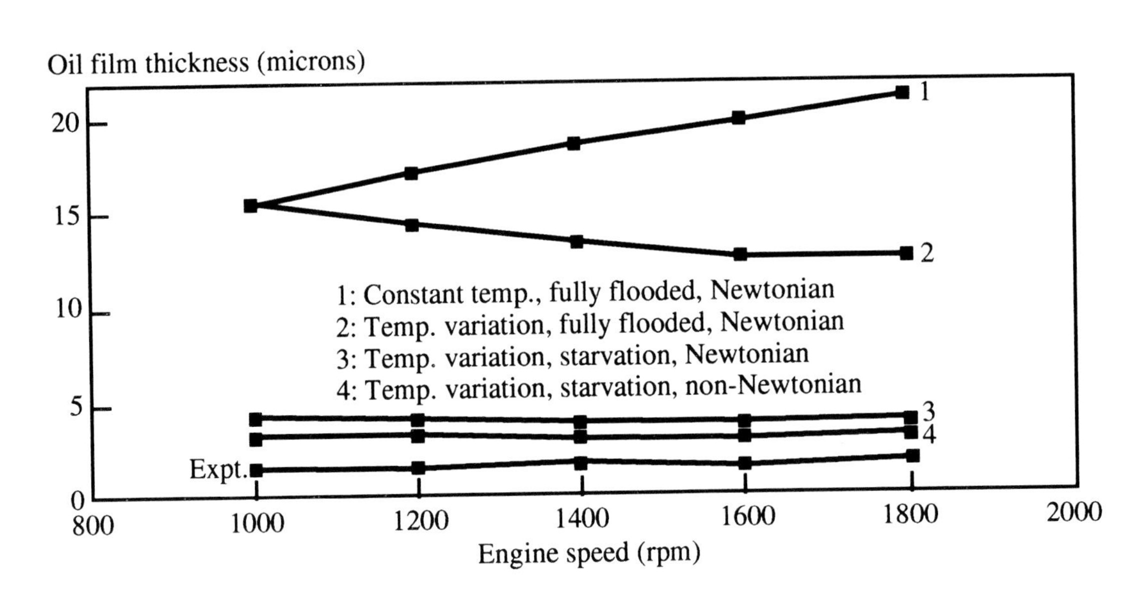

Fig. 12 Improved piston ring oil film thickness modelling.

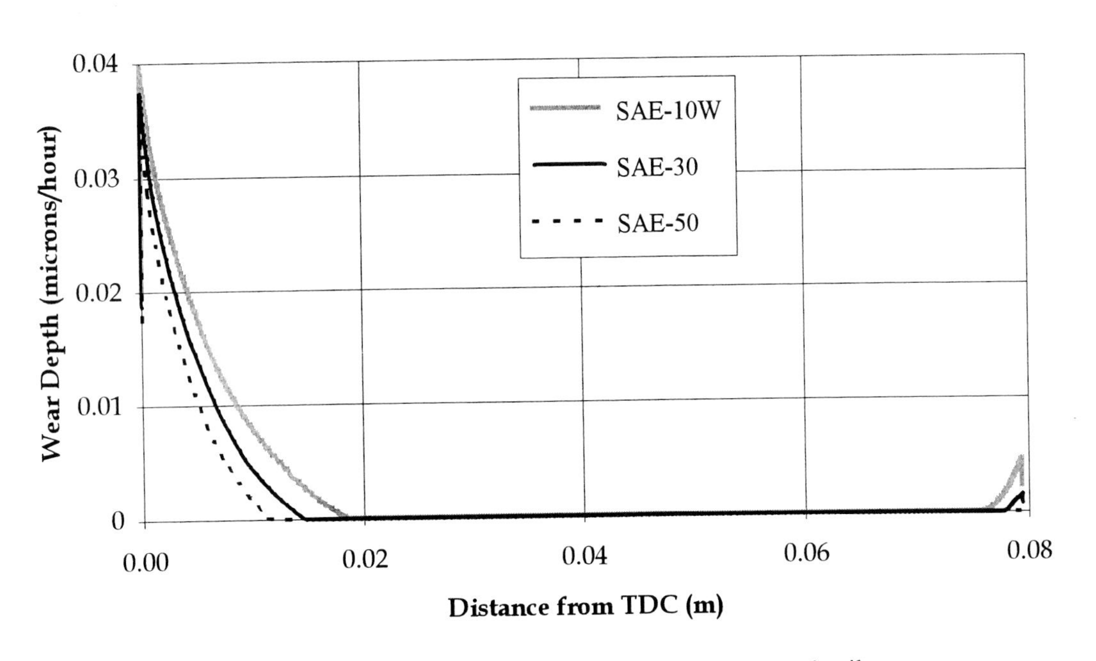

Fig. 13 Modelled cylinder linear wear for three single grade oils.

is concentrated around top of ring travel where velocities are low. The greatest wear occurs with the lowest viscosity single grade oil, 10W. Priest (**48**) has used the same approach to obtain ring wear and has included ring twist to obtain good agreement between modelled and experimental ring wear profiles for both diesel and gasoline engines.

The piston ring-pack lubrication model described above calculates the OFT under each ring and since the lubricant viscosity is known at each point along the liner (the temperature of the liner is assumed to vary linearly from top dead centre to bottom dead centre), the instantaneous friction force acting on each ring can be calculated. Using the friction transition model (solid line in Figure 5 bottom dead centre), the instantaneous friction force acting on each ring can be calculated. Using the friction transition model (solid line in Figure 5) the proportion of hydrodynamic/boundary friction for each ring can be calculated. If these instantaneous friction forces are multiplied by the piston velocity instantaneous power loss is obtained and this can be averaged over one engine cycle. Tables 1 (**9**) shows the variation of peak friction force, F_m, and friction mean effective pressure, P_f, for the two 10W50 oils shown in Figure 1 for a mean temperature of 70°C. The friction results rank the two oils in the same order as the experimental results and clearly show that a full rheological model is necessary to explain the friction performance of two oils nominally of the same viscosity. Table 2 shows the average power loss is lower for oil A than oil B The boundary terms in Table 2 are negligible apart from for the top ring where the most severe conditions occur. These results also confirm the need for a detailed rheological model.

temperature variation etc. at these extreme conditions. Thus for this severe, non-steady state failure mechanism, it does not appear to be appropriate to use the steady wear model. This also applies to friction where boundary lubrication can be neglected as the λ ratio is still above 1 even for catastrophic failure. These bearings can be considered to operate entirely under hydrodynamic conditions.

Using a full 2-D lubrication model (**7,11**) the effect of varying viscosity for both single and multigrade lubricants can be investigated. Tables 3 and 4 summarise the modelling data applied to the Mercedes Benz M111 for a fully warmed up engine. In these tables the valve train friction is entirely boundary and the bearing friction entirely hydrodynamic. The bearing data show as expected that the friction loss increases with increasing viscosity for the single grade (Newtonian) oils. However, the increase in friction losses for the multigrade oils is much less as the effective viscosities, at the high shear rates encountered in the bearings, are similar due to shear thinning. There is also an effect due to viscoelasticity which has been shown to give enhanced load bearing capacity experimentally (**51**). Full lubrication analysis (**11**) predicts that viscoelastic oils should not only give enhanced load bearing capacity at very high eccentricities (high pressures) but also reduced friction (**52**).

Table 1 Comparison of measured and predicted peak friction force F_m, and friction mean effective pressure P_f, for 1000rev/s and one quarter full load.

Oil	$\eta_\infty(70^0C)$	P_f (expt)	P_f (model)	F_m (expt)	F_m (model)
	(cSt)	(kPa)	(kPa)	(N)	(N)
SAE-10W50A	21.09	40.8	23.1	460	295.1
SAE-10W50B	24.06	43.4	25.0	440	264.6

Table 2 Average power loss of each ring for two 10W50 multigrades (H indicates hydrodynamic losses, B boundary losses)

10W50 Oils	Average Power Loss (Watts)							
	Oil control ring		2nd ring		top ring		total	
	H	B	H	B	H	B	H	B
A	47.54	0.02	68.40	0	80.43	3.27	196.37	3.29
B	52.07	0.01	74.12	0	87.53	2.47	213.72	2.48

5 JOURNAL BEARING WEAR AND FRICTION

Main and big end bearing wear does not appear to be amenable to this form of analysis. Because the surfaces of journals and bearings have low roughness and the surfaces are conformal the premise that there is a steady mild wear process taking place does not appear to hold. In a bench durability test (**49, 50**) big-end bearing failure was catastrophic and the computed minimum OFT was ~ 0.2 μm still above the mean roughness. Failure is thus dependent on the effects of distortion/deflection,

Table 3 Friction loss predictions for single grade oils in Mercedes Benz M111 gasoline engine (P indicates piston assembly, VT valve train and B bearing losses respectively).

GRADE	TOTAL, W	P ,W (%)	VT,W (%)	B ,W (%)
SAE-10W	1420.7	461.5 (33)	413.3 (29)	545.9 (38)
SAE-30	1472.7	620.1 (42)	276.8 (19)	575.8 (39)
SAE-50	1584.4	850.8 (54)	88.1 (6)	645.5 (41)

Table 4 Friction loss predictions for multigrade oils in Mercedes Benz M111 gasoline engine (P indicates piston assembly, VT valve train and B bearing losses respectively).

SAE grade	TOTAL (W)	P,W (%)	VT,W (%)	B,W (%)
10W/30	1455.1	528.4 (36)	371.4 (26)	555.3 (38)
15W/40	1513.3	638.9 (42)	286.9 (19)	587.5 (39)
20W/50	1577.6	807.4 (51)	140.5 (9)	629.7 (40)

6 TOTAL ENGINE FRICTION

Finally the friction models for valve trains, pistons and bearings can be combined to give a total engine friction model. In Tables 3 and 4 the total friction is tabulated on the assumptions given above, namely that valve train and bearing frictions are entirely boundary and hydrodynamic respectively. It can be seen that the boundary friction from the valve train reduces as viscosity increases whereas for piston assembly and bearings the friction increases with increasing viscosity.The multigrade oils give lower friction than their equivalent single grades due to shear thinning effects. Another approach is to apply the transition friction model to all three components . In this model it has been assumed that there is boundary friction in the bearings (ie that there is a steady wear condition for low λ ratios) so that as the viscosity reduces in all components of the model there comes a point when friction starts to increase. Also EHD friction in the valve trains has been included. Figure 14 schematically plots friction versus viscosity for each component and their sum for a given engine operating condition. There is thus an optimum viscosity for this set of engine conditions which gives a minimum friction loss. This optimum will change with engine conditions but for example the friction losses for a limited set of conditions, as in a fuel economy test, can be modelled and an overall optimum explored. Sensitivity analyses to viscosity, boundary coefficient of friction, transition model parameters, etc. will then provide options for the formulator and engine designer to explore.

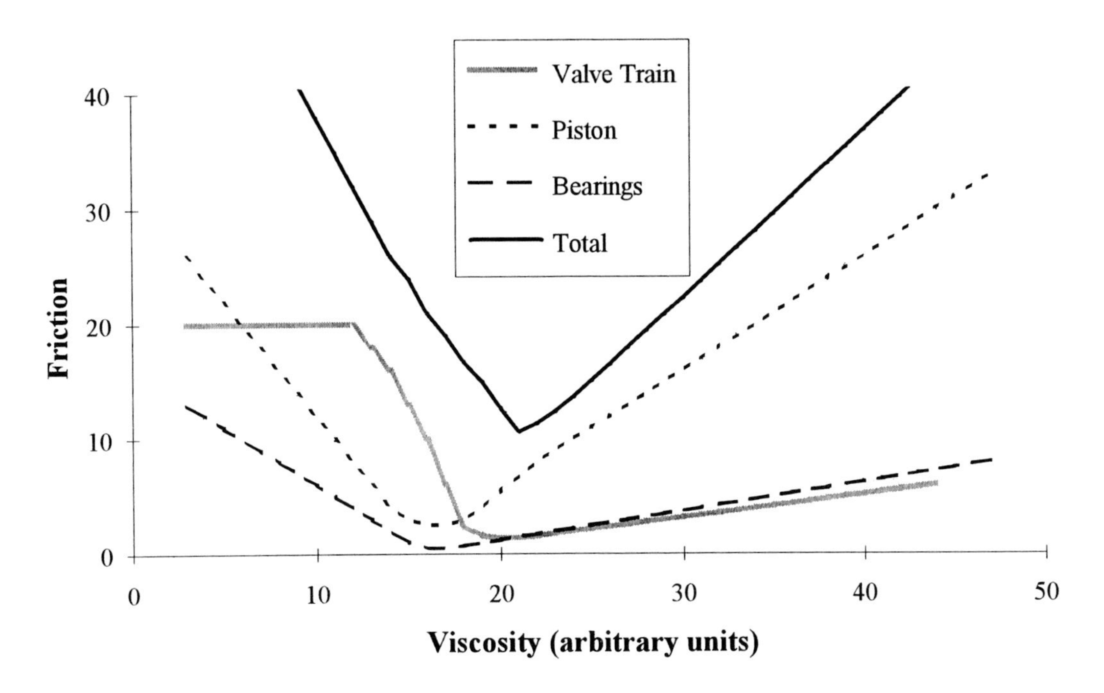

Fig. 14 Schematic plot of total engine friction, sum of different Stibeck curves.

7 CONCLUSIONS

This paper has reviewed the approaches to modelling friction and wear in engines. It is very much a 'Shell' view. Over many years researchers in Shell and elsewhere have improved and optimised these models as our knowledge has grown and ever more sophisticated mathematical and modelling tools have become available. They have demonstrated that detailed rheological models are required where the effects of temperature, pressure and shear rate all need to be taken into account if realistic models of friction and wear are to be developed. For polymer containing multigrade oils the effect of shear thinning under hydrodynamic conditions is both of a viscometric and elastic nature where the later provides an added bonus under extremes of pressure and shear rate. However, these benefits are difficult to substantiate experimentally as they occur under such extreme conditions; nevertheless substantial progress has been made in this area. These non-Newtonian effects are also present in base oils under even more severe elastohydrodynamic conditions and are crucial to modelling friction. Again progress has been made in our understanding and measurement of these complex phenomena.

The use of a simple linear wear model has been shown to be effective in reproducing qualitatively the wear patterns seen in engine valve trains and piston assemblies. This has been used to explain many failures observed in the field and has been used with engine designers to explore valve train optimisation. To get more quantitative results, non-linear, time dependent wear models involving wear transitions (**53**) combined with dynamic EHD models are currently being developed.

The wear model has not been so successful for bearing wear. However, great progress has been made in developing a complete lubrication bearing model that has been validated experimentally and which explains why, under severe transient conditions, viscoelastic properties are important.

Finally, combining these models to get a picture of total engine friction and wear provides insight for both lubricant formulator and engine designer which will lead to better engines and lubricants. Sophisticated lubrication models are clearly required to explore the detailed parameters needed to enhance the design of critical engine components. Appropriate levels of sophistication can be selected dependent on the nature of the problem under investigation.

8 ACKNOWLEDGEMENTS

The author wishes to thank many colleagues who have contributed over the years to the body of work described here. In particular to J.C. Bell, S. Chynoweth, G.W. Roper, L.E. Scales A.G. Schlijper, R.I. Taylor and B.P. Williamson. Acknowledgement is also due to Shell Research Ltd for permission to publish.

9 REFERENCES

1 **Korcek S.** and **Nakada, M.,** Engine oil Performance Requirements and reformulation for Future Engines and Systems *Proceedings of the International Tribology Conference,* Yokohama, 1995, pp 783-788.

2 **Gairing, M., Frend M.,** and **Reglitzky, A.,** Environmental Needs and New Automotive Technologies Drive Lubricants Quality, *14th World Petroleum Congress,* Norway, 1994.

3 **Dohner B.R.** and **Wilk, M.A.**, *Formulating for ILSAC-GF/2 – Part 2: Obtaining Fuel Economy Enhancement from a Motor Oil in a Modern Low Friction Engine*, SAE 952343.

4 **Tseregounis S.I.** and **McMillan, M.L.**, *Engine Oil Effects on Fuel Economy in GM Vehicles – Comparison with the ASTM Sequence VI A Engine Dynamometer Test,* SAE 952347.

5 American Petroleum Institute Publication 1509, *Engine Oil Licencing and Certification System,* 1996; European Association des Constructeurs Europeens d'Automobiles Engine Oil Sequences, 1996.

6 **Bell, J.C.,** Effects of Valve train Design Evolution on Motor Oil Anti-wear Requirements, *Fifth CEC International Symposium on the Performance Evaluation of Automotive Fuels and Lubricants,* Gothenberg, May 1997.

7 **Taylor, R.I.,** Engine Friction: The Influence of Lubricant rheology, *Fifth CEC International Symposium on the Performance Evaluation of Automotive Fuels and Lubricants,* Gothenberg, May 1997.

8 **Taylor, R.I., Brown, M.A., Thompson D.M.,** and **Bell, J.C.,** The Influence of Lubricant Rheology on Friction in the Piston Ring Pack, SAE 941981.

9 **Taylor, R.I., Kitahara, T., Saito T.**, and **Coy, R.C.,** Piston Assembly Friction and Wear: The Influence of Lubricant Viscometry, *Procs. of the International Tribology Conf.,* Yokohama, 1995, pp1423-1428.

10 **Colgan T.** and **Bell, J.C.,** *A Predictive Model for Wear in Automotive Valve Train Systems*, SAE 892145.

11 **Davies A.R.** and **Li, X.K.**, Numerical Modelling of Pressure and Temperature Effects in Viscoelastic Flow Between Eccentrically Rotating Cylinders, *J. Non-Newtonian Fluid Mech.,* **54**, pp 331-350, 1994.

12 **Goenka, P.K., Paranjpe R.S.** and **Jeng, Y.-R.,** *FLARE : An Integrated Software Package for Friction and Lubrication Analysis of Automotive Engines – Part I: Overview and Applications*, SAE 920487.

13 **Bovington C.** and **Spikes, H.,** Predictions of the Influence of Lubricant Formulations on Fuel Economy, from Laboratory Bench Tests, *Proceedings of the International Tribology Conference,* Yokohama, 1995, pp817-822.

14 **Bates, T.W., Williamson, B.P., Sperot, J.A.** and **Murphy, C.K.,** *A Correlation Between Engine Oil Rheology and Oil Film Thickness in Engine Journal Bearings.* SAE Paper No. 860376 (1986).

15 **Bates T.W., Roberts G.W., Oliver D.R.** and **Milton A.L.,** *A Journal bearing Simulator Bench test for Ranking Lubricant Load-Bearing Capacity: A Theoretical Analysis,* SAE Paper No. 922352 (1992).

16 **Scales, L.E., Rycroft, J.E. Horswill N.R.,** and **Williamson, B.P.,** *Simulation and observation of transient effects in elastohydrodynamic lubrication, SP-1182,* SAE International Spring Fuels and Lubricants Meeting, Dearborn, Michigan, (1996) 23-34.

17 **Cameron, A.,** *Basic Lubrication Theory,* Third Edition, (published by Ellis Horwood Ltd., 1983)

18 **Cross, M.M.,** Rheology of non-Newtonian fluids: A new flow equation for pseudoplastic systems. *J. Colloid Sci.,* **20** (1965) 417.

19 **Okrent, E.H.,** The Effect of Lubricant Viscosity and Composition on Engine Friction and Bearing Wear, *ASLE Trans.,* **4** (1961) 97.

20 **Okrent, E.H.,** The Effect of Lubricant Viscosity and Composition on Engine Friction and Bearing Wear. II, *ASLE Trans.,* **4** (1961) 257.

21 **Okrent, E.H.,** Engine Friction and Bearing Wear. III. The Role of Elasticity in Bearing Performance, *ASLE Trans.,* **4** (1964) 147.

22 **Olson, D.H.,** *Relationship of Engine Bearing Wear and Oil Rheology* SAE Paper No. 872128 (1987).

23 **Hutton, J.F., Jackson, K.P.** and **Williamson, B.P.**, *The Effects of Lubricant Rheology on the Performance of Journal Bearings.* ASLE Preprint No. 84-LC-1C-1 (1984).

24 **Roberts, G.W.** and **Walters, K.,** Oil Viscoelastic Effects in Journal Bearing Lubrication, *Rheol. Acta.,* **31** (1992) 55.

25 **Berker, A., Bouldin, M.G., Kleis S.J.,**and **Van Arsdale, W.E.,** Effect of Polymer Flow in Journal Bearings, *J. Non-Newtonian Fluid Mech.,* **56,** (1995) 333-347.

26 **Rastogi, A.** and **Gupta, R.K.,** Lubricant elasticity and the performance of dynamically loaded short journal bearings, *J. Rheo.,* **34(8)** (1990), 1337 - 1356

27 **Gwynllyw, D.Rh., Davies, A.R.,** and **Phillips, T.N.,** *On the Effects of a Non-Newtonian Lubricant on the Dynamics of a Journal Bearing,* Paper to be published in *J. Rheology.*

28 **Gwynllyw, D.Rh., Davies, A.R.,** and **Phillips, T.N.,** A Moving Spectral Element Approach to ther Dynamically Loaded Journal Bearing Problem, *J. Computational Phys.* **123**,1996, pp476-494.

29 **Bair S.** and **Winer, W.O.**,The high pressure, high shear stress rheology of liquid lubricants, *ASME J. Tribology* **114** (1992) 1.

30 **Bair S.** and **Winer, W.O.,** The high shear stress rheology of liquid lubricants at pressures of 2 to 200 mPa, *ASME J. Tribology,* **112** (1990) 246.

31 **Chynoweth, S. Coy R.C.** and **Michopoulos, Y.** Generic properties of rheological flow curves, *Proc. Inst. Mech. Engrs.,* **209(J4)** (1995) 243-254.

32 **Chynoweth, S., Coy R.C.,** and **Michopoulos, Y.** Simulated non-Newtonian Lubricant behaviour under Extreme Conditions *Proc. Instn. Mech. Engrs., Part J: J. Eng. Tribology*, vol. **208,** pp. 243-254, (1995).

33 **Chynoweth, S., Coy, R.C., Michopoulos, Y.,** and **Scales, L.E.,** Simulated lubricant behaviour under elastohydrodynamic conditions, *Proc. of the International Tribology Conference,* Yokohama, 2 (1995) 663-668.

34 **Chynoweth, S., Coy, R.C., Holmes A.J.,** and **Scales, L.E.**, Simulated Lubricant non-Newtonian Behaviour under Elastohydrynamic Conditions *Proc. of the 23rd, Leeds/Lyon Symposium on Tribology*, 1996.

35 **Rouse Jr., P.E.,** A theory of the linear viscoelastic properties of dilute solutions of coiling polymers, *J. Chem. Phys.,* **21** (1953) 1272-1280.

36 **Dowson. D.** and **Higginson. G.R.,** A numerical solution to the elastohydrodynamic problem, *J. Mech. Eng. Sci.,* **1** (1959) 7-15.

37 **Roper G.W.** and **Bell, J.C.,** *Review and Evaluation of Lubricated Wear in Simulated Valve Train Contact Conditions,* SAE 952473.

38 **Dyson, A.** Kinematics and Wear Patterns of Cam and Finger Follower Automotive Valve Gear, *Tribology International*, Vol. **13,** No. 3, 1980.

39 **Coy R.C.** and **Dyson, A.,** A Rig to Simulate the Kinematics of the Contact Between Cam and Finger Follower, *Lubrication Engineering*, Vol. **39**, No. 3, 1983, pp143-152.

40 **Bell J.C.** and **Colgan, T.A.**, Critical Physical Conditions in the Lubrication of Automotive Valve Train Systems, *Tribology International,* Vol. **24,** No. 2, 1991, pp 77-84.

41 **Bell, J.C. Davies P.T.,** and **Fu, W.B.,** Prediction of Automotive Valve train Wear Patterns with a Simple Mathematical Model, *Proceedings 12th Leeds-Lyon Symposium on Tribology*, Lyon, 1985.

42 **Archard, J.F.,** Contact and Rubbing of Flat Surfaces, *J. App. Phys.*, Vol **24,** 1953, p 981.

43 **Williamson, B.P.** and **Perkins, H.N.** *The effects of engine oil rheology on the oil film thickness between a cam and rocker follower,* SAE 92, 2346.

44 **Williamson B.P.** and **Bell, J.C.** *The Effects of Engine Oil Rheology on the Oil Film Thickness and wear between a Cam and Rocker Follower*, SAE 962031.

45 **Staron J.T.** and **Willermet, P.A.** An Analysis of Valve Train Friction in Terms of Lubrication Principles, SAE 830165

46 **Rycroft, J.E. Taylor R.I.** and **Scales, L.E.** Elastohydrodynamic Effects in Piston Ring Lubrication in Modern Gasoline and Diesel Engines, *Proceedings 23rd Leeds-Lyon Symposium on Tribology*, Leeds, 1996.

47 **Taylor, R.I.. Brown, M.A Thompson D.M.** and. **Bell, J.C** *The Influence of Lubricant Rheology on Friction in the Piston Ring Pack*, SAE 941981.

48 **Priest, M.** The Wear and lubrication of Piston Rings, Ph.D. Thesis, University of Leeds, Department of Mechanical Engineering, October, 1996.

49 **Bates, T.W.** and **Toft, G.B.,** *Effect of Oil Rheology on Journal Bearing Performance: Part 4 – Bearing Durability and Oil Film Thickness* SAE Paper No. 892154 (1989).

50 **Coy, R.C. Kirsch, L.J. Bates T.W.** and **Burnett, P.J.** Automotive Lubrication Studies, *Proceedings of the 4th International Tribology Conference*, 'Austrib 94', Vol. 2, 1994, pp751-759.

51 **Williamson, B.P.** *Enhancement of Journal Load Bearing Capacity by Polymer Containing Oils*, SAE 971697

52 **Scales, A.R. Davies, D.Rh. Gwynllyw, T.N. Phillips, T. Tasche** and **B.P. Williamson**, *The Effect of Lubricant Rheology on the Stability and Performance of Dynamically Loaded Journal Bearings,* SAE 971697.

53 **Bell J.C.,** and **Willemse P.J.,** The development of scuffing failure in an innovative valve train system, *World Tribology Congress, London,* 1997.

Tribochemistry of ceramics: science and applications

T E FISCHER
Stevens Institute of Technology, Hoboken, NJ, USA

SYNOPSIS

It is now well established that friction modifies the kinetics of chemical reactions of solids with each other, with the gaseous environment or with liquids to the extent that reactions, which ordinarily need high temperatures to be observable, occur at moderate temperatures during sliding. There are several mechanisms for this interaction besides the obvious frictional heat, namely: removal of product scale, acceleration of diffusion, and direct mechano-chemical excitation. Depending on the solid and the reaction, this interaction can lead to accelerated fracture and wear, or to the decrease of contact stress concentration and reduced wear. The reactions occur only at the frictional contact and can therefore produce very smooth surfaces. This phenomenon has applications in the functioning of lubricant additives, in the operation of seals and bearings and in the polishing of ceramics. When the material removal is entirely chemical, the polishing method avoids surface defects such as microcracks and produces surfaces with excellent mechanical and electrical integrity. With a simple machine, we have produced silicon nitride, silicon,and silicon carbide surfaces with R_a = 0.5 nm at 50 μm cut-off and 4 nm at 8 mm cut-off length.

1 INTRODUCTION

Since friction, in its simplest form, involves the making and breaking of adhesive bonds between the sliding bodies and is always influenced by surface films that modify this adhesion, it is almost obvious that chemical reactions of the bodies with each other, with the environment or with liquid lubricants play a major role in tribology. It is widely observed that chemical reaction rates are strongly modified, usually accelerated, by the simultaneous occurrence of friction. This phenomenon bears the name of tribochemistry and has been extensively studied in the Institute of Peter-Adolf Thiessen in Berlin (**1,2**). Their work was primarily focused on the exploration of the tribochemical reactions produced in ball mills. Thiessen and coworkers describe many industrial uses of this phenomenon for the processing of materials.

We will concentrate our attention on the tribochemical reactions that occur in sliding (**3**). Modifications of lubricants in sliding service are well known. Polymer molecules dissolved in the oils to decrease the temperature dependence of the viscosity are straightened and aligned or cracked by the high shear, leading to temporary or permanent loss of viscosity. The action of antiwear and 'extreme pressure' additives in lubricants is based on tribochemical reactions. In oxidative wear it has been shown that the oxidation reaction is accelerated by friction and that this creates an oxide surface film that shears preferentially to the substrate with consequent decrease in wear (**4-7**). Tribochemical reactions of silicon nitride in humid ambient lead to a decrease in wear (**8**), Diamond-like carbon films likewise are known to interact with the environment (**9**).

We shall concern ourselves here with the kinetics of tribochemical reactions; namely, with the modification of the reaction rates by simultaneous friction (**3**). We first describe various mechanisms by which friction modifies reaction kinetics, we then review the tribochemistry of ceramics to gain an overview of the different phenomena that are encountered, finally we examine the application of these phenomena to the polishing of ceramics and semiconductors.

2 OBSERVATIONS OF TRIBOCHEMISTRY

Tribochemical reactions during sliding manifest themselves in various ways. Generally, the wear particles are a chemical reaction product: in oxidative wear of steels, the wear particles are iron oxide (**4-7**); in the tribochemical wear of silicon nitride, an amorphous oxide is found on the surface even when sliding occurred at low temperatures (**8**). Often, also, the kinetics of wear are modified by the reaction. In the oxidative wear of steel, a specific dependence of the wear rate (in mm^3/Nm) on the sliding speed allows comparism it with a chemical kinetic model that provides information on the mechanisms by which friction modifies the reaction rate (**4-7**). In silicon nitride, the generation of a smooth wear surface and the absence of solid wear particles reveals the presence of tribochemical dissolution in water (**10**).

3 MECHANISMS OF TRIBOCHEMISTRY

The most obvious mechanism by which friction increases the rate of chemical reactions is frictional heat which has been used since the earliest times for the production of fire. One distinguishes between general increase in temperature and flash temperature (**11-13**); the latter consists of the temperature flashes that occur at contacting asperities. The increases in reaction rates caused by frictional heat are no different from those caused by other increases in temperature.

When the sliding velocity and load are kept low to avoid frictional heating, one observes that other mechanisms operate by which friction increases reaction rates. One such mechanism is the exposure of clean surfaces. It is well known that the rates of chemical reactions are controlled by diffusion of reagents through the layer of reaction product that is formed on the surfaces. Wear of these scales exposes fresh surfaces and accelerates the reaction; in steady state, the reaction rate equals the rate of removal of the reaction product. Another mechanism is the modification of the material surfaces. Since wear constitutes a severe deformation of material on a small scale, it can introduce defects in the surface layer which serve as high-energy sites with increased reactivity on the surface and as diffusion paths that cause large increases in diffusion-limited reaction rates in the subsurface region. Razavizadeh and Eyre, for example, observed relatively thick layers of oxide on the surface of friction tracks in aluminum (**14**).

Rubbing and fracture cause charge separations in ionic materials and set up large electrostatic potentials that can lead to discharge in the surrounding gas. With metallic bodies separated by a lubricant, electrochemical potentials are often observed. In addition, exoelectron emission is often quoted as a cause of tribochemical reactions despite the very low emission currents usually observed.

Finally, there is the possibility of direct mechanical stimulation of chemical reaction. Consider two atoms that are separated by large mechanical stresses. As the distance of these atoms increases, the energy splitting between the highest occupied molecular orbital (HOMO, usually bonding) and the lowest unoccupied molecular orbital (LUMO, usually antibonding) decreases. This diminishes the activation energy of the electron transfer taking part in a chemical reaction. Such phenomena occur only where atoms are separated; namely, in friction and in fracture. In solids, such a mechanism participates in the phenomenon of adsorption-induced fracture or stress corrosion cracking (**15, 16**); it is responsible for the wear increase of oxide ceramics by water and polar hydrocarbons.

The experimental identification of tribochemical mechanisms, namely the details of the interaction of friction and chemical reaction to modify the kinetics of both, is difficult. As often done in chemistry, the mechanisms can be inferred from details of the reaction kinetics. This has been possible in the oxidative wear of steel (**4–7**). The temperature dependence of tribochemical reactions can provide some information on the mechanisms. Heinicke and coworkers (**1, 2**) have shown that the temperature-dependence of tribochemical reactions can exhibit Arrhenius-type behavior with activation energies different from simple chemical reaction;, they can also exhibit a rate that does not change or decrease with increasing temperatures. Figure 1 shows, as an example, the temperature dependence of the tribochemical polishing of silicon nitride (**17**). We see that in water, the reaction shows complex temperature dependence: increasing at low temperature and abruptly decreasing above 65°C. In the presence of chromic acid, the same reaction exhibits an Arrhenius behavior with a small activation energy of 20 kJ/mole.

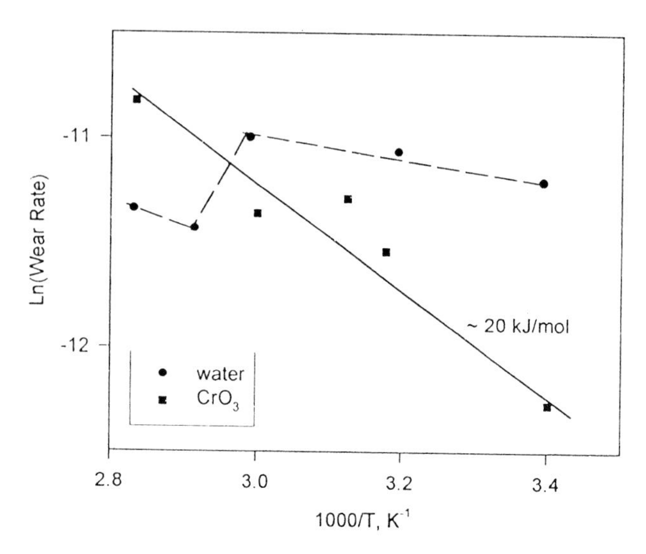

Fig. 1 Temperature dependence of the tribochemical removal rate of silicon nitride when polished in water and in chromic acid. The data of chromic acid follow an Arrhenius plot corresponding to an activation energy of 20 kJ/mole.

We are presently attempting to observe the occurrence or the results of tribochemical reactions by FTIR in situ (**17**). One of the rubbing surfaces is deposited on to a germanium substrate through which frustrated total internal reflection spectroscopy is possible. Figure 2 shows such an example of the polishing of silicon oxide in water: during friction, we observe a modification of the concentration of Si-H on the surface. These are early observations and their interpretation is not complete.

4 TRIBOCHEMISTRY IN CERAMICS

The tribochemistry of ceramics takes several forms, depending on the materials, the environment and the mechanical conditions of rubbing; it can consist in chemically induced cracking that increases wear rates (**18**), in modifications of surface composition and topology that decrease wear (**8,19**), in a purely chemical form of wear (by dissolution in the liquid environment) (**10, 20**).

The ambient humidity has a pronounced effect on the wear of silicon nitride and other ceramics (**18–24**): not only the amount of wear, but the wear mechanism itself is modified by humidity. Silicon nitride wears rapidly in dry argon, but if the environment contains various amounts of water vapor, the wear rate decreases by as much as two orders of magnitude. Under these conditions, the wear scar is much smoother than after sliding in dry gases and is covered with an amorphous silicon oxide which is probably strongly

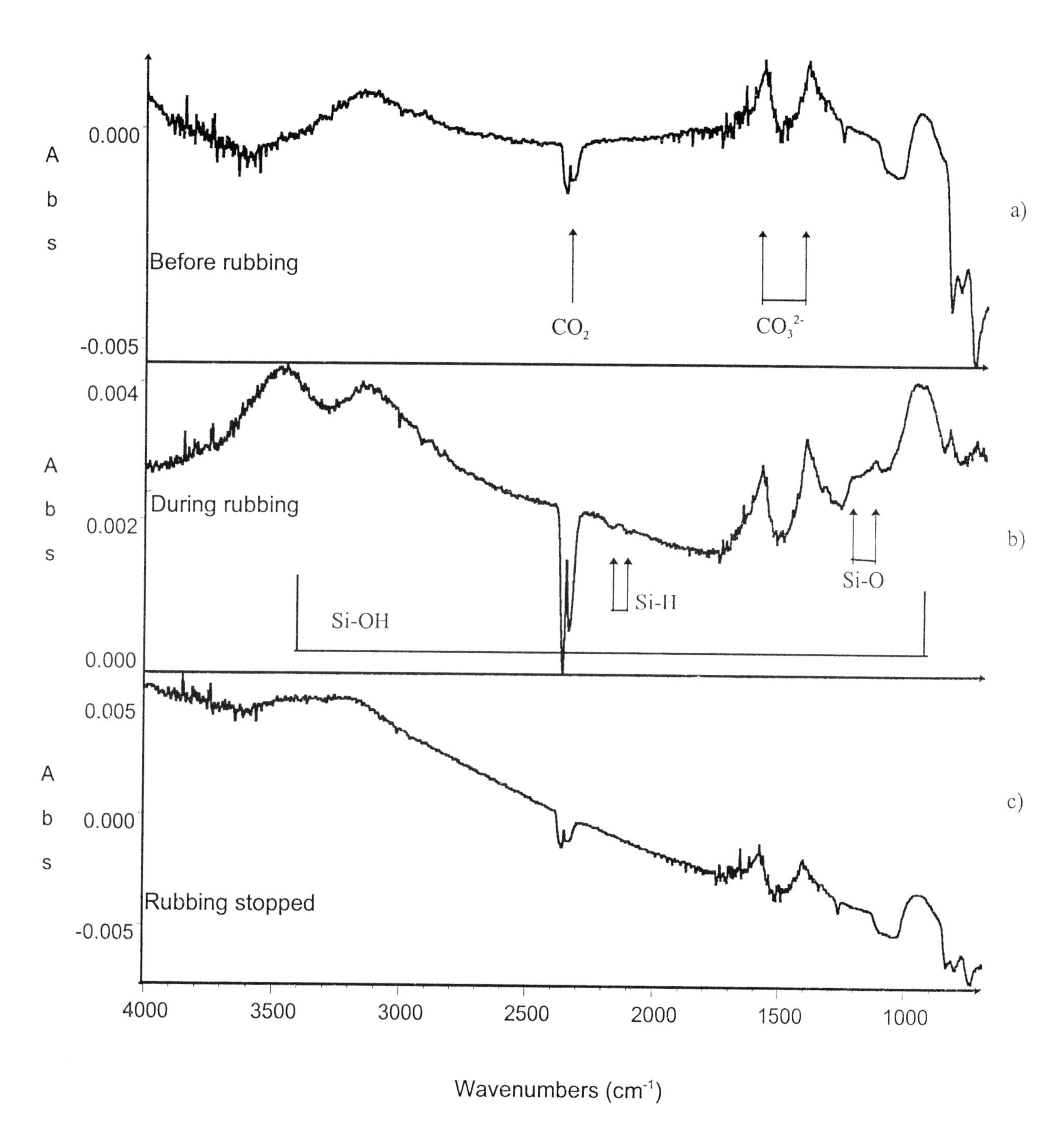

Fig. 2 ATR-FTIR spectra of silicon in contact with the buffer solution with pH 10. a) The absorption spectrum taken before rubbing started. b) A ratio of the spectrum during rubbing and before rubbing to reveal the changes in surface composition occurring during rubbing: the concentration of Si-H, Si-OH sites is modified. c) A spectrum taken after rubbing stopped.

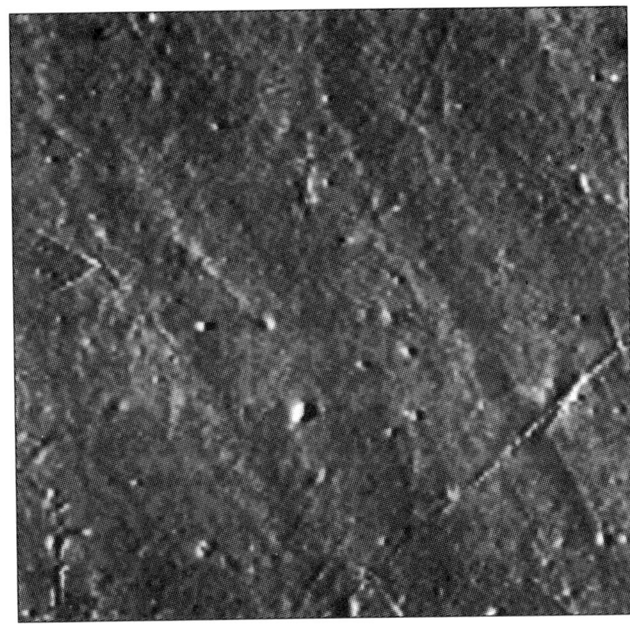

Fig. 3 a) Atomic Force micrograph of a tribochemically polished surface of silicon nitride. The micrograph is entirely featureless and the roughness is $R_a < 0.5$ nm. The graininess of the picture is not due to surface roughness but is an electronic artifact. b) Atomic force micrograph of an abrasively polished surface of a high-performance silicon nitride ball bearing.

hydrated (**24, 25**). The result is a reduction of the local stresses responsible for the mechanical wear.

In the absence of friction, measurable oxidation rates of silicon nitride are obtained only above 1000 K (**26**); they are increased one thousandfold by the presence of humidity in the air (**27**). During friction, massive oxidation is obtained even at room temperature. How this occurs exactly has not been determined yet. One can speculate that the reaction is accelerated because the hydroxide formed on the surface is continuously removed and a fresh surface is exposed by friction, but clear experimental evidence on the mechanisms is still lacking.

When silicon nitride slides in water, the tribological reaction is a dissolution (**20**) of the material at the contacting asperities. The resultant surfaces are so flat that hydrodynamic lubrication (**10**) is obtained even in water at low sliding speeds. This modified topography of the surfaces also reduces the contact stress concentrations and therefor suppresses microcracking and other mechanical wear phenomena that introduce defects in the material.

5 TRIBOCHEMICAL POLISHING

Since the tribochemical dissolution of silicon nitride in water occurs only at contacting asperities that are removed, and results in extremely smooth surfaces, this phenomenon can be used as the basis for a novel technology for the polishing of ceramics and the planarization of wafers in the manufacture of integrated circuits (**28**). If the material removal is purely tribochemical, no microfracture or plastic deformation takes place. The smooth surface produces a large real contact area and small local contact stresses. As a consequence, the polished surface is free of microcracks and other defects. By judicious choice of fluids and mechanical parameters, it is possible to produce surfaces of high quality at competitive polishing rates. In addition, the process is simple and easily controlled.

For polishing to be effective and entirely tribochemical, a number of conditions must be satisfied (**29**).

1) Friction must be sufficient to stimulate a tribochemical reaction. When the surfaces are very smooth, hydrodynamic lubrication will occur even at low sliding speed of the tool, consequently friction will be very low and the stimulation of the reaction will stop.
2) Local contact stresses must remain below the threshold for mechanical wear, otherwise the surface is roughened by the microfracture and plastic deformation.
3) The reactivity of the material towards the polishing reagent must be low enough so that the reaction does not occur generally but only where stimulated by friction at the asperities. If the reactivity is too high, general corrosion will occur. The latter is often fastest at surface defects (such as emerging dislocation or grain boundaries) with the consequence that pitting and roughening of the surface occurs.

Tribochemical polishing thus occurs in a limited window of mechanical and chemical parameters. In this respect, it is fundamentally similar to stress corrosion cracking where the stresses and the corrosion are too slow to take place individually and occur only in concert.

The surface quality of tribochemically polished silicon nitride was explored by Nomarski differential interference contrast optical, electron and atomic force microscopy (**29**). Surface residual stresses were measured by X-ray diffraction, and the fracture strength by the biaxial stress method. The surface roughness measured by atomic force microscopy is Ra $\leq$ 0.5 nm at 50 µm cut-off length (see Figure 3), $R_a \leq 2$ nm at 150 µm. On a large scale, the surface roughness is $R_a \leq 4$ nm at 2.5 mm cut-off length and $R_a \leq 6$ nm at 8 mm; the R_a value at 8 mm is the measurement of a slight macroscopic curvature with a radius of 94 m and not of real roughness. The surface roughness (R_a) was compared with that of a silicon wafer polished by chemomechanical polishing (CMP) in the semiconductor industry, of the highest quality, suitable for IC chip processing. The surface roughness of the latter is $R_a \leq 4$ nm and 6 nm over the cut-off lengths of 0.8 and 2.5 mm. In the cut-off length of 8 mm, it is $R_a \leq 10$ nm. Tribochemical polishing of silicon nitride not only produces a very low roughness, but stress-free surfaces: surface residual stress measurements by X-ray diffraction indicate the presence of compressive stress at 50 MPa. No degradation of the fracture strength is found. The average fracture strength measured by the biaxial stress method is about 770 MPa.

In the production of integrated circuits, the polishing or planarization of 'damascene', a pattern of oxide, silicon and metal on an IC chip, is a critical step. To this end, we have investigated the tribochemical polishing of tungsten, copper and aluminum (**30**). With tungsten, we were able to obtain surfaces with a roughness $R_a < 5$ nm. The attempts at polishing aluminum met with failure, but taught us some principles of tribochemical polishing. Aluminum is naturally covered with a thin layer of Al_2O_3 which is mechanically and chemically more stable than aluminum. Thus any chemical or mechanical process that removes aluminum oxide will remove aluminum much more rapidly. Chemical reactions dissolve aluminum below pin holes in the oxide and produce rough surfaces. Stresses causing mechanical removal deform the substrate much more than the oxide. We must conclude that the polishing of aluminum on a scale of one nanometer, by mechanical or chemical means, is impossible.

6 ACKNOWLEDGEMENTS

This overview is based on work supported by ARPA under contract #F33625-92-C-59178814312, by the National Science Foundation under Grant CMS-9414976 and the State of New Jersey under an AIMS grant.

7 REFERENCES

1 **Heinicke, G.**, 1984, *Tribochemistry,* Munich: Carl Hanser Verlag.
2 **Thiessen, P.A., Meyer, K.,** and **Heinicke, G.,** eds., 1967, Grundlagen der Tribochemie, Berlin, Akademie Verlag.
3 **Fischer, T.E.,** *Ann. Rev. Mater. Science,* **18** (1988) 303.
4 **Quinn, T.F.J.,** *Tribol. Int.* **16** (1983) 305.
5 **Quinn, T.F.J.,** *Tribol. Int.* **16** (1983) 257.
6 **Sexton, M.D.** and **Fischer, T.E.**, *Wear* **96** (1984) 17-30.
7 **Fischer, T.E.** and **Sexton, M.D.,** 1984, *Physical Chemistry of the Solid State; Applications to Metals and their Compounds,* ed. P. Lacome, p. 97, Amsterdam: Elsevier (1984).
8 **Fischer, T.E.** and **Tomizawa, H.**, *Wear,* **105** (1985) 29.
9 **Kim, D.S. Fischer, T.E.** and **Gallois, B.,** *Surf. & Coatings Technol.,* **49** (1991) 537.
10 **Tomizawa, H.** and **Fischer, T.E.,** *ASLE Transactions,* **30** (1987) 41.
11 **Archard, J.P.** *Tribology* **7** (1974) 213.
12 **Tabor, D.** *Proc. R. Soc. London Ser.* **A251** (1959) 378.
13 **Ashby, M. F.. Abulawi, J** and **Kong, H-S.** *STLE Tribol. Trans.* October 1991.
14 **Razavizadeh, K.** and **Eyre, T. S.,** *Wear* **79** (1982) 325.
15 **Wiederhorn, S.M. Freiman, S.W., Fuller, E.R.,** and **Simmons, C.J.,** *J. Materials Sci.,* **27** (1982) 3460.
16 **Michalske T.A.** and **Bunker, B.C.,** *J. Appl. Phys.,* **56** (1984) 2686.
17 **Muratov, V.A.** and **Fischer, T.E.** to be published.
18 **Fischer, T.E. Anderson, M.P. Jahanmir, S.** and **Salher, R.,** *Wear,* **124** (1988) 133.
19 **Jahanmir, S.** and **Fischer, T.E.,** *Tribology Trans.,* **31** (1988) 32.
20 **Sugita, T., Ueda, K.,**and **Kanemura, Y.,** *Wear,* **97** (1984) 41.
21 **Shimura, H.** and **Tsuya, Y.,** *Wear of Materials,* 1977, K. Ludema, ed. ASME, New York (1977) 452.
22 **Gates, R.S., Hsu, S.M.,** and **Klaus, E.E.,** *Tribology Trans.,* **32** (1989) 357.
23 **Wallbridge, N., Dowson D.,** and **Roberts, E.W.,** *Wear of Materials,* 1983, K. C. Ludema ed,., ASME, New York 202.
24 **Tomizawa H.** and **Fischer, T.E.,** *ASLE Transactions,* **29** (1986) 481.
25 **Fischer, T.E. Liang, H., Mullins, W.M.**, *Mat. Res. Soc. Symp. Proc.,* **140** (1989) 339.
26 **Kiehle, A.J., Heung, L.K., Gielisse P.J.,** and **Rockett T.J.,** *J Am. Ceram. Soc.,* **58** (1975) 17.
27 **Singhal, S. C.** *J. Am. Ceram. Soc.,* **59** (1976) 82.
28 **Hah, S.R.** and **Fischer, T.E.**, Patent applied for.
29 **Hah, S.R.** and **Fischer, T.E.**, to be published.
30 **Wei, J.J.** and **Fischer, T.E.,** patent applied for and to be published.

Tribology in the space environment

P D FLEISCHAUER
Technology Operations, The Aerospace Corporation, El Segundo, CA, USA

SYNOPSIS

The environments of spacecraft – be they launch vehicles, orbiting satellites, or exploration vehicles (for the moon or other planets) – are definitely extreme. The major condition with which space tribology must be concerned is the vacuum environment; i.e., the absence of atmospheric gases that often provide protective coatings to minimize wear and reduce friction. Apparatus can be forced to operate in extreme heat or cold, in radiation environments, or under severe conditions of load, speed, and direction of motion. Moving mechanical assemblies can include rolling element bearings, sliding latches and actuators, gears (including harmonic drives), and sliding electrical contacts. Tribological surfaces for space applications include steels, both high carbon and corrosion resistant, other metals such as titanium and aluminum, ceramics and ceramic thin films, and polymers and composites. Perhaps surprisingly, spacecraft manufacturers in the United States and in many other countries favor the use of oil and grease lubricants, although for many applications solid films are preferred. The latter are limited to relatively short distances of travel, less than ten million passes for rolling or sliding contacts, and relatively low speeds. Fluids, on the other hand, must be confined to reduce evaporative loss over long lifetimes of operation at moderate to high speeds. Some newer synthetic oils with outstanding lubricating properties have made sealing requirements less severe in recent designs. Finally, there is a variety of performance requirements for space mechanisms, ranging from high-speed, highly loaded ball bearings in controlled moment gyroscopes to cryogenically cooled bearings in some liquid rocket engines that obviously become very hot during operation, as evidenced by discoloration of steel bearings in these devices. In this paper, numerous examples of spacecraft mechanisms will be cited with discussions of the lubricants and tribologies associated with each. Observations and opinions of the author, based on experience and laboratory studies, concerning best practices for a number of specific applications will be presented.

1 INTRODUCTION

The original invitation for this paper was for a presentation on tribology in extreme environments. The definition of an extreme environment is very much a function of the intended application. Thus, what is extreme for automotive applications is not necessarily extreme for space applications. This paper will present information concerning tribology in the space environment, specifically concerning moving mechanical assemblies (MMAs), structural materials and surfaces of bearings and other tribological components, and lubricants for these components and MMAs. The environments of spacecraft – including launch vehicles (rockets), orbiting satellites, and exploration vehicles (for lunar or planetary surfaces) – can be extreme. Typical environmental factors considered by space vehicle manufactures include pyrotechnic shock, random (3-axis) vibration, natural and man-made radiation, thermal cycling, orientation effects (gravitational effects in ground testing), and electromagnetic interference/charging. These factors can all be of concern for the tribologist, along with variable conditions of load, speed, and direction of motion. However, the single most significant environmental factor with which space tribologists must be concerned is the vacuum environment, i.e., the absence of atmospheric gases that often provide protective coatings to minimize wear and reduce friction. These same gases can be destructive to space-qualified lubricants during ground storage and testing.

The mechanisms utilized on spacecraft encompass various sliding and rolling elements; they have surfaces of variable composition, hardness, and toughness; and they operate under grossly different conditions of speed and load. In the latter case, speeds can vary from the extremely slow rotation of a solar array on a geosynchronous, three-axis stabilized satellite (approximately one revolution per day) to moderately high-speed spin bearings in gyroscopes (up to 15,000 rpm) and momentum wheels (usually less than 6,000 rpm). For all known conditions of operation, the contact mechanics are relatively benign, so that it is safe to say that any space mechanism will operate to its requirements as long as sufficient lubricant is maintained in the contact regions. [One possible exception to this statement involves the turbopump bearings on rocket engines such as those used on the space shuttle main engine (SSME). In this case, it would probably also hold true if it were possible to have lubricant, but because of the extremes of heat and pressure, material substitution (the use of silicon nitride balls) was needed to improve reliability.] This observation means that most designs and structural materials will meet most of the objectives of most missions, providing that the lubricant has been engineered into the system in the correct fashion.

The lubricant is the single most significant part of any

spacecraft moving mechanical mechanism. Engineering cultural differences have led to different philosophies in the approach to lubricant selection, depending on which side of the Atlantic the designers and manufacturers live. Simply stated, building apparatus to operate for years in space vacuum with no servicing requires methods of confinement of lubricant to the pertinent area of contact. Europeans have exploited the advantages of solid-film lubrication for appropriate applications (**1**), while in the United States much more emphasis has been on engineering systems to accommodate the use of oils and greases (**2**). Both solids and fluids have their advantages, and there are tribological conditions that demand one or the other (**3**). New preparations of both types of lubricant have emerged during the ten years since the above-mentioned assessments were prepared (**1,2**). Doped and multilayered MoS_2 thin films were developed specifically for spacecraft applications (**4, 5**), and synthetic hydrocarbon oils have been adopted from terrestrial industrial and automotive applications for use on spacecraft (**6, 7**).

This paper is based on the author's experience working on research of solid and fluid lubricants and a wide variety of practical, operational problems associated with spacecraft hardware. An exhaustive review of the literature is not intended. Instead, specific examples of space-environment-related problems and our approach to their solutions will be presented. After a discussion of various types of MMAs used on spacecraft (here I will concentrate on orbiting satellites since my experience is in this area), I will cover materials and surfaces of concern in these devices and then concentrate on lubrication. As indicated in the previous paragraph, aspects of solid and fluid lubricants will be discussed with emphasis on environmental factors that can determine proper material selection and ultimate performance.

2 MOVING MECHANICAL ASSEMBLIES (MMAs)

Early satellites (Sputnik 1, Oct. 4, 1957; Explorer 1, Jan 31, 1958) were very simple by today's standards and had almost no moving parts. If they had communications antennas, they were omnidirectional so that pointing was not an issue. As time progressed designs evolved and the level of sophistication increased. Two basic techniques were employed for stabilizing satellites in their orbits: (**1**) spinning the main body of the satellite to create gyroscopic stabilization and (**2**) three-axis stabilization with the use of spinning fly wheels to control momentum in all directions or by means of small rocket motors (known as thrusters) to continuously adjust position. The complexity and evolution of MMAs on satellites were functions, in part, of the type of stabilization process employed. Thus, for spin-stabilized systems (mostly communications satellites), the principal moving parts consisted of the deployment latches and actuators and a device known generally as either a 'de-spin mechanical assembly' (DMA) or a 'bearing and power transfer assembly' (BAPTA). This latter device served as the interface between the spinning satellite and the de-spun or pointed antenna. Besides providing the structural rotation between the two parts of the satellite, this device also had to provide for conduction of power and signals across the rotating interface.

Figure 1 depicts a Defense Satellite Communications System II (DSCS II) satellite from the mid and late 1970s and early 1980s (**8**). Though eventually replaced by the 3-axis stabilized DSCS III, the DSCS II satellites had an exceptional record of success with typical operational lifetimes of eight to ten years. The DMA (BAPTA) for this satellite is characteristic of many that are used for spin-stabilized systems and for solar array drives on 3-axis systems. It contains relatively large (90 to 110 mm bore) bearings to support the rotating structures and a slip-ring assembly (also known as an 'electrical contact ring assembly') consisting of gold wire 'brushes' sliding against gold (plated) rings (**9**). The slip-ring shaft is supported by a set of smaller ball bearings. Both sets of bearings and the sliding surfaces are lubricated with a mineral oil formulated with an antiwear additive to minimize wear in the boundary lubrication regime generated by the 60-rpm rotation rate. Two real-time life tests of the flight hardware were conducted for seven and ten years, respectively, and showed outstanding results with no evidence of abnormal or anomalous performance. The bearings and lubricants were in excellent condition after the tests, and oil migration studies indicated that lubricant remained in the regions of initial application over the entire seven-year life test.

The design philosophy embodied in DSCS II employed oil lubrication with care taken to provide labyrinth seals to prevent evaporative loss of oil during operation in the space vacuum (**10**). This system worked very well; however, for the case of solar array drives involving even slower rotation rates, solid lubrication of the bearings and slip-ring surfaces also has proven to be very effective with perhaps less engineering complexity because of the absence of concern over evaporative losses. Both bonded and sputtered MoS_2 films have been used on the bearings (**11**). The slip-ring assemblies consist of silver (85%)–MoS_2 (12%)–graphite (3%) composite brushes sliding against silver (plated) rings. These dry-lubricated systems are quite robust in the space environment, but the moisture in normal air (of say 50% relative humidity) can cause reactions of the MoS_2 and silver that result in electrical noise in signal circuits. Thus, standard atmospheric conditions constitute an 'extreme environment' in this case. The degradation mechanism, during periods of storage, that leads to the generation of electrical noise is believed to be reaction of the MoS_2 with moisture in the ambient to liberate H_2S. The H_2S then reacts with the silver on the ring to form Ag_2S directly beneath the brush. Ag_2S is a semiconductor, so the films cause electrical noise spikes when brushes pass over the affected areas of the ring. The films generally wear away in time, providing a healing or recovery of nominal behavior, but the recovery process may be rapid or may persist throughout the intended mission, depending on the rotation rate of the assembly. In extreme cases, if the bulk of the MoS_2 reacts to form MoO_3, there is no lubrication, and severe wear of the rings occurs upon operation. Consequently, it is essential that such systems be protected from the atmosphere during storage, which usually is accomplished by enclosing the hardware (in a bag or box) and then purging the enclosure with a dry, inert gas.

One other configuration, which is used for BAPTAs with relatively high spin rates (30 rpm), is to have oil-lubricated bearings with composite silver-on-silver slip-ring assemblies. In these applications, the units experience more than 150 million revolutions in a ten-year life, so dry-film lubricated bearings are impractical.

Figure 2 depicts other United States Air Force Satellites encompassing communications, surveillance, weather (meteorological), and navigation missions, and provides a minimal indication of significant MMAs. These satellites are all fixed in stable positions in orbit. (The surveillance satellite shown actually rotates slowly, but the entire vehicle is pointed in a given direction.) The mechanical subsystems indicated are ones for which we have provided the greatest amount of tribological consultation over the years in order to increase reliability and performance characteristics of the overall missions. Historically, operational problems with these devices have caused the greatest concern, and we have conducted laboratory tests of lubricants and surface treatment procedures to minimize torque disturbances and maximize lifetimes during operation. The types of problems encountered with these devices concerned ball bearings and sliding electrical contacts. In all cases, the problems were solved by either lubricant substitution, control over manufacturing conditions, environmental control after fabrication, or a combination of all three. Other types of tribological contacts are encountered in latches used to secure solar panels, sensors and their covers, extendible booms (for scientific measurements or probes), and any other apparatus that are stowed during launch and then deployed after reaching orbit. Typical latches are lubricated with solid films based on MoS_2 or perfluoropolyalkylether greases. Since the normal operational mode is for only one pass in the space-vacuum environment, there are no real lifetime considerations, except that during ground testing the devices can be operated for 10 to 100 passes, often in air! Again, there are potential 'extreme environment' problems because the lubricants are designed to perform in the absence of potentially deleterious oxygen and water. It is imperative to guarantee that the lubricant is not removed or reacted during testing so that proper function is not compromised once the satellite reaches its orbit.

Another set of MMAs that have required considerable attention so that proper lubrication was provided encompasses actuators, including harmonic drives, that are used to steer antennas, solar arrays, and sensors on stabilized platforms. Typically, stepper motors drive a gear train that provides very fine adjustments to the orientation (or pointing) of the device. The gears and the associated ball bearings operate in the boundary lubrication regime and can undergo changes in direction of motion. Ideally, these devices can be lubricated with MoS_2-based, solid-film lubricants, as long as the gear ratios are not such that the number of revolutions or passes exceeds the limit for these films (a maximum of about 10 to 20 million revolutions). In practice, in the United States, fluorocarbon-based or, more recently, synthetic hydrocarbon-based greases are used. Both life tests and flight experience have shown that the synthetic hydrocarbon greases are superior in terms of torque stability, wear, and contamination considerations (**6**). In combination with some of these actuators, position indicators, such as potentiometers, are also used. Potentiometers are another form of sliding electrical contact. The metal wipers slide against metallized polymers and are usually lubricated with a mineral or synthetic hydrocarbon oil. Fluorocarbon oils have been found to create highly resistive films and to degrade under the boundary conditions.

The mechanical devices employed on satellites, that have provided some of the greatest challenges for tribologists, consist of the various types of fly wheels used for momentum compensation and stabilization. By convention, such wheels are called *momentum wheels* when the rotation is in one continuous direction at a constant angular velocity; *reaction wheels* when angular velocity is changed during a mission with the possibility of reversals in direction; and *controlled moment gyroscopes* (CMGs) when the rotating wheel, also running at constant angular velocity, is gimbaled to change the direction of the momentum vector and thus the interaction with the main body of the satellite. Typically, wheels are arranged in clusters in a satellite so that, in conjunction with appropriate control systems, all axes of motion can be stabilized. Frequently, designs will include spare wheels to provide redundancy in the event of an anomaly or failure during the mission. Wheel systems, specifically the spin bearing arrangements, vary somewhat depending on the manufacturer, but normally one or more pairs of very high-precision bearings supports the spinning elements. Oils lubricate the bearings during operation in either elastohydrodynamic (momentum wheels and CMGs) or boundary regimes (often for reaction wheels). Highly refined mineral oils (sometimes in grease formulations) have been used, primarily with phosphate ester additives, in practically all wheel applications. For the most part, wheel operation was very successful with lifetimes of operation in excess of ten years. However, designers have recently increased the rotation speeds of wheels to over 5,000 rpm to achieve greater moments of inertia and momentum capabilities, and numerous anomalies and failures have

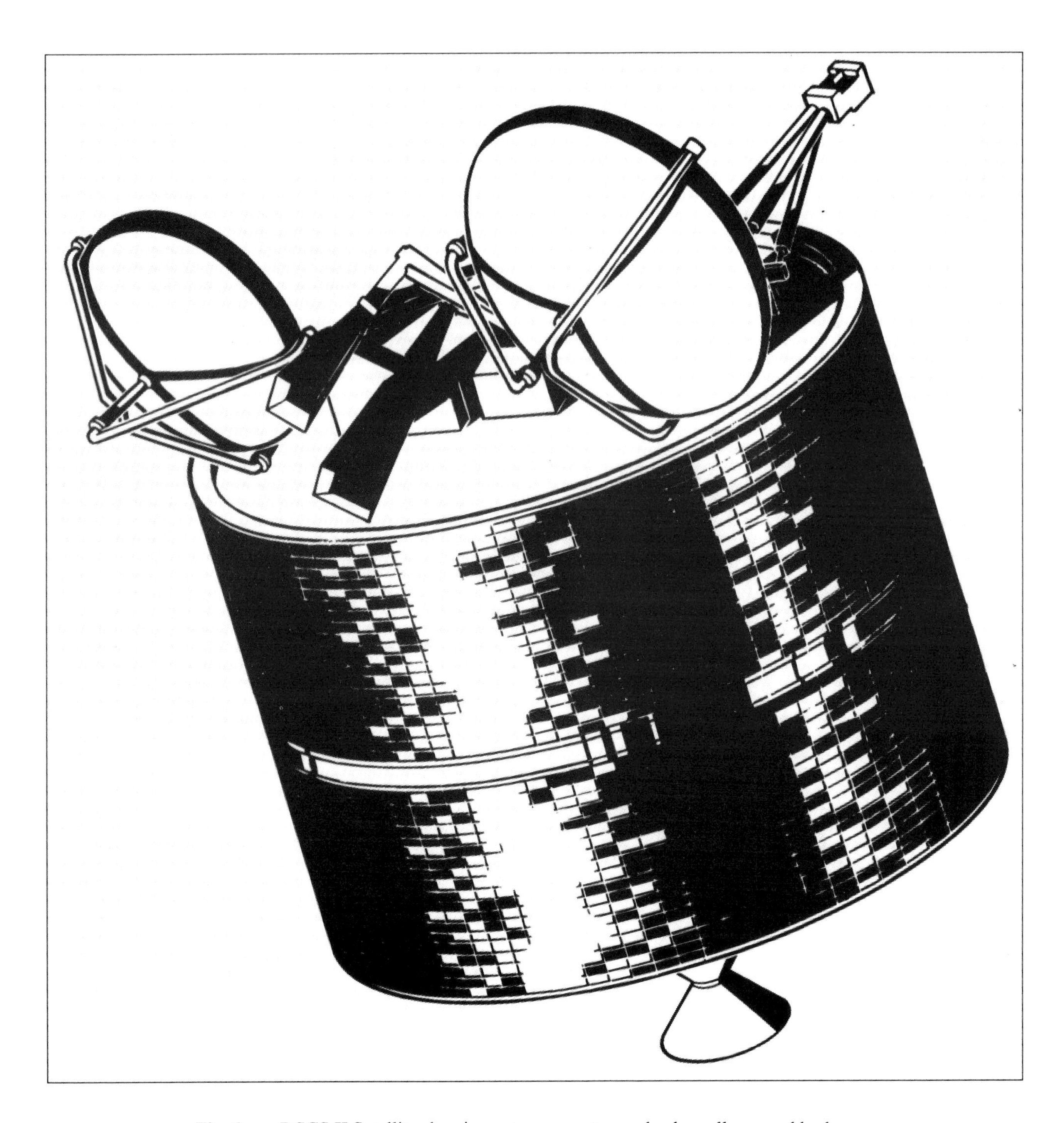

Fig. 1 DSCS II Satellite showing antennas on top and solar cells around body.

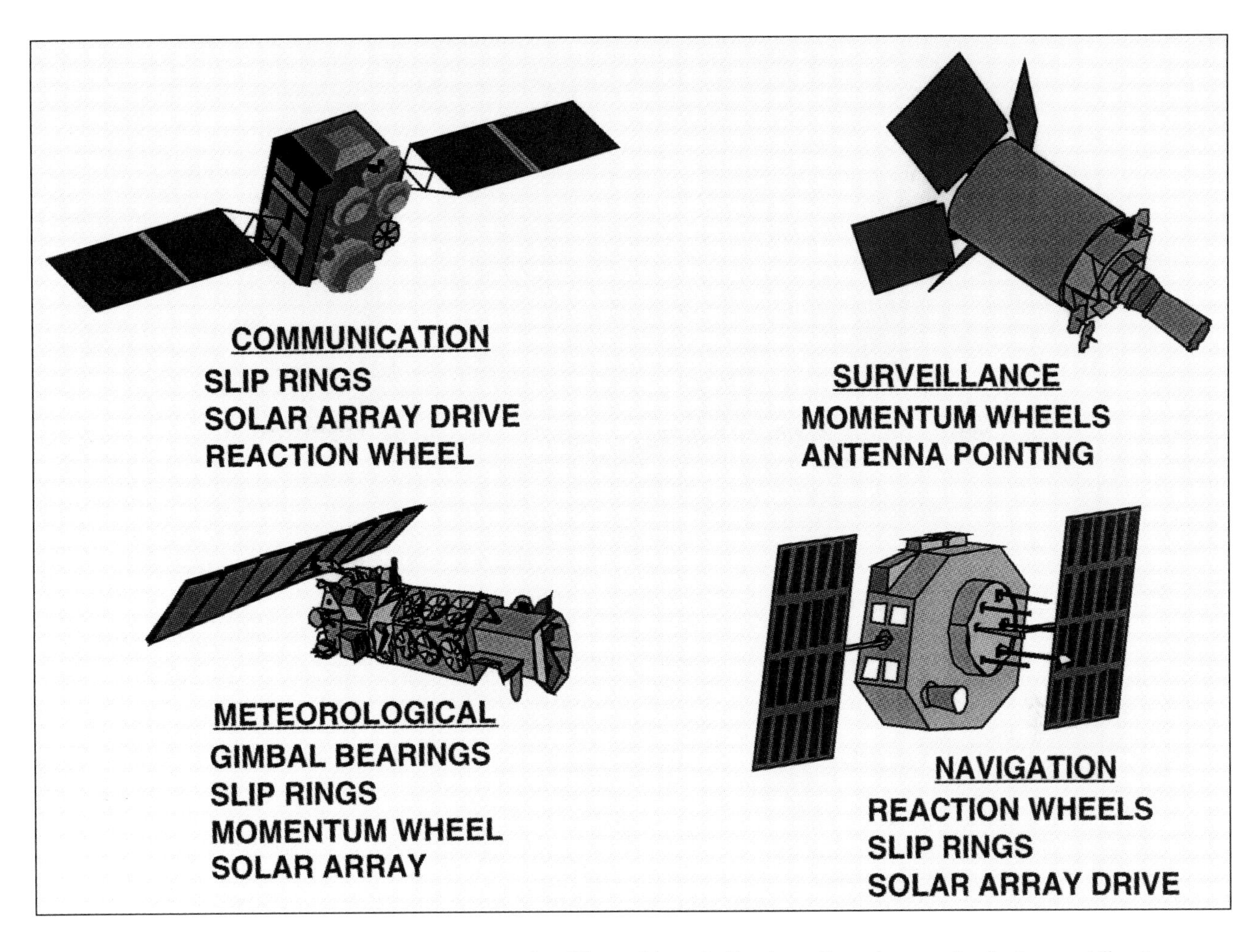

Fig. 2 Four United States Air Force Satellites with an indication of moving mechanical assemblies in each for which tribological consultation was required.

occurred. (Other, slower-speed systems have failed, but it has usually been because of inadequate, initial oil supplies or extended, on-the-ground storage times prior to launch.)

A schematic torque trace for a wheel in the process of exhibiting major anomalies and ultimate failure is shown in Figure 3A. This type of behavior is very characteristic of many failures that we have studied. Initial periods of high torque are followed by recovery and nominal operation for significant time. Gradually, the periods of unstable, high torque become more frequent and increase in amplitude. Ultimately, the system torque exceeds the capability of the motor and the system shuts down. Disassembly and analysis of ground-test systems that have experienced this type of failure have shown that the ball-retainer pockets have been 'burned,' and that, in general, the ball pathways and the balls are lacking in oil. The failure mechanism has been attributed to (a) loss of oil on the critical surfaces (the balls, the ball pockets, and the ball tracks on the raceways); (b) increased friction between the balls and the retainers or between the retainers and the controlling land; and (c) eventual retainer instability that exacerbates the entire cycle. Laboratory simulations (Fig. 3B) have shown that bearings lubricated with limited initial quantities of oil exhibit the same type of torque behavior, and this torque run-away is cured by the addition (injection) of oil directly into the bearing raceway. Another observation was that the use of synthetic hydrocarbon oils [poly-alpha-olefins or multiply alkylated cyclopentanes (**12**) delayed the onset of instability substantially and generally provided for much smoother and longer operation.

During the investigation of these retainer instability events, exhaustive studies of retainer properties and oil absorption characteristics were conducted in The Aerospace Corporation's tribology laboratory (**13**). Figure 4 depicts the type of behavior that is characteristic of the classical cotton-phenolic type of retainer. The major conclusion of this very comprehensive study was that oil is absorbed into the retainers in a two-step process, one that is rapid (within a few days), and a second that is very slow (taking months and even years). The rapid process is believed to occur via capillary action along the reinforcement threads of the composite material. The slow process probably corresponds to diffusion of oil into the phenolic matrix. Most retainers need to be immersed in oil

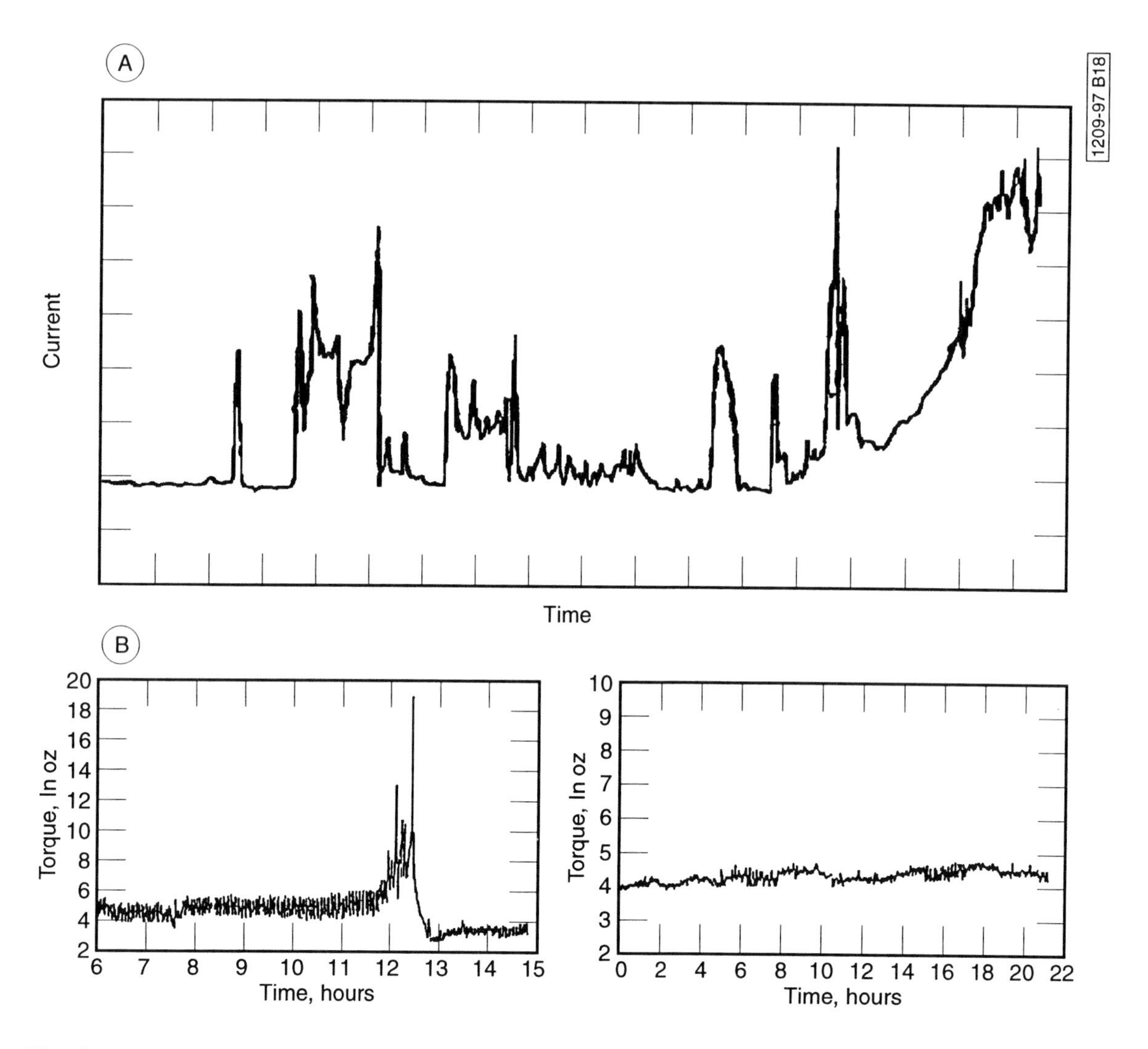

Fig. 3 (a) Schematic torque (motor current) trace for a momentum wheel experiencing bearing retainer instability and eventual failure.
(b) Torque traces from bearing tests for which the initial quantity of oil was a very thin film applied from a 50:1 solvent to oil solution. On the left is a trace for KG-80 oil showing the onset of retainer instability which is quieted by the injection of 100 μL of fresh oil. The right trace is for the same bearing with NYE 2001 (Pennzane) oil, showing no instability for twice the running time. (Note the change in torque scale.) A 100:1 solvent to oil solution treatment was necessary to obtain instability within a reasonable operating time for NYE 2001.

for a minimum of seven days to complete the capillary process. It is not necessary to conduct the impregnation process in a vacuum, but it is essential to dry the retainer prior to putting it into the oil and for the environment surrounding the immersed retainer to be rigorously moisture free. Also, exposure of an impregnated retainer to moisture can force oil out of the cotton-phenolic matrix. Fully impregnated retainers assist in maintaining oil in bearings because they do not absorb oil from the lubricated metal parts. There is no net delivery of oil from the retainer to a bearing, even if the metal is depleted (**14**), but impregnated retainers tend to provide lower friction surfaces for contact with metal parts and thus help minimize retainer instability events. Additional information regarding the superior performance of synthetic lubricants will be presented in the section on lubricants.

3 MATERIALS AND SURFACES

The materials used to fabricate most spacecraft MMAs are ones adapted from terrestrial applications and are known to most tribologists. There is very little about the space environment that makes material selection, apart from lubricants, very unusual. Instead, standard criteria such as predicted loads and stresses, speeds of operation, and fatigue properties are used to select most materials. One criterion, potentially unique to space applications, concerns the outgassing properties of any material, but again this normally relates to the lubricants (see the next section) rather than to the structural materials. This lack of uniqueness is not to say that there is not a variety of different materials and surface treatments used for space applications. Selection depends entirely on the type of apparatus and the conditions of use and performance.

Standard bearing steels, mostly 52100, are used in ball bearings for space mechanisms. In some instances, because of concerns about corrosion during ground operations, stainless steels (440C) are utilized. To my knowledge, no space ball bearing has failed because of fatigue problems within the bulk material. Failures result from depletion or failure of lubrication and subsequent wear of metal or retainer parts. However, recent projections of fatigue requirements of a wheel bearing system have concluded that steel alloys with higher fatigue ratings (e.g., Rex 20) will be required for the specific application. Once the materials are selected to meet the load and stress predictions, one can expect that the ultimate determination of lifetime of operation will again be the effectiveness of the lubrication system.

The preparation of steel surfaces for ball bearing use has received renewed attention during the recent past because of the restrictions placed on chlorofluorocarbon solvents and the solvents' effects on the Earth's environment. Studies of steel surfaces have revealed that the oxide structure is complex and that cleaning can alter this structure (**15**). The composition and structure of the surface of a steel is extremely important because of the effects on protective film formation and lubrication. The formation or growth of antiwear additive layers and films [i.e., tricresylphosphate (TCP)] are particularly sensitive to surface chemical compositions (**16**). In the past, solvent rinsing (with Freon) of TCP-coated parts left the films intact; but the newer, environmentally friendly, detergent-based cleaners remove most of the phosphate-containing material on test surfaces. On the other hand, wear tests with The Aerospace Corporation's eccentric bearing tester have shown that TCP pretreatment of bearings is not nearly as important as having the additive formulated with the base stock (**17–19**) (see Figure 5). The chemical interaction(s) of the additive with the steel still determine the degree of surface protection and the overall performance, in spite of their complexity. Such interactions can be thermally induced, as with the TCP pretreatment, or they can be tribochemically induced. Data in Figure 5 show that a

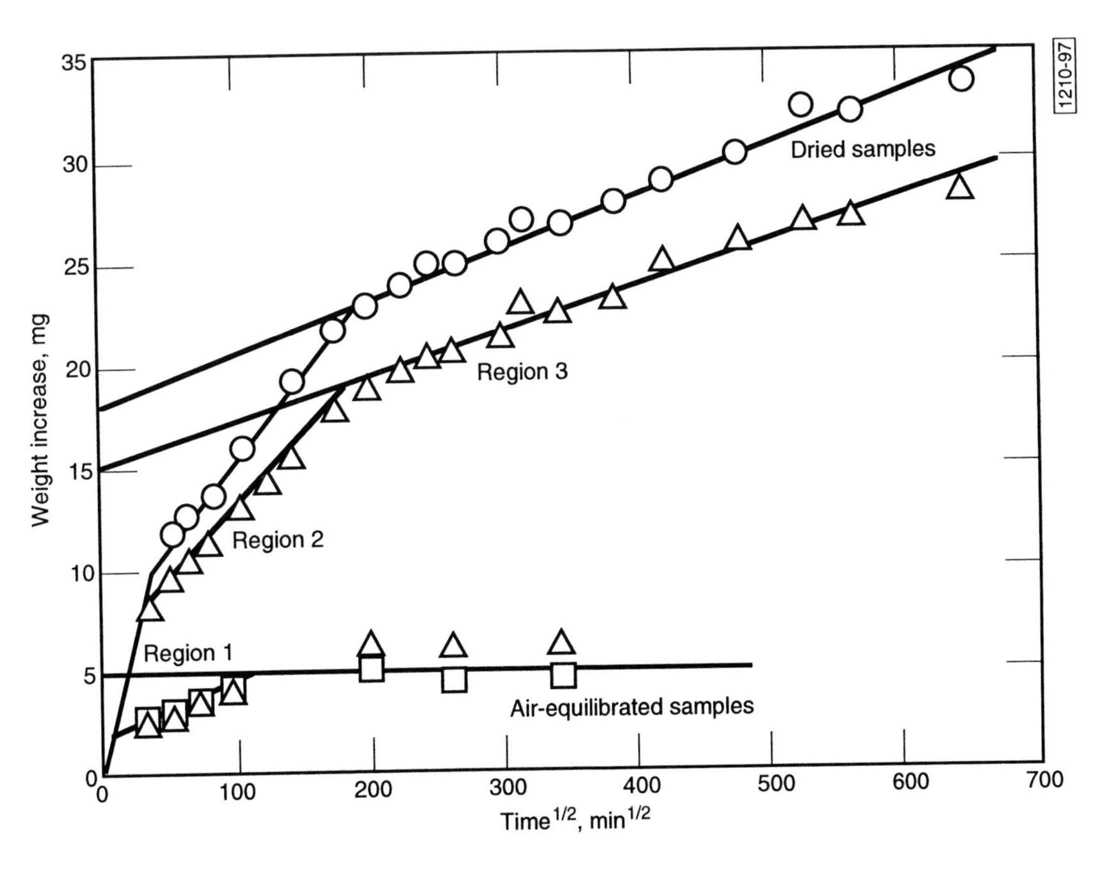

Fig. 4 Oil Absorption into cotton-phenolic material showing a two-step absorption mechanism and the drastic effect of ambient moisture on absorption. Regions 1 and 2 correspond to capillary fill associated with the threads and region 3 is fit by diffusion from the capillaries into the resin matrix. (Adopted from ref. 13.)

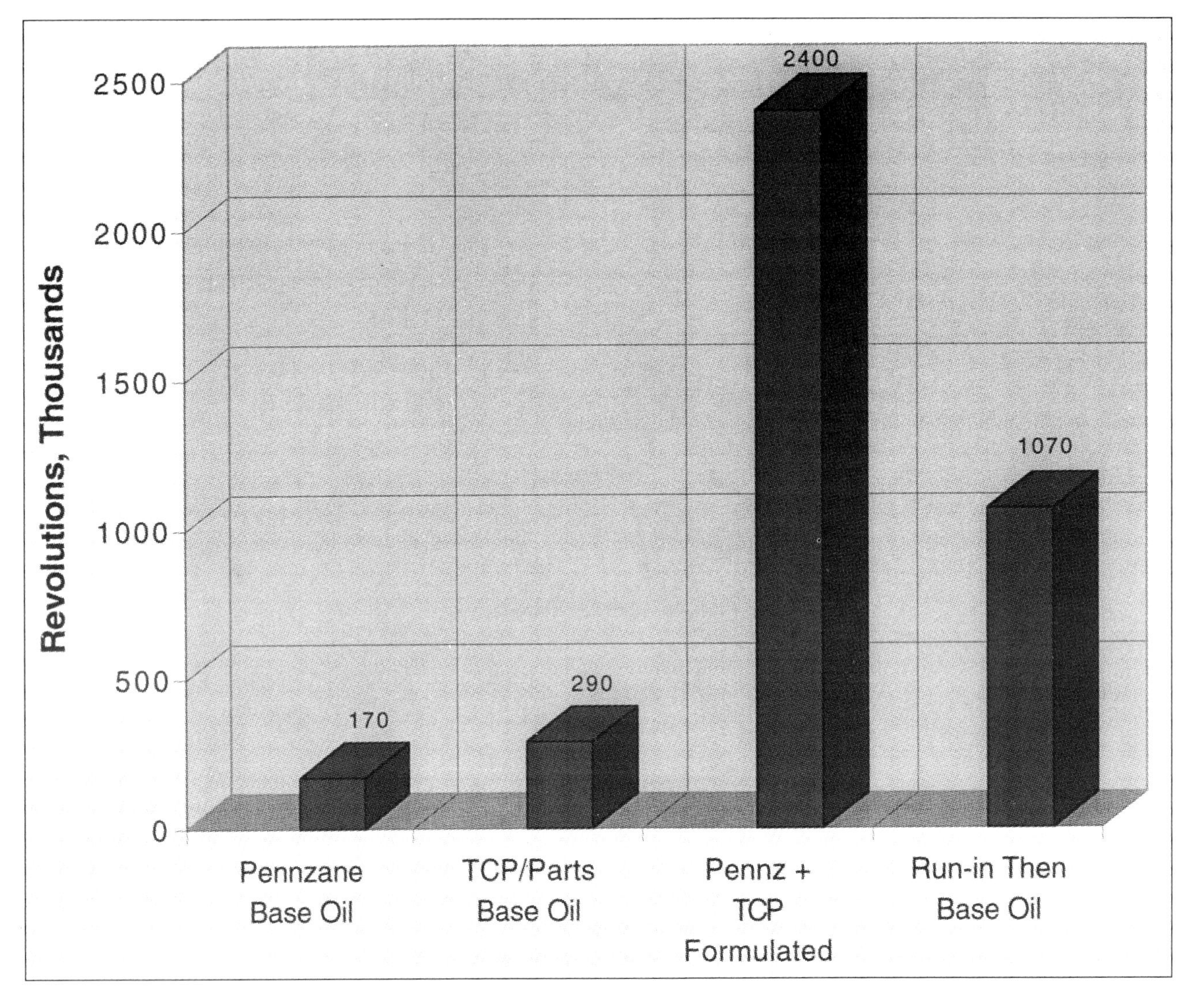

Fig. 5 Wear-test results from the Aerospace Corporation eccentric bearing tester. The formulated oil ran nearly 20 times longer than the base oil and almost 10 times longer than the pretreated sample with base oil. When the bearing was run-in for 180,000 revolutions and then rinsed and run with base oil, it lasted 3 times longer than the pretreated sample with base oil.

bearing initially run-in for approximately 10% of total normal life with formulated oil, then operates two to three times longer than one pretreated with the standard thermal, TCP process. Both tests were run with pure base oil after the respective pretreatments (run-in or heating).

The coating of ball and raceway surfaces with hard, thin films has gained significant popularity with the designers of European space hardware but is only just beginning to be explored in the United States space business (**20, 21**). TiC-coated balls have been used in BAPTA and momentum wheel bearings and in gyroscope bearings for aircraft applications with great success. Tests of such bearings have run for more than twice the life of uncoated bearings with identical lubrication and shown no evidence of degradation (**22**). Though all of these results are very supportive of hard-coated surfaces, it is important to realize that the surface compositions of such parts are grossly different from those of standard steel bearings, and that these differences could significantly impact the reactions and performance of friction-reducing additives in lubricants. So far, the evidence seems to indicate that surface chemical differences are of little consequence, but investigations are in progress to better understand possible implications for other device applications and for the development of more space (vacuum) friendly additives.

The materials and surfaces of other MMAs present some variations as, for example, with the use of polymers in some latches and in potentiometers. Also, titanium is a material of choice for as many parts as possible because of its light weight. Care is required with titanium surfaces, however, because of its tendency to gall, so that coatings or surface oxidation and minimization of contact stresses are essential. In general, various steels are used for most devices, including latches, actuators, and gears. TiN coating of gears has been considered, but is not widely used at present. Finally, as mentioned previously, gold and silver are used for slip-ring assemblies to provide good electrical conductivity and relative chemical inertness.

4 LUBRICANTS

The selection of a solid or fluid lubricant for a particular space application depends primarily on the conditions of load and speed (life) of the contacting surfaces. Solid lubricants provide many advantages in terms of the lack of susceptibility to space vacuum and radiation, but they are limited to applications involving relatively benign stress and low number of passes (rotations). The most common solid-film lubricants are based on MoS_2 or Teflon. They are applied to the critical surfaces in the form of bonded, burnished, or sputtered films (**11**). Sputter-deposited MoS_2-lubricated ball bearings have obtained a maximum of 20 million rotations before failure (**5**) and should not be designated for contact stresses in excess of approximately 1 GPa. Sputter-deposited MoS_2 films are particularly sensitive to moisture in the operating and storage environments (**23**). Figure 6 shows data for sliding tests of rf sputter-deposited films of MoS_2 of different thicknesses in moist [50% relative humidity (RH)] and dry air and for doped (Ni) and undoped films after storage in humid (85% RH) and dry environments. The most obvious observation is that the life in moist air is a small fraction of that in dry air. These data are for older films made in our laboratories, and some improvements for multilayered materials have been observed, but the basic conclusion that moisture threatens lifetime remains. Storage conditions also can have a dramatic effect on the ultimate life of MoS_2 films. It is imperative that ground testing and storage of devices lubricated with MoS_2 films be effected in a protected environment, either vacuum or inert gas. [There is some concern that inert gas does not reproduce the operational environment of space vacuum, but there is no disagreement that moisture is extremely detrimental (**24**).]

Certain doped and ion irradiated MoS_2 sputter-deposited films and multilayered materials have provided increased resistance to severe environmental conditions and are much more durable in conventional sliding and rolling wear tests (**5, 25, 26**). The significant structural attribute of these films seems to be that they have a parallel, layered configuration; the basal planes of MoS_2 are aligned with the plane of the substrate surface (**27**). Measurements of the susceptibility of parallel and perpendicular films to storage-induced oxidation indicated that the parallel films have considerably greater resistance (**28**). Controlled methods of production of films with parallel orientation should provide some measure of control over the critical sensitivity of sputter-deposited MoS_2 films to the 'extreme' Earth environment. MoS_2 films prepared under ultra-high vacuum conditions have exhibited extremely low friction coefficients, but when exposed to laboratory air they incorporate oxygen, which is believed to raise the friction (**29**). An obvious conclusion with respect to the preceding rhetoric is that for sputter-deposited MoS_2 films and MoS_2 lubricants in general, air is an extreme environment!

Fluid lubricants (oils and greases) are employed in most spacecraft devices that require millions of cycles of operation. These materials require special design and engineering procedures to minimize loss to space vacuum. Labyrinth (also known as molecular) seals are fabricated into the devices so that oil molecules have to follow a torturous pathway to escape the bearing cartridges (**2, 10**). The useful lifetimes of spin-bearing systems are normally estimated by calculating the free molecular loss (pumping) of oil through the labyrinths (**10, 30**) and then incorporating at least two times the total quantity of oil needed for the respective mission. Manufacturers installed porous reservoirs within the bearing cavities in an attempt to provide replacement oil to the bearings during operation. These reservoirs do provide vapor within the cavity that can exit the labyrinth and therefore reduce the rate of loss from the bearings, but they do not add oil to the bearings because the latter have the highest temperature surfaces in the closed system. Recently, various schemes for adding oil to operating bearings have been developed and tested. Some are operating successfully in orbiting satellites. However, the most significant development to impact the design and lifetimes of spin bearing systems for space applications has been the development of synthetic hydrocarbon oils to replace the standard mineral oils (**12, 31, 32**).

The implementation of synthetic hydrocarbon oils, in general, was slowed by the infrastructure surrounding the refining of crude oil and the tremendous economic impact of oil products. Arguments against synthetics, such as that they are not better lubricants but they do have excellent volatility and viscosity properties (**31**), were exactly what the space industry needed. Both viscosities and viscosity indexes can be tailored by selection of appropriate synthesis conditions, and synthetic materials are pure compounds that have low reactivities and measurable vapor pressures (**12**), two facts that make them ideal for use in spacecraft devices. No claim is made that synthetic oils were developed for the space industry; the market could not support the costs. However, their availability and exceptional fit (specifically their low volatilities) to the needs of space mechanisms is very fortunate. Formulated synthetic oils, be they poly-alpha-olefins or multiply alkylated cyclopentanes, have outperformed analogous mineral oils in all tests performed in our laboratories (**6, 33**) and in most other studies know to us (**7, 34**). The low evaporation rates of synthetic oils make the need for effective resupply systems less severe, though such systems are preferred for maximum reliability. The next challenge

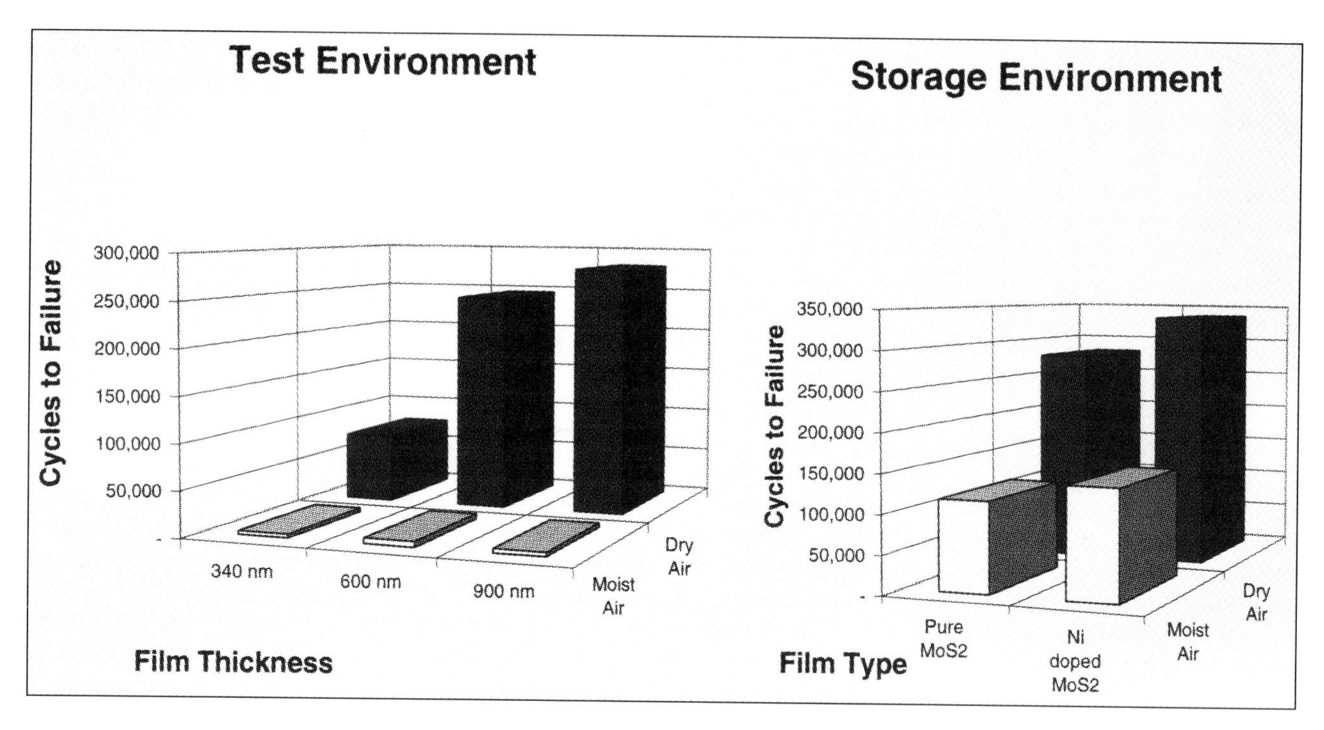

Fig. 6 Sliding wear-test results showing the negative effects of moisture during testing and storage of sputter-deposited MoS_2 films. Test life in 50% humid air was barely measurable. Life increased with increasing film thickness but seemed to reach a maximum value. Nine-months storage in 85%-RH air reduced wear life by about a factor of three compared to storage in 0%-RH air; Ni doping had little effect on life or resistance to moisture.

is to provide antiwear and antifriction additives with comparable low volatilities. A U.S. Air Force, Wright Laboratory study to develop and test such materials is in progress and should produce results in the very near future.

5 SUMMARY

The space environment actually poses few unusual challenges for the tribologist, except for the most obvious factor, the low pressure (vacuum) of the surroundings of any spacecraft. In specialized cases, very low or high temperatures are experienced, and specific designs must be developed. Shock and vibration associated with launch must be considered in any MMA design to avoid denting of contacting surfaces. Natural or man-made radiation can impinge on exposed tribological surfaces, such as some latches and actuators, but most devices are shielded by the materials of construction and housings surrounding the moving parts. For those devices that can be exposed to radiation, solid-films of MoS_2 or perfluoropolyalkylether greases are quite robust. Gravitational effects during ground storage or testing can cause oils to migrate out of bearings or other contact areas, but normally orientations that mitigate such losses can be employed. Electromagnetic interferences or charging events rarely compromise tribological surfaces unless, for example, currents flow through oil lubricated bearings for long periods of operation. For space-qualified MoS_2 solid lubricants, exposure to humid air prior to launch can be one of the most severe environments of all. Proper care must be taken to minimize quiescent exposure and eliminate exposure during operation (testing). New developments in synthetic hydrocarbon oils and additives are providing the satellite designer with much greater latitude than ever before. The low volatilities of the oils and acceptable viscosities are making significantly longer missions possible. In the case of existing programs, this life extension is often possible with no hardware changes, a condition that is very easy to justify to skeptical program managers.

6 ACKNOWLEDGEMENTS

This work was supported by the United States Air Force Materiel Command, Space and Missile Systems Center under Contract No. F04701-93-C-0094. I would like to thank the members of the Aerospace Corporation Tribology Section, specifically Drs P.A. Bertrand, D.J. Carré, S.V. Didziulis, and J.R. Lince, and Mr A.R. Leveille of our Vehicle Engineering Division, for their contributions to this effort and many helpful discussions. I'd also like to thank Dr. M.N. Gardos of Hughes Aircraft Co. for his many significant contributions over the years.

7 REFERENCES

1 **Rowntree, R. A.** and **Todd, M. J.,** A Review of European Trends in Space Tribology and its Application to Spacecraft

Mechanism Design, *Mater. Res. Soc. Symp. Proc.*, Vol **140**, 1989, pp 21-34.

2 **Fleischauer, P. D.** and **Hilton, M. R.**, Assessment of the Tribological Requirements of Advanced Spacecraft Mechanisms, *Mater. Res. Soc. Symp. Proc.*, Vol **140**, 1989, pp 9-20.

3 **Hilton, M. R.** and **Fleischauer, P. D.**, Lubricants for High-Vacuum Applications, Metals Handbook, vol **18**: *Friction, Lubrication, and Wear Technology*, ASM International, Materials Park, OH, 1992, pp 150-161.

4 **Hilton, M. R.** and **Fleischauer, P. D.**, Applications of Solid Lubricant Films in Spacecraft, *Surf. and Coat. Technol.* **54/55** (1992) 435-441.

5 **Roberts, E. W.** and **Price, W. B.**, Advances in Molybdenum Disulphide Film Technology for Space Applications, *Proc. Sixth European Space Mechanisms & Tribology Symp.*, Technopark, Zürich, 4-6 October 1995, pp. 273-278.

6 **Carré, D. J., Kalogeras, C. G. Didziulis, S. V. Fleischauer, P. D.** and **Bauer, R.**, Recent Experience with Synthetic Hydrocarbon Lubricants for Spacecraft Applications, *Proc. Sixth European Space Mechanisms & Tribology Symp.*, Technopark, Zürich, 4-6 October 1995, pp. 177-183.

7 **Gill, S.** and **Rowntree, R. A.**, Interim Results from ESTL Studies on Static Adhesion & the Performance of Pennzane SHF X-2000 in Ball Bearings, *Proc. Sixth European Space Mechanisms & Tribology Symp.*, Technopark, Zürich, 4-6 October 1995, pp. 279-284.

8 TRW Space Log, 1977, TRW Space & Electronics Group, Redondo Beach, CA.

9 **Forster, A. S.** and **Feuerstein, S.**, Preliminary Results of an Advanced-Technology Despin Mechnical Assembly Life Test Programme, *Proc. First European Space Tribology Symp.*, Frascati, Italy, 9-11 April 1975, pp 123-141.

10 **Gardos, M. N.**, Labyrinth Sealing of Aerospace Mechanisms-Theory and Practice, *ASLE Trans.* **17** (1974) 237-250.

11 **Lince, J. R.** and **Fleischauer, P. D.**, Solid Lubricants, Chapter 12 in *Space Vehicle Mechanisms: Elements of Successful Design*, P. Conley, ed., John Wiley & Sons, Inc. New York, NY, in press (1997).

12 **Venier, C. G.**, Multiply-Alkylated Cyclopentanes (MACs): A New Class of Synthesized Hydrocarbon Fluids, *Lubrication Engineering*, **47** (1991) 586-591.

13 **Bertrand, P. A.**, Oil Absorption into Cotton-Phenolic Material, *J. Mater. Res.* **8** (1993) 1749-1757.

14 **Bertrand, P. A. Carré, D. J.**, and **Bauer, R.**, Oil Exchange between Ball Bearings and Cotton-phenolic Ball-bearing Retainers, *Tribol. Trans.*, in press (1997).

15 **Didziulis, S. V., Hilton, M. R.**, and **Fleischauer, P. D.**, The Influence of Steel Surface Chemistry on the Bonding of Lubricant Films, *Surface Science Investigations in Tribology*, Y. W. Chung, A. M. Homola, and G. B. Street, eds, ACS Symp. Ser. **485**, (1992) 43-57.

16 **Arezzo, F.**, Oil-aging Mechanism on 52100 Steel with Hydrocarbon Oils Containing Tricresyl Phosphate (TCP), *ASLE Trans.*, **28** (1984) 203-212.

17 **Fleischauer, P. D.**, Performance of Fluid and Solid Lubricants in spacecraft Applications, *J. Synthetic Lubrication*, **12** (1995) 3-12.

18 **Didziulis, S. V.** and **Bauer, R.**, *Volatility and Performance Studies of Phosphate Ester Boundary Additives with a Synthetic Hydrocarbon Lubricant*, Aerospace Corporation TR-95(5935)-6, 20 December 1995.

19 **Kalogeras C. G.** and **Didziulis, S. V.**, *Bearing Tests of Lubricant Additive Formulation and Pretreatment Processes, Aerospace Corporation* TR-95(5935)-5, 20 December 1995.

20 **Boving H. J.** and **Hintermann, H. E.**, Wear-resistant Hard Titanium Carbide Coatings for Space Applications, *Tribol. Int.*, **23** (1990) 129-133.

21 **Sproul, W. D.**, Reactive Sputtering of TiN, ZrN, and HfN, *Thin Solid Films*, **107** (1983) 141-147.

22 **McKee, F. B.**, Technical Note: Gyro Spin Axis Bearing Performance Using Titanium Carbide Coated Balls *Surf. and Coat. Technol.*, **33** (1987) 401-404.

23 **Fleischauer, P. D.** and **Bauer, R.**, Chemical and Structural Effects on the Lubrication Properties of Sputtered MoS_2 Films, *Tribol. Trans.*, **31** (1988) 239-248.

24 **Gardos, M. N.**, Anomalous Wear Behavior of MoS_2 Films in Moderate Vacuum and Dry Nitrogen, *Tribol. Letters*, **1** (1995) 67-85.

25 **Seitzman, L. E., Singer, I., Bolster, R. N.**, and **Gossett, C. R.**, Effect of Titanium Nitride Interlayer on the Endurance and Composition of a Molybdenum Disulfide Coating Prepared by Ion-beam-assisted Deposition, *Surf. and Coat. Technol.*, **51** (1992) 232-236.

26 **Hilton, M. R., Bauer, R., Didziulis, S. V., Dugger, M. T., Keem, J.**, and **Scholhamer, J.**, Structural and Tribological Studies of MoS_2 Solid Lubricant Films Having Metal-Multilayer Nanostructures, *Surf. and Coat. Technol.*, **53** (1992) 13.

27 **Fleischauer, P. D.**, The Effect of Crystallite Orientation on the Environmental Stability and Lubrication Properties of Sputtered MoS_2 Thin Films, *ASLE Trans.*, **27** (1984) 82-88.

28 (a) **Fleischauer, P. D.** and **Tolentino, L. U.**, Structural Studies of Sputtered MoS_2 Films by Angle Resolved Photoelectron Spectroscopy, *Proc. 3rd ASLE Int. Solid Lubrication Conf.*, 1984, *ASLE SP-14*, Denver, CO, pp. 223-229.

(b) **Fleischauer, P. D.** and **Stewart, T. B.**, Effects of Crystallite Orientation on the Oxidation of MoS_2 Thin Films, Aerospace Corporation TR-0084A(5945-03)-2, 9 September 1985.

29 **Donnet, C., Mogne, M. Belin, Le, T.** and **Martin, J. M.**, Solid Lubricant Studies in High Vacuum. *Proc. Sixth European Space Mechanisms & Tribology Symp.*, Technopark, Zürich, 4-6 October 1995, pp. 259-264.

30 **Carré D. J.** and **Bertrand, P. A.**, A Model to Calculate Evaporative Oil Loss in Spacecraft Mechanisms, submitted to *Tribol. Trans.* (1996).

31 **Beerbower, A.**, What's So Hot About Formulated Synthetics? *Tribology in the 80's*, Vol. **1**, NASA Conf. Publ. 2300, Cleveland, OH, 18-21 April 1983, pp. 477-499.

32 **Klaus, E. E.**, Status of New Direction of Liquid Lubricants, *Tribology in the 80's*, Vol. *1*, NASA Conf. Publ. 2300, Cleveland, OH, 18-21 April 1983, pp.367-389.

33 **Kalogeras, C. G,. Hilton, M. R., Carré, D. J., Didziulis, S. V.**, and **Fleischauer, P. D.**, The Use of Screening Tests in Spacecraft Lubricant Evaluation, *Proc. 27th Aerospace Mechanisms Symp.*, Sunnyvale, CA, 12-14 May 1993, NASA Conf. Publ. 3205, p. 197.

34 **Bialke, B.**, Space-Flight Experience and Test Performance of a Synthetic Hydrocarbon Lubricant, *Proc. Sixth European Space Mechanisms & Tribology Symp.*, Technopark, Zürich, 4-6 October 1995, pp. 285-291.

The problem-solving role of basic science in solid lubrication

M N GARDOS
Components & Materials Laboratory, Hughes Aircraft Company, El Segundo, CA, USA

SYNOPSIS

Lamellar solid lubricants such as graphite, layered transition metal dichalcogenides and lubricious oxides are examined, and the limits of manipulating their properties by doping are reviewed. The electronic structures of the host and the dopants interacting within a layered lattice are correlated with the characteristically anisotropic electrical and thermal conductivities, shear modulus, shear strength and chemical reactivity. Factors controlling the critical resolved shear stress of a single crystallite and those influencing the more global behavior of particles sheared within a lubricant film are compared, highlighting the effects of adsorbates chemisorbed on the grains during interparticle sliding. The adhesion of these films to their underlay is explained in terms of activated bonding between the basal planes or the edge sites of the grains and the load-bearing substrate. The gamut of knowledge transcending the customary regime of engineering tribology down to the atomic-level fundamentals is a highly effective diagnostic tool that helps solve tough problems by better selection or additive-induced improvement of layered-lattice-type solid lubricants.

1 INTRODUCTION

The greatest challenge facing the designers of moving mechanical assemblies (MMAs) and the tribologists who are called upon to lubricate these machine elements, is the ability to embrace more science-oriented methods of problem solving in place of brute-force engineering. Traditional approaches increasingly fail to resolve tough extreme-environment MMA problems where solid lubricants offer the best approach (e.g., in air and vacuum, especially at high temperatures). The only hope of solution lies in clarifying the atomic-molecular level interaction within the lubricant layer, as well as between this layer and its substrate exposed to some environmental stress. Only then can extraction of the universal rules of basic material behavior be applied to a wide range of specifics.

A schematic of solid lubricated surfaces under load and tangential shear subjected to a variety of thermal-atmospheric environments is shown in Fig. 1. With respect to behavior at the macroscopic scale responsible for engineering performance, the following axiomatic assumptions may be advanced (**1**):

1. A stepwise reduction in shear strength (τ_S) should exist, as the path normal to the plane of shear from the substrate toward the sliding counterface is traversed. This sequence of reduction must be preserved under any and all conditions.
2. The solid lubricant layer must adhere to its substrate but not adhere to itself. These simultaneous requirements appear to be mutually exclusive, because lubricants are designed to be inherently low surface energy materials.
3. The chemistry and crystal structure of a lubricating film must be engineered not only to resist *adhesive delamination* from its substrate, but also to yield a low *net* interfacial shear strength at the sliding counterface. Ideally and in line with 1 above, the film itself should contain cross-sectionally graded regions with progressively decreasing τ_S going from the substrate toward the sliding interface to prevent premature *cohesive* failure. Achieving this latter requirement via compositional and/or density variation by various physical and chemical vapor deposition techniques is just as difficult as forcing a low surface energy lubricant to adhere to its substrate. The adhesion versus cohesion balance is often precarious.

A low *net* interfacial τ_S may be attained by coating both bearing surfaces with the same sacrificial or replenished (transferred) solid lubricant film. Alternatively, one may be coated, then mated with the counterface treated with another low surface energy layer. The key objective is to preferentially shear (wear) the low energy, low τ_S but attrition-resistant layer(s) at the interface instead of wearing the load-bearing substrates.

The main goal of this paper is to help tribologists accurately predict the performance of MMAs solid lubricated with layer-lattice-type lubricants such as graphite, layered transition metal dichalcogenides and certain lubricating oxides by showing that conformance to engineering requirements is governed by the atomic-level interaction within the lubricant and between the lubricant and its underlay. The respective interactions can be made more synergistic by molecular engineering. A lubricant tailored with dopant(s) occasionally fills a need of which the pristine host is not capable, and surface activation enhances bonding between the lubricant and its substrate.

However, most doped lamellae can play only limited roles; they are not panaceas. Their range of environmental stability is limited: they tend to work best at moderate

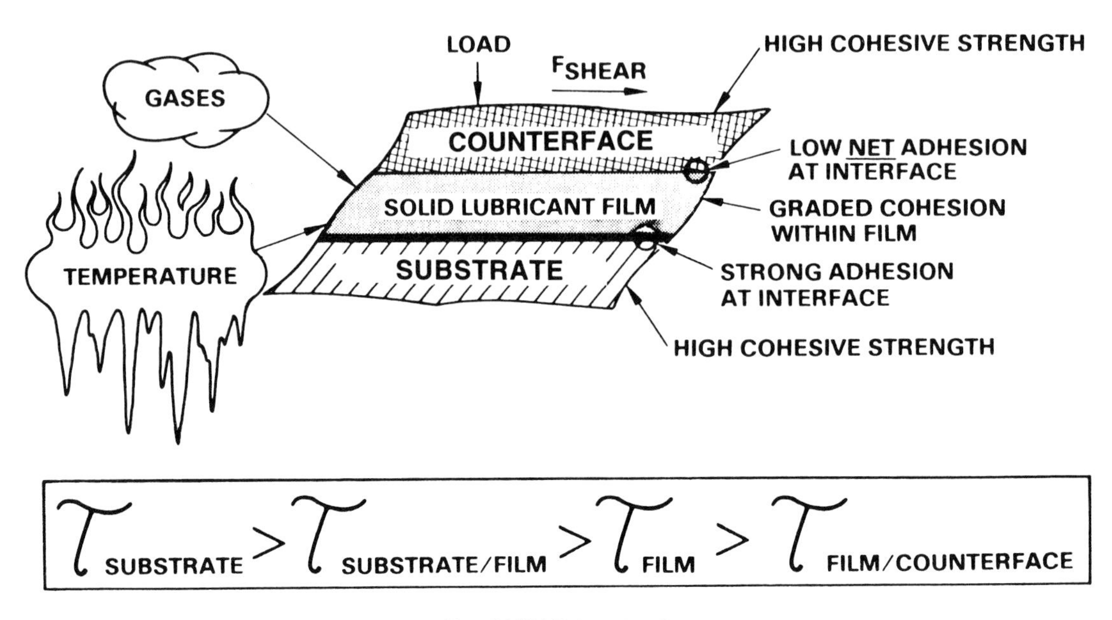

Fig. 1. A schematic model of solid lubricated bearing surfaces under normal load, tangential shear and environmental stresses; after (**1**).

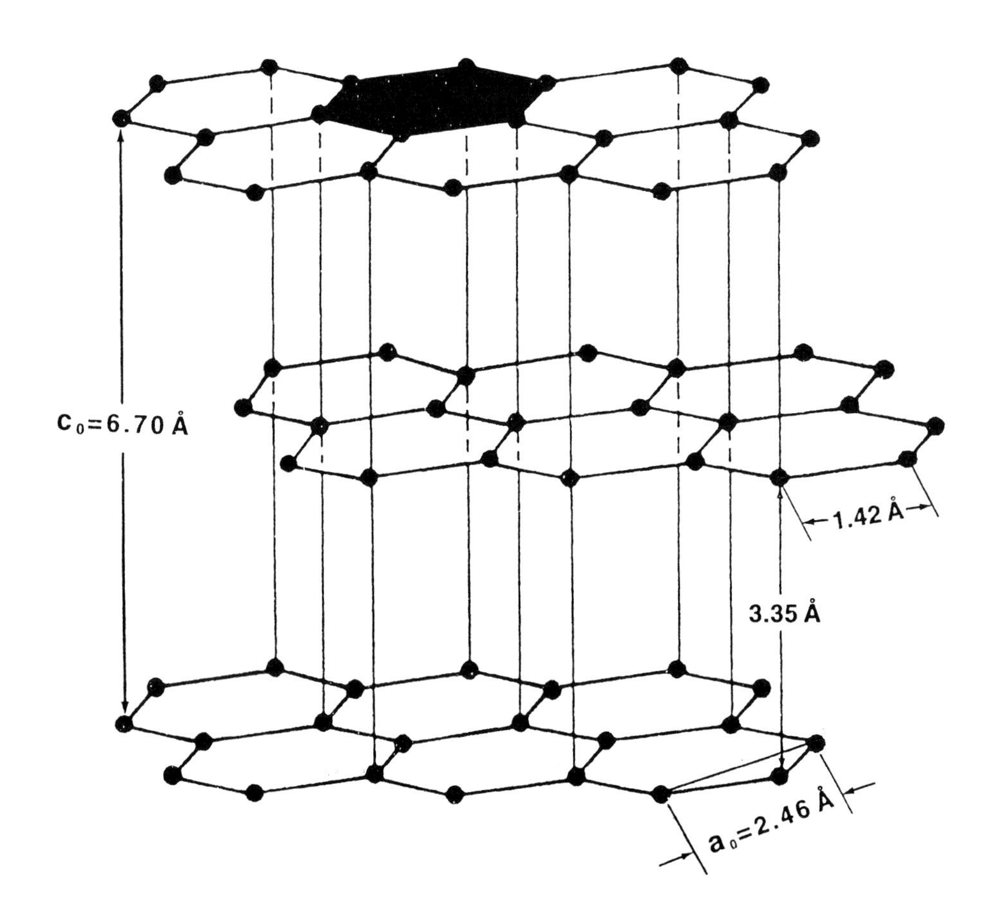

Fig. 2. The crystal structure of graphite.

temperatures, preferably in the absence of atmospheric moisture and oxygen. The fundamental laws of nature, which bestow favorable lubricating properties to certain layered crystallinities with or without dopants are also unfortunately responsible for degradation at elevated temperatures and in reactive atmospheres. Familiarity with these inviolable rules is indispensable for providing hardware with the lowest possible friction and the longest possible wear life, especially where the success or failure of extreme-environment applications is at stake.

2 BACKGROUND

The most often used sacrificial solid lubricants or those replenished from a binder have lamellar crystal structures. Examples of these commonly employed materials are the crystallites of highly ordered pyrolytic graphite (HOPG) in Fig. 2, the similarly layered-hexagonal transition metal dichalcogenides (LTMDs) such as MoS_2 in Fig. 3, or lubricious oxides (LOs) which may be considered somewhat lubricating due to their anisotropic structure, e.g., certain oxides containing grains with tailorable critical resolved shear stress (τ_{cr}) cleavage planes (Fig. 4). The ability of the lubricant particles to serve as building blocks of low-τ_S surface layers is manifested by two simultaneously exhibited properties: low *inter-planar* attraction accompanied by high-strength *intraplanar* chemical bonding. The former attribute leads to low τ_{cr} and thus to the low τ_S of a complete film, whereas the latter helps resist asperity penetration of the more-or-less aligned basal planes under the enormously high micro-Hertzian stresses that develop between contacting asperities.

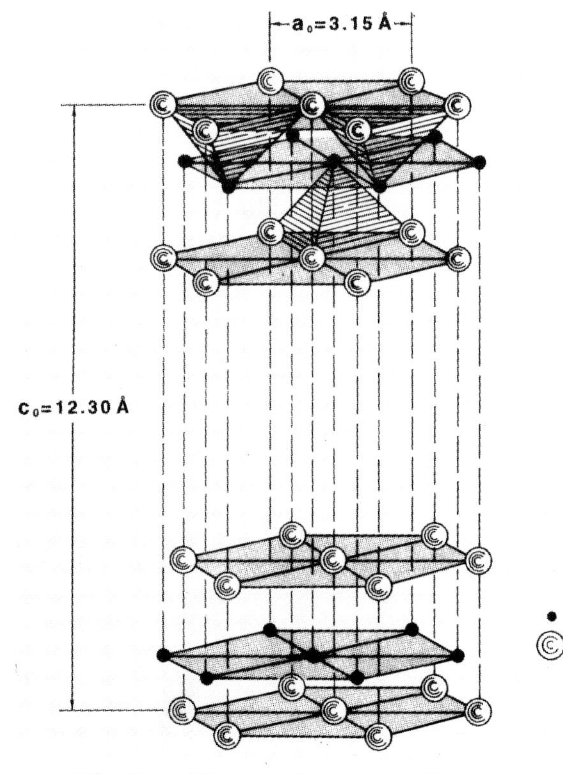

Fig. 3. The crystal structure of MoS_2.

The directional variation in the strength of the chemical bonds determines not only the anisotropic mechanical behavior, but the equally directional electrical and thermal properties of the crystallites as well. Thin films assembled from particles of acicular morphology, favorably aligned by the preparation process and the ensuing burnishing and/or run-in, are equally affected. The semimetallic, semiconducting or insulating nature and the τ_{cr}, τ_S, electrical-thermal conductivities and thermal-atmospheric stability of HOPG, LTMD and LO grains are intimately related to and dependent on their anisotropic interatomic bonding and their lay within the film.

However, lamellae suitable for solid lubrication do not always satisfy a given application completely. The electronic structure-induced electrical-thermal conductivities and the τ_{cr} (and τ_S) of naturally layered compounds often offer only limited solutions. The best measure of understanding the basic laws of nature is our ability to improve less-than-ideal crystal structures by molecular engineering as well as possible.

Generally, tribologists employ one of the techniques of (interstitial) intercalation or substitutional doping to reduce interplanar bonding within a crystallite. As shown later in this paper, the two general classes of altered lamellar structures are very different chemically. The most effective form of interstitial doping is the insertion of additional layers of electron donor or acceptor atoms or molecules (intercalants) between the sheets of the host material to alter its properties in a controlled manner. Dopants can also be substitutional, either introduced randomly into the lattice or in ordered distributions within the crystal structure. As also described later, the oxygen-deficient rutile polymorph of TiO_{2-x} was made to form layered Ti_nO_{2n-1}. Magnéli phase crystallographic shear structures (Fig. 4) with tailorable τ_S and electrical conductivity. This was attained not only by removing oxygen from the crystal structure (heating in vacuum is one way), but by doping with certain cations lower in valency than the host Ti^{4+}.

Unavoidably, lamellar lubricants are never in the form of infinitely large, single-crystal sheets. The individual particles of ground-up lubricant powders or the grains in sputtered-polycrystalline films have a variety of broken bonds, vacancies and other defects even on their stable basal planes and, especially, at the highly reactive edge sites. How these sites interact with (a) each other, (b) their substrate, and (c) the thermal-atmospheric environment during intra- and inter-particle slip determine the friction and wear life of a solid lubricant layer. It is not generally recognized that an unfortunate byproduct of inherent or intercalation-induced electrical conductivity (e.g., to provide an electrically conductive solid lubricant layer on a slip-ring, connector or a switch) is an enhanced sensitivity to environmental degradation. The higher conductivity is often synonymous not just with a higher-than-desired τ_{cr}, but with reduced oxidative and/or hydrolytic stability.

There are cases, where the environmental influences on the τ_S of a solid lubricant film outweigh the reduction in the

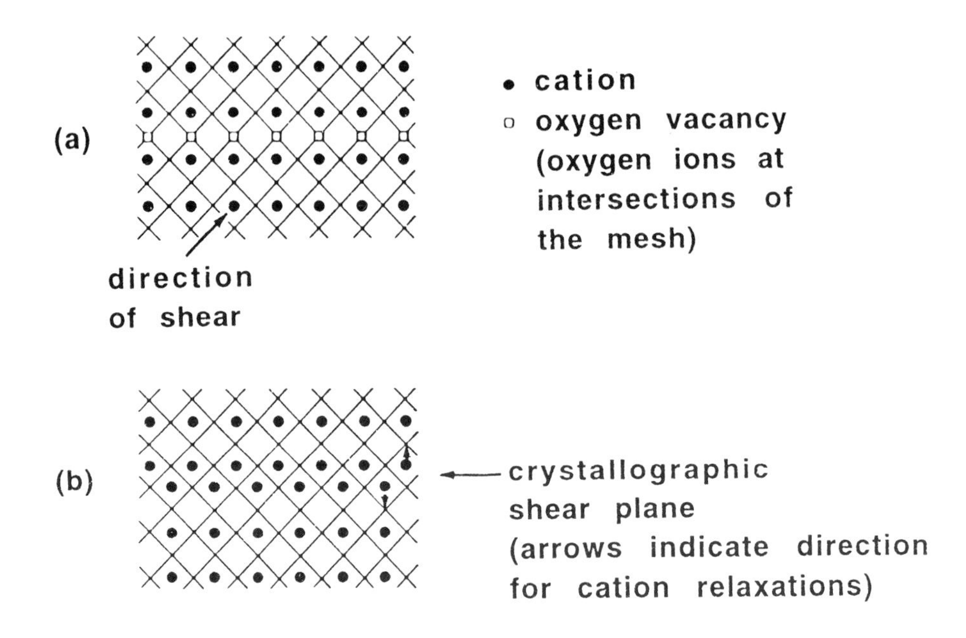

Fig. 4. Schematic representation of forming a crystallographic shear plane in a hypo-thetically reduced oxide structure capable of forming Magnèli phases: (a) aligned oxygen vacancies in the lattice prior to shear displacement in the direc-tion of the arrow, yielding the structure in (b). In the resulting structure the vacancies have been eliminated, but a fault is generated in the cation sub-lattice. The planes of faults, depending on the cation-to-cation distances and the resulting repulsive or attractive interactions, become altered on prog-ressive reduction to produce grains with different shear strengths; after (**2**).

τ_{cr} of the pristine (or doped) crystallite building blocks. The surface and even the bulk of each conductive crystallite sliding against its sister particles becomes detrimentally altered by thermal-atmospheric reactions. There may be a resultant increase in inter- and intraparticle attraction, because the global τ_S reduction aimed for by molecular engineering has been lost due to exposure to the mundane (but unavoidable and often outright harmful) environment of ambient air and moisture. Film failure then occurs by rapid delamination or excessive wear of the oxidized-hydrated or otherwise reacted grains, or both.

Advancing the argument that science fundamentals have problem-solving roles, the contents here deal with the author's interpretation of what, where, why and how the best compromises might be reached by using pure or doped (intercalated) HOPG, LTMDs and LOs as solid lubricants for a variety of conventional and special applications. An attempt is made to explain the pertinent atomic-level issues in terms of relevance to an engineering tribologist, and set the general limits of doping for product improvement.

3 PURE AND INTERCALATED GRAPHITES

A large portion of this section's contents was excerpted from (**3**), complemented by other references where the information in the main citation was found insufficient.

In this section, it is shown that pristine HOPG is a poor lubricant in its un-intercalated form. Even when intercalated, the doped versions are troublesome enough not to equally satisfy the multifaceted requirements of a tribological application. The solid (and for that matter, liquid) lubrication of electrical contacts is especially rife with compromises. The enhanced environmental sensitivity of graphite intercalation compounds (GICs) dictates that their gainful employment be restricted to certain special applications, with the mitigation of friction and wear playing only a minor role.

3.1 Highly Ordered Pyrolytic Graphite

To set the stage for understanding the concept of intercalation, let us examine pure HOPG first. In general, it is a good lubricant only in air, where it is permeated with sufficient amount of atmospheric moisture. Unintercalated, the π-π^* orbitals (i.e., sub-bands of the graphite band structure) overlap at selected sites of the Brillouin zone (Fig. 5). According to the data in Table 4.1 of (**3**), the overlap of these sub-bands can result in attractive interactions ranging from as low as 0.39 eV (8.97 kcal/mol = 37 kJ/mol), to as high as 0.8 eV (16 kcal/mole = 66 kJ/mol). The actual interaction force is somewhere within this range, depending on how the band parameters influence the net inter- and intra-planar attraction of a *family* of imperfect graphitic planes of particles interacting in a solid lubricant layer. The ~9 to 16 kcal/mol binding energy between the basal planes depicted in Fig. 2 is enough to render pristine HOPG a high friction and wear material.

The H_2O molecules are weakly amphoteric donor-acceptors. Water can transfer and/or receive charge to/from the neighboring polar species, depending on their chemistry. In the case of graphite, the oxygen in H_2O acts as a weak donor to the high electron density of the aromatic π clouds. As a result, the electron-hole attraction (and τ_{cr}) between the basal planes is reduced. This is why HOPG is an excellent lubricant in sufficiently humid air at temperatures low enough where water does not desorb, but a poor one in vacuum and/or at high temperatures where desorption does occur.

In pure HOPG, the large anisotropy in properties normal to the c-axis ($\perp c$ or within the basal planes) relative

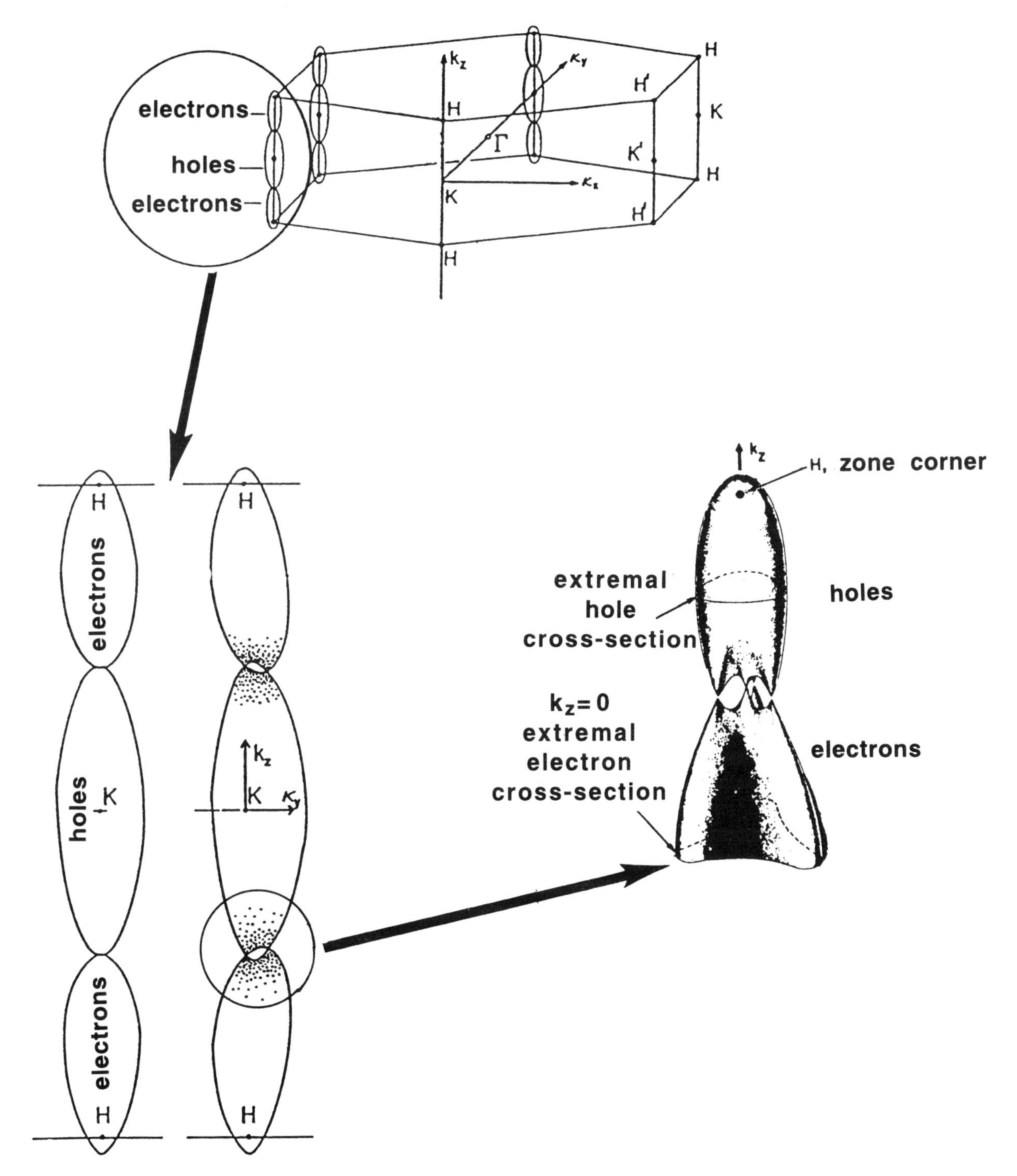

Fig. 5. Schematic of the Brillouin zone and Fermi surface cross-section of HOPG; after (**3**), (**4**) and (**5**).

to parallel to the c-axis (∥c or perpendicular to the planes) is manifested by a high directionality in the elastic constants and the electrical-thermal conductivities. The π-π* sub-band overlap responsible for the increased ∥c chemical bonding of the high-τ_{cr} (unintercalated) material is the cause of some electrical and thermal conductivity in the same direction. However, this mechanism cannot match the high magnitude of the inherently large ⊥c conductivities enabled by the delocalized π electrons moving freely between (within) the graphitic sheets. Since heat is conducted by both electrons and phonons, the ⊥c action of the delocalized π electrons is complemented by the hard phonon modes developing between the strongly-bonded intraplanar carbon atoms. The ⊥c C_{11} (in-plane) elastic constant (as large as that of diamond!) is about 30-times higher than the across-the-plane (∥c) C_{33} equivalent, because of the reduced chemical bonding and the resulting softer phonon modes in the ∥c direction (**6**).

The overall result is that the *intraplanar* electrical and thermal conductivities and elastic stiffness of HOPG are always higher than the *interplanar* equivalents. This dichotomy does matter where (a) the solid lubricant layer must allow removal and/or transfer of heat through the Hertzian contacts of a ball bearing, or (b) high-current, high heat-generating electrical contacts must be lubricated without suffering thermally-induced desorption of an intercalant (e.g., the loss of H_2O at temperatures not much higher than its boiling point).

3.2 Intercalation with donors and acceptors

The most characteristic similarity GICs share with those of the HOPG host and the intercalant within is manifested by the characteristically strong *intralayer* bonding in the alternating graphitic and dopant layers and the relatively weak *interlayer* bonding between graphite-intercalant and certain (but not all) graphite-graphite sheets.

The intercalation reaction with either donors (Lewis bases) or acceptors (Lewis acids) is generally accompanied by charge transfer between the intercalant species and the parent graphitic layers (a.k.a. *galleries*). This transfer is the driving force of the intercalation reaction. The perturbation introduced by each dopant layer is largely confined to the neighboring graphite bounding layers, while the interior graphite galleries remain essentially unaffected.

Let us suppose that the *donor* is potassium (or another alkali metal). Each K atom positioned in the intercalant layer has one valence electron which, when totally transferred to the neighboring graphite sheet, renders the dopant layer positively charged. With only a typical K^+ concentration of 10% or less, this attraction between the positively charged K^+ layer and the charge carrier electrons results in a 10^3 increase in carrier concentration as compared to pure HOPG. For *acceptors* such as the halogens or certain metal halides, the intercalant layer becomes negatively charged by extracting electrons from the graphite bounding layers. As a consequence, a high hole concentration is found in these neighboring graphene sheets, with the holes acting as the charge carriers.

Trouble arises where the electrical conductivity of a lubricant must be preserved without degrading its shear strength (e.g., for dry-lubricating electrical contacts). Since a compromise must somehow be struck between competing (and mutually exclusive) τ_{cr} versus electrical conductivity properties in the ∥c versus ⊥c directions, it matters to both an electrical engineer and the tribologist whether the intercalant is a donor or an acceptor.

Donor intercalation (mostly with alkali metals) tends to increase the ∥c electrical and thermal conductivity. Because of the extensive charge transfer between these metals and the graphitic layers, there is greater coupling between the intercalant and graphene planes in donor-type GICs than with the acceptor-type (e.g., metal halide- or halogen-intercalated) compounds. The donor-doped materials thus become more 'three-dimensional' (**3**). A more isotropic electrical conductivity arising from the enhanced sub-band overlap is commensurate with a higher τ_{cr} of the lamellae.

In contrast, doping with the inherently planar metal halide (mostly metal chloride) *acceptors* such as $FeCl_3$, $CuCl_2$, $NiCl_2$, $CdCl_2$, $CrCl_3$ or $SbCl_5$ tends to *decrease* the ∥c conductivities of GICs. The diminished flow of electrons is attributed to the high electrical impedance across the intercalant layer, a result of the small overlap between the graphite p_z (π) orbitals and the molecular orbitals of the chlorides. Owing to the orders-of-magnitude increases in ⊥c conductivities and an equal decrease in ∥c conductivities for typical acceptor compounds upon intercalation, an anisotropy ratio in excess of 10^6 can be achieved (**3**). We as tribologists benefit in terms of lower τ_{cr} and τ_s, but only where high ∥c electrical and thermal conductivities through Hertzian contacts are not required.

Because the planar metal chloride molecules are not divisible, they remain on either the upper or the lower graphite bounding layer on cleaving due to the weak binding force between the graphite layers and the intercalants (**7**). The weak force is the results of the fact that these intercalants have no empty states near the Fermi level, i.e., they have no dangling bonds [that is why they are planar in the first place, see (**8**)]. Why can these chemically stable compounds still act as acceptors? Wertheim, et al (**9, 10**) have shown that the intercalant species form ~10 nm diameter islands in the graphene galleries. The charge removed from the graphite π band is localized on excess chlorine at the edges of these characteristically packed islands to maintain full halogen coordination of the cations, as evidenced by the detection of inequivalent chloride ions in XPS spectra.

The weak interaction between the metal halide and graphite sheets is both a blessing and a curse, depending on the problem at hand. As far as the tribologist is concerned, it should not matter where the cleavage occurs, as long as the low τ_s of the solid lubricant film is maintained and high ∥c conductivity across the tribocontact is not required. A low τ_S is the main reason why only metal chloride-GICs (at least some of them) have been found to be good solid lubricants. However, the number and distribution of the exposed intercalant molecules should matter, because these dopants are extremely sensitive to environmental degradation (section **3.4**). Yet, at the same time, they are probably the only active sites for bonding directly (without using a separate binder) to the bearing substrates, as discussed in Section **3.5**.

Even the tribologically poor donors, where the increased bonding between the electron-donating intercalant layers and their neighboring graphite sheets translates into higher τ_S, may be useful for certain friction and wear applications where high electrical and thermal conductivity of the contacts is required. The charge distribution in the neighboring donor-doped sheets shifts toward the positively charged intercalant layer and thus the chemical bonding (the π-π^* overlap) between the graphite bounding layer and its immediately neighboring interior layer is decreased. However, this mechanism is not active enough to render alkali-intercalated GICs lubricants as good as the metal chloride-doped versions. Although the metal chloride (and other kinds of) acceptors exhibit the opposite effect, namely an increase in bonding between the bounding HOPG layer and the interior layers once-removed, the substantial net reduction in τ_S still favors acceptor-GICs (Section **3.6**).

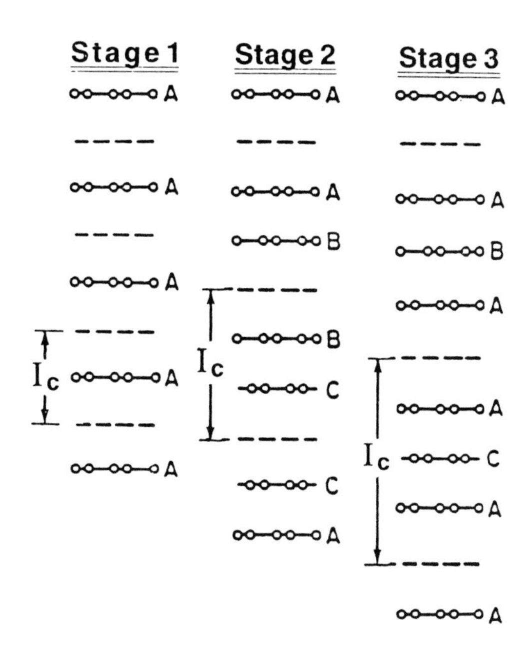

Fig. 6. Schematic of staging in GICs; after (**1**).

3.3 The staging phenomenon during intercalation

Among the many structural parameters of GICs, the degree of intercalation by a variety of donors and acceptors is best defined by the stage index I_C. It designates the number of graphene sheets between two successive layers of intercalant (Fig. 6). The transitions between stages are abrupt, because they depend on the lateral motion of the intercalant within the layer kinks of the imperfect graphitic planes (Fig. 7). The propagation of these kinks and the associated transition energetics occur in quantum jumps. Since energy is required in quanta to induce these transitions, intercalation and de-intercalation occur stepwise, e.g., on heating to various temperatures (**11**). This is an important in terms of thermal stability limits of GICs.

For example, the inherently planar metal chloride molecules readily enter into the galleries of HOPG. Using iron chloride as a model, $FeCl_3$ crystallizes into a layered solid composed of $Cl_3Fe_2Cl_3$ intercalant sandwiches that are incommensurate with the graphite host. The intercalant layer tends to have the same sandwiched configuration as in the molecular solid, and maintains the same structure as found in solid $FeCl_3$. The behavior of other similar compounds (e.g., $CuCl_2$, $NiCl_2$, $CdCl_2$, $CrCl_3$ and $SbCl_5$) is analogous.

The sandwiched dimerization of the intercalant species is a manifestation of the physical basis for the staging phenomenon. There is strong interatomic intercalant-versus-intercalant binding compared to intercalant-graphite bond, favoring a more or less close-packed planar arrangement of a chloride dopant between the graphitic sheets. Introducing each new intercalant layer adds substantial strain energy to the crystal as it expands to accommodate the chloride molecules. The insertion of a minimum number of intercalant layers with the largest possible layer separation is favored as a function of a given average dopant concentration. Therefore, the most stable stacking arrangement is the one which minimizes the volume needed to accommodate the available intercalant in the interlayer space.

Larger elastic constants imply higher thermal conductivity by phonons, commensurate with faster propagation of acoustic energy. The sound velocities of the Stage 1 and Stage 2 $FeCl_3$-GICs were studied to determine their $\|c$ (C_{33}) elastic constants. These constants (*ergo* the stiffness along the *c*-axis) increase with an increasing stage index I_C, because higher-stage GICs have more vacant galleries which possess strong interlayer attraction between the graphite layers (**12**). More environmentally stable, low-I_C metal chloride-GICs are, therefore, better lubricants.

3.4 Environmental stability

Donor-doped (e.g., with Li, Na or K) GICs are easily oxidized and hydrated (**13**), so they should be kept under

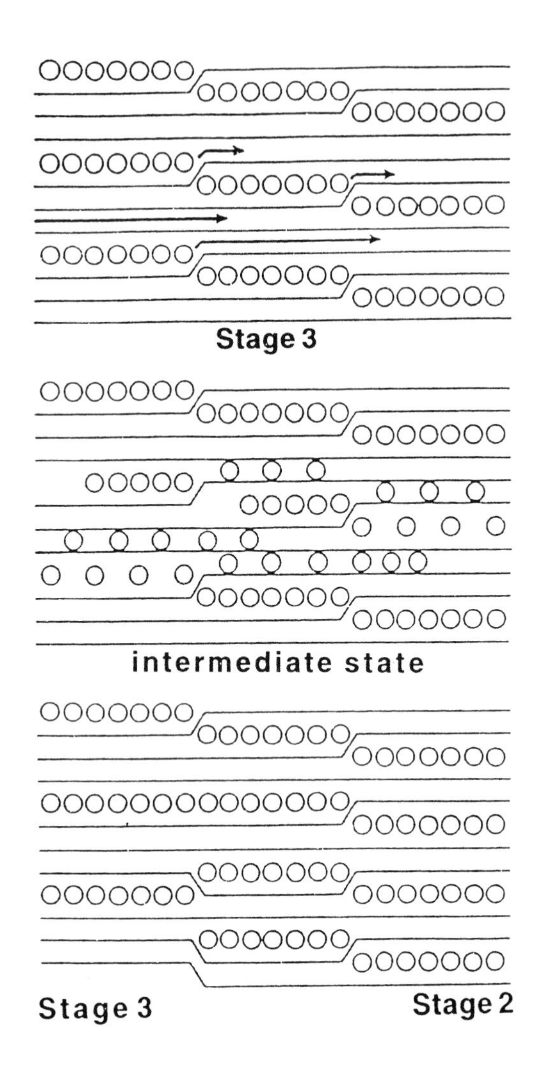

Fig. 7. The mechanism for the transition between GIC stages according to the Daumas-Herold domain model; after (**11**).

an inert atmosphere or in vacuum. Acceptors such as the metal halides and the halogens are also moisture-sensitive. In addition, they become thermally desorbed from the intercalated HOPG structures at elevated temperatures (**1**). The temperature and the degree of desorption depends on the type of intercalant, the particular stage number (I_C) and the film thickness.

The thickness plays a role, because the sequential intrusion of the dopant species into the interlayer galleries of the HOPG (Fig. 7) proceeds more readily from the edge of the crystallite at sites closest to its outermost basal-planar surface. Then, they go into the other subsurface layers in a similarly sequential order. In this way, the elastic energy required to cause *c*-axis expansion of the lattice is minimized. It also explains why, because of the reduced elastic impedance to swelling, thin samples intercalate (or de-intercalate) more readily than thick samples. Since solid lubricating GIC layers are generally thin, they are especially vulnerable to heat-induced degradation. However, Gardos (**1**) did show that relatively thick samples of both $CdCl_2$- and $NiCl_2$-GICs maintained their low τ_S in both air and in vacuum, even after heating to ~700°C. Surprisingly low friction (but not low wear) were observed for isostatically pressed, 3 mm-thick GIC samples sliding against α-SiC, despite heat-induced sublimation of $CdCl_2$ and possible decomposition of the $NiCl_2$ at elevated temperatures in vacuum, and a most probable increase in I_C.

Since the carrier density for both donor (acceptor) GICs is dominated by the respective charge transfer from (to) the intercalant layers, the density is expected to be essentially *temperature independent*. In contrast with the HOPG host, the thermal dependence of the electrical conductivity in GICs is directly associated not just with the carrier density but with carrier mobility also. The $\perp c$ electrical conductivity increases with reduced temperatures, because there is less scattering of the electrons by phonons. This effect is larger for donors than acceptors. Although intercalation causes a decrease in carrier mobility because the dopant species are in fact 'defects' which scatter electrons, this decrease is much less than the very large increase in the carrier density. Compared to pristine HOPG, the combined *net* effect is an order-of-magnitude or higher electrical conductivity on intercalation.

3.5 Adhesion to the substrate

So far, GIC powders have been used as burnished films or have been bonded to the substrate by polymers. Expanding beyond this limited application depends on the atomic-level manipulation of the GIC and/or its substrates for direct bonding of GICs to the bearing surface. For example, could GIC layers be grown directly on metals or ceramics with high adherence but low τ_S?

As advocated and outlined in (**1**), *abinitio* and molecular dynamics calculations of idealized substrates and lubricant molecules can predict the type of surface engineering needed for better adhesion. A recent example of this methodology may be found in (**8**), where bonding metal halides to the surface oxygens of oxides was achieved via $TlCl_3^{2+}$, $HfCl_2^{2+}$ species and $Al_2Cl_4^{2+}$ dimers. As described in section 4.2.3, the halides normally acting as acceptors can change into donors, if the mating surface (such as an oxide) exhibits strong Lewis acid (acceptor) properties. The same material can act amphoterically, depending on the chemistry (weak/strong acid/base nature) of its counterface. For example, each aluminum chloride is converted as $AlCl_3 \rightarrow AlOCl \rightarrow Al_2O_3$ by the application of heat, in the presence of water (**14**). By the use of this interfacial reaction, one might be able to coat GIC directly to an oxidized bearing surface by some chemical vapor deposition method. There are ab-initio and molecular dynamics methods readily available for predicting the kind of surface modifications needed for good adhesion (**15**).

3.6 The history of intercalated graphite use

Graphite intercalation compounds have been the subject of considerable interest for the past several decades, because of the rich variety of chemical species which can be inserted between the graphene galleries, and because of the extensive ability to tailor the electronic and mechanical properties of the HOPG host for a variety of practical applications.

To date, HOPG (a semimetal) has been converted to a true metal by intercalation, mostly for use by the electronics industry. For example, a Boeing 727 aircraft carries ~2000 lbs of Cu wire, with larger planes carrying considerably more. GICs, especially in fiber form, are potential competitors because of their 3.3-times lower density, 10-times the tensile strength and 5-times the elastic modulus of Cu (**16, 17**). Using (resistivity × density)$^{-1}$ = 6.3 × 10^5 cm^2/Ω for Cu as the customary figure of merit, a GIC must exhibit an electrical resistivity lower than 6 × 10^{-6} Ω cm to compete with Cu. Over 10 years ago resistivities of about 2×10^{-6} Ω cm have already been achieved (**16**), close to that of Cu = 1.67×10^{-6}Ω cm.

The unique properties of GICs are also useful for upgrading the performance of GIC fiber-reinforced composites (**18–20**). The improved fiber properties arise because of the special alignment and strengthening of the intercalated graphitic basal planes parallel with the axis of each fiber by certain acceptors, e.g., with bromine. GICs have also seen applications in EMI shielding, as electrode materials in batteries, catalysts for organic synthesis as well as a material for hydrogen storage or the separation of its isotopes (**21**).

Although the early 1950s and 60s brought forth some sporadic exploratory research on metal chloride-doped GICs as lubricants for a variety of extreme environment applications (**22–24**), it was Conti's award-winning STLE paper (**25**) which called renewed attention to the tribological utility of metal chloride-GICs. In his experiments, the $CoCl_2$-GIC pigment in a resin-bonded solid lubricant exhibited over a fivefold wear life increase, while the $NiCl_2$-intercalated version provided greater than a twofold increase in load-carrying capacity relative to pristine HOPG. Conti's work inspired Gardos (**1**) to show, as previously mentioned, high tribothermal stability and low friction (but not necessarily low wear) for certain isostatically-pressed GICs sliding against α-SiC in moderate vacuum, to temperatures as high as 700°C. More recently, the capacity for being a good lubricant in vacuum combined with the tailored $\|c$ versus $\perp c$ electrical conductivity have led to the use of GICs as lubricants for satellite potentiometers operated with resin-bonded carbon resistive elements. Due to the high friction and wear (and the resulting electrical noise) of hard carbons and graphites in vacuum, these potentionometers widely used in terrestrial applications are not suitable for space use (**26**).

Graphite fluoride (CF_X, where x = 1.1 in the ideal stoichiometry) has also seen use as a halogen (acceptor)-intercalated solid lubricant with some limited success, mostly as a powder additive in self-lubricating composites and bonded solid lubricant films (**27**). Today, $CF_{1.1}$ is employed more extensively as a battery electrode material (**28**).

4 PURE AND INTERCALATED DICHALCOGENIDES

Most of the fundamentals relevant to tribologists and lubrication engineers on these subjects have been excerpted from (**29**). Complementary atomic-level information on the most widely used MoS_2 was taken from the work of Fleischauer and his colleagues (**30–32**) and other appropriate references, as cited. The main intercalation-related papers are those of Friend and Yoffe (**33**) expanding on the general basics, and Jamison's classic work (**34**) dealing with intercalated LTMDs prepared specifically as lubricants.

The main topic of this section is that similar to HOPG and GICs, the layered structure of LTMDs renders them mechanically, electronically and environmentally anisotropic. By choosing an LTMD according to the filling of their transition metal d-orbitals which are either empty or partially occupied, the electronic (and thus the physical) properties of LTMDs can be tailored to certain given requirements. It is possible to select the right kind of semiconductor-to-metal or metal-to-semiconductor transitions, and the associated τ_S, electrical conductivity and environmental stability trade-offs even without intercalation. The paucity of practical intercalated LTMD lubricants to date may have been caused by this relatively wide range of undoped candidates, with the kind of specialized attributes suitable for most tribological applications. However, manipulation by doping has confirmed our understanding of the chemical bonding-induced changes in crystal structure and properties.

4.1 Pure layered transition metal dichalcogenides

4.1.1 Fundamental bonding considerations

Similar to HOPG and GICs, the basic atomic structure of loosely coupled X-M-X packets, where X = a chalcogen such as S, Se, Te and M = a transition metal such as Nb, Mo and W, renders LTMDs mechanically and electronically anisotropic. For example, the $\perp c$ electrical conductivity of the semiconductor MoS_2 (Fig. 3) is a factor of at least 100 higher than the $\|c$ conductivity (Table 1). Similarly, the thermal conductivity of LTMDs is ~10-times higher in the $\perp c$ direction; the velocity of sound and the elastic modulus in the $\perp c$ direction are also higher than those in $\|c$ (**6**). In contrast, the coefficient of thermal expansion is about a factor of 10 greater $\|c$ than $\perp c$.

The lowest unoccupied molecular orbitals in the LTMD

host layers are the transition metal d-bands, which are either empty or partially filled. An increased filling of these bands predictably fine-tunes the electronic (and thus the physical) properties of LTMDs. By choosing the LTMD according to the partial filling of the d_{z^2} sub-band and the type of chalcogen, it is possible to select the desired properties even without intercalation.

The highest occupied molecular orbital of the MX_2 layer structures is full with 16 electrons (ZrS_2). Subsequent electron additions for NbS_2, MoS_2 and TeS_2 go into the non-bonding, d-based d_{z^2} sub-band. This gradually narrows the band gap, progressively increasing the electrical conductivity (reducing the resistivity). In the highest resistivity but lowest τ_{cr} MoS_2, all the accessible Mo and S orbitals are involved in strong *intralayer* bonding, leaving only high energy anti-bonding orbitals available for *interlayer* bonding. The d_{z^2} is completely filled with both electrons in MoS_2, but only half-filled (with an unpaired electron) in $NbSe_2$. The long range attraction between these singly-occupied orbitals (dangling bonds) is high enough to align the Nb atoms over each other between the Se-Nb-Se layers, as depicted in Fig. 8. In contrast, the Mo atoms are staggered between the S-Mo-S layers, because there is repulsion between the completely filled orbitals (**30**). The end-result is a significantly higher τ_{cr} for $NbSe_2$ than for MoS_2. For the same reasons, the nature of bonding in $NbSe_2$ leads to larger elastic constants: a larger C_{11} for $\perp c$, C_{33} for $\| c$ and C_{44} (the shear modulus) than those of MoS_2 (**6**). In qualitative analogy with pristine HOPG and donor-GICs, the chemical bonding-cum-crystal structure-dependent τ_{cr} and τ_s are again at odds with the electrical (and thermal) conductivity across the planes and thus through a Hertzian contact.

Much has been made in (**34**) about using the *c/na* ratios of unintercalated and intercalated LTMDs as a measure of the τ_s, by virtue of a larger van der Waals gap indicated by a larger *c/na*. This seems to be correct when the height of the MoS_2 coordination unit ($C_o/2a_o$ = 1.95 in Fig. 3), commensurate with a wider band-gap and van der Waals gap as well as a lower τ_{cr}, is compared to the $C_o/2a_o$ ratio for $NbSe_2$ (1.82). The equivalent $NbSe_2$ parameters are semimetallic in character, with a narrower van der Waals gap and higher τ_{cr}.

However, the $\| c$ unit cell elongation of the Group VIb LTMDs is also *decreased* from MoS_2 to $MoSe_2$ and, in turn, to $MoTe_2$. It is clear that this shrinkage cannot be attributed to the action of the chemically stable, doubly occupied (highly satisfied) d_{z^2} orbital. We have seen that this action is repulsive, not attractive. The causes lie elsewhere: the reduction in the $\| c$ unit cell elongation stems from the small size of the Mo atom (it is one of the smallest atoms to form LTMDs) and the decreasing ionicity (S>Se>Te) accompanied by an inversely increasing size (Te>Se>S) of the chalcogens. The sulfide is the most ionic, with the sequence of MoS_2>$MoSe_2$>$MoTe_2$. The X-X (chalcogen-to-chalcogen) core distances in the $\| c$ direction (an indicator of the van der Waals gap size) are *inversely* proportial to their respective *c/na* ratios, translating into a progressively more compacted character of the neighboring X-M-X packets going from MoS_2 to $MoTe_2$ (**29**). A

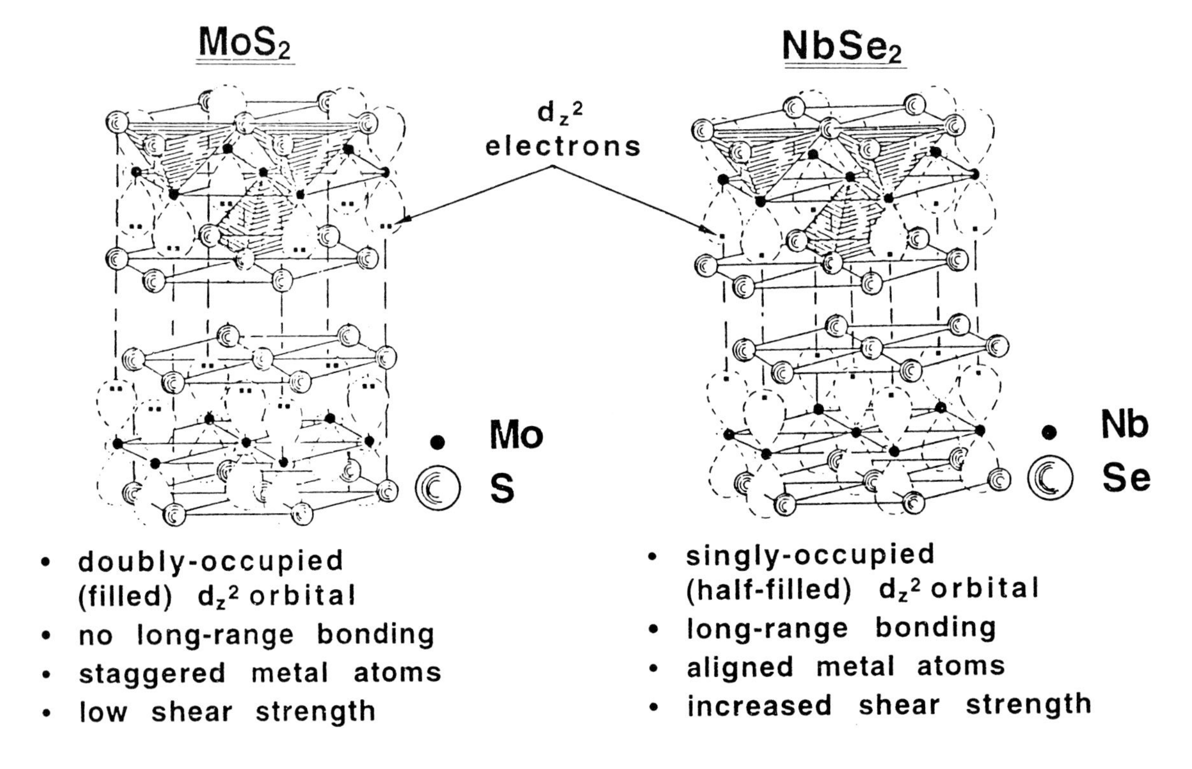

Fig. 8. Influence of filling the the d_{z^2} orbitals of MoS_2 and $NbSe_2$ on their structural and lubrication properties (courtesy of Dr. P.D. Fleischauer of The Aerospace Corporation; also see (**30**).

diminishing van der Waals gap translates into an equally diminished capacity to act as a solid lubricant, with MoS_2 being the best and $MoTe_2$ the worst in that family of LTMDs. In parallel, there is an increasing ionic character of the Group IV metals (Ti>Zr>Hf) as well as the Group V and Group VI equivalents (V>Nb>Ta and Cr>Mo>W, respectively). One must be careful, however, in defining the degree of ionic versus covalent bonding nature of one LTMD to another *between* groups (**6**).

4.1.2 Electrical Conductivity

Electrical conduction in LTMDs is accomplished by excitation of electrons out of the d_{z^2} valence sub-band into the d/p conduction sub-band. The charge carriers in LTMDs are created by thermally or electrically forcing electrons to hop across the band gap (if any). One can also view this mechanism as sending electrons across the van der Waals gap: a wider gap comes with a lower τ_{cr}, but the conductivity is also reduced. The size of the band gaps and the associated resistivities are predictably different, as shown in Table 1. The 'wider gap, lower electrical conductivity and higher volume resistivity' rule is always followed. In the more strongly interlayer-bonded Group Vb LTMDs the hopping barrier to the electrons is much lower. Therefore, the Group VIb LTMDs are semiconductors, while the Group Vb LTMDs are semimetals. For example, the volume resistivity of NbS_2 and $NbSe_2$ is about five-orders-of magnitude less than that of MoS_2, because the lone dangling bond electron in the half-filled d_{z^2} orbitals of NbS_2 (or $NbSe_2$) is a charge carrier. On application of a sufficiently large electric potential commensurate with the necessary activation energy, the conduction electrons become easily trapped and detrapped at these sites to provide the flow of current.

The chalcogen-caused property changes include the highest electrical conductivity of $MoTe_2$ in Group VIb (Table 1), diminishing in the order of $MoTe_2$>³$MoSe_2$>MoS_2. Typically, ligand field strength decreases as one goes down each group of the periodic table. This means that splitting the d-levels in the central metal of the octahedral MX_6 units decreases going from S to Se to Te. The energy of the lowest unoccupied molecular orbital (the d/p conduction sub-band) is thus brought closer to that of the d_{z^2} orbital. Such an effect leads to a narrower band gap and higher electrical conductivity, depending on the temperature and the current-voltage characteristics of the electrical contact. On the whole, $MoSe_2$ may be the best compromise for an electrically conductive solid lubricant, simultaneously providing acceptable τ_{cr} and τ_s characteristics. Although its shear properties are somewhat worse than that of the semiconducting MoS_2, the environmental stability is greater than that of the semimetallic but relatively high-τ_{cr} $NbSe_2$ (see 4.1.3 below).

Table 1 Band gap and near-room-temperature volume resistivities of selected Group Vb and VIb LTMDs. Values followed by the particular reference source. Variation in data attributed to differences in sample type (see footnotes) and the measurement method.

Group no.	**LTMD**	**Band gap (eV)**	**ρ (Ω·cm)**
Vb	NbS_2		8×10^{-4} (35)[(1)] 5.35×10^{-4} (36)
	$NbSe_2$	none (semi-metals)	1.5×10^{-1} (29,36)[(2)] 3.5×10^{-4} (29) [(2)] 5.55×10^{-4} (37)
	$NbTe_2$		$\sim 1 \times 10^{-4}$ (29)
VIb	MoS_2	1.6 (38)	$\parallel c = 2 \times 10^3$ (29) $\perp c = 12$ (29) [(1)] 3 to 100 (29) [(1)] 851 (36,37)
	$MoSe_2$	1.1 (38)	1.5 (29)[(1)] 3.4 to 5 (29)[(2)] 1.86×10^{-2} (36,37)
	$MoTe_2$	1.0 (38)	0.4 (29) [(1)] 8.5 (29) [(1)]

[(1)]Single crystal.
[(2)] Pressed compact.

Another molecular engineering method of preparing more stable electrical contact lubricants is to formulate a quasi-binary alloy of the more electrically conductive but less environmentally stable and lubricating (Group Vb) $NbSe_2$ and the less electrically conductive, but more stable and better lubricant (Group VIb) WSe_2. Replacing only 3% of the W atoms with Nb atoms in the MoS_2-like WSe_2 lattice lowers the electrical resistivity from 3.6×10^{-1} Ω.cm to 1.6×10^{-3} Ω.cm (**39**).

There are caveats, however, where the electrical conductivities of burnished and sputtered LTMD films compacted in concentrated contacts need to be predicted by the behavior of the individual crystallite building blocks. Waghray, *et al.* (**40**) showed that the sliding contact resistances of powder-burnished and sputtered sulfide/selenide/telluride films of Mo, W, Nb and Ta were all about the same and did not show any correlation with static resistivities. They did, however, differ in tribological performance. All LTMDs formed adherent burnished layers except NbS_2 and $TaTe_2$ (MoS_2 performed best both electrically and tribologically). Only the sputtered films of disulfides and diselenides of Nb, Ta, Mo and W adhered well to the stainless steel substrate, each performing satisfactorily in all respects except NbS_2, $NbSe_2$ and TaS_2. Regula, et al (**41**) found the static contact (sheet) resistance of sputtered WS_X (x = 1.6 to 1.8) to be controlled by the deposition temperature, the H_2S pressure of the sputtering atmosphere and the resultant grain alignment in the films. The measured activation energy for the sheet resistances depended more on grain boundary effects (intergrain potential barriers) than on the intrinsic properties of the grains.

4.1.3 Environmental stability

The environmental sensitivity of LTMDs is inherent with the lamellar nature of their crystal structure. Point-defect-free basal planes are orders-of-magnitude more inert and stable than their dangling bonds-terminated edge sites exposed by tribological action. The freshly generated active sites readily react with gases, liquids or solids, the stability of the basal planes notwithstanding. The edge sites of MoS_2 are $\sim 10^{11}$-times more reactive than a defect-free basal plane (**42**), therefore they must be protected from unwanted environmental effects by certain additives (see section 4.2.2 below).

Another degradative factor is the inherently higher environmental sensitivity of the Group Vb LTMDs attributed to their singly occupied $d_Z{}^2$ orbitals. For example, the unpaired d-electron of Nb is readily donated to oxygen (i.e., $NbSe_2$ is easily oxidized), especially at planar and edge defects and step or ledge sites. This renders unintercalated $NbSe_2$ less oxidatively stable than MoS_2. From early Russian experiments (**39**) to generating and moving Nb_2O_5 and MoO_3 islands about on their respective $NbSe_2$ and MoS_2 substrates (**43**), $NbSe_2$ was shown to suffer greater tribo-oxidative degradation than MoS_2. The thermal-atmospheric sensitivity of LTMDs is typically manifested by the resultant increase in friction and a reduction in wear life.

As previously mentioned, the data in Table 1 hints at a possible compromise. Since (a) the filled $d_Z{}^2$ orbitals of Mo in the Group VIb MoS_2 and $MoSe_2$ imply greater oxidative stability than the half-filled orbitals of the Group Vb Nb-based counterparts, and (b) the larger Se atom provides narrower band-gaps (thus the generally higher electrical conductivity of the selenides compared to the sulfides), $MoSe_2$ might be a more oxidatively stable choice among these conductive solid lubricants for certain electrical contact applications in ambient air.

As far as the consequences of elevated temperatures are concerned (i.e., without any danger of oxidation in an inert environment), the charge carriers and phonons couple in a unique fashion in the LTMD layer structure. In an apparent paradox, the electrical conductivity increases (**33**) and τ_S decreases with increasing temperature. The reduced τ_S may be ascribed to H-bonded moisture desorbed from the oxidized sites of the crystallites (**44**). It appears that increased $\|c$ interplanar binding *within* a crystallite by greater thermal excitation of the electrons through the band-gap is less than the binding effects of water molecules *between* crystallites, as discussed below.

4.2 The special case of MoS_2

4.2.1 Environmental sensitivity to degradation

Over the past several decades, polymer-bonded or sputtered LTMD films, especially those based on the lowest-τ_S and relatively stable MoS_2, have become the coating of choice for solid lubricated MMAs operated in a variety of conventional and extreme environments. However, even the most recent overview papers promoting the low friction, long wear life and volatile condensible-free nature of these thin films for vacuum (aero-space) applications call attention to the well-known sensitivity of MoS_2 to air and atmospheric moisture (**45–50**).

This sensitivity manifests itself both in terms of increased friction (0.15 to 0.3 in humid air vs. 0.01 to 0.05 in inert gas or vacuum) and drastically reduced life of burnished, bonded or sputtered MoS_2 films, as well as water vapor-assisted oxidation of the deposited layers during storage (**49, 50**). For this reason, it is common aerospace industry practice to tribotest or use MoS_2 in vacuum or inert gas atmospheres, and store or ground-test MoS_2-lubricated space mechanisms under a blanket of dry nitrogen or argon. Aerospace technologists consider vacuum and inert gas environments virtually interchangeable in terms of keeping moisture away from MoS_2, expecting the same (good) tribological performance under both atmospheric conditions.

To help understand how the electronic structure-induced morphology and chemical stability affect the oxidation of MoS_2, high temperature experiments were conducted in 1 torr of O_2 using a small piece of natural molybdenite, inside a Knudsen cell-like hot-stage of an environmental SEM (**1**). Molybdenum trioxide (MoO_{3-x}) was generated at the reactive edge, step and defective basal plane sites of the crystallite at 850°C (Fig. 9). The MoO_{3-x} is <010> textured on the MoS_2 basal surfaces, being the lowest energy habit plane of the oxide (**1, 41**). As described in (**43**), a completely oxidized island (similar to the one marked by the arrow on the top of Fig. 9b) can be moved around on (i.e., sheared against) the yet-unoxidized substrate with the application of a very small tangential force. The easy shear of $MoO_3(010)$-on-$MoS_2(0001)$ should not be surprising, given the low surface energy γ = 20 to 50 dynes/cm of the basal plane of MoS_2 (**51**) and the reported γ = 70 dynes/cm for MoO_3 (**52**).

Not until recently did Roberts (**45, 46**) raise the question about the possibly harmful effect of not just oxygen but oxidation-enhancing residual H_2O remaining behind in the moderate ($\sim 10^{-4}$ Pa $\approx 10^{-6}$ torr) vacuum of certain spacecraft atmospheres. The gases remaining there are reported to be about 80% water vapor and 20% CO.

Spacecraft internal atmospheres right after launch can indeed be as high as 10^{-3} to 10^{-4} torr, and it can take as long as 1000 hours for the pressure to become reduced to the 10^{-5} to 10^{-6} torr range. In some spacecraft enclosures (e.g., an open Space Shuttle bay) the water content of the desorbed species can be as high as 48% after 12 hours (**53**). The gaseous environment of the Shuttle bay is, in fact, predominantly water. With all the non-metallics outgasing, especially the Al-coated polyimide thermal control films wrapped on exposed structures everywhere (the only major component emanating from these films is H_2O, see Ref. **54**), the high moisture content is not surprising (**55**). In general, the atmosphere surrounding a spacecraft may contain fluctuating partial pressures of certain gases where (a) the return flux of the outgased species is either high (**56**) or low (**57–59**) depending on the different vehicle orientations on orbit, and (b) the gases are trapped within hermetic enclosures or labyrinth-sealed regions (**60**).

If the MoS_2-lubricated hardware is located at an ultrahigh vacuum site of a satellite, there is no problem. If not, it is not only possible but probable that the residual gas composition in the working environment of a MoS_2-lubricated satellite bearing or gear remains detrimental enough to reduce the life of the mechanism, especially if it is exercised continuously soon after launch. This probability also presents the specter of ground-testing MMAs in an atmosphere not representative of their microclimate on orbit. Indeed, Gardos (**44**) recently showed unexpectedly rapid MoS_2 film wear in moderate (~$1x10^{-5}$ torr) vacuum containing over 90% residual water vapor. It was proposed that high-τ_S (5 to 6 kcal/mol) hydrogen-bonded H_2O bridges developing in such an environment translates into faster generation and removal of MoO_3 from the underlying MoS_2 films by tribological action (Fig. 10).

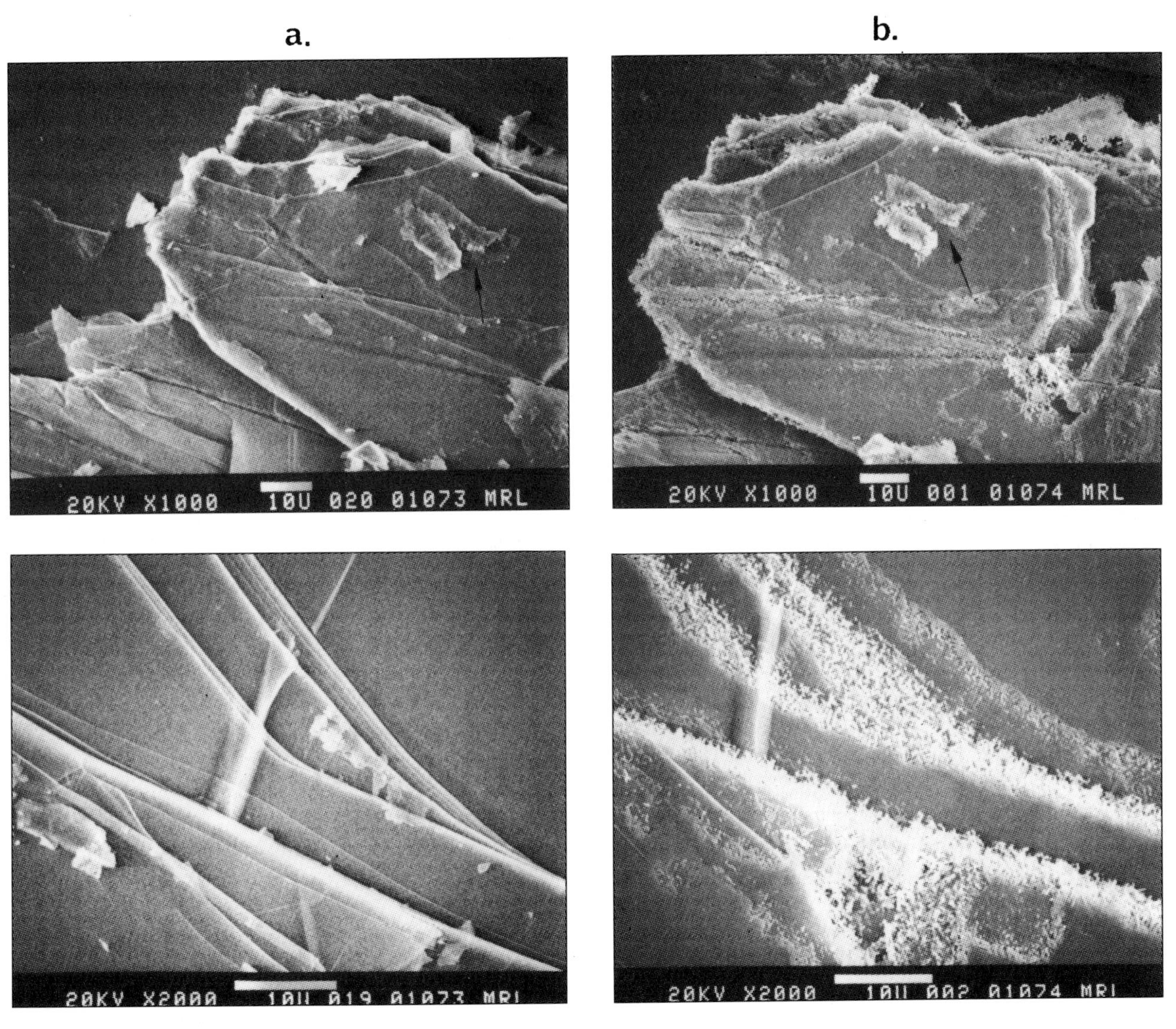

Fig.9. SEM photomicrographs of a nature molybdenite (MoS_2) (a) before and (b) after partial oxidation for 5 minutes in 133 Pa = 1.0 torr O_2; from (**1**).

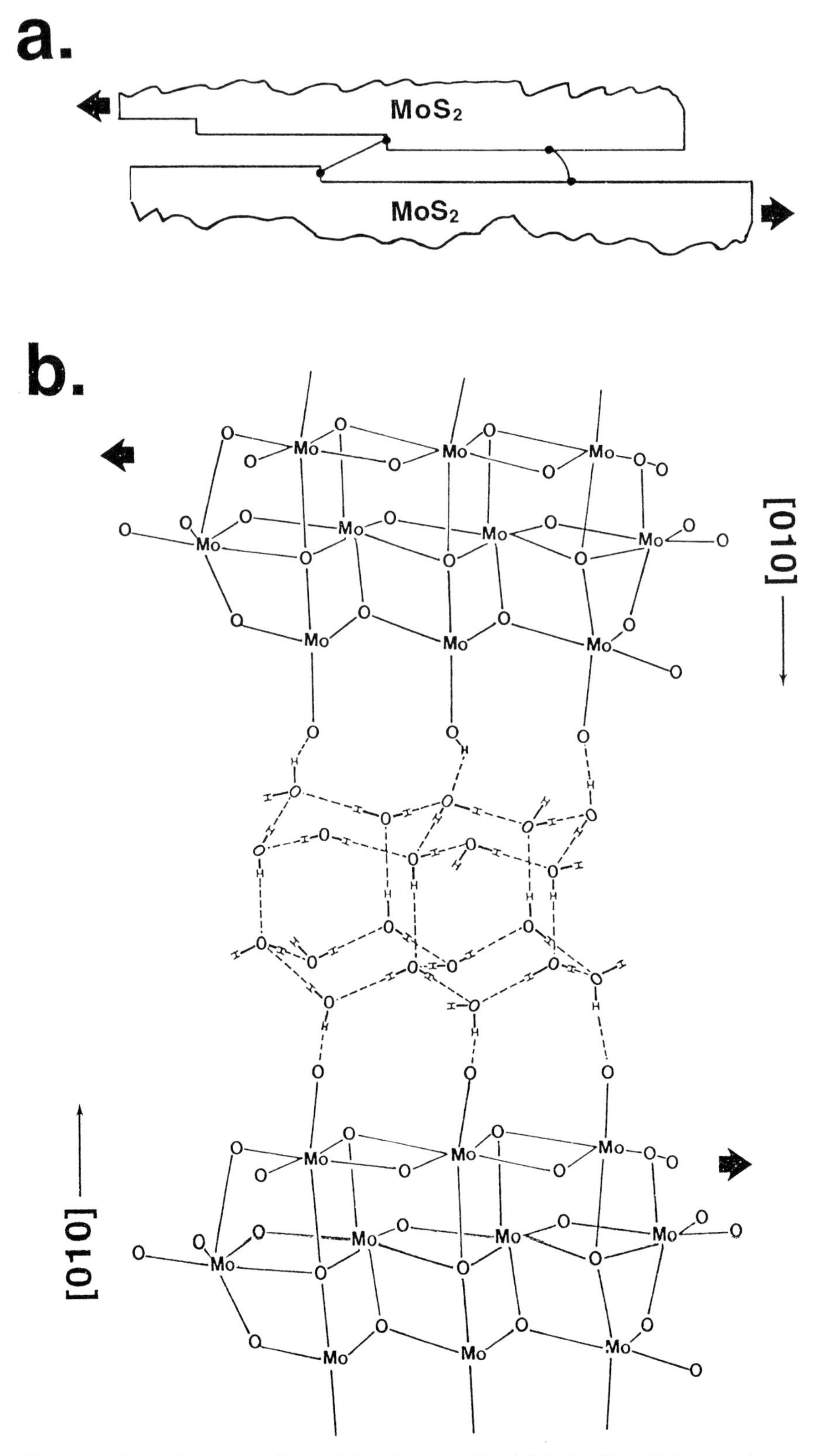

Fig. 10. Microscopic and nanometric models of an oxidized MoS_2 film sliding against itself. Higher shear stresses are provided by hydrogen-bonded water bridges anchored to to the oxidized and/or hydrated moieties, linking the active coun-terface sites together: (a) cross-sectional schematic at the grain structure level, and (b) atomic-level schematic of mating MoO_3(010) planes bonded together by an idealized, ice-like network of H-bonded water molecules; from (**44**).

4.2.2 Increasing the tribo-oxidative stability of sputtered MoS_2

The wear and friction sensitivity of sputtered films to moisture and oxygen may be mitigated in several ways:

1. Operate a solid lubricated mechanical assembly in high-to-ultrahigh vacuum or in inert gas, where the residual H_2O and O_2 is below some yet undetermined (but very low) limit. However, shielding dry-lubed MMAs from ambient air and humidity in more conventional applications is not practical.
2. Design a deposition method which yields the most defect-free crystal structure and test it in an entirely moisture- and oxygen-free environment. Martin and his staff (**61–64**) employed a surface analytical tribometer operated in an ultrahigh vacuum test atmosphere and a unique coating technique to show unmeasurably small friction ('superlubricity') of fully stoichiometric MoS_2. This unique effect manifested itself under incommensurate interfacial orientation of the perfect, oxide-free crystallites. The considerable scientific significance of the findings notwithstanding, the special deposition rate is too low and thus the process is too expensive for cost-effective industrial production of coatings. Furthermore, there is no proof yet that the ultra-low friction films also give sufficiently long lives.
3. Protect the dangling (or oxygen-capped) molybdenum bonds at the chalcogen vacancies with clusters of moisture-blocking additives.

The latter technique has become the most widely used palliative, with selected metals co-deposited mainly with MoS_2. The metallic additives are either simultaneously co-sputtered with MoS_2 from a homogeneously mixed, composite target effectively yielding a composite-like film (**65**), or sequentially sputtered from multiple targets by rotating the substrate underneath to form nanometer-thick, alternating metal/lubricant films with a sandwiched structure (**66**). Recent work by Jayaram, et al (**67**) revealed island-like growth of the metal, instead of forming evenly formed interlayers. For all practical purposes, the resulting structure is similar to that of the more conventionally co-sputtered films. In either case, the co-deposited metal poisons (**68, 69**) the otherwise friable and edge-plane-columnar (less oxidatively stable) structure of the MoS_2 sputtered by itself. The microcrystalline composite layers are more dense, exhibiting lower wear rates and higher resistance to oxidation and hydration, with only a small penalty of somewhat increased friction. The most commonly employed additives for MoS_2 are Ni and Sb_2O_3, while the 'layered' technique uses Au-20%Pd at the present time (**66, 70**).

Gardos suggested in (**41**) that vacuum-evaporated Au, which preferentially decorates the heated and oxidized edge and basal plane sites of MoS_2, might also protect the same sites from *additional* hydro-oxidative degradation. Decoration comes about from the attraction of Au (or other metal) atoms to the dangling, and even the already oxygen-capped, Mo bonds.

The degree of attraction may be modeled by the differences in the respective Schottky barrier heights between the perfect-basal versus the defective edge, step and ledge sites of MoS_2 and the atomic clusters of the metal. A hypothesis was advanced in (**71**), whereby the degree of adhesion and friction of metals to diamond, Si and GaAs is attributed to the magnitude of the respective Schottky barrier heights. A high (low) one-directional barrier to electron flow across the metal-semiconductor junction is commensurate with low (high) adhesion and friction. The size of the barrier, in turn, was interpreted in terms of (a) the presence or absence of dangling or passivated (saturated) bonds on the semiconductor surface, and (b) the chemistry and defect structure of the contacting metal. While the barrier height between Au and the inert (perfect) basal plane of MoS_2 is high signifying no electronic (*ergo* chemical) interaction (**72**), the barrier between Au and the prismatic plane (the edge sites) should be lower. Certain metallic dopants which coordinate sufficiently with the unsaturated Mo atoms should also help protect these active MoS_2 sites from oxygen and water molecules. In the author's opinion, the metallic clusters do double duty: they (a) act as gas barriers at the atomic level, and (b) produce denser and more basal-plane-aligned films to reduce gas-permeability of the lubricant layers.

There is a trade-off between the moisture-protecting role of a metallic additive and its contribution to toughness-induced increases in wear life. Hertzian indentation-induced spallation is less with the retention of some porosity (at the cost of some reduction in density), and increases where the metal content is too high and the pore-free nature tends to disappear due to the higher densities (**70**). Even the right amount of metal that offers the necessary toughening of the coating structure may not serve well as a water vapor barrier in increasingly moisture-laden environments. The case in point is Ni, which readily becomes oxidized during the co-sputtering process (**73**). Water molecules hydrogen-bond just as willingly to nickel oxide-hydroxide sites as to molybdenum oxide-hydroxide sites. Furthermore, Ni causes catalytic degasification of MoS_2 in hydrogen by promoting the removal of sulfur in the form of H_2S. The result is an increase in the number of reactive edge sites (**74**). Similar action with reactive atmospheric gases (such as O_2) is also possible. Gold is inert, although the Pd additive in the sputtered Au 'multilayers' is a well-known catalyst. It is reasonable, therefore, to advance the hypothesis that the electronic structure of the interface should become progressively more favorable to degradative reactions during intrusion of moisture. The Schottky barrier height would increase and the adhesion between MoS_2 and a readily oxidizable metal (e.g., Ni) would decrease within a co-sputtered film. The overall result might be an increase in friction and a severe reduction in wear life.

4.2.3 Adhesion to the substrate

It has been proposed that the degree of donor-acceptor charge transfer also referred to as Lewis base-Lewis acid interaction or the analogous Schottky barrier versus ohmic contact formation between a metal and a semiconductor defines the chemical bonding-induced part of interfacial react with strong acids (acceptors), while weak base-donors prefer to react with weak acid-acceptors (**75**). For example, in a metal versus n-type semiconductor contact the metal is the weak acid (acceptor), while the semiconductor is the weak base (donor). This happens because the chemical potential of a clean n-doped semi-conductor is higher than that of the metal. When they come in intimate contact with each other, the electrons tunnel from the semiconductor (the weak base-donor) to the metal (the weak acid-acceptor) until the chemical potentials become the same. There will, however, be a step from the highest occupied metal state (the chemical potential) to the lowest unoccupied state of the semiconductor at the interface. This step is called the Schottky barrier, which is a rectifying (low adhesion, low shear strength) junction. According to the hypothesis in (**71**), to get a low-barrier or ohmic (conductive, high adhesion and high shear strength) contact one needs a sufficiently large number of interface states (dangling bonds) on an adsorbate-free semiconductor surface and an ultraclean, unoxidized metal counterface.

Industrial experience with structural and pressure-sensitive *adhesives* is in agreement with the fundamental principles of charge transfer at the interface. Basic adhesives are best for clean acidic surfaces, whereas acidic adhesives are best for clean basic sub-strates (**76**, **77**). Of late, the same principles have been applied to tailoring both the bearing surface and the immediately attached portions of a *lubricant* film for maxium adhesion (**32, 71, 78**).

The atomic-level fundamentals of sputtering the semiconductor (weak base) MoS_2 for maximum adhesion to bearing steels (weak acids when clean, stronger acids when oxidized) have been most extensively addressed by the surface scientists and tribologists of The Aerospace Corporation in the United States (**32, 79**). It was shown that the best bond between the counterfaces can indeed be achieved by scrupulously cleaning the stainless steel bearing surfaces by argon ion etching prior to deposition. The loose Fe_2O_3 top layer to which, on both physical and chemical basis, MoS_2 would be poorly anchored must be removed from the more adherent Cr_2O_3 underneath. Further ion bombardment then exposes nascent Fe and some of the more desirable Cr component (the primary corrosion resistant constituent of bearing steels such as the widely used 440C) for the needed chemical interaction with MoS_2. The prerequisite for this interaction is the activation of MoS_2 into an n-type semiconductor by the sputtering-induced removal of some sulfur. Where they are missing (mostly from the edge sites), the fully coordinated Mo^{6+} ions change to the reactive Mo^{5+}

Unfortunately, activating both the steel substrate and the lubricant by the plasma treatment leads to the undesirable edge-island interface and edge-site-rich, columnar growth of the sputtered MoS_2. This is not the best crystallite lay in a film for the lowest friction, longest wear life and highest environmental stability. Metal-co-sputtered films provide a better alignment, with the basal planes more or less parallel with the substrate surface (**73**). The preferred lay may be attributed to the fact that Au or Ni, deposited first on the steel substrate to act as the initial bonding interlayer and then periodically between the lubricant grains, adheres well enough to the steel substrate but interacts with the dangling Mo^{5+} bonds less than clean Fe, Cr or even their oxides. Gold bonds to MoS_2. sufficiently, but not well enough for inducing edge-columnar growth of the sputtered lubricant (**80**).

4.3 Intercalated layered transition metal dichalcogenides

There has been more work done on enhancing the adhesion and environmental stability of sputtered MoS_2 films than to tailor the interplanar properties of any other pristine LTMD via doping. One would think that systemic improvement of the grains within a film by intercalation, combined with the advantages of co-sputtering with an edge site-protecting metal, would lead to the best overall product.

The paucity of molecular-engineered LTMD lubricants to date may have been caused by the relatively wide range of undoped candidates with the kind of τ_{cr} versus electrical conductivity trade-offs suitable for most tribological applications. Nevertheless, manipulation of the electronic structure (therefore the physical properties) of LTMDs is possible by intercalation. As with HOPG, the reaction is accompanied by charge transfer between the dopant specie(s) and the host layers. However, all LTMD intercalants are electron donors only. No one has yet succeeded in intercalating with acceptors, as is possible with HOPG. This may be explained by the nature of Coulomb repulsions involved between the ionized acceptor and the chalcogen layers carrying an effective negative charge. The ensuing repulsion is probably greater than any electronic energy gained by intercalation (**33**).

The electronically stable MoS_2 is willing to accept only very strong donors (e.g., alkali metals) as intercalants to turn this semiconductor into a metal. At the same time, it is capable of retaining those intercalants in the crystal structure only at very low temperatures. At higher temperatures the dopants become unstable and come out of the lattice (**81**), often accompanied by an explosively defoliating chemical reaction with such common environmental components as water (**82**).

There is not much else that can be done to MoS_2 to improve its lubricating properties besides passivating the environmentally sensitive, reactive edge sites. At the same time, the intercalation of LTMDs have not led to products better than MoS_2. In the author's opinion, the results to date are useful to a tribologist only in demonstrating our knowledge of the atomic-level interactions involved. A case in point is altering the crystal structure and properties of $NbSe_2$by intercalation with Au and Cu (**34**).

Both Ag and Cu have one valence electron on their respective bonding orbitals. By incorporating a certain amount of these metal atoms into the van der Waals gap, a part of the intercalant's valence electron charge becomes donated to the half-filled d_z^2 orbital of Nb. The attraction between the Nb atoms depicted in Fig. 8 is sufficiently reduced to turn the relatively poor lubricant $NbSe_2$ into the substantially lower-τ_S $Ag_{0.33}NbSe_2$ and $Cu_{0.33}NbSe_2$ (**34**). However, the electrical conductivity of the intercalated versions becomes lost, because the barrier height to electron hopping (the band gap) was simultaneously increased. In an ironic parallel with pristine HOPG and donor-GICs, by achieving a lower τ_{cr}, the $\| c$ conductivity becomes reduced. Jamison (**34**) argued, however, that preferential oxidation of the intercalant may possibly reduce the oxidation kinetics of the host, improving the overall environmental stability of doped LTMDs. No work has been done yet to prove this point.

Inadvertent intercalation with Ag or Cu could destroy the electrical conductivity of certain precious metal electrical contact lubricated with pristine semimetallic LTMDs such as $NbSe_2$. For example, electrocatalytic leaching of some of the 30% Ag and 14% Cu from Paliney No. 7 [a commonly used precious metal electrical contact alloy, see (**83**)] into the mating contact lubricated with a thin $NbSe_2$ film may sufficiently alter both the crystal structure and electrical conductivity of the lubricant layer in the long term. Diffusional leaching of Ag and Cu out of the precious metal contact and their intercalation into the lubricant film is conceivable, rendering the film more lubricative but eventually less conductive. Polarity changes in the electrical contact, especially with the precious metal wiper negatively biased, may enhance current-induced electromigration (**84**) of Ag and Cu into the intercalant-hungry $NbSe_2$ film. An unacceptable increase in electrical noise would follow.

This may well be the reason why the tribologically handicapped $NbSe_2$ worked well only in lubricating the short-design-life potentiometers of the Surveyor Moon-landing vehicles. The pots served to ascertain the direction to which the TV camera was pointing. The bare Paliney No. 7 wipers of the first Moon lander, sliding against bare Ni-Cr wire-wound resistive elements, failed badly due to metal-to-metal seizure and the accompanying electrical noise. The $NbSe_2$-lubed versions operated satisfactorily (**83, 85**). In contrast, $NbSe_2$ consistently failed as the lubricant additive in Ag-containing slip ring brushes mated against coin-silver sliprings for long-life applications. MoS_2-containing brush materials were a lot more successful and are now routinely used in satellites (**86, 87**).

5 LUBRICIOUS OXIDES

Stable metal oxides would be ideal high temperature lubricants for critical applications in air. Unfortunately, most of these materials are abrasive with poor shear properties. They remain unacceptable even at elevated temperatures, where some softening would be expected. Those that are not excessively abrasive or high-wearing are still high in friction, tribochemically unstable or volatile (**88**). Based on the realization that certain oxides such as MoO_3 have reasonably low shear strength planes (**42, 44, 51**), Gardos (**88–91**) investigated the possibility of finding other layered metal oxide structures of low τ_{cr} and thus low τ_S that persist in a wide environmental regime. In particular, the rutile polymorph of titania (TiO_{2-x}) was used as a model compound.

5.1 Pure oxygen-deficient rutile (TiO_{2-x})

Anything that is done to an oxide surface can, and often will, change its composition (**88**). In almost all cases, the material becomes oxygen-deficient. Rutile is no exception to this general rule. In oxygen-poor or reducing environments, especially at high temperatures, a progressively larger number of anion vacancies are generated both on the surface and in the bulk of the rutile lattice. Vacancy generation leads to the creation of new crystallographic shear systems also known as the Ti_nO_{2n-1} Magnèli phases schematically represented in Fig. 4. The high index planes of these phases shift with oxygen content and dominate the actions of the natural (low index) planes. They also determine the τ_{cr} of the crystallites and thus the τ_S of a film or a bulk oxide surface. It was hypothesized, then experimentally demonstrated, that the predictably changing Magnèli phases either reduce or increase the lattice strain energy (E_l) and the commensurately smaller and larger τ_S-controlled friction. In particular, two low E_l and τ_S regimes were predicted and found near $TiO_{1.98}$ to $TiO_{1.93}$ and at $TiO_{1.70}$, surrounding a four-to-five times higher peak of those values at $TiO_{1.80}$ (Fig. 11a). Independent research by NASA Lewis Research Center tribologists found that the hardness also follows the same trend (**93**).

The generation of oxygen vacancies starts on the surface, but will not necessarily lead to the formation of Magnèli phases. Some of the anion removal processes are not energetic enough to induce diffusion of the vacancies into the bulk. Diffusion is important, because the self-arrangement of the clustered cations into ordered, planar arrays (see Fig. 4) is the dominant mechanism for creating the low- and high-τ_S shear planes. The gradients in τ_S depicted in Fig. 11a change, because of the lattice-distortive interaction of charge density waves produced by corner-, edge- and face-sharing TiO_6 octahedra on progressive reduction. The layer-to-layer attraction or repulsion of the shear planes can be decomposed into contributions from different sets of one-dimensional chains of the variously assembled octahedral building blocks. For example, those derived from edge- and face-shared chains have shorter Ti-Ti distances and stronger interaction (thus higher τ_{cr} and hardness) than layers of corner-sharing chains only.

The changing Ti-Ti distances in the lattice, which control shear plane attraction or repulsion, also determine the electrical conductivity. When an oxygen is lost from reduced rutile, two electrons are left in the vacancy to maintain charge balance. Initially, the electrons may occupy the vacancies formed. As the reduction proceeds, a neighboring Ti^{4+} may change its charge state to Ti^{3+} ; the other electron is essentially free to act as a charge carrier. The color of the oxide changes from the ivory-yellow-white of the fully stoichiometric version to the progressively darker, blue-black coloration of the more anion-deficient sample. The color change is caused by the absorption of red light by the conduction electrons, signaling the transformation from a large band gap insulator to an n-type semi-conductor and, on further reduction, to the near-metallic $TiO_{1.80}$. In line with the bonding fundamentals of HOPG, GICs or pristine and intercalated LTMDs, a higher (lower) τ_{cr} and τ_S here is commensurate with higher (lower) electrical conductivity (Fig. 11a and b).

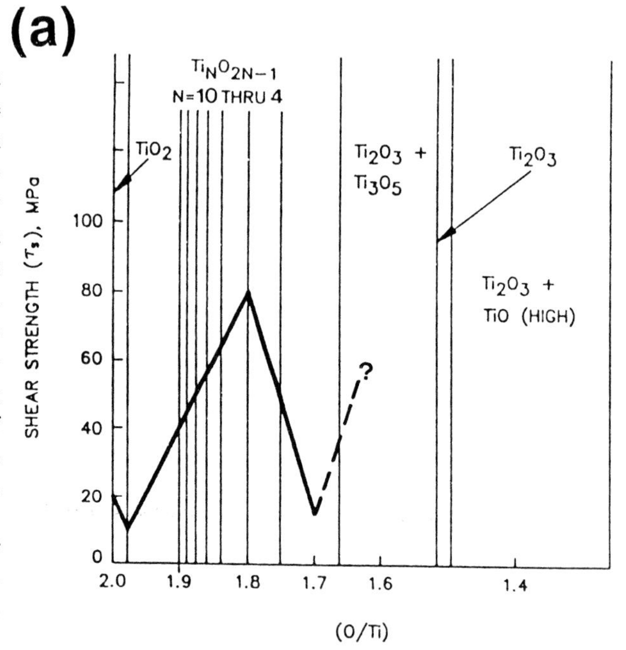

5.2 Molecular engineering of rutile by doping

More recently, an attempt was made to generate stable Magnèli phases of the desired kind(s) by doping rutile with cations similar in size and polarizability to the Ti^{4+} , but with lower valencies (**93**). Doping was predicted to cause chemical expulsion of oxygen from the lattice to maintain charge balance and to generate specific shear structures. The same basic concept is used to create high temperature superconductive oxides. Additional tribometric experiments, similar to those completed with rutile reduced by a thermal-vacuum treatment, were done in vacuum and in various partial pressures of 99.995% pure O_2 gas at temperatures to 1000°C. Polycrystalline rutile blended and hot-pressed with CuO to a purposely induced stoichiometry of $(Ti+Cu)O_{1.80}$ served as the model compound. Copper was selected as the dopant of choice, because both the theoretical predictions and preliminary experiments with other likely cations indicated that Cu^{1+} and/or Cu^{2+} had the best chance of yielding the expected results.

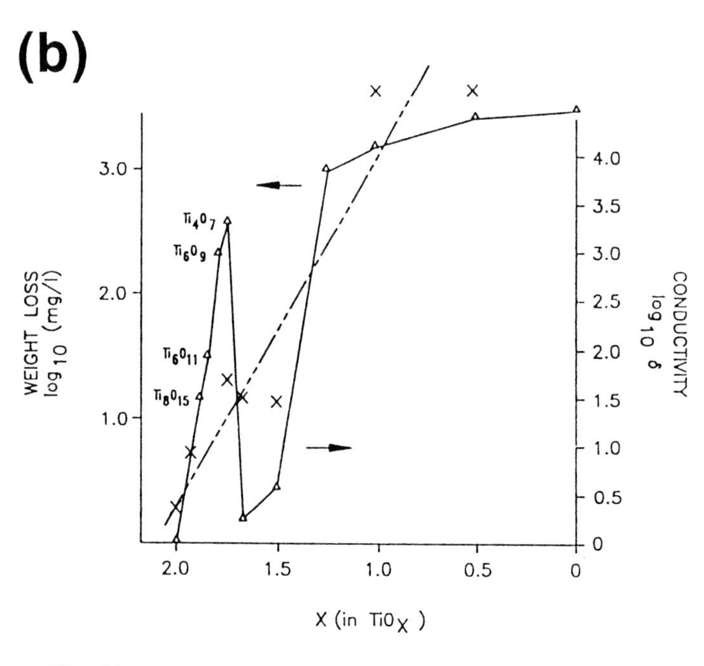

Fig. 11. The effect of oxygen stoichiometry on the shear strength (a) and electrical conductivity and weight loss (b) of rutile; from (**92**).

The analyses of the hot-pressed molds and the tribometric data indicated that copper entered the rutile lattice, even though the preparation technique was far from ideal. The reaction was not stoichiometric, however. While some of the residual dopant evaporated (or remained) as Cu_2O or was left behind as metallic copper (Cu^o), a small portion did react to generate Magnèli phases. A Ti-Cu oxide previously not reported in the literature was formed, equivalent to an analogous compound resembling the X-ray diffractogram of $V_3Ti_6O_{17}$. The Ti-V analog is a known electrode and catalyst equivalent in oxygen stoichiometry to the rutile-structured $(Ti+Cu)O_{1.89}$. The new Ti-Cu oxide also exhibited the electrical conductivity of undoped rutile electrodes (Fig. 11b) reduced to $TiO_{1.89}$. The dopant-induced removal and environmental stabilization of oxygen content mitigated the characteristic rise in τ_S as the oxygen stoichiometry was changed to and beyond the $O_{0.18}$ peak by heating and sliding in vacuum (Fig. 11a).

The Cu-doped rutile compound may be considered only as a model for the initial experimental confirmation of the hypothesis. It is not considered as a practical solid lubricant. Nevertheless, in view of other nonstoichiometric oxides that also form Magnéli phases (e.g., VO_{2-x}, WO_{3-x} and even MoO_{3-x}), doping-induced molecular engineering of lubricative oxides other than rutile may be possible.

More work is needed to establish the practical usefulness of this molecular engineering concept.

5.3 Grain orientation

In bulk-polycrystalline LOs, notwithstanding a doping process, the mostly random orientation of the single-crystal-like (but tailored-τ_{cr}) grains will impart less favorable properties than the more favorably aligned grains in HOPG or LTMD films. With these latter lubricants and morphologies, purposely induced lay of the plate-like crystallites parallel with the bearing surface is more attainable. Obviously, growing doped LO films the same way, with low-τ_s crystallographic textures, would also be the ideal solution. However, this configuration has not yet been realized.

6 CONCLUSIONS

The performance of MMAs solid lubricated by the most commonly employed pristine and doped lamellar HOPGs, LTMDs and, in the future, by layer-lattice LOs, is governed by the atomic-level interaction within the grain structure of each lubricant crystallite, and between the grains and the load-carrying bearing substrate. The ability of these lamellar particles to serve as the building blocks of low-τ_s engineering films is manifested by low interplanar and interparticle attraction simultaneously accompanied by high-strength intraplanar chemical bonding. The former attributes lead to low τ_{cr} and thus a low τ_s of a thin film. The latter helps resist asperity penetration of the strong basal planes aligned by the deposition process and run-in.

The anisotropic strength of the chemical bonds results in equally anisotropic mechanical, electrical and thermal behavior of the crystallites and the films assembled therefrom. The conductivities across the lamellae of a single grain cannot be increased without some degradation of its τ_{cr} and the resulting increase in the τ_s of an entire lubricant film. A compromise must usually be struck between shear properties competing with electrical-thermal conductivity needs.

Unfortunately, the higher conductivities sometimes required in concentrated Hertzian contacts are synonymous not just with a high τ_{cr} and τ_s, but with reduced oxidative and/or hydrolytic stability as well. Even where the electrical and thermal aspects of the lubricant are not factors, the degenerative environmental influences on the global shear properties of a film often outweigh the beneficially low τ_{cr} of its constituent grains. The surface and even the bulk of each lubricant crystallite sliding against its sister particles can become detrimentally altered by commonly encountered thermal-atmospheric reactions. Antagonistic inter and intra-particle effects emerge via exposure to such inevitable environments as ambient air and moisture.

To a certain limited extent, one can mitigate the mutually exclusive τ_s versus conductivity conundrum and the associated chemical stability problems by molecular engineering. A lubricant tailored with dopant(s) and additives occasionally fulfills a need better than the pristine host. In general, the main tribological goal of intercalation or substitutional doping is to reduce interplanar bonding (the τ_{cr}) within the crystallites comprising a thin film. So far, the most effective form of chemical manipulation of lamellar structures has been the insertion of electron donor or acceptor intercalant layers between the sheets of the host compound. Ideally, this should be done without exacerbating environmental sensitivity, a goal seldom reached. At least in the case of MoS_2, certain non-intercalating metal additives simply co-deposited with the inherently oxidation-hydration-susceptible lamellae serve better by enhancing film-to-substrate bonding and protecting the defect sites from thermal-atmospheric degradation.

The days of high-cost and time-consuming problem resolution by trial-and-error are over. Technical communities cannot afford not to embrace more fundamentals-oriented methods in place of brute-force engineering. The information given in this paper may help diagnose and eliminate tough friction and wear problems by clarifying the atomic-molecular level interactions of layered-lattice-type solid lubricants and their substrates stressed by the environment. The next essential step is to extend the basic elements of science into the realm of mechanical systems engineering. Only such broad understanding will give us the full quiver of capabilities we need to face the technical challenges of the coming century.

7 ACKNOWLEDGEMENTS

The author would like to thank Mr Alf Conte of the US Naval Air Warfare Center (Warminster, PA, USA) and Drs Paul Fleischauer, Jeff Lince and Mike Hilton of The Aerospace Corporation (El Segundo, CA, USA) for invaluable discussions and pertinent references on intercalated graphites and layered transition metal dichalcogenides. The section on lubricating oxides could not have been written without the experimental help of Prof. Kevin Kendall (U. of Keele, England), Drs Vaughan White of the Industrial Research and Development (Lower Hutt, New Zealand), Andre Ezis and James Shih of CERCOM Inc. (Vista, CA, USA) and Robert Clarke (Orinda, CA, USA). The management of Hughes Aircraft Company is also gratefully recognized for permission to publish this paper.

8 REFERENCES

1 **Gardos, M.N.,** Determination of the tribological fundamentals of solid lubricated ceramics, Volume 1: Summary. *WRDC-TR-90-4096*, Nov. 1990, Hughes Aircraft Co., El Segundo, CA, USA.

2 **Catlow, C.R.A.,** Defect clustering in nonstoichiometric oxides, Ch. 2 in *Nonstoichiometric oxides*, (Ed. O.T. Sorensen), 1981, (Materials Science Series, Academic Press

Inc., New York), pp. 61–98.

3 **Dresselhouse, M.S.** and **Dresselhouse, G.**, Intercalation compounds of graphite. *Advances in Physics*, 1981, **30** (2), 139–326.

4 **McClure, J.W.**, Energy band structure of graphite. *IBM J.*, 1964, **8**, 255–261.

5 **Gardos, M.N.**, Determination of the tribological fundamentals of solid lubricated ceramics, Part 2: Formulation of improved tribomaterials for advanced cryogenic turbopump applications, Volume 1: Summary and design guide addendum. *WL-TR-93-4089*, Aug. 1993, Hughes Aircraft Co., El Segundo, CA, USA.

6 **Gardos, M.N.**, On the elastic constants of thin solid lubricant films. in *Proc. 16th Leeds-Lyon Symp. on Tribology, Mechanics of Coatings*, Elsevier Trib. Ser. 17, (Eds D. Dowson, C.M. Taylor, and M. Godet), Elsevier Science Publ.B.V., 1990, pp.3–13.

7 **Ikemiya, N., Shimazu, E., Hara S., Shioyama, H.** and **Sawada, Y.**, Atomic force microscopy and scanning tunneling microscopy studies of $AlCl_3$-graphite intercalation compound. *Carbon*, 1996, **34**, 277–279.

8 **Ylilammi, M.**, Monolayer thickness in atomic layer deposition. *Thin Solid Films*, 1996, **279**, 124–130.

9 **Wertheim, G.K., van Attekum, P.M.Th.M., Guggenheim, H.J.** and **Clements, K.E.**, Acceptor site in metal halide intercalated graphite. *Solid State Commun.*, 1980, **33**, 809–811.

10 **Wertheim, G.K.**, Island formation in metal halide intercalation compounds. *Solid State Commun.*, 1981, **38**, 633–635.

11 **Kamimura, H.**, Graphite intercalation compounds. *Physics Today*, Dec. 1987, **40**, 64–71.

12 **Abe, T., Mizutani, Y., Shinoda, N., Ihara, E., Asano, M., Harada, T., Inaba, M.** and **Ogumi, Z.**, Debye-Waller factors of $FeCl_3$- and ICl-graphite intercalation compounds. *Carbon*, 1995, **33**, 1789–1793.

13 **Oh, W.-C., Cho, S.-J.**, and **Ko, Y.-S.**, The Stability of potassium-graphite deintercalation compounds. *Carbon*, 1996, **34**, 209–215.

14 **Kaae,J., Dresselhouse, M., MacKenzie, J., McQuillan, B., Price, R., Reynolds, G., Steckert, H.** and **Tilley, D.**, Studies in support of oxidation-resistant composite materials. *AFOSR-TR-88-0284*, Feb. 1988, General Atomics, San Diego, CA, USA.

15 **Goddard, W.A.**, *et al*, in Advances at MSC in biology, chemistry, and material sciences. *MSC'96, Proc. Ann. Conf. Materials and Process Simulation Center*, California Institute of Technology, March 21–22, 1996, Pasadena, CA, USA.

16 **Oshima, H.** and **Woolam, J.A.**, Metallic conductivity and air stability in copper chloride intercalated carbon fibers. *J. Appl. Phys.*, 1982, **53**, 9220–9223.

17 **Oshima, H., Natarajan, V., Woolam, J.A., Yavrouian, A., Haugland, E.J.** and **Tsuzuku, T.**, Electronic properties of carbon fibers intercalated with copper chloride. *Jap. J. Appl. Phys.*, 1984, **23**, 40–43.

18 **Dziemianowicz, T.S.** and **Forsman, W.C.**, Acceptor intercalated graphite fibers: kinetics, product stability and applications to composites. *Synthetic Metals*, 1983, **8**, 297–304.

19 **Gaier, J.R.** and **Jaworske, D.A.**, Environmental stability of intercalated graphite fibers. *Synthetic Metals*, 1985, **12**, 525–532.

20 **Chung, D.D.L., Li, P.** and **Li, X.**, Strengthening graphite-polymer composites by using intercalated graphite. in: *Interfaces in polymer, ceramic, and metal matrix composites*, Proc. Second Int. Conf. on Composite Interfaces (ICCI-II), Cleveland, OH, June 13–17, 1988, New York, Elsevier, 1988, pp. 101–105.

21 **Akuzawa, N., Amari, Y., Nakajima, T.** and **Takahashi, Y.**, Electrical resistivity and hydrogen-physisorption behavior of potassium-graphite intercalation compounds in the course of reactions with ammonia, water, and oxygen. *J. Mater. Res.*, 1990, **5**, 2849–2853.

22 **Hughes, R.I.** and **Daniel, S.G.**, The influence of an intercalated inorganic salt on the frictional characteristics of graphite. *Proc. Inst. Mech. Eng.*, 1963–64, **178**, 273–282.

23 **Lipp, L.C.** and **Klemgard, E.N.**, Graphite lubricant combinations for high temperature applications. *Lubr. Eng.*, 1966, **22**, 187–195.

24 **Lipp, L.C.** and **Stern, H.**, Effect of radiation and additives on graphite lubricants for high temperature applications, in *Bearing and seal design in nuclear power machinery, ASME Proc. Symp. Lubr. in Nuclear Applications*, June 5–7, 1967, Miami Beach, FL, pp. 288–303.

25 **Conte, A.A. Jr.**, Graphite intercalation compounds as solid lubricants. *ASLE Trans.*, 1983, **26**, 26–33.

26 **Gardos, M.N., Lipp, L.C.** and **Griffin, W.S.**, Solid lubricated resistive ink for potentiometers. *U.S. Patent No. 5,035,836*, July 30, 1991.

27 **Gardos, M.N.** and **McConnell B.D.**, Development of a high load, high temperature, self-lubricating composite – Part I: Polymer matrix selection; Part II: Reinforcement selection; Part III: Additives selection; Part IV: Formulation and performance of the best composites. *ASLE SP-9*, 1982, and the references therein.

28 **Watanabe, N., Nakajima, T.** and **Touhara, H.**, Studies in Inorganic Chemistry 8 – Graphite Fluorides. 1988, (Elsevier Science Publishers B.V., New York).

29 **Wilson, J.A.** and **Yoffe, A.D.**, The transition metal dichalcogenides – discussion and interpretation of the observed optical, electrical and structural properties. *Advances in Physics*, 1969, **18**, 193–335.

30 **Fleischauer, P.D.**, Fundamental aspects of the electronic structure, materials properties and lubrication performance of sputtered MoS_2 films. *Thin Solid Films*, 1987, **154**, 309–322.

31 **Fleischauer, P.D., Lince, J.R., Bertrand, P.A.** and **Bauer, R.**, Electronic structure and lubrication properties of MoS_2: a qualitative molecular orbital approach. *Langmuir*, 1989, **5**, 1009–1015.

32 **Didziulis, S.V.** and **Fleischauer, P.D.**, Application of Surface Science to Solid Lubricants. in *Surface Diagnostics in Tribology*, Eds. K. Miyoshi and Y.-W. Chung, 1994, p. 135 (World Scientific Publ. Co. Ptc. Ltd).

33 **Friend, R.H.** and **Yoffe, A.D.**, Electronic properties of intercalation complexes of the transition metal dichalcogenides. *Advances in Physics*, 1987, **6**, 1–94.

34 **Jamison, W.E.**, Intercalated dichalcogenide solid lubricants. in *Proc. 3rd Int. Conf. on Solid Lubrication*, Aug. 7–10, 1984, ASLE SP–14, pp.73–87.

35 **Golubnichnaya, A.A.** and **Kalikhman, V.L.**, Antifriction properties of alloys of the system WS_2-NbS_2. *Isvestiya Academiya Nauk. Neorg. Mater. (English translation)*, 1978, **14**, 228–231.

36 **Magie, P.M.**, A review of the properties and potentials of the new heavy metal derivative solid lubricants. *Lubrication Engineering*, 1966, **22**, 262.

37 **Boes, D.J.**, New solid lubricants: preparation, properties and potentials. *IEEE Trans. Aerospace,* 1964, **2**(2), 457–466.

38 **Grant, A.J.**, **Griffiths, T.M.**, **Pitt, G.D.** and **Yoffe, A.D.**, The electrical properties and the magnitude of the indirect gap in the semiconducting transition metal dichalcogenide layer crystals, *J. Phys. C: Solid State Phys.*, 1975, **8**, L17–L23.

39 **Kalikhman, V.L.** and **Pravoverova, L.L.**, Antifriction and electrical properties of quasi-binary WSe_2 – $NbSe_2$ alloys. *Poroshkovaya Metallurgiya (English translation from Soviet Powder Metallurgy)*, 1972, **7** (115), 93–96.

40 **Waghray, H.**, **Lee, T.-S.** and **Tatarchuk, B.J.**, A Study of the tribological and electrical properties of sputtered and burnished transition metal dichalcogenide films. *Surf. Coat. Technol.*, 1995, **77**, 415–420.

41 **Regula, M.**, **Ballif, C.**, **Moser, J.H.** and **Lévy, F.**, Structural, chemical, and electrical characterization of reactively sputtered WS_x films, *Thin Solid Films*, 1996, **280**, 67–75.

42 **Gardos, M.N.**, The synergistic effects of graphite on the friction and wear of MoS_2 films in air. *Tribol. Trans.*, 1987, **31**, 214–227.

43 **Kim, Y.**, **Huang, J.-L.** and **Lieber, C.M.**, Characterization of nanometer scale wear and oxidation of transition metal dichalcogenide lubricants by atomic force microscopy. *Appl. Phys Lett.*, 1991, **59**, 3404–3406; **Lieber, C.M.** and **Kim, Y.**, Characterization of the structural, electronic and tribological properties of metal dichalcogenides by scanning probe microscopies. *Thin Solid Films*, 1991, **206**, 355–359.

44 **Gardos, M.N.**, Anomalous wear behavior of MoS_2 films in moderate vacuum and dry nitrogen. *Tribol. Lett.*, 1995, **1**, 67–85.

45 **Roberts, E.W.**, The advantages and limitations of sputtered molybdenum disulfide as a space lubricant. *Proc. Fourth European Symp. On 'Space Mechanisms and Tribology'*, ESA SP-299, March 1990, pp. 59–65.

46 **Roberts, E.W.**, Ultralow friction films of MoS_2 for space applications. *Thin Solid Films*, 1989, **181**, 461–473.

47 **Spalvins, T.**, Lubrication with sputtered MoS_2 films: principles, operation, limitations. *NASA TM-105292*, 1991.

48 **Fleischauer, P.D.** and **Hilton, M.R.**, Assessment of the tribological requirements of advanced spacecraft mechanisms. *Aerospace Rept. No. TOR-0090(5064)-1*, 30 Sept. 1991, The Aerospace Corp., El Segundo, CA, USA.

49 **Hilton, M.R.** and **Fleischauer, P.D.**, Applications of solid lubricant films in spacecraft. *Surf. Coat. Technol.*, 1992, **54/55**, 435–441.

50 **Peebles, D.E.** and **Ohlhousen, J.A.**, Surface chemistry of MoS_2 lubricant films: effects of high humidity storage on sputtered films. *Sandia Rept. No. SAND-92-0052*, May 1992, Sandia National Labs, Albuquerque, NM,USA.

51 **Brudnyi, A.I.** and **Karmadonov, A.F.**, Structure of molybdenum disulfide lubricant film. *Wear*, 1975, **33**, 243–249.

52 **Hayden, T.F.**, **Dumesic, J.A.**, **Sherwood, R.D.** and **Baker, R.T.K.**, Direct observation by controlled atmosphere electron microscopy of the changes in morphology of molybdenum oxide and sulfide supported on alumina and graphite. *J. Catal.*, 1987, **105**, 299–318.

53 **Scialdone, J.J.**, An estimate of the outgassing of space payloads and its gaseous influence on the environment. *J. Spacecraft*, 1986, **23**, 373–378.

54 **Wood, B.E.**, **Bertrand, W.T.**, **Bryson, R.J.**, **Seiber, B.L.**, **Falco, P.M.** and **Cull, R.A.**, Surface effects of satellite material outgassing products. *AIAA 22nd Thermophysics Conf.*, June 8–10, 1987, Honolulu, HI, Paper No. AIAA-87-1583.

55 **Pickett, J.S.**, **Murphy, G.B.** and **Kurth, W.S.**, Gaseous environment of the Shuttle early in the Spacelab 2 mission. *J. Spacecraft and Rockets*, 1988, **25**, 169–174.

56 **Bird, G.A.**, Spacecraft outgas ambient flow interaction. *J. Spacecraft*, 1981, **18**, 31–35.

57 **Neumann, R.J.**, Analysis of the performance of the Space Ultravacuum Research Facility in attached and free-flyer mode. *NASA-TM-100325*, March 1988.

58 **Sega, R.M.** and **Ignatiev, A.**, A Space ultra-vacuum experiment – application to material processing. *Proc. 1st AIAA/IKI Microgravity Science Symp.*, May 13–17, 1991, Moskow, USSR, AIAA, Wash. DC, pp. 172–178.

59 **Ignatiev, A.**, **Sterling, M.** and **Sega, R.M.**, Use of ultra-vacuum facility for high quality semiconductor film growth. 43rd Int. Astronautical Congress, Aug. 28–Sept. 5, 1992, Washington DC, IAF Paper No. 92- 0931.

60 **Gardos, M.N.**, Labyrinth sealing of aerospace mechanisms - theory and practice. *ASLE Trans.*, 1974, **17**, 237–250.

61 **Martin, J.-M.**, **Donnet, C.** and **LeMogne, Th.**, Superlubricity of molybdenum disulfide. *Phys. Rev. B.*, 1993, **48**, 10,583–10,586.

62 **Donnet, C.**, **Le Mogne, T.** and **Martin, J.-M.**, Superlow friction of oxygen-free MoS_2 in ultrahigh vacuum. *Surf. Coat. Technol.*, 1993, **62**, 406–411.

63 **Martin. J.-M.**, **Pascal, H.**, **Donnet, C.**, **Le Mogne, T. Loubet, J.L.** and **Épicier, T.**, Superlubricity of MoS_2: crystal orientation mechanisms. *Surf. Coat. Technol.*, 1994, **68/69**, 427–432.

64 **Le Mogne, T.**, **Donnet, C.**, **Martin, J.-M.**, **Tonck, A.**, **Millard-Pinard, N.**, **Fayeulle, S.** and **Moncoffre, N.**, Nature of superlubricating MoS_2 physical vapor deposition coatings. *J. Vac. Sci. Technol.*, 1994, **A12** (4), 1998–2004.

65 **Stupp, B.C.**, Synergistic effects of metal co-sputtered with MoS_2. *Thin Solid Films*, 1981, **84**, 257–266; **Stupp, B.C.**, Performance of conventionally sputtered MoS_2 versus co-sputtered MoS_2 and nickel. *Proc. 3rd Int. Conf. on Solid Lubrication*, ASLE SP-14, 1984, pp. 217–222.

66 **Hilton, M.R.**, **Bauer, R.**, **Didziulis, S.V.**, **Dugger, M.T.**, **Keem, J.M.** and **Scholhamer, J.**, Structural and tribological studies of MoS_2 solid lubricant films having tailored metal-multilayer nanostructures. *Surf. Coat. Technol.*, 1992, **53**, 13–23.

67 **Jayaram, G.**, **Marks, L.D.** and **Hilton, M.R.**, Nanostructure of Au-20%Pd layers in MoS_2 multilayer solid lubricant films. *Surf. Coat. Technol.*, 1995, **76/77**, 393–399.

68 **Ronay, M.** and **Serrano, C.M.**, Reducing the grain size of polycrystalline lead films by the use of barriers to grain growth. *J. Appl. Phys.*, 1983, **54**, 652–659; **Bright, A.A.** and **Ronay, M.**, Reducing the grain size of polycrystalline lead films by the use of $AuIn_2$ barriers to grain growth. *J. Appl. Phys.*, 1984, **55**, 810–816.

69 **Barna, P.B.** and **Adamik, M.**, Growth mechanisms of polycrystalline thin films. in *Science and technology of thin*

films, Ed. F.C. Matacotta and G. Ottavini, 1995, pp. 1–28 (World Scientific Publ.).

70 **Hilton, M.R.,** Fracture in MoS_2 solid lubricant films. *Surf. Coat. Technol.*, 1994, **68/69**, 407–415.

71 **Gardos, M.N.,** Tribology and Wear Behavior of Diamond. Ch. 12 in *Synthetic Diamond: Emerging CVD Science and Technology* (Electrochem. Soc. Monogr.), Eds K.E. Spear and J.P. Dismukes, 1994, pp. 419–504, (John Wiley & Sons, New York).

72 **Lince, J.R., Carré, D.J. and Fleischauer, P.D.,** Schottky-barrier formation on a covalent semiconductor without Fermi-level pinning: the metal-MoS_2 (0001) interface. *Phys. Rev. B.*, 1987, **36**, 1647–1656.

73 **Zabinski, J.S., Donley, M.S., Walck, S.D., Schneider, T.R.** and **McDevitt, N.T.,** The effects of dopants on the chemistry and tribology of sputter-deposited MoS_2 films. *Tribol. Trans.*, 1995, **38**, 894–904; **Hilton, M.R., Jayaram, G.** and **Marks, L.D.,** Microstructure of co-sputter-deposited metal-and oxide-MoS_2 solid lubricant thin films. *J. Mater. Res.*, in press.

74 **Chu, X.** and **Schmidt, L.D.,** Processes in MoS_2 gasification. *J. Catal.*,1993, **144**, 77–92.

75 **Lee, L.-H.,** Hard-soft acid-base (HSAB) principle for solid ahesion and surface triboInteractions. Ch. 2.5 in *Tribology in the USA and the former Soviet Union -- studies and applications*, Eds V.A. Belyi, K.C. Ludema and N.K. Myshkin, 1994, pp.89–101, (Allerton Press, Inc., New York).

76 **Good, R.J.** and **Chaudhury, M.K.,** Theory of adhesive forces across interfaces, Part 1. The Lifshitz-van der Waal component of interaction and adhesion; Good, R.J., Chaudhury, M.K. and van Oss, C.J., Part 2. Interfacial hydrogen bonds as acid-base phenomena and as factors enhancing adhesion. in *Fundamentals of adhesion*, Ed. L.H. Lee, 1991, pp. 137–172, (Plenum Press, New York).

77 **Mittal, K.L** and **Anderson, H.R.,** Eds, *Acid-base interactions: relevance to adhesion science and technology*, 1991, (VSP, Zeist, The Netherlands).

78 **Donnet, C.,** Problem-solving with surface analysis in tribology, in *Problem-solving methods for surfaces and interfaces*, Eds. J.C. Rivière and S. Myhra, (Marcel Dekker Inc.), in press.

79 **Didziulis, S.V., Hilton, M.R.** and **Fleischauer, P.D.,** The Influence of steel surface chemistry on the bonding of lubricant films. Ch. 3 in *Surface science investigations in tribology*, Eds. Y.-W. Chung, A.M. Homola and G. Bryan Street, 1992, pp. 43–57, ACS Symp. Ser. No. 485, (Am. Chem. Soc.).

80 **Jayaram, G., Doraiswamy, N., Marks, L.D.** and **Hilton, M.R.**, Utrahigh vacuum high resolution transmission electron microscopy of sputter-deposited MoS_2 thin films. *Surf. Coat. Technol.*, 1994, **68/69**, 439–445.

81 **Woolam, J.A.** and **Somoano, R.B.,** Physics and chemistry of MoS_2 intercalation compounds. *Mat. Sci. Eng.*, 1977, **1**, 289–295.

82 **Frindt, R.F., Arrott, A.S., Curzon, A.E., Heinrich, B., Morrison S.R., Templeton, T.L., Divigalpitiya, R., Gee, M.A., Joensen, P., Schurer, P.J.** and **LaCombe, J.L.,** Exfoliated MoS_2 monolayers as substrates for magnetic materials. *J. Appl. Phys.*, 1991, **70**, 6224–6226.

83 **Jones, J.R.,** Friction and wear of potentiometer contacts in vacuum. *Wear*, 1968, **2**, 355–367.

84 **Ho, P.S.,** and **Kwok, T.,** Electromigration in metals. *Rep. Prog. Phys.*, 1989, **52**, 301–348.

85 **Heckel, D.T., Jones, J.R.** and **Davidson, R.S.,** Improving potentiometer life. *The Electr. Engineer*, May 1967, pp. 71–72.

86 **Nielsen, L.E.** and **Small, R.A.,** Electrical sliprings for space simulation testing. *Proc. AIAA/ASTM/IES 4th Space Simul. Conf.*, Sept. 8–10, 1969, Los Angeles, California, USA, AIAA Paper No. 69-1035.

87 **Lewis, N.E., Cole, S.R., Glossbrenner, E.W.** and **Vest, C.E.,** Friction, wear, and noise of slip ring and brush contacts for synchronous satellite use. *IEEE Trans. on Parts, Hybrids, and Packaging*, **Vol. PHP-9**, No. 1, March 1973, pp. 15–22, and the references therein.

88 **Gardos, M.N.,** The effect of anion vacancies on the tribological properties of rutile (TiO_{2-x}). *Tribol. Trans.*, 1988, **31**, 427–436; also see discussions on this paper, *Tribol. Trans.*, 1989, **32**, 30–31.

89 **Gardos, M.N., Hong, H.-S.** and **Winer, W.O.,** The effect of anion vacancies on the tribological properties of rutile (TiO_{2-x}), Part II: Experimental evidence. *Tribol. Trans.*, 1990, **32**, 209–220.

90 **Gardos, M.N.,** The Triboxidative behavior of rutile-forming substrates. *Mat. Res. Soc. Symp. Proc. Vol. 140, New Materials Approaches to Tribology: Theory and Applications*, Eds. L.E. Pope, L.L. Fehrenbacher, and W.O. Winer, 1989, pp. 325–338.

91 **Gardos, M.N.,** The effect of Magnéli phases on the tribological properties of polycrystalline rutile (TiO_{2-x}). in *Proc. 6th Int. Congr. on Tribology, Eurotrib '93*, Ed. M. Kozma, **Vol. 3**, Aug. 30–Sept. 2, 1993, Budapest, Hungary, pp. 201–206.

92 **Gardos, M.N.,** Determination of the tribological fundamentals of solid lubricated ceramics, Part III: Molecular engineeering of rutile (TiO_{2-x}) as a lubricious oxide. *WL-TR-94-4108*, Oct. 1994, Hughes Aircraft Company, El Segundo, CA, USA.

93 **DellaCorte, C.** and **Deadmore, D.L.,** Vickers indentation hardness of stoichiometric and reduced single crystal TiO_2 (rutile) from 25 to 800°C. *NASA TM-105959,* Apr. 1993.

Coatings tribology – contact mechanisms and surface design

K HOLMBERG and **H RONKAINEN**
VTT Manufacturing Technology, Espoo, Finland
A MATTHEWS
RCSE, Hull University, UK

SYNOPSIS

The fundamentals of coatings tribology are presented using a generalised holistic approach to the friction and wear mechanisms of coated surfaces in dry sliding contacts. It is based on a classification of the tribological contact process into macromechanical, micromechanical, nanomechanical, and tribochemical contact mechanisms and material transfer. The important influence of thin tribo- and transfer layers formed during the sliding action is shown. Optimal surface design regarding both friction and wear can be achieved by new multilayer techniques which can provide properties such as reduced stresses, improved adhesion to the substrate, more flexible coatings and harder and smoother surfaces. The differences in contact mechanisms in dry, water- and oil-lubricated contacts with coated surfaces are illustrated by experimental results from diamond-like carbon coatings sliding against a steel and an alumina ball. The mechanisms of the formation of dry transfer layers, tribo-layers and lubricated boundary and reaction films are discussed.

1 INTRODUCTION

Recent advances in coatings technologies now permit the deposition of films with properties which were unachievable even a decade ago. Examples are multilayered and metastable coatings with extreme mechanical and chemical properties. The plasma and ion-based vacuum coating techniques have been at the forefront of these new developments. They allow the coating/substrate system to be designed in such a way that the combination performs in an optimal manner. This objective is further aided by improvements in our fundamental understanding of contact mechanisms between surfaces, at the macro, micro and nano level.

This paper therefore begins with an overview of the present level of understanding of contact mechanisms, especially for coated surfaces. This includes an appreciation of stress states, mechanical properties such as hardness, and chemical influences such as oxidation.

The possibilities opened up by using multiple layers of coatings are then discussed; this includes the possibility to grade the functional properties from the surface to the interface between coating and substrate. Special mention is made of recent developments in carbon-based coatings which combine excellent frictional properties with good wear resistance.

2 CONTACT MECHANISMS

The tribology of a contact involving surfaces in relative motion can be understood as a process with certain input and output data (**1**). Input data that are used as a starting point for the analysis of a tribological contact are the geometry of the contact, both on a macro- and micro-scale, the material properties based on the chemical composition and structure of the different parts involved and the environmental parameters, as shown in Figure 1. Other input data are the energy parameters such as normal load, velocity, tangential force and temperature.

The tribological process takes place as the two surfaces are moving in relation to each other, and both physical and chemical changes occur in accordance with the physical and chemical laws with respect to the input data. As a function of time the tribological process causes changes in both the geometry and the material composition and results in energy related output effects: friction, wear, velocity, temperature, sound and dynamic behaviour.

2.1 Tribological contact mechanisms

The complete tribological process in a contact in relative motion is very complex because it involves simultaneously friction, wear and deformation mechanisms at different scale levels and of different types. To achieve a holistic understanding of the complete tribological process taking place and to understand the interactions it is useful to separately analyse the tribological changes of four different types: the macro- and micro-scale mechanical effects, the chemical effects and the material transfer taking place, as shown in Figure 2. In addition there has recently been an increasing interest in studying tribological behaviour on a molecular level; i.e. nanomechanical effects (**2**).

A better and more systematic understanding of the mechanisms involved in a tribological contact is necessary for the optimisation of the properties of the two contacting surfaces in order to achieve the required friction and wear performance. Approaches to the tribological optimisation of surfaces have been presented by Matthews, *et al.* (**3**) and

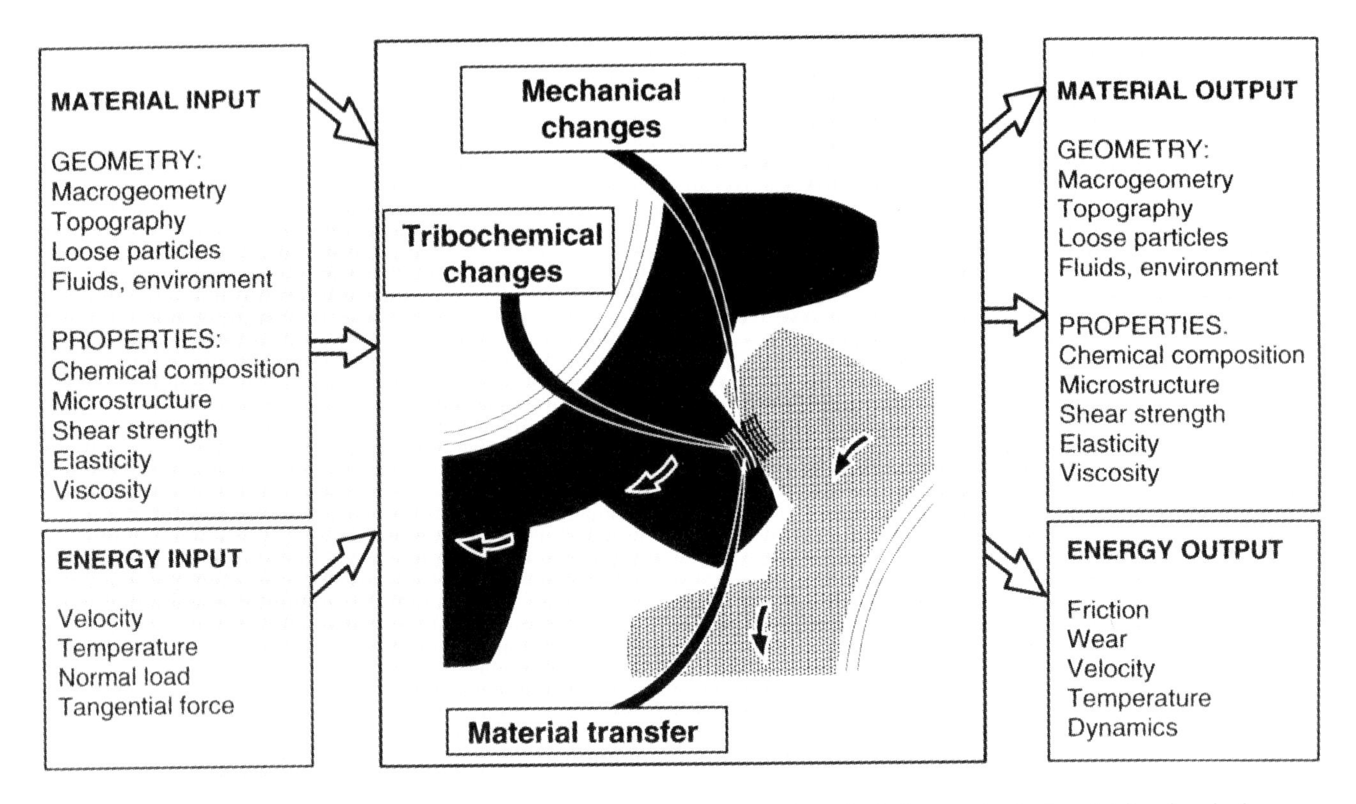

Fig. 1 The tribological process in a contact between two surfaces includes mechanical and tribochemical changes as well as material transfer.

Franklin and Dijkman (**4**) who introduce eight wear design rules in an expert system for assisting the selection of metallic materials, surface treatments and coatings during the initial stages of engineering design.

2.2 Stresses and strain in the surfaces

The phenomena that take place in a tribological contact are influenced by the force pressing the two surfaces together. Calculation methods for the stress fields and deformations in a coated surface have been reviewed by Holmberg and Matthews (**1**). A useful approach to calculating the coating/substrate interface stresses has more recently been used by Ramalingam and Zheng (**5**) to assess the problems that may be encountered when hard coatings are applied on substrates with lower or higher compliance than the coating. Numerical solutions are obtained for different stress distributions on the coated surface. They show how tensile stresses in the wake of the contact tend to separate the coating from the substrate and how this can be solved by adjusting the film thickness and choice of coating conditions to ensure that the coating to substrate adhesion strength can withstand the stresses.

The beneficial effect of high compressive stresses in hard and smooth diamond coatings has been shown by Gunnars and Alaheliste, (**6**) and Gåhlin, *et al.* (**7**). They found that highly stressed coatings may exhibit smoother surfaces and wear rates which were only 5–20 % of those of the stress-free coatings. The internal stresses influenced the crack propagation direction in diamond coatings.

Finite element methods have been applied to evaluate the stress field in the hard coating and the substrate under frictional loads (**8**, **9**). Diao and Kato (**10**) analysed the von Mises stress distributions in hard coatings and in elastic sliding. An elliptical distribution of normal and traction contact pressure was assumed for the analysis of von Mises stress for different coating thicknesses, friction coefficients and elastic moduli of the coating and the substrate. The position of yield could be calculated and they introduced a local yield map showing the yield strength ratio in relation to the ratio of the coating thickness to the contact half-width. The local yield maps showed that yield at the coating-substrate interface on the substrate side is the most common case under a wide range of contact conditions.

A new parameter for the prediction of the onset of the spalling of a ceramic coating under sliding contact was developed by Diao and Kato (**11**) and equations for calculating the critical normal load for the generation of cracks in ceramic coatings under indentation were developed by Diao, *et al.* (**12**). Eberhardt and Peri (**13**) determined the stresses near the tip of a crack by using a finite element method and their results support the use of a layer thickness with a large contact-radius to layer-thickness ratio and an appropriate lubricant to reduce friction and prevent surface crack propagation.

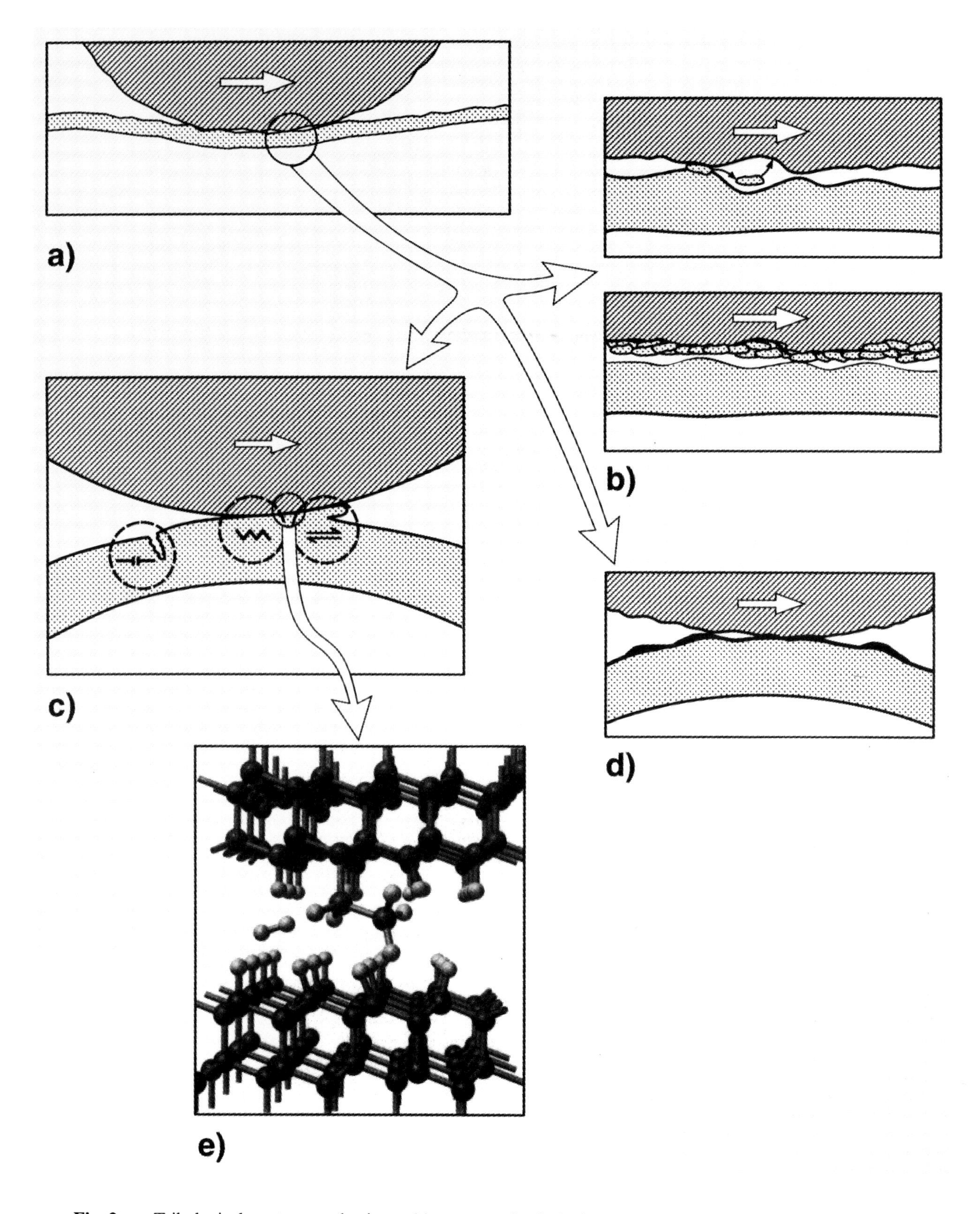

Fig. 2 Tribological contact mechanisms: (a) macromechanical, (b) material transfer, (c) micromechanical, (d) tribochemical, (e) nanomechanical contact mechanisms.

The surface roughness is an important parameter influencing the contact stresses, especially at the coating/substrate interface. This was first shown by Sainsot, *et al.* (**14**), who also concluded that in hard coatings with thicknesses less than 15 μm on softer substrates the maximum von Mises stresses both in the coating and in the substrate will be located just at their interface. This clearly points out the importance of analysing the stresses at the coating/substrate interface and comparing them to the coating adhesion strength.

More recently Mao, *et al.* (**15**) have developed a numerical technique including integral transform and finite element methods for solving contact problems in layered rough surfaces. They apply their numerical model (**16**) for the two-dimensional dry sliding contact of two elastic bodies on real rough surfaces, where an elastic body contacts with a multilayer surface under both normal and tangential forces. Of special interest is that their model uses surface profile data directly recorded with a stylus measuring instrument. They analyse the contact pressure distribution for layers with different coefficients of friction, thicknesses and elastic moduli.

Similar work has been carried out for the contact of gear teeth by Varadi, *et al.* (**17**). They analyse the contact of spur gears, taking into account real surface topography, by using numerical algorithms. The contact pressure distribution and contact area are calculated for different types of gear surfaces resulting from shaping, grinding or milling. They also evaluate the sub-surface stresses for different pressure distributions by a finite element method.

2.3 Macromechanical friction and wear mechanisms

The macromechanical tribological mechanisms describe the friction and wear phenomena by considering the stress and strain distribution in the whole contact, the total elastic and plastic deformations they result in, and the total wear particle formation process and its dynamics. In contacts between two surfaces of which one or both are coated, four main parameters can be defined which control the tribological contact behaviour. They are the coating-to-substrate hardness relationship, the thickness of the coating, the surface roughness, and the size and hardness of any debris in the contact which may originate from external sources or be produced by the surface wear interactions themselves.

The relationship between these four parameters will result in a number of different contact conditions characterised by specific tribological contact mechanisms. Fig. 3 shows schematically twelve such very typical tribological contacts, with different mechanisms influencing friction, when a hard spherical slider moves on a coated flat surface (**18**). The corresponding wear mechanisms have been described in a similar way (**19**).

2.3.1 Hardness of the coating

An important parameter is the coating hardness and its relationship to the substrate hardness. It is common to consider hard coatings and soft coatings separately (**5, 20, 21**). The advantages of using a soft coating to reduce friction are well known, owing to the work of Bowden and Tabor (**22**) and others. Soft coatings such as Ag and Au may also have the function of reducing sliding-originated surface tensile stresses, that contribute to undesirable subsurface cracking and subsequently to severe wear (**23**). A hard coating on a softer substrate can decrease friction and wear by preventing ploughing both on a macro-scale and a micro-scale (**19, 24–26**). These coatings typically exhibit residual compressive stresses which can prevent the likelihood of tensile forces occurring.

Further decreases in friction and wear can be achieved by improving the load support, that is by increasing the hardness of the substrate to inhibit deflections and ploughing due to counterpart load. This effect has been shown for hard coatings on polymers (**27**) but multilayer or gradient layer structures especially, offer excellent possibilities in this respect (**28–30**). Hard coatings are thus particularly useful in abrasive environments. Low friction can be achieved with hard coatings if a low shear strength microfilm is formed on the top of coating. Thus the shear takes place within the microfilm and the load is well supported by the hard coating. This mechanism is discussed in more detail in section 2.5.

2.3.2 Thickness of the coating and roughness of the surfaces

For soft coatings the thickness of the coating influences the ploughing component of friction, while for rough surfaces it affects the degree of asperity penetration through the coating into the substrate as shown in (a), (b), (e) and (f) in Fig. 3 (**23, 31–34**). A thick hard coating can assist a softer substrate in carrying the load and thus decrease the contact area and the friction. Thin hard coatings on soft substrates are susceptible to coating fracture because of stresses caused by substrate deformation. Rough surfaces will reduce the real contact area, although the asperities may be subject to abrasive or fatigue wear, as shown in (c), (d), (g) and (h) in Fig. 3 (**20, 26, 35, 37–41**).

2.3.3 Debris at the interface

Loose particles or debris are quite often present in sliding contacts. They can either originate from the surrounding environment or be generated by different wear mechanisms. Their influence on friction and wear may be considerable in some contact conditions, depending on the particle diameter, coating thickness and surface roughness relationship and the particle, coating and substrate hardness relationship. Particle embedding, entrapping, hiding and crushing represent typical contact conditions involving the influence of debris, as shown in Fig. 3 (i)–(l) (**18**).

2.4 Micromechanical tribological mechanisms

The origin of the friction and wear phenomena that we

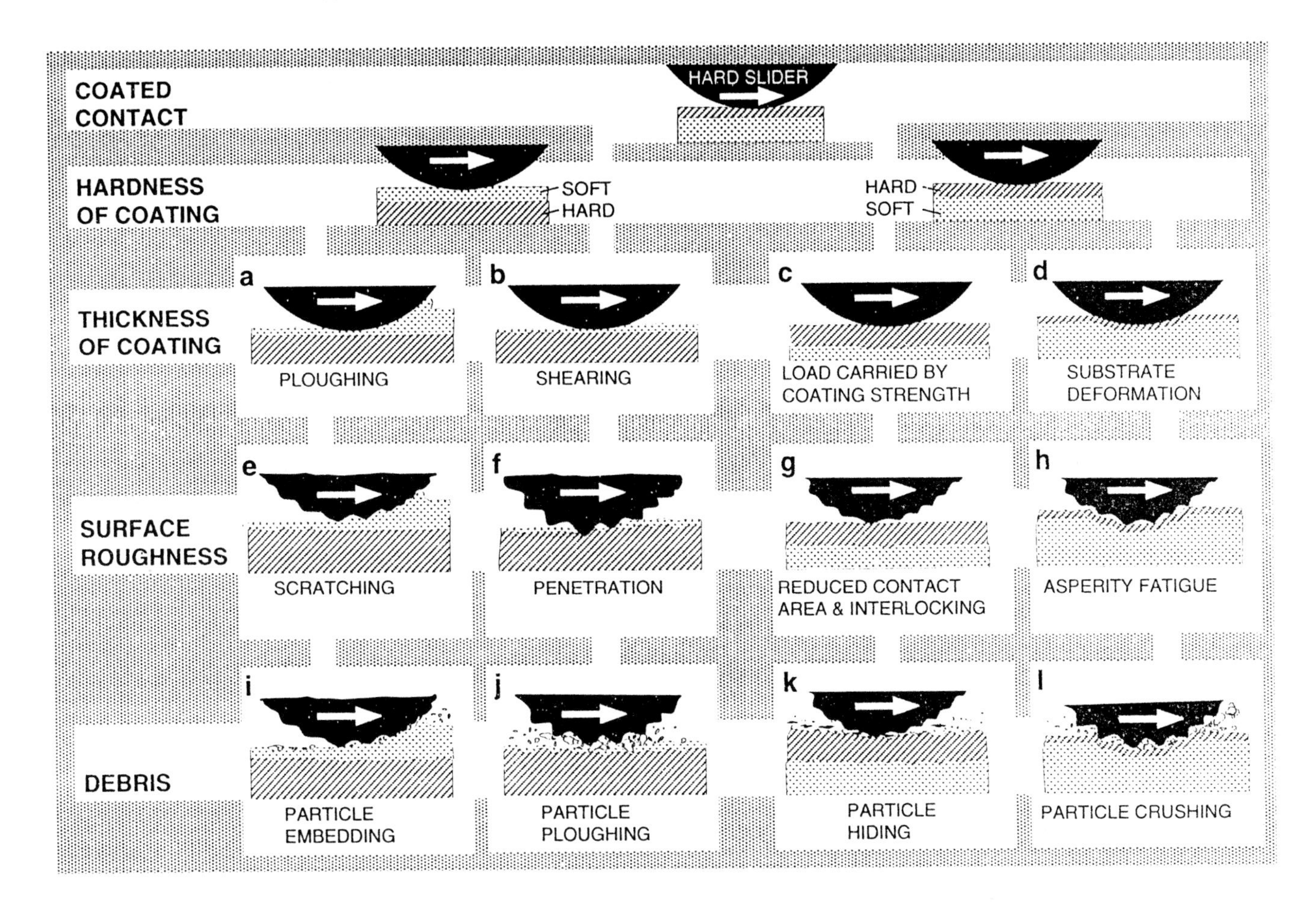

Fig. 3 Macromechanical contact conditions for different mechanisms which influence friction when a hard spherical slider moves on a coated flat surface.

observe on the macro-level is found in the mechanisms that take place at the micro-level. The micromechanical tribological mechanisms describe the stress and strain formation at an asperity-to-asperity level, the crack generation and propagation, material liberation and particle formation. In typical engineering contacts these phenomena are at a size level of about 1 μm or less down to the nanometre range.

Shear and fracture are two basic mechanisms for the first nucleation of a crack and for its propagation until it results in material liberation and formation of a wear scar and a wear particle. These mechanisms have been discussed by, for example, Argon (**42**) and Suh (**20**), but there is still today only a very poor understanding of these quite fundamental phenomena. Another approach is to study the tribological micromechanical mechanisms by using the velocity accommodation mode concept developed by Berthier, *et al.* (**43**).

Recent studies of the stress intensity factor at the tip of a surface-breaking crack in a layered surface indicate that compressive stresses may tend to prevent crack propagation (**13**). The crack pattern of ceramic coatings in indentation have been analysed and correlated to coating thickness and load, and the critical normal loads have been estimated by considering the deformation of the substrate (**12**). There is a strong influence on both the fracture load and the fracture pattern by the mismatch of elastic properties between the layer and the substrate (**44**). Variations in coating properties and coating thickness may change the fracture loads by up to a factor of ten.

2.5 Tribochemical mechanisms of coated surfaces

The chemical reactions taking place at the surfaces during sliding contact, and also during the periods between repeated contacts, change the composition of the outermost surface layer and its mechanical properties. This has a considerable influence on both friction and wear because they are determined to a great extent by the properties of the surface, where phenomena such as shear, cracking and asperity ploughing take place (**45**). The chemical reactions on the surfaces are strongly influenced by the high local pressures and the flash temperatures, which can be over 1000°C, at spots where asperities collide.

2.5.1 Formation of thin microfilms on hard coatings

Very low coefficients of friction (down to $\mu = 0.1$) have been reported for a hard titanium nitride coating sliding against itself (**46**) and even lower values (down to about $\mu = 0.01$ but more typically 0.05) have been measured for diamond-like hard carbon (DLC) coatings sliding against different counter-materials (reviewed by Donnet, **21**) and diamond coatings sliding against diamond and ceramics (**26**, **37**, **47**, **48**). This can be explained by the formation of low-shear microfilms on the hard coating or perhaps only on the asperity tips of the coating. Thus, if we consider the contact on a micro-scale, there is effectively a soft coating on a hard substrate, although now the coating (e.g. diamond) plays the role of hard substrate and the soft microfilm formed plays the role of a coating. It is obviously advantageous if the substrate under the hard coating is as hard as possible, to avoid fracture of the brittle coating by deformation, to improve the load support and to decrease the real area of contact. The very low coefficients of friction of polished diamond and diamond-like coatings are further explained by the extreme smoothness of the surface excluding effects such as interlocking and asperity ploughing, as well as by the hard coating reducing the ploughing component of friction.

It has been shown that when a ceramic counterface slides against a titanium nitride coating an oxide layer is often formed. The wear rate and the friction is highly affected by this reaction product. This process has been described in an oscillating fretting wear contact by Schouterden, *et al.* (**49**), Fouvry, *et al.* (**50**) and Mohrbacher, *et al.* (**51**) and in rotational sliding by Vanvoille, *et al.* (**52**).

A transfer layer is soon built up on the counterface when a steel or ceramic surface slides against a diamond-like hard carbon coating (**53, 54**). The explanation for the low shear strength between the two surfaces is however the formation of an extremely slippery microfilm between the surfaces. Different explanations for the structure of the microfilms have been presented and it is very difficult to analyse them conclusively (**55**).

Recently Erdemir, *et al.* (**56, 57**), Liu, *et al.* (**58**) and Schouterden, *et al.* (**49**) have published convincing evidence, in the form of Raman spectra of the surfaces and the wear products, which indicate that graphitization is taking place and a low shear strength graphitic microfilm is formed between the surfaces. Perhaps the difficulty in finding the low shear strength film is associated with it occurring mainly in the form of graphitic wear products or third bodies present in the moment of sliding but partly disappearing thereafter. Erdemir, *et al.* (**56**) showed that long term sliding is needed before a more distinct graphitic layer can be detected and they measured coefficients of friction as low as $\mu = 0.02$. The low friction and low wear behaviour of DLC coatings has been confirmed by several authors (**24, 48, 59, 60, 62–64**). The lowest value for the coefficient of friction, $\mu = 0.006$, has been measured by Donnet, *et al.* (**64**) for a diamond-like carbon film deposited on a silicon wafer sliding against a steel sphere in vacuum below 10^{-1} Pa. Higher sliding velocities and loads improve the graphitization process and result in decreased friction and wear (**53, 58**). The DLC coatings can show low friction at temperatures up to 300°C (**65**). Decreased friction and wear properties are generally reported for decreasing environmental humidity (**21, 64, 66, 67**).

The friction and wear behaviour of polycrystalline diamond films, as influenced by complex physical and chemical effects, has been reviewed by Gardos (**55**). The tribo-conditions under which one of the controlling factors dominates over the others are not yet clearly understood. In experiments carried out in dry N_2 with ceramic surfaces sliding on diamond coatings Erdemir, *et al.* (**26**) have shown that micrographitization takes place and results in extremely low friction down to $\mu = 0.04$ (**47, 68**). In open air, evidence on interface graphitization has not yet been found. Experimental results have been reported for aluminium oxide sliding against a diamond coating where lower humidity results in decreased wear but increased friction (**69**). The coefficient of friction of a spheroidal fullerene C_{70} surface in the form of a pure C_{70} film on silicon sliding against different pin materials such as alumina and stainless steel is high, in the range of $\mu = 0.5–0.9$ (**70**).

2.5.2 Oxidation of soft coatings

In environments containing oxygen, such as air, a thin (about 1–10 nm thick) oxide layer is formed very quickly on most metal surfaces. Some oxide layers, like copper oxide, are sheared more easily than the metal, while others, such as aluminium oxide, form a very hard layer. Improved lubricity caused by the oxidation of the top layer of the coating has been observed for lead coatings (**31**) and for molybdenum disulphide coatings (**71–73**). When sliding molybdenum disulphide coatings, a low shear transfer film is rapidly built up on the counter-face which can result in ultra-low coefficients of friction down to $\mu = 0.002$ (**54, 74**). Low friction and wear properties are achieved in vacuum or dry environments while increased air pressure or humidity results in increased friction (**21, 54, 75**).

2.6 Nanomechanical contact mechanisms

Recently emerging technologies such as the atomic force microscope and the surface force apparatus (**76**) have opened the possibility to study friction and wear phenomena on a molecular scale and to measure frictional forces between contacting molecules at the nano-newton level. Increased computational power has made it possible to study friction and associated phenomena by molecular dynamic simulations of sliding surfaces and to investigate the atomic scale contact mechanisms. The friction that arises from slippage between solid-to-solid interfaces (**77**) and between closely packed films in sliding contact (**78**) have been investigated. The atomic scale mechanisms of friction when two hydrogen-terminated diamond surfaces are in sliding contact have been studied, and the dependence of the coefficient of friction on load, crystallographic sliding direction and roughness have been investigated (**79**). In another study, the atomic-scale roughness effect on friction when two diamond surfaces are placed in sliding contact was examined (**80**).

The increased understanding of the origin of friction at the atomic-scale and even why friction exists, has resulted in an examination of the relationship between the commonly used laws of friction at a macro-scale and the molecular frictional behaviour on a nano-scale. There have been suggestions that friction arises from atomic lattice vibrations occurring when atoms close to one surface are set into motion by the sliding action of atoms in the opposing surface. Thus some of the mechanical energy needed to slide one surface over the other would be converted to sound energy, which is then eventually transformed into heat (**81**). Today we are only at the very beginning of the understanding of the nanomechanical tribological contact effects that explain the origin of friction and wear and there is no doubt that in the near future many new theories and explanations for the origin of tribological phenomena will become available.

The scaling up of the nanomechanical explanations of contact mechanisms to practically useful conclusions on a macro-scale is a most challenging and complex task and will take many years. Already there are practical applications on a nano scale where the increasing knowledge of tribological nanomechansims can be used. This has resulted in the development of Micro Electro Mechanical Systems (MEMS) such as motors, transducers, gears and bearings of sizes in the micrometer range. For these extremely small components silicon has been used in the early applications for production reasons but studies have shown that tribological improvements can be achieved by using polycrystalline diamond or MoS_2 thin coatings or hydrogenated diamond-like carbon coatings (**54, 82, 83**).

2.7 Mechanisms of material transfer

When a wear particle has been liberated from the surface, it can influence the tribological behaviour of the contacts in two ways. Loose wear debris in the contact may influence friction and wear as discussed above, although the wear debris may also attach to the counterface to form a transfer layer on it which changes significantly the tribological properties of the counterface. It can be said that a new counterface is formed, resulting in a new material pair.

Observations of the considerable influence of transfer films on the tribological behaviour of the contact have been reported for, example for polytetrafluoroethylene and polyimide coatings (**84**), titanium nitride coatings (**38, 85, 86**), diamond-like carbon coatings (**53, 54–57, 60**), diamond coatings (**87**) and molybdenum disulphide coatings (**54, 88**).

3 MULTILAYER AND MULTICOMPONENT COATINGS

3.1 Multilayer Coatings

In recent years, the use of multilayers has often been cited as the way forward to improve the mechanical, tribological and chemical properties of coatings. This has been argued from a number of standpoints. Firstly, taking the CVD coating industry as a guide, the move there has been to use many layers of different ceramics (e.g. **89–91**). Each layer can impart a specific property, such as the ability to act as a thermal barrier or diffusion barrier, or to impart abrasion resistance in metal cutting. Each layer in the coatings then has a discrete functional purpose. It has also been found that multi-layer CVD coatings can enhance the life of punch tools and injection moulding dies (**92**).

In the PVD coating sector, multilayering work was carried out in the late 1970s and early 1980s by researchers at the Los Alamos laboratories in the USA (**93, 94**). They attempted to build on the earlier models of Koehler (**95**) which predicted that high yield strength materials could be fabricated by alternating thin layers of high shear modulus material with thin layers of low shear modulus. The model was based on the inhibiting of dislocation formation and mobility. The Los Alamos work investigated Al/Al_xO_y films deposited by a pulsed gas process in an electron beam gun system, and showed that a Hall-Petch type relationship was obeyed for yield stress, based on the layer spacings. Bunshah and his co-workers also deposited microlaminate coatings by evaporation techniques; both metal/metal (**96**) and metal/ceramic (**97**) couples being studied. In general, an improvement in mechanical properties was observed for decreasing layer thicknesses.

In recent years, the literature, and at a certain level the theoretical understanding of multilayer, films has grown, in particular with regard to compositionally modulated 'superlattice' thin films (e.g. **98, 99**). However, many of the theories developed are applicable to highly epitaxial or single crystal layers, and whilst considerable progress has been made in understanding the theoretical aspects of superlattice hardening and strengthening mechanisms, their value in practical tribological systems is somewhat limited.

A more practical approach is to consider how the requirements of a surface differ at different locations within it – i.e. at the interface with the substrate, within the coating itself and at its surface as shown by Figure 4. Multilayer coatings offer a possibility to design the surface according to such requirements (**1, 100–102**).

Holleck tended to use multilayers of different ceramic materials, such as TiC and TiB_2, selected on the basis of their dominant bonding mechanism, i.e. metallic, covalent or ionic. He demonstrated improvements in hardness, indentation toughness, adhesion and wear performance under optimised layer thickness conditions (**103**). He attributes this, in part, to the crack deflection and stress relaxation mechanisms for the TiC/TiB_2 system.

This will pertain to many kinds of contact, especially where cyclic, fatigue-inducing, conditions prevail. Note that the benefits are no longer defined in terms of the early Koehler-type arguments based on the maximisation of yield strength. Rather they relate to the macro-behaviour of the multi-layer stack. However, the concept of the alternation

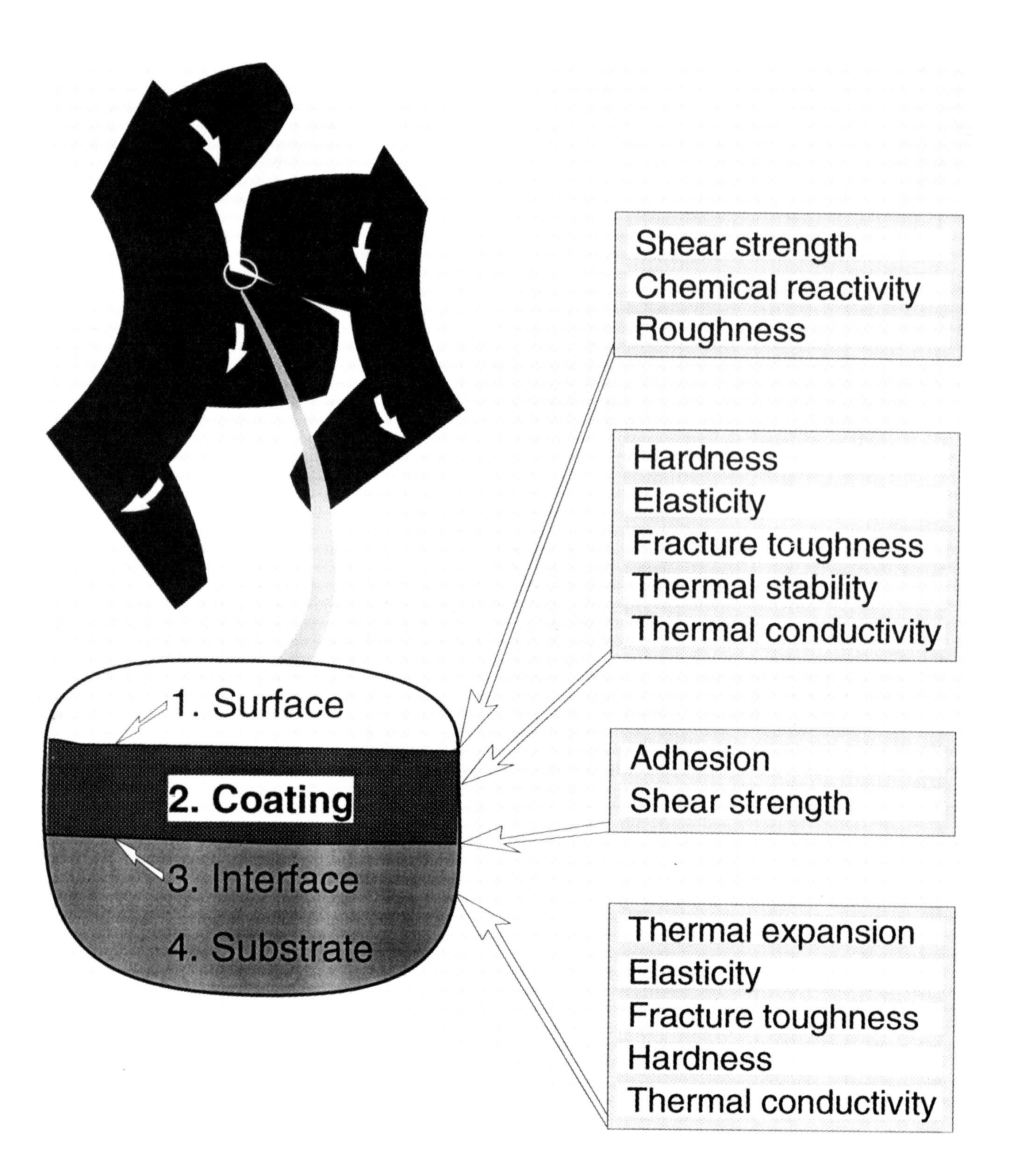

Fig. 4 Tribologically important properties in different zones of the coated surface.

of high and low shear modulus layers does seem to provide benefits beyond those described even by Holleck. We can cite, for example, the considerable benefits exhibited by multi-layer diamond-like-carbon and metal carbide films (**104**) or the multi-layer TiN/Ti films (**105–107**). In these cases a different argument has to be developed in order to fully explain the benefits observed.

One way to consider the behaviour of these films is to model the coating under a normal force which can be a point or distributed load, causing the coating to deflect. This will be accompanied by deformation of the substrate, which ideally should remain elastic.

If we consider the bending stresses induced in the coating under such a normal load, the level of maximum stress will increase with thickness for a given deflection and radius of curvature, see Fig. 5.

The diagram is highly schematic, and treats the coating in two dimensions as a beam, taking no account of the influence of the substrate on the stress distribution. Nevertheless it is illustrative since it shows that many multilayers, if bent to a similar radius, would individually see a much lower stress than one thick layer. This concept only works if the layers can effectively slide over each other, i.e. in the manner of a sheaf of paper, or paperback book being bent. In effect, we see another benefit of multilayer coatings, that is that the role of one of the sets of alternate layers can be to offer a shear zone to permit the harder (more brittle) layers to deflect under load without fracture. Thus a straight line scribed on such a cross-section would deflect as shown in Fig. 6.

This ensures that each of the hard layers will be subject to a much lower maximum bending stress than would be the case for a thick film bent to the same radius, although clearly the deflection under the same load would be greater, due to the lower 'stiffness' of the surface coating. That is not necessarily an issue, since for many tribological contacts with thin surface coatings it is the substrate which provides the main load support; the coating is present to provide a hard, low friction outer layer to reduce abrasive and/or adhesive wear.

Thus we can now see why coatings based on DLC/WC and Ti/TiN, for example, have proved successful in many practical contact conditions. The former are now used on gears and bearing surfaces which see cyclic contact conditions (**104**), whilst the latter have shown promise in erosive conditions (**99**). Since this concept is relatively new, the development of models to permit layer optimisation in terms of relative thickness and mechanical properties has been limited. The constraint on both the hard and soft layers may be that the elastic limit should not be exceeded, although this criterion seems more important for the hard layers, since if they yielded then rapid fracture would ensue.

The softer layers will be present to give an easy shear, i.e. they should have a low shear modulus, which equates to a low elastic modulus. Also these layers should have a long elastic strain to failure, meaning that yield should not occur. Although in this case plastic yielding is to be avoided, especially in contacts subject to cyclic loading and fatigue, the main purpose will be to prevent the harder layers from exceeding their yield stress. There are some interdependent variables in the optimisation of layer thicknesses. For example, since the layers also carry a load normal to the surface, they will be subject to compressive stresses, which will tend to squash the layers and in particular the softer ones. This will have the effect of limiting their minimum thickness – otherwise they may collapse and provide inadequate support for the harder layers within the coating. Note that whilst an increase in the thickness of the layers will increase the relative sliding distance of one hard layer over another, it will not necessarily increase the shear strain – which is a ratio between the sliding distance of one hard layer over the next and the thickness. The ideal seems to be to incorporate a large number of thin layers, thereby ensuring that the load support is not compromised. In the case of DLC/WC coatings produced commercially, layer thicknesses of 20 nm are typical, for sliding and rolling contacts, whilst for Ti/TiN films, layers 1 μm and 2μm thick have proved effective in erosion conditions. The optimisation of layer thickness will thus depend on the application and the loading conditions. In erosive or abrasive contacts, for example, a greater overall coating thickness is often needed, especially for coarse and hard third bodies.

The benefits of the multi-layering of relatively hard and relatively elastic layers have recently been demonstrated in a specially developed cyclic impact test (**108, 109**). In that work a relatively soft substrate (316 stainless steel) was

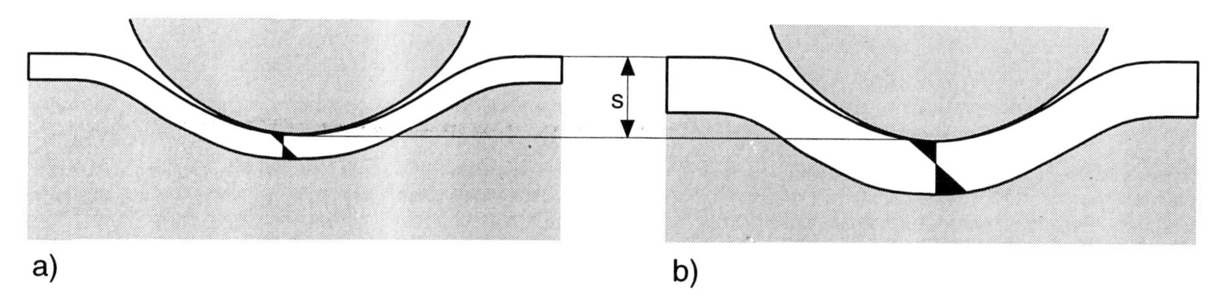

Fig. 5 (a) Thin hard coatings on a soft substrate generate lower stresses in the coating and at the coating to substrate interface compared to (b) thick hard coatings with the same deflection.

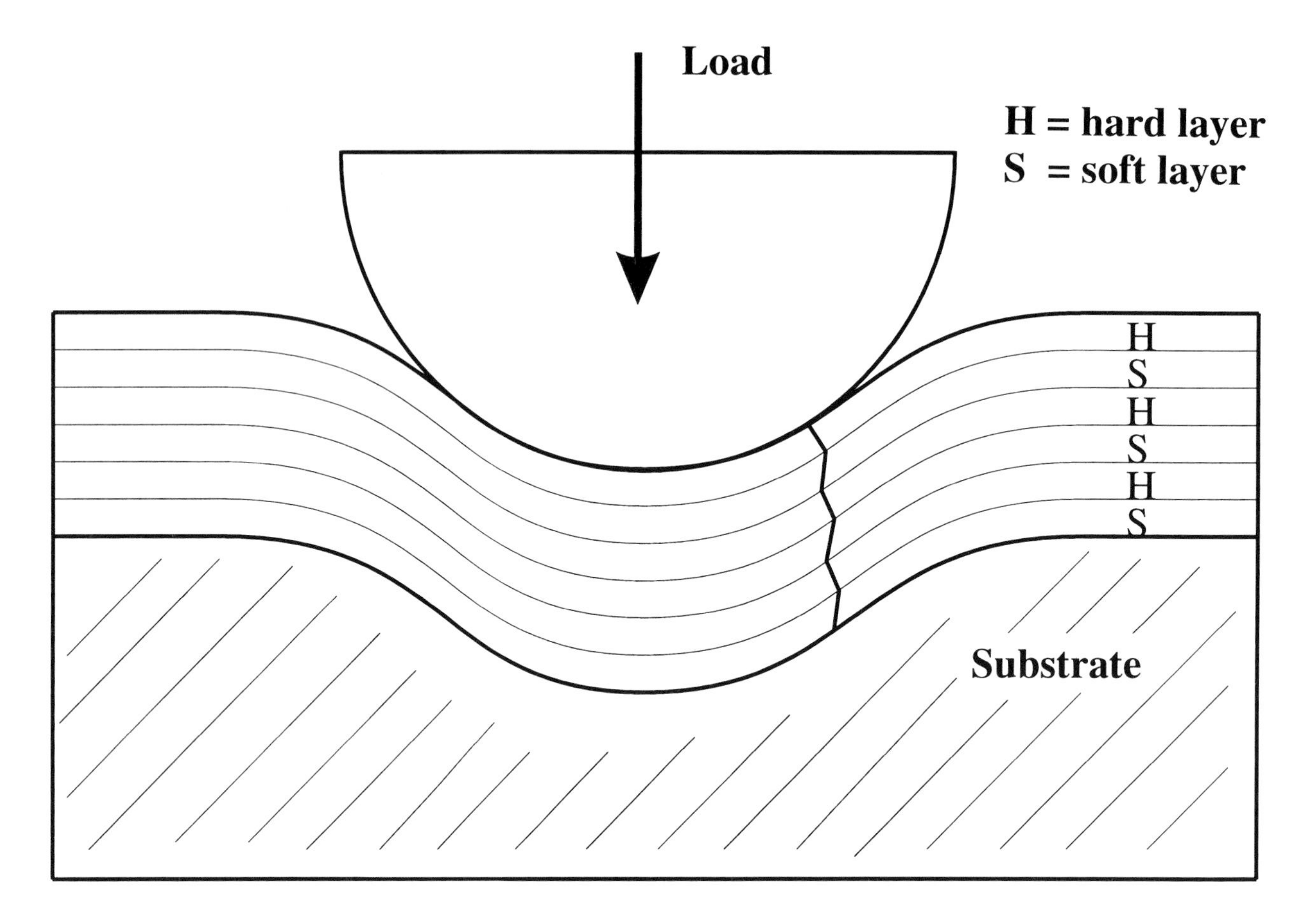

Fig. 6 A multilayer coating with alternate hard and soft layers can allow deflection to occur under load without yielding of the hard layers. They effectively slide over each other, with shear occurring in the soft layer. The pattern of shear is illustrated by the line through the film which was initially straight in the unloaded condition.

used. This deformed plastically, under repeated loads of 900 N applied at a frequency of 8 Hz by a 6 mm diameter tungsten carbide ball, coatings comprising multilayer stacks of Ti, TiN, TiCN, TiC, Ti/DLC. TiC, and Ti/DLC showed good wear performance, though not optimised for adhesion. Further tests using machines such as this, which apply the kinds of impact loads often encountered in industry, will allow the full benefit of multi-layered coatings to be achieved through further optimisation.

In effect the above represents a functionally graded coating designed to utilise specific layers for the provision of distinct properties (Fig. 7) (**25, 101, 110**). Such a concept can even be achieved with just two layers; for example an electroless nickel coating can be used as a load supporting layer for a hard PVD TiN or CrN coating, to improve the abrasive wear performance and enhance the corrosion resistance (**111, 112**). It is even possible to include within this approach the concept of pre-treating a substrate surface to produce a hardened outer layer, which can then support the coating more effectively. Such approaches have been termed *hybrid* or *duplex* coatings (**113–115**).

3.2 Multicomponent Coatings

Although multilayer films are all effectively 'multicomponent', this title is usually reserved for coatings in which there is a mixing of phases but not in a layered sense. For many years it has been known, for example, that mixed phases of ceramics, such as HfN and TiN could exhibit higher hardnesses than each phase on its own. Now there is intense research being carried out into other mixed ceramics, especially TiAlN (**116–118**), TiCN (**118–120**), Ti,B,N (**121, 122**) and others (e.g. **123, 124**). These coatings can exhibit much improved properties – e.g. in terms of high temperature hardness and oxidation resistance or suitability for interrupted cutting applications due to improved impact resistance. In some processes these mixed phases are deposited using an approach which is identical to that for multilayered films – e.g. by rotating samples in front of two or more sources or targets of each material. Whether this leads to multilayering or complete mixing depends largely on the speed of rotation and/or the degree of scattering and mixing of the vapour. The boundary between the definitions of multi-layered and

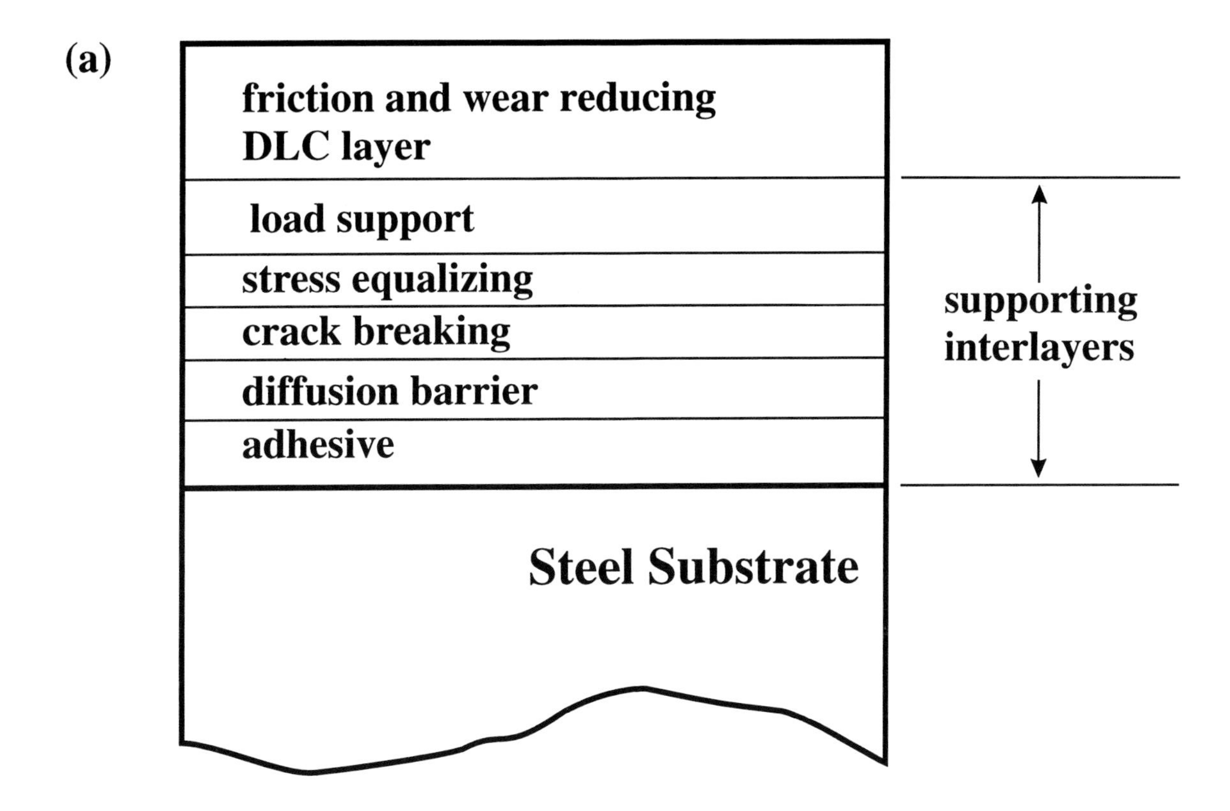

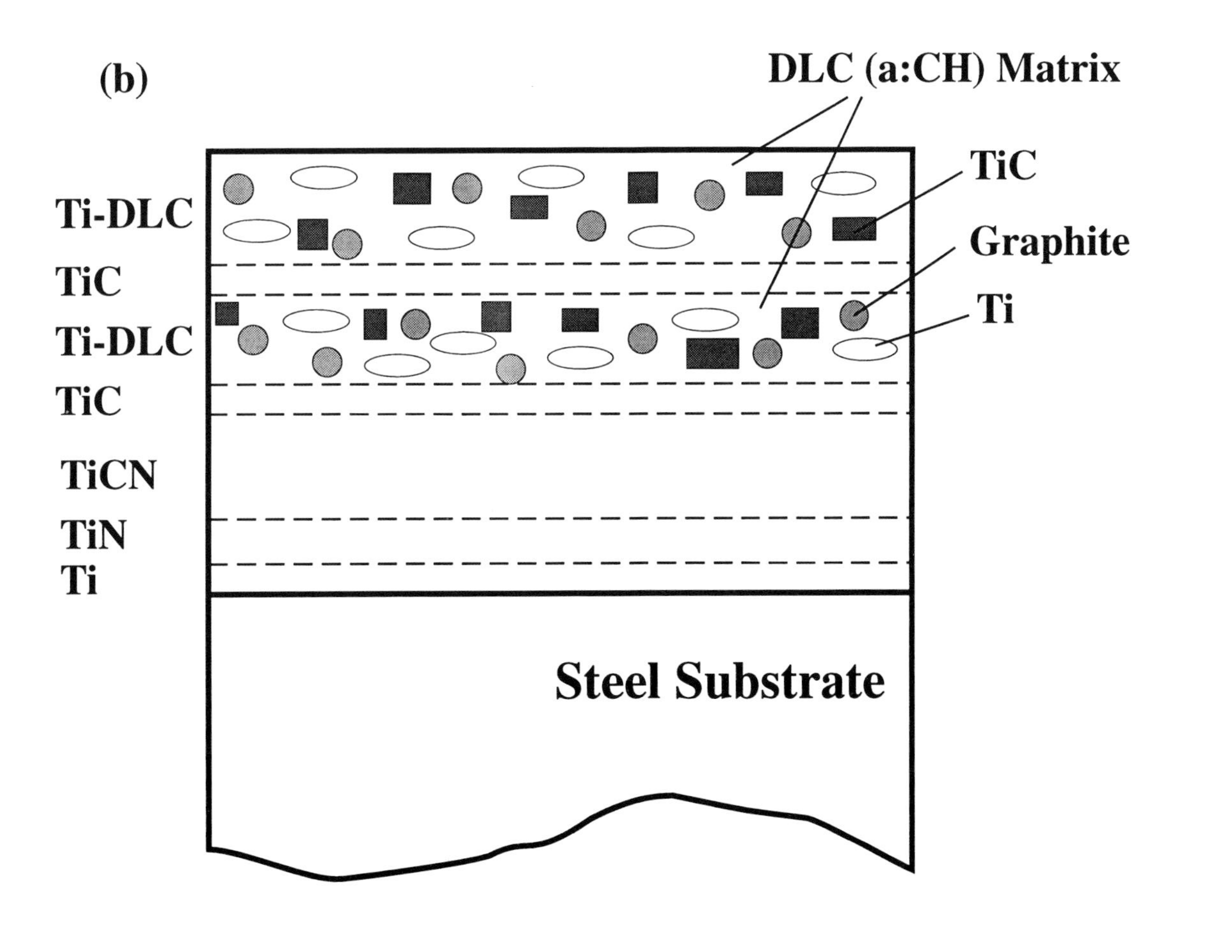

Fig. 7 (a) A functionally graded multilayer coating design to utilise specific layers for distinct properties. (b) A particular multilayer coating tested by Voevodin, *et al.* (**25**).

multi-component is thus somewhat blurred. Subramanian and Strafford (**91**) add further types 'multi-phase' and 'composite', depending upon the size and distribution of any secondary phase(s) in the primary matrix. Terminology thus presents a further complication, and there is no doubt that, on these definitions, even many multi-layered coatings could be defined as composites, or at least as mixed multilayer-composite structures. In a sense this illustrates the enormous potential of these coatings; at the same time the requirement to coat complex shapes in a controlled way with such 'fine-tuned' coatings can present problems (**125**).

As mentioned in Section 2, hard coatings can be further improved if a relatively low shear strength film is deposited at the outer surface. This represents a further modification to the multilayering concept. In effect, the gradient coatings described above, with DLC as the outer layer, are an example of this approach. Another suitable low friction layer is MoS_2 (**21**), although at the present time the greatest research interest lies in carbon-based layers. As explained in the next section, when suitably deposited these can even provide benefits in lubricated contacts, thus providing a wider range of operating conditions. It is likely that further developments will see other solid lubricant outer layers being developed for specific applications – for example films for high temperature sliding at 700°C and above are already under test (**126, 127**).

With developments such as those described above, advanced coating processes are now allowing tribologists to specify exactly the optimised surface properties which they need.

4 LUBRICATED COATED SURFACES

We have seen that thin diamond and diamond-like carbon coatings deposited in low pressure conditions, such as with the PVD and CVD techniques, can provide excellent low friction and wear protection properties in dry sliding conditions (**1**). The typical coefficient of friction values measured against steel are lower than typical values for boundary lubricated sliding with good oil lubrication, and thus considerably lower than for any other hard coated surfaces in dry sliding conditions. Only thin MoS_2 coatings can provide a similarly low friction. But at the same time diamond and diamond-like carbon coatings also provide excellent wear resistance, of about one order of magnitude lower wear than for any other wear resistant coatings. The diamond-like carbon coatings are of special interest because of their considerably lower deposition temperature compared to diamond films, which enables a larger variety of substrate materials to be coated.

The excellent tribological properties of diamond-like carbon coatings have been verified in many tests in dry conditions. However, the tribological performance of coatings in industrial applications may also be influenced by water or oil from the surrounding environment. In practical applications the coatings will at least be exposed to humidity. Earlier, the tribological performance of amorphous hydrogenated carbon films (a-C:H) and the hydrogen-free hard carbon (a-C) films in dry, unlubricated conditions was studied (**63, 128, 129**). The results verify the low friction and high wear resistance properties of a-C:H and a-C coatings referred to by several authors. To ensure the functionality of a-C:H and a-C coatings in water and oil environments, their tribological performance was studied in such lubricated conditions (**130**).

As a general trend the hydrogen-free hard carbon coatings, deposited by vacuum arc, are more wear resistant compared to hydrogenated carbon films deposited in a capacitively coupled r.f. plasma. On the other hand the counterparts sliding against a-C films suffer from more severe wear compared to counterparts sliding against the a-C:H films. The coefficient of friction is also generally lower for the a-C films.

4.1 The effect of water lubrication

When the coatings are exposed to water lubrication, the water is a very aggressive medium for the a-C:H coatings and causes catastrophic failure of the coating (**129**). The coating is peeled off in an early stage of the test, after only about 1.5 hours sliding. The same type of behaviour has also been reported by Drees, *et al.* (**131**). The a-C:H(Ti) film, with a titanium-alloyed surface layer, endures the test even though the water lubrication increases the wear of the coating (Fig. 8). Since titanium alloying of the a-C:H coating improves the performance of the coating in the water, it is suggested that the hydrogenated structure itself is vulnerable in a water environment. Conversely, the a-C coating performed well with water lubrication, since no measurable wear could be detected even after 21 hours of sliding in the pin-on-disc test. Obviously the hydrogen-free structure of the a-C coatings is preferable compared to a-C:H coatings providing lower wear and friction, even when water is introduced in to the system.

The pin wear rates increased against both a-C and a-C:H(Ti) coatings in water-lubricated conditions. The tribolayers formed on the wear surfaces in dry sliding conditions usually reduce the pin wear and the friction coefficient. In a water environment the formation of protective tribolayers is prohibited and thus the pin wear is increased. However, the coefficient of friction is significantly lower for the water-lubricated coatings, particularly for the a-C coating.

4.2 The effect of oil lubrication

Oil lubrication reduces the friction coefficients in uncoated, as well as in coated tribosystems. The mineral base oil (oil 1) has only a minor effect on friction behaviour of a-C and a-C:H coatings, but the hydraulic oil with EP additives has a greater effect on the friction coefficient as shown in Figure 9 (**129**). For the titanium alloyed a-C:H(Ti) coating the friction coefficient is increased when oil is introduced in the contact. This shows that in dry conditions a

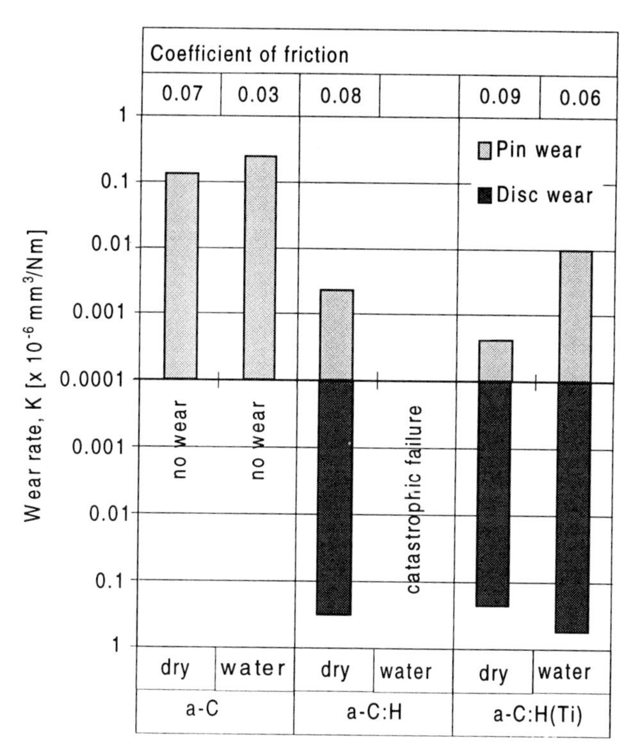

Fig. 8 The friction and wear results of the water lubricated reciprocating wear tests. The counter face material was an alumina ball with 10 mm diameter. Sliding speed 4 mm/s and normal load 5 N.

tribologically beneficial tribolayer formed on the wear surfaces, which provides a low coefficient of friction. Actually the a-C:H(Ti) coating shows a lower friction coefficient in dry conditions than the uncoated contacts in oil-lubricated conditions. This low friction behaviour of alloyed a-C:H films has also been reported also by several other authors (e.g. **104, 132, 133**). By the addition of oil the formation of this tribolayer is prevented, thus causing an increase in the friction coefficient. The frictional behaviour of the coatings in oil-lubricated conditions is governed by the lubricative properties of the oil, particularly when oil with EP additives is used. Surprisingly in tests with oil containing EP additives, the a-C coated contacts had the lowest friction, which suggests that the hydrogen-free coating can have a beneficial effect when used with EP additives. No wear of the a-C and a-C:H coating is detected when oil lubrication is used. However, the oil lubrication prevented the formation of the tribolayer on the pin wear surface in the same manner as water lubrication, which increased the pin wear rates to some extent.

The lubricated tests showed that the DLC coatings give low friction and high wear resistance performance in dry conditions, which can still be improved with oil lubrication. The a-C coatings also operated well in water-lubricated conditions, but for the a-C:H coating the water-lubricated conditions were detrimental. Considering the utilisation of DLC films in practical components, it is suggested that a-C coated components can also be used in lubricated conditions to give additional safety for friction and wear control in e.g. poorly lubricated conditions.

5 CONCLUSIONS

The fast development of a large variety of different advanced surface modification and coating deposition techniques during the last few decades offers remarkable possibilities to improve the tribological properties of surfaces and thus improve the functional reliability and lifetime of tools, components and other surfaces in sliding contact. Full advantage of these developments can only be achieved when appropriate methods for surface design and optimisation are available. To support this a systematic approach to the analysis of tribological coated surfaces has been outlined. It is based on a classification of the tribological contact process into macro-, micro- and nanomechanical and tribochemical contact mechanisms and material transfer.

The use of thin multilayers offers an excellent possibility for surface design to achieve the required properties at the surface. Increased coating/substrate adhesion, improved load support, surface stress reduction and improved crack propagation resistance can be achieved by different concepts of multilayer surface design.

The diamond and diamond-like hard carbon coatings represent from a tribological point of view the most dramatic new coating concept introduced during the last few decades. In dry sliding the friction can be extremely low with a coefficient of friction less than 0.01 and a wear resistance more than one order of magnitude better than for any other hard coating. The diamond-like coatings can also be used with good tribological performance in oil or water lubricated conditions with some care e.g. hydrogenated

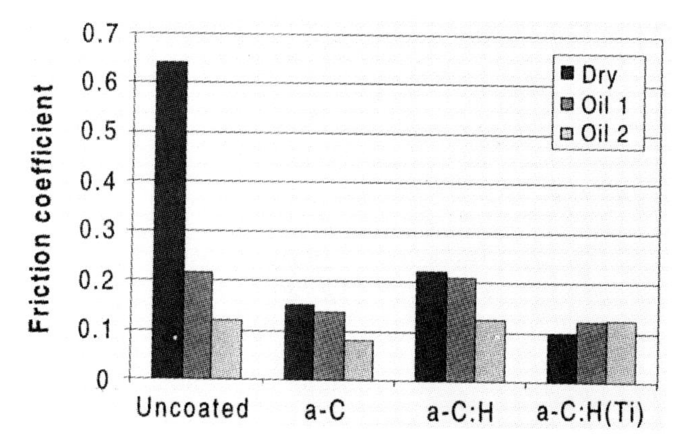

Fig. 9 The coefficient of friction from the oil lubricated reciprocating sliding tests. The counter face material was a steel ball with 10 mm diameter. Sliding speed 4 mm/s and normal load 10 N.

diamond-like coatings may show catastrophical failure when sliding in a water environment while hydrogen-free DLC coatings perform well.

6 REFERENCES

1 **Holmberg, K.** and **Matthews, A.**, *Coatings tribology – properties, techniques and applications in surface engineering*. Elsevier Tribology Series, **28**, 1994, 442 p. (Elsevier Science B.V., The Netherlands).

2 **Bhushan, B.**, (Ed.) *Handbook of Micro/Nanotribology*, 1995 (CRC Press).

3 **Matthews, A.**, **Holmberg, K.** and **Franklin, S.**, A methodology for coating selection. Thin Films in Tribology. Proc. *19th Leeds-Lyon Symp. on Tribology*, 8–11 Sept. 1992, Leeds, UK, Dowson, D., *et al.* (Eds), Elsevier Tribology Series, Vol. **25**, 1993, pp. 429–439 (Elsevier, Amsterdam).

4 **Franklin, S.E.** and **Dijkman, J.A.**, The implementation of tribological principles in an expert-system ('PERCEPT') for the selection of metallic materials, surface treatments and coatings in engineering design. *Wear*, 1995, **181–183**, 1–10.

5 **Ramalingam, S.** and **Zheng, L.**, Film-substrate interface stresses and their role in the tribological performance of surface coatings. *Tribology International*, 1995, **28**, (3), 145–161.

6 **Gunnars, J.** and **Alahelisten, A.**, Thermal stresses in diamond coatings and their influence on coating wear and failure. *Surface and Coatings Technology*, 1996, **80**, 303–312.

7 **Gåhlin, R.**, **Alahelisten, A.** and **Jacobson, S.**, The effects of compressive stresses on the abrasion of diamond coatings. *Wear*, 1996, **196**, 226–233.

8 **Diao, D.F.**, **Kato, K.** and **Hayashi, K.**, The maximum tensile stress on a hard coating under sliding friction. *Tribology International*, 1994, **27**, (4), 267–272.

9 **Wong, S.K.** and **Kapoor, A.**, Effect of hard and stiff overlay coatings on the strength of surfaces in repeated sliding. *Tribology International*, 1996, **29**, (8), 695–702.

10 **Diao, D.F.** and **Kato, K.**, Interface yield map of a hard coating under sliding contact. *Thin Solid Films*, 1994, **245**, 115–121.

11 **Diao, D.F.** and **Kato, K.**, Spalling mechanism of Al_2O_3 ceramic coating coated WC-Co substrate under sliding contact. *Thin Solid Films*, 1994, **245**, 104–108.

12 **Diao, D.F.**, **Kato, K.** and **Hokkirigawa, K.**, Fracture mechanics of ceramic coatings in indentation. Trans. ASME, *J. of Tribology*, 1994, **116**, 860–869.

13 **Eberhardt, A.W.** and **Peri, S.**, Surface cracks in layered Hertzian contact with friction. *Tribology Transactions*, 1994, **38**, (2), 299–304.

14 **Sainsot, P.**, **Leroy, J.M.** and **Villechaise, B.**, Effect of surface coatings in a rough normally loaded contact. In: *Mechanics of coatings*, Dowson, D., *et al.* (Eds), Tribology series **17**, 1990, pp. 151–156 (Elsevier, Amsterdam).

15 **Mao, K.**, **Sun, Y.** and **Bell, T.**, Contact mechanisms of engineering surfaces: State of the art. *Surface Engineering*, 1994, **10**, (4), 297–306.

16 **Mao, K.**, **Sun, Y.** and **Bell, T.**, A numerical model for the dry sliding contact of layered elastic bodies with rough surfaces. *Tribology Transactions*, 1995, **39**, (2), 416–424.

17 **Varadi, K.**, **Kozma, M.** and **Poller, R.**, The effect of surface roughness on contact and stress states of spur gears. *Tribotest Journal*, 1995, **2**, 25–35.

18 **Holmberg, K.**, A concept for friction mechanisms of coated surfaces. *Surface and Coatings Technology*, 1992, 56, 1–10.

19 **Holmberg, K.**, **Matthews, A.** and **Ronkainen, H.**, Wear mechanisms of coated sliding surfaces. *Thin Films in Tribology*. Proc. 19th Leeds-Lyon Symp. on Tribology, 8–11 Sept. 1992, Leeds, UK, Dowson, D., *et al.* (Eds), Elsevier Tribology Series, Vol. 25, pp. 399–407 (Elsevier, Amsterdam).

20 **Suh, N.P.**, *Tribophysics*, 1986, 489 p. (Prentice-Hall, Englewood Cliffs, New Jersey, USA.)

21 **Donnet, C.**, Tribology of solid lubricant coatings. *Condensed Matter News*, 1995, **4**, (6), 9–24.

22 **Bowden, F.P.** and **Tabor, D.**, *Friction and Lubrication of Solids*, 1950, 374 p. (Claredon Press, Oxford).

23 **Spalvins, T.** and **Sliney, H.E.**, Frictional behavior and adhesion of Ag and Au films applied to aluminium oxide by oxygen-ion assisted screen cage ion plating. *Surface and Coatings Technology*, 1994, **68/69**, 482–488.

24 **Voevodin, A.A.**, **Donley, M.S.**, **Zabinsky, J.S.** and **Bultman, J.E.**, Mechanical and tribological properties of diamond-like carbon coatings prepared by pulsed laser deposition. *Surface and Coatings Technology*, 1995, **76–77**, 534–539.

25 **Voevodin, A.A.**, **Schneider, J.M**, **Rebholz, C.** and **Matthews, A.**, Multilayer composite ceramic-metal-DLC coatings for sliding wear applications. *Tribology International*, 1996, **29**, (7), 559–570.

26 **Erdemir, A.**, **Bindal, C.**, **Fenske, G.R.**, **Zuiker, C.**, **Csencsits, R.**, **Krauss, A.R.** and **Gruen, D.M.**, Tribological characterization of smooth diamond films grown in Ar-C_{60} and Ar-CH_4 plasmas. *Diamond Films and Technology*, 1996, **6**, (1), 31–47.

27 **Bull, S.J.**, **McCabe, A.R.** and **Jones, A.M.**, Mechanical and tribological performance of ion-beam deposited diamond-like carbon on polymers. *Surface and Coatings Technology*, 1994, **64**, 87–91.

28 **Holleck, H.** and **Schier, V.**, Multilayer PVD coatings for wear protection. *Surface and Coatings Technology*, 1995, **76–77**, 328–336.

29 **Voevodin, A.A.**, **Rebholz, C.**, and **Matthews, A.**, Comparative tribology studies of hard ceramic and composite metal-DLC coatings in sliding friction conditions. *Tribology Transactions*, 1995, **38**, (4), 829–836.

30 **Voevodin, A.A.** and **Donley, M.S.**, Preparation of amorphous diamond-like carbon by pulsed laser deposition: a critical review. *Surface and Coatings Technology*, 1996, **82**, 199–213.

31 **Sherbiney, M.A.** and **Halling, J.**, Friction and wear of ion-plated soft metallic films. *Wear*, 1977, **45**, 211–220.

32 **Aubert, A.**, **Nabot, J.P.**, **Ernoult, J.** and **Renaux, P.**, Preparation and properties of MoS_2 films grown by d.c. magnetron sputtering. *Surface and Coatings Technology*, 1990, 41, 127–134.

33 **Finkin, E.F.**, A theory for the effect of film thickness and normal load in the friction of thin films. *J. Lubr. Tech.*, 1969, July, 551–556.

34 **Dayson, C.**, The friction of very thin solid film lubricants on surfaces of finite roughness. *ASLE Trans.*, 1971, **14**, 105–115.

35 **Burnett, P.J.** and **Rickerby, D.S.**, The mechanical properties of wear-resistant coatings I: Modelling of hardness behaviour. *Thin Solid Films*, 1987, **148**, 41–50.

36 **Jahanmir, S., Deckman, D.E., Ives, L.K, Feldman, A.** and **Farabaugh, E.**, Tribological characteristics of synthesized diamond films on silicon carbide. *Wear,* 1989, **133**, 73–81.

37 **Hayward, I.P., Singer, I.L.** and **Seizman, L.E.**, Effect of roughness on the friction of diamond on CVD diamond coatings. *Wear,* 1992, **157**, 215–227.

38 **Sue, J.A.** and **Troue, H.H.**, Friction and wear properties of titanium nitride coating in sliding contact with AISI 01 steel. *Surface and Coatings Technology,* 1990, **43–44**, 709–720.

39 **Guu, Y.Y., Lin, J.F.** and **Ai, C.-F.**, The tribological characteristics of titanium nitride coatings Part I. Coating thickness effects. *Wear,* 1996, **194**, 12–21.

40 **Guu, Y.Y.** and **Lin, J.F.**, The tribological charasteristics of titanium nitride coatings Part II. Comparison of two deposition processes. *Wear,* 1996, **194**, 22–29.

41 **Miyoshi, K.**, Friction and wear of plasma-deposited diamond films. *J. Appl. Phys.*, 1993, **74**, 4446–4454.

42 **Argon, A.S.**, Mechanical properties of near-surface material in friction and wear. In: *Fundamentals of Tribology.* Suh, N.P. & Saka, N. (Eds.), 1980, pp. 103–125 (MIT Press, London).

43 **Berthier, Y., Godet, M.** and **Brendle, M.**, Velocity accommodation in friction. *Tribology Trans.,* 1989, **32**, (4), 490–496.

44 **Oliveira, S.A.G.** and **Bower, A.F.**, An analysis of fracture and delamination in thin coatings subject to contact loading. *Wear,* 1996, **198**, 15–32.

45 **Gee, M.** and **Jennett, N.M.**, High resolution characterization of tribochemical films on alumina. *Wear,* 1995, **193**, 133–145.

46 **Mäkelä, U.** and **Valli, J.**, Tribological properties of PVD TiN and TiC coatings. *Finnish Journal of Tribology,* 1985, **4**, (2),74–85.

47 **Zuiker, C., Krauss, A.R., Gruen, D.M., Pan, X., Li, J.C., Csencsits, R., Erdemir, A., Binal, C.** and **Fenske, G.**, Physical and tribological properties of diamond films grown in argon-carbon plasmas. *Thin Solid Films,* 1995, **270**, 154–159.

48 **Habig, K.-H.**, Fundamentals of tribological behaviour of diamond, diamond-like carbon and cubic boron nitride coatings. *Surface and Coatings Technology,* 1995, **76–77**, 540–547.

49 **Schouterden, K., Blanpain, B., Celis, J.-P.** and **Vingsbo, O.**, Fretting of titanium nitride and diamond-like carbon coatings at high frequencies and low amplitude. *Wear,* 1995, **181–183**, 86–93.

50 **Fouvry, S., Kapsa, P.** and **Vincent, L.**, Fretting behaviour of hard coatings under high normal load. *Surface and Coatings Technology,* 1994, **68/69**, 494–499.

51 **Morbacher, H., Blanpain, B., Celis, J.-P. and Roos, J.R.**, The influence of humidity of fretting behavour of PVD TiN coatings. *Wear,* 1995, **180**, 43–52.

52 **Vancoille, E., Celis, J.P.** and **Roos, J.R.**, Tribological and structural characterization of a physical vapour deposited TiC/Ti(C,N)/TiN multilayer. *Tribology International,* 1993, **26**, (2), 115–119.

53 **Ronkainen, H., Likonen, J., Koskinen, J.** and **Varjus, S.**, Effect of tribofilm formation on the tribological performance of hydrogenated carbon coatings. Int. Conf. on *Metallurgical Coatings and Thin Films,* ICMCTF'94, 25–29 April 1994, San Diego, 14 p.

54 **Donnet, C., Martin, J.-M., Le Mogne, T.** and **Belin, M.**, How to reduce friction in the millirange by solid lubrication. Proc. *Int. Tribology Conf.,* Yokohama, 1995, 6 p.

55 **Gardos, M.N.**, Tribology and wear behavor of diamond. *Synthetic Diamond: Emerging CVD Science and Technology,* K.E. Spear & J.P. Dismukes (Eds), 1994, pp. 419–502 (John Wiley & Sons).

56 **Erdemir, A., Bindal, C., Pagan, J.** and **Wilbur, P.**, Characterization of transfer layers on steel surfaces sliding against diamond-like hydrocarbon films in dry nitrogen. *Surface and Coatings Technology,* 1995, **76–77**, 559–563.

57 **Erdemir, A., Bindal, C., Fenske, G.R., Zuikner, C.,** and **Wilbur, P.**, Characterization of transfer layers forming on surfaces sliding against diamondlike carbon. *ICMCTF96 Conf.,* 22–26 April, 1996, San Diego, 11 p.

58 **Liu, Y., Erdemir, A.** and **Meletis, E.I.**, An investigation of the relationship between graphitization and frictional behavior of DLC coatings. *ICMCTF96 Conf.,* 22–26 April, 1996, San Diego, 11 p.

59 **Neernick, D., Persoone, P., Sercu, M., Goel, A., Venkatraman, C., Kester, D., Halter, C., Swab, P.,** and **Bray D.** Diamond-like nanocomposite coatings fro low-wear and low-friction applications in hunid environments. 10th Int. Conf. on *Thin Films,* 23–27 Sept. 1996, Salamanca, Spain, 3 p.

60 **Meletis, E.I., Erdemir, A.,** and **Fenske, G.R.**, Tribological characteristics of DLC films and duplex plasma nitriding/DLC coating treatment. *Surface and Coatings Technology,* 1995, 73, 39–45.

61 **Watanabe, S., Miyake, S.,** and **Murakawa, M.**, Tribological behaviour of cubic boron nitride film sliding against diamond. Trans. ASME, *J. of Tribology,* 1995, **117**, 629–633.

62 **Zaidi, H., Le Huu, T., Robert, F., Bedri, R., Kadiri, E.K.** and **Paulmier, D.**, Influence of defects and morphology of diamond-like carbon coatings on their tribological behaviour. *Surface and Coatings Technology,* 1995, **76–77**, 564–571.

63 **Holmberg, K., Koskinen, J., Ronkainen, H., Vihersalo, J., Hirvonen, J.-P.** and **Likonen, J.**, Tribological characteristics of hydrogenated and hydrogen-free diamond-like carbon coatings. *Diamond Films and Technology,* 1994, **4**, (2), 113–129.

64 **Donnet, C., Belin, M., Auge, J.C., Martin, J.M., Grill, A.** and **Patel, V.**, Tribochemistry of diamond-like carbon coatings in various environments. *Surface and Coatings Technology,* 1994, **68/69**, 626–631.

65 **Erdemir, A.** and **Fenske, G.R.**, Tribological performance of diamond and diamondlike carbon films at evaluated temperatures. *Tribology Transactions,* 1996, **39**, 4, 787–794.

66 **Bhushan, B.** and **Ruan, J.**, Tribological performance of thin film amorphous carbon overcoats for magnetic recording rigid disks in various environments. *Surface and Coatings Technology,* 1994, **68/69**, 644–650.

67 **Maillat, M.** and **Hintermann, H.E.**, Tribological properties of amorphous diamond-like carbon coatings. *Surface and Coatings Technology,* 1994, 68/69, 638–643.

68 **Pimenov, S.M., Smolin, A.A., Obraztsova, E.D., Konov, V.I., Bögli, U., Blatter, A., Maillat, M., Leijala, A., Burger, J., Hintermann, H.E., Loubnin, E.N.**, Tribological behaviour of smooth diamond films. *Surface and Coatings Technology,* 1995, **76–77**, 572–578.

69 **Skopp, A., Klafke, D., Buchkremer-Herrmanns, H., Ren, H.** and **Weiss, H.**, Oscillating sliding behaviour of MW PACVD diamond coatings. 10th Int. Colloquium on *Tribology – Solving Friction and Wear Problems,* 9–12

January 1996, Technische Academie Esslingen, Ostfildern, 1996, pp. 1943–1959.

70 **Zhao,W., Tang, J., Li, Y.** and **Chen, L.**, High friction coefficient of fullerene C_{70} film. *Wear*, 1996, **198**, 165–168.

71 **Fayeulle, S., Ehni, P.D.** and **Singer, I.L.**, Role of transfer films in wear of MoS_2 coatings. In: *Mechanics of coatings*. Dowson D., *et al.* (Eds), Elsevier Tribology Series, Vol. 17, 1990, pp. 129–138. (Elsevier, Amsterdam).

72 **Singer, I.L.**, A thermomechanical model for analyzing low wear rate materials. *Surface and Coatings Technology,* 1991, **49**, 474–481.

73 **Bhushan, B.** and **Gupta, B.K.** *Handbook of Tribology – Materials, Coatings and Surface Treatments,* 1991 (MCGraw-Hill, New York).

74 **Martin, J.M., Donnet, C.** and **Le Monge, T.** Superlubricity of molybdenum disulphide. The American Phys. Soc., *Physical Review B,* 1993, **48**, (14), 10583–10586.

75 **Singer, I.L., Fayeulle, S.** and **Ehni, P.D.** Wear behavior of triode-sputtered MoS_2 coatings in dry sliding contact with steel and ceramics. *Wear*, 1996, **195**, 7–20.

76 **Israelachvili, J.N.** and **Tabor, D.** The measurement of van der Walas dispersion forces in the range 1.5 to 130 nm. *Proceedings of the Royal Society*, 1972, **A331,** 19–38.

77 **Thompson, P.A.** and **Robbins, M.O.** Simulations of contact-line motion: slip and the dynamic contact angle. The American Physical Society, *Physical Review Letters,* 1989, **63**, (7), 766–769.

78 **McClelland, G.M.** and **Glosli, J.N.** In: Singer, I.L. and Pollock, H.M. (eds), NATO ASI *Proc. on Fundamentals of Friction,* 1992, pp. 405–426 (Kluwer, Dordrecht).

79 **Harrison, J.A., White, C.T., Colton, R.J.** and **Brenner, D.W.** Molecular-dynamics simulations of atomic-scale friction of diamond surfaces. The American Physical Society. *Physical Review B,* 1992, **46**, (15), 9700–9708.

80 **Harrison, J.A., Colton, R.J., White, C.T.** and **Brenner, D.W.** Effect of atomic scale surface roughness on friction: a molecular dynamics study of diamond surfaces. *Wear*, 1993, **168,** 127–133.

81 **Krim, J.** Friction at the atomic scale. *Scientific American*, October, 1996, 48–56.

82 **Beerschwinger, U., Albrecht, T., Mathieson, D., Reuben, R.L., Yang, S.J.** and **Taghizadeh, M.** Wear at microscopic scales and light loads for MEMS applications. *Wear*, 1995, **181–183**, 426–435.

83 **Gardos, M.N.** Tribological behaviour of polycristalline diamond films. NATO Advanced Research Workshop *Protective Coatings and Thin Films*, 30.5. – 5.6.1996, Portimao, Algarve, Portugal, 12 p.

84 **Yamada, Y., Tanaka, K.** and **Saito, K.** Friction and damage of coatings formed by sputtering polytetrafluoroethylene and polyimide. *Surface and Coatings Technology*, 1990, **43–44**, 618–628.

85 **Ronkainen, H., Likonen, J.** and **Koskinen, J.** Tribological properties of hard carbon films produced by the pulsed vacuum arc discharge method. Int. Conf. on *Metallurgical Coatings and Thin Films* ICMCTF-1992, San Diego, CA, 6–10 April 1992, 20 p.

86 **Huang, Z.P., Sun, Y.** and **Bell, T.**, Friction behaviour of TiN, CrN and (TiAl)N coatings. *Wear*, 1994, **173**, 13–20.

87 **Bull, S.J., Chalkner, P.R., Johnston, C.** and **Moore, V.**, The effect of roughness on the friction and wear of diamond thin films. *Surface and Coatings Technology*, 1994, **68/69**, 603–610.

88 **Hopple, G.B.** and **Loewenthal, S.H.**, Development, testing and characterization of MoS_2 thin film bearings. *Surface and Coatings Technology*, 1994, **68/69**, 398–406.

89 **Schintlmeister, W., Wallgram, W** and **Kranz, J.**, Properties, applications and manufacture of wear-resistant hard material coatings for tools, *Thin Solid Films*, 1983, **107**, 117–127.

90 **AB Sandvik Coromant,,** *Modern Metal Cutting: A Practical Handbook*, Sandvik Coromant, Sweden, 1994.

91 **Subramanian, C.** and **Strafford, K.N.** Review of multicomponent and multilayer coatings for tribological applications, *Wear*, 1993, **165**, 85–95.

92 **Franklin, S.E.** Wear behaviour of CVD and PVD coated tools in metal stamping and plastics injection moulding, in *Surface Engineering Practice*. K. N. Strafford, P. K. Datta and J. S. Gray (Eds.), 1990 (Ellis Horwood, Chichester).

93 **Springer, R.W.**, and **Catlett, D.S.** Structure and Mechanical Properties of Al/Al_xO_y Vacuum Deposited Laminates, *Thin Solid Films*, 1978, **54**, 197–205.

94 **Springer, R.W.** and **Harford, C.D.** Characterisation of aluminium-aluminium nitride coatings sputter-deposited using the pulsed gas process, *J. Vac. Sci. Technol.*, 1982, **20**, 462–464.

95 **Koehler, J. S.** Attempt to design a strong solid, *Phys. Rev. Sect. B2,* 1970, 547–551.

96 **Bunshah, R.F., Nimmagadda, R., Doerr, H.J., Movchan, B.A., Grechanuk, N.I.** and **Didkin, G.G.** Structure-property relationships in Cr/Cu and Ti/Ni microlaminate composites, *Thin Solid Films*, 1984, **112**, 227–236.

97 **Sans, C., Deshpandey, C., Doerr, H.J., Bundshah, R.F., Movchan, B.A.** and **Denschishin, A.V.** Preparation and Properties of TiC/Ni Microlaminates, *Thin Solid Films*, 1983, **107**, 345–351.

98 **Barnett, S.A.** *Deposition and mechanical properties of superlattic thin films,* 1993 (Academic Press).

99 **Barnett, S.A.** and **Shinn, M.** Plastic and elastic properties of compositionally modulated thin films, *Annu. Rev. Mater. Sci.*, 1994, **24**, 481–511.

100 **Holleck, H.**, Material selection for hard coatings, *J. Vac. Sci. Technol.*, A4, 1986, 2661–2669.

101 **Holleck, H.**, Designing advanced coatings for wear protection, *Proc. Surface Engineering and Heat Treatment, Past, Present and Future*, Conf. Dec. '89, London. P.H. Morton (ed.), Inst. of Materials, 1990.

102 **Holleck, H.** and **Schier, V.**, Multilayer PVD coatings for wear protection, *Surf. Coat. Technol.*, 1995, **76–77**, 328–336.

103 **Holleck, H., Lahres, M.** and **Woll, P.**, Multilayer coatings – influence of fabrication parameters on constitution and properties, *Surf. Coat. Technol.*, 1990, **41**, 179–190.

104 **Matthews, A.** and **Eskildsen, S.S.** Engineering applications for diamond-like carbon. *Diamond and Related Materials*, 1994, **3**, 902–911.

105 **A Leyland** and **A Matthews,** Thick Ti/TiN multilayered coatings for abrasive and erosive wear resistance, *Surf. Coat. Technol.*, 1994, 70, 19–25.

106 **Bull, S.J.** and **Jones, A.M.** Multilayer coatings for improved performance, *Surf. Coat. Technol.*, 1996, **78**, 173–184.

107 **Ensinger, W., Enders, B., Emmerich, R., Martin, H., Hubler, R., Wolf, G.K., Alberts, L., Schroer, A., Schreiner, W.H., Stedile, F.C.,** and **Baumol, I.J.R.** Properties of multilayered hard coatings deposited by ion

beam and plasma techniques, *Le Vide*, Les Couches Minces, Supplement No. 261, Mars-Avril 1992, 355–357.

108 **Bantle, R.** and **Matthews, A.** Investigation into the impact wear behaviour of ceramic coatings, *Surf. Coat. Technol.*, 1995, **74–75**, 857–868.

109 **Voevodin, A.A., Bantle, R.** and **Matthews, A.** Dynamic impact wear of TiC_xN_y and Ti-DLC composite coatings, *Wear*, 1995, **185**, 151–157.

110 **Vetter, J., Burgmer, W., Dederichs, H.G.** and **Perry, A.J.** The architecture and performance of multilayer and compositionally gradient coatings made by cathodic arc evaporation, *Surf. Coat. Technol.*, 1993, **61**, 209–214.

111 **Leyland, A., Bin-Sudin, M., James, A.S., Kalantary, M.R., Wells, P.B., Housden, J., Garside, B.** and **Matthews, A.** TiN and CrN PVD coatings on electroless nickel coated steel substrates, *Surf. Coat. Technol.*, 1993, **60**, 474–479.

112 **Bin-Sudin, M., Leyland, A., James, A.S., Matthews, A., Housden, J.** and **Garside, B.** Substrate surface finish effects in duplex coatings of PAPVD TiN and CrN with electroless nickel-phosphorus interlayers, *Surf. Coat. Technol.*, 1996, **81**, 215–224.

113 **Matthews, A., Leyland, A., Dorn, B., Stevenson, P., Bin-Sudin, M., Rebholz, C., Voevodin, A.A.** and **Matthews, A.** Plasma based surface engineering processes for wear and corrosion protection, *J. Vac. Sci. Technol.*, 1995, A13, 1202–1207.

114 **Matthews, A.** and **Leyland, A.** Hybrid techniques in surface engineering, *Surf. Coat. Technol.*, 1995, **71**, 88–92.

115 **Matthews, A.** Plasma-based vacuum deposition processes to enhance wear and corrosion performance. NATO Advanced Research Workshop on *Protective Coatings and Thin Films*, Alvor, Portugal, May 1996. (To be published by Kluwer Academic Publishers).

116 **Knotek, O., Bohmer, M.** and **Leyendecker, T.** On structure and properties of sputtered Ti and Al based hard compound films, *J. Vac. Sci. Technol.*, 1986, A4, 2695–2700.

117 **Jehn, H.A., Hofmann, S., Ruckborn, V.-E.** and **Munz, W.-D.** Morphology and properties of sputtered (Ti, Al)N layers on high speed steel substrates as a function of deposition temperature and sputtering atmosphere, *J. Vac. Sci. Technol.*, 1986, A4, 2701–2705.

118 **Bienk, E.J., Reitz, H.** and **Mikkelsen, N.J.** Wear and Friction Properties of Hard PVD Coatings, *Surf. Coat. Technol.*, 1995, **76–77**, 475–480.

119 **Sproul, W.D.** Multilayer, multicomponent and multiphase physical vapor deposition coatings for enhanced performance, *J. Vac. Sci. Technol.*, 1994, A12, 1595–1601.

120 **Vancoille, E., Celis, J.P.** and **Roos, J.R.** Dry sliding wear of TiN-based ternary PVD coatings, *Wear*, 1993, **165**, 41–49.

121 **Gissler, W.** Structure and properties of Ti-B-N coatings, *Surf. Coat. Technol.*, 1994, **68/69**, 556–563.

122 **Ronkainen, H., Holmberg, K., Fancey, K., Matthews, A., Matthes, B.** and **Broszeit, E.** Comparative tribological and adhesion studies of some titanium-based ceramic coatings, *Surf. Coat. Technol.*, 1990, **43/44**, 888–897.

123 **Knotek, O., Loffler, F.** and **Kramer G.** Multicomponent and multilayer physical vapour deposited coatings for cutting tools, *Surf. Coat. Technol.*, 1992, 54/55, 241–248.

124 **Sproul, W.D.** Physical vapour deposition tool coatings, *Surf. Coat. Technol.*, 1996, **81**, 1–7.

125 **Jehn, H.A., Rother, B.** and **Hoffman, S.** Preparation and homogeneity of multicomponent thin films deposited by vacuum and plasma-assisted processes. 12th Italian Conference on *Vacuum Science and Technology*, March 1993, Bolzane, Italy, Vuoto, 1994, 23, pp. 10–17.

126 **Pauleau, Y., Marechal, N., Juliek, P., Rouzaud, A., Zimmermann, C.** and **Gras, R.** Sputter deposited lubricant thin films for high temperature applications, J. Soc. Tribologists and Lubrication Engineers, *STLE Lubrication Engineering*, 1995, **52**, 481–487.

127 **Zabinski, J.S., Day, A.E., Donley, M.S., Dellacorte, C.** and **McDevitt, N.T.** Synthesis and characterisation of high temperature oxide lubricant, *J. Mat. Sci.*, 1994, **29**, 5675–5879.

128 **Ronkainen, H., Koskinen, J., Likonen, J., Varjus, J.** and **Vihersalo, J.** Characterization of wear surfaces in dry sliding of steel and alumina on hydroginated and hydrogen-free carbon films. *Diamond and Related Materials*, 1994, **3**, 1329–1336.

129 **Neuville, S.** and **Matthews, A.** Hard carbon coatings: the way forward. To be published.

130 **Ronkainen, H.** and **Holmberg, K.** Friction and wear properties on dry, water and oil lubricated DLC against alumina and DLC against steel contacts. To be published.

131 **Drees, D., Celis, J.P., Dekempeneer, E.** and **Meneve, J.** The electrochemical and wear behaviour of amorphous diamondlike carbon coatings and multilayered coatings in aqueous environments. To be published.

132 **Miyake, S., Miyamoto, T.** and **Kaneko, R.** Microtribological improvement of carbon film by silicon inclusion and fluorination. *Wear*, 1993, **168**, 155–159.

133 **Grischke, M., Bewilogua, K., Trojan, K.** and **Dimigen, H.** Application-oriented modifications of deposition processes for diamond-like-carbon-based coatings. *Surface and Coatings Technology*, 1995, **74–75**, 739–745.

Abrasive and erosive wear tests for thin coatings: a unified approach

I M HUTCHINGS
Department of Materials Science and Metallurgy, University of Cambridge, UK

SYNOPSIS

Abrasive and erosive wear tests are increasingly applied to surface engineered components, since they have the potential for use as quality assurance methods as well as for more fundamental tribological characterisation. A simple theoretical treatment of such tests is outlined, in which the concept of the 'tribological intensity' of the test conditions is introduced. Available test methods are reviewed and their suitability for thinly coated samples is discussed. There is considerable scope for further development of tests since only a few are satisfactory for these important applications.

NOTATION

A	apparent area of contact in abrasive wear
b	chordal radius of spherical wear scar (see Figure 8)
E	mass removed by erosion per unit mass of erodent particles
n	total mass of erodent particles passing along nozzle
$\dot{m}$	mass flow per unit time along erosion nozzle
N	normal load applied to sliding contact
N_C	normal load carried by coating
N_S	normal load carried by substrate
$\overline{N}_C$	mean load carried by coating over test duration
$\overline{N}_S$	mean load carried by substrate over test duration
P	nominal contact pressure (= N/A)
q	volume lost by wear per unit sliding distance
Q_C	'erosion durability' of coating (see equation 18)
R	radius of spherical counterbody
$\mathbf{r}$	position vector (see Figure 1)
r_C	radius of eroded area from which coating has been removed
S	total sliding distance in abrasion test
t	time
T	test duration
T_C	time elapsed at which coating is penetrated
U	relative sliding velocity in abrasive wear
v	impact velocity of erodent particles
V	volume of material removed by abrasive wear
V_C	volume worn from coating
V_S	volume worn from substrate
x	displacement of surface due to wear (see Figure 1)
α	defines divergence of particle flux in erosion test
θ	impact angle of erodent particles
ϕ	flux of erodent particles (mass per unit area per unit time)
κ	specific wear rate or dimensional wear coefficient
κ_C	κ for coating
κ_S	κ for substrate
ρ	density of material removed by erosion
ψ	'tribological intensity' of test conditions

1 INTRODUCTION

In many engineering applications, surface coatings and other methods of surface treatment are used to increase the lives of components exposed to abrasion or erosion. Reproducible and well-characterised methods are therefore needed to determine the resistance of surface engineered materials to these types of wear (**1**).

Reliable methods are also required to assess the mechanical properties of engineered surfaces, in order to provide quality assurance in the manufacturing process and quality control in finished products.

For both these purposes, there is thus a considerable incentive to develop innovative methods for abrasion and erosion testing, and much progress has been made in this area over the past two decades. It is the aim of this paper to review these advances, to highlight problem areas which remain to be addressed, and to provide some pointers to future progress.

Coatings and surface treatments vary greatly in the depth to which they affect material properties. Depths or coating thicknesses range from less than 0.1 μm in the case of ion implantation, to tens or even hundreds of millimetres for some weld hardfacings (**2–4**). In terms of wear testing, and also service performance, a thick homogeneous coating can be treated as if it were effectively a bulk material. Thin coatings, in contrast, pose particular problems since an accelerated wear test is likely to lead to coating penetration. The terms 'thick' and 'thin' used here must be defined. A thick coating (or deep surface treatment) is one which is not penetrated during the service life of the component or during the wear test, and in which the properties of the substrate do not significantly influence the tribological performance. A thin coating, or shallow surface treatment, will often be penetrated in testing even if not in service, and the tribologist will then be faced with the problem of extracting information about the performance of the coating from data which may relate at the beginning of the test to the coating alone, but later, as the coating wears, to some composite performance of both coating and substrate. This problem is central to the test methods we shall consider below.

A further problem results from the very small mass changes associated with removal of a thin surface layer. Complete removal of a typical coating a few micrometres thick over an area of 1 cm^2 will cause a mass change only of the order of 1 mg. Many wear test methods intended for bulk samples rely on weighing to measure the extent of wear, but for thin coatings there are clear advantages in designing a test method in which wear is detected in some other, potentially more accurate, way.

2 THEORETICAL BACKGROUND

In comparing the various types of abrasion and erosion test which can be applied to thin coatings, it is useful to start from a common theoretical basis, which will be developed here.

The rate of recession of the specimen surface due to wear, dx/dt as defined in Figure 1, can be expected to depend on two factors: the tribological conditions to which the surface is subjected, and the response of the material to these conditions. Although it is recognised that the rate of wear in general depends on all aspects of the tribological system (**5,6**), under the well-defined conditions of a wear test it may be justifiable to treat the influences of the test conditions and the material as independent, and write:

$$\frac{dx}{dt} = \psi(r)\kappa(r) \tag{1}$$

Here $\psi(\mathbf{r})$ describes the conditions causing wear at the point on the surface defined by the position vector $\mathbf{r}$ (Figure 1), and $\kappa(\mathbf{r})$describes the response of the material at that point.

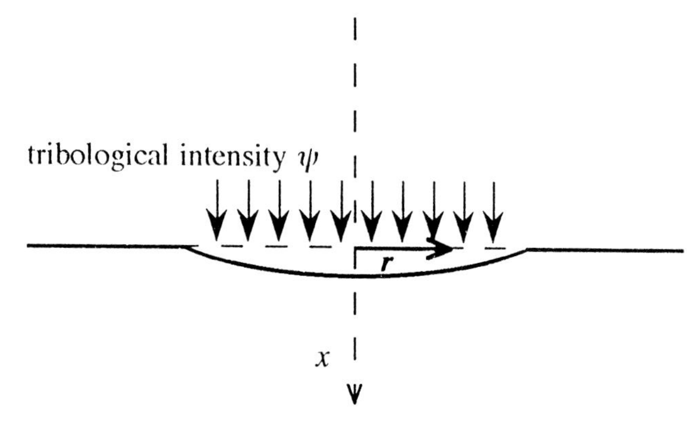

Fig. 1 Schematic illustration of the recession of the specimen surface during a wear test, in which it is subjected to abrasion or erosion with tribological intensity ψ.

For consistency with previous treatments of wear the quantity κ, which is associated with the behaviour of the material under the test conditions, will be taken to be the 'specific wear rate' or 'dimensional wear coefficient', commonly expressed in the context of sliding abrasion as volume loss per unit sliding distance, per unit normal load on the contact (**2**). κ thus has dimensions $[\text{mass}]^{-1} \times [\text{length}] \times [\text{time}]^2$, and its value is often quoted in units of $mm^3m^{-1}N^{-1}$. The 'wear resistance' of the material under these test conditions can conveniently be described by $1/\kappa$.

The quantity ψ describes the severity of the mechanical stimulus applied to the surface which results in wear; we will call it the 'tribological intensity' of the test. ψ has dimensions $[\text{mass}] \times [\text{time}]^{-3}$.

The assumption implicit in equation (1), that the material's 'wear resistance' and the 'tribological intensity' of the test can be combined as independent factors in producing wear, is of course a gross simplification, but is nevertheless commonly made in theoretical models of wear. κ is often effectively constant over quite large ranges of experimental conditions, although any change in wear mechanism will usually cause a substantial change in κ and invalidate the assumptions of equation (1). We shall examine below how this equation can be further developed and used to analyse various abrasive and erosive wear tests.

The most straightforward type of test to interpret involves subjecting the whole of the specimen surface, or a well-defined part of it, to uniform abrasion or erosion under conditions of constant tribological intensity. ψ is then constant over the whole wearing area, and for a homogeneous specimen equation (1) can be integrated to give the depth of material removed after a test of duration T:

$$x(T) = \psi \int_0^T \kappa dt = \psi\kappa T \tag{2}$$

This depth, which in an idealised case will be uniform over the whole worn area, can in principle either be measured directly or deduced from a measurement of mass loss, and the specific wear rate κ can then be readily calculated.

If the test is performed on a sample carrying a coating of thickness x_C and the coating is penetrated during the test, then the total wear depth x at the end of the test will be related to the wear coefficients of the coating κ_C and the substrate κ_S by:

$$x(T) = \psi\left(\int_0^{T_c} \kappa_c dt + \int_{T_c}^{T} \kappa_s dt\right) = x_c + \psi \int_{T_c}^{T} \kappa_s dt \tag{3}$$

where T_c is the time elapsed at the moment at which the coating has just been worn through. Interpretation of the test results is then more complex than in the former case, since the wear coefficient of either the coating κ_c, or the substrate κ_s, or the time at which the coating is penetrated

T_c must be known independently in order to determine κ_c from measurement of the final depth of wear or the mass loss.

3 ABRASIVE WEAR

3.1 Tribological intensity in abrasive wear

The simplest model for abrasive wear under sliding conditions, as derived for example by Rabinowicz (**7**), suggests that the volume lost from the surface per unit sliding distance, q, is linearly proportional to the normal load, N:

$$q = \kappa N \tag{4}$$

where the constant of proportionality κ has the same meaning as κ in equation (1). Simple manipulation leads to an expression for the rate of linear recession of the surface, dx/dt:

$$\frac{dx}{dt} = \kappa \frac{UN}{A} \tag{5}$$

where A is the (apparent) area of contact between the sliding surfaces and U is the relative sliding velocity. Equation (5) can be compared with equation (1) to show that:

$$\psi = UP \tag{6}$$

where P is the nominal contact pressure ($P = N/A$). For sliding abrasion, the 'tribological intensity' y can thus be interpreted in physical terms as the product of sliding speed and contact pressure.

Several practical abrasion tests are available in which both ψ and the worn area are nominally constant during the whole test. These include the Taber abraser method and various polishing and grinding tests. In other methods, such as the dry sand or wet sand rubber wheel abrasion tests and the ball-cratering method, both ψ and the worn area change continuously as wear progresses. We shall examine these classes of test in turn.

3.2 Abrasion tests with nominally constant ψ

Various polishing and grinding tests constitute a group of methods in which the nominal tribological intensity ψ remains constant over the worn area. In these tests, illustrated schematically in Figure 2, a plane specimen is loaded against the flat surface of a much larger counterface in the presence of abrasive particles, and is subjected to a polishing or grinding action. The method has been used as a wear test for bulk materials and also for coated samples for at least 20 years (**8, 9**). In most cases, the test uses a conventional metallographic polishing machine, and has been promoted by at least one manufacturer of such equipment (**10**). Both rotary (**8, 11–13**)) and vibratory polishing equipment (**11, 14, 15**) have been used. Abrasive particles have included silica sand (**12**), silicon carbide and alumina abrasives (**13, 14**) and diamond polishing grits with a wide range of sizes (**11, 16**).

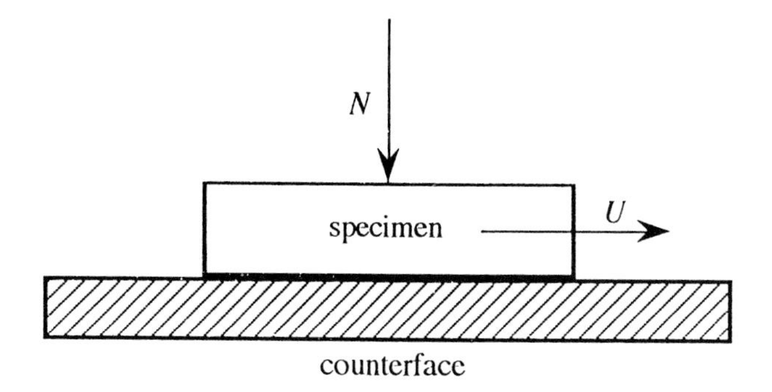

Fig. 2 Polishing wear test, in which the specimen moves with velocity U over a plane counterface under normal load N in the presence of abrasive particles.

The progress of wear is usually determined by periodically interrupting the test and weighing the sample, and with sufficiently gentle abrasion and accurate weighing this method can detect the removal of very small depths of material. For example, the use of a microbalance with a reproducibility of 5 μg can result in measurement errors corresponding to a depth of only 5 nm in steel, although the depth of penetration of the abrasive particles may be considerably greater than this and provide the practical limit to sensitivity (**11**). Nevertheless, changes in wear resistance with depth over a few tens of nanometres have been reliably reported in polishing wear studies on ion-implanted metals, in which peak wear resistance was achieved at a depth of about 100 nm (**11, 17, 18**).

Ideally, a polishing wear test carried out on an evenly coated (or surface-modified) sample should produce progressive wear of the coating, leading eventually to complete and even removal of the coating over the whole specimen area. Several critical studies of the polishing method for ion-implanted metals (**17, 18**), hard coatings (**16**) and bulk metals (**15**) have revealed that this is unfortunately not the case, and that wear is generally not uniform over the specimen surface. In some cases, wear is most rapid at the specimen edge, while in other work most wear has been observed at the centre. These variations in wear rate may have several origins. It has been suggested that in the case of a coated specimen, residual stress variation across the surface may influence the local wear resistance (**16**), but the fact that non-uniform wear is also seen in ion-implanted and bulk samples suggests that it must also be due to a variation in the tribological intensity over the specimen. This in turn may be ascribed to uneven contact pressure on the sample (**16, 18**), while the complexities of fluid-particle flow through the gap between polishing disc and specimen are also likely to play a role.

When the wear is uneven, the mass loss after the coating has been first penetrated will contain contributions from both substrate and coating, as illustrated schematically in Figure 3. Interpretation of the resulting data is then difficult, unless the distribution of contact pressure or wear rate between the coating and substrate is known (**19**).

The Taber abraser, shown schematically in Figure 4, provides a further test method in which the tribological intensity ψ and the worn area both remain nominally

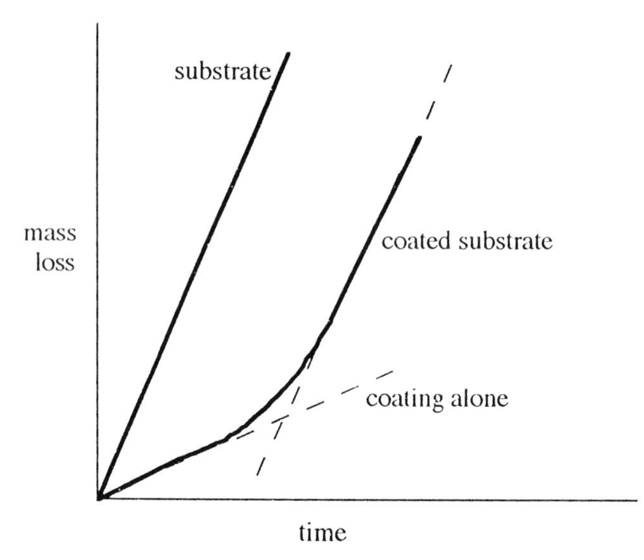

Fig. 3 Comparison of the response of a coated specimen, and an uncoated substrate, to an abrasive wear test. The coating is penetrated unevenly and the subsequent mass loss contains contributions from both substrate and coating.

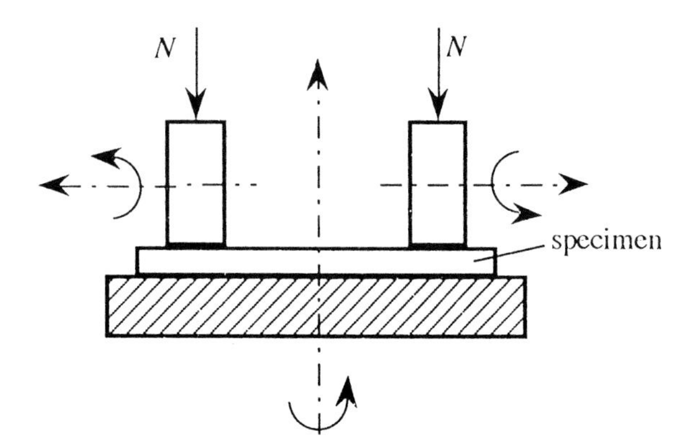

Fig. 4 The proprietary Taber abraser test, in which a coated specimen is rotated about a vertical axis beneath two counter-rotating abrasive wheels. The specimen is a sheet typically 100 mm across (not to scale).

constant during the test. In this procedure a plane specimen is rotated at a speed of 1 revolution per second about a vertical axis beneath a pair of composite abrasive wheels which rotate freely and independently about a horizontal axis which does not intersect the axis about which the specimen is driven. Various types of wheels are available, containing abrasive grit particles (usually silicon carbide) dispersed in (most commonly) an elastomeric binder; the different grades of wheel contain different binder materials and sizes of grit. The abrasive wheels, which are pressed against the specimen surface under a dead load (typically in the range 1 to 10 N on each wheel), roll and slide against the specimen and produce an abraded annular wear track some 12 mm wide (**20**).

The use of the Taber tester has been standardised by the American Society for Testing and Materials (ASTM) for organic paint coatings (**21**), but it is also quite widely applied to other coated systems. For example, it has been used for relatively thick plasma-sprayed alumina coatings (**20**) and electroplated and electroless plated metals (**22**), as well as for much thinner PVD titanium nitride (**23, 24**) and more complex nitride and carbide coatings (**25**). The extent of wear is usually assessed by weighing.

The Taber test results in two-body abrasion, and in theory at least, constant tribological intensity. In practice the reproducibility of the test between laboratories is disappointing (**21**). This appears to be largely because of variability in the properties of the abrasive wheels, the hardness of which changes with time as the binder compound ages. Added variability may result from the need to dress the surfaces of the wheels in a standard way before each test in order to expose fresh abrasive; from frictional heating of the wheels and specimen during the test; and from the influence of the wear debris which is transferred to some extent to the wheel surface, despite measures to remove it continually from the specimen. Although the abraded area remains constant, it is thus clear that ψ will almost certainly change as the test progresses, in a manner which depends to some extent on the properties of the specimen material.

A final method in which ψ would be expected to remain effectively constant has been proposed recently by Axén, *et al.* (**19**), and is illustrated in Figure 5. A hollow cylinder is rotated with its axis normal to the specimen surface, in the presence of a slurry of abrasive particles. As in the Taber test, an annular wear scar results, but the constant supply of fresh abrasive particles to the contact zone should avoid some of the problems which cause ψ to change in that method. The cylinder diameter can be small, so that the wear test can be performed on a specimen area only a few millimetres across. Only preliminary results have so far been published, but the method appears to offer significant potential for further development.

3.3 Abrasion tests with variable ψ and abraded area

Several practical abrasive wear tests involve continual and

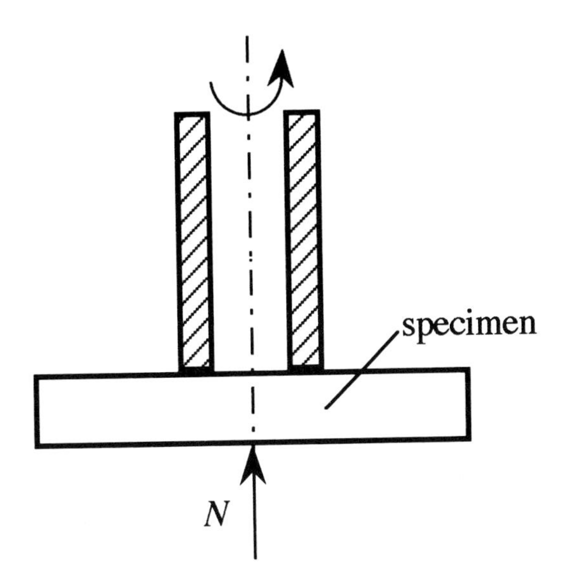

Fig. 5 Abrasive wear test proposed by Axén, *et al.* (19) in which a hollow flat-ended cylinder is pressed and rotated against a plane specimen surface in the presence of an abrasive slurry.

progressive changes in both the tribological intensity during the test and the area exposed to wear.

One important example is the rubber wheel abrasion test, in which a rubber-rimmed wheel slides against the surface of a plane test sample in the presence of abrasive particles. Although the applied load is held constant, as the sample wears the contact becomes more conformal and the area of contact increases; the pressure P and the value of ψ therefore fall. The test, which is shown schematically in Fig. 6, has been standardised by ASTM with dry silica sand as the abrasive (**26**), and is widely used as a standard method for evaluating low-stress abrasion in bulk materials. A short-duration test is specified for use on coatings (Procedure C), although non-standard conditions have also been used by several investigators. A similar test can be performed with an aqueous slurry of silica sand, and has also been standardised (**27, 28**).

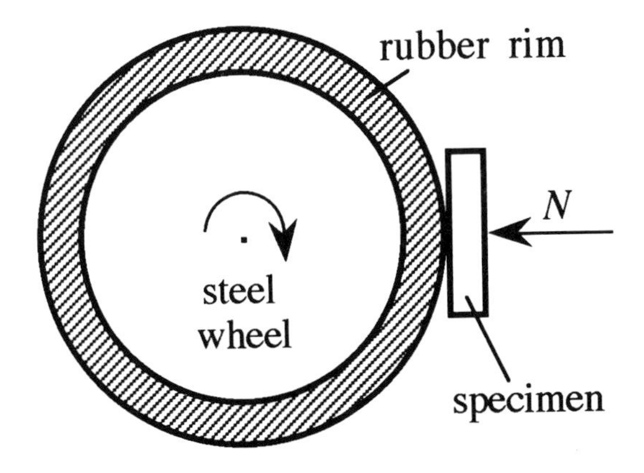

Fig. 6 The dry sand rubber wheel abrasion test (not to scale). A plane specimen is pressed against a rotating rubber-rimmed wheel, 230 mm in diameter. Abrasive particles (silica sand) are fed into the gap between specimen and wheel and are dragged through the contact zone.

When a thick coating is tested by rubber wheel abrasion, wear may be confined to the coating alone. The ASTM test has been used, for example, to provide useful information on the relative abrasion resistance of various electroplated and electroless plated metallic coatings, all at least 40 μm thick, as well as anodized aluminium (**29**). When the test is applied to thin coatings, however, the coating is rapidly penetrated and the specimen mass loss involves contributions from both substrate and coating (**30**). These contributions cannot be separated unless the variation of tribological intensity ψ over the worn area is known. This in turn will depend on the distribution of contact pressure over the wear scar. Although in principle this might be derived from an analysis of the elastic deformation of the rubber wheel in contact with the worn profile of the specimen surface, such an analysis has not, as far as is known, been attempted.

It must be concluded that if the coating is penetrated during the test, interpretation of the mass loss data from a rubber wheel abrasion test is not straightforward, and that for this reason such methods are not suitable for use on thin coatings.

An important group of tests consists of those in which ψ and the worn area change during the test, but in which the geometry of the wear scar is imposed by the test apparatus. An example of such a method is the 'ball-cratering' test illustrated in Figure 7a, in which a hard sphere is pressed and rotated against the specimen surface in the presence of an abrasive slurry. In a typical implementation of the test a steel ball about 25 mm in diameter rests against the surface of the specimen and is driven by friction from a rotating shaft. The normal force acting on the specimen derives from the weight of the ball, and can be varied to a limited extent by moving the drive shaft relative to the specimen. Other test methods can be devised which employ the same principle, such as the use of a rotating cylindrical disc (Figure 7b). A rotating disc with a spherically domed rim can also be used, in combination with rotation of the specimen about an intersecting perpendicular axis (Figure 7c). With the correct rim profile this then produces a wear crater in the form of a spherical cap, of the same form as that produced by the ball-cratering method. This method is often referred to as a 'dimple grinder' test, since it can be carried out with commercially available equipment more commonly used for the mechanical thinning ('dimpling') of specimens for transmission electron microscopy.

The important distinction between all these methods and the rubber wheel abrasion test lies in their use of a much more rigid counterbody. Elastic deformation of a rubber wheel results in displacements of the wheel rim which are comparable to the depth of the wear scar in the

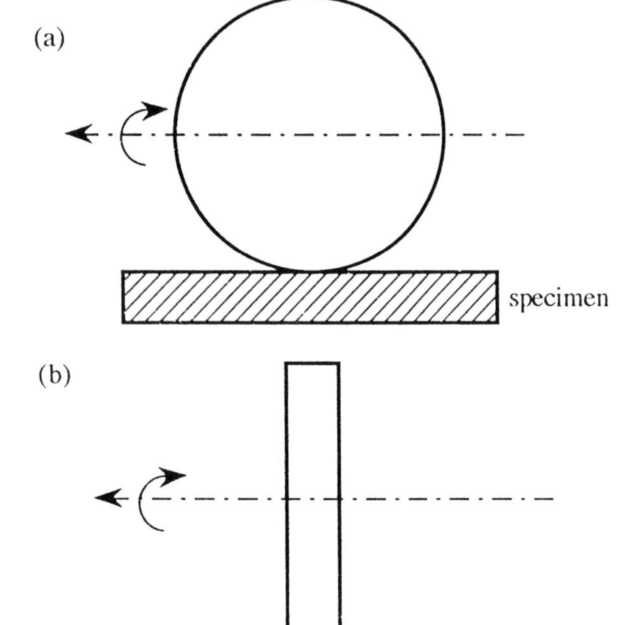

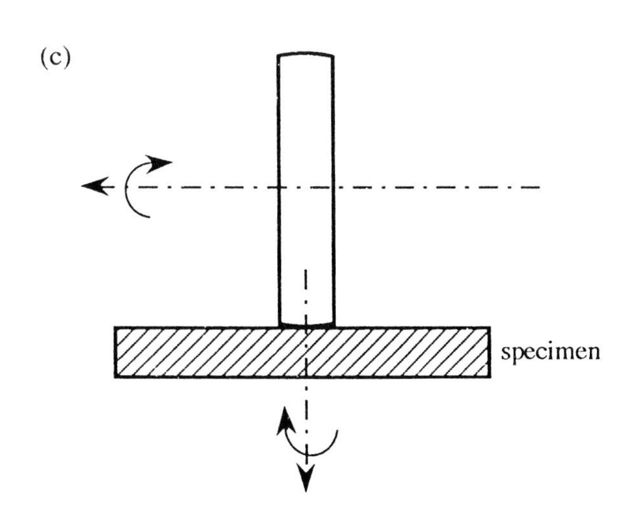

Fig. 7 Examples of abrasion tests which produce imposed wear scar geometries. In each case a hard counterbody is pressed and rotated against the specimen surface in the presence of a slurry of abrasive particles. (a) Spherical counterbody: 'ball-cratering' test. (b) Cylindrical counterbody. (c) Domed cylindrical counterbody combined with rotation of the specimen about a perpendicular axis: 'dimple grinder' test.

specimen. The shape of the scar is thus determined by both the distortion of the wheel and the wear resistance of the specimen. With a rigid counterbody, in contrast, any local reduction in wear resistance in the specimen will lead to a local increase in wear rate, which will however increase the gap between the specimen and counterbody at that point, reducing the contact pressure and thus reducing the wear rate. This mechanical feedback results in a wear scar with a conformal geometry which almost precisely replicates the shape of the counterbody, as illustrated in Figure 8.

The use of a cylindrical steel wheel for abrasion testing (Figure 7b) dates back to the work of Brinell in 1921 (**31**), and the principle of an imposed wear scar geometry has been used since at least 1956 to investigate depth profiles in modified surfaces and to measure the thicknesses of thin coatings (**32, 33**); it has been standardised by ASTM for the latter purpose (**34**). The use of a spherical counterbody was first reported in 1979 (**35**), and commercial ball-cratering equipment is now widely available for the determination of coating thickness. It is only recently, however, following the important work of Kassman, *et al.* (**36**), that abrasive wear tests which utilise the principle of imposed wear scars have become established.

For thick coatings it is possible to perform abrasive wear tests by the imposed wear scar method without penetrating the coating. In that case, or for bulk uncoated samples, it is straightforward to measure the dimensions of the wear scar, and to note that equation (5) can be recast in the following form:

$$A\frac{dx}{dt} = \frac{dV}{dt} = UN\kappa \tag{7}$$

where V is the volume of material removed by wear. With an imposed wear scar geometry, V can be calculated from either the scar width or its depth, and the value of κ simply derived provided that N remains constant during the test and is known.

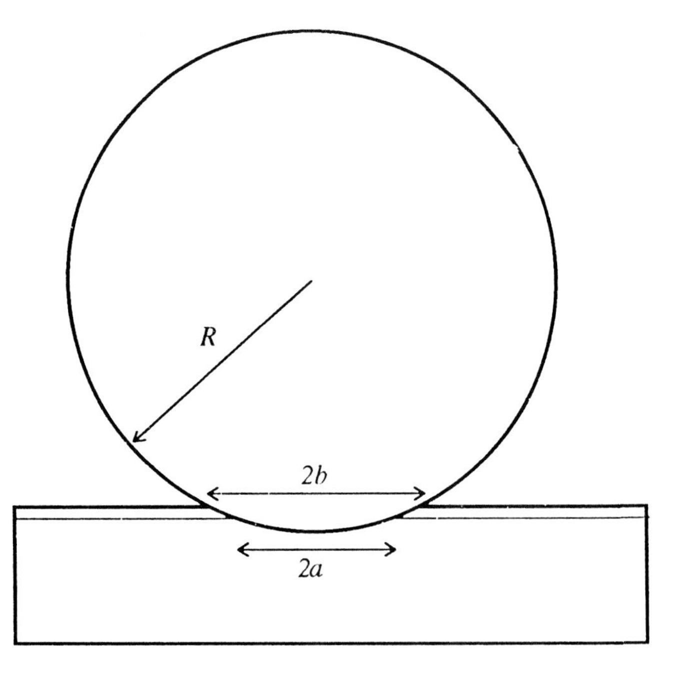

Fig. 8 Geometry of an imposed spherical wear scar on a coated sample, as produced by the ball-cratering or dimple grinder test.

For example, the volume of a spherical scar with chordal diameter $2b$ (Figure 8) is given to a close approximation by:

$$V \approx \frac{\pi b^4}{64R} \tag{8}$$

for $b<<R$, where R is the radius of the spherical counterbody. If the total sliding distance during the test is S, equation (7) can be integrated over the test duration and combined with equation (8) to give:

$$SN \approx \frac{1}{\kappa}\left(\frac{\pi b^4}{64R}\right) \quad \text{for } b<<R \tag{9}$$

A plot of b^4 against S should therefore be linear, and the value of κ can be readily deduced from its slope. Results which confirm the validity of this approach have been reported from tests with both the dimple-grinder and the ball-cratering apparatus on a wide range of bulk specimens and thick coatings: materials have included silicon single crystals, high speed steel and cemented carbide (**36**); soda-lime glass, various engineering ceramics, copper and aluminium (**37**); ultra-high molecular weight polyethylene for orthopaedic prostheses (**38**); and relatively thick polymeric paint films (**39**).

When thinly coated samples are tested, the substrate is often revealed at an early stage of the test unless particularly low loads and/or short test durations are employed. The fact that the load applied by the rotating counterbody is then partitioned between coating and substrate, and that this distribution changes during the test, would seem to suggest that derivation of the wear coefficients for the coating and substrate via equation (5) would not be straightforward. However, equation (7) will apply to both the coating and the substrate separately, so that we can write:

$$\frac{dV_s}{dt} = UN_s\kappa_s \tag{10a}$$

$$\frac{dV_c}{dt} = UN_c\kappa_c \tag{10b}$$

which can be integrated over the test duration to yield:

$$V_s = S\overline{N}_s\kappa_s \tag{11a}$$

$$V_c = S\overline{N}_c\kappa_c \tag{11b}$$

where $V_{s,c}$ are the total volumes removed by wear from the substrate and coating respectively, and $N_{s,c}$ are the mean loads supported by each component over the duration of the test. Since the total normal load, N is equal to ($N_c = N_s$) at all times, equations (11a) and (11b) can be combined to give:

$$SN = \frac{V_c}{\kappa_c} + \frac{V_s}{\kappa_s} \tag{12}$$

This important equation was first derived by Kassman, *et al.* (**36**) who used it to deduce values of κ_C and κ_S from abrasive wear experiments performed with a dimple grinder (Figure 7c) on TiN and TiC coated steel and cemented carbide substrates. For a spherical wear scar geometry the volumes V_C and V_S can be readily computed from measurements of scar diameter made on the surface of the specimen. Although Kassman, *et al.* performed a supplementary test on an uncoated substrate to measure κ_S, an alternative method of data reduction can be used by which the values of both κ_C and κ_S can be determined from a single test on a coated sample (**37**).

The ball-cratering and dimple-grinder methods have been further developed and used extensively over the past five years to assess the abrasive wear resistance of thin hard coatings (**40–46**); it should be noted that these methods are not restricted to use on plane specimens, but can be applied successfully to any coated surface, even one with compound curvature (**47**). Both methods, and recent refinements, have been comprehensively reviewed by Rutherford and Hutchings (**48**) and Gåhlin, *et al.* (**49**). In particular, important advances have been made in understanding the influences of imperfections in the wear scar geometry, which are a disadvantage of the dimple-grinder method (**49**), and in optimising the method of data analysis to reduce errors (**37, 48**).

There is still further scope for development of this type of abrasive test, however. For example, in the most common implementation of the ball-cratering method, the normal force acting on the specimen is linked to the weight of the ball and can be varied within only a small range; it must be measured during the test since it is influenced by the friction between the specimen and the ball (**37**). For some purposes it may be useful to vary the load more widely, and alternative implementations of the test are possible in which the ball is driven positively through a shaft and the load is applied through a lever arm. Such an arrangement also makes it possible to measure the depth of penetration of the ball continuously during the test (**50**). Although throughout their original work on the dimpler test Kassman, *et al.* expressed their theoretical analysis in terms of wear scar depth, they and subsequent investigators of both the dimple-grinder and the ball-cratering test actually measured the surface diameter of the scar. There may be significant advantages in modifying the test so that depth is the primary measurement, and recent work has shown that the practical problems involved in making such a delicate measurement continuously during the test can be solved (**51**).

4 EROSIVE WEAR

4.1 Tribological intensity in erosive wear

In order to derive an expression for the tribological intensity ψ associated with an erosive wear test, we shall assume a simple model for E, the mass lost from a surface due to the impact of unit mass of abrasive particles at velocity v (**2**):

$$E = \frac{\kappa}{2}\rho v^2 \tag{13}$$

where ρ is the density of the surface material. Here κ has the same interpretation as in equation (1). If the flux of erodent particles ϕ represents the mass of particles striking unit area in unit time, the rate of linear recession of the surface, dx/dt, at any point on the surface can be expressed as:

$$\frac{dx}{dt} = \frac{E\phi}{\rho} \tag{14}$$

On substituting for E from equation (13), and comparing the resulting expression for dx/dt with equation (1), we deduce that for erosive wear the tribological intensity ψ is given by:

$$\psi = \tfrac{1}{2}\phi v^2 \tag{15}$$

ψ is thus equal to the kinetic energy of the particles striking unit area in unit time.

The model used to derive equation (13) is very simple; in practice, E is usually found to depend on a higher power of v (typically about 2.4), and also on the impact angle θ. A more general expression for ψ is therefore:

$$\psi = \phi f(v, \theta) \tag{16}$$

For the purposes of the present discussion, it is important to note that the tribological intensity ψ is linearly proportional to the particle flux ϕ for constant v and θ, and that it is a strong function of the particle velocity v.

4.2 Erosion tests

Although several other types of erosive wear test have been devised, only two basic designs are in widespread use. In the gas-blast rig (Figure 9), erodent particles are accelerated along a nozzle in a flowing stream of gas, usually air. The nozzle may be a parallel cylinder, as specified for example in an ASTM standard (**52**), or may have a more complex shape. The specimen is mounted at a fixed stand-off distance from the end of the nozzle. The

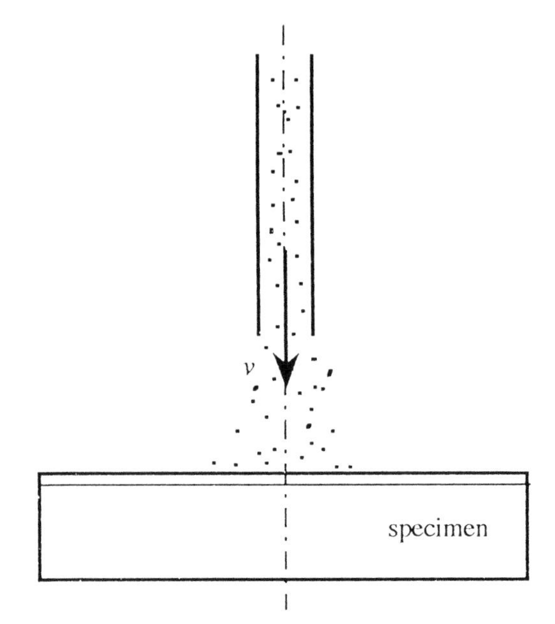

Fig. 9 Gas-blast erosion test in which particles are accelerated along a parallel cylindrical nozzle and strike a plane coated specimen.

centrifugal accelerator, shown schematically in Figure 10, uses rotation to accelerate particles fed into the centre of the rotor, and produce a stream of particles in a (usually) horizontal plane. The apparatus can be used to test more than one specimen simultaneously, held at suitable angles around the periphery of the rotor.

Both methods produce divergent streams of particles, so that the particle flux ϕ will in general vary over the

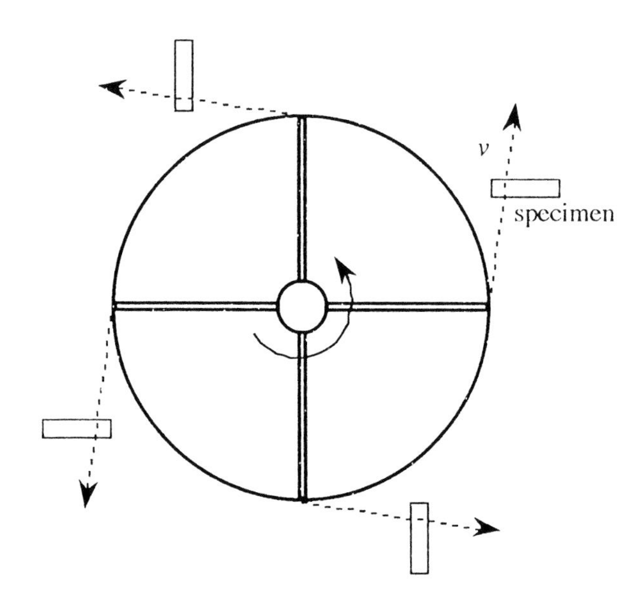

Fig. 10 Centrifugal erosion accelerator in which particles are fed into the centre of the rotor and move outwards along radial channels. On leaving the rim of the rotor they strike fixed specimens (of which four are shown) mounted around the circumference.

surface of the specimen. The impact velocity and angle may also vary over the specimen, so that significant variation of the tribological intensity ψ is to be expected. Some investigators have masked the specimen area so that ψ varies little over the exposed area (**23, 53–55**), and very few have attempted to quantify the distribution of flux over the eroded area (**56, 57**); most, however, appear to have ignored the effect.

Shipway and Hutchings (**57**) have recently reviewed the application of solid particle erosion testing to coated samples. As in abrasive wear testing, by far the most common method of quantifying the extent of wear is by mass loss. Sufficiently thick coatings can be tested in the same way as bulk samples, but as for abrasion, problems arise in the interpretation of the results once the coating is penetrated. Attempts have been made to separate the mass loss due to the coating from that of the substrate, but the method depends on detecting a change in the rate of mass loss as the substrate is revealed, and can be used successfully only if the erosion rate of the substrate is significantly greater than that of the coating, or if there is little variation of ψ over the specimen surface (**58**).

Some investigators have accepted penetration of the coating as inevitable, and used the mass of erodent particles, or the time of exposure to the particle stream, required to remove the coating completely, as a measure of its erosion resistance (**59–61**). Alternative approaches, which do not demand continuous observation of the specimen during the test and are thus less prone to operator error, can also be used to assess the durability of the coating in terms of the dose of particles required to penetrate it. Both energy-dispersive X-ray microanalysis (EDX) and back-scattered SEM imaging have been applied to a small area of the eroded specimen, over which ψ can be assumed to be effectively constant, to quantify the extent of coating removal (**56, 62**); the erosion resistance of a coating has also been described by the particle dose just required to expose 50% by area of the underlying substrate (**42**).

If the durability of a coating under erosion conditions is of interest, however, it is better to measure the dose required to remove the coating completely; Hedenqvist and Olsson (**63**) demonstrated in principle how this could be done by exposing a specimen to a spatially varying flux of erodent particles, and noting the position on the surface at which the coating has been fully penetrated. Hedenqvist and Olsson used a centrifugal accelerator and inferred the distribution of particle flux over the specimen surface by eroding glass specimens. An essentially similar approach has been further developed by Shipway and Hutchings using a gas-blast apparatus (**57**).

The particles leaving a long parallel nozzle in a gas-blast erosion apparatus form a divergent plume, the properties of which can be deduced from erosion experiments on coated specimens (**64**). To a good approximation, the flux of particles $\phi(r)$ striking the specimen at a radial distance r from the nozzle axis is:

$$\phi(r) \approx \frac{\dot{m}\alpha^2 \exp(-\alpha r)}{2\pi} \tag{17}$$

where $\dot{m}$ is the mass feed rate of particles along the nozzle, and α defines the divergence of the particle flux. The divergence of the particle stream is radially symmetrical, and as the coating is penetrated there is a circular boundary on the specimen between the central region over which the substrate is revealed and the outer region over which coating material is still present. It can be shown that the radius of this circular wear scar, r_C, after a total mass m of particles have passed down the nozzle is given by (**57**):

$$r_c = \frac{1}{\alpha} \ln m - \frac{1}{\alpha} \ln\left(\frac{2\pi Q_c}{\alpha^2}\right) \tag{18}$$

where Q_C is the dose of particles (mass per unit area) required to remove the coating. A plot of r_C against the logarithm of the mass of erodent particles should therefore be linear, with slope $1/\alpha$, and Q_C can be derived from the intercept of the line. Equation (18) has been found to be valid for many different coating systems tested by the gas-blast erosion method with a parallel cylindrical nozzle; these include thin hard PVD coatings, anodized films on aluminium, diamond-like carbon films and polymeric paint coatings (**39, 45, 57, 65, 66**). The value of Q_C, which has been termed the 'erosion durability' of the coated system, has been shown to be a sensitive indicator of coating adhesion for thin hard coatings, and to vary in a systematic way with impact velocity, coating thickness and erodent particle size (**67**).

5 GENERAL DISCUSSION AND CONCLUSIONS

To be useful, a laboratory wear test method must involve a reproducible and well-characterised means of subjecting the sample to abrasion or erosion, coupled with an accurate method of quantifying the extent of wear. In addition, tests which are to be applied to coated or surface engineered materials should allow the wear rate of the coating to be determined independently of the substrate, and this condition imposes significant constraints on the test.

It is important also to acknowledge the purpose of the test. As outlined above, samples may be subjected to abrasion or erosion testing in order to assess their durability when exposed to this type of wear. In this case, it is important that the conditions of the test reproduce, as far as possible, those of the practical application for which the component is intended (**1**). In particular, the influence on wear rate and wear mechanism of the size, shape and most importantly, hardness of the abrasive or erosive particles must be recognised (**2**). In the case of brittle materials, transitions in wear mechanism can occur as the test conditions are varied, which in the extreme case may render the results from a particular wear test completely irrelevant to the behaviour of a material in a given application (**68**).

Wear tests may also be employed in a more general way as part of the process of characterising a material; for example, tribological tests of various kinds combined with measurements of scratch response and information from indentation testing can be used to generate a 'tribological profile' which is particularly relevant to coated systems (**41**). Abrasion and erosion can also be used as controlled methods of mechanically loading the surface of a material, and can be used to generate information on the adhesion of a coating (**56, 57, 63, 66**). In these respects, such tests may have a valuable role to play in the quality assurance of coated or other surface engineered components, although there is still much progress to be made in understanding the complex interaction between test conditions and tribological response, and in relating this response to more fundamental properties such as interfacial adhesion.

Any test which is intended for quality assurance on surface engineered components should ideally be applied directly to those components, and not to samples specially produced for test purposes, even if these have been made from the same material as the production components and surface treated in the same batch. Of the abrasion and erosion tests described above, only a few can readily be used to test sufficiently small areas to be applicable to most coated components; these include the imposed crater geometry abrasion tests (e.g. by ball-cratering), the rotating cylinder abrasion test (**19**) and the erosion test described by Shipway and Hutchings (**57**).

It is clear that there has been significant progress over the past twenty years in devising and developing new approaches to the wear testing of surface engineered materials, and that these can make a valuable contribution to our understanding and use of these economically important technologies. There is, however, still ample opportunity for further development of versatile and robust wear test methods, and for building on our current understanding of material response to abrasion and erosion.

6 REFERENCES

1 **Bull, S.J.**, Tribological and microtribological phenomena in coatings, *Materials Science Forum*, 1997, **246**, 105–152.

2 **Hutchings, I.M.**, *Tribology: friction and wear of engineering materials*, 1992, (Edward Arnold, London)

3 **Holmberg, K.** and **Matthews, A.**, *Coatings tribology*, Tribology Series **28**, 1994 (Elsevier, Amsterdam).

4 **Bhushan, B.** and **Gupta, B.K.**, *Handbook of tribology*, 1991 (McGraw Hill, New York).

5 **Czichos, H.**, *Tribology – a systems approach to the science and technology of friction, lubrication and wear*, Tribology series **1**, 1978, (Elsevier, Amsterdam).

6 **Celis, J.P.**, A systems approach to the tribological testing of coated materials, *Surface and Coatings Technology*, 1995, **74–75**, 15–22.

7 **Rabinowicz, E.**, *Friction and wear of materials*, 1964 (John Wiley, New York).

8 **Rabinowicz, E.**, Abrasive wear resistance as a materials test, *Lubrication Engineering*, 1977, **33**, 378–381.

9 **Rabinowicz E.**, **Doherty, P.**, and **Boyd, D. M.**, Measurement of the abrasive wear resistance of hard coatings, *Thin Solid Films*, 1978, **53**, 301–302.

10 **Muller, K.** and **Fundal, E.**, Description of the micro wear test and application studies, *Proc. 4th Nordic Symposium on Tribology*, J. Jakobsen, M. Klarskov and M. Eis (eds.), Technical University of Lyngby, Denmark, 1990, pp. 499–515.

11 **Bolster, R.N.** and **Singer I.L.**, Polishing wear studies of coating materials, in *Mechanical properties, performance and failure modes of coatings*, T.R. Shives and M. B. Peterson, eds., 1984, pp. 201–207, (Cambridge University Press, Cambridge).

12 **Leyland, A.** and **Matthews, A.**, Thick Ti/TiN multilayered coatings for abrasive and erosive wear resistance, *Surface and Coatings Technology*, 1994, **70**, 19–25.

13 **Olsson, M., Kahlman L.** and **Nyberg, B.**, Abrasive wear of structural ceramics, *Bull. American Ceramic Soc.*, 1995, **74**, 48–52.

14 **Bunshah, R.F.**, Selection and use of wear tests for coatings, in *Selection and use of wear tests for coatings*, ASTM STP769, R. G. Bayer, ed., American Society for Testing and Materials, 1982, pp. 3–15.

15 **Blau, P.J.**, Needs and challenges in precision wear measurement, *J. Testing and Evaluation*, 1997, **25**, 216–225.

16 **Bromark M.**, A novel abrasive wear test for thin hard coatings, *Tribologia* (submitted for publication), reproduced in Bromark, M., Dissertation for PhD degree, Uppsala University, Sweden, 1996.

17 **Bolster, R.N., Singer I.L.** and **Vardiman, R. G.**, Composition, structure and wear resistance of Ti-6Al-4V implanted with carbon or boron to high doses, *Surface and Coatings Technology*, 1987, **33** 469–477.

18 **Singer, I.L., Bolster, R.N., Pollock, H. M.,** and **Ross, J.D.J.**, Polishing wear behaviour and surface hardness of ion-beam modified Ti-6Al -4V, *Surface and Coatings Technology*, 1988, **36**, 531–540.

19 **Axén, N., Jacobson, S.** and **Hogmark, S.**, Principles for the tribological evaluation of intrinsic coating properties, *Wear*, 1997, **203–204**, 637–641.

20 **Koutsomichalis, A.**, Wear behaviour of alumina plasma-sprayed coating on a Al-Cu alloy, *Materials Letters*, 1993, **18**, 19–24.

21 **American Society for Testing and Materials**, Standard test method for abrasion resistance of organic coatings by the Taber abraser, ASTM D 4060–84, 1984.

22 **Kelley, J.E., Stiglich, J.J.** and **Sheldon, G.L.**, Methods of characterisation of tribological properties of coatings, in *Surface modification technologies*, 1988, eds. T. S., Sudarshan and D.G. Bhat, The Metallurgical Society, pp. 169–187.

23 **Rickerby, D. S.** and **Burnett, P. J.**, The wear and erosion resistance of hard PVD coatings, *Surface and Coatings Technology*, 1987, **33**, 191–211.

24 **Bull, S.J., Rickerby, D.S., Robertson, T.** and **Hendry, A.**, The abrasive wear resistance of sputter ion plated titanium nitride coatings, *Surface and Coatings Technology*, 1988, **36**, 743–754.

25 **Knotek, O., Lugscheider, E., Löffler F., Krämer, G.** and **Zimmermann, H.**, Abrasive wear resistance and cutting

performance of complex PVD coatings, *Surface and Coatings Technology*, 1994, **68–69**, 489–493.

26 **American Society for Testing and Materials**, Standard practice for conducting dry sand/rubber wheel abrasion tests, G65–85, 1985.

27 **American Society for Testing and Materials**, Test method for conducting wet sand/rubber wheel abrasion tests, G105–89, 1989.

28 **Saltzman, G.A.**, Wet-sand rubber-wheel abrasion test for thin coatings, in *Selection and use of wear tests for coatings*, ASTM STP769, R. G. Bayer, ed., American Society for Testing and Materials, 1982, pp. 71–91.

29 **Budinski, K.G.**, Wear characteristics of industrial platings, in *Selection and use of wear tests for coatings*, ASTM STP769, R. G. Bayer, ed., American Society for Testing and Materials, 1982, pp. 118–133.

30 **Sirvio, E. H., Sulonen, M.** and **Sundquist H.**, Abrasive wear of ion-plated titanium nitride coatings on plasma-nitrided steel surfaces, *Thin Solid Films*, 1982, **96**, 93–101.

31 **Brinell, J.A.**, Undersöknig rörande järns och stals samt en del andra kroppars förmåga att motstå nötnig, *Jernkontorets Annaler*, 1921, **76**, 347–398.

32 **Happ, W.** and **Shockley, W.**, Diffusion depths in silicon measured using cylindrical grooves, *Bull. American Physical Soc.*, 1956, **1**, 382.

33 **McDonald, B.** and **Goetzberger, A.**, Technique for the determination of the depth of diffused layers, *Journal of the Electrochemical Society*, 1962, **109**, 141–144.

34 **American Society for Standards and Materials**, Standard test method for measurement of surface layer thickness by radial sectioning, ASTM Standard E 1182–87, 1987.

35 **Thomson, V., Hintermann, H.E.** and **Chollet, L.**, The determination of composition depth profiles using spherical erosion and scanning Auger electron spectroscopy, *Surface Technology*, 1979, **8**, 421–428.

36 **Kassman, Å, Jacobson, S., Erickson, L., Hedenqvist, P.** and **Olsson, M.**, A new test method for the intrinsic abrasion resistance of thin coatings, *Surface and Coatings Technology*, 1991, **50**, 75–84.

37 **Rutherford, K. L.** and **Hutchings, I. M.**, A micro-abrasive wear test, with particular application to coated systems, *Surface and Coatings Technology*, 1996, **79**, 231–239.

38 **Choudhury, M** and **Hutchings, I.M.**, The effects of irradiation and ageing on the abrasive wear resistance of ultra high molecular weight polyethylene, *Wear*, 1997, **203–204**, 335-340.

39 **Rutherford, K.L., Trezona, R.I., Ramamurthy, A.C.** and **Hutchings, I.M.**, The abrasive and erosive wear of polymeric paint films, *Wear*, 1997, *203-204*, 325–334.

40 **Nothnagel, G.**, Wear resistance determination of coatings from cross-section measurements of ball-ground craters, *Surface and Coatings Technology*, 1993, **57**, 151–154.

41 **Hogmark, S.** and **Hedenqvist, P.**, Tribological characterisation of thin hard coatings, *Proc. 6th Nordic Symposium on Tribology, Nordtrib 1994*, 735–747.

42 **Bromark, M., Larsson, M., Hedenqvist, P., Olsson, M., Hogmark, S.,** and **Bergmann, E.**, PVD coatings for tool applications, *Surface Engineering*, 1994, **10**, 205–214.

43 **Hedenqvist, P., Bromark, M., Olsson, M., Hogmark, S.** and **Bergmann, E.**, Mechanical and tribological characterization of low-temperature deposited PVD TiN coatings, *Surface and Coatings Technology*, 1994, **63**, 115–122.

44 **Bromark, M., Gåhlin, R., Hedenqvist, P., Hogmark, S., Håkansson, G.** and **Hansson, G.**, Influence of recoating on the mechanical and tribological performance of TiN-coated HSS, *Surface and Coatings Technology*, 1995, **76–77**, 481–496.

45 **Rutherford, K.L., Bull, S.J., Doyle, E.D.** and **Hutchings, I.M.**, Laboratory characterisation of the wear behaviour of PVD-coated tool steels and correlation with cutting tool performance, *Surface and Coatings Technology*, 1966, **80**, 176–180.

46 **Rutherford, K.L., Hatto, P.W., Davies, C.** and **Hutchings, I.M.**, Abrasive wear resistance of TiN/NbN multi-layers: measurement and neural network modelling, *Surface and Coatings Technology*, 1996, **86–87**, 472–479.

47 **Rutherford, K.L.** and **Hutchings, I.M.**, Micro-scale abrasive wear testing of PVD coatings on curved substarates, *Tribology Letters*, 1996, **2**, 1–11.

48 **Rutherford, K.L.** and **Hutchings, I.M.**, Theory and application of a micro-scale abrasive wear test, *Journal of Testing and Evaluation*, 1997, **25**, 250–260.

49 **Gåhlin, R., Larsson, M., Hedenqvist, P., Jacobson S.** and **Hogmark, S.**, The crater grinder method as a means for coating wear evaluation – state of the art, *Surface and Coatings Technology*, 1997, submitted for publication.

50 **Klümper-Westkamp, H.** and **Mayr, P.**, Mikrotribometer zur Qualitätssicherung von Randschichten, *Prakt. Met.*, 1995, **Sonderbd. 26**, 433–442.

51 **Rutherford, K.L.** and **Hutchings, I.M.**, 1996, unpublished work.

52 **American Society for Testing and Materials**, Practice for conducting erosion tests by solid particle impingement using gas jets, ASTM standard G76–83, 1983.

53 **Wert, J.J.** and **McKechnie, T.N.**, The effect of composition and process parameters on the erosion resistance of sputtered Ni-TiB2 coatings, *Wear*, 1987, **116**, 181–200.

54 **Kral, M.V., Davidson, J.L.** and **Wert, J.J.**, Erosion resistance of diamond coatings, *Wear*, 1993, **166**, 7–16.

55 **Burnett, P.J.** and **Rickerby, D.S.**, The erosion behaviour of TiN coatings on steels, *Journal of Materials Science*, 1988, **23**, 2429–2443.

56 **Jönsson, B., Akre, L., Johansson, S.** and **Hogmark, S.**, Evaluation of hard coatings on steel by particle erosion, *Thin Solid Films*, 1986, **137**, 65–77.

57 **Shipway, P.H.** and **Hutchings, I.M.**, Measurement of coating durability by solid particle erosion, *Surface and Coatings Technology*, 1995, **71**, 1–8.

58 **Bromark, M., Larsson, M., Hedenqvist, P.** and **Hogmark, S.**, Determination of coating erosion resistance using the mass-loss technique, *Proc. 6th Nordic Symposium on Tribology, Nordtrib '94*, 1994, 207–214.

59 **Shanov, V., Tabakoff, W.** and **Metwally, M.**, Erosive wear of CVD coatings exposed to particulate flow, *Surface and Coatings Technology*, 1992, **54–55**, 25–31.

60 **Garg, D., Dyer, P.N., Dimos, D.B., Sunder, S., Hintermann, H.E.** and **Maillat, M.**, Low-temperature chemical vapor deposition tungsten carbide coatings for wear/erosion resistance, *Journal of the American Ceramic Society*, 1992, **75**, 1008–1011.

61 **Garg, D.** and **Dyer, P.N.**, Erosive wear behaviour of chemical vapor deposited multilayer tungsten carbide coating, *Wear*, 1993, **162–164**, 552–557.

62 **Olsson, M.**, **Hedenqvist, P.**, **Stridh, B.** and **Söderberg, S.**, Solid particle erosion of hard chemically vapour-deposited coatings, *Surface and Coatings Technology*, 1989, **37**, 321–337.

63 **Hedenqvist, P.** and **Olsson, M.**, Solid particle erosion of titanium nitride coated high speed steel, *Tribology International*, 1990, **23**, 173–181.

64 **Shipway, P.H.** and **Hutchings, I.M.**, Influence of nozzle roughness on conditions in a gas-blast erosion rig, *Wear*, 1993, **162–164**, 148–158.

65 **Hutchings, I.M.** and **Shipway, P.H.**, Improved method of testing durability and adhesion of paints and other coatings, *Proc. ISATA Conference on New and Alternative Materials for the Automotive Industries*, 1993, 165–170.

66 **Chandra, L.**, **Allen, M.**, **Butter, R.**, **Rushton, N.**, **Hutchings, I.M.** and **Clyne, T.W.**, The effect of biological fluids on the response of DLC films to a novel erosion durability test, *Diamond and Related Materials*, 1996, **5**, 410–414.

67 **Rutherford, K.L.** and **Hutchings, I.M.**, Development of the erosion durability technique for thin coatings, *Surface and Coatings Technology*, 1996, **86–87**, 542–548.

68 **Hutchings, I.M.**, Development of wear maps for the abrasion and erosion of ceramic materials, in *Advanced ceramics for structural and tribological applications*, H.M. Hawthorne and T. Troczynski, Eds., 1995, pp. 127–138, (The Metallurgical Society of The Canadian Institute of Mining, Metallurgy and Petroleum, Montreal).

Life prediction in rolling element bearings

E IOANNIDES
SKF Engineering & Research Centre, Nieuwegein, The Netherlands, and Imperial College of Science, Technology and Medicine, London, UK

SYNOPSIS

Rolling elements were the first componentsto be subjected to stochastic life prediction methods. The Weibull life model was adopted by Lundberg and Palmgren as early as 1947 and since then it has been used as the basis of National and International Standards and has gradually been modified to incorporate emerging knowledge of lubrication and to predict the life increases of the constantly improving bearings. It focuses on the most appropriate failure mode of those times, the subsurface-originated fatigue. Bearings have however, been continuously improved in terms of design, manufacturing precision, steel integrity and heat treatment and their calculated lives, according to the above methodology, are now seriously under-predicted. Moreover, bearing failures now predominantly occur on the surface.

This paper reviews all levels of modern modelling which increasingly use realistic stress distributions in the steel, generalised fatigue criteria and an endurance limit. It is therefore now possible to calculate the life of rolling bearings under more realistic application conditions where the site of failure shifts to the surface and effects of roughness, contamination, edge and internal (hoop and residual) stresses can be included in the life calculations.

NOTATION

a_1	life factor for reliability
a_2	life adjustment factor for material
a_3	lubrication – life modification factor for lubrication
a_{23}	adjustment factor for material and lubrication
C	bearing basic load rating, N
L_{10}	rating life (90% reliability), rev. $\times 10^6$
L_{na}	rating life 100-n%, rev. $\times 10^6$
N	number of revolutions $\times 10^6$
P	bearing equivalent radial load, N
p	exponent of the life equation
S	probability of survival
V	volume under stress, mm^3
τ_o	maximum orthogonal shear stress below raceway contact, MPa
τ_u	fatigue limit shear stress, MPa
z_o	depth where τ_o occurs, mm
η_c	contamination factor
κ	viscosity ratio

1 INTRODUCTION

The rolling element bearing is the most widespread moving machine element, with some fifty billion bearings in operation throughout the world. The origin of rolling bearings can be traced in antiquity where an example of an unearthed Roman ship had bearings supporting platforms with rolling elements made of bronze and wood. Even earlier, a Celtic cart was found with hubs, having spacing for supporting wooden elements, rolling or possibly sliding (**1**). Despite these early origins, rolling bearings were introduced industrially only during the latter part of the nineteenth century after a flurry of patents during the previous century. The chief reason for this is that only then could the manufacturing processes of the day achieve the low surface roughness of the rolling elements and the contacting raceways needed to ensure the satisfactory separation of these components necessary for satisfactory operation. This, together with the use of fatigue-resisting high alloy steels, was essential to ensure long and trouble-free operation under the high contact pressures that develop in the minute contact areas between the relatively small rolling elements and the raceways, as it can be seen in the photoelastic model of Fig. 1.

As a consequence the modern rolling element bearing is a precision machine element, usually mass-produced, at high speed, with very exacting dimensional and working contact surface requirements. The high demands of dimensional accuracy of the different bearing components originate from the requirements for running accuracy of machinery whilst, as indicated above, the need for low roughness originates from the need to build a separating lubricating film.

Life prediction is the essential tool in optimising the selection of rolling bearings in machinery. Thus it is an important issue, not only because of the large numbers of bearings involved, but also because of the critical role these components play in many applications, including safety, for example in transport systems. Today high performance and reliability requirements, which are the result of keen competition for all manufactured equipment, are combined with the need for high speed development from concept to finished product. This in turn implies that numerical modelling is usually preferable to testing because of both speed and cost. Considerable effort has, therefore, been put into the modelling of bearing performance in increasingly realistic terms by including, for example, effects of roughness and lubricant contamination. This is not an entirely new trend, as rolling elements were the first component to be subjected directly to stochastic life prediction methods. The Weibull life model was adopted by Lundberg and Palmgren as early as 1947 and since then it has been used as the basis of National and International Standards and has gradually been modified to incorporate emerging knowledge of lubrication and to predict the life increases of the constantly improving bearings. The Lundberg-Palmgren theory was an important step in bearing technology in the 1940s and it was the first calculation method to satisfy the need for quantification of bearing performance. It focuses on the most appropriate failure mode of those times, the subsurface originated fatigue. With subsequent modifications, in the form of life factors a_2 and a_3, for material and lubrication effects respectively, it is still the basis of national and international calculation methods (ISO 281).

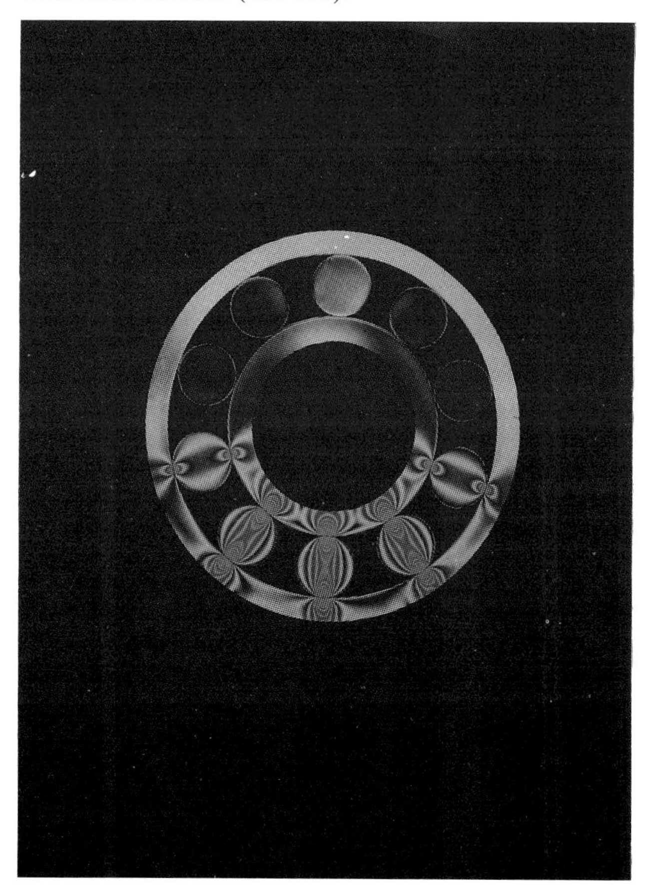

Fig. 1 Photoelastic model of a radially loaded bearing.

These broad issues, lubrication and contamination of rolling bearings, are also dominant as causes of failure, Fig. 2, (**2**). It should be noted, however, in this connection, that the vast majority of bearings, 99.5%, are either scrapped with the original equipment or replaced during maintenance without having failed.

Considerable effort is being put into modelling bearing performance in realistic terms. Bearings are selected (by type and size) such that the percentage probability (n) to failure in an application within a prescribed number of revolutions L_{na}, i.e. the expected life of the bearing, is known. The operational parameters of the applied load, bearing steel quality and lubrication are included in almost all bearing life L_{na} calculations. However, as indicated above, conditions such as surface dents, internal stress etc. must also be taken into account in the bearing life prediction models to provide the accuracy needed (**3**).

The paper discusses the range of methodologies available for life prediction to the bearing user. These generally fall into two categories: simplified calculations which can be performed manually and which include the standardized methods, advanced methods, embodied in computer programs, both quasi-static and dynamic methods capable of handling detailed bearing internal geometry description, of using advanced contact mechanics to account for roller profiles, ring distortions etc. and detailed stress fields from real (lubricated) contacts that are in turn used in life predictions.

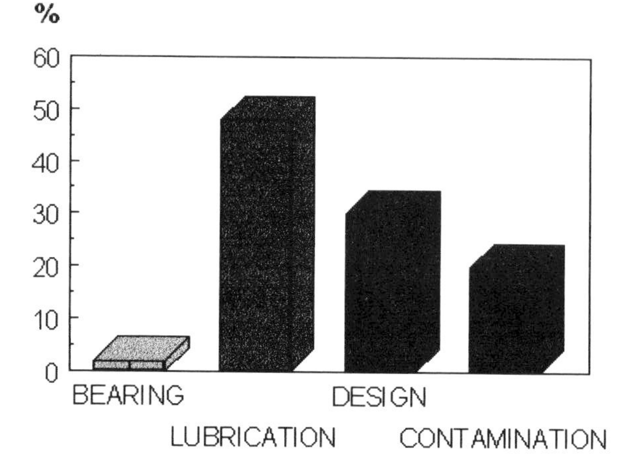

Fig. 2 Main Causes of Bearing Failure.

Clearly, the more complicated methods require the acquisition of more detailed input, and the selection of the appropriate method for a particular bearing life calculation depends on the accuracy required and to some extent on the availability of the appropriate input.

2 BEARING LIFE MODELS

2.1 General

Central to all bearing life models is fatigue life modelling. However two different approaches have been traditionally followed, initiation and propagation modelling. The traditional initiation model has its origins in Weibull (**4**), was extended by Lundberg and Palmgren (**5, 6**) and invokes a stress power law to account for the portion of the life spent in the initiation of a crack, which dominates the complete lifetime in brittle bearing steels. This method evolved from using a single stress value (the maximum orthogonal shear stress) to ascribe the danger of fatigue, which is appropriate for strictly Hertzian stress fields because of their self-similarity, to a model which uses a 3-D grid in the material with a multi-axial fatigue criterion (**3**).

In contrast, the propagation models evoke a form of the Paris law to calculate the propagation rate of an embryonic crack to a pre-defined position or length and from this the fatigue life is calculated (**7–9**).

Both formulations need to account for a Weibull distribution of the bearing lives, and also for the presence of a fatigue limit. Localized features of metallurgical significance, such as precipitates and inclusions, or other localized stress concentrations such as surface damage (e.g. debris dents) act as initiation sites for cracks which subsequently may be arrested or propagate with speeds that depend on the local properties of the microstructure. Thus both types of model require sets of experimentally adjustable constants so that engineering predictive models can be constructed from either model.

2.2 Simplified models

2.2.1 Standardized method

Lundberg and Palmgren (**5, 6**), as indicated above, extended Weibull's theory (**4**) to predict bearing life. Their bearing life prediction is based on a fundamental equation, Eqn. (1) below:

$$\ln\frac{1}{S} \approx N^e \frac{\tau_o^{\,c}}{z_o^{\,h}} a z_o \tag{1}$$

By substituting in this equation the Hertzian contact parameters and in terms of the applied load and the contact geometry, they obtained the load-life relationship for the bearings. This is of elegant simplicity and can be expressed in its final form as:

$$L_{10} = \left(\frac{C}{P}\right)^p \tag{2}$$

where C is the bearing dynamic capacity, a constant which depends on the bearing geometry and material, and P the equivalent load. Later, Eqn. (2) was adopted by ISO (**10**) and multiplicative constants were introduced, ISO (**11**), to account for different reliability, material fatigue properties and lubrication, resulting in today's form:

$$L = a_1 a_2 a_3 \left(\frac{C}{P}\right)^p \tag{3}$$

or:

$$L = a_1 a_{23} \left(\frac{C}{P}\right)^p \tag{4}$$

as is used by many bearing manufacturers who have recognised the inter-relationship of material and lubrication effects.

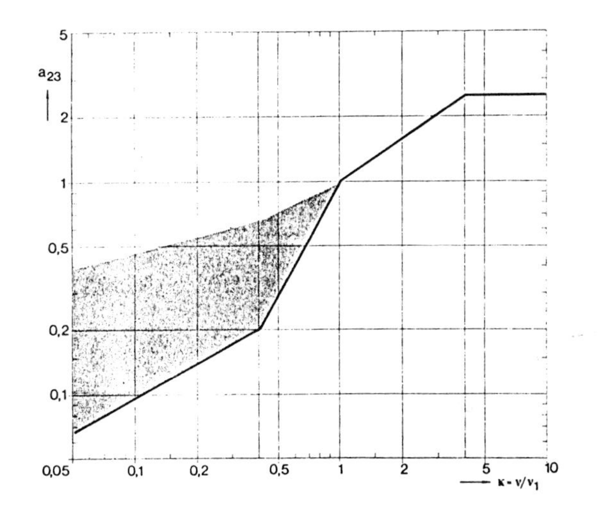

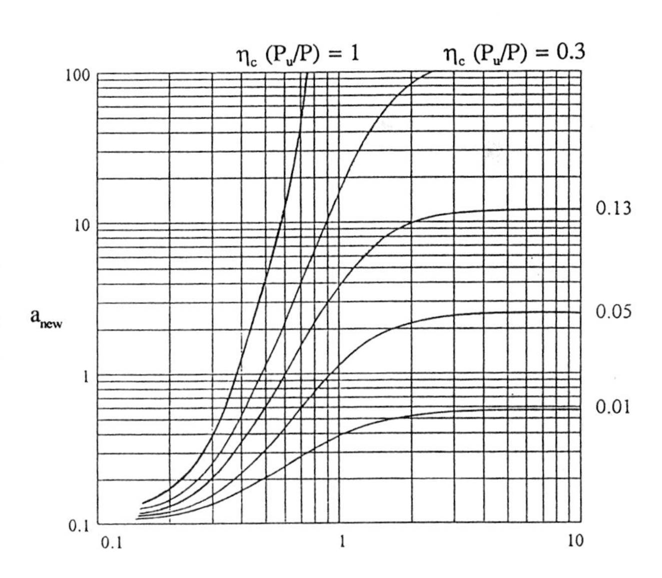

Fig. 3 a) Life adjustment factor a_{23} dependence on the lubrication parameter κ.

b) Life adjustment factor a_{new} dependence on the lubrication parameter κ and on the contamination parameter η_c.

2.2.2 Other methods

Experimental work in the 1970s and early 80s (e.g. (**12**)) indicated the strong dependency of the bearing life on particulate contamination in the lubricant. At the same time bearing tests were also indicating much longer bearing lives than the ones predicted according to ISO under clean lubricant conditions. In addition the load/life curve exhibited a 'fatigue limit knee' at lower loads (**13**). Some bearing manufacturers have now incorporated these effects in the life calculation methods of their catalogues (**14–17**). The life equation in this approach remains similar to Eqn. 2, but now a new life factor is required to account for the presence of the fatigue limit and the complex interactions of the material, lubrication and contamination. This method is an extension of the ISO method as can be seen in Fig. 3a, and b where the a_{23} factor of Eqn. 4 is shown, Fig. 3a, together with a similar new factor a_{new} which contains the effects of contamination η_c, and of the steel fatigue limit, Fig. 3b. It can be seen in this figure that a_{new} is a generalisation of a_{23} allowing for the possibility of longer lives for clean conditionsm, when η_c approaches 1, and shorter lives for contaminated lubrication, when η_c approaches 0, with a corresponding variation with the lubrication parameter κ.

The use of a multiplicity of independent life parameters a_i accounting for the various conditions that affect bearing life, e.g. contamination, hoop stress, residual stresses, is advocated in (**18**) whilst in (**19**) two new *a* factors are introduced for the fatigue limit and contamination which however are not independent.

2.3 Advanced models

2.3.1 Quasi-static models

The continuous improvement of bearing design has led to the development of unique roller and raceway profiles and conformities of the ball-raceway contacts. These will result in quite different bearing lives when actual contact stresses are considered. No provisions exist in the ISO life calculation method for non-uniform load distributions nor for flexible bearing rings. Thus, while the life calculated according to the ISO method has served industry well over the years, the modern down-sized bearing is required to have a longer actual service life and this necessitates greater sophistication in the calculation methods to account for actual operating conditions.

To remedy this situation bearing companies have developed dedicated, and usually proprietary, computer programs to enhance the support provided to customers, e.g. (**20**) where a computer program provides analysis of bearings in an environment in which both housing and bearing rings may be considered as flexible, with the flexibility of the housing and the surrounding structures usually provided from FEM calculations. The main results of such analysis with this program are the loads and the displacements. The life of the bearings is calculated with a method which takes the load for each rolling element into account, but makes a detailed contact stress analysis of the most heavily loaded element in each row. Life is then calculated according to Eqn 5:

$$\ln\frac{l}{S} \approx \frac{N^e \{H(\tau_o - \tau_u)\}^c \, az_o l}{z'^h} \tag{5}$$

that utilizes a fatigue stress limit and takes contamination into account through evaluation of τ_o. The elasticity of bearings, shafts and housings as indicated above is also considered. The bearing model considers specific internal geometry such as the raceway groove radius of ball bearings, and the profiles of rollers and roller bearing raceways. Life calculations take into account operating conditions of clearance, true contact under load, the influence of interference fits, temperature differential between the rings, misalignment and edge stresses in roller bearings as well as in ball bearings where the contact ellipse is truncated. In comparing such calculated life results with the classical ISO method, the advanced method tends to yield: 1) shorter lives at high *C/P* values because the ISO method tends to over-estimate the contact areas 2) longer lives under normal loading conditions because of the modern bearing's actual contact geometries and 3) shorter lives under heavy loads due to the effects of edge stresses, demonstrating the advantage of special profiles, see Fig. 4 and Fig. 5.

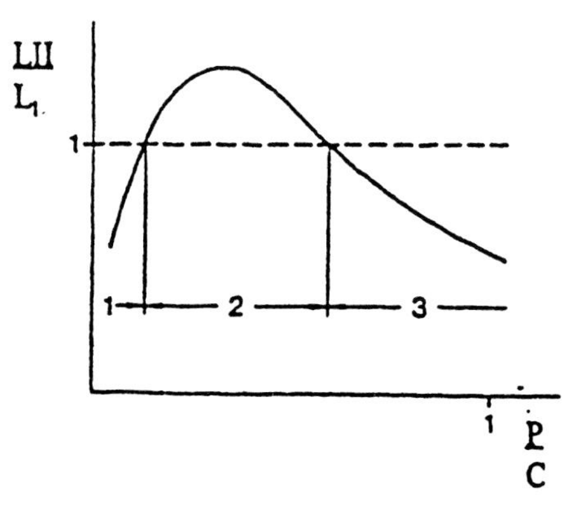

Fig. 4 Comparison between bearing life as calculated by the ISO method and the method in (**20**) as a function of load. In region 1 only a part of the roller length is utilized. The roller is optimally utilized in region 2. In region 3 edge stresses occur. See also Fig. 5.

Contamination

Surface damage can arise from a number of different sources, such as damage occurring during transport, damage from incorrect mounting procedures, and damage

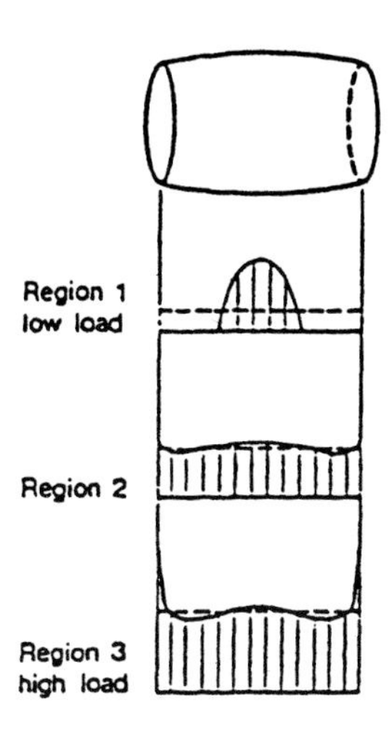

Fig. 5 Load distribution as a function of load under conditions of zero misalignment. Level of ideal line load is indicated by a broken line.

from the over-rolling of large contaminant particles in the lubricant. Of all these possible sources the over-rolling of contaminant particles is probably the most commonly neglected source of surface damage. In the past, a considerable amount of experimental work has been reported in which contaminated lubricants were used in rolling element bearings and rolling contacts. An overview of this work is given in (**21**). In general, such experimental work points to a wide range of life reductions, usually severe, indicating the need for a predictive model which can account for the effects of such realistic conditions. For this reason theoretical studies have recently been emerging in which an effort is made to systematize the many parameters that govern the effects of particulate contamination. These studies have concentrated on two main aspects: the mechanisms and the prediction of damage by particles to the raceways and rolling elements (**21, 22**), and a theoretical confirmation of how such localised damage can réduce the fatigue life of a bearing (**23, 24**). For the latter, fatigue models such as the one reported in (**16**) have been used to predict the life reductions associated with damaged raceways and rolling elements. An alternative approach has been the use of Fracture Mechanics and Micromechanics modelling, e.g. (**9**).

In many cases the starting points for life predictions are the dry contact stress calculations. These calculations themselves are quite complex and require a great deal of computer memory and time if sufficient detail of the real damaged surface is modelled during the passage of the rolling element over the damaged ring. Early examples are given in (**22**) where FE methods were used to simulate the effect of a certain level of filtration in helicopter gearboxes. Again with the advent of multilevel-multi-integration techniques it is now possible to calculate realistic contacts in bearings requiring many thousands of grid points. Fig. 6a shows the ball/raceway dry contact pressure distribution in a deep groove ball bearing with a large circular dent (600 µm diameter) in the centre of the contact (**24**), whilst Fig. 6b shows the corresponding fatigue criterion distribution in the subsurface region of the ring. The brighter areas in this Figure signify higher values of the criterion and consequently higher risk of fatigue. It can further be seen in this Figure that the risk of fatigue is higher in the rim area of the dent and the distribution is symmetrical with respect to the centre of the dent. This implies that the risk for initiation of fatigue spalls is also evenly distributed around the rim of the dent. This contradicts the experimental evidence that spalls invariably start near the trailing edge of the trailing side of the dent. With a full EHL calculation, however, it is possible to show that the film pressure and the corresponding value of the criterion are higher in the trailing edge of the dent and that therefore this is the predicted preferential site for fatigue spall initiation in agreement with the experimental evidence, (**24**).

The predicted contact pressures and the values of the

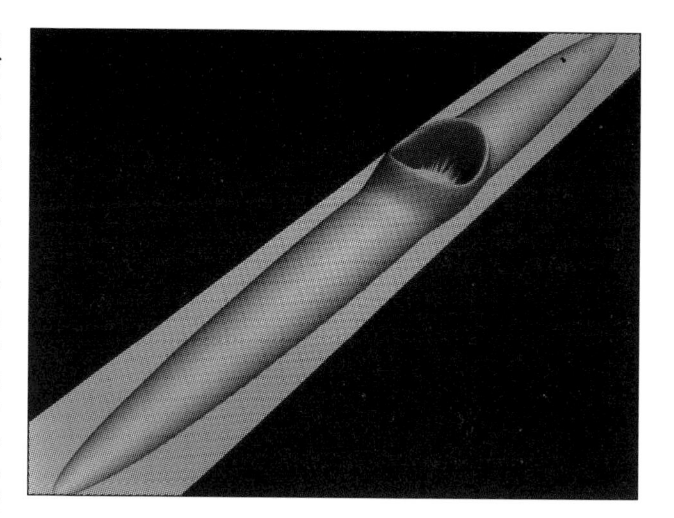

Fig. 6 a) Pressure distribution in deep groove ball bearing with a dent of 600 µm diameter.

b) Subsurface fatigue criterion distribution.

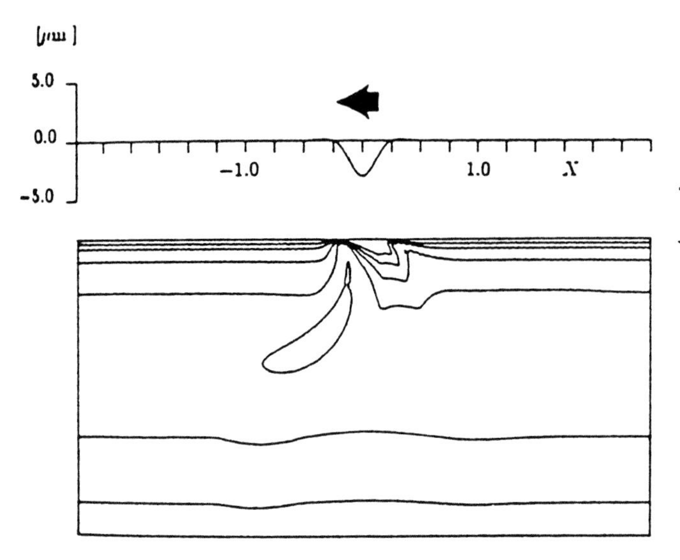

Fig. 7 Predicted risk of fatigue in the neighbourhood of a dent using EHL calculations.

fatigue criterion according to (**16**) are asymmetric, as shown in Fig 7. When effects of sliding are introduced, additional detrimental effects are suggested. (**25**).

In addition, theoretical predictions have been compared with endurance tests with artificially indented bearings. The indentations were made with a small hard ball and the calculations used the actual spherical shape of the dent. Fig. 8 shows the theoretical life prediction compared with experimental lives (90% confidence intervals) obtained for different indentation sizes. Fig. 9 shows this comparison of life as a function of load for one particular dent size. From both Figures it can be concluded that the life reduction is significant and that the theoretical predictions match the experimental lives quite well. In contrast, the use of simple life reduction factors, which are only related to filter rating as suggested in (**18**) overlook the influence of bearing load, which is clearly important (Fig. 9).

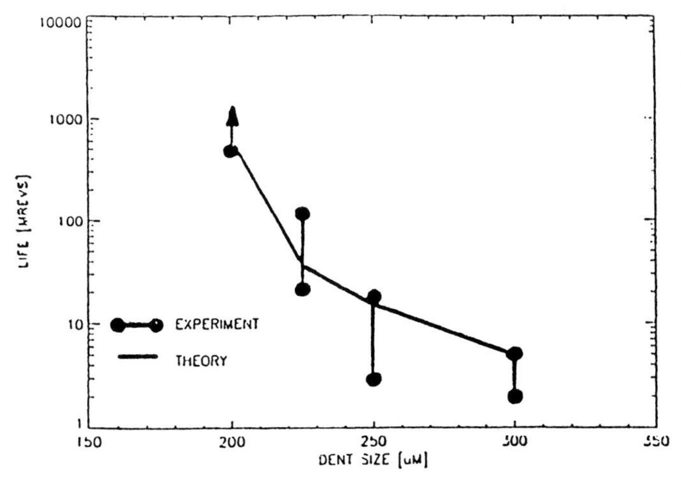

Fig. 8 Life as function of dent size, 3 dents, 6309, $C/P = 2.8$.

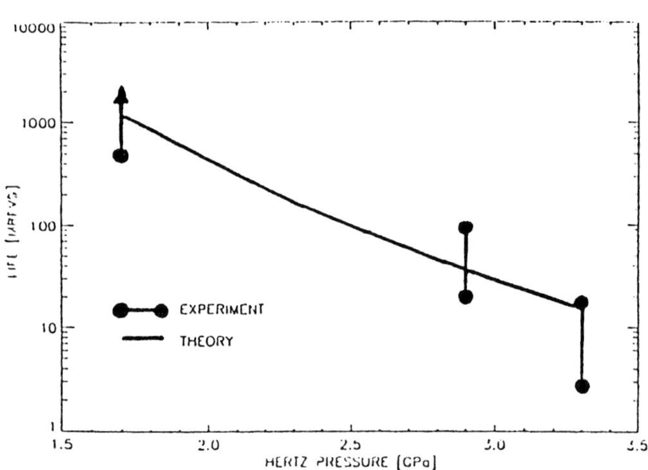

Fig. 9 Life as a function of load, 3 dents, 6309, dent size 250 μm.

Roughness

Estimating the life of real rough surfaces is of primary interest in the effort to predict the life of rolling bearings. The most direct consequence of roughness comes through its effect on surface stresses. In dry or thin film contact, there are large contact stress fluctuations associated with the random changes of surface slope. For bearing steels, the normal surface stress fluctuation amount to as much as 1.4 GPa/degree, giving a maximum shear stress just beneath the asperities of 0.35 GPa/degree.

Thus, for surfaces with slopes greater than about 1 degree, the fatigue threshold of $\tau_u = 350$ MPa is exceeded and the contact life is consequently reduced. The amount of reduction depends not only on the surface slope but also on the proportion of the surface where the threshold stress is exceeded, determined by the skewness of the surface height distribution. Surface fatigue thus depends principally on surface slope Δ_q and skewness R_{sk} and only to a lesser extent on the mean asperity height R_q.

To calculate the surface life reduction, given the present day calculation limitations, one has to use expressions or design curves derived from tests (**13**) or utilise stress fields obtained from dry contact solutions to estimate the expected life. The latter approximation is valid for thin films where the pressure distribution approaches asymptotically the dry contact one as the film thickness diminishes. In the absence of pressure fluctuation attenuation by the lubricant film, the dry contact pressure distributions provide a conservative limit for the purposes of life calculation. Such dry contact calculations have been reported by a number of authors, (**26–27**), and more recently the extension of the methods to layered contacts has been reported (**28–29**). The dry contact subsurface stress fields were used in (**30**) to assess the relative effect of roughness parameters on life. In Fig. 10, from this reference, the computed surfaces show that averaged L_{10} values increase dramatically when the slope Δ_q of the

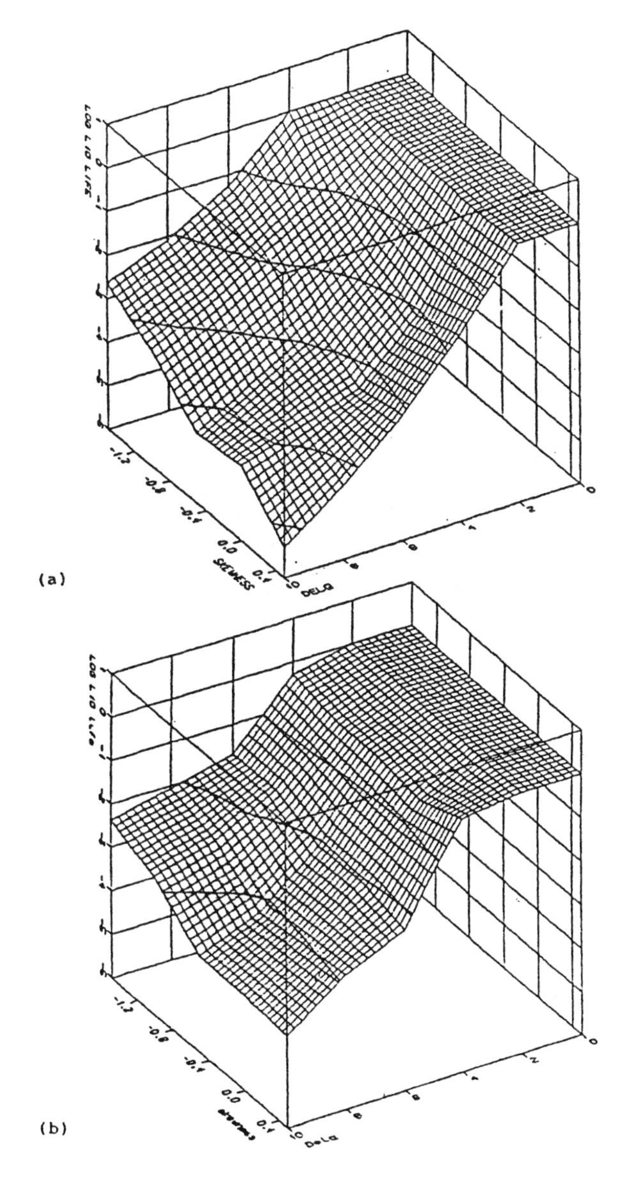

Fig. 10 Relative life L_{10} variation versus slope, Δ_q (deg.) and skewness R_{sk}, for two different average surface wavelengths, $\lambda_q = 360(R_q/\Delta_q)$:
a) $\lambda_q = 72$ mm; b) $\lambda_q = 36$ mm.

surface is reduced or when the skewness R_{sk} is made more negative. The benefit deriving from a reduction in the roughness amplitude R_q itself is, by contrast, more modest.

Residual and hoop stresses

Internal stresses (residual and hoop) influence the fatigue life of bearings. Residual stresses are mostly determined by heat treatment and manufacturing processes. Hoop stress is caused by the interference fit between inner ring bore and shaft to prevent the inner ring rotating on its seating and to avoid fretting corrosion on the interface. The sum of such internal stresses and the load over-rolling stresses defines fatigue life.

For normal interference fits, the influence on life is already included by the comparison with life tests; however, in some applications a heavy interference fit is required. Under such conditions the life of the bearing can be reduced. This life reduction can be evaluated by employing the model described in (**3**), adding the hoop stress to the hydrostatic stress component. Figure 11, a comparison of the theoretical life prediction with experimental results, shows a satisfactory agreement between the predictions and the experimentally obtained life. Since life is a function of contact stress, hoop stress and the fatigue limit of the material, it will also depend on the combinations of bearing type and size, the load, and the applied interference fit.

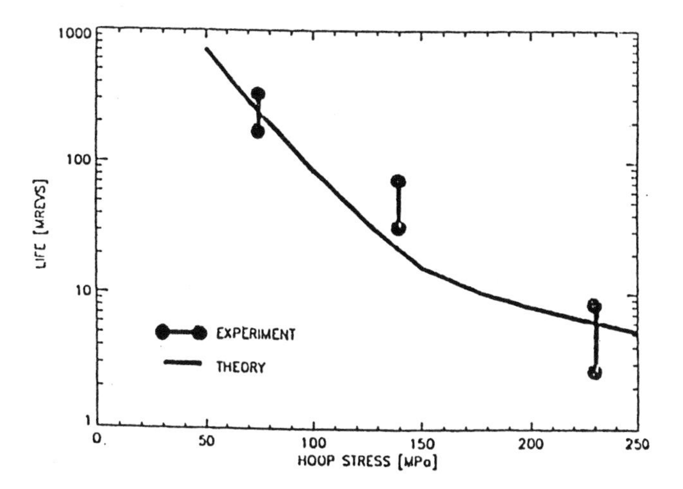

Fig. 11 Life as a function of hoop stress, 6309, high oxygen steel, $C/P = 2.8$.

2.3.2 Dynamic models

The rolling bearing is a fully dynamic system with moving parts, the rolling elements, the cage and at least one of the rings. As the rotational speed of the application increases the inertia forces increase in relative magnitude to the contact forces and it becomes necessary to describe the bearing as a dynamic mechanism.

This is by necessity a very complex approach because the complete bearing is a 'system' where the interaction of its components, through EHL and other lubricated contacts, is responsible for the overall performance. In this, the connection with the environment, i.e. the bearing housing, the shaft and rest of the machinery can also be important. Furthermore, while environmental trends require reduction of noise and vibration, rolling bearings can be both a source and a transmitter of vibration in rotating machinery. Thus description of the dynamic behaviour of a bearing fulfils two purposes: the understanding, modelling and optimisation of vibration and noise, together with more accurate friction and in certain cases life prediction. For the latter, analysis of long term behaviour is required which, because of the enormous amounts of calculations and computer storage required, is possible today only when a steady state condition is quickly reached. (**31**) and (**32**) are examples of programs that offer such a possibility.

3 DISCUSSION AND CONCLUSIONS

Bearing life predictions for modern applications require consideration of application and environmental conditions not currently covered in the calculation methods embodied in the national and international standards (e.g. ABMA, ISO). Down-sizing trends and increased reliability requirements necessitate more exacting calculation methods to meet the demands of today's bearing users. This is further highlighted by the change in the failure mode away from the classical subsurface fatigue, upon which the standards are based, to surface failure. As indicated before, contamination has become an important failure mode and as such must be accounted for in today's calculations. In an attempt to consider more realistic application conditions, not covered in today's standards, various bearing manufactures and users are proposing changes in the standardized calculation methods by introducing new life adjustment factors that are similar to the current a_2 and a_3. Such factors must be based on test results and field experience. This is needed because the interactions of materials, stresses, lubrication and environmental conditions are complex. However, progress has been made with calculation methods that include a fatigue stress limit and which are supported by the bearing manufacturers who possess most of the available test data. In addition to this, a more realistic modelling approach to the environmental conditions as they relate to the stresses in the contact areas provides a means to calculate the effects of contamination and internal stresses on bearing performance. These new methodologies have been used for some time in research and in computer programs of considerable complexity but they are now finding their way into the simpler and possibly standardized methods, as the recent effort to extend ISO 281 indicates.

The effect of all this is manifold; it requires the consideration of the bearing as a system which includes the effects of the lubrication and of the machinery around the bearing, especially in the case of open bearings, and it emphasizes the need to protect the bearing from the contamination in the environment with increasing use of sealed (and greased for life) bearings. It is also leading bearing designers away from traditional design rules that maximize load carrying capacity based only on knowledge of the bearing, to new design rules that optimize the life of a bearing, involving its lubricant, its seal and the environment as important parts of the system.

4 ACKNOWLEDGEMENT

The author wishes to thank Dr H.H. Wittmeyer, Managing Director of SKF Engineering and Research Centre B.V., for his kind permission to publish this paper.

5 REFERENCES

1 **Dowson D.**, *History of Tribology*, Longman Group Ltd., London, 1979.
2 **Engel L.**, and **Winter**, Walzlagerschaden, *Antriebs Technik* **18**, (1979), Nr. 3.
3 **Ioannides, E.**, **Jacobson, B.** and **Tripp, J.**, Prediction of rolling bearing life under practical operating conditions, *Proceedings of the 15th Leeds-Lyon Symposium on Tribology*, (Leeds, 1966) 181–187, (1989).
4 **Weibull W.**, A Statistical Representation of Fatigue Failures in Solids, Acta Polytechnica, Mechanical Engineering Series 1, *Royal Swedish Academy of Engineering Sciences*, **No.9**, 49, (1949).
5 **Lundberg, G.** and **Palmgren, A.**, Dynamic capacity of rolling bearings, Acta Polytechnica, Mechanical Engineering Series, *Royal Swedish Academy of Engineering Sciences*, Vol **1**, No. 3,7, (1947).
6 **Lundberg, G.** and **Palmgren, A.**, Dynamic capacity of rolling bearings, Acta Polytechnica, Mechanical Engineering Series, *Royal Swedish Academy of Engineering Sciences*, Vol **2**, No. 4, 96, (1952).
7 **Tallian, T.E.**, Simplified contact fatigue life predictions model – Part II: New model, *ASME J. Trib.*, **114**, 214–222, (1992).
8 **Zhou, R.S.**, **Cheng, H.S.**, and **Mura, T.**, Micropitting in Rolling and Sliding Contact under Mixed Lubrication, *ASME J. Trib.*, **111**, 605–613, (1989).
9 **Cheng, W.**, **Cheng, H.S.**, **Mura, T.**, and **Keer, L.M.**, Micromechanics Modeling of Crack Initiation under Contact Fatigue, *ASME J. Trib.*, **116**, 2–8, (1994).
10 International Standards Organization, Roller bearings – dynamic load ratings and rating life – Part 1: Calculation methods, *International Standard*, 1962.
11 International Standards Organization, Roller bearings – dynamic load ratings and rating life – Part 1: Calculation methods, *International Standard, 281*, (1990).
12 **Sayles, R.S.** and **Macpherson, P.B.**, *Influence of wear debris on rolling contact fatigue, Rolling Contact Fatigue Testing of Bearing Steels*, (ed. J.J.C. Hoo) ASTM STP **771** 225–274, (1982).
13 **Lorösch, H. K.**, *Influence of Load on the Magnitude of the Life Exponent for Rolling Bearings, Rolling Contact Fatigue Testing of Bearing Steel*, ASTM STP **771**, (ed. J.J.C. Hoo) 275–292, (1982).
14 FAG Rolling Bearings, Catalogue WL 41520 ED, FAG, ed., Schweinfurt, Germany, (1995).
15 SKF General Catalogue 4000, AB SKF, Gothenburg, Sweden, (1989).
16 **Ioannides, E.** and **Harris, T.A.**, A new fatigue life model for rolling bearings, *ASME J. Trib.*, **107**, 367–378, (1985).
17 **Lösche, T.**, New Aspects in the Realistic Prediction of the Fatigue Life of Rolling Bearings, *Wear* **134**, 357–375, (1989).
18 **Zaretsky, E. V.**, *STLE Life Factors for Rolling Bearings*, STLE Publication SP-34, (1992).
19 **Takata, H.**, Possibility of a New Method for Calculating Fatigue Life for Rolling Element Bearings, *Japanese Journal of Tribology*, **36**, 707–718, (1994).

20 **Tengroth, S.** and **Haggblad, H.A.,** A General Computer Program for Optimized Analysis and Calculation of Bearing Arrangements, *SKF Ball Bearing Journal* **#236**, Newport Print Limited, England, 2–8, (1990).

21 **Sayles, R.S., Hamer, J.C.,** and **Ioannides, E.,** The effects of particulate contamination in rolling bearings – a state of the art review, the *Proceedings of IMechE,* **204**, 29–36, (1990).

22 **Ko, C.N.** and **Ioannides, E.,** Debris Denting – The Associated Residual Stresses and their Effect on the Fatigue Life of Rolling Bearings: A FEM Analysis, *Proceedings 15th Leeds-Lyon Symposium on Tribology,* (Leeds 1988) 199–207, (1989).

23 **Webster, M.N., Ioannides, E.,** and **Sayles, R.S.,** The effects of topographical defects on the contact stress and fatigue life in rolling element bearings, *Proceedings 12th Leeds-Lyon Symposium on Tribology* (Lyon 1985), 207–221, (1986).

24 **Lubrecht, A.A., Venner, C.H., Lane, S., Jacobson, B.,** and **Ioannides, E.,** Surface damage – Comparison of theoretical and experimental endurance lives of rolling bearings, *Proceedings Japan International Tribology Conference,* Nagoya, 185–190, (1990).

25 **Ai, X.,** and **Cheng, H.S.,** The Influence of Moving Dent on Point EHL Contacts, *STLE Trib. Trans.,* **37,** 323–335, (1994).

26 **West, M.A.** and **Sayles, R.S.,** A 3-dimensional method of studying 3-body contract geometry and stress on real rough surfaces, *Proceedings 14th Leeds-Lyon Symposium on Tribology* (Lyon 1987) 195–200, (1988).

27 **Bailey, D.M.** and **Sayles, R.S.,** Effect of roughness and sliding friction on contact stress, *ASME J. Trib.,* **113,** 729–738, (1991).

28 **Cole, S.J.** and **Sayles, R.S.,** A numerical model for the contact of layered elastic bodies with real rough surfaces, *ASME J. Trib.* **114,** 334–340, (1992).

29 **Olver, A.V., Cole, S.J.,** and **Sayles, R.S.,** Contact stresses in nitrided steels, *Proceedings 19th Leeds-Lyon Symposium on Tribology* (Leeds, 1992), 71–80, (1993).

30 **Tripp, J.H.** and **Ioannides, E.,** Effects of surface roughness on rolling bearing life, *Proceedings Japan International Tribology Conference,* Nagoya, 797–802 (1990).

31 **Kleckner, R.J.** and **Pirvies, J.,** Spherical Roller Bearing Analysis, *J. Lub. Tech.* **104**/99, (1982).

32 **Gupta, P.K.,** ADORE – *Advanced Dynamics of Rolling Elements User Manual,* PKG Clifton Park, New York 12065, 7–4, 7–5, (1992).

At the boundary between lubrication and wear

B JACOBSON
SKF Engineering & Research Centre B.V., Nieuwegein, The Netherlands

SYNOPSIS

For more than 100 years it has been possible to predict the oil film thickness, pressure build-up and power loss in plain hydrodynamic bearings assuming the surfaces to be smooth. The calculated oil film thickness for smooth surfaces was then compared with the statistical root mean square roughness of the surfaces, and if the film thickness to roughness ratio, lambda, was larger than 10, experience showed that the bearings worked well.

Since 1939 it has been possible to calculate and predict elastohydrodynamic oil film thicknesses for smooth surfaces. When these calculated oil film thicknesses were compared with the combined surface roughness of lubricated gears and rolling bearings it was found that if lambda was about 2–3 the gears and rolling bearings worked well despite the fact that peak to valley surface roughness heights could be 3 to 7 times larger than the calculated film thickness. Still the lubrication worked well and the surfaces were totally separated by a continuous oil film.

Recent experiments with very smooth ball bearings have shown that a separating oil film is present between each of the balls and the rings even when the calculated oil film thickness is much lower than the combined root mean square roughness of the lubricated surfaces. This indicates that the calculated oil film thickness can be 20 times smaller than the peak to valley roughness of the surfaces, and still the surface roughness peaks do not penetrate the oil film.

This separation is possible if the local shear stress in the oil is low enough at the local temperature and pressure not to remove the last molecular layer of oil from the lubricated surface. As the shear stress in the lubricant for a given film thickness is proportional to the surface roughness slope for rolling contacts, thin film lubrication is only possible for surfaces with low surface roughness slopes.

The surface finish of elastohydrodynamically lubricated gears and bearings has so far been rough enough to show the effect of running-in. The surface structure is changed by running and if that change is slow and well controlled, the surfaces will run in until they become smooth enough to no longer wear. This running-in thus determines the running conditions and the functionality of the surfaces. If the lubricant contains particles, and these are large and hard enough, the running-in of the surfaces can be destroyed and they will never become smooth enough to be separated by an oil film. This leads to wear and/or spalling failure of the surfaces. Whether the surfaces will wear or spall depends on the properties and the concentration of contaminant particles. For some moderate levels of contamination the surfaces can first wear and then spall when the load distribution over the contact area has been changed by the wear.

To ensure that the surfaces can work in a stable mode without wear or spalling, the running conditions have to be ideal for each point on the lubricated surfaces all the time. This means that the lubrication analysis has to be made at an asperity size level. It is not possible to predict the behaviour of a lubricated, elastically deformed asperity just by knowledge of the mean oil film thickness and the mean contact pressure. Besides the lubricant film thickness, it is also necessary to know the local pressure, temperature, shear stress and lubricant rheology, as well as the form and and elasticity of the asperity.

1 INTRODUCTION

Before Beauchamp Tower's publication 1883 of his findings regarding the pressure distribution in journal bearings (**1**), it had been assumed that the pressure was constant and equal to the load divided by the projected bearing area, just as if the journal had been floating on a 180 degree hydrostatic bearing. After the publication of Reynolds' equation (**2**) it was possible to calculate the real pressure distribution for simple bearing geometries. The first attempt to analyse the lubrication of gear teeth was made by Martin in 1916 (**3**). He assumed that the surfaces were smooth and that the lubricant retained its atmospheric properties. The calculations resulted in predicted oil film thicknesses much smaller than the typical surface roughness features, so he concluded that something was incorrect in his analysis. Twenty-five years later Meldahl (**4**) repeated Martin's calculations but also included the elastic deformations of the lubricated surfaces caused by the pressure distribution in the lubricant. This resulted in thicker predicted films, but still too thin to explain how gears and rolling bearings could work as well as they did. Not until Ertel and Grubin (**5, 6**) included both the elastic deformations of the surfaces and the Barus' type exponential increase of the viscosity with the pressure in the lubricant film was it possible to get calculated oil film thicknesses of the same order as typical surface roughnesses. This type of analysis was later refined to give accurate calculation results, both for complex geometries and different types of mathematically described rheologies for the lubricants (**7–12**).

At least two different size scales are important to the behaviour of the lubricated contact: the size of the macro-Hertzian contact and the size of the asperities on the surface. The size of the macro-Hertzian contact combined with the surface velocities gives the time available for compression of the asperities and their re-emergence into the oil film during the passage through the Hertzian contact. The size and height of each single asperity gives information about the local pressure variation and pressure gradients, and thus about the local shear stresses in the lubricant when the motion of the surfaces and the lubricant film thickness is known. When bearings and gears are run, occasional surface asperities are penetrating the oil film and make direct metal-to-metal contact. This results in high local stresses, and controlled wear or plastic deformation on the asperity level takes place until the surfaces are run-in and separated by a continuous lubricant film. This run-in state is only valid for one specific running condition, where the highest asperity tops just stop touching each other through the oil film. If the load or speed is changed new asperities will come into contact and require some more running-in before the surfaces are totally separated by a continuous oil film again.

2 LUBRICATION OF NON-SMOOTH SURFACES

Recently, when computer modelling of EHL films and dry contacts was made possible, all the way down to the level of single surface roughness asperities, using powerful computers and modern numerical techniques, it became evident that, just as for dry contacts, the Newtonian lubricant models produced very spiky pressure distributions. In fact, the variation in pressure became almost a mirror image of the local roughness pattern, magnified by some orders of magnitude. The resulting surface roughness on the deformed lubricated surfaces amounted to only a few percent of the oil film thickness, reflecting the variation in the compression of the lubricant in the high pressure region. For lubricating oils compressed into the solid glassy state, the compressibility is only 3–5 % per gigapascal, giving an oil film thickness variation of the order of ±4% for a pressure variation of ±1 GPa superimposed on the normal elastohydrodynamic pressure distribution.

Such a spiky pressure distribution induces shear stresses in the lubricant, which are proportional to the pressure gradients and proportional to the local oil film thickness for pure rolling contacts. When some sliding motion is superimposed on the rolling, such as the kinematic sliding in deep groove ball bearings and spherical roller bearings, high stresses in the lubricant are also induced in the direction of slip.

Elastohydrodynamically lubricated gears and bearings have until now shown the effect of running-in. The surface structure is changed by the running, and if that change is slow and well controlled, the surfaces will run-in until they have become smooth enough not to wear any more. This light wear thus determines to a large extent the functionality of the surfaces as long as the running conditions do not change. If the load, speed or temperature are changed so that asperity tops again touch each other through the oil film, running-in will start again. The running-in continues until there is no more direct asperity contact for a given running condition, indicating that the real minimum oil film thickness between the highest roughness peaks will be very close to zero.

If the forces, stresses and/or generated temperatures during the running-in are too large the surface contacts on the asperity level will damage the surface and increase the roughness. This leads to higher local contact pressures between the asperities and thus to a progressively more severe contact condition, making the surfaces increasingly rough while the wear rate remains high. There is thus a bifurcation point at which the surfaces can either run-in and become smooth or become rougher and rougher depending on some small variation in one or a few of the contact parameters.

When the the peak to valley surface roughness is very much larger than the existing oil film thickness, a major part of the load on the contact will be carried by direct solid contact between the asperity tops and the opposite surface. As the reduced radius of curvature of the asperity contact is usually very small, the local asperity contact pressures will be very high, and plastic deformation takes place both as a result of the high normal pressure and as a result of the large tangential stresses caused by tangential motion of the two surfaces relative to each other. The sharper the contact points, the lower the force needed to reach the limit for plastic deformation and thereby the limit for break-up of the protective oxide layers on the contacting surfaces. The steeper the asperity slopes, the more difficult it is to generate a separating oil film, so large asperity slopes are bad for cooperating surfaces, both from a stress pont of view and from a lubrication point of view.

3 NON-NEWTONIAN LUBRICANT BEHAVIOUR

For heavily loaded lubricated contacts the total shear deformation of the oil in the high pressure region gives stresses well above the limit for Newtonian behaviour of the lubricant. This means that stresses and lubricant flows in two perpendicular directions in the oil film will be coupled to each other via the the limited shear strength of the oil. A pressure gradient in one direction will also influence the flow in the perpendicular direction, much the same as for dry friction between solid bodies.

As early as 1941, P.W. Bridgeman (**13**) showed experimentally that liquids of the same type as lubricating oils transformed to solids at pressures much below the pressures found in lubricant films between the load-carrying parts of a ball bearing. Bridgeman's experiments were all static, and as the high viscosity/solid behaviour at high pressure inside an EHL contact is only present for a millisecond or a fraction of a millisecond, the behaviour on that time scale is even more solid-like. This solid type of behaviour for the oil in the central high pressure part of an elastohydrodynamically lubricated point contact was used by Jacobson (**9**) in his calculation of the oil film thickness distribution in a circular contact. To get input data for the computer calculations, Jacobson measured the shear strength of the solidified oil and the pressure at which it solidified. Later more accurate measurements were published (**14**), which show that typical liquid lubricants convert to glassy solids at pressures ranging from 0.5 GPa to above 2.2 GPa depending on the temperature and molecular structure of the lubricant. As soon as the lubricant is compressed into the solid state, even very low sliding speeds superimposed on the rolling speed will introduce high shear stresses in the direction of the surface sliding. These stresses, superimposed on the stresses emanating from the pressure distribution in the contact drive, the lubricant into a strongly non-Newtonian state. When the limiting shear strength of the lubricant is approached, a motion of the surfaces in one direction will directly influence the lubricant flow perpendicular to it.

This had already been experimentally shown in reference (**15**), when high speed photography was used to study the behaviour of an oil film in a sapphire disc machine where a sliding vibration perpendicular to the rolling entrainment velocity had a major influence on the formation of the oil film, see Figure 1.

For a Newtonian lubricant the small sideways vibration with an amplitude of 0.1 mm should not have any influence on the oil film build-up, but the experiments in Figure 1 show that the central film thickness in the inlet decreased almost to the same level as the film thickness in the side-lobes of the elastohydrodynamic contacts. The oil was obviously already non-Newtonian far out in the inlet region, causing a back-flow to occur when the sideways sliding was superimposed on the rolling motion. This means that not only the stresses but also the flow of lubricant inside the high pressure region of the EHL contact are strongly affected by the the non-Newtonian

a)

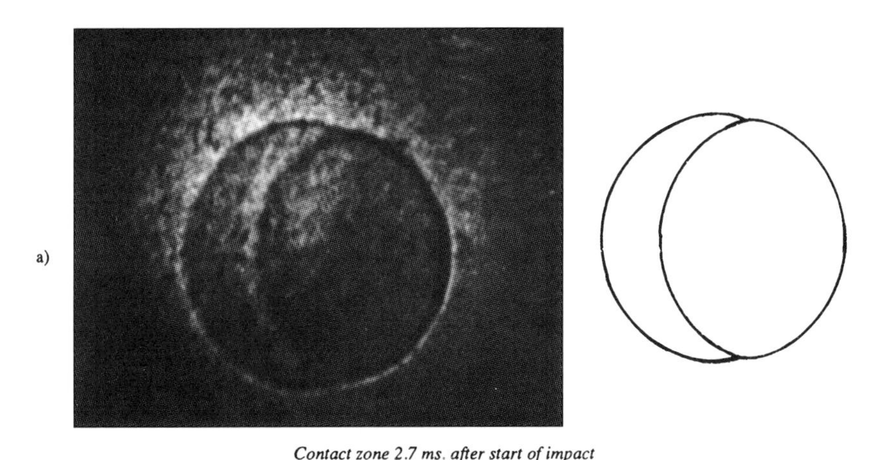

Contact zone 2.7 ms. after start of impact

b)

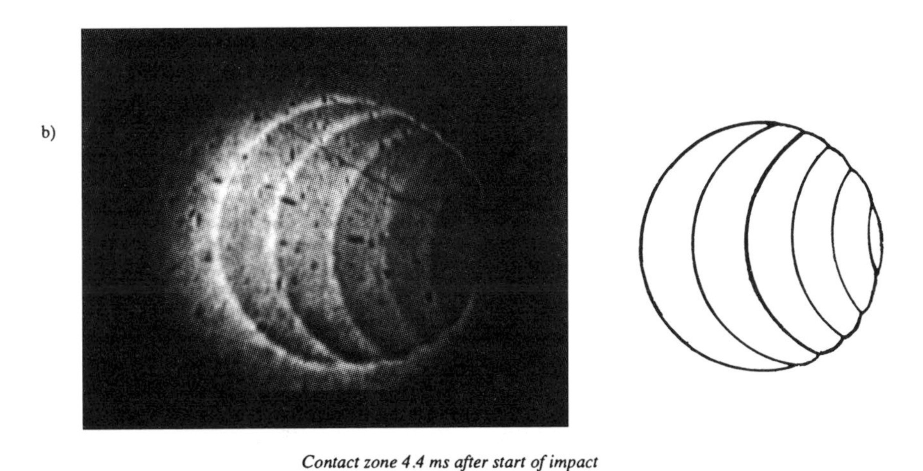

Contact zone 4.4 ms after start of impact

Fig. 1 Disturbed oil film build-up due to slight sideways vibration.

behaviour of the lubricant. Therefore the asperity behaviour in a heavily loaded lubricated contact will be quite dependent on the presence or absence of a sliding motion component superimposed on the main rolling velocity component. If no sliding is present the oil can behave like a Newtonian liquid, building up high local pressures on the asperity tops by squeeze motion which pushes them down and makes the lubricated surfaces conform much more than outside the Hertzian contact. The composite surface roughness of the two contacting surfaces becomes almost zero and the lambda value becomes large.

This ideal condition is quickly destroyed if some sliding is introduced between the load-carrying surfaces. The sliding motion in one direction uses all of the available strength of the oil in that direction which makes it very easy for local pressure variations within the contact to push the oil down into the valleys of the surface structure, allowing the asperity tops to reach through the oil film and collide. The interaction between the surface asperity behaviour and the lubricant rheology determines the state of lubrication.

4 MIXED LUBRICATION

Depending on the local rheology of the lubricant in the EHL contact and the surface structure of the contacting surfaces, the asperities will be more or less elastically deformed in the high pressure zone by the local pressure variations. For pure squeeze or rolling motion, the shear strength of the oil is high enough to deform the surfaces elastically so that they become totally conforming and thus need only a very thin oil film to avoid metallic contact. For pure sliding motion, on the other hand, the shear stresses in the high pressure zone of the contact reach the lubricant shear strength and thus no strength is left to keep the oil from flowing out sideways from the asperity contacts, and a mean oil film thickness large enough to prevent the high tops of the asperities from colliding is necessary.

Thus, depending on the details of the surface structures of the cooperating surfaces and the kinematics of the contact, a very large variation can be found in the lambda value needed to lubricate the contact without breaking through the oil film. Lambda values as high as 20 can give occasional asperity contacts, while for well run-in surfaces, lambda as low as 0.3 can stop all metallic contact. It all depends on the lubrication of the asperities and on whether the local pressure hills can elastically make the contacting surfaces more conformable inside the Hertzian contact area compared with the roughness they have outside the contact.

As shown in references (**16–18**), the lubricated asperity behaviour inside the Hertzian contact zone is mainly a function of the surface roughness slope, wavelength, contact pressure, rolling and sliding velocities and the time it takes for an asperity to move through the Hertzian contact. Numerical calculations of the oil film thickness over an asperity top in an elastohydrodynamic contact (**18**) showed that even when the asperity was 10 times higher than the central oil film thickness in the contact, virtually no trace of it could be seen in the converged numerical solution. The asperity built up its own pressure spike so that it was almost totally flattened and pushed into the surface, see Figure 2. For a non-Newtonian lubricant model in sliding, the asperity was scraping away the lubricant and left a zero film-thickness trace behind it, see Figure 3.

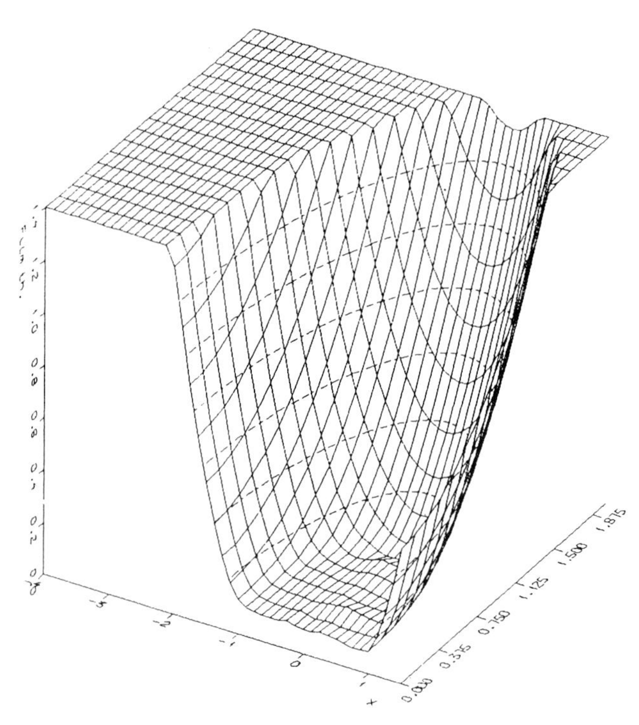

Fig. 2 Asperity lubricated with Newtonian oil.

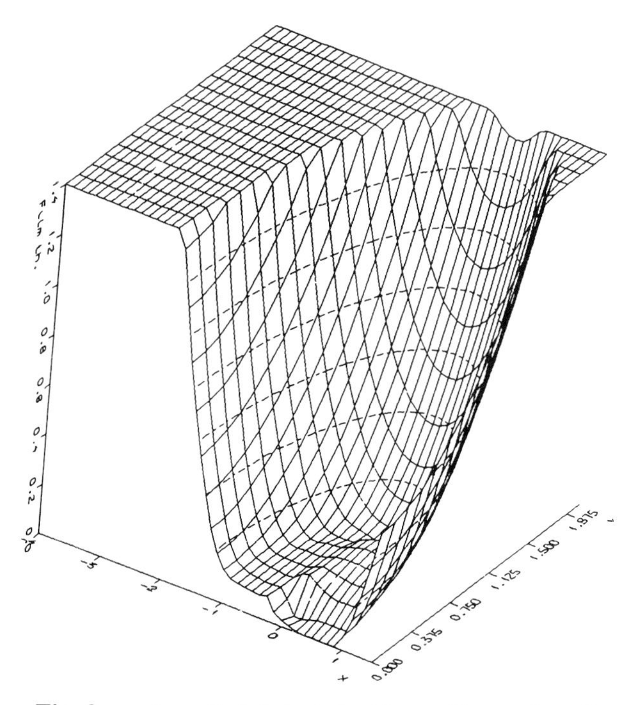

Fig. 3 Asperity lubricated with non-Newtonian oil.

A simple steady-state calculation of the minimum oil film thickness above an asperity top showed that, if a rough surface is stationary and a smooth lubricated surface is rolling and sliding over it with local contact pressures above the glass transition pressure for the lubricant, the oil film thickness will fall to zero at each asperity top if enough time is available. A simple model of the lubricant in the glassy state is in the form of a solid with a given shear strength τ_L which varies with pressure and temperature. If the smooth surface then slides over the asperity, the film thickness will fall to:

$$h_{min} = 2\tau_L \left|\frac{\partial p}{\partial x}\right| / \left\{\left(\frac{\partial p}{\partial x}\right)^2 + \left(\frac{\partial p}{\partial y}\right)^2\right\}$$

if enough time is available. This leads to oil film collapse on all the asperity tops with contact pressures above the glass transition pressure. The dynamic analysis of the compression of the asperities in the inlet zone and their re-appearance during the transport through the contact showed that the important parameters determining whether metallic contact takes place or not are, in addition to the contact pressure, the surface roughness, the number of asperities inside the Hertzian contact area, the sliding speed, the transport time for an asperity through the lubricated contact, and the mean oil film thickness.

This could be summarised in the equation:

$$n\frac{u_1 - u_2}{u_2} \bullet h_{min} < \Delta h \qquad \text{to avoid film collapse}$$

where:

n is the number of asperities from the inlet to the outlet of the lubricated contact.

U_1 and U_2 are the surface speeds.

h_{min} is the mean oil fim thickness.

Δh is the allowable elastic spring-back of the asperity during its passage through the contact.

This means that a surface with 5 wavelengths within the Hertzian contact can have the relative sliding speed: $(U_1 - U_2)/U_1 = 0.44 = 44\%$ for a given asperity spring-back, while a surface with 36 asperities needs less than 10% slip for the same spring-back.

The analysis of these parameters shows that using the ratio between the calculated oil film thickness and the composite surface roughness as a measure of how well the surfaces are separated by an oil film will underestimate how well smooth surfaces are lubricated compared with rough surfaces. Smooth surfaces need not only proportionally thinner oil films for good lubrication, but also the ratio of oil film thickness to surface roughness can be decreased.

The non-Newtonian behaviour of the oil, manifested by a limiting shear strength, can lead to oil film collapse, as discussed above. This collapse can be studied by considering the solidified oil as a plastic solid and applying the extrusion theory (19). As expected, increasing sliding speeds and decreasing asperity wavelengths increase the likelihood of film collapse. Figure 4 shows the minimum film thickness reduction with time for different asperity wavelengths.

5 STARVATION

Lubrication analysis generally assumes a sufficient supply of lubricant to ensure that the inlet to the contact is fully flooded. If this condition is not met and the contact is operating in the starved lubrication regime, then significant deviation from the predicted film thickness will occur. Many machine elements operate in the starved regime; the conditions are particularly severe for grease lubricated bearings, and the accurate prediction of lubricating film thicknesses under these conditions is almost impossible.

In the starved condition there is insufficient lubricant to fill the inlet and as a result the inlet does not experience the full hydrodynamic pressure build-up required to generate the high viscosity necessary to maintain full contact separation. This inevitably leads to lower film thicknesses than in the fully-flooded condition. It is the inlet lubricant supply condition that determines the degree of film reduction, and it is not possible to predict this for many practical applications.

In early experimental studies (**20**), starved film thicknesses were measured using optical interferometry. As

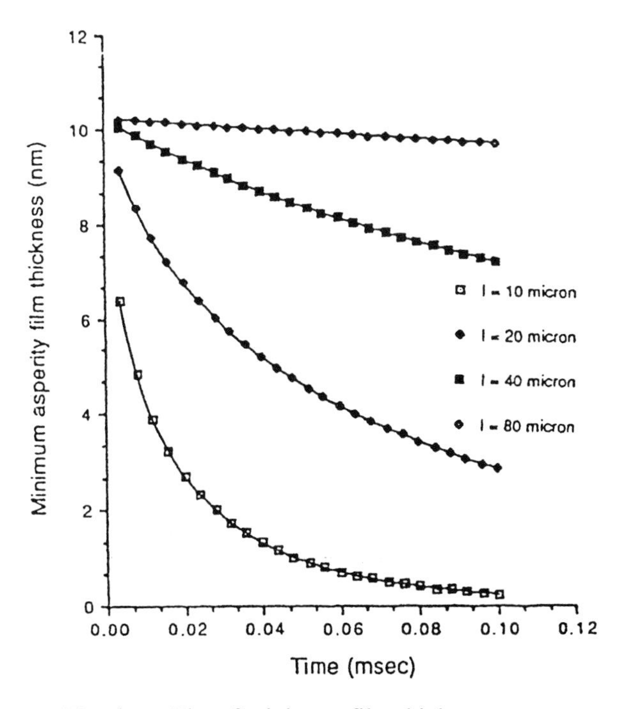

Fig. 4 Plot of minimum film thickness duringcollapse of different asperity wavelengths (**19**).

starvation increases and the inlet meniscus approaches the Hertzian radius, the film thickness drops below the lower detection limit of 100 nm and it is impossible to quantify the fully-starved condition. More recent experimental studies (**21, 22**) of the fully-starved or parched regime have shown that a thin film does persist, maintaining contact separation, also when the meniscus reaches the Hertzian contact edge. The regime of parched lubrication was first proposed by Kingsbury (**21**) to denote the fully starved condition where there is no inlet meniscus. In the parched regime no free oil is present and only a thin oil film remains on the surface. It was shown that instrument bearings could run successfully for several hours with only 80–200 nm of deposited oil as lubricant. More direct evidence of residual film separation in the parched regime has been obtained with thin film optical interferometry measurements of heavily-starved contacts (**22**).

Experimental studies of film formation by greases in a model contact (**23**) have shown that, unless an external supply is used to replenish grease constantly within the raceway, it is rapidly pushed to either side of the raceway, thus starving the contact. Grease lubrication can therefore be considered to be guided by a starvation phenomenon where the development of the EHL film is governed by the supply of lubricant from the surrounding grease reservoir. This is seen in Figure 5 where EHL film thickness is plotted against increasing rolling speed (**24**). Initially the thickness increases with speed and then drops rapidly as the system passes through the speed/viscosity starvation boundary. The final speed-independent film thickness for the fully-starved condition is approximately 40 nm. This behaviour is characteristic for greases although the starvation speed and final film thickness level vary greatly with different types of grease and operating conditions. The grease lubrication film has two components: a residual film of deposited gellant particles and a speed-dependent EHL contribution. Both components show a speed and time dependence making prediction of film thickness in rolling contacts for a grease very difficult. In all cases the film thickness is dependent on test history, and isolated results need to be interpreted with care.

One example of how limited the ability is to replenish the inlet zone of a starved contact by redistributing the lubricant sideways is shown in Figure 6 where both theoretical calculations and experiments give very stable lubrication over half a point contact while the other half is severely starved (**25**). Theoretical results in the same thesis, see Figure 7 (**25**), show that side-flow is virtually non-existent in the high pressure zone of pure rolling contacts.

6 CONCLUSIONS ABOUT THIN FILM LUBRICATION

The lambda criterion provides a broad guide to the safe/unsafe operating limits for a lubricated contact. It cannot be used to assess operation at low lambda values; here the criterion is invalid. Under these conditions the lambda value is far too crude a tool as paradoxically many systems are known to run successfully at low lambda ratios (**21**). This is particularly true for thin film lubrication where the estimated film thickness is less than the undeformed root mean square roughness value. Again the

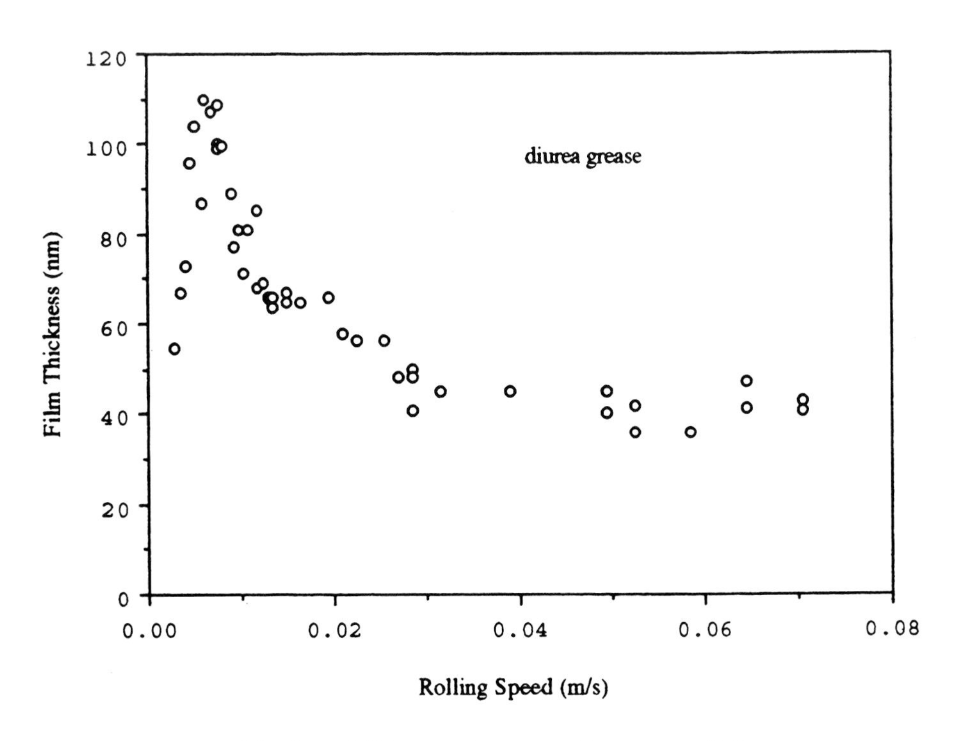

Fig. 5 EHL starvation curve for a grease (**24**).

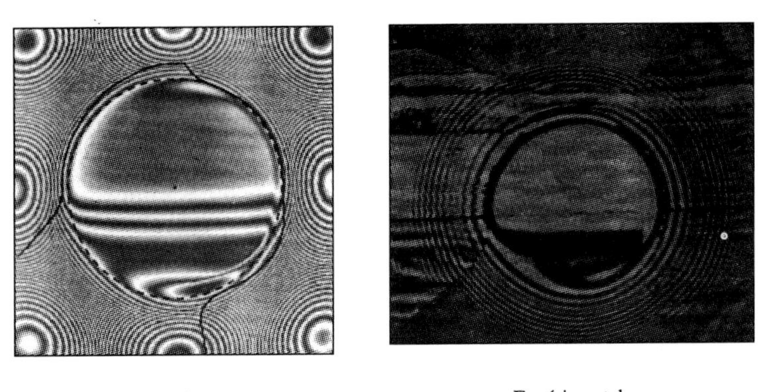

Fig. 6 Comparison between numerical solution and experiment for asymmetrical starvation of a point contact (**25, 20**).

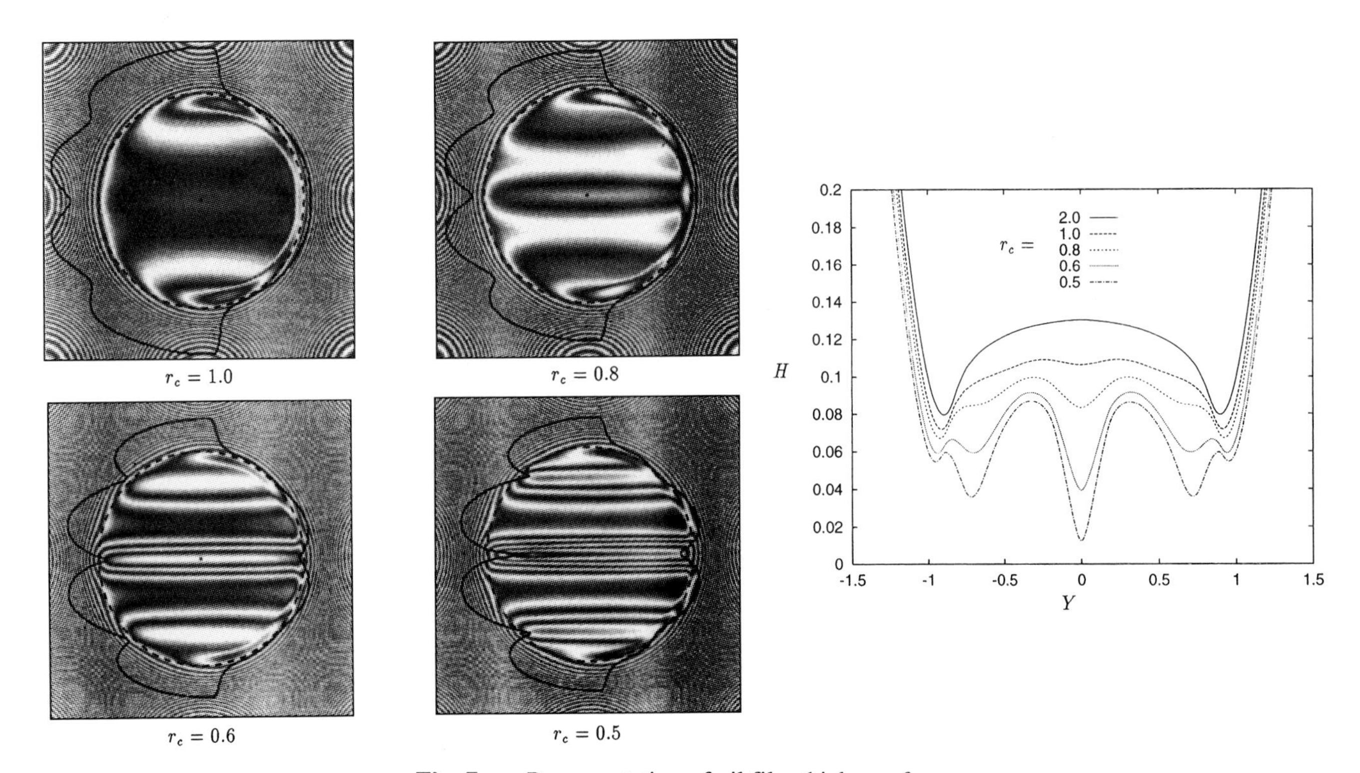

Fig. 7 Representation of oil film thickness for uneven starvation of the contact inlet (**25**).

problem lies in being unable to predict with confidence the separating film thickness arising from the local contact conditions.

The above analyses show that using the basic bulk properties of the lubricant, such as viscosity and viscosity-pressure coefficients, and the surface roughness properties as measured outside the heavily loaded contacts and comparing the measured roughness with the theoretically calculated lubricant film thickness, does not allow prediction of the contact behaviour for low lambda applications. Only when the calculated oil film-thickness is large compared with the surface-roughness, and no starvation effects are present, can the smooth surface approximation be used to predict the lubricant film thickness. As soon as the lubricated surfaces are not mathematically smooth and the roughness heights are not negligible compared with the mean oil film thickness, the local pressure fluctuations caused by the asperities will have an influence on the elastic deformations of the surfaces. This leads to sensitivity for roughness structure and directionality by the lubricated contact, as well as sensitivity to the amount of oil available for lubrication.

If the contact is starved because too little active lubricant is present, the normal EHL theory overestimates the film thickness if fully flooded results are used, but at the same time the local asperity pressure variation within the heavily loaded contact can partly flatten the roughness asperities, leading to smoother surfaces in the contact area.

For very smooth surfaces, the elastic deformations lead to an almost total conformity within the EHL contact and thus to a very high ratio of oil film thickness to roughness, even for very thin lubricant films. Thus the calculated oil film thickness compared with the measured roughness of the surfaces cannot directly determine if a separating oil film is present in the contact or not. It depends both on the details of the surface structure and on the amount of lubricant available.

The mechanism describing how the last molecular layer of lubricant is removed from the surface tops and moved into the valleys in the structure or out from the contact still needs to be determined.

7 ACKNOWLEDGEMENT

The author wishes to thank Dr H.H. Wittmeyer, Managing Director of SKF Engineering & Research Centre B.V., for his kind permission to publish this paper.

8 REFERENCES

1 **Tower, B.,** First Report on Friction Experiments (Friction of Lubricated Bearings). *Proc. Institution of Mechanical Engineers,* London (1883) pp. 632–659.

2 **Reynolds, O.,** On the Theory of Lubrication and its Application to Mr Beauchamp Tower's Experiments, including an Experimental Determination of the Viscosity of Olive Oil. *Phil. Trans. Roy. Soc.,* vol. **177,** p 157. London, (1886).

3 **Martin, H.M.,** The Lubrication of Gear Teeth. *Engineering,* London, **119**, Aug. 11, 1916.

4 **Meldahl, A.,** *Beitrag zur Theorie der Gertiebe-Schmierung und der Beanspruchung geschmierter Zahnflanken.* Brown Boveri Mitteilungen, Nov. 1941, s 374 ff.

5 **Ertel, A.M.,** Hydrodynamic lubrication based on new principles, *Akad. Nauk SSSR Prikadnaya Math. Mekh.,* **3**(2), (1939) 41–52.

6 **Grubin, A.N.** and **Vinogradova, I.E.,** *Central Scientific Research Institute for Technology and Mechanical Engineering.* Book No.30, Moscow 1949, (D.I.S.R. translation No. 337).

7 **Dowson, D.** and **Higginson, G.R.**, A Numerical Solution to the Elastohydrodynamic Problem. *J. Mech. Eng. Sc.* **1,** (1959) p 1, 6.

8 **Dowson, D.**, *History of Tribology.* Longman Group Limited, London, (1979).

9 **Jacobson, B.O.,** *On the Lubrication of Heavily Loaded Spherical Surfaces Considering Surface Deformations and Solidification of the Lubricant.* Acta Polytechnica Scandinavica, Mech. Eng. Series No. 54 (1970).

10 **Jacobson, B.O.,** Elasto-Solidifying Lubrication of Spherical Surfaces. *ASME* Paper No. **72**–Lub–7, (1972).

11 **Hamrock, B.J.** and **Dowson, D.**, Isothermal Elastohydrodynamic Lubrication of Point Contacts, Part II – Ellipticity Parameter Results. *Trans. ASME, Journal of Lubrication Technology,* Vol. **98**, July 1976, pp. 375–383.

12 **Jacobson, B.O.** and **Hamrock, B.J.**, *Non-Newtonian Model Incorporated into Elastohydrodynamic Lubrication of Rectangular Contacts.* NASA TM 83318, (1983).

13 **Bridgeman, P.W.,** *Proc. Am. Acad.* **74,** (12) (1942) pp. 399–424.

14 **Höglund, E.** and **Jacobson, B.**, Experimental Investigation of the Shear Strength of Lubricants Subjected to High Pressure and Temperature. *Trans. ASME, Journal of Tribology,* **108,** (1986), pp. 571–578.

15 **Jacobson, B.**, *Solid Oil in Rolling Contacts.* Intertribo, (1981), pp. 92–98. Çeskoslovenská Vedeckotechnická Spoloçnost, Vysoké Tatry.

16 **Jacobson, B., Ioannides, E.** and **Tripp, J.H.**, Redistribution of Solidified Films in Rough Hertzian Contacts, Part I: Theory, *Proc. 14th Leeds-Lyon Symp.,* Lyon (1987). In Interface Dynamics, Tribolgy Series 12, Elsevier, Amsterdam (1988), pp. 51–57.

17 **Jacobson, B.,** Redistribution of Solidified Films in Rough Hertzian Contacts, Part II: Experimental. *Proc. 14th Leeds-Lyon Symp., Lyon* (1987). In Interface Dynamics, Tribology Series 12, Elsevier, Amsterdam (1988), pp. 59–63.

18 **Jacobson, B.** Mixed Lubrication, *Wear,* **136,** (1) (1990), pp. 99–116.

19 **Hamer, J.C., Sayles, R.S.** and **Ioannides, E.,** The Collapse of Sliding Micro-EHL Films by Plastic Extrusion. *Proc. Joint ASME/STLE Tribology Conf.,* Toronto, Canada (1990), Paper 90–Trib–47.

20 **Wedeven, L.D., Evans, D.** and **Cameron, A.,** Optical Analysis of Ball Bearing Starvation. *Trans. ASME, Journal of Lubrication Technology,* **93,** (1971), p. 349.

21 **Kingsbury, E.,** Parched Elastohydrodynamic Lubrication. Trans. *ASME, Journal of Tribology* **107,** (1985), p. 229.

22 **Guangteng, G., Cann, P.** and **Spikes, H.A.,** A Study of Parched Lubrication. Wear, 153, (1992), p. 91.

23 **Cann, P.M.** and **Spikes, H.A.,** Film Thickness Measurements of Lubricating Greases under Normally Starved Conditions. *NLGI Spokesman,* **56,** (1992), p. 21.

24 **Cann, P.M., Ioannides, E., Jacobson, B.** and **Lubrecht, A.A.,** The Lambda Ratio – A Critical Re-Examination. *Wear,* **175,** (1994), pp. 177–188.

25 **Chevalier, F.,** Modélisation des Conditions d'Alimentation dans les Contacts Élastohydrodynamiques Ponctuels. *These présentée devant l'INSA de Lyon,* 96 INSAL 0124, 20 décembre 1996.

Tribology as a maintenance tool

Y KIMURA
Faculty of Engineering, Kagawa University, Japan

SYNOPSIS

Tribology plays an important role in maintenance engineering on the one hand, while a major contribution of tribology is achieved through maintenance on the other. This paper describes the significance and framework of maintenance engineering and introduces some work on maintenance tribology.

1 INTRODUCTION

Engineering can be divided into two areas: engineering for production and engineering for maintenance. Production in industry starts from the perception of the need for a commodity or item which should have a certain required function. The process of defining a specific attribute for the required function is called *design,* and the process of giving substance to the defined attribute is called *manufacturing*.

Once what has been manufactured comes into the possession of its user and is put into operation, sooner or later its substance will be changed physically or chemically. In many cases these changes will impair its function, and are called *deterioration*. Maintenance can be defined to control the condition of components, systems, plants etc., so as to maintain their function above the required levels for the whole of their life. It includes various processes, e.g. monitoring the changes in the substance, diagnosing its function, defining and implementing practical maintenance actions like repairing, replacing and scrapping.

It can be said that the major part of engineering has so far been concerned with the processes of production. At the present time, however, there are reasons for us to consider that engineering for maintenance described above has attained a significance comparable to or even greater than that for production. First, recognition that resources are limited urges us to realize the importance of making full use of those facilities already in our possession. Secondly, the increase in the scale and complexity of systems for manufacturing, energy supply, transportation, communication etc. has opened a possibility that simple failure of a component can cause a breakdown of the total system or even a disaster.

In the present paper, the significance of maintenance as an economic sector and the important concern of tribology are first outlined. Secondly, the framework of maintenance engineering is discussed. Finally, some examples of contributions to maintenance tribology from the author's laboratory are introduced.

2 ECONOMICS OF MAINTENANCE

Extensive surveys of the economical significance of maintenance in Japan were made in the 1980s (**1, 2**). The estimated total maintenance cost of mechanical systems in Japan in 1979 amounted 8.3×10^{12} yen, which occupied 3.75% of the gross national product (**1**). Here, the mechanical systems were defined as the equipment and machinery for business use in the mining, construction, manufacturing, transportation, communication and energy industries and building equipments. These figures seem to increase with the maturity of society.

More recent data on the estimated maintenance cost of production plants in Japan are shown in Fig. 1 (**3**) comparing it with her total product sales. Although the economical recession reduced those amounts in 1992, the maintenance cost amounts to around 10 trillion yen occupying some 3% of the total product sales. This percentage is different between industries as shown in Fig. 2 (**3**) for 1994; it is as high as 5% in the textile, paper

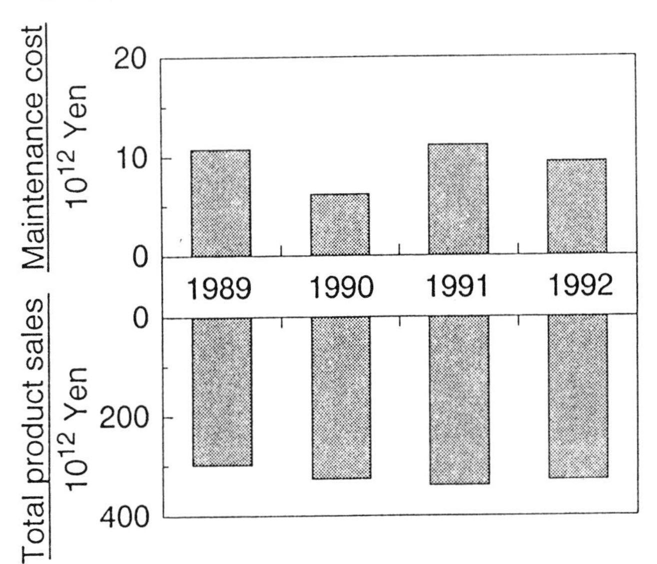

Fig. 1 Maintenance cost of production plants and total product sales in Japan (**3**).

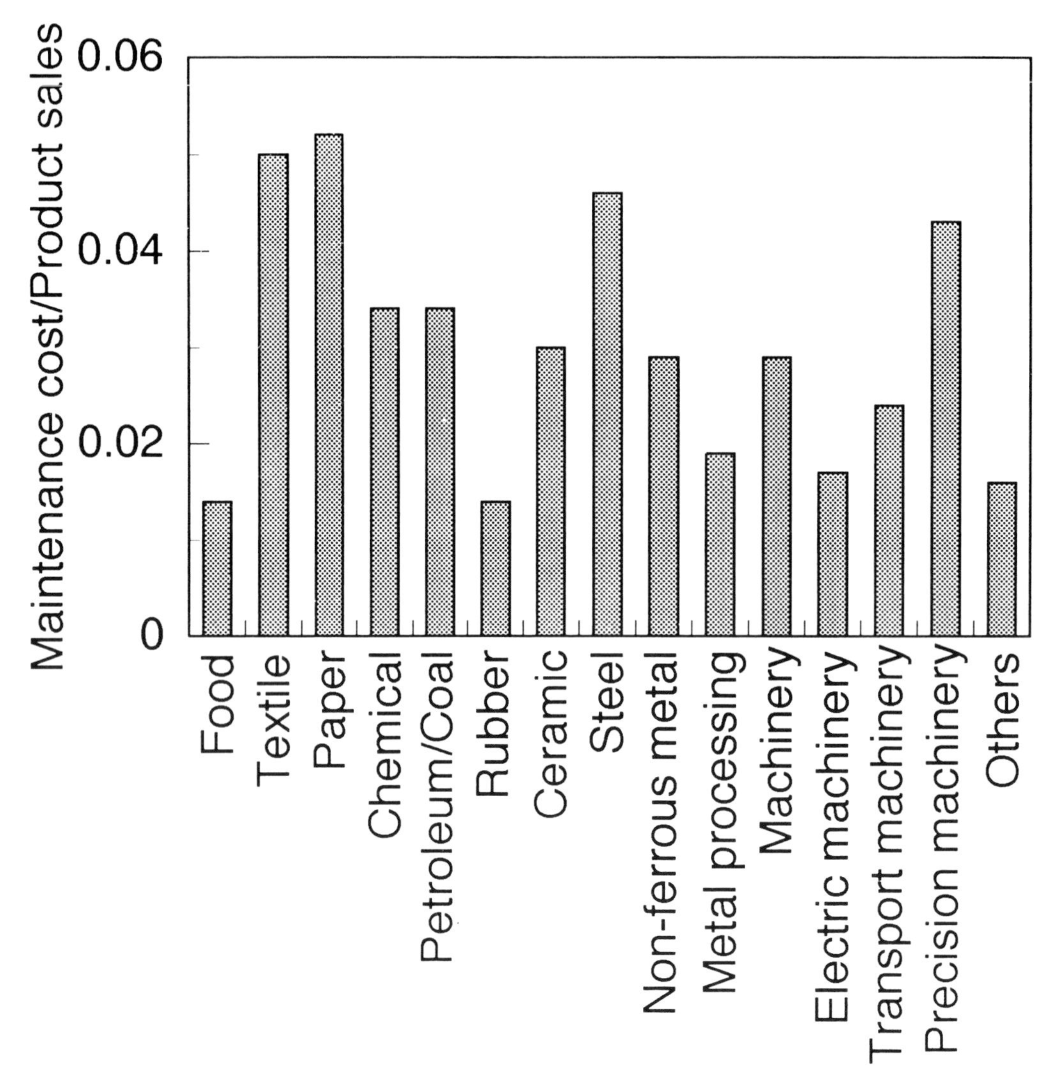

Fig. 2 The ratio of maintenance cost to product sales in individual industries in Japan, 1994 (**3**).

and steel industries, while less than 2% in the food, rubber, metal and electromechanical industries. Although the maintenance cost amounts to such a large sum, maintenance industry does not form a big independent economic sector. Its total sales are roughly a half of the total maintenance cost, and the remaining half is the cost of in-house maintenance.

No definite data are available about the economics of maintenance tribology. However, thirty years ago when the term 'tribology' was born, estimation was made of savings reasonably obtainable by British industry through tribology as shown in Table 1 (**4**). It should be noted that the four items (d) to (g) in the table are directly or indirectly concerned with maintenance. That is, more than 90% of the economic contribution of tribology could be achieved through maintenance, evidently showing the deep concern of tribology in maintenance.

3 FRAMEWORK OF MAINTENANCE ENGINEERING

Maintenance tribology has not yet been systematized as an engineering field. However, this is not a particular feature of maintenance tribology, but also holds true of maintenance engineering as a whole. In what follows, major research areas of maintenance engineering are introduced.

3.1 System analysis

If we consider maintenance of a plant, for example, it is composed of thousands of mechanical systems comprising millions of components. However, it is reasonable to select components or systems as the objects of maintenance which have critical significance for operation of the plant. Analytical tools are available for this task (**5**):

Table 1 Estimated figures of savings reasonably obtainable by British industry (**4**).

		10^6 £s
(a)	Reduction in energy consumption through lower friction	20
(b)	Reduction in manpower	10
(c)	Savings in lubricant costs	10
(d)	Savings in maintenance and replacement costs	230
(e)	Savings of losses consequential upon breakdowns	115
(f)	Savings in investment due to higher utilization and greater mechanical efficiency of machinery	22
(g)	Savings in investment through increased life of machinery	100
	1965 Totals	515

1. FMECA (Failure mode, effects and criticality analysis). A failure mode is first assumed for each component, and its consequences are assessed by following chains of causes and effects. The criticality of the failure of the component is then determined.
2. FTA (Fault tree analysis). A well-known analysis which schematically forms a logic tree connecting events from possible basic events to a top event with 'and gates' and 'or gates'. This gives a structure for the propagation of failures and a quantitative prediction of a critical failure through failure statistics of components.
3. ETA (Event tree analysis). An event tree is formed starting from a basic event through multiple stages of forks of 'success' and 'failure' with definite probabilities up to top events, failure of a system, through various paths. This gives the possible contribution of basic events to certain critical failures.

3.2 Failure physics

This is to study the failure modes and failure patterns of particular components. Fatigue, corrosion, wear and other tribological failures are typical failure modes of mechanical components. Prediction of the possibility of their occurrence in particular systems or components under their operating conditions is a primary issue in failure mode studies, for which we can learn from traditional fields of engineering for production, of which tribology provides a typical example.

On the other hand, a generalized failure pattern can be defined by the change in a function level with operating time, operating cycles, or other time-like variables as illustrated in Fig. 3 (**6**). The essential phenomena and mechanisms bringing about the failure modes have been understood to certain extents and, as is typically the case with wear, we have a huge mass of data on tribological failures. However, it is a feature of engineering for production that those data have been compiled so that they can be fed back to the initial conditions, like design of tribocomponents and selection of lubricants and tribomaterials. What is needed in the present context of maintenance engineering is to feed forward the data to predict the future behaviour of tribosystems or tribocomponents. This means that new compilation of available data is required for them to be used as reference in maintenance engineering.

3.3 Condition monitoring technology

This is the most extensively studied area in maintenance engineering, and no detailed description seems necessary. In the present terminology, this is the technology of monitoring the deterioration, i.e. the decrease in the function level, of a system or a component, continuously or intermittently. A survey of the maintenance technology in production plants in Japan (**2**) revealed that bearings

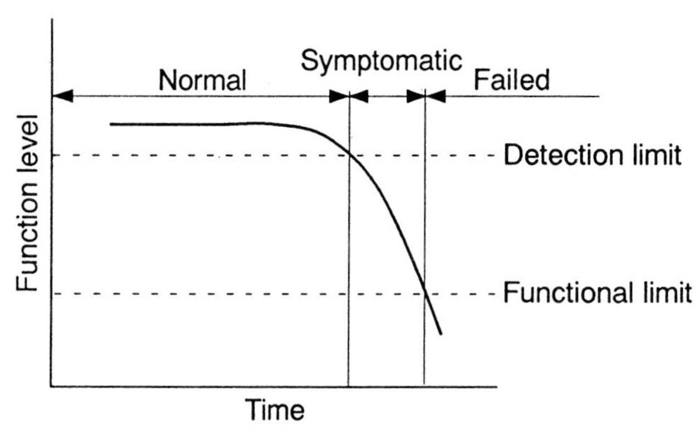

Fig. 3 A typical failure pattern showing lowering of functional level with operating time (**6**).

occupied about 50% of the total number of components being monitored, showing the importance of tribology in maintenance.

A number of monitoring technologies has been developed which are based on various principles. With bearings and other tribological components, vibration, temperature and oil contamination are the quantities commonly monitored. Although these technologies have been working successfully, high reliability, long life times and low prices of bearings have, paradoxically, made monitoring impractical in many cases; that is, monitoring instruments are less reliable, have shorter lives and are much more expensive. Further, those quantities being monitored have been selected by the ease of monitoring and, in most cases, interpretation of the data has been made according to experience with particular components or systems. Subjects for future study in this area include planning reasonable instrumentation for monitoring and developing technology for monitoring quantities which convey direct information about tribological deterioration.

3.4 Diagnosis

For condition monitoring technology to be significant, its outputs should be interpreted properly to give prediction of the future evolution of deterioration in specific components or systems in specific failure modes. Forming a contrast to the development of monitoring technology, however, this is still in a primitive stage.

In some cases, simple extrapolation of data can be used for prediction, when direct quantitative determination of deterioration is possible. For example, least square approximation of past wear-time data by a polynomial function is used to predict the time when the wear will reach a critical value. However, this needs to bring a system to a halt, which is not always allowed, and then the quantities as listed above, which can be monitored without causing down-time, must be used to indicate deterioration, though indirectly.

In many cases when those indirect indices are monitored, it is necessary to make a rule based on a knowledge base about the deterioration of specific components or systems. For example, a combination of the vibration speed at specific frequencies, direct-reading ferrography data and the trend of the increase in iron content in the lubricating oil is used to predict failure of gears in DIXPERT diagnosis logic (**7**). In another example with bearings in a steam turbine generator (**8**), necessary maintenance measures are given for specific levels of oil contamination with wear debris.

Either way, these diagnosis methodologies are based on our empirical knowledge of those deterioration processes. If we realize that one of the main roles of a theory is prediction, theories of tribological failures should provide means of predicting their future evolution. Model-based diagnostics of mechanical systems (**9**) may be interesting in this connection. At the present time, however, it is still unclear whether we can have a generalized common knowledge base for tribological failure modes.

3.5 Maintenance strategy

With simple equipment like a desk lamp, it would be enough to go out to purchase a new bulb to replace one which has failed. This is typical 'breakdown maintenance' and it might cause little inconvenience except for a night just before the deadline for submission of a manuscript. However, the increasing scale and complexity of systems no longer allow such unexpected down-time for safety, reliability and economical reasons, and the concept of 'preventive maintenance' was introduced in the 1950s.

A number of confusing terms are used to represent methods of maintenance, but a broad classification into 'time-based maintenance' and 'condition-based maintenance' seems reasonable. Time-based maintenance includes 'overhaul' and 'preventive maintenance', while condition-based maintenance includes 'breakdown maintenance' and 'predictive maintenance'. Time-based maintenance is based on the 'bathtub curve', which represents a time-reliability relationship. That is, the probability of failure is high at the outset of operation because of deficiencies in production; it is then reduced to a practically constant low level where only chance failures are expected; and finally it increases again showing the wear-out failures. Once a bathtub curve is obtained for a system or a component from statistics, repair or replacement before the onset of the wear-out failure can minimize the risk of its failure.

Although large-scale overhaul is being performed for systems of critical importance, marked trend is that condition-based maintenance is taking over for some reasons. First, system analysis has succeeded in identifying non-essential components for which the most economical breakdown maintenance would suffice. Second, perfect implementation of time-based maintenance eliminates failure statistics on which it should be based. Third, time-based maintenance is essentially impractical for novel systems which lack those statistics. Figure 4 shows a result of a recent survey the on maintenance of rolling bearings (**10**).

The concept of predictive maintenance is shown in Fig. 5. It is dependent on the technology of condition monitoring. Changes in substance are detected mainly through certain changes in its function. Diagnosis is then made to determine necessary maintenance actions. For the predictive maintenance to be practical, it is essential that the deterioration being monitored has a sufficient duration of the symptomatic period, Fig. 3. This depends on the nature of the deterioration, detection limit of the condition monitoring technology, the diagnosis system etc., and these factors limit the application of predictive maintenance.

A maintenance strategy for a particular system or component must be developed by comparing the advantages and disadvantages of the maintenance methods

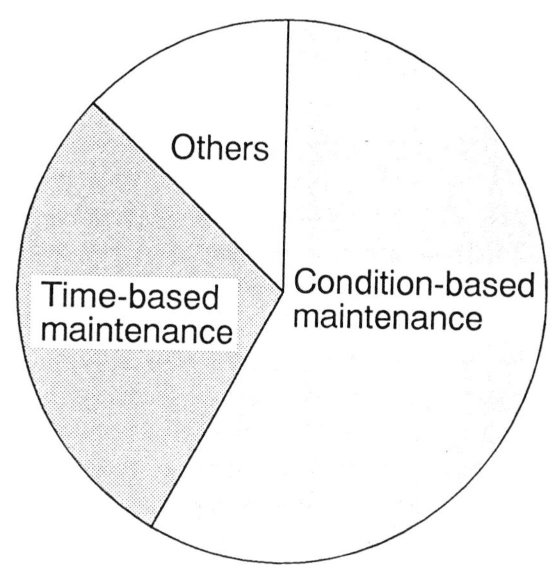

Fig. 4 Maintenance of rolling bearings (**10**).

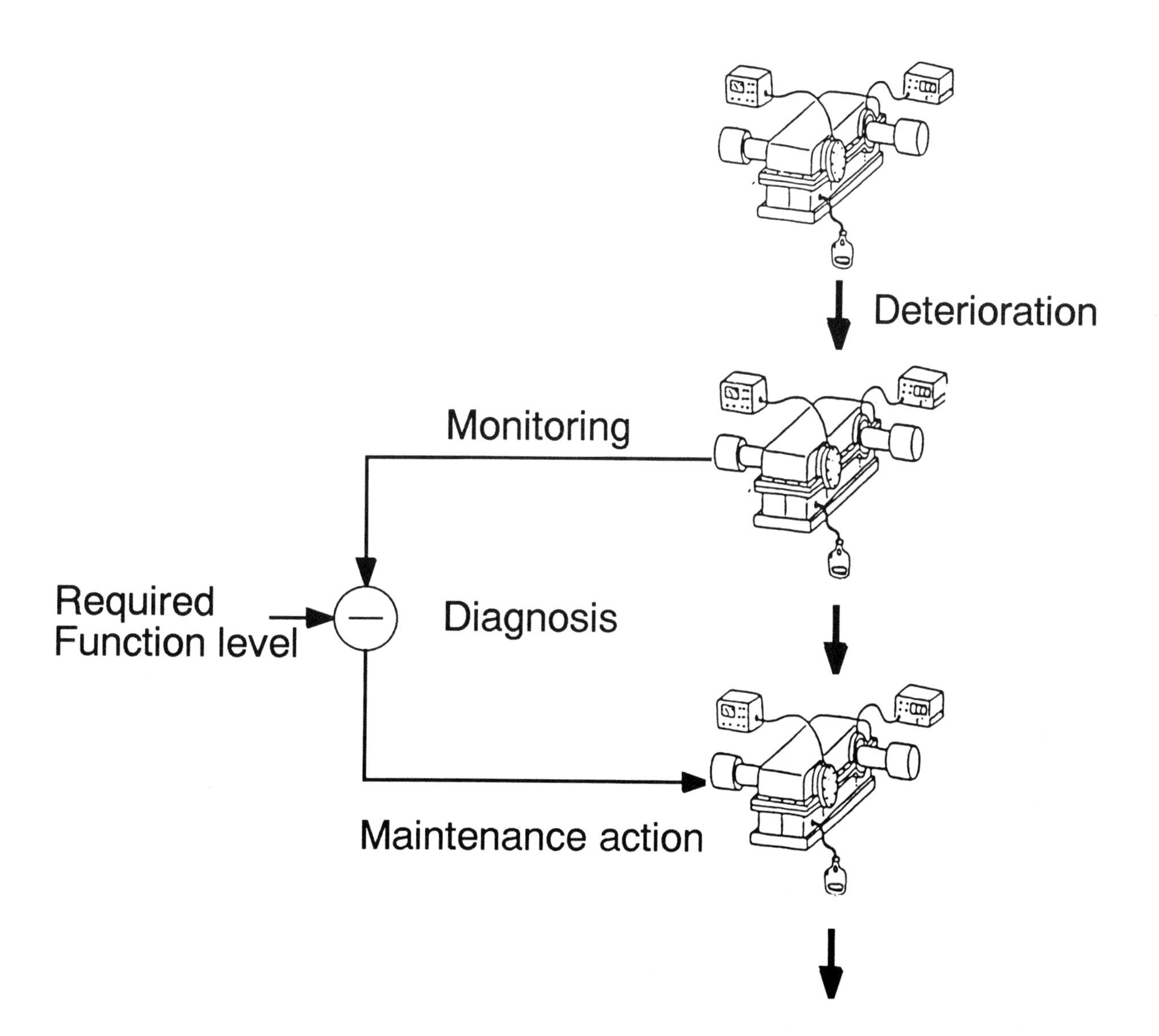

Fig. 5 Concept of the predictive maintenance.

based on the required levels of safety, reliability and economy. Computerized programs have been developed to determine and implement maintenance strategy for individual mechanical systems in a whole plant, which are called 'maintenance systems' (**11**).

4 STUDIES ON MAINTENANCE TRIBOLOGY

As mentioned earlier, the majority of possible contributions of tribology must be made through maintenance, and most tribological work can be said to have some relevance to maintenance engineering. In what follows, some works are introduced which were conducted with the explicit intention of contributing to maintenance tribology.

4.1 Effects of deterioration of gasoline engine oils on wear

4.1.1 Objective

Engine oils deteriorate during their use in engines. The deterioration is commonly expressed in terms of the changes in characteristic physical and chemical parameters, e.g. viscosity, total acid number, total base number, quantity of insolubles and absorbance in infrared spectra. These changes represent various aspects of the fall in lubrication performance of engine oils in a number of tribological contacts which are operating under various conditions.

This work (**12**) tried to determine changes in wear of steel lubricated with engine oils which had deteriorated through use in real gasoline engines, and to correlate the changes with those of the characteristic parameters.

4.1.2 Experimental

Engine oils used for the experiments were of SAE 10W/30 viscosity grade and of SF or SG performance grades with different additive formulations. Two types of gasoline engines were used: 550 cm^3 two-cylinder and 1400 cm^3 four-cylinder overhead camshaft engines. Their operation modes included constant-speed endurance tests and stop-and-go type tests up to 40000 km in equivalent running distance at the maximum. Oil samples were taken from the engines at predetermined intervals, followed by filling with the identical fresh oil.

Wear experiments were conducted on a three-roller-on-ring machine, in which a flat end of a rotating ring, 34 mm in outer and 26 mm in inner diameters, was made to slide against the cylindrical surface of three bearing rollers of 10 mm diameter, at right angle to their axis. The rings and the rollers were made from AISI 52100 bearing steel hardened to 750 HV. The sliding was made at a speed of 47.1 mm/s under a static load of 785 N; the initial maximum Hertzian pressure was 0.7 GPa which simulated cam-follower contacts. A gimbal mechanism supporting the sliding system allowed uniform contact loads at the three contacts. Oil temperature was kept at 40°C, and volumetric wear on the three rollers was determined after sliding for 1500 m in total.

4.1.3 Results and discussion

It was found that the oil deterioration through its use in the engines caused marked changes in wear. With an increase in the equivalent running distance of the engines, wear first decreased and then increased, exceeding that in the fresh oil.

Effects of various characteristic parameters of deterioration on wear were extensively studied, and the most definite correlation was observed when wear was plotted against the increase in the total acid number (TAN) above the initial value, as seen in Fig. 6(a). With the increase in TAN, wear first decreased slightly, and then tended to increase after achieving a minimum; this increase became evident when TAN exceeded 1 mgKOH/g.

This finding indicates the presence of opposing effects of acidic components in the oils. The initial value of TAN, 1.35–2.61 mgKOH/g in this case, is caused by acidic constituents of the additives, which should protect steel surfaces by forming adsorbed films. On the other hand, the increase in TAN during use represents the increase in active acidic materials produced in engines. Once deterioration makes the latter effect predominate, it accelerates corrosive wear resulting in the marked increase in the amount of wear.

Changes in the total base number (TBN) were also correlated with wear, Fig. 6(b). It was found that wear was minimized while TBN was larger than 2 mgKOH/g as determined by the HCl method. The TBN for fresh oils is dependent on the species and amount of additives such as detergents or dispersants. With used oils, it represents remaining basicity, which should neutralize the active acidic materials mentioned above. Acidic attack is activated when TBN becomes smaller than some critical value, 2 mgKOH/g in this case.

It is interesting to note that an on-line engine oil deterioration sensor recently developed (**13**) makes use of the change in infrared absorbance at 1630 cm^{-1}. This wave number was chosen to give an indication of the presence of nitric esters in used oils. This absorbance was reported to be related to the increase in TAN; the absorption shows an abrupt increase when TAN becomes 1–2 mgKOH/g.

In conclusion, if we take a wear-preventing property as a function of gasoline engine oils, it can be represented by a change in TAN, with which on-line sensing through the change in the infrared absorbance is possible.

4.2 Debris size distribution and wearing conditions

4.2.1 Objective

Analyses of wear debris in lubricating oils in use are established technologies condition monitoring such as

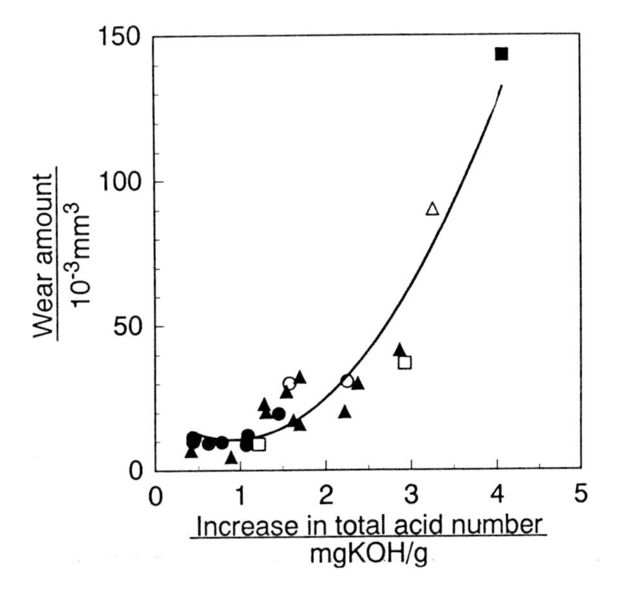

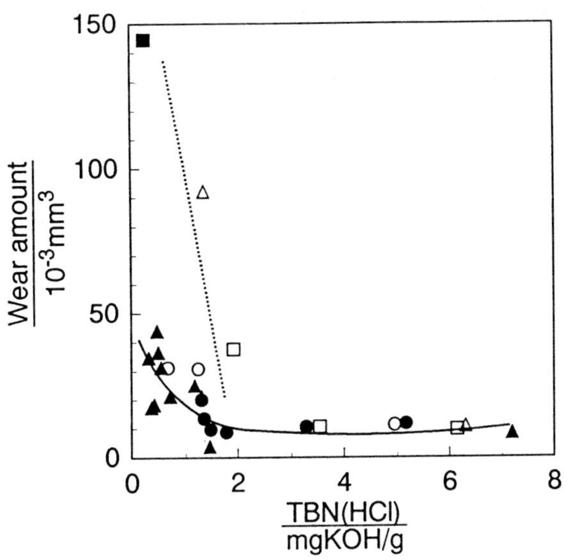

Fig. 6 Changes in wear with deterioration of engine oil; (a) change with the increase in TAN, (b) change with TBN (**12**).

ferrography and spectrometric oil analysis program. In direct-reading ferrography, it is suggested that both the total amount of debris and the relative concentration of larger debris will increase when the wearing condition approaches a catastrophic stage, or the region of the wear-out failure in the bathtub curve. In laboratory tests on a wear tester, however, simplified sliding conditions seem to make appearance of wear-out failure unlikely. This work (**14**) tried to find out what changes in wearing conditions had effects on the debris size distribution.

4.2.2 Experimental

Sliding experiments were conducted in which the lubricated wear of steel was determined on a three-pin-on-disk machine under varied conditions. A flat end of each pin of 4 mm in diameter was made to slide against a disk at a speed of 47.1 mm/s. The pins and the disks were made of 0.35% carbon steel having 160 HV. A standard sliding condition (STD) was defined; normal load was 392 N, temperature was 40°C, and the lubricant was a paraffinic base oil having viscosity of 13.3 mm^2/s at 40°C and TAN of 0.01 mgKOH/g.

Changes in sliding conditions were made in five ways. That is: the lubricant was contaminated with 1 vol% water to form an emulsion (WCL); the lubricant was previously deteriorated by oxidizing to have TAN of 0.3 mgKOH/g (DET); the load was increased to 588 N (HL1); and 784 N (HL2); and the temperature was raised to 80°C (HT). In each of these the other conditions were left unchanged.

Sliding was interrupted after certain periods to collect wear debris in the lubricant, and started again with the identical lubricant. Wear debris were collected on a membrane filter having 0.45 μm pores to form a sample. A CCD camera mounted on an optical microscope fed images of the debris to an image analyzer. Since the debris had generally flat shapes, an equivalent diameter was defined on their projected figure from which a debris size was represented.

4.2.3 Results and discussion

For each sample, 15-20 frames were observed under the microscope, which formed about 1% of the total area of the filter. Results showed that the size of the debris practically followed an exponential distribution in any case. An average equivalent diameter was determined for each sliding condition where small erratic deviations for large debris size were ignored.

The change in the average equivalent diameter with sliding time duration for STD was given by a dashed curve in Fig. 7, which shows a gradual decrease to a steady value. This was compared with the data for WCL, DET and HT in Fig. 7(a). Although initial values of the average equivalent diameter were different depending on the conditions, they tended to converge to a steady value which was practically the same as that for STD. In contrast, the normal load was found to affect the steady value of the average equivalent diameter as shown in Fig. 7(b). A higher load resulted in a larger steady value. Further, the convergence to the steady value seemed to take place more rapidly for a higher load.

It was concluded that, under the conditions examined in this series of experiments, only the normal load affected the steady value of the size of wear debris. This implies a possibility that a cause of anomaly in wear behavior can be identified by monitoring the debris size if it is a change in the actual load supported by solid contact. It can take place in various cases. For example, misalignment in assembly or reassembly and uneven wear can cause localized high contact loads, and these accordingly necessitate particular maintenance actions.

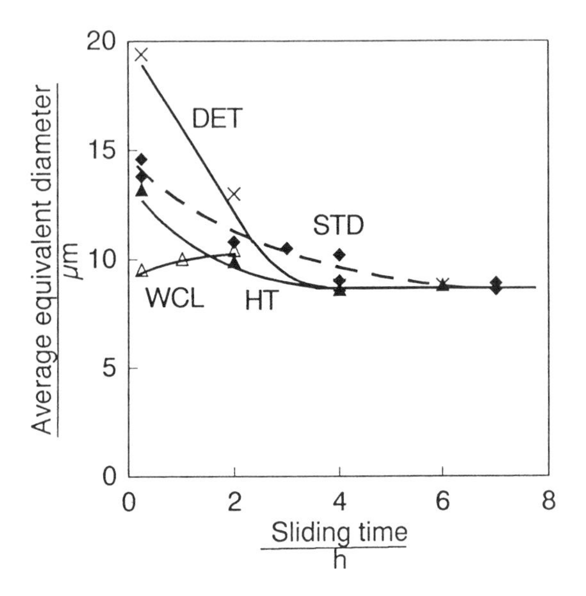

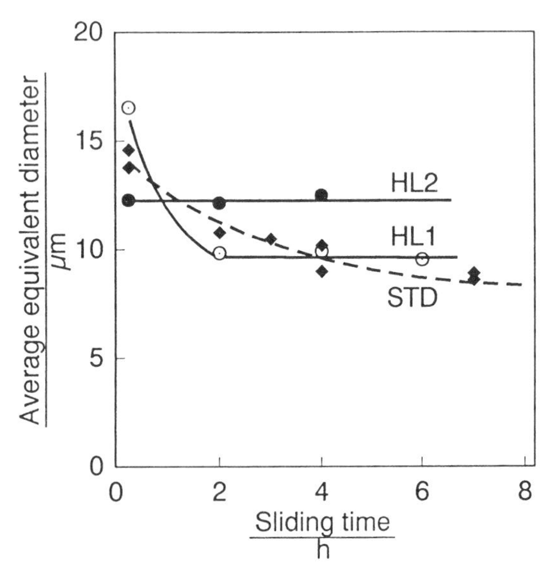

Fig. 7 Changes in the size of wear debris with sliding time under varied sliding conditions; (a) in contaminated lubricant and at high temperature, (b) under different loads (**14**).

4.3 Deterioration-tree analysis of tribocomponents

4.3.1 Objective

Normally, mechanical systems include a number of tribological components, and deterioration of one of them can propagate in the system in various ways to cause that of others and an eventual system failure. This work proposes an analytical method by which propagation of basic tribological deterioration to a system failure can be described and the possibility of condition monitoring in the system can be assessed.

4.3.2 Deterioration transfer function

Let us consider a rolling element bearing, for example. Basic tribological deterioration modes can be wear, flaking etc. which take place at the interfaces between the rolling elements and the races. The deterioration pattern i.e. the evolution with time of wear in a dimensional change may be represented as the value *X1* at the top left in Fig. 8, and that of flaking in a pitted area may be represented as *X1* at the center left in the same figure; the former is progressive while the latter starts abruptly after a certain period of operation.

As mentioned earlier, deterioration of bearings is most commonly monitored by the change in the vibration level. The increase in the vibration acceleration caused by the evolution of wear and flaking will typically be represented by the graphs at the right in Fig. 8. The vibration level X2 due to wear will also grow progressively, while that due to flaking will jump from nil to a certain level.

A deterioration transfer function can be defined as the relation between the basic tribological deterioration *X1* and the deterioration in a function of a component *X2*. As suggested in the figure, this function can have various shapes which are characteristic to particular tribocomponents and particular deterioration modes.

4.3.3 Deterioration tree

Basic tribological deterioration like wear and flaking can cause secondary deterioration in the component or in the system. For example, vibration can increase instantaneous loads leading to increased friction, wear can cause adverse shape changes leading to deficiency of hydrodynamic lubrication and increase in friction, increase in friction can cause temperature rise leading to material deterioration or decrease in viscosity of lubricant, and so forth.

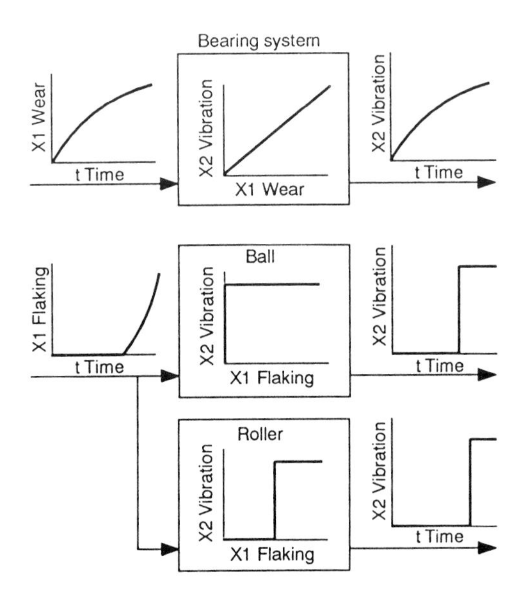

Fig. 8 Examples of the deterioration transfer function.

How the propagation of deterioration takes place in a components or a system can be described by a deterioration tree as illustrated in Fig. 9. Similarly to the fault tree, it starts from basic events representing basic deterioration at individual interfaces making tribological contact. The deterioration transfer function defined above, together with 'and gates', or 'or gates' connect them to secondary and successive deterioration modes, eventually up to a failure of the whole system.

Advantages of the use of the deterioration tree can be expected as follows. First, this can provide quantitative prediction of failures if a database is furnished with tribological deterioration of individual elements and the deterioration transfer functions of components or systems. Secondly, adequate sites for the condition monitoring can be indicated in the system. They are shown by small open triangles in Fig. 9, while solid triangles show monitoring does not make sense at those sites.

5 CONCLUDING REMARKS

Although the importance of maintenance engineering has recently been stressed in many occasions, its systematization seems to be proceeding very slowly. This review has tried to emphasize the economical significance of maintenance, to describe fields of maintenance engineering and to illustrate some researches, putting stresses on the relation with tribology.

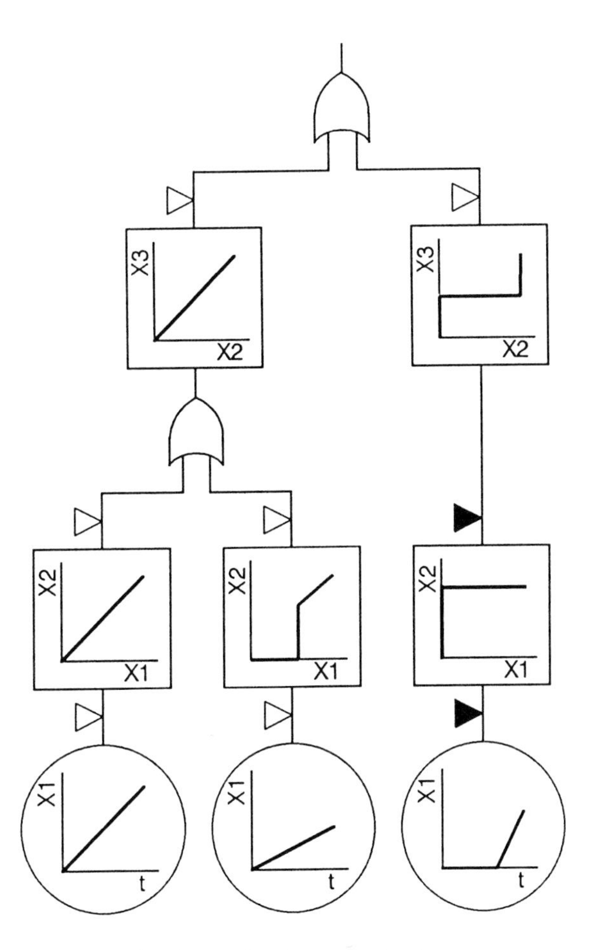

Fig. 9 The deterioration tree.

Tribology plays an important role in maintenance engineering on the one hand, while a major contribution of tribology is achieved through maintenance on the other. The following is a partial list of subjects of tribological study in this field.

1. Recompilation of experimental data and development of theories for the prediction of the evolution of tribological deterioration.
2. Development of a methodology for processing data provided by condition monitoring of a tremendous number of tribocomponents.
3. Development of a logic to describe the propagation of deterioration of tribocomponents: from basic events to the failure of a system.
4. Establishment of the whole system of maintenance engineering.

6 REFERENCES

1 **Watanabe, Y.,** Economical aspect of industrial maintenance, *J. Japan Soc. Precision Eng.,* 1982, **48,** pp. 11–15, in Japanese.

2 Report on a survey of maintenance technology for plants, *Japan Inst. Plant Maintenance*, 1985.

3 *Study on present status and future trend of maintenance works in Japan*, Japan Soc. for the Promotion of Machine Industry, 1996.

4 **Jost, H. P.,** Lubrication (Tribology) – *A report on the present position and industry's needs,* Department of Education and Science, 1966, H. M. Stationery Office. Quoted from: Some economic factors of tribology, *Proc. JSLE-ASLE Intern. Tribology Conf.* (Sakurai, T. ed.) Tokyo, Japan, 9–11 June, 1975, pp. 2–19 (Elsevier Sci. Publ. Co., Amsterdam).

5 **Anderson, R.T.** and **Neri, L.,** *Reliability-centered maintenance: Management and engineering methods,* 1990, (Elsevier Applied Science, London).

6 **Takata, S.** and **O'shima, E.,** *Framework for computer assisted life-cycle Maintenance,* Selected Papers from IFAC-MACS Symp. (Isermann, R. and Freyermuth, B. eds.) Baden-Baden, Germany, 10–13 Sept., 1991, pp. 451–456.

7 **Holmberg, K., Kuoppala, R.,** and **Vuoti, A.,** Expert system for wear failure prediction, *Proc. 5th Intern. Congress on Tribology* (Holmberg, K. and Nieminen, I. eds.) Espoo, Finland, 12–15, June, 1989 pp. 32–37.

8 **Ahn, H.S.,** *et al.,* Practical contaminant analysis of lubricating oil in a steam turbine-generator, *Tribology International,* 1996, **29**, pp. 161–168.

9 **Natke, H.G.** and **Cempel, C.,** *Model-aided diagnosis of mechanical systems,* 1997, (Springer, Berlin).

10 *Report of the Research Group on Diagnosis and Evaluation of Machinery,* 1996, in Japanese.

11 **Okazaki, E.** and **Oozasa, K.,** Development of a computerized maintenance system 'ADAMS', *Plant Engineer,* Dec., 1986, pp. 63–71, in Japanese.

12 **Moon, W-S.** and **Kimura, Y.,** Wear-preventive property of used gasoline engine oils, *Wear,* 1990, **139**, pp. 351–365.

13 **Murakami, Y., Matsumoto, E.,** and **Nogami, Y.,** Research and development of an oil deterioration sensor, *Nissan Tech. Review,* 1991, **30**, pp. 92–97, in Japanese.

14 **Kimura, Y.** and **Hisamura, K.,** Changes in debris size distribution and wearing conditions, *Proc. 10th International Congress on Condition Monitoring and Diagnostic Engineering Management,* Jantunen, E., Holmberg, K., and Rao, R.B.K.N., eds. 1997, **1,** pp. 313–332.

Recent developments in wear-mechanism maps

S C LIM
Department of Mechanical and Production Engineering, National University of Singapore, Singapore

SYNOPSIS

This paper presents a summary of the author's personal view of the development of wear-mechanism maps, culminating in the presentation of some recently proposed maps. These maps, which present wear data in a graphical manner, are able to provide a more global picture of how materials in relative motion behave when different sliding conditions are encountered; they also provide the relationships between various dominant mechanisms of wear that are observed to occur under different sliding conditions as well as the anticipated rates of wear. Some thoughts on future directions for research in this area are also presented.

1 INTRODUCTION

Wear is a complex phenomenon. It occurs whenever surfaces come into sliding contact, even in the presence of a lubricant. To the designers and engineers who have to make optimal decisions in situations where tribological considerations are significant, it is important for them to have ready access to information pertaining to the fundamental understanding of the wear processes of interest. Some kind of user-friendly databases would be most helpful here. These databases should be able to provide the appropriate information for materials selection, choice of the suitable (optimal) operating condition – such as contact geometry, speed and environment – for a particular pair of materials in tribological contact.

There are many ways of presenting wear data. The more common modes of presentation include the tabulation of wear rates and the elucidation of the dominant mechanisms of wear observed under the sliding conditions of interest, and the latter is usually accomplished through the presentation of micrographs showing features on the worn surfaces. However, these presentations tend to be restrictive in the sense that they usually cover a relatively narrow (localised) range of sliding conditions. This can be inadequate, and a more complete approach is perhaps through the linking of the wear rates and wear mechanisms over a much wider range of sliding conditions in the form of a wear-mechanism map as Tabor had suggested earlier (**1**). Such a map not only provides a multi- (most of the time, two-) dimensional graphical presentation of wear data, it also provides an overall framework for the wear behaviour of a particular sliding system into which individual wear mechanisms observed under various sliding (operating) conditions may be fitted.

Many names have been given to diagrams which describe the overall behaviour of wear; the more commonly used ones include *wear-mechanism map*, *wear-mode map*, *wear-transition map* and *wear-regime map*; sometimes, the word 'diagram' is used instead of 'map'. Generally, wear-mode, wear-transition and wear-regime maps tend to focus on the description of the mode of wear, namely, *mild wear*, *severe wear* and the transition between them. In the case of wear-mechanism maps, details of the dominant wear-mechanisms are given and the regions of their dominance are indicated; often, predicted rates of wear are also included in the maps. Names such as *fretting map* and *erosion map* have also been used for more specific wearing conditions.

In the following sections, the general methodology for the construction of a wear map will first be described. This will be followed by an almost chronological description of the development of mostly sliding wear maps for metals, ceramics, metal-matrix composites and polymers. Maps describing the wear of coatings, tool wear, fretting wear, erosion and time-dependent wear transitions are then introduced. The paper will conclude by suggesting some future directions for research in wear mapping.

2 METHODOLOGY OF WEAR MAPPING

Researchers involved in the construction of wear-mechanism maps each have their individually favoured approach; the choice of these slightly different approaches is almost always a personal one. The following describe briefly the steps adopted by the author:

1. For the pair of materials of interest, their mode of contact (for example, uni-directional sliding), contact geometry (for example, pin-on-disk), the environment in which they are to interact (for example, atmospheric condition) and lubrication condition will first have to be decided.
2. Gather experimental data from the literature on wear rates and wear mechanisms pertaining to this sliding pair measured in conditions exactly like or very close to those specified in step 1. In-house tests will have to be carried out if data are lacking. mathematical models describing wear behaviour of this pair should be

gathered as well.

3. The parameters to be used as axes for the map will be decided. One can construct a two- or three-dimensional wear diagram; so far, the majority of wear maps are of the two-dimensional type. The range of sliding conditions to be included in the map will also have to be decided. It is desirable to select as wide as possible a range. For situations such as machining, the range should preferably be similar to that recommended for that particular group of tools whose wear behaviour is to be mapped.
4. Construct the empirical wear maps. This is done first by grouping the wear data according to the mode and mechanism of wear. The wear-rate and wear-mechanism data, appropriately classified, are then plotted into the (usually) two-dimensional space defining the map. The field of dominance of each mechanism is then demarcated using field boundaries and the approximate locations of the contours of constant wear-rate are located. At this stage, the wear map is sufficiently informative and it should provide a summary of the global wear behaviour of the sliding pair of interest.
5. This final step is to introduce the appropriate mathematical models available to describe the wear behaviour of this sliding system. When these are not available, new models will have to be developed. The calibrated model for each field is then used to calculate the projected wear rates for conditions in the field where no experimental data are available. These wear-rate contours are then superimposed onto the map. A complete wear-mechanism map is thus generated.

3 DEVELOPMENTS IN VARIOUS GROUPS OF WEAR MAPS

3.1 Wear maps for metals

The concept of creating wear maps of one form or another for metals is not new: attempts were made as early as 1941 to present wear data in this fashion (**2**). In their work on the wear of cast iron and steel, Okoshi and Sakai (**2**) presented wear rates as a surface in a three-dimensional space (Fig. 1), with the sliding conditions (load and speed) as the two horizontal axes and the wear rate as the vertical (third) axis. It is clear from this wear-rate surface that wear rate depends on the two sliding conditions in a slightly non-linear fashion. As far as the author is aware, the next wear diagram of some significance was to arrive more than two decades later, when Welsh (**3**) presented a diagram summarising the sliding conditions corresponding to the mild-wear/severe-wear transitions observed in the wear of steels.

In the early 1980s, a series of diagrams, mostly for the unlubricated wear of steels with different test configurations, were proposed. These include the works of

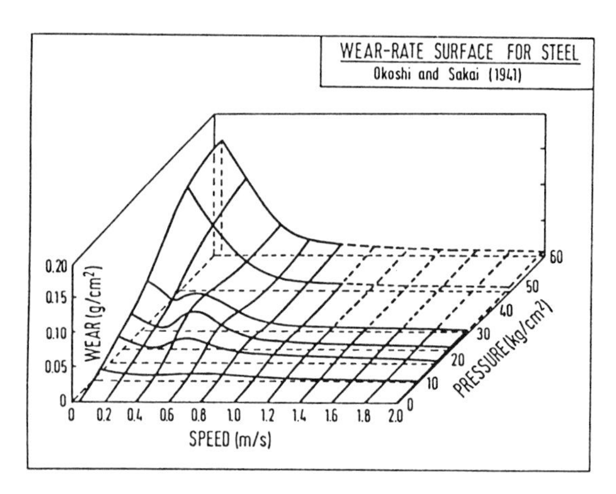

Fig. 1 The wear-rate surface for steel proposed by Okoshi and Sakai (**2**).

Childs (**4**), Eyre (**5**), Marciniak and Otimianowski (**6**) and Egawa (**7**). Apart from the diagram of Marciniak and Otimianowski which gives a wear-rate surface similar to the work of Okoshi and Sakai (Fig. 1), the other three show the boundaries between mild- and severe-wear behaviour in their respective sliding systems and within the range of sliding conditions investigated. The only diagram that is significantly different in nature from the rest is the wear-regime map for soft steels sliding on soft steels presented by Childs (**4**). In this map (shown in Fig. 2), not only is the mode of wear described: in this case severe metallic wear, but the expected dominant wear-mechanisms in five different regions (A to E) in the map are also described although this was done in the text and not represented on the map. Later, the mode of mild oxide-protected wear was added in (**8**). Furthermore, it is the only diagram that has a wide range of sliding conditions (in this case, the sliding speed covers more than four decades of values), fulfilling an important requirement of a useful wear map, that is, to cover a wide range of sliding conditions. It undoubtedly

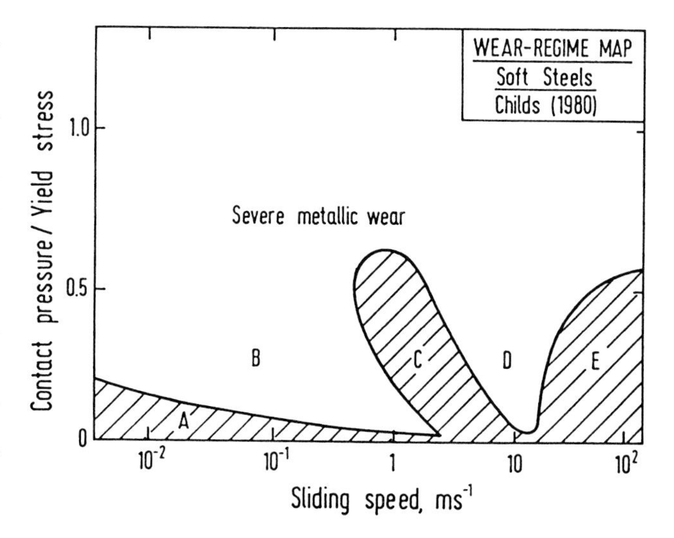

Fig. 2 The wear-regime map for soft steels proposed by Childs (**4**).

suggested how subsequent wear maps were to be constructed.

Another line of approach to generate wear maps for metals was taken by Kato and Hokkirigawa (**9**) who summarised their SEM observations of in-situ abrasive wear into an abrasive-wear diagram, and this is shown in Fig. 3. They found from tests carried out on brass, 0.45%C-steel and 18-8 austenitic stainless steel samples that three wear types, namely, cutting, wedge and ploughing were operative under different conditions. They described the mechanisms of wear observed rather than the wear mode as had been done mostly till that point in time. They introduced an index, the degree of penetration (D_p), to describe the severity of contact and this was used as one of the axes of the diagram. The abrasive-wear diagram shows the possible region for each type of wear defined by D_p and the shearing strength at the contact interface. Similar abrasive-wear diagrams for aluminium under both dry and lubricated conditions were subsequently presented, showing the occurrence of another wear mode: cleaving (**10**). More recently, in their investigation of the effect of tilted contact between sliding surfaces, simulating misaligned contacts found in machine elements which generally lead to abrasive wear, Hokkirigawa, *et al.* (**11**) proposed a wear map describing the three modes of wear of a tilted steel pin sliding with base-oil lubrication, against a steel or aluminium-alloy plate, with the latter sometimes covered with a layer of hard anodic oxide film. The three modes include cutting, ploughing and plastic deformation, with the last mode generally occurring when the plate is harder.

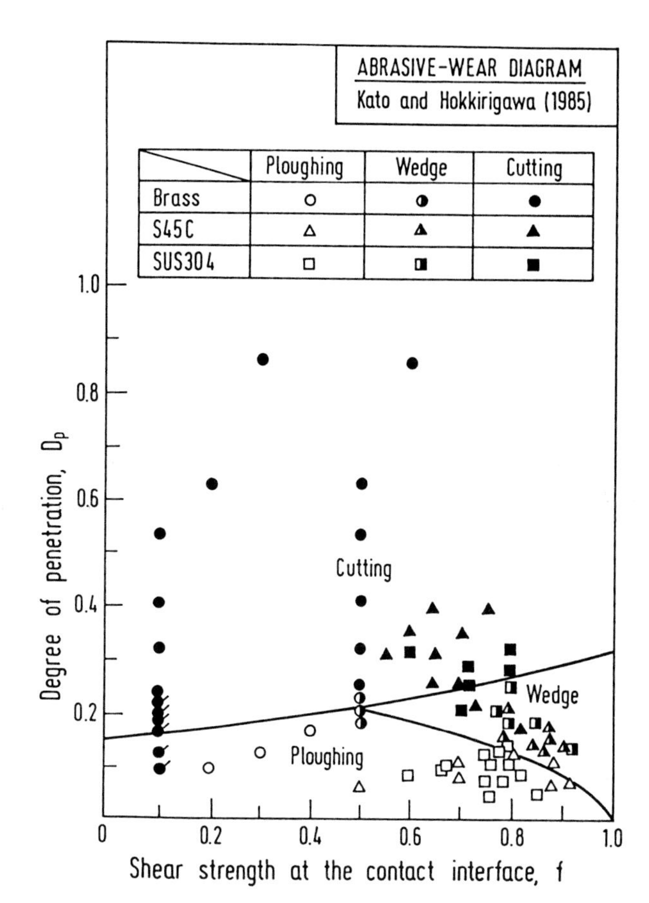

Fig. 3 The abrasive-wear diagram for three different metals proposed by Kato and Hokkirigawa (**9**).

Turning back to sliding wear, the wear maps available so far, with the exception of Childs' wear-regime map (**4**), had two major limitations, namely, the limited range of operating conditions covered and the lack of information on the dominant mechanisms of wear. The information provided was limited to whether mild wear, severe wear, or a transition between them was observed. These limitations were addressed by the construction of the wear-mechanism map for steels (**12**). This map (Fig. 4) describes the unlubricated pin-on-disk wear behaviour of steels over a wide range of sliding conditions: some seven decades of values for speed and five decades of values for contact pressure. The map predicts the field of dominance of one wear mechanism and when its contribution becomes less important; contours of predicted normalised wear-rates are superimposed over these fields. For completeness, a companion wear-mode map and a wear-transition map were later proposed (**13**) and one of them (the wear-transition map) is shown in Fig. 5. These two maps summarise the sliding conditions associated with mild and severe wear as well as how the various wear-transitions reported in the literature could be related: information which the wear-mechanism map for steels could not conveniently present. With such a wear-transition map, the operating conditions under which a mild-wear condition exists for a steel sliding component are clearly demarcated.

Several refinements to make these maps more accurate have also been suggested (**14**). More recently, Kato, *et al.* (**15**), using the same methodology, constructed wear-mechanism maps to clearly illustrate the effects of nitriding on the global wear behaviour of steels, providing an additional dimension of information not possible to be included in the wear-mechanism map for steels (Fig. 4). Following the same methodology used in constructing the maps for steels, a wear-mechanism map for the unlubricated sliding of aluminium and aluminium alloys on steel was later proposed by Liu, *et al.* (**16**). This map is a considerable improvement over the earlier empirical wear-map for the same group of alloys presented by Antoniou and Subramanian (**17**).

One of the important considerations in the construction of wear maps is the temperature generated at the sliding interface. At the onset of the mapping of steel wear, it was recognised that the interfacial temperature would control the dominance of different mechanisms of wear in a significant manner. When the interfacial temperature is high (and this occurs when the sliding speed is higher than a certain critical value), mechanisms involving oxidation, phase transformation and melting would be important; below this speed of sliding, the sliding interface can be considered as 'cold' and the dominant mechanisms are essentially controlled by the plasticity of the materials

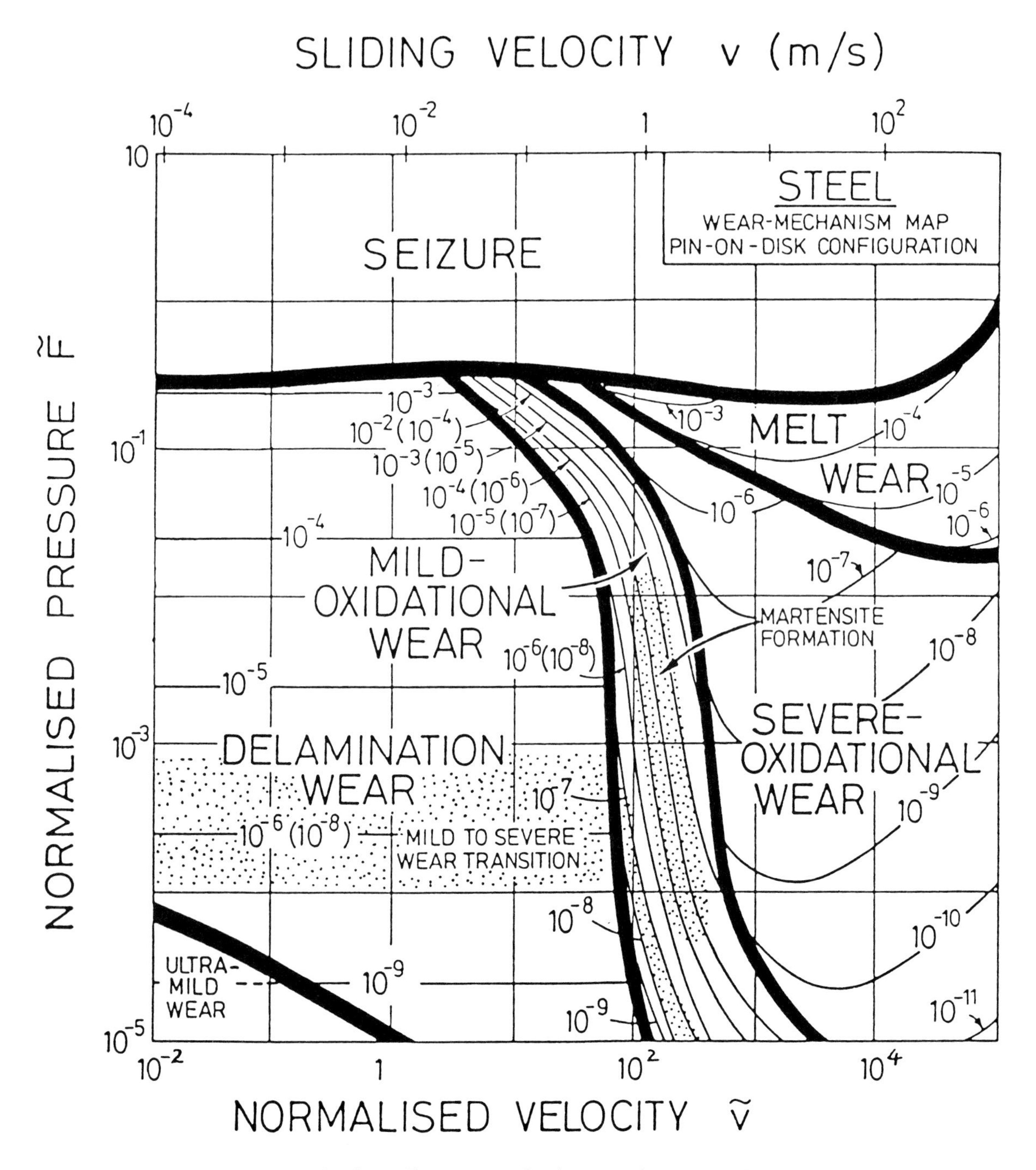

Fig. 4 The wear-mechanism map for steels (**12**).

forming the interface. The mathematical models used in the prediction of wear rates would then reflect how the wear processes are dependent on the interfacial temperature. Attempts were therefore made to compute, based on one-dimensional equivalents of the more complex three-dimensional patterns of heat flow, both the flash and bulk temperatures. This gave rise to a temperature map for the dry sliding of steels (**12**). This method of temperature computation has since been further extended and refined (**18–21**), resulting in the creation of the T-MAPS PC-based software (**22, 23**). Refinements of this methodology to construct temperature maps for other material pairs in sliding contacts have also been carried out, for example, by Wang and Rodkiewicz (**24**) who proposed minor changes to the temperature maps for steels and some ceramic materials.

There is another group of maps which one might consider as 'failure maps'. The most notable example is the transition diagram of the International Research Group on Wear of Engineering Materials (IRG-OECD) which presents the critical load-velocity curves for the failure of thin-film-lubricated sliding concentrated contacts (**25, 26**). In this diagram, illustrated in Fig. 6, the three regimes of different tribological behaviour during lubricated sliding

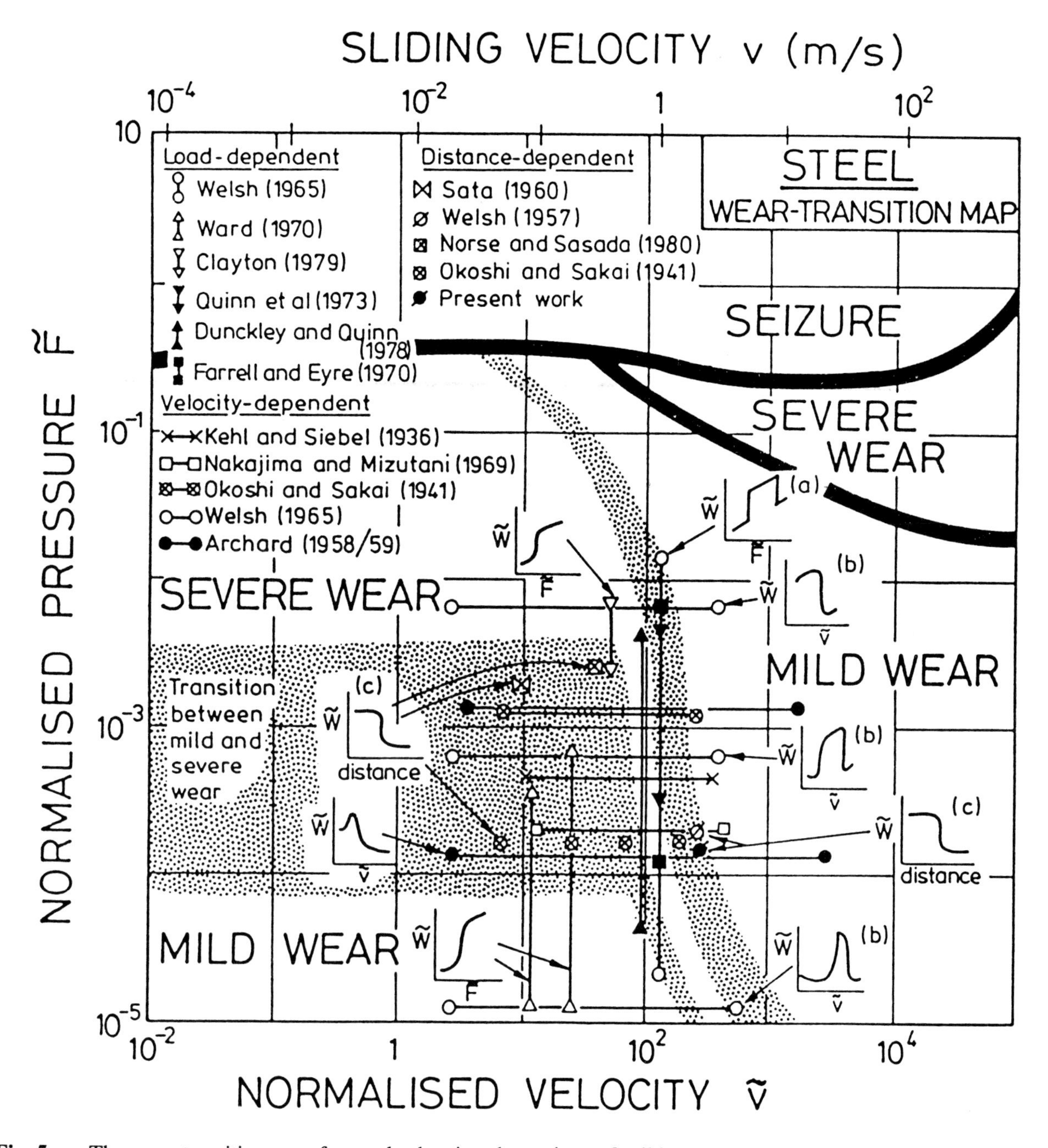

Fig. 5 The wear-transition map for steels showing the regions of mild wear and severe wear. The sliding conditions corresponding to the different types of wear transitions observed are also indicated (**13**).

are demarcated. Sliding contacts within region I will theoretically suffer no wear, while those operating in region III will experience severe wear. This is useful not only for the selection of proper sliding conditions, it can also be very useful for failure analysis. Actually, diagrams showing the 'safe' operating conditions for various machine elements, such as bearings operated under different kinds of lubrication conditions; are already available in the literature, for example, ref (**27**). These diagrams have helped designers and engineers select the correct machine elements and operating conditions to meet the design requirements. Such an approach was also used during the construction of wear maps for cutting tools, where safety zones and least-wear regions are identified in which the rate of wear of tools would be a minimum or sufficiently low (see below).

This concept of defining regions of safe operation was brought one step further when Landheer, *et al.* (**28**) proposed a theoretical wear-mechanism map for plain journal bearings based on the IRG transition diagram. They found that their diagram agreed well with data taken from actual bearing practices presented by Neale (**27**). From such a map, it is possible to predict the state of lubrication experienced by the journal bearing if the loading condition

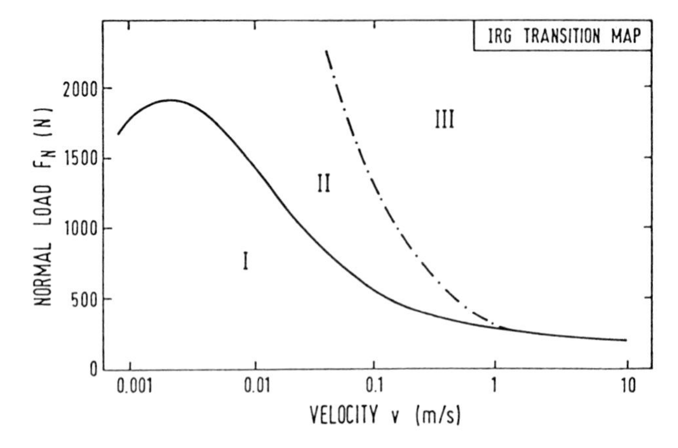

Fig. 6 The IRG transition map showing the critical load-velocity curves for the failure of thin-film-lubricated sliding concentrated contacts, demarcating three regimes of different tribological behaviour (**25, 26**).

is known. The 'safe zone' of bearing operation then corresponds to the region where a no-wear condition is predicted. Qualitative information on bearing failure could also be obtained from this map. Having identified pitting, abrasive wear and scuffing as the three major processes of tooth distress in gears, Tweedale (**29**) proposed a map showing the various zones of tooth distress: this is another form of a 'failure map'. It is only a qualitative one because no values were suggested for the limits of the two operating parameters, namely, load and speed, defining the various distress zones. This diagram would be an excellent tool for the diagnosis of gear teeth failure if details of the limiting values of load and speed that would lead to the operation of undesirable wear mechanisms could be provided. It is the wish of the author to see more maps and diagrams of such diagnostic nature being generated to help designers and engineers in their continual combat against wear.

3.2 Wear maps for ceramics

The concept of wear mapping has also been extended to include ceramic materials (**30**). To date, substantial progress has been achieved in developing wear maps for ceramics, see for example the works of Hsu, *et al.* (**31, 32**), Kato (**33**), Lee, *et al.* (**34**), Hokkirigawa (**35**), Kong and Ashby (**36**), Dong and Jahanmir (**37**), Gautier and Kato (**38**), Blomberg, *et al.* (**39**). In these diagrams, regimes of different dominant wear-mechanisms are demarcated and, in some of them, the sliding conditions leading to transitions between mild and severe wear are also indicated.

Fig. 7 shows the wear-mechanism map for an alumina ball sliding on an alumina disk proposed by Kong and Ashby (**36**). In this map, the locations of the dominance of seven different wear mechanisms are indicated together with contours of constant flash and bulk temperatures generated during sliding. The map clearly shows the inter-relationships of these mechanisms with each other as well as their dependence on the temperatures generated at the sliding interface. The different field boundaries on the map suggest where transitions from one dominant wear-mechanism to another may take place. Although the wear models developed were unable to accurately describe the wear rates, they nevertheless provided an overall framework for the wear characteristics of alumina. Because of the wide range of sliding velocity and contact pressure covered by the map, it should enable the designer to intelligently decide whether alumina will be able to meet the set of requirements for a particular tribological application. To be useful during the design process, wear maps should ideally encompass as wide as possible a range of sliding conditions, so that global wear characteristics can be clearly displayed.

3.3 Wear maps for metal-matrix composites

The emergence of composite materials, especially metal-matrix composites with different reinforcement phases, as a group of advanced tailor-made materials for tribological application has created the need to summarise the wear mechanism and wear transition information of some of them to optimise design. Unlike many monolithic metallic materials, understanding of the processes by which these composites wear during dry sliding is still limited, and in some cases, controversial (**40, 41**). Such a situation renders the construction and calibration of wear-mechanism maps for composite materials much more difficult than for monolithic metals such as steels or aluminium alloys. The way to avoid the difficulties associated with the limited understanding of the wear processes involved is to construct empirical wear-mechanism maps by carefully integrating the wear-rate and wear-mechanism data observed during sliding. This has been done for the Al(6061)-SiC_w composite (**42**) where over a small range of normal load and speed, the transition boundaries separating different modes of wear are drawn; an indication of the dominant mechanism is also given in each regime. This map is shown in Fig. 8.

More recently, an attempt was made to examine whether a map could be constructed for Al-SiC_p composites over a larger range of sliding conditions. This was done based on extensive experimentation as well as data from the literature, and the empirical wear-mechanism map is shown in Fig. 9 (**43**). The extensive experimentation enabled a sufficiently large amount of data, especially high-speed data, to be generated, thereby extending the range of sliding conditions covered by this map. In the course of developing this map, the amount of SiC reinforcement particles in the composites was considered to have a greater influence on their wear behaviour than the matrix material (made of different grades of aluminium alloys). As a result, this map was constructed based on data from different Al-SiC_p composites with nearly the same volume

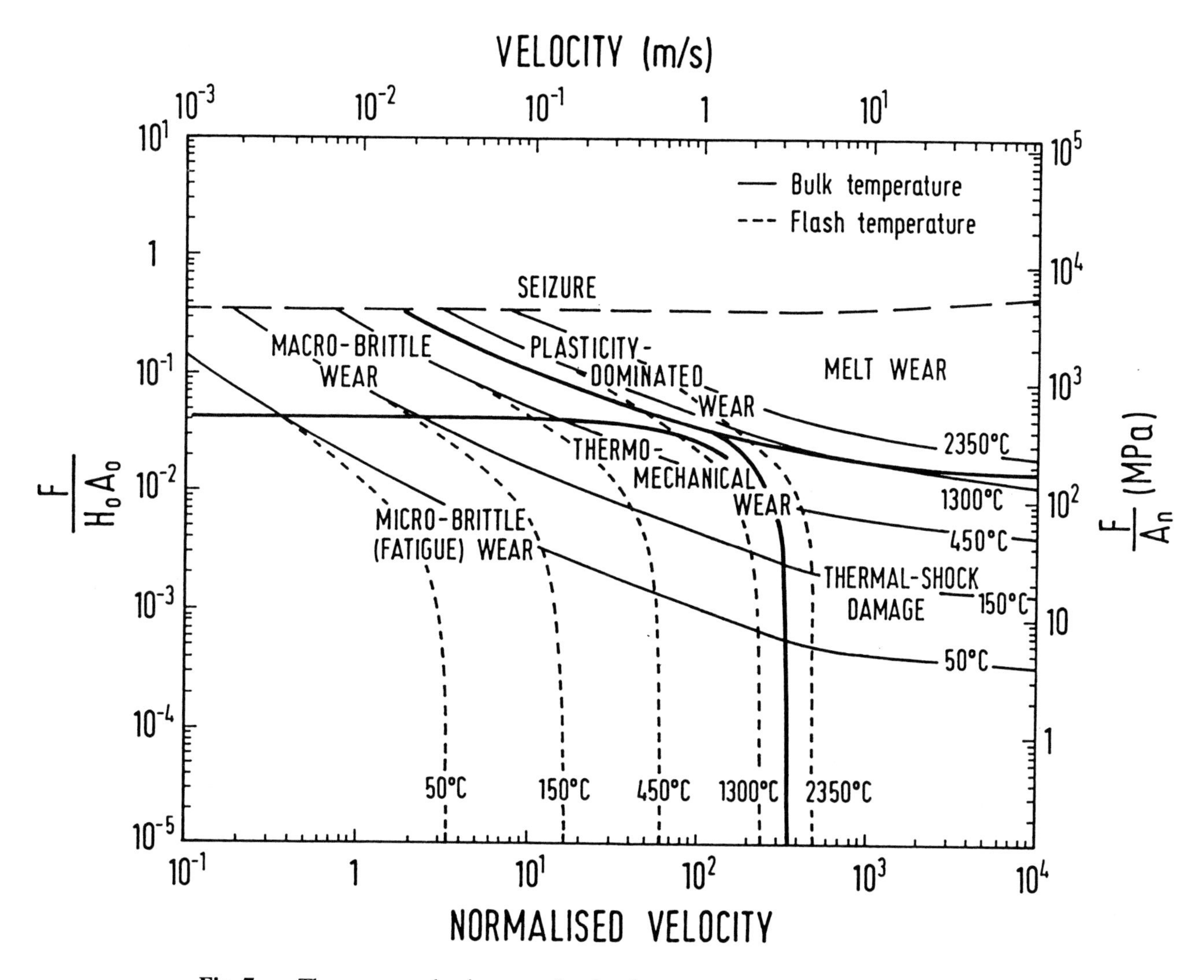

Fig. 7 The wear-mechanism map for alumina proposed by Kong and Ashby (**36**).

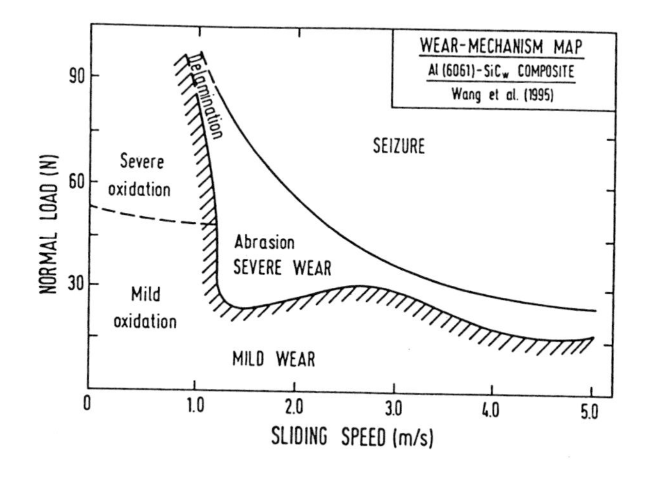

Fig. 8 The empirical wear-mechanism for Al(6061)-SiC_w composite proposed by Wang, *et al.* (**42**). The range of sliding condition covered in this map is however rather limited.

fraction of SiC particles (about 20%). This map shows that thermal effects play an increasingly important role in the wear behaviour of this group of composites when the sliding speed exceeds about 3 m/s. Some attempts were made to develop physical models to explain the observed responses of these composites in terms of changes to their reinforcement phase under different sliding conditions, especially during higher-speed sliding (**43**). This map (Fig. 9) provides the framework for future work to better understand the overall wear behaviour of these Al-SiC_p composites.

3.4 Wear maps for polymers

It is interesting to note that diagrams were proposed a long time ago for the purpose of design and the *initial* selection of suitable bearing materials. In a 1973 paper, Crease (**44**) commented that the data available then for most bearing materials on wear performance under the intended

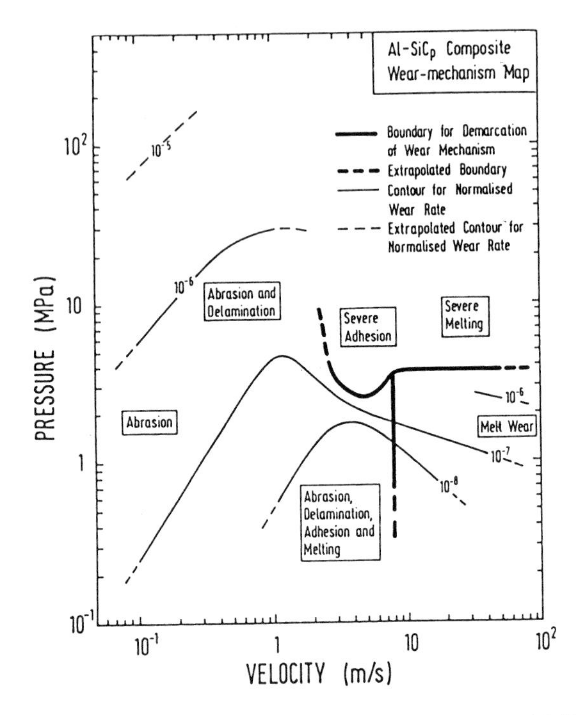

Fig. 9 The empirical wear-mechanism map for Al-SiC$_p$ composites (**43**). The regions of dominance of six different wear mechanisms are demarcated with the contours of constant normalised wear rates superimposed over them.

operating conditions was largely inadequate. He stressed that it is important to know how the wear factor (a measure of the wear rate) varies with parameters such as bearing pressure and surface temperature over the possible range of these variables likely to be met in practice. Crease proposed diagrams, based on available performance data of a series of polymeric-based bearings, relating the wear factors to some of the operating variables. An example of such a wear-factor diagram is given in Fig. 10. In this diagram, the range of bearing pressure within which three different polymeric-based bearing materials could safely operate and the corresponding expected values of wear factor are given. The next significant wear map for polymers came many years later in the form of the deformation map for the unlubricated sliding of PTFE on steel (**45**). Much effort is needed to address the paucity of polymeric wear maps in the technical literature.

3.5 Wear maps for coatings

The usefulness of wear-mechanism maps is not restricted to bulk materials. Borel, *et al.* (**46**) have suggested that it is also meaningful to employ these maps to understand the wear of abradable coatings deposited on certain gas-turbine components. They proposed wear-mechanism maps fc. ωwo different AlSi-plastic coatings tested in a high-temperature environment. They concluded that these maps can be usec for modeling wear mechanisms and the design of coating systems to provide enhanced performance of gas turbines at elevated temperatures. Further work along this line supported the earlier findings that these wear maps enabled the influence of coating microstructure variations on the abradability to be determined quickly, leading to the formulation of a general abradability model for aero-engine coatings (**47**). Wear characteristics of TiN and TiC coatings on tool inserts were also examined using mapping techniques (**48**), and these will be detailed in the next section.

3.6 Wear maps for cutting tools

The originator of graphical representation of tool wear can be traced to Trent who in the late 1950s produced a series of machining charts (**49**). The idea of a diagram to describe tool wear did not catch on again until Yen and Wright (**50**) proposed a qualitative wear map for cutting tools, and this was later taken up by Kendall (**51**) who made an attempt to relate qualitatively the observed tool wear with the wear-mechanism map for steels (**12**), and proposed that a safe zone exists within which excessive tool wear would not occur: the same concept first suggested by Yen and Wright

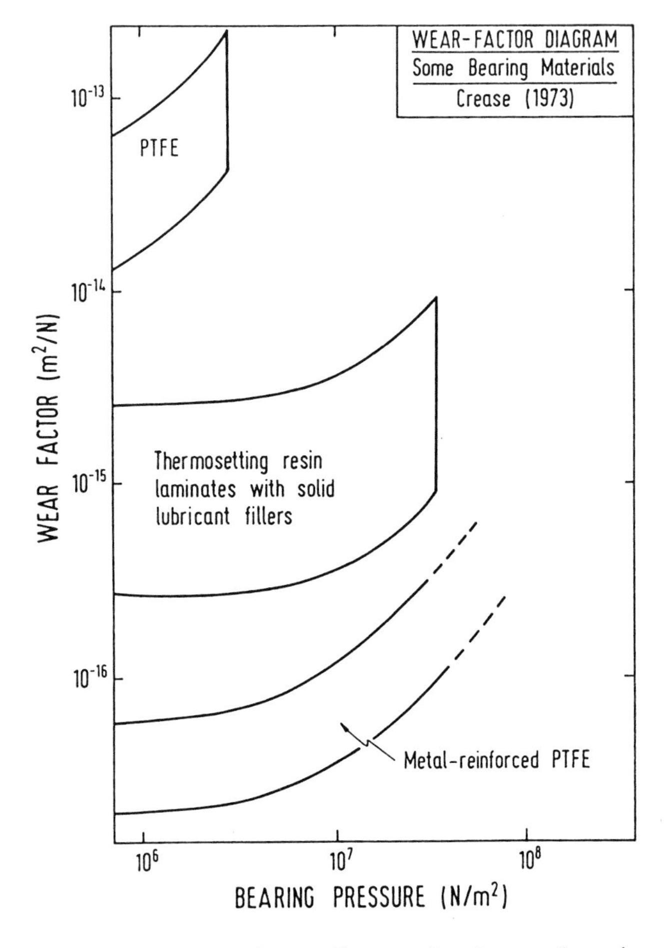

Fig. 10 The wear-factor diagram for three polymeric-based bearing materials (after Crease (**44**)).

(**50**). With such a background, a proposition was made by Lim, *et al.* (**52**) to construct a series of empirical wear maps for different cutting tools. These maps should enable machinists to select the machining conditions in which the desired productivity (in terms of material removal rate) could be attained at an acceptable rate of tool wear. A similar qualitative approach was taken independently by Quinto and he recently presented a tool failure mode diagram (**53**) derived from Kendall's qualitative wear map (**51**), Trent's machining chart (**49**) and the wear-mechanism map for steels (**12**). In this map, Quinto proposed that the safe zone is a region of gradual wear associated with predictable and reliable tool performance. All these maps are intended to inform the machinist of the machining conditions which would give rise to the least amount of tool wear.

The effort to construct empirical wear maps for cutting tools has resulted in several maps for two groups of uncoated tools. These maps display the global characteristics of flank and crater wear (the two major forms of tool wear during turning operations) over the recommended range of machining conditions for uncoated high-speed-steel (HSS) and carbide tools (**54**). One key difference between these maps and the earlier qualitative ones is that data (both wear rates and wear mechanisms) drawn from actual machining operations were used, thereby allowing the optimisation of actual machining operations. In parallel, the same methodology was extended to coated tools, and the flank-wear map first constructed for TiN-coated HSS tool inserts shows the significant enlargement of the safety zone from that found in the uncoated case (**55**). More importantly, this map shows that the amount of tool wear reduction which such a coating might provide is critically dependent on the machining conditions employed. Such a map will help end-users to employ these coated inserts in a cost-effective manner. An example is shown in Fig. 11 (**48**) which superimposes the crater-wear map of the uncoated HSS tools on to the corresponding one for the TiN-coated HSS inserts. The expansion of the safety zone and least-wear regime as a result of the application of the TiN coating is clear.

3.7 Fretting maps

When tribo surfaces come into oscillatory contacts with displacement of small amplitudes, they are often damaged by fretting wear. It will be useful from the design point of view to know when transition to reciprocating sliding would take place. Addressing this issue, Vingsbo and Söderberg (**56**) proposed fretting maps summarising the fretting wear behaviour of three metallic materials, namely low-carbon steel, austenitic stainless steel and pure niobium. They believed these maps would help clarify some of the confusion concerning the distinction between different types of fretting and in practical service life

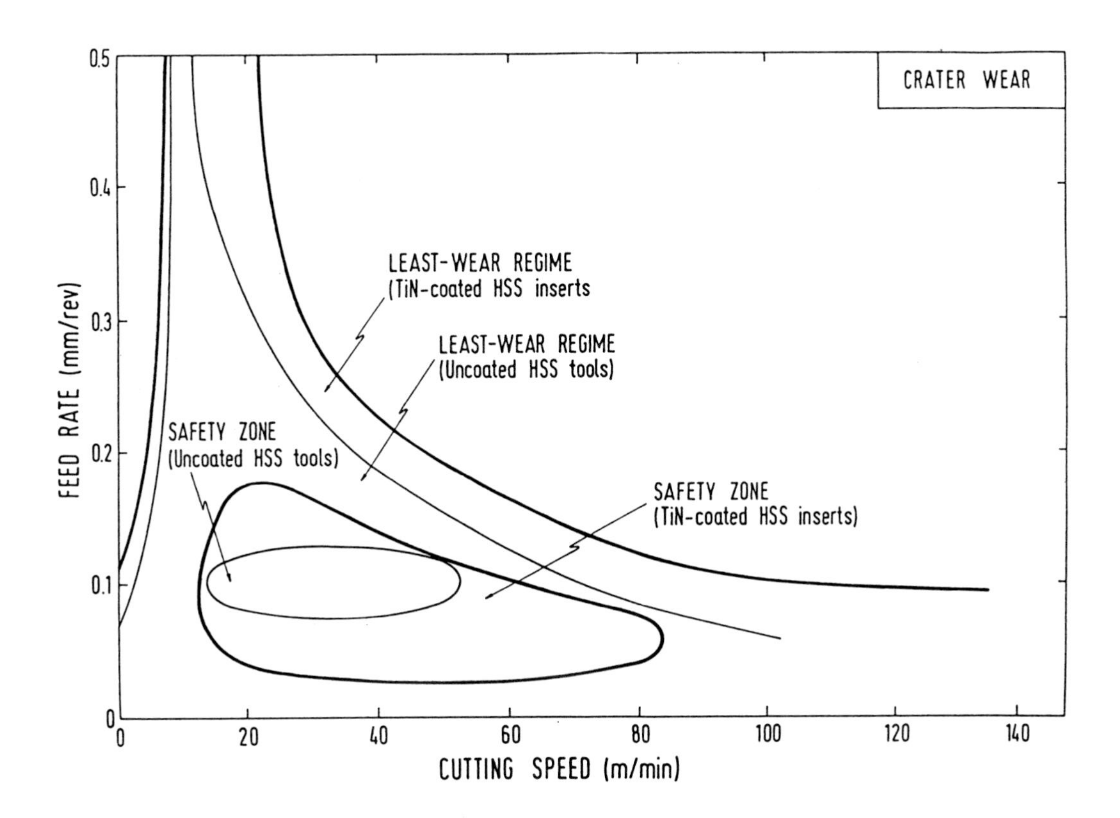

Fig. 11 Map showing the expansion of safety zone and least-wear regime as a result of the application of TiN coatings on the crater wear of high-speed-steel (HSS) tools during dry turning operations (**48**).

prediction. A map detailing the various damage mechanisms sustained by silver-plated copper contacts was later proposed (**57**); such a map is not only useful in design but in failure analysis as well. Running condition fretting maps for WC-Co and TiN coatings have also been constructed (**58, 59**).

3.8 Erosion maps

When hard particles, either carried by a gas stream or contained in a flowing liquid, strike a surface, erosion takes place. In the case of ceramics and brittle materials, a relatively small change in conditions such as impact velocity or angle past a certain threshold (critical) value can result in a significant change in the mechanism of wear: a wear transition. Hutchings (**60**) has argued that such transitions are best understood through erosion maps and he has provided examples of such maps which display the regimes of particle size and impact velocity over which different mechanisms of erosion dominate. When erosion takes place in a corrosive environment, Stack, *et al.* (**61**) suggested that the total wastage and degradation of material as a result of the synergistic effects of erosion and corrosion may in some cases be greater than that which would be observed from these processes operating separately. They proposed an aqueous erosion-corrosion map showing the transitions between the various regimes of aqueous erosion-corrosion in terms of erodent velocity and potential. It is interesting to note that there is a regime in this schematic map within which no corrosion or erosion is expected to occur. This is the concept of 'safety zones' (**62**). The ability to locate such safety zones where damage to the surface or the rate of degradation (wastage) of materials would be at a minimum is one of the major strengths of wear maps. A similar approach has been adopted in the wear maps for cutting tools discussed earlier.

3.9 Maps for time-dependent wear transitions

While addressing the durability issue of ceramic machine components, Yust (**63**) observed that because ceramics are generally susceptible to fatigue effects, wear data obtained from short-term wear tests may be misleading when long-term service requirements are considered. Fatigue effects may considerably influence the anticipated lifetime before transition into severe wear from the designed mild-wear condition. The same may be true for other materials used as machine elements operating in an environment exposed to cyclic stresses. Yust (**63**) argued that with the introduction of a third 'time' axis into a two-dimensional wear map, the wear-transition surface could provide a basis for materials selection which includes anticipated lifetime as a factor. He illustrated his concept using the IRG transition map, and showed hypothetically the possibility that wear testing at a constant load, for example, may eventually result in a transition from mild to severe wear due to time-dependent failure of the tribo-surface. Similar comments were also made by Luan, *et al.* (**64**) who observed from their experiments on high-carbon steels under boundary lubrication that both the coefficient of friction and the mechanism of wear changed with increasing testing time under the same load and speed. They went on to propose a three-dimensional wear map for boundary-lubricated steels incorporating a third (time) axis.

What then are the implications of these findings? Two questions concerning materials selection for tribological applications immediately follow. First, how important is such time-dependent wear behaviour? Second, if it is important, how long will the components maintain their pre-transition mode of wear? This second question points directly to the availability of relevant wear data. If time-dependent wear-transition maps for different materials are available, designers will be able to incorporate some form of safety factor into the design using the 'rate of progression' toward the severe-wear transition provided in the relevant map. This may ensure mild wear for the components throughout their expected service life.

4 FUTURE DIRECTIONS FOR WEAR MAPPING

The following are some personal views on future directions for research in wear mapping:

1. Wear maps for different tribological conditions should be constructed; they should not be limited only to describing wear in uni-directional sliding. Effort to construct time-dependent wear-transition maps should be encouraged.
2. More wear maps designed to serve primarily as diagnostic tools, such as 'failure maps' and 'operating-conditions maps', should be prepared.
3. Wear maps should be constructed with an aim to serve the end-users.
4. Wear maps should have a sufficiently wide range of sliding (or operating) conditions. Even if these maps are of an empirical nature, they will still be able to provide an overall framework for the better understanding of wear behaviour.
5. More wear map should use as axes (whether in a two- or three-dimensional map) parameters that can be easily controlled in practice. These parameters include contact pressure (which is often closely related to the contact geometry) and speed; other material-related parameters could be integrated into them if necessary. Designers may find it difficult to use maps having parameters that they cannot readily link to other design parameters.
6. There is a need to address the paucity of wear maps describing lubricated conditions as real-life tribo systems almost always operate in a lubricated condition.
7. Companion friction maps may be useful for designers too. Earlier work on steels showed that when the sliding becomes more severe, the measured coefficient of

friction depends on the sliding condition much more than on the surface properties (such as surface roughness) which were found to be more important during slower sliding (**65**). It may be possible to generate friction maps from the corresponding wear maps.

8. One pre-requisite as well as an outcome of a wear mapping exercise is the ordering of wear data. Often, much thought has to be put in to understanding the cause of scatter of wear data reported by different research groups, even though very similar test conditions and environments were used. Such ordering of data and the ensuing production of physical and mathematical models to describe the observed wear behaviour will contribute to the increased understanding of the underlying wear processes. These are often difficult and time-consuming tasks. The challenge to tribologists remains.

5 CONCLUDING REMARKS

This review is not intended to be an exhaustive one and the omission of some pieces of work is therefore unavoidable. The views expressed above are entirely personal; they are based on the author's experience in constructing wear-mechanism maps as well as his understanding of this subject matter gained through interaction with colleagues engaged in similar endeavours.

In the author's opinion, wear maps are useful to designers and engineers when they have to make engineering decisions where wear is one of the major considerations. Wear maps can also play the role of a diagnostic tool during failure analysis. These maps combine wear rates and wear mechanisms observed under a certain set of sliding (or operating) conditions, and at the same time provide a framework for the overall wear behaviour of the materials in relative motions. Wear mapping is slowly gaining acceptance as a user-friendly approach to the presentation of wear-related information; this can be seen from the increasing number of wear maps presented during recent years. Notwithstanding these achievements, a greater effort could be channeled into the construction of new wear-maps for the many materials used in tribological applications as well as the continual refinement of existing ones. Such an endeavour may contribute toward fulfilling Ludema's hope for the publication of a Complete Handbook for Tribological Design (**66**) in the not-too-distant future.

6 ACKNOWLEDGEMENTS

I would like to thank Dr J.K.M. Kwok and Ms C.Y.H. Lim for permission to use some of their results.

7 REFERENCES

1 **Tabor, D.**, *Proc. Int. Conf. Tribology in the 80's*, April 18–21, 1983, NASA Lewis Research Centre, Cleveland, OH, NASA, pp. 1–17.
2 **Okoshi, M.** and **Sakai, H.**, *Trans. JSME*, Vol. **7**, 1941, pp. 29–47.
3 **Welsh, N.C.**, *Proc. Roy. Soc.*, Ser. **A**, Vol. **257**, 1965, pp. 31–50.
4 **Childs, T.H.C.**, *Tribo. Int.*, Vol. **13**, 1980, pp. 285–293.
5 **Eyre, T.S.**, *Powder Metall.*, Vol. **24**, 1981, pp. 57–63.
6 **Marciniak, J.** and **Otimianowski, T.**, *Proc. Eurotrib '81*, Warsaw, Poland, September 21–24, 1981, The Polish Tribology Council, 1981, Vol. **1/A**, pp. 241–256.
7 **Egawa, K.**, *J. JSLE Int.* Edition, No. **3**, 1982, pp. 27–30.
8 **Childs, T.H.C.**, *Proc. I.Mech.E.*, Vol. **202**, No. C6, 1988, pp. 379–395.
9 **Kato, K.** and **Hokkirigawa, K.**, *Proc. Eurotrib '85*, Ecully, France, September 9–12, 1985, Elsevier, 1985, Vol. 4, Section 5.3, pp. 1–5.
10 **Kato, K.**, **Hokkirigawa, K.**, **Kayaba, T.**, and **Endo, Y.**, *Trans. ASME, J. Tribo.*, Vol. **108**, 1986, pp. 346–351
11 **Hokkirigawa, K.**, **Kato, T.**, **Fukuda, T.**, **Shinooka, M.**, and **Takahashi, J.**, *presented 10th Int. Conf. on Surface Modification Technologies*, SMT10, Singapore, September 2–4, 1996.
12 **Lim, S.C.** and **Ashby, M.F.**, *Acta Metall.*, Vol. **35**, 1987, pp. 1–24.
13 **Lim, S.C.**, **Ashby, M.F.**, and **Brunton, J.H.**, *Acta Metall.*, Vol. **35**, 1987, pp. 1343–1348.
14 **Ashby, M.F.** and **Lim, S.C.**, *Scripta Metall.*, Vol. **24**, 1990, pp. 805–810.
15 **Kato, H.**, **Eyre, T.S.**, and **Ralph, B.**, *Acta Metall.* Mater., Vol. **42**, 1994, pp. 1703–1713.
16 **Liu, Y.**, **Asthana, R.**, and **Rohatgi, P.**, *J. Mater. Sci.*, Vol. **26**, 1991, pp. 99–102.
17 **Antoniou, R.**, and **Subramanian, C.**, *Scripta Metall.*, Vol. **22**, 1988, pp. 809–814.
18 **Ashby, M.F.**, **Abulawi, J.**, and **Kong, H.S.**, *Tribo. Trans.*, Vol. **34**, 1991, pp. 577–587.
19 **Kong, H.S.** and **Ashby, M.F.**, *MRS Bulletin*, October 1991, pp. 41–48.
20 **Kong, H.S.** and **Ashby, M.F.**, Cambridge University Engineering Department Report, *CUED/C-MATS/TR.*186, 1991.
21 **Nichols, F.A.** and **Ashby, M.F.**, in *Contact Problems and Surface Interactions in Manufacturing and Tribological Systems*, M.H. Attia and R. Komanduri (ed.), PED-Vol. 67, Book No. H00895, ASME, 1993, pp. 75–86.
22 **Ashby, M.F.**, **Kong, H.S.**, and **Abulawi, J.**, *T-MAPS 4.0 Background Reading: Frictional Heating at Dry Sliding Surfaces*, Cambridge University Engineering Department, 1992.
23 **Ashby, M.F.**, **Abulawi, J.**, and **Kong, H.S.**, *Operating Manual for T-MAPS 4.0'*, Cambridge University Engineering Department, 1992.
24 **Wang, Y.** and **Rodkiewicz, C.M.**, *Tribo. Int.*, Vol. **27**, 1994, pp. 259–266.
25 **Salomon, G.**, *Wear*, Vol. **36**, 1976, pp. 1–6.
26 **de Gee, A.W.J.**, **Begelinger, A.**, and **Salomon, G.**, *Proc. 11th Leeds-Lyon Symp. on Tribology*, Leeds, September 4–7, 1984, D. Dowson and C.M. Taylor (ed.), Butterworths, London, 1985, pp. 105–116.
27 **Neale, M.J.**, (ed.), *Tribology Handbook*, Butterworths, 1973.
28 **Landheer, D.**, J.P.M. Faessen and A.W.J. de Gee, *Tribo.*

Trans. Vol. **33,** 1990, pp. 418–424.

29 **Tweedale, P.,** *Prof. Engrg,* Vol. **4,** 1991, pp. 25–26.

30 **Ramsey, P.M.** and **Page, T.F.,** *Br. Ceram. Trans. J.,* Vol. **87,** 1988, pp. 74–80.

31 **Hsu, S.M., Wang, Y.S.,** and **Munro, R.G.,** *Wear,* Vol. **134,** 1989, pp. 1–11.

32 **Hsu, S.M., Lim, D.S., Wang, Y.S.,** and **Munro, R.G.,** *J. STLE,* Vol. **47,** 1991, pp. 49–54.

33 **Kato, K.,** *Wear,* Vol. **136,** 1990, pp. 117–133.

34 **Lee, S.W., Hsu, S.M.,** and **Munro, R.G.,** *Proc. Conf. on Tribology of Composite Materials,* Oak Ridge, TN, May 1–3, 1990, P.K. Rohatgi, C.S. Yust and P.J. Blau (ed.), ASM International, 1990, pp. 35–41.

35 **Hokkirigawa, K.,** *Wear,* Vol. **151,** 1991, pp. 219–228.

36 **Kong, H.S.** and **Ashby, M.F.,** *Acta Metall Mater.,* Vol. **40,** 1992, pp. 2907–2920.

37 **Dong, X.** and **Jahanmir, S.,** *Wear,* Vol. **165,** 1993, pp. 169–180.

38 **Gautier, P.** and **Kato, K.,** *Wear,* Vol. **162–164,** 1993, pp. 305–313.

39 **Blomberg, A., Olsson, M.,** and **Hogmark, S.,** *Wear,* Vol. **170,** 1994, pp. 77–89.

40 **Rohatgi, P.K., Liu, Y.,** and **Asthana, R.,** *Proc. Conf. on Tribology of Composite Materials,* Oak Ridge, TN, USA, May 1–3, 1990, P.K. Rohatgi, C.S. Yust and P.J. Blau (ed.), ASM International, 1990, pp. 69–79.

41 **Rohatgi, P.K., Liu, Y.,** and **Lim, S.C.,** in *Advances in Composite Tribology,* K. Friedrich (ed.), Elsevier, 1993, pp. 291–309.

42 **Wang, D.Z., Peng, H.X., Liu, J.,** and **Yao, C.K.,** *Wear,* Vol. **184,** 1995, pp. 187–192.

43 **Kwok, J.K.M.,** PhD thesis, National University of Singapore, 1996.

44 **Crease, A.B.,** *Tribo.,* Vol. **6,** 1973, pp. 15–20.

45 **Briscoe, B.J.** and **Evans, P.D.,** *Proc. Int. Conf. on Wear of Materials* 1989, April 9–13, 1989, Denver, CO, K.C. Ludema (ed.), ASME, 1989, pp. 449–457.

46 **Borel, M.O., Nicoll, A.R., Schläpfer, H.W.,** and **Schmid, R.K.,** *Surf. Coat. Technol.,* Vol. **39/40,** 1989, pp. 117–126.

47 **Barbezat, G., Clarke, R., Nicoll, A.R.,** and **Schmid, R.,** *Int. Symp. on Tribology,* Beijing, China, October 19–22, 1993, Y.S. Jin (ed.), International Academic Publishers, Vol. 2, pp. 756–766.

48 **Lim, C.Y.H.,** PhD thesis, National University of Singapore, 1996.

49 **Trent, E.M.,** *J. Inst. Prod. Engrs,* Vol. **38,** 1959, pp. 105–130.

50 **Yen, D.W.** and **Wright, P.K.,** *Trans. ASME, J. Engrg for Industry,* Vol. **105,** 1983, pp. 31–38.

51 **Kendall, L.A.,** in *Metals Handbook,* 9th edition, Vol. **16:** Machining, J.R. Davies (ed.), ASM International, 1989, pp. 37–48.

52 **Lim, S.C., Liu, Y.B., Lee, S.H.,** and **Seah, K.H.W.,** *Wear,* Vol. **162–164,** 1993, pp. 971–974.

53 **Quinto, D.T.,** *Int. J. Refrac. Met. Hard Mater.,* Vol. **14,** 1996, pp. 7–20.

54 **Lim, S.C., Lee, S.H., Liu, Y.B.,** and **Seah, K.H.W.,** Tribotest, Vol. **3,** 1996, pp. 67–87.

55 **Lim, S.C., Lim C.Y.H.,** and **Lee, K.S.,** *Wear,* Vol. **181–183,** 1995, pp. 901–912.

56 **Vingsbo, O.** and **Söderberg, S.,** *Wear,* Vol. **126,** 1988, pp. 131–147.

57 **Kassman, A.** and **Jacobson, S.,** *Wear,* Vol. **165,** 1993, pp. 227–230.

58 **Carton, J.F., Vannes, A.B., Vincent, L., Berthier, Y., Dubourg, M.C.,** and **Godet, M.,** *Proc. Int. Symp. on Tribology,* Beijing, China, October 19–22, 1993, Y.S. Jin (ed.), International Academic Publishers, Vol. **2,** pp. 1110–1119.

59 **Wei, J., Kapsa, Ph.,** and **Vincent, J.,** presented *10th Int. Conf. on Surface Modification Technologies,* Singapore, September 2–4, 1996.

60 **Hutchings, I.M.,** in *Erosion of Ceramic Materials* J.E. Ritter (ed.), Trans Tech Publications, 1992, Chapter 4.

61 **Stack, M.M., Zhou, S.,** and **Newman, R.C.,** *Wear,* Vol. **186–187,** 1995, pp. 523–532.

62 **Stack, M.M.,** *Proc. 1st Int. Symp. on Process Industry Piping,* NACE, Houston, TX, 1993, paper 37.

63 **Yust, C.S.,** in *Tribological Modeling for Mechanical Designers,* ASTM STP 1105, K.C. Ludema and R.G. Bayer (ed.), ASTM, 1991, pp. 153–161.

64 **Luan, D.C., Qu, J.X.,** and **Shao, H.S.,** *Proc. Int. Symp. on Tribology,* Beijing, China, October 19–22, 1993, Y.S. Jin (ed.), International Academic Publishers, Vol. 1, pp. 629–636.

65 **Lim, S.C., Ashby, M.F.,** and **Brunton, J.H.,** *Acta Metall.,* Vol. **37,** 1989, pp. 767–772.

66 **Ludema, K.C.,** in *Achievements in Tribology,* L.B. Sibley and F.E. Kennedy (ed.), ASME, New York, 1990, pp. 111–127.

Teaching tribology

K C LUDEMA
University of Michigan, Ann Arbor, MI, USA

1 INTRODUCTION

This paper is a status report on what is available, what needs to be done, and who would benefit most directly from instruction in tribology. Few of the thousands of designers in industry are well informed on the broad principles of tribology and as a consequence many consumer products are not well designed. The problem lies mostly in the inadequate teaching of tribology in engineering schools, an omission that academia is not equipped to resolve in the near future.

To begin, it may be useful to define the term 'tribology'. The 'Jost Report' (**1**) ascribes a very broad scope to the term, namely, 'the science and technology of interacting surfaces in relative motion, and of associated subjects and practice. The term is taken to imply that motion will be resisted by friction, that continued motion may produce wear in some form, and that both might be alleviated by lubrication of some type. The complexity lies in the details.

The following sections explore the major issues related to tribology and the teaching thereof, outlined as:

1. How do we know there is a need?
2. What is available now?
3. The disciplinary range of tribology.
4. The education of the Complete Tribologist.
5. Who might be interested in a broad course in tribology?
6. What should be the form and content of a comprehensive course?
7. Who will develop such a course?

2 HOW DO WE KNOW THERE IS A NEED FOR TEACHING TRIBOLOGY?

2.1 Evidence of insufficient education in tribology

Though some segments of our broad technical community have achieved great sophistication in tribology research, significant educating still needs to be done. Engineers and scientists seem to be no less likely than the general public to purchase the bogus, 'cure-all' lubricants that abound in our service centres and shopping malls. The endorsements are so sincere and polished, particularly when a popular race car driver is the central figure in the endorsement. Bogus products appear faster than tribologists can learn of their capabilities (or incapabilities).

Various authors have provided some estimates of the annual cost of inadequate attention to tribology. The costs are of various kinds, including the cost of energy that is lost because of unnecessarily high friction, the cost of materials that are wasted by high wear, and the cost of reduced functioning of machinery because the load carrying capacity of sliding surfaces is less than the 'designed' load. Figures of loss to economies range from 2.5% to 6% of the world gross national product (GNP) (**1, 2**). This works out to several hundred dollars for each of us each year.

Direct evidence for inadequate knowledge of tribology among designers is apparent in very many consumer products. Unsatisfactory performance is seen in such products as automobile brakes that make distracting noises, hard disks in computers that fail and obliterate your work, and artificial skeletal joints that fail before adjacent parts do. 'Interacting surfaces in relative motion' are apparently not designed with the same confidence, finesse and dispatch as are most other components in the consumer product industry. Rather, the choices of materials, lubricants and contact stresses are often taken from past experience in the hope that this latest update will not push a design beyond the limits of the system.

2.2 Common occurrence in the consumer product industry

Between 75 and 100 students in my summer classes, most of whom are engineers in the consumer product industry have, over the years, related tales of confusion in the tribological aspect of designing products. There are differences from one company to another, but a fairly accurate composite of comments can readily be related. Frequently a junior engineer is assigned to work out a tribology problem, often a problem of long standing. The problem is assigned late in the design stage of a product though it had been known to be a problem much earlier. As lead time to production diminishes, several more people of various disciplines may be assigned to help, and each offers a different solution. Eventually the problem simply 'goes away' without anyone knowing why, only to return in the next iteration of product development. By the time of the next iteration, most if not all of those who worked on the problem previously have been moved to a different product or promoted out of engineering, and the (futile) search is renewed. Usually there is no written record of what was done before or why it was done, only an oral tradition. Often the only concrete evidence of a previous study of the problem is a bench test or simulator device, perhaps still in use because it once had a high priority.

Bench tests themselves are often the major problem. A prototype product may fail in bench tests, but miraculously the production items may perform well in use! The opposite also occurs, probably more often. Good bench tests cannot be designed without understanding tribology.

It is clear that tribology is the most poorly understood of all issues in mechanical design. By contrast, the methods of finite element analysis, control systems, estimation of fatigue life and many other engineering arts are well taught and well supported by validation test methods. Proof of this disparity may be seen in the differences in warranty costs: that for tribological inadequacies is many times greater than for all other defects combined in many products. It is amazing therefore that there is so little call in industry for more and better instruction in tribology in universities.

2.3 Some significant successes in tribology

Very likely the economic value of many successful tribological applications has gone unnoticed, but they are important to people nonetheless. Many products would not be available today had there not been some attention devoted to tribology. These include:

1. Hair 'conditioning' substances to make combing easy,
2. Slip resisting floor materials,
3. Tubing that can be inserted long distances into blood vessels,
4. Woven fabrics that fold, hang and stretch well,
5. Cook-ware that cleans easily,
6. Fluid seals that prevent dripping and spilling of liquids,
7. Paint that adheres firmly to plastics.

Most people working with these products would probably not classify their work as tribology, but rather as an application of science. Still, their insights should be gathered into the general wisdom in tribology.

3 WHERE IS TRIBOLOGY TAUGHT TODAY AND BY WHOM?

3.1 Course offerings

Only about 20 'full length' courses on some aspect of tribology are offered in academic institutions in North America each year. The majority of these courses focus on either fluid film lubrication or on the mechanics of elastic contact, largely because these topics are simpler and more highly developed than are most others. Minority topics include abrasion, erosion, specific wear mechanisms in various materials, etc. Very few cover the basics of friction, wear in general or chemical boundary lubrication to any great extent.

The driving force behind these offerings are the academic faculty members who are engaged in research in one or another of the divisions of tribology. The students in tribology courses are usually fourth year undergraduates and graduate students, who need a few units to graduate. Undergraduate students often look for a course that is a small extension beyond their favourite required courses, whereas the graduate students are more inclined to take an unusual type of course in order to broaden their experience. Historically, few of these students have been employed as tribologists in industry, however.

In addition to full academic courses, some colleges offer one or two clock hours of instruction in tribology as part of design or manufacturing courses. Most of this instruction is taken from narrowly focused books on tribology and it is mysterious that so few faculties in design and manufacturing consider any part of tribology to be important. Perhaps one reason is that design courses in universities have traditionally ignored cost or customer satisfaction as a factor in optimizing product design.

About five or six courses of various duration, two to five days, are also offered in the summer in universities and attended mostly by people from industry. Short courses are fairly popular with industry. Their 'shortness' appeals to managers, and advertisements for such courses imply that great quantities of knowledge can be imparted in a short time, if well done.

Outside academia, there are approximately 15 tutorial sessions offered each year in conjunction with the meetings of professional societies. The Tribology Division of the ASME–I (The American Society of Mechanical Engineers – International) and the STLE (The Society of Tribologists and Lubrication Engineers) are more active than others in this area in that they have education committees that regularly organize tutorials in some aspect of tribology. These tutorials usually consist of lectures by speakers from industry who are active in several specialized fields.

3.2 Available printed information

A restricted library on tribology would require about 100 feet (30 meters) of shelf length, of which perhaps 90% would be journals and conference volumes. About 10% of the shelf length would be required for various monographs, handbooks and textbooks. A serious scholar should be able to assemble a working knowledge of tribology from these books, but it would require a great amount of study. Most books are rather narrow and report little more than research results, but some can serve as text books. The best books for academic purposes are those in the specialized areas, in particular, hydrodynamic lubrication and contact mechanics. The least useful are those that attempt to explain friction and wear from the perspective of a single discipline, or from studies of single mechanisms of friction or wear in steady state conditions. Few of them help a reader understand friction or wear in the real world of time varying combinations of friction and wear mechanisms. Perhaps most authors assume that all (or most) practical situations are simply summations of 'components' of friction or wear and the responsibility of the practising engineer is to determine which components are present. There is a great cultural gap between practising engineers and authors of books and research papers.

3.3 Overview

Overall there are several venues for teaching and learning tribology, but they are evidently not sufficient to meet the great need in the product design function. The need has been known, or certainly discussed for many years. The Jost Report (**1**) pointed to the great need to improve the tribological aspect of product design 30 years ago, and called for the development of interdisciplinary training in tribology. Then 18 years ago Czichos wrote a textbook (**3**) and offered a "systems approach" to tribology, which showed the value of interdisciplinary study. Countless other papers and discussions have repeated these themes over the years, but tribology is not yet recognized as a vital topic in engineering education. Why not? I submit that there are three major reasons:

1. Good understanding of tribology requires large scale interdisciplinary co-operation and few technical people are prepared devote much time to such an effort,
2. Good teaching of tribology requires a central theme to rescue tribology from being seen as an arbitrary collection of disjointed topics, and there is yet no central theme,
3. The succession of specialized themes and methods of research in tribology over the years has diluted efforts to view the broad 'picture' of tribology. Each 'new' theme adds new elements to tribology, recent examples being the Scanning Tunnelling and related microscopes and probes, which have become isolated topics in themselves rather than adding to the understanding of old topics.

4 THE DISCIPLINARY RANGE OF TRIBOLOGY

The proper focus in teaching a broad course in tribology is toward developing methods or *guidelines for designing* 'interacting surfaces in relative motion'. This is rarely emphasized in seminars, tutorials, conferences and symposia on tribology. Rather, conferences are most often organized around narrow and specialized research areas, and some are on such popular topics as 'manufacturing tribology', 'automotive tribology', 'computer tribology', and most recently 'nano-tribology'. The contents of these programs, in the aggregate, do not define the bounds of the field, nor do they help much in the consumer-product design activity.

The elements in a broad course must surely parallel the considerations in the design process, some of which are:

1. Type of materials (various metals, polymers, ceramics, coatings and all composites), condition of the material, mechanical as well as tribological properties,
2. Lubricants and lubrication methods, lubricant handling, additives in process fluids (for hydraulic systems, computer hard disks, shampoos, etc.),
3. Surface roughness, and methods of producing surfaces,
4. Tolerances and clearances, estimated through analytical methods,
5. Break-in methods, accommodation of start-up (initiation of sliding) and intermittent overload,
6. Testing and simulation, and methods of ranking of friction and life performance of candidate materials, lubricants and methods of manufacture,
7. Prediction of friction and wear life of the product in regular use, and expected change of function before total failure of the component occurs.

The above list may be seen to derive from several very different academic disciplines. The disciplines and some of the topics relevant to tribology in each discipline are:

1. Chemistry and chemical engineering – molecular structure and chemical pathways in the formulation of lubricants, the surface state of solids,
2. Solid mechanics – contact stresses, fracture mechanics, temperature rise in sliding, etc.
3. Fluid mechanics – designing mechanical systems that slide or roll on thick fluid films, etc.
4. Mechanical dynamics – analysis of vibration modes, analysis of experimental data, dynamic modelling, etc.
5. Materials engineering – the properties of materials and how they behave in sliding and wearing contact, etc.
6. Physics and other sciences – characterizing molecularly thin films, both theoretically and experimentally.

An adequate knowledge of tribology actually does not require a full education in all of the above disciplines, but there is hardly a way to pick and choose only the most relevant courses in each academic unit. Every disciplinary course of studies is carefully crafted to channel students through a sequence of increasingly more sophisticated teachings to produce specialists in the discipline. Further,

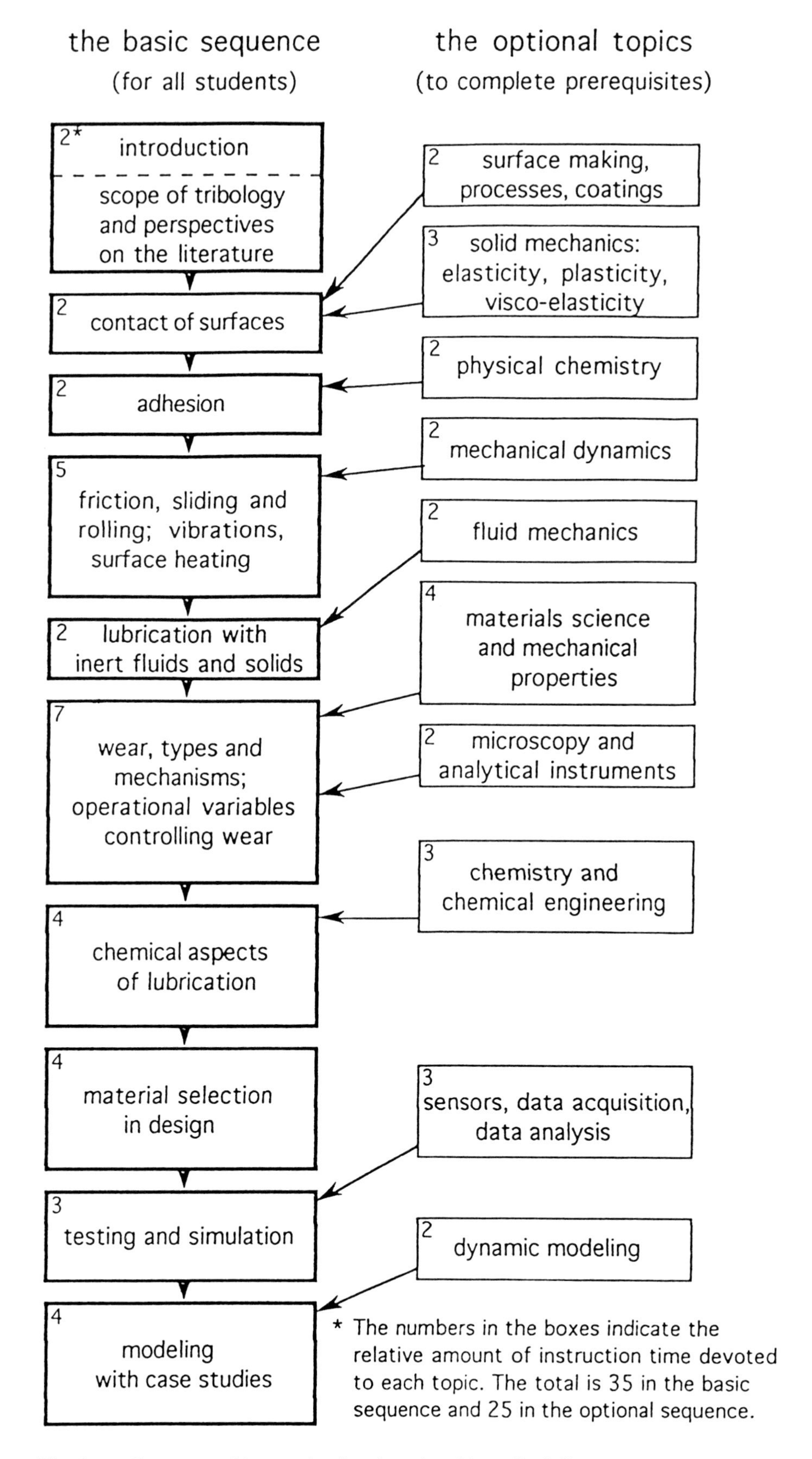

Fig. 1 Sequence of instruction in a broad and interdisciplinary course in tribology.

the methods in the several disciplines differ considerably from each other, and few temperaments are compatible with the full range of these methods. The topics based on fluid and solid mechanics are usually presented with extensive mathematics. The materials related topics, by contrast, are often presented in the form of disjointed litanies of types or mechanisms of friction and wear, supported by micrographs and spectra from various analytical instruments. Teaching in the chemical areas covers many molecular structures and reaction paths from one structure to another in lubricants, and industrially based courses often convey the simple but vital formulas for lubricants that one should use to keep a particular machine in operation at the lowest cost. Much teaching involves subjective terms and definitions, conventional and seriously outdated descriptions of the causes of and cures for tribological problems, often ignoring the time varying chemical and physical events on surfaces that control friction and wear rate. It may be time to explore another approach, and that would be to craft a special course of studies for tribologists. However, teaching only those historical topics, in divided units as listed above is not sufficient: this teaching should be 'seamless' in that few if any of the traditional and partisan views should be taught as 'settled' or complete knowledge. The next section suggests some additional philosophical positions in the more profitable type of teaching, using the 'complete tribologist' as the end product of the new teaching.

5 EDUCATING THE COMPLETE TRIBOLOGIST

Friction, wear and lubrication involve chemical, mechanical and materials variables, working in concert. Further virtually every tribological system operates in a transient state (except perhaps machinery that is lubricated by thick fluid films). The Complete Tribologist should be conversant across the entire range of tribological knowledge, and should acquire the habit of:

1. Describing all tribological systems as continua of disciplinary phenomena, rather than breaking up problems into specialized and disciplinary bits,
2. Defining terms and concepts such that they have the same meaning in every aspect of tribology,
3. Embedding useful models, however primitive, into computer-based design work stations.

5.1 Tribological system description

Real systems embody mechanisms and details that are described and understood in different ways by materials experts, solid mechanics experts, chemists, etc. But, a complete description should take account of oxides (on metals) and adsorbed gases, chemical reactivity, physical surface change, composition and size distribution of loosened particles (if any), state of degradation of lubricants (if any), changes in contact stress distribution, temperature distribution on the asperity scale, etc. To describe problems with any of these elements missing is to miss most of the possible solutions to the problem. Now and then some aspect of a tribological problem can be solved by a specialist in a single discipline, just as some medical problems can be solved by a medical specialist. However, efficient design, as well as medical practice, requires a first view by someone with a general perspective.

5.2 Terms and concepts

The diverse meaning of many terms and definitions in tribology serves to limit clear thinking. Scuffing, scoring, galling and seizing are four confused (in the literature) terms that refer generally to damage to lubricated surfaces. In some places the terms are synonymous, in other places these terms may describe a succession of damage severity. There have been several attempts to define each of the terms by reference to photos of magnified surfaces, but even these are not convincing because of the very many ways of preparing and viewing surfaces: each person is accustomed to a different scheme or magnification of images. It would be useful if teachers of tribology were to work in concert to standardize terminology, and it should be done in conjunction with engineers in industry.

Another term on which agreement should be developed is 'boundary lubrication'. It has been defined in at least four different ways:

1. The state of (inadequate) lubrication when a fluid film is so thin that some load is carried by 'solid-solid contact' of asperities on opposing surfaces, or,
2. When the coefficient of friction of a lubricated system exceeds some particular value (e.g., 0.1, 0.15, 0.2, etc.), often stated as absolute definitions each of which ultimately applies to specific machinery, or,
3. When sliding surfaces are separated by *physically* adsorbed substances (Hardy's original definition), or,
4. When sliding surfaces are at least partially separated by substances that are formed in situ by chemical conversion (as with lubricants containing active *boundary* additives).

Each definition is used (and apparently understood) within different disciplinary groups, but misunderstood or dismissed by others. It may seem unreasonable to suggest that everyone agree on a single definition, and yet it is useful to do so: good solutions to problems are more likely to arise where everyone uses the same terms, with the same meanings.

Perhaps the most widely misused or improperly understood term is the term 'adhesive' as a type of wear. There can be no dispute that adhesion is operative in sliding systems since all matter will bond together naturally. Adhesion is said to be operative when sliding surfaces roughen without the agency of abrasives, though direct tearing of primary bonds is rarely demonstrated. However, the term 'adhesive wear' is also invoked when a plot of wear rate is seen to be linear against hardness, or against sliding speed or contact pressure (the same linear plot is seen by others as proof of abrasive wear!). Adhesive wear is therefore taken to mean a type of behaviour and a fundamental force of nature, neither concept conveying any suggestion of solutions to wear problems.

5.3 Modelling – a new imperative

The design process is increasingly oriented toward optimization between and among many variables. The reason is simple: an optimum design cannot be accomplished without consideration of all relevant variables. The methods of optimization not only require full characterization of all variables, but also require some method of quantifying the phenomena involved: this exercise may be called *modelling*.

The practice of teaching and learning engineering by manipulation of models, vital as it is in general technical education presents a problem in teaching tribology. Too few tribological phenomena have been modelled analytically. Teaching a broad course in tribology therefore requires a new approach. It is tempting to spend the most time on the well-modelled topics, hydrodynamics and contact mechanics as mentioned earlier. But it is necessary to make a best effort to model friction and wear as well.

For perspective, it is probably fair to state that 'under-modelled' fields are seen as 'less worthy' of consideration in some curricula, which attitude may be one reason for the under-representation of tribology in curricula. In essence until 'someone else' develops models for some phenomena, many academics prefer not to consider the topic. However, scholarliness consists in using models where they are available and developing even crude models where none exist at this time. At the very minimum, tribological phenomena can be described in 'word models', but it must be carefully done.

There are 'more and less' profitable approaches to modelling. Most of the existing equations for friction and wear have been assembled semi-empirically, which is justifiable, but built up as if both friction and the wearing process involve steady-state events (**4**). Perhaps the assumptions of mechanics are too deeply embedded in technical minds to accept an inhomogeneous or discontinuous world. Friction and wear are usually not steady-state events and the best analogy is the modelling of human life: the progression toward the end of life is fraught with many contingencies. In the same way, the wear life of a sliding pair is 'under threat' by several simultaneous possibilities, one or two of which may become prominent depending on some intermittent overload, change in surrounding chemistry or other phenomena. Teachers of tribology might profitably look away from the methods of solid mechanics to those events closer to their own existence when framing models for friction and wear.

On the range of complexity of models, those for fluid film lubrication and elastic contact mechanics are the simplest because they involve relatively few variables. Liquid impingement erosion involves more variables, solid particle erosion still more, and dry sliding wear involves a great number of variables. Topics for which no assessment of the number of controlling variables has been made so far include chemical boundary lubrication, start-stop effects and the myriad of other transient events. Descriptions of these phenomena still require some empiricisms and anecdotal treatment.

The great number of variables makes modelling rather difficult, especially for those of the technical backgrounds that usually populate the field of tribology. The easiest (and historically most common) approach is to examine the influence of one variable on wear rate at a time, holding all other variables at some 'standard' value. The next logical step is to combine all variables into one equation, either retaining the original coefficients (of proportionality) on each variable or combining all individual coefficients into one grand coefficient. There may well be some sliding systems for which this procedure works, perhaps in deep abrasion with durable abrasives, but not in more complicated systems. The problem is the usual interdependence of variables. In such cases, unless all relevant variables have been identified and characterized, the coefficients on the individual variables are not robust, that is, a missing variable will have the effect of changing coefficients in unexpected places when one identified variable is changed during an experiment. An example of apparently having identified all of the relevant variables is a paper on the friction of oil lubricated lip seals, in this conference (**5**). An important point in establishing the robustness of coefficients is that all variables must be varied over wide ranges in the experiments, which may explain why it is not often done.

A serious impediment to overall and broad modelling is that successful modellers in the narrow specialities see little challenge in other areas. However, chemical boundary lubrication is a clear example of the value of interdisciplinary co-ordination. The success of chemical boundary lubrication lies in the transporting of active chemical species to regions of distress on sliding surfaces. That transport rate depends on the flow of fluid carrier through the contact region (as done by fluid mechanics), and the concentration of chemical compounds in that fluid (as known by chemists). the loss of reaction film materials depends on the size of, and contact stresses in and type of motion of the sliding members (solid mechanics) and the mechanical properties of the reaction film (material science). The problem cannot be fully described and no progress can be made with any one of the dimensions omitted.

5.4 Perspective

The complete tribologist is a new entity in technology, conversant in relevant parts of several disciplines, well versed in design methods, uses terms with broad and precise meaning, and is inclined to model tribological systems so that proper solutions can be applied. In modelling there are yet few guidelines for systems as diverse as tribological systems so the complete tribologist must also be inventive beyond what is usually taught in formal education.

6 WHO MIGHT BE INTERESTED IN LEARNING TRIBOLOGY?

In developing a course of instruction in tribology it is useful to keep the potential audience in mind. The most readily suggested target or potential recipient for instruction in tribology is the college student, usually in Mechanical Engineering. However, the greatest need and the greatest potential for direct application of the teaching is with the engineers in industry who are engaged in the design and development of consumer products. They have the greatest incentive to learn the subject and often have case studies in hand. In distinction to the consumer product industry, the capital products industry also has tribology problems but these are usually long term problems and specialized in nature. Designers of capital products often 'grow into the problem' or have experts available to help them.

Most engineers, by far, who work on tribology problems have the, first degrees in mechanical engineering, though there are many with degrees in other disciplines as well. Few will refer to themselves as 'tribologists' just as they do not single out any of the other topics in their engineering curriculum as their speciality.

Engineers will often attempt to solve tribology problems with their resident skills as if they are design problems. For example, when a sliding pair vibrates, most mechanical engineers will view the problem as one of mechanical dynamics and overlook the key role of frictional behaviour as the initiator of vibration. Doubtless this is the result of most mechanical engineers having taken a (required) course in mechanical dynamics and none in tribology.

7 THE OPTIMAL INSTRUCTION SEQUENCE IN TRIBOLOGY

Tribology spans several disciplines and few people will have had previous instruction in many of them. An exposure to tribology within one of the sciences or branches of engineering appears not to be sufficient for much of industrial practice. Thus some preparation in several disciplines is useful but in limited amounts. Few people would devote more than 100 instructional hours (plus computational practice, outside reading, etc.) in preparation to becoming a complete tribologist. Mechanical engineers will have had the most of these 100 hours already (perhaps half), materials engineers somewhat less, and others still less.

Several course structures could be envisioned according to some assumed level and breadth of preparation beyond sophomore physics, math and chemistry. One possibility is that relevant topics from materials science, mechanics and chemistry, etc., could be woven into one continuous course along with specifically tribology topics. In this course many students would be 'covering the same ground' as in their degree programs (though review can sometimes be enlightening). Alternatively, there could be a 'mainstream' of tribology instruction, with modules available in background topics for students to study before the need arises in the mainstream instruction. A topical listing of such a course is shown in the following block diagram. In any case, there should be a single, coherent ending package of instruction, which brings all students to the same understanding of the principles of tribology. Common understanding is vital for discussions on tribology, and is, as well, the starting point in the co-ordination of the several specialities and skills in solving tribology problems.

8 WHO SHOULD DEVELOP THE DESIRABLE COMPREHENSIVE COURSE?

The key to the development of a successful course, finally, is that it be developed by an interdisciplinary coalition of tribologists, from industry and from academia. Each will make the greatest contribution by learning how little of their own field is required to fully educate the Complete Tribologist. Otherwise the coalition will naturally suggest 2 or 3 years of study.

Perhaps the most profitable action can be taken by the education committees of the professional societies, particularly those that also sponsor scholarly journals. These committees are usually composed of people from industry, academia and government facilities. The benefits of providing comprehensive education in tribology surely would be to improve consumer products, both for the satisfaction of the consumer and to enhance world economies.

9 REFERENCES

1 *Lubrication (Tribology) Education and Research: A Report on the Present and Industry's Needs*, commonly known as the Jost Report, Dept. of Education and Science, HMSO, London, 1966.

2 **Barwell, F.T.**, Wear of Metals, *Wear*, **1**, pp. 317–332, 1957.

3 **Czichos, H.,** *Tribology*, Elsevier Tribology Series, v.1, 1978.
4 **Meng, H.C.** and **Ludema, K.C.,** Wear Models and Predictive Equations: Their Form and Content, *Wear,* **181–183** (1995), pp. 443–457.
5 **Wassink, D.A., Lenss, V.G., Levitt J.A.,** and **Ludema, K.C.,** Physically Based Modeling of Sliding Lip Seal Friction, *First World Tribology Conference*, London, September 1997.

Simulation of real contact in tribology

N K MYSHKIN, M I PETROKOVETS, and **S A CHIZHIK**
Metal-Polymer Research Institute of Belarus Academy of Sciences, Gomel, Belarus

SYNOPSIS

Measuring techniques for topography of very smooth surfaces are discussed. Simulation methods of real contact are reviewed. A new version of Archard's model is proposed. The model is developed for real surfaces involving two levels of roughness, micro-roughness and submicro-roughness.

NOTATION

$\tilde{A}$	dimensionless area of contact
A_a	nominal contact area
A_r	real contact area
D	density of summits
E	reduced modulus of elasticity
F_i	Lennard–Jones potential
F_s	specific adhesive force
h	separation
H	deformed layer thickness, normalizing parameter
I_1, I_2	table functions
K	reduced stiffness of contacting materials
l_i	clearance between flat body and rough surface
$\tilde{p}$	dimensionless load
p	contact pressure
S_i	area element
z_i	surface relief height relative to the lowest point
z_{max}	maximum of a surface relief height
α	bandwidth parameter
β	curvature radius of asperity summit
δ_i	compression
ε	interatomic distance
γ	interface energy
Δ_c	adhesional parameter considering topography parameters and material properties = $\frac{3}{4}\frac{1}{\sigma\beta^{1/3}}\left(\frac{\pi\beta\gamma}{E}\right)^{2/3}$
σ	mean-root-square deviation of asperity height
ν	Poisson's ratio

1 INTRODUCTION

The point that the friction of solids is substantially conditioned by the roughness of rubbing surfaces and their free energy (adhesion) has been perceived adequately by the tribology founders. Thus, Coulomb considered interlocking asperities, while Desaguliers called researchers' attention to the molecular interaction of surfaces. Yet only in modern times have tribologists become cognizant of the difficulty and beauty of the inheritance received by them. Here, the case is not only the well-known discussion of the relation between mechanical (deformation) and molecular (adhesion) components of friction, although this discussion is continuing with a wide range of opinions from the flat denial of adhesion in friction to the nearly full neglect of the mechanical component. The matter is that these subjects of investigation (roughness and adhesion) have turned out to be far more complex than the founders of tribology thought. There was something behind that mechanicians avoided solving contact problem with account for roughness and adhesion. Considerable advances have been made in the area of mechanics. This was not without impact of tribology, but in their turn the advances allowed tribologists to refine their understanding of friction, including the real contact area, one of basic elements involved in friction.

The aim of the present review is to trace briefly the development of some notions in the area of simulation of real contact, having regard to a considerable step forward in surface topography measurement.

2 MEASUREMENT AND DESCRIPTION OF SURFACE TOPOGRAPHY

2.1 Measuring techniques for topography of very smooth surfaces

An analysis of tendencies during the last quarter of the century reveals that the dimensional range in technologies and research has shifted to the submicro- and nanometre scales (**1**). High precision rubbing parts and other mated components for precise mechanics (e.g., magnetic recording devices, high-precision lathes, sealing systems, etc.) can be produced using only very smooth surfaces. Modern technologies can produce working surfaces with standard deviations under 10 nm.

New requirements for roughness measurement techniques and new means are needed to check the quality of these parts. The accuracy of traditional contact (probing) as well non-contact techniques has been

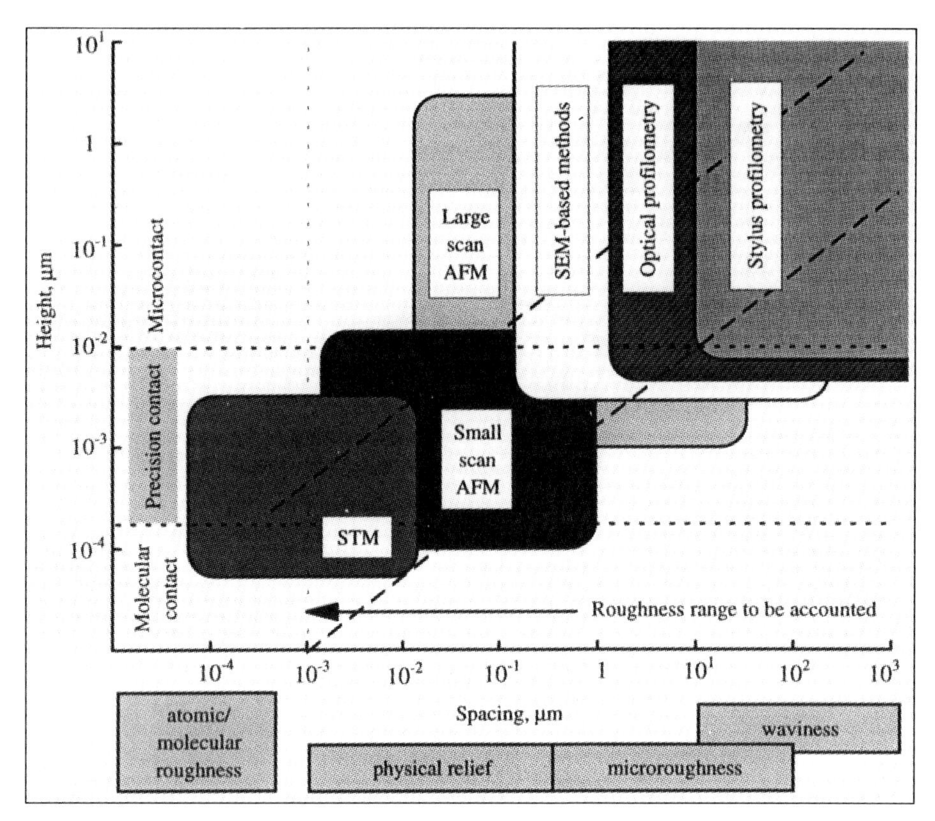

Fig. 1 Diagram of the height and spacing parameters and ranges of vertical–lateral resolution for different methods of roughness measurement.

perfected to a level allowing measurement of roughness in the nanometre range (Fig. 1). The most accurate profilometer probes allow measurement of summit heights of several Ångstroms (**2, 3**); optical instruments have the same ultimate vertical resolutions (**4, 5**). Yet, the comparatively poor lateral (horizontal) resolution significantly limits application of these techniques to nanometre topographies when the distance between asperities is much less than there solution or 0.1–1 µm.

The development of techniques using probes smaller than the radius of the probing needle or the light wavelength makes it possible to extend the spectrum of surfaces studied. The electron beam in the scanning electron microscope (SEM) is an example of such a probe. By interpreting the emission intensity of the secondary electrons the topographic pattern can be restored, and the SEM technique can be used to gauge topography with a comparable resolution both vertically and laterally (**6**).

A still finer probe is the electron flux tunneled between the target surface and the needle tip in the scanning tunneling microscope (STM). In this case the surface topography resolution is 0.01 and 0.1 nm in the vertical and lateral directions, respectively (**7**). Hence, the STM technique and others resulting from its progress make it possible to use this approach for more accurate topographic investigations of solids on the nanoscale.

Significant prospects are connected with the application of atomic force microscope (AFM) (**8**) in which atomic-molecular surface effects are detected. In our studies we have used an atomic force microscope with probes in the form of a platinum blade cut to a sharp angle (1.75 × 0.5 × 0.005 nm) with rigidity 1 N/m. It was arranged at an angle to the target surface and vibrated at a frequency of 50–100 kHz and an amplitude of about 1-2 nm. When the probe tip approaches the surface (as close as 1 nm) a force of attraction (repulsion) is experienced by the surface, causing changes in the amplitude of vibrations which, in their turn, are registered by an optical interferometer. A constant clearance was maintained during scanning by the electronic feedback system and drivers using piezoceramic members. The maximum scanning field was 20 × 20 µm. The vertical resolution was not less than 0.2 nm and the horizontal resolution was 2 nm. The scanning site was selected using an optical microscope either manually or with the help of a stepper device with indexing of not more than 3 µm along the X and Y axes.

Though scanning probe microscopy is widely used for studying the surfaces of solids, the problem of measuring the average parameters on the nanoscale needs more thorough investigation. This is essential, in the first place, for solving tribological problems. It is difficult to compare surfaces, to reveal the role of fine roughness during friction and to model contacts mathematically unless the problem is solved.

The problem to determine the average dimensions of

Fig. 2 Image processing procedure for separation of roughness scale levels. A=B+C, where A is initial image, B is its long-wave component, C is the subroughness image.

topographical elements (asperities and valleys) forming the surfaces of solids at the molecular level can be solved only providing a sufficient bank of statistically valid representative data is available. As has been noted, the STM and AFM can produce the necessary degree of resolution. Still, the problem remains how to accumulate data based on the maximum possible topographical resolution of the available instruments since it requires extended scanning and creating large banks of numbers with the problem of storing and processing them. In this situation it is natural to use the traditional tribological analysis of the profile data which makes it possible to obviate the above problems when determining the fine roughness parameters.

The natural quest for inspection of rubbing surfaces in minute detail and advanced methods of roughness measurement gave rise to establishment of three fundamental facts: erratic arrangement of asperities in surface and the random distribution of their parameters (height, slope, summit curvature); the multiscale character of roughness (that is, the significant size of asperities can vary from the length of the sample to the atomic scale); and non-stationarity which manifests itself as a dependence of the numerical values of roughness parameters on the scale of measurement. These properties are not independent; for example, non-stationarity owes its origin to the multiscale topography, all other things being the same.

2.2 Some features of surface topography

The random character of roughness has became platitudinous, but its role in friction processes appears to be underestimated. It is because of erratic roughness that friction shows as a stochastic process with increased dissipation of energy. It is not without reason that surfaces with regular roughness may exhibit better tribological behavior.

The multiscale topography appears to be due to the action of numerous factors including chemical and physical ones (atomic-molecular and supermolecular structures) as well as processing factors (type of surface processing, inaccuracy in the tool, etc.). It is convenient to recognize five major levels of topography in accordance with length scale: errors in form, waviness, (micro) roughness or technological roughness, subroughness or physical relief, and atomic-molecular roughness. Whilst strict lines of demarcation between these levels are absent, the first three levels have been standardized. A demand for the last two levels arose owing to the growing field of nanotechnology.

Real surfaces, as a rule, contain at least two levels of asperities, such as waviness plus roughness or roughness plus subroughness. The important implication of this fact for tribology was first perceived by Archard who proposed the well-known multilevel model (**9**). He pioneered in the application the multiscale topography to deduction of the friction laws for the contact conditions when deformation is entirely elastic. This model will be discussed below at greater length.

The non-stationarity of surface topography was first identified by Sayles and Thomas (**10**) and gives investigators a lot of trouble. This fundamental problem is beyond the scope of the present review. Yet it unfailingly accompanies any multilevel contact model. In this connection we propose a crude method to separate a real topography measured by AFM into two levels, roughness and subroughness. This method provides the fulfillment of at least one of two conditions for stationarity, namely, zero mean of random process. The procedure is as follows. Preliminary analysis of an AFM-image of a rough surface (Fig. 2, A) revealed that the profile contains long-wave component inherent microroughness rather than subroughness. To eliminate the component, we used the procedure of repeated median filtration of the image. After each application of the procedure the utmost (on perimeter) points of the images which heights were deformed most of all as a result of edge effects were cropped. The filtration procedure was carried out up to seven times. Computational experiment has shown that, for the images, it was enough to filter out subroughness components of the relief. A greater number of filtration repetitions does not significantly change the spectral function for appropriate profile sections of the image. At the following stages the filtered image (B) was subtracted from the initial image the (A) to the obtain resultant image (C) reflecting surface subroughness (Fig. 2). Parameters for the submicro-relief were determined on the basis of the

C-type images (10 × 10 μm, 128 × 128 points).

The similarity of the spacing parameters obtained from image C (Fig. 2) and those measured by Talystep on the same sample may be good indirect evidence that the above procedure is acceptable.

The mathematical description of rough surfaces must consider the main features found from measurement; in so doing it must be applied to contact problems in tribology. There is a great diversity of mathematical theories of roughness from simplest the deterministic models (which describe a rough surface as a regular array of bodies with simple geometry) to complex models based on theory the of probability and fractal geometry. Here it is pertinent to note that the computer simulation of roughness with prescribed properties may be useful (**11, 12**).

3 SIMULATION OF REAL CONTACT

3.1 Deterministic models of contact

Coincidentally with the evolution of theories and measurement of roughness, contact models of rough surfaces were developed. The models may be categorized according to the theories of roughness used primarily as deterministic and nondeterministic. The former stand out because of their utmost simplicity. In essence, they deal with the sole feature of contact between rough surfaces, its patchy (discrete) pattern. In some cases, this would suffice to describe the contact behavior of tribological interest, as demonstrated by numerous publications on contact mechanics and friction (e.g., (**13, 14**)).

3.2 The Greenwood and Williamson model (GW model)

An elegant model of elastic contact advanced by Greenwood and Williamson (**15**) awakened great interest in the simulation of real contact by statistical methods. Their approach is rather straightforward and involves three main steps:

- determination of the probability distribution for asperity parameters using measurement and/or theories of roughness;
- solution of the governing equation for mechanical or another interaction of a single asperity with the counterbody; in doing so the asperity is generally replaced with the a simple geometrical shape (e.g; cone, sphere, elliptic paraboloid, etc.);
- ensemble average of asperities.

The first two steps open a wide avenue to attack the simulation of real contact. There is no way of listing every modification of the GW model. It should be mentioned that this work was concerned with using various statistics of surface topography, deformation modes, models of single asperities, and other factors. For example, a joint probability density of profile peak heights and curvatures has been exploited by Onions and Archard (**16**). Bush and his co-workers (**17**) replaced asperities with elliptic paraboloids and used the statistical geometry theory. Plastic contact was examined by Nayak (**18**). The transition region between elastic and ideal plastic contact, including the work-hardening effect, was studied by Francis (**19**).

Two additional examples can be given. The capillary condensation of moisture at the interface of smooth surfaces' contact (e.g., head-disc interface) can produce a significant effect on parameters of the contact. The effect has been estimated employing the approach under discussion in (**20, 21**).

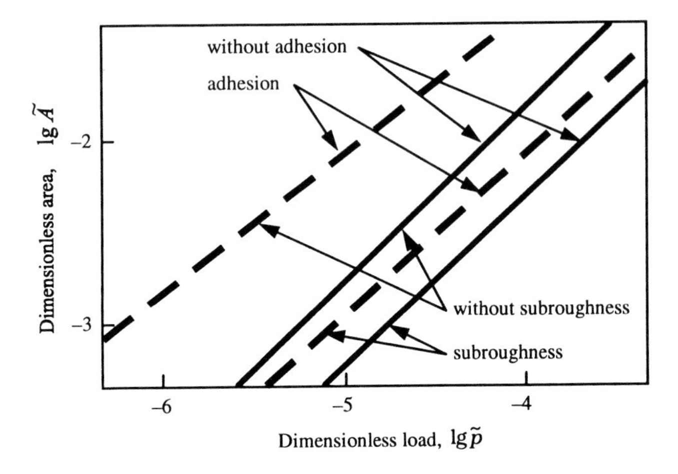

Fig. 3 Effect of load on discrete contact area.

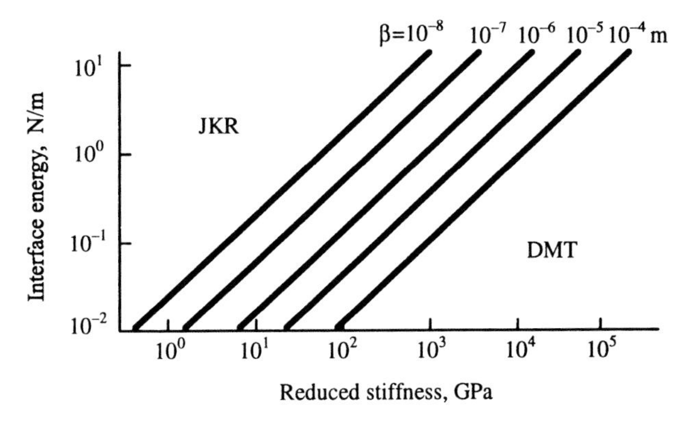

Fig. 4 Domains of parameters (K, γ) where either of two models (JKR or DMT) is correct.

The GW approach proved to be efficient at developing the so-called two-level model (model of Archard-type) (**22–24**). The combination of two levels, roughness and subroughness, was examined. To take into account subroughness on the large asperities (roughness), the solution of Greenwood and Tripp (**25**) for the contact of rough spheres was used as the governing equation for a single asperity. Analysis of the two-level model showed that the highest asperities of the first level (roughness) come into contact and form the individual contact spots. Yet, contrary to the traditional view, the spots are not continuous but multiply connected; that is, each of the spots consists of a set of subspots the total area of which was conditionally named the 'physical contact area'. This area is due to contact of submicroasperities (fine or nanoscale roughness) and is less that the real contact area by an order of magnitude (Fig. 3). The strong interaction between the mating surfaces may occur within the subspots of the physical contact. The contribution of this physical, and maybe chemical-interaction to total resistance to relative displacement of the rubbing surfaces may be very significant.

3.3 Simulation of real contact with adhesion

The first successful theories of adhesive contact between sphere and a flat surface were formulated by Johnson, Kendall and Roberts (JKR model) (**26**) and Derjaguin, Muller and Toporov (DMT model) (**27**). The former considers the adhesion as a change in surface energy only where the two bodies are in contact (that is, the attractive forces are infinitely short-range). The latter takes into account the interaction outside the contact zone (the Lennard-Jones potential). Analysis shows that each of these models is true for certain combinations of physico-mechanical and geometric characteristics (Fig. 4). The DMT theory is applied to the materials for which the point with coordinates (K, γ) is below the corresponding line plotted with constant radius of asperity tip β. Here K is the reduced stiffness of the contacting materials and γ is the interface energy. That is, the DMT theory is applied to bodies of micron sizes having the properties of metals or to bodies of submicron sizes having the properties of polymers.

Both of the theories were used in the simulation of rough contact with adhesion (**28–32**) based on the GW model. It was shown that surface roughness markedly reduces the adhesive force. For example, if roughness increases by a factor of two, adhesion drops by about two orders of magnitude. On the other hand, for a two-level model (roughness plus subroughness), calculation of the so-called physical contact area revealed that the area may increase by several times due to surface force (Fig. 3).

3.4 Fractal models

Considerable recent attention is focused on fractal characterization of surface topography and its application to contact and friction problems. This stems from the fact that the fractal approach can aid in overcoming the difficulties with which the conventional statistical methods are faced, in particular the multi-scale character of roughness and its non-stationarity. Majumdar and Bhushan (**33**) pioneered in applying fractal geometry to surface contact mechanics. They gave a fractal representation of rough the surface and developed a fractal model of contact. The number of papers using the fractal theory is increasing, e.g. (**34, 35**). It is reasonable that the approach under discussion is applied only to fractal surfaces. There is an opinion that many real surfaces, both natural and man-made, appear to be fractal with a very simple square-law power spectrum (**36**). Yet, to prove (or disprove) this suggestion for each specific case is a challenging task.

3.5 Computer simulation

Application of crude computing power allows the simulating rough contact based on 3 D images of surface topography. The topography is represented as a matrix of the surface and may be generated by special-purpose algorithms or measured by a scanning technique (stylus and optical profilers as well as scanning tunnelling and atomic force microscopies). Aside from pictorial rendition, this approach has a number of other advantages. It offers freedom the of replacing asperities by a simple geometry (as done by models of the GW-type) and does not set limits on the statistics of the surface (in contrast to the fractal approach). Computer simulation permits one to consider any factor of interest to tribologists (the surface forces, boundary films, anisotropy of roughness and mechanical properties). So far the stationary contact has been modeled. Hopefully, computer simulation will provide a way of studying movable contacts or at least transition from static to dynamic friction.

Some progress toward using computer simulation of real contact has been achieved (**37–42**).

Computer simulation is a very adaptable method in that it can be combined with other methods. Such an example is presented below where computer simulation is applied in conjunction with the GW model to the examination of a two-level contact model (roughness plus subroughness).

4 NEW VERSION OF ARCHARD'S MODEL

Let us consider the contact of a rough surface with an ideal smooth surface. This contact simulates the magnetic head-disc interface. Real topography was measured by AFM, and subsequent analysis showed that the topography involves two levels (scales) of roughness (microroughness, level 1, and submicroroughness, level 2) (see Fig. 2). It is believed that the contact is formed from deformation of both the levels of roughness due to external the load and surface forces. As an approximation, the problem is solved in two stages. At the first stage the contact is examined under the assumption that level 2 is absent. This stage is intended for evaluation of the load supported by subroughness (level 2). Solution of the problem follows the GW scheme taking account of the adhesion. Here we use the random process theory and the JKR model. The real contact area A_r and load p are functions of the dimensionless separation h (**23**):

$$A_r(h) = 3^{1/3}\pi D\sigma\beta\Delta_c A_a I_2(h,\alpha,\Delta_c)$$

$$p(h) = \sqrt{3}\,D\sigma\beta(\sigma/\beta)^{1/2}\Delta_c^{\ 3/2} E A_a I_1(h,\alpha,\Delta_c)$$

where $I_1(h,\alpha,\Delta_c)$ and $I_2(h,\alpha,\Delta_c)$ are the table functions presented in (**23**).

At the second stage the calculated load is distributed over the asperities of level 2 (Fig. 2, C) and the subroughness contact is examined by computer simulation. Material deformation is assumed to follow the Winkler foundation model.

For a volume element of height z_i and area $\Delta x_i \times \Delta y_i$ the force of resistance to deformation is a function of compression δ_i (Fig. 5):

$$p_i(\delta_i) = \frac{E}{kH}\delta_i$$

where $k = (1+\nu)(1-2\nu)/(1-\nu)$.

Molecular interaction is assumed to follow the Lennard–Jones potential $F_i = f(l_i)$ where l_i is clearance. The deformed surface takes on its form by compression at the contact points and by a tension out of contact Δz_i which is derived from the force balance on an area element:

$$\Delta z_i = \frac{8}{3}\gamma k H \varepsilon^2 E^{-1} l^{-3}$$

For the deformed surface $z_i' = z_i + \Delta z_i$ and it is possible to calculate compressive (contact) and tensile (out of contact) stresses at each point. After summation over all the points of the image we obtain the nominal contact pressure p and the specific adhesive force F_s:

$$p(h) = -\frac{E}{kH}\Delta x \Delta y \sum_i (h - z_i)$$

$$F_s(h) = \frac{8}{3}\gamma \varepsilon^2 \Delta x \Delta y \sum_i (h - z_i')^{-3}$$

In the general case the separation does not correspond to an equilibrium state of the surfaces. Then the resultant force between contact and the out-of-contact force:

$$N = F_s(h) + p(h)$$

is not equal to zero. This force is a compressive load or a pull-off force depending on its sign and is necessary to balance the surfaces at separation h. In this case the contact area is $A_r = n\Delta x\Delta y$, where n is amount of the image points of for which $z_i > h$.

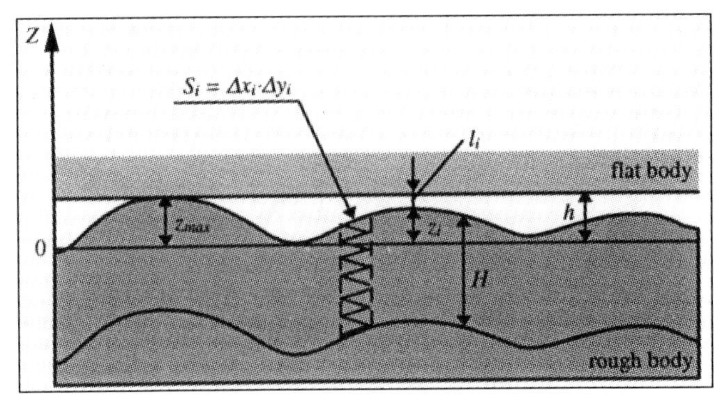

Fig. 5 The geometric parameters of real contact.

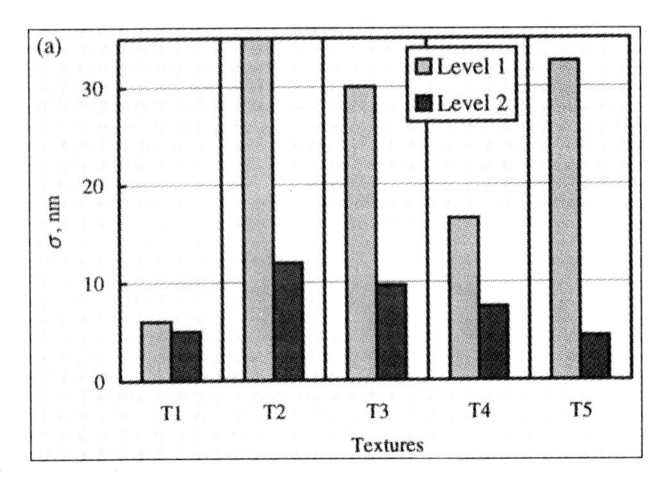

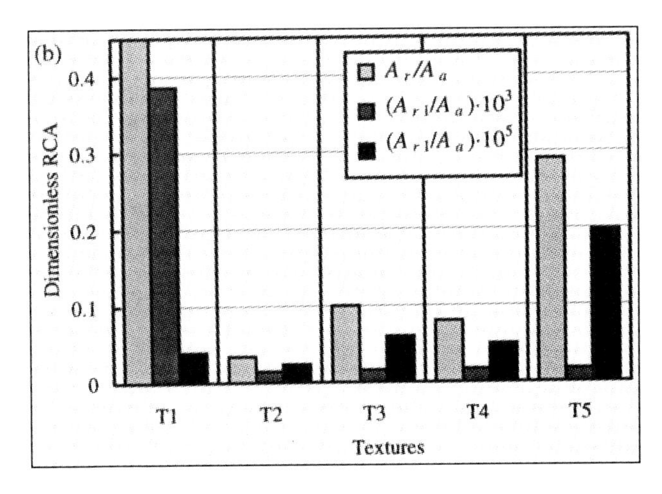

Fig. 6 Comparative diagrams for height parameter σ (a) and real contact area (b).

The algorithm was tested by modeling a sphere–plane contact.

The results obtained correlate well with the JKR and DMT theories. This two-level model was used to analyze head–disc interface. Hard magnetic disc substrates is of special-purpose aluminum alloy with 10 μm NiP coating. Original texture T1 was made by fine polishing with a diamond tool. Textures T2–T5 were obtained by grinding with free abrasive. The height parameter σ and the real contact area (RCA) calculated for textures T1–T5 are presented in Fig. 6. Visualization of contact spots at micro- and submicrolevels for T1 shown in Fig. 7. From this Figure we notice that spots of real contact (Fig. 7b) break up into more small-sized spots (Fig. 7c) which make up that we term the physical contact area.

5 CONCLUSION

At the present time there is a strong trend towards a transition from macro- to micro- and nano-scale that may give a new insight into the basic problems of tribology,

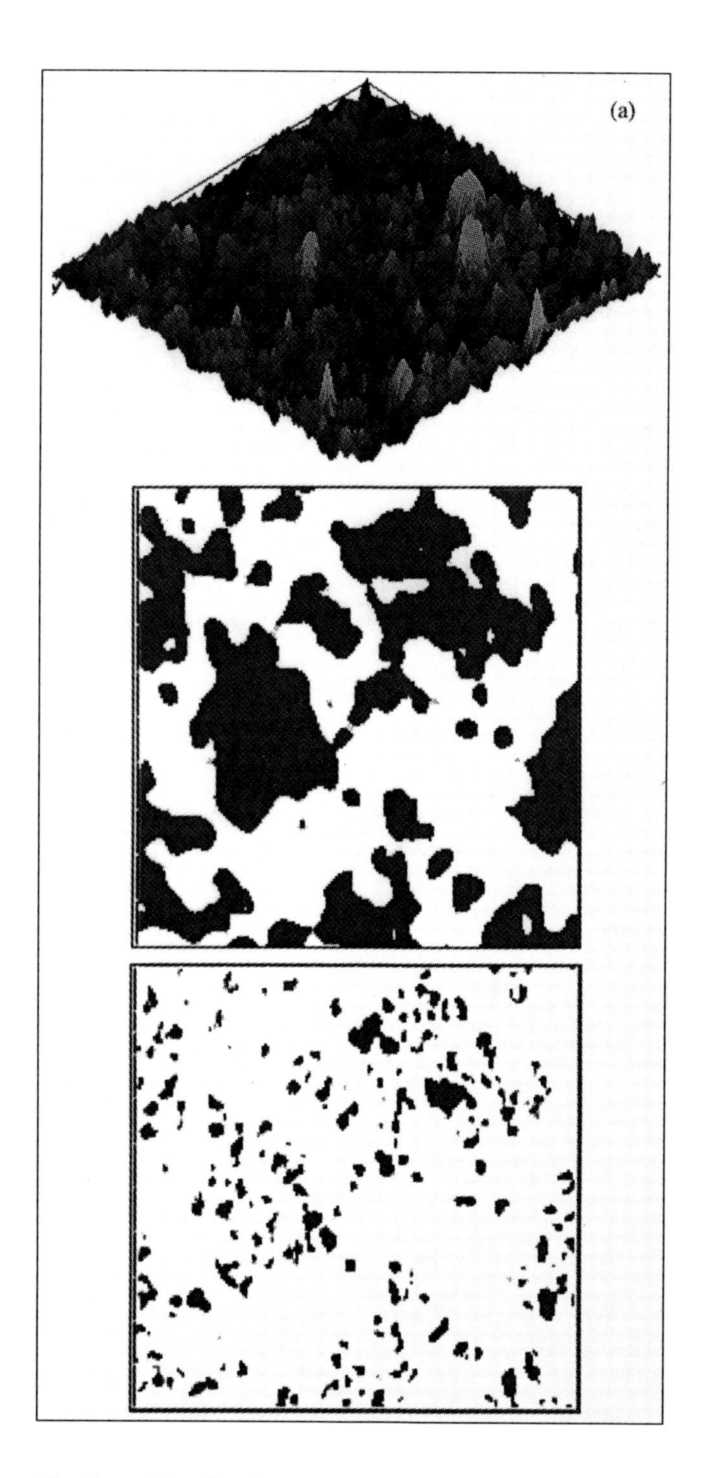

Fig. 7 Vizualization of contact spots at various scale levels: (a), AFM-image of an analyzed area (texture 1, scan size 15 × 15 μm); (b), actual contact spots at microlevel; (c), actual contact spots at submicrolevel.

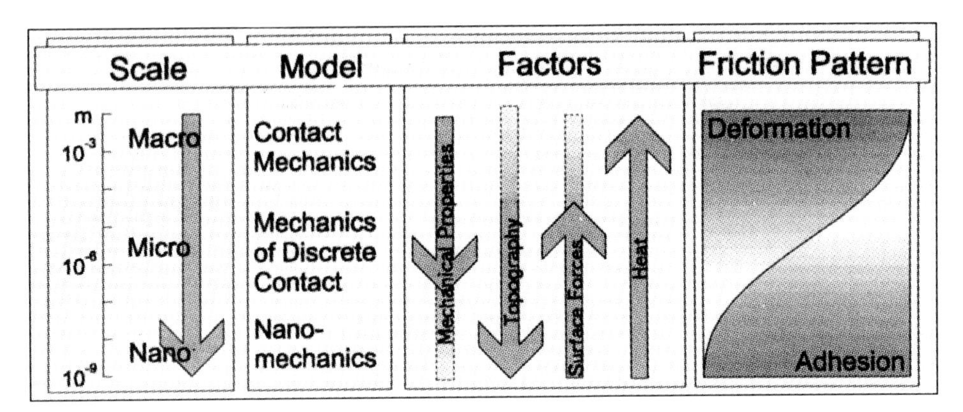

Fig. 8 Combination of factors affecting friction.

e.g., correlation of deformation and adhesion mechanisms of friction. This trend, along with a combination of factors with variable impact, is presented in Fig. 8.

Mechanical properties of contact materials should be taken into account at any scale level, but depending on this level such parameters as Young's modulus and hardness can differ not only in magnitude but also in their physical interpretation. Scaling from macro- to nano-scale can be related to the transition from bulk properties of material to surface layer properties, local Young's modulus and nanoindentation data (**43–45**). New promising results are expected from the application of SPM methods to evaluation of mechanical properties. The basic problem here is the physical interpretation of experimental data and their self-affinity in changing the scale factor.

In simulating real contacts these circumstances along with conventional factors such as temperature, sliding velocity, to name but a few, should be taken properly into account.

6 ACKNOWLEDGMENT

This work was supported by the Foundation of Fundamental Research of the Republic of Belarus (Grants T94-350 and T94-228).

7 REFERENCES

1 **Rohrer, H.,** 10 years of STM. *Proc. Inf. Conf. on Scanning Tunneling Microscopy*, North-Holland 1992, 1–6.

2 **Bennet, J.M.** and **Dancy, J.H.,** Stylus profiling instrument for measuring statistical properties of smooth optical surfaces. *Appl. Opt.*, **20,** 1785–1802.

3 **Bennet, J.M., Bristow, T.C., Arackelian, K.,** and **Wyant, J.C.,** Surface profiling with optical and mechanical instruments. Presented at the *Optical Fabrication and Testing Workshop*, Seattle, October, 1986.

4 **Kohno, T., Osawa, N., Miyamoto, K.** and **Musha, T.,** Practical non-contact surface measuring instrument with one nanometric resolution. *Proc. Eng.* 1985, **7**, 231–232.

5 **Bhushan, B., Wyant, J.C.,** and **Meiling, J.,** A new three-dimensional non-contact digital optical profiler. *Wear*, 1988, **122**, 301–312.

6 **Myshkin, N.K., Grigoryev, A.Ya.,** and **Kholodilov, O.V.,** Quantitative analysis of surface topography using scanning electron microscopy. *Wear*, 1992, **153**, 1, 119–133.

7 **Binnig, G.** and **Rohrer, H.,** Scanning tunneling microscopy. *Helv. Physica Acta*, **55**, 1982, 726-735.

8 **Sarid, D.,** *Scanning force microscopy*, 1991 (New York, Oxford University Press).

9 **Archard, J.F.,** Elastic deformation and the laws of friction. *Proc. Roy. Soc. Lond.*, **A243**, 190–205.

10 **Sayles, R.S.** and **Thomas, T.R.,** Surface topography as a non-stationary random process. *Nature*, 1978, **271**, 5644, 431–434.

11 **Patir, N.,** A numerical procedure for random generation of rough surfaces. *Wear*, 1978, **47**, 2, 263–267.

12 **Hu, Y.Z.** and **Tonder, K.,** Simulation of 3-D random surface by 2-D digital filter and Fourier analysis. *Int. J. of Mach. Tool Manufact.*, 1992, **32**, 82–90.

13 **Bowden, F.P.** and **Tabor, D.,** *The friction and lubrication of solids*, Part II, 1964 (Oxford University Press).

14 **Johnson, K.L.,** *Contact mechanics.* 1985 (Cambridge University Press).

15 **Greenwood, J.A.** and **Williamson, J.B.P.,** Contact of nominally flat surfaces. *Proc. Roy. Soc., Lond.*, **A295**, 300–319.

16 **Onions, R.A.** and **Archard, J.F.,** The contact of surfaces having a random structure. *J. Phys. D.: Appl. Phys.*, 1973, **6**, 289–304.

17 **Bush, A.W., Gibson, R.D.,** and **Thomas, T.R.,** The elastic contact of a rough surface. *Wear*, 1975, **35**, 87–111.

18 **Nayak, P.R.,** Random process model of rough surfaces in plastic contact. *Wear*, 1973, **26**, 334–340.

19 **Francis, H.A.,** Application of spherical indentation mechanics to reversible and irreversible contact between rough surfaces. *Wear*, 1977, **45**, 222–269.

20 **Tian, H.** and **Matsudaira, T.,** The role of relative humidity, surface roughness and liquid build-up on static friction behavior of the head/disc interface. *ASME J. of Tribology*, 1993, **115**, 28–35.

21 **Chizhik, S.A.,** Capillary mechanism of friction and adhesion for rough surfaces. *J. Frict. Wear*, 1994, **15**, 1, 11–26.

22 **Sviridenok, A.I., Chizhik, S.A.,** and **Petrokovets, M.I.,** Statistical model of elastic contact accounting subroughness. *Sov. J. Frict. Wear*, 1985, **6,** 6, 57–65.

23 **Sviridenok, A.I., Chizhik, S.A.,** and **Petrokovets, M.I.,** *Mechanics of discrete friction contact* (in Russian), 1990 (Nauka i technika, Minsk).

24 **Sviridenok, A.I., Chizhik, S.A.,** and **Sveklo, I.F.,** Microtribological problems of physics and mechanics in smooth surface contact. *Tribology International*, 1996, **29,** 5, 377–384.

25 **Greenwood, J.F.** and **Tripp, J.H.,** The elastic contact of rough spheres. *ASME J. of Appl. Mech.*, 1967, **34,** 153–159.

26 **Johnson, K.L., Kendall, K.,** and **Roberts, A.D.,** Surface energy and the contact of elastic solids. *Proc. Roy. Soc Lond.*, **A324,** 301–313.

27 **Derjaguin, B.V., Muller, V.M.,** and **Toporov, Yu.P.,** Effect of contact deformation on the adhesion of particles. *J. Colloid Interface Sci.*, 1975, **53,** 314–326.

28 **Fuller, K.N.G.** and **Tabor, D.,** The effect of surface roughness on the adhesion of elastic solids. *Proc. Roy. Soc.Lond*, 1975, **A345,** 327–342.

29 **Bush, A.W., Gibson, R.D.,** and **Keogh, G.R.,** Calculation of rough surfaces based on statistical geometry. *Wear*, 1976, **40,** 399–403.

30 **Chang, W.R., Etsion, I.,** and **Bogy, D.B.,** Adhesion model for metallic rough surfaces. *ASME J. of Tribology*, 1988, **110,** 50–58.

31 **Sviridenok, A.I., Petrokovets, M.I.,** and **Chizhik, S.A.,** Estimating the adherence force for elastic rough spheres (in Russian). *BSSR Academy Reports* (*Doklady Akademii Nauk BSSR*), 1989, **33,** 5, 422–425.

32 **Maugis, D.,** On the contact and adhesion of rough surfaces. *J. Adhesion Sci. Technol.*, 1996, **10,** 2, 161–175.

33 **Majumdar, A.** and **Bhushan, B.,** Role of fractal geometry in roughness characterization and contact mechanics of surfaces. *ASME J. of Tribology*, 1990, **112,** 205–216.

34 **Borodich, F.M.** and **Onishchenko, D.A.,** Fractal roughness in contact and friction problems (the simplest models). *J. Frict. Wear*, 1993, **14,** 3, 14–19.

35 **Zhou, G.Y., Leu, M.C.** and **Blackmore, D.,** Fractal geometry model for wear prediction. *Wear*, 1993, **170,** 1–14.

36 **Thomas, T.R.,** Surface topography in tribology. *Physics Bulletin,* 1982, **9,** 326–328.

37 **Sayles, R.S.** and **Webster, M.N.,** Pattern recognition in tribology. *J. Appl. Math.,* 1982, **81,** 19–31.

38 **Webster, M.N.** and **Sayles, R.S.,** A numerical model for the elastic frictionless contact of real rough surfaces. *ASME J. Tribology,* 1986, **108,** 314–320.

39 **Ren, N.** and **Lee, Si.C.,** Contact simulation of three-dimensional rough surfaces using moving grid method. *ASME J. Tribology,* 1993, **115,** 597–601.

40 **Liang, X., Kaiyuan, J., Yongging, J.** and **Darong, Ch.,** Variation in contact stress distribution of real rough surfaces during running-in. *ASME J. Tribology,* 1993, **115,** 602–606.

41 **Chizhik, S.A., Gorbunov, V.V.** and **Myshkin, N.K.,** The analysis of molecular scale roughness effect on contact of solids based on computer modeling. *Precision Eng.,* 1995, **17,** 3, 186–191.

42 **Tian, X.** and **Bhushan, B.,** A numerical three-dimensional model for the contact of rough surfaces by variational principle. *ASME J. Tribology,* 1996, **118,** 33–42.

43 **Burnham, N.A.** and **Colton, R.J.,** Measuring the nanomechanical properties and surface force microscope. *J. Vac. Sci. Technol.,* 1989, **A7**(4), 2906–2913.

44 **Lu, C.J., Bogy, D.,** and **Kaneko, R.,** Nanoindentation hardness tests using a point contact microscope. *ASME J. Tribology,* 1994, **116,** 175–180.

45 **Bhushan, B.** and **Koinkar, V.N.,** Nanoindentation hardness measurements using atomic force microscope. *Appl. Phys. Lett.,* 1994, **64,** 1653–1655.

New probe methods for nanoscale asperities and thin films

S A ASIF and **J B PETHICA**
Department of Materials, University of Oxford, UK

SYNOPSIS

This paper considers some of the changes in mechanical properties of materials that may occur as sample volume is decreased towards molecular dimensions. Such very small regions are often important in tribology, and may be sampled in Nanoindentation. A brief outline is given of some simple scaling processes which are partly responsible for the changes. Two new capabilities of nanoindentation testing are then introduced for determination of rate-sensitive mechanical property measurement in very small volumes and thin films. Application examples are given of creep in metals, and of loss and storage modulus determination in polymers.

1 INTRODUCTION

Numerous tribological problems have as an essential component the behaviour of very thin surface films and asperities, often at the sub-micrometres scale. In an ideal world, tribologists would only have to ensure that device designs gave perfectly lubricated contacts. Unfortunately, in practice, the lifetime of tribological systems is determined by the rare but crucial moments where lubricating films become very thin, and asperity contacts give permanent material deformation. There is thus some interest in measuring the mechanical properties of asperities and thin films at the nanometre scale, so that effects of contact can be assessed, and minimised by observable surface treatments or processing.

In this paper a brief overview is given of the use of nanoscale probe testing for mechanical property measurement well below the micrometre scale, and of some reasons why nanoscale material deformation may differ from that of bulk. Two new measurement methods are described, specifically aimed at determining the time-dependent strength properties of surface and thin film materials.

Nanoindentation is the technique whereby a finely controlled force is applied to a sharp tip, normally diamond, and the resulting penetration of the tip into the test surface is measured with resolution typically better than 0.1 nm. The tip load is programmed to execute a chosen ramp pattern with time, typically load up, hold at fixed load, unload to separation. More complex control patterns can be made to check for specific processes of interest. The overall force-displacement data can then be processed by techniques which are now well established (**1–3**) to give for example elastic modulus from the unloading response, and hardness from the residual impression depth. The resolution is sufficient to allow characterisation of films less than 10 nm thick. The size of probe also means high lateral resolution is available, so that spatial features such as wear tracks, particles, distinct phases etc. can be individually probed.

The addition of a small modulation to the force, with synchronous detection of the resulting displacement, gives even greater sensitivity (**4, 5**) and this forms the basis of the new results described here. Applied to AFM it has recently been possible to measure the strength of a single molecular bond across a contact (**6**).

Modelling methods have been advancing in parallel with experimental techniques. The advent of near atomic-scale experiments allows direct theory-to-experiment comparisons with molecular dynamics simulations (**7**). Great insight has been obtained and will continue to be gained into mechanisms responsible for tribological processes at the molecular scale. In fact one of the surprises has been how good continuum modelling methods remain as the molecular scale is approached. Down to, say 1 nm, classical continuum concepts of contact mechanics (**8**) are remarkably useful and successful.

2 SCALE DEPENDENCE OF MATERIALS PROPERTIES

It is generally observed that as the zone deformed in an indentation test is reduced very much below 1 μm, hardness is found to steadily increase (**9**). In other words, hardness is not a scale-independent material property. Approaching the 10 nm or so scale, hardness, or yield strength, can approach that of an ideal, theoretical defect-free lattice (**4, 10, 14**). This phenomenon is also observed in molecular dynamics simulations (**7**). Even liquid films cease to behave as pure liquids below a few nm thickness, and start to exhibit solid-like properties such as an ability to sustain finite static shear stresses (**11**). To compound the problems at the nanoscale, the onset of plasticity becomes extremely sensitive to interfacial chemical conditions (**12, 13**). Clearly all these effects will be crucial in tribological contacts.

Whilst it might be thought obvious that mechanical properties at the molecular scale might not resemble those of continua, it is interesting to note that the changes actually start well before the molecular scale is reached, and in fact at the scale where continuum modelling is still quite valid. Two simple examples serve to illustrate the importance of scaling, even before atomic and molecular structure is taken into account.

First consider the idealised punching of a single prismatic dislocation loop of radius R, as in Fig. 1a. The resulting 'indent' is a single atom depth disc.

The line energy of the loop is given roughly by $2\pi RK$. K is the energy per unit length, which is itself only very weakly (logarithmically) dependent on R (**14**). The work done by the punch in creating the surface depression is $\sigma\pi R^2 b$, where σ is the mean stress on the face of the punch and b is the Burgers vector. Therefore the loop will be stable (i.e. can be created and exist) if $\sigma R^2 b > 2\pi RK$, that is, if $\sigma > K/Rb$. This argument is along the lines suggested by Gane many years ago (**15**). Note the inverse dependence on R, suggesting a drastic rise in stress for very small radius loops. In fact the stress will exceed the lattice limit (~10% strain) when R<20 nm for many metals, which means that complete dislocation loops will not be formed in such compressive contact geometries. Plastic deformation takes place by a less correlated motion of atoms, and crucially, occurs at very high lattice strains and hence lattice elastic energies. In molecular dynamics simulations, the slip coincides with periods of disorder (**7**). Clearly, all this does not resemble macroscopic material behaviour, and there are thus ranges of length scales around tens of nm where it is reasonable to expect significant changes in mechanical properties.

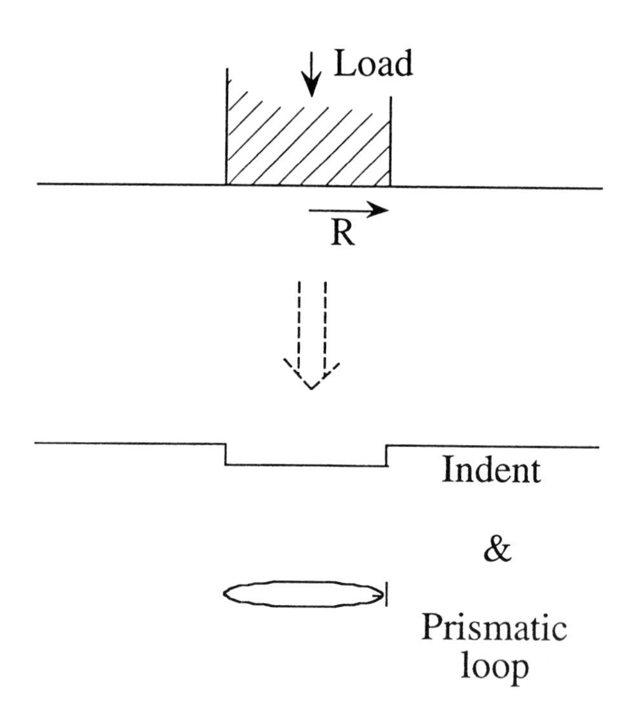

Fig. 1a Simple indent geometry, punching a prismatic loop.

A second, also extremely simplified example, illustrates the influence of adhesive forces. In the geometry of Fig. 1b, there will be a force acting between tip and surface which is of the order of $2\pi R\gamma$ where γ is a surface or interfacial energy. This may be due to long-range attractive forces (DMT) (**16**), to capillary liquid present in the tip surface gap (JEI) (**11**), or to interfacial energy of the solid-solid contact (JKR) (**17**). Whatever the mechanism, this attractive force will be balanced when in equilibrium by elastic compressive forces F in the contact. If we take simply the Hertz model for contact with E^* the reduced elastic modulus of the contacting surfaces (see below, and (**8**)), the mean stress in the compressive region of the interface will be:

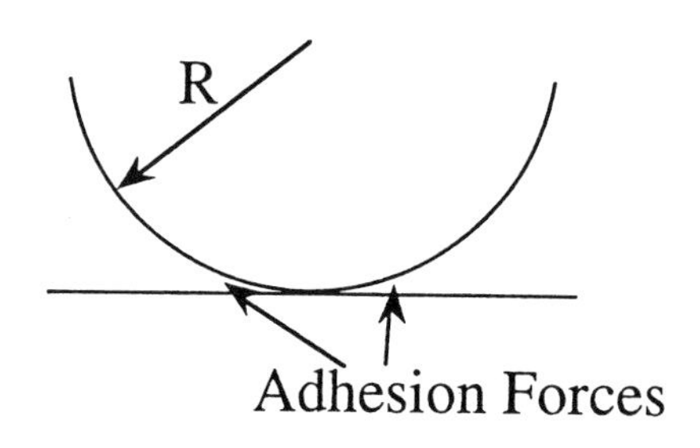

Fig. 1b Surface (adhesion) forces act between spherical tip and surface.

$$\sigma \propto F\frac{1}{3}\left(\frac{E^*}{R}\right)^{\frac{2}{3}} \quad \text{whence:} \quad \sigma \propto E^{*\frac{2}{3}}\left(\frac{\gamma}{R}\right)^{\frac{1}{3}}$$

Again there is a clear increase of contact stress with decreasing size of contact. A fuller calculation (**18**) suggests that asperities having tip radius much below 5 nm, may not actually be mechanically stable against plastic deformation in such a contact.

In summary, we can see that the actual size of a contact is important, and there may be quite drastic changes of behaviour at the smallest scales. This scaling is fundamentally caused by the different dimensionality of the 3-D elastic energy storage in solids and the 1-D or 2-D nature of the energy sources and defect processes.

Now this is all fine, but the process of transition from macroscopic to nanometre is not sudden, and transitional changes in materials properties will occur. Modelling in this intermediate length scale is very challenging. Experiments also show there are a variety of parameters such as environment and chemistry (**13**) that need to be controlled. Strain rate sensitivity of mechanical properties such as hardness must be allowed for in experiments at the nanoscale, where (as in tribology) dimension changes may range over some orders of magnitude (**19**). Clearly, what is needed is reliable and as comprehensive experimental data as possible. To this end we have been expanding the capabilities of nanoindentation testing, and in what follows we show how it can be used to characterise two aspects of time dependent deformation.

3 EXPERIMENTAL

The experimental set-up is based on a Nanoindenter II (**20**), with Berkovitch diamond indenter tip. The indenter shaft is suspended by a system of leaf springs inside a housing head, and the load is provided by passing a current through a coil and magnet assembly at the top of the shaft. The displacement of the indenter is measured by a capacitance gauge. The load resolution of the system is about 0.3 μN and the displacement resolution is < 0.05 nm. Parameters such as approach speed, load, and the rate of loading and unloading, are controlled by computer, and can be set over very wide ranges. Our apparatus has been also modified so that the temperature of the indenter tip and the specimen sample can be varied between about 10°C and 70°C. This is done by means of two separate thermoelectric peltier cells, one supporting the sample, and the other holding the tip. The temperature of both indenter and sample are monitored by thermocouple, and can be kept stable and equal to within a few millidegrees. More details are given elsewhere (**21**).

The stiffness S of the contact zone is monitored continuously by AC modulation δF of the force, and lock-in detection of the resulting displacement δz; that is, $S = \delta F/\delta z$. Now S is related to the contact area A by:

$$S = 2E^*\sqrt{\frac{A}{\pi}} \quad \text{where:} \quad \frac{1}{E^*} = \frac{\left(1-\nu_T^2\right)}{E_T} + \frac{\left(1-\nu_s^2\right)}{E_s}$$

and E and ν are Young's modulus and Poisson's ratio, suffices referring to tip and surface (**8**). This leads to a number of interesting possible measurements.

First, if E^* is known, or can be taken as constant, S can be used to monitor changes in contact area during indentation. Since it is a modulation (AC) measurement, and so is insensitive to DC drifts, it provides an accurate way of measuring the very small dimensional changes associated with creep. The effective strain rate can be defined as (**19, 22**):

$$\dot{\varepsilon} = \frac{1}{S}\frac{dS}{dt} \tag{1}$$

and the mean stress or Hardness σ can evaluated from:

$$\sigma = \frac{P}{\pi}\frac{4}{S^2}E^{*2} \tag{2}$$

where P is the applied load. Both of these quantities can therefore be obtained directly from measured values of S during an experiment.

Alternatively, if the contact area A is known, or can for example be estimated from the DC force-displacement curve with a known tip shape (**1–3**), it becomes possible to measure E directly during all stages of an indentation by measuring S. The analysis to obtain E from stiffness is described in detail by Oliver and Pharr (**3**). Here we note that particularly for polymeric materials, the elastic response may have a loss as well as storage term, giving an additional phase shift between the AC force and displacement. Analysis of the AC response of the indent and apparatus shows how both of these may be determined. The indentation and associated apparatus may be represented by the components in Fig. 2 (**4, 21**) K_m is the machine frame stiffness, which may be taken as very large (infinite) for the results discussed here. K_i is the stiffness of the leaf springs that hold the indenter shaft and C_i is the damping coefficient of the air gap in the displacement sensing capacitor. These parameters are all known and their accurate determination is described elsewhere (**21**). It is thus possible to evaluate K_S and C_S, the spring constant and damping coefficient respectively of the contact region under test. The following standard equations for system oscillatory motion (**2, 21**) give the displacement AC amplitude X_O resulting from modulating force amplitude F_O, and the phase shift F between force and displacement:

$$X_o = \frac{F_o}{\sqrt{\left(K_s + K_i - m\omega^2\right)^2 + \left(\left(C_i + C_s\right)\omega\right)^2}} \tag{3}$$

$$\phi = \tan^{-1}\frac{\left(C_i + C_s\right)\omega}{K_i + K_s - m\omega^2} \tag{4}$$

K_S and C_S can thereby be found. The storage modulus G' and the loss modulus G' are then given by:

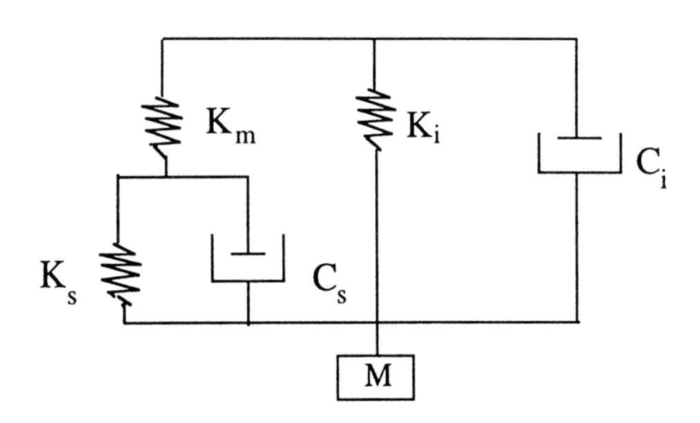

Fig. 2 Dynamic model of indenter system and sample.

$$K = \alpha\sqrt{A}E' = \alpha\sqrt{A}\frac{G'}{2(1+\upsilon)} \quad \text{and}$$

$$\omega C_s = \alpha\sqrt{A}E'' = \alpha\sqrt{A}\frac{G''}{2(1+\upsilon)}$$

where ω is the angular frequency of the AC modulation and $\alpha = 2/\sqrt{\pi}$.

4 NANOINDENTATION CREEP

The creep properties of two contrasting metals, indium and tungsten, are described here. Indium was investigated at three different temperatures. After the system reached thermal equilibrium at each temperature, the tip was approached to the specimen surface at a velocity of 4 nm/s. After the tip contacted the specimen surface, the load was held constant and the thermal drift monitored. Experiments were carried out only when the DC thermal drift was less than 0.05 nm/s. To measure the creep properties, the load was then ramped up to maximum load of 5 mN in one second. After reaching the maximum load, the load was held constant for 1600 seconds. Fig. 3 shows the resulting variation of stiffness S with time whilst the load is held constant. Using the equations (1) and (2) above, the effective strain rate can be plotted against mean stress, as shown in Fig. 4. A stress exponent of 7.7 is observed at 17°C, in good agreement with conventional creep tests (**23**). At 45°C the exponent is 4.5 and at 60°C it is 4.8. It is perhaps surprising that indentation creep can give results comparable to uniaxial testing, since the strain geometries are so dissimilar. In fact the first 50–100 seconds of data in Fig. 3 is not used in producing Fig. 4, as there are significant deviations from a simple power law. However it is clear that despite the complications, realistic materials testing is possible, and unlike conventional testing, is possible at the nanometre scale.

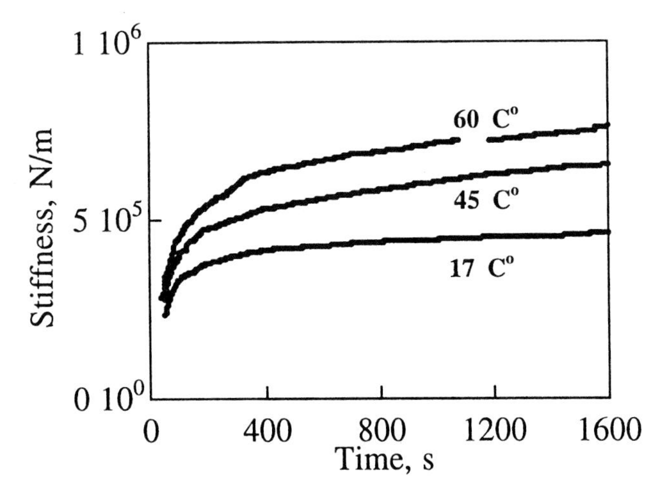

Fig. 3 Time behaviour of AC stiffness following step increase in load on indium.

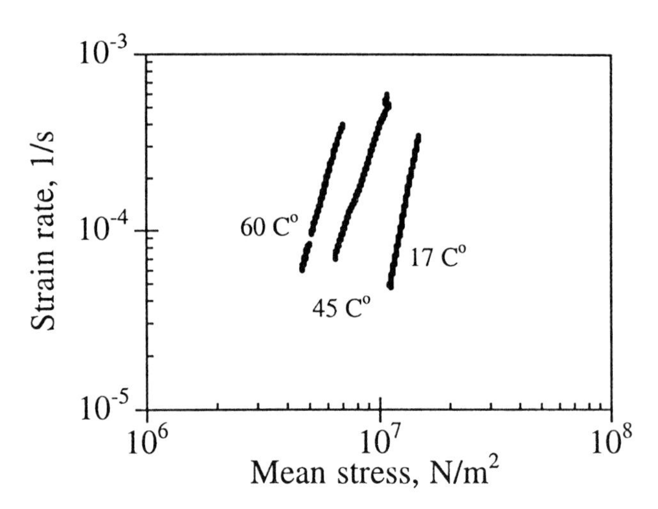

Fig. 4 Relation between strain rate and mean stress for indium from indentation data.

Creep at temperatures above 0.5 T_m is expected to follow an Arrhenius type of behaviour, and the strain rate can be written as:

$$\dot{\varepsilon} = \dot{\varepsilon}_o \exp(-Q/RT)$$

where Q is the activation energy, R is the gas constant, and T is the absolute temperature. The activation energy Q of the creep process can therefore be obtained from the slope of the semi-log plot of strain rate versus $1/RT$ at a chosen stress. It is obvious from Fig. 4 that the real behaviour is probably more complex than a single activation process. However, if we simply extrapolate the lines for each temperature to give a data point at a single stress around 9×10^6 N/m^2 we find a value of Q of 77 kJ/mol, which is in surprisingly good agreement with the activation energy for self diffusion in pure indium (in the range of 75–78 kJ/mol (**24**). Again, caution is required in inferring too great an accuracy in this comparison, but clearly, meaningful measurements can be made at the very small scale.

In view of the great sensitivity of the AC modulation method, creep behaviour of tungsten was also investigated. Fig. 5 shows the overall variation of stiffness with time during a complete indentation. The data commences at the point of contact (zero of time). The stiffness rises as the load is increased, and at a critical load, sudden indentation (pop-in) occurs. This is the point at which significant numbers of dislocations are suddenly created in the tungsten, and the phenomenon is described in detail elsewhere (**4, 10, 13, 25**). The load is increased beyond the pop-in point, and then is held constant. The continuing creep is easily seen. The load is then stepped down by a defined percentage and the subsequent creep monitored. Fig. 5 shows the superposition of results from three separate tests, in which the step unloads were set to 10%, 50% and 70%. It is clear that at about 50% unload, there is zero creep. Less unload continues the positive creep, and more causes reverse creep. This is clear evidence of the action of back stresses. There are likely to be (at least) two origins of such back stress. The first is microstructural, the stored energy in strained dislocation networks and this is the source of reverse creep seen in traditional stress reduction creep WWW tests. The second will be geometric within the indentation – the elastically strained region beyond the plastic indent will apply a reverse stress to the indent once the load is reduced. Clearly, useful information characterising quite complex material behaviour, in geometries directly relevant to tribological contacts, can be obtained from the creep properties of indents only a few tens of nm deep.

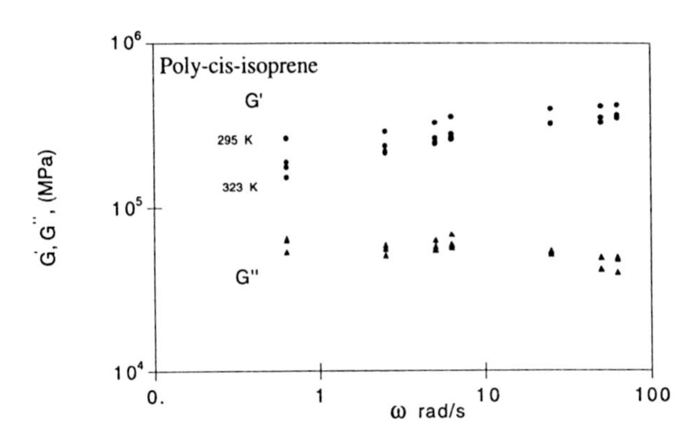

Fig. 6 Storage and loss moduli for poly-cis-isoprene determined from nanoindentation.

5 STORAGE AND LOSS MODULUS OF THIN FILMS

We describe here the use of the AC modulation method described above to measure G' and G" on two bulk polymers. The first is poly-cis-isoprene, which is extremely soft. Even at very low loads <100 μN, indents were many microns deep. The results are therefore not a test of small volumes, but rather of materials whose plastic deformation requires extremely small stresses. The second sample used was poly-trans-isoprene, a considerably harder natural rubber. The results on this material were obtained from indents around 300 nm deep and at a DC load of 300 μN.

Fig. 6 shows the variation of G' and G" with angular frequency of the test modulation, for poly-cis-isoprene. The values are in quite good agreement with macroscopic test values (**26**). For this material there is only a weak dependence on frequency in the range accessible with present Nanoindenter configurations. Note that these results are obtained for total loads on the sample of around 300 μN, which indicates the great sensitivity of the technique.

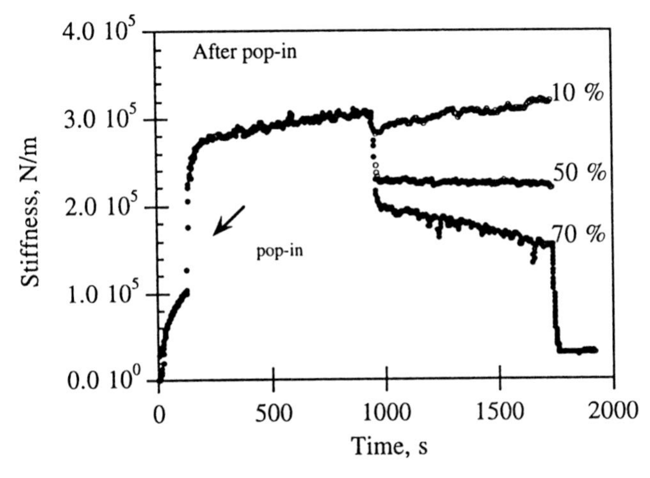

Fig. 5 Evolution of contact stiffness with time during indentation into electropolished tungsten. Data for three indents superimposed on each other. The creep relaxation response depends on the magnitude (percentage) of the step unload. Maximum load in all cases was 2 mN.

Figs. 7 and 8 show the behaviour of poly-trans-isoprene, for a number of different temperatures and frequencies. Poly-trans-isoprene is normally crystalline at room temperature, flows around 60°C and the modulus drops with temperature (**27**). This behaviour is clearly seen in the Figures. It can also be seen that if the curve for one particular frequency is displaced horizontally, it approximately coincides with the next frequency. Thus a master curve can be obtained for a reference frequency by shifting the curve of a particular frequency to coincide with the reference frequency. We describe in more detail elsewhere (**21**, and to be published) how time-temperature equivalence can thereby be assessed in this system.

It is interesting to see that these are similar to macroscopic results although the measurements are taken from volumes somewhat below a micron across. Localisation of mechanical loss mechanisms to individual sites or molecular structures, which must occur eventually, has to be at rather small length scales. We are investigating improvements in the technique to detect such localisation of losses.

We conclude this section by remarking that to obtain results such as those in Figs. 6 to 8, very accurate knowledge of the dynamic response of the measuring apparatus is required. Not only must the parameters in equations (3) and (4) be accurately known, but any other ancillaries such as electronics which might, for example produce frequency-dependent offsets of phase, must be clearly investigated and allowed for. Greater detail is provided in (**21**).

6 CONCLUSIONS

We have demonstrated some new uses of the indentation technique in studying time-dependent deformation processes in nanoscale volumes. Quite reasonable comparison is shown with traditional uniaxial measurements and expected material behaviour. Nanoscale

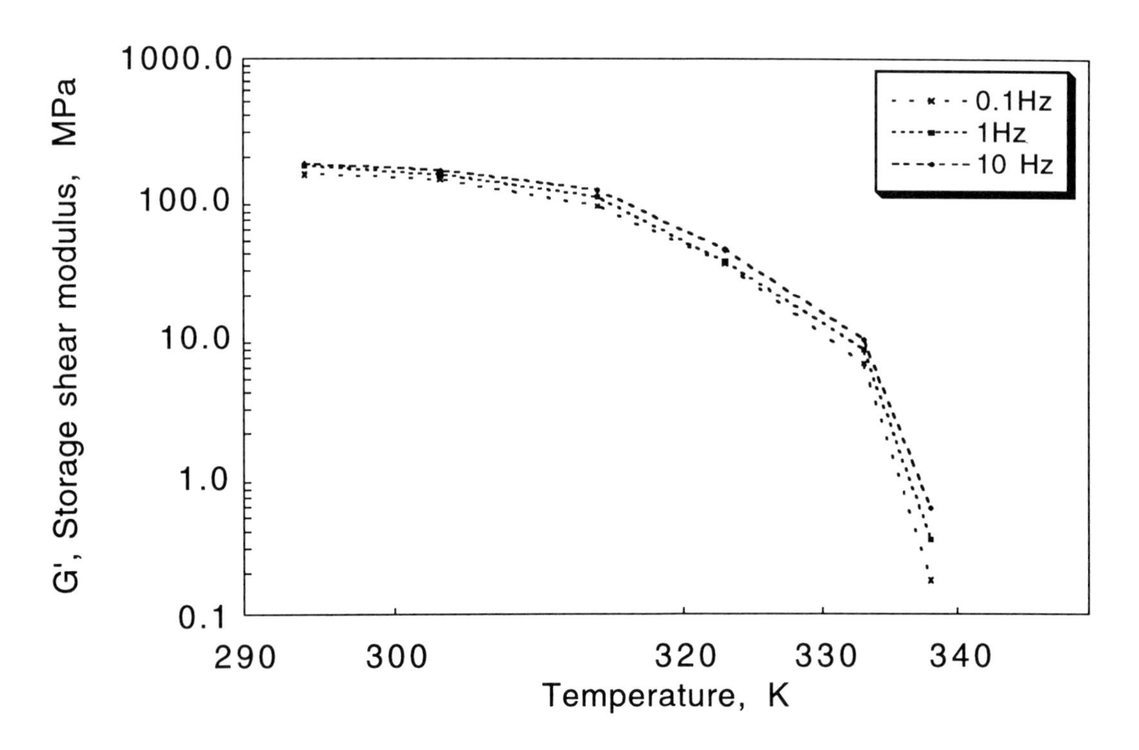

Fig. 7 Storage modulus for poly-trans-isoprene as a function of temperature and frequency.

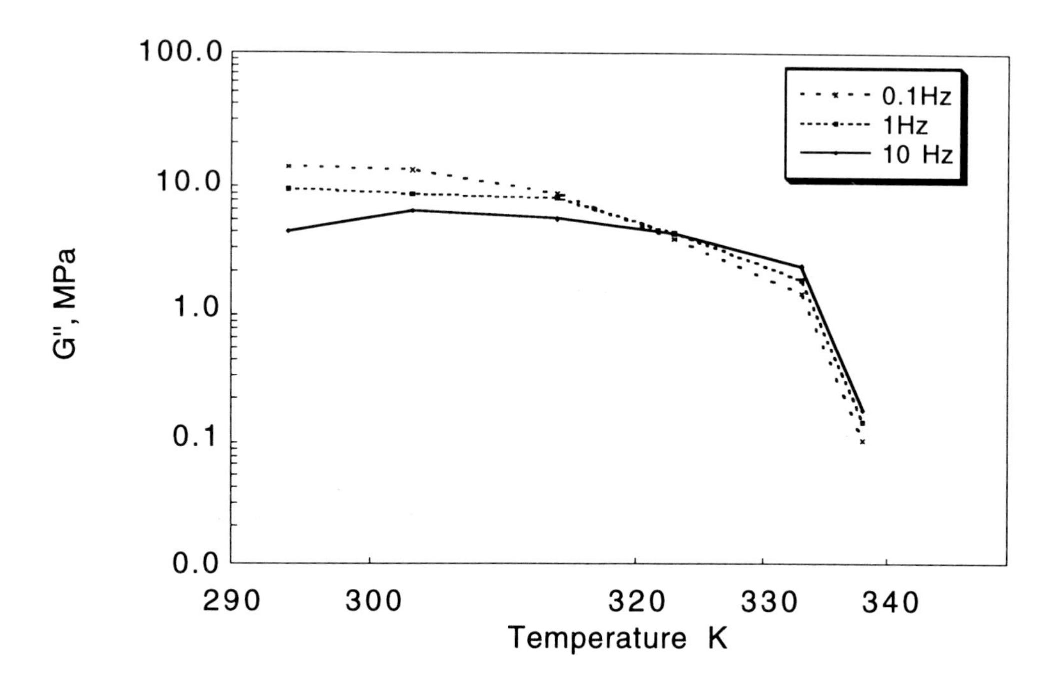

Fig. 8 Loss modulus for poly-trans-isoprene as a function of temperature and frequency.

measurements are likely to become important in observing and understanding the changes in materials properties as the molecular scale is approached. These are an essential component of many tribological critical systems.

7 REFERENCES

1 **Pethica, J.B., Hutchings, R.,** and **Oliver, W.C.,** *Phil. Mag.* A **48**, 593, (1983).

2 **Doerner, M.F.** and **Nix, W.D.,** *J. Mater. Res.*, **1**, (4), 601–609, (1986).

3 **Oliver, W.C.** and **Pharr, G.M.,** *J. Mater. Res.*, **7** (6), 1654, (1992).

4 **Pethica, J.B.** and **Oliver, W.C.,** *Mat. Res. Soc. Symp. Proc.*, **130**, 13, (1989).

5 **Weihs, T.P.** and **Pethica, J.B.,** *Mat. Res. Symp. Proc.*, **108**, 325, (1992). **Pethica J.B.** and **Oliver W.C.,** *Physica Scripta*, **T19**, 61, (1987).

6 **Jarvis, S.P., Yamada, H., Yamamoto, S-I., Tokumoto, H.,** and **Pethica, J.B.,** *Nature*, **384**, 247–249, (1996).

7 **Sutton, A.P., Pethica, J.B., Rafii-Tabar, H., and Nieminen, J.A.,** (1992), Mechanical properties of metals at the nanometre scale, Ch.7 of *Electron Theory in Alloy Design*, (pp. 191–233), Eds. D.G. Pettifor and A.H. Cottrell, Institute of Materials, London. (Book #534). Also **Landman, U., Luedtke W.D.,** and **Ringer, E.M.,** Molecular dynamics simulations of adhesive contact formation and friction, (1992) in *Fundamentals of Friction*, 463–508, eds. I.L. Singer & H.M.Pollock, Kluwer Academic Publishers, Dordrecht.

8 **Johnson, K.L.,** Contact mechanics, *Cambridge University Press*, 1985.

9 **Pethica, J.B., Oliver, W.,** and **Hutchings, R.,** pp. 90–108 in *Microindentation techniques*, eds. P. Blau & B. Lawn, ASTM, Philadelphia, (1985).

10 **Page, T.F., Oliver, W.C.,** and **McHargue, C.J.,** *J. Mater. Res.* **7**, 450 , (1992).

11 **Israelachvili, J.N.,** Intermolecular and surface forces with applications to colloidal and biological systems, *Academic Press Ltd.* (1985).

12 Surface Effects in Crystal Plasticity, eds. R.M. Latanision and J.T. Fourie, (1977), Noordhoff, Leyden.

13 **Mann, A.B.** and **Pethica, J.B.,** *Langmuir*, **12**, 4583–4586, (1996), also *Appl. Phys. Lett.*, **69**, 907–909, (1996).

14 **Kelly, A.,** (1980) *Strong Solids*, Clarendon Press, Oxford.

15 **Gane, N.,** *Proc. Roy. Soc. Lond.* A. **317**, 367 (1970).

16 **Derjaguin, B.V., Muller, V.M.,** and **Toporov, Yu.P.,** *J. Colloid Interface Sci.* **73**, 293, (1980).

17 **Johnson, K.L., Kendall, K.,** and **Roberts, A.D.,** *Proc. R. Soc. Lond.* **A324**, 301, (1971).

18 **Roy Chowdhury, S.K.** and **Pollock, H.M.,** (1981) *Wear*, **66**, 307. Also **Maugis, D.** and **Pollock, H.M.,** (1984), *Acta. Metall*, **32**, 1323, and **Pashley, M.D., Pethica, J.B.,** and **Tabor, D.,** (1984) *Wear*, **100**, 7–31, (1984).

19 **Lukas, B.N.** and **Oliver, W.C.,** *Mat. Res. Symp. Proc.*, **356**, 137 and 645, (1995) and **Lucas, B.N., Oliver, W.C.,** and **Pharr, G.M.,** *Mat. Res. Symp. Proc*, Spring meeting, (1996).

20 Nano Instruments Inc., Knoxville, Tennessee, U.S.A.

21 **Asif, S.A.,** D.Phil. thesis, University of Oxford, 1997.

22 **Atkins, A.G., Silverio, A.,** and **Tabor, D.,** *J. Inst. Metals*, **94**, 369, (1966). **Mayo, M.J.** and **Nix, W.D.,** *Acta Metall. Mater.*, **6**, 2183, (1988).

23 **Weertman, J.,** *Trans. AIME*, **218**, 207–218, (1960).

24 **Eckert, R.E.** and **Drickamer, H.G.,** *J. Chem. Phys.*, **20**, 13, (1952). Also **Powell, G.W.** and **Braun, J.D.,** *Trans. ASME, J. Engr. Mat. Tech.*, **101**, 387, (1979).

25 **Gerberich, W.W., Venkataraman, S.K., Huang, H., Harvey, S.E.,** and **Kohlstedt, D.L.,** *Acta Metall. Mater.*, **43**, 1569, (1995).

26 **Pyne, A.R.,** Rheology of Elastomers, Eds. P. Mason, N. Wookey, Pergamon, London, 1958.

27 **McCrum, N.G., Read, B.E.,** and **Williams, G.,** Anelastic and Dielectric effects in Polymer solids, John Wiley, London, 1967.

Surface forces, surface chemistry, and tribology

K FELDMAN, M FRITZ, G HAHNER, A MARTI, and **N D SPENCER**
Laboratory for Surface Science and Technology, Department of Materials, Zürich, Switzerland

SYNOPSIS

Scanning probe methods have been applied to the investigation of tribological phenomena on the nanometer and nanonewton scale. The systems studied have included parallel investigation of identical tribosystems on the macro and nano scales, where the inherent differences in the AFM/LFM and flat-on-disk experiments have been compared; oxide-covered surfaces in contact under electrolytes, where the adhesion hysteresis and frictional behaviour were shown to be strongly dependent on the solution pH; and polymer surfaces, where advantage can be taken of variations in the interactions between the scanning tip and different polymers, to perform chemically sensitive, high-resolution surface imaging of polymer blends.

1 SCANNING PROBE METHODS IN TRIBOLOGY

The invention of the atomic force microscope (AFM) in 1986 (**1**) was a crucial moment in the history of tribology. Although many tribologists interested in fundamental phenomena had speculated about such issues as forces acting on single asperities and friction on the atomic scale, the development of the AFM and its frictional counterpart, the lateral force microscope (LFM) (**2**), actually allowed these quantities to be measured and imaged across a sample surface.

The principle of operation for an AFM/LFM (Figure 1) is related to that of the scanning tunnelling microscope (**3**) (STM), for which Binnig and Rohrer won the 1986 Nobel Prize.

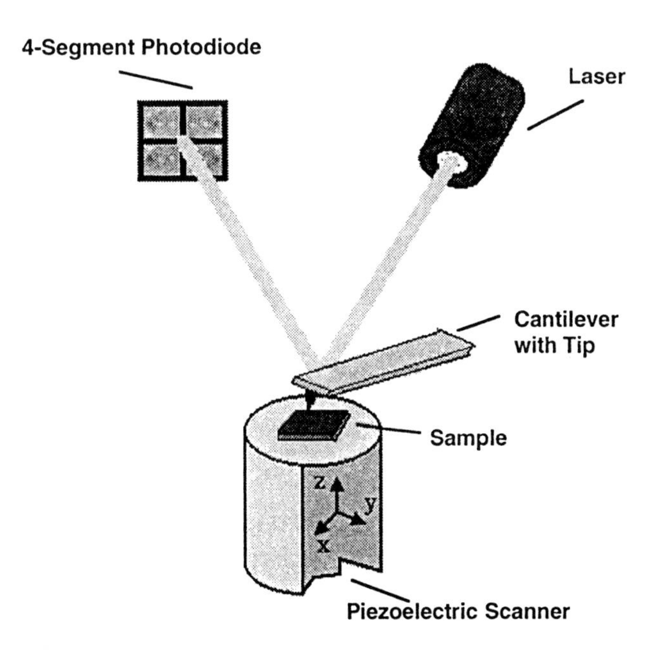

Fig. 1 Schematic of an atomic force microscope showing the optical lever principle.

In short, a very fine tip (radius 30 nm) positioned on the end of a flexible cantilever is rastered over the surface of the sample in a direction perpendicular to the longest axis of the cantilever. The tip and cantilever are displaced vertically upon encountering surface morphology, and this movement is detected by a laser beam, which is bounced off the rear of the cantilever (optical lever principle) (**4**). The sample, which is supported on a piezoelectric displacement system, is moved up and down via a servo control mechanism that maintains the deflection of the cantilever (and thus the applied load) at a constant value. The z-displacement of the sample can thus be displayed as a function of x- and y-position to provide a morphological image of the sample. At the same time, the laser beam can be used to detect lateral (torsional) movements of the tip and cantilever as the sample is rastered. These movements are a convolution of morphological effects and tip-sample friction, which may readily be deconvoluted via a friction loop measurement (Figure 2).

Thus, by monitoring both vertical and lateral tip movements, the surface morphology and friction coefficient can be simultaneously mapped across a sample surface. Another type of measurement of importance in tribological studies is the so-called 'force curve'. In this experiment, the tip is moved up and down over a single location, moving in and out of contact with the sample. During these cycles the deflection of the cantilever is monitored, providing an indication of the tip-sample interaction forces as a function of separation (Figure 3).

As the tip is withdrawn from the surface, tip-sample adhesion leads to a minimum in the force curve before the tip jumps back away from the surface. Of particular interest to tribologists is the hysteresis in this measurement, which (providing the mechanical parameters are appropriately chosen) corresponds to the hysteresis in tip-sample adhesion and in many cases to friction (**5**).

A particular strength of AFM/LFM is its ability to

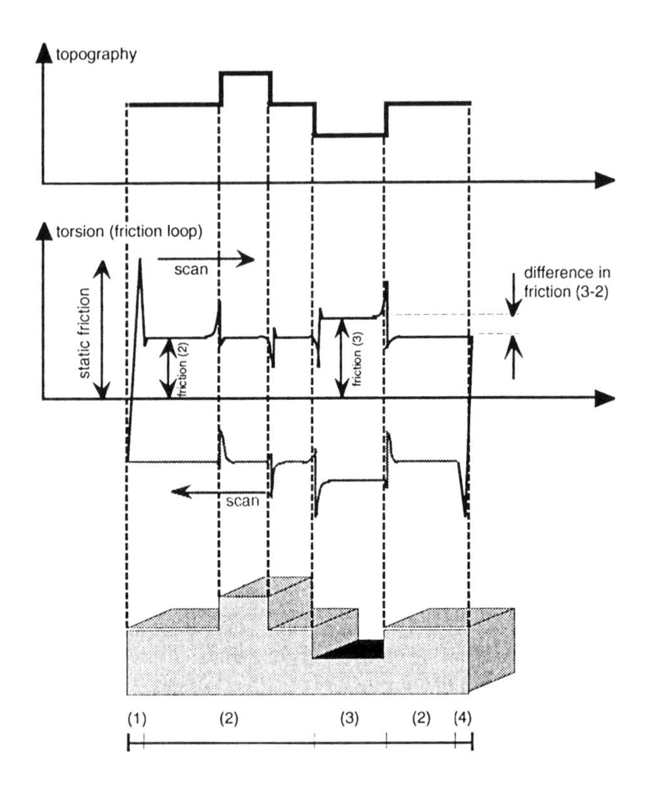

Fig. 2 Friction loop and topography on a heterogeneous stepped surface. Terraces (2) and (3) are composed of different materials. In regions (1) and (4), the cantilever sticks to the sample surface because of static friction F_{ST}. The sliding friction is t_1 on part (2) and t_3 on part 3. In a torsional force image, the contrast difference is caused by the relative sliding friction, $\Delta F_{SL} = t_1 \quad t_3$. Morphological effects may be distinguished from frictional ones by their non-inverted behavior upon scanning in the opposite direction. After R. Overney and E. Meyer, *MRS Bulletin,* May 1993, 26 Fig. 5.

operate under liquids. In this way, the effects of liquid medium can be observed on friction – a very promising approach for studying lubrication on a nanometer scale.

2 NANOTRIBOLOGY AND MACROTRIBOLOGY

While nanotribological measurements with the AFM/LFM are clearly of interest for the understanding of the fundamentals of friction and lubrication, little has been done to actually correlate nano- and macrotribological studies. The problem lies partly with the type and scale of systems that are typically investigated by the different approaches. In our laboratory, we have tried to measure identical systems (silicon-silicon and gold-gold, where the sample is produced by evaporating gold on to tips and on to silicon wafers) using both flat-on-disk and LFM methods. As can be seen in Figure 4 and Table 1, the normal load used in the two methods differs typically by at least seven orders of magnitude. The contact area and velocities are also in somewhat different ranges. The results show that, as expected, the macro experiment demonstrates Amon tons' law, displaying a linear relationship between friction and normal load. In the case of the nano experiment, Amon tons' law is not obeyed, and the system is dominated by adhesion-influenced contact mechanics, with the normal load at zero applied load being due to the adhesion contribution. 'JKR' theory (**6**) has been found to be an appropriate formalism with which to describe the observed behavior. This work is in progress, and has the ultimate aim of distinguishing between the mechanical and chemical causes of friction.

Table 1 Comparison of important tribological parameters in nano- and macro-tribological experiments.

	Nano (LFM)	**Macro (Pin-on-Disk)**
Contact Area	10^{-16} m^2	10^{-5} m^2
Load	0-200 x 10^{-9} N	1-10 N
Pressure	0-2 GPa	0.1-1 MPa
Velocity	0.05-12μm s^{-1}	10μm s^{-1}- 260 mm s^{-1}

3 SURFACE FORCES UNDER ELECTROLYTES

The study of surface forces under electrolytes is particularly interesting, since, providing the surfaces are amphoteric, the charges on the surfaces, and thus the forces between the surfaces can be controlled by means of pH

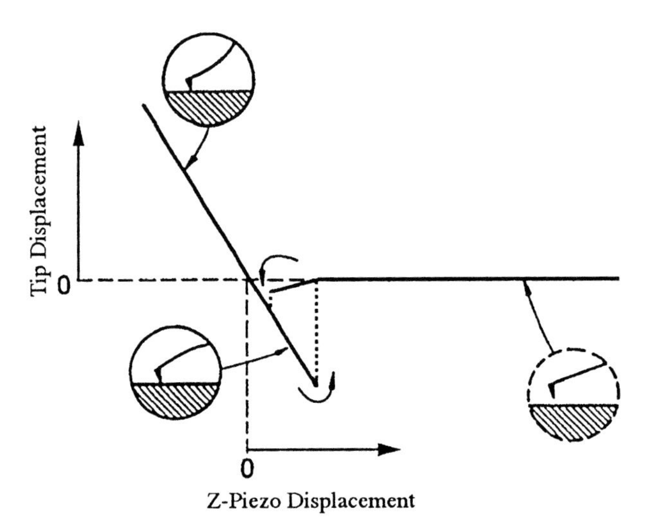

Fig. 3 'Force' curve generated by an AFM showing tip-sample approach (right), contact (top), and adhesion (bottom left). Adapted from Butt, H-J., *Biophys.* J. 60 (1991), 1438–1444.

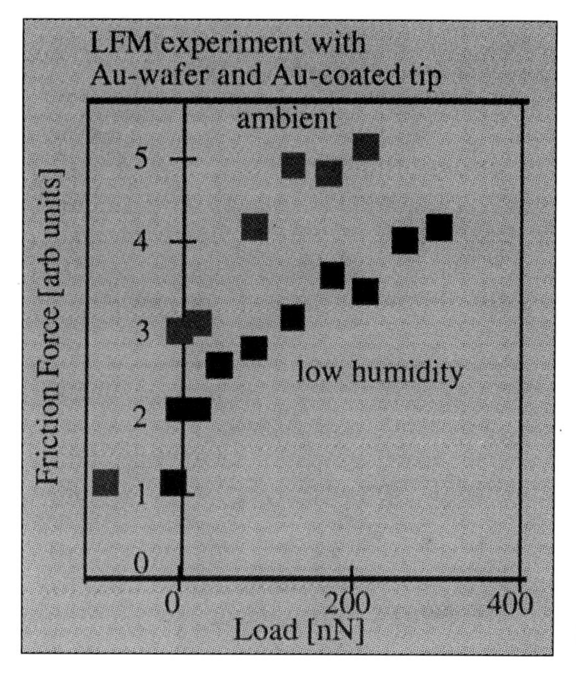

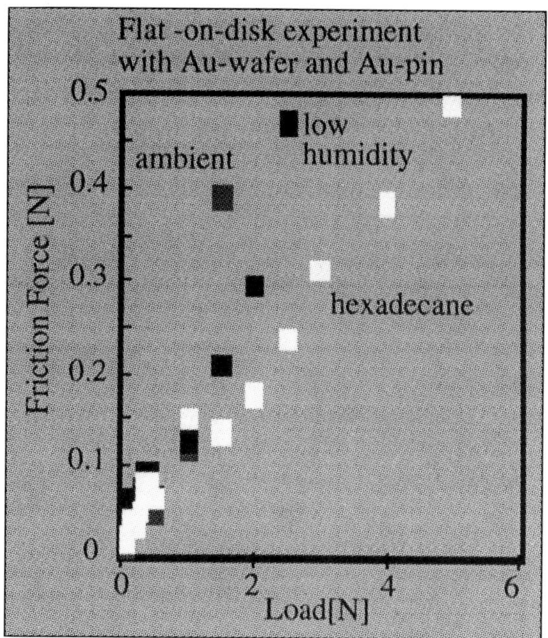

Fig. 4 Comparison between tribological measurements in the nano and macrotribological regimes for the Au-Au system.

variation. In the case of a silicon nitride AFM/LFM tip, the isoelectric point (IEP) of the surface is pH≈5.5-7 (depending on the degree of oxide coverage), above which it is negatively charged and below which it is positively charged. Silica, for example, has an IEP of 2, so that its interactions with a Si_3N_4 tip under electrolyte can be either repulsive (below pH=2 or above pH=5) or attractive (between pH 2 and 5). This results in characteristic behaviour that can be seen in AFM force curves (**7**) (Figure 5).

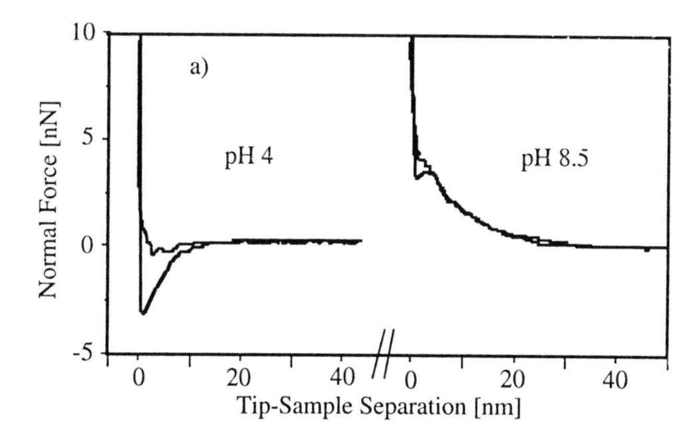

Fig. 5 Force versus tip-sample separation curves at pH 4 and pH 8.5 (1mM NaCl) for the system SiO_2-Si_3N_4. In each case, the upper curve corresponds to decreasing and lower curve to increasing tip-sample separation. The enclosed areas correspond mostly to adhesion hysteresis (**7**).

In this figure, not only is the attractive and repulsive behaviour evident but also a marked difference in the amount of hysteresis can be observed in the force curves. To a certain extent this can be accounted for by mechanical hysteresis in the system (**8**), but the largest contributor to this effect is the adhesion hysteresis itself, which corresponds to a dissipation taking place between making and breaking contact. Such adhesion hysteresis has been correlated by Israelachvilis with frictional behaviour, since the latter can be thought of as sequential making and breaking of adhesive contacts as one body slides over another. Indeed in our AFM studies of aluminium oxide (sapphire, IEP = 9), the adhesion hysteresis (measured from force curves) and the lateral (frictional) force (measured by LFM) seem to be highly correlated (Figure 6) (**9**).

While the fact that an interaction between charged surfaces occurs is conceptually simple, hysteresis in such a system is more difficult to understand. One possible explanation is based on the mixing of the charges as the charged surfaces are brought into contact. This could be thought of as leading to an increase in the entropy of the system, and a corresponding free energy change, which is presumably released as heat (Figure 7). Upon separation of the surfaces, the adiabaticity of the system requires that work be done upon it, corresponding to the *F*d*x* work done by the AFM as the tip is pulled out of adhesive contact with the surface. This and related models are currently under intensive investigation in our group.

Since our work (**7, 9**) has shown that lateral forces of surfaces in frictional contact under electrolytes are dependent both on pH and on the isoelectric points of the

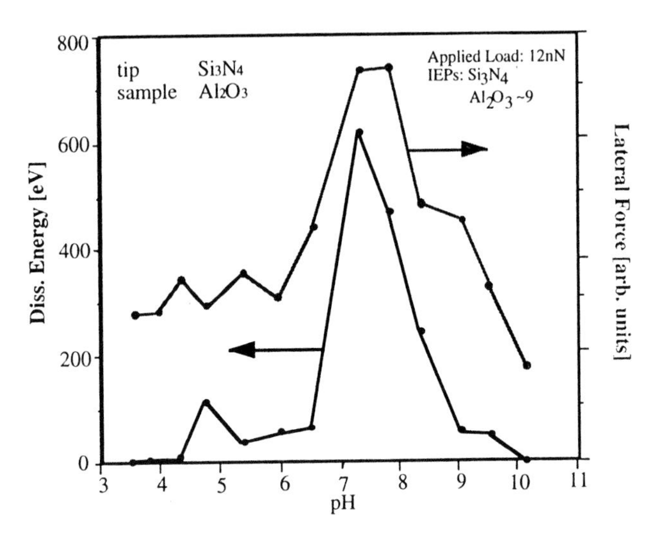

Fig. 6 Correlation of Dissipated Energy (the adhesion hysteresis measured by force curves resembling those shown in Fig. 5) and Lateral Force (measured by LFM) for the Si_3N_4-Al_2O_3 system under NaCl solution (1mM) (**9**).

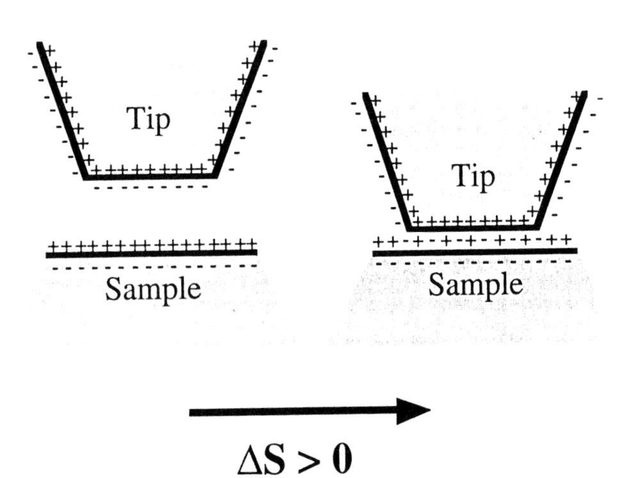

Fig. 7 Mixing of surface charges on tip and sample leading to entropy increase upon tip-sample contact.

tip and sample materials, it then follows that this phenomenon can form the basis of an imaging method for, among others, oxide surfaces. Figure 8 shows that, for alumina and silica surfaces measured with a Si_3N_4 tip, the relative magnitude of frictional forces is reversed between pH≈4 and pH≈7.5.

This means that on a mixed surface consisting of both SiO_2 and Al_2O_3 regions, the frictional contrast, as seen by LFM images, will reverse on moving from pH≈4 to pH≈7.5. Indeed this is the case (Figure 9) and allows LFM to be used to extract chemical information from a surface with high spatial resolution (**10**).

This approach for surface chemical characterization via pH-dependent frictional measurements shows great promise for the analysis of oxide mixtures, and is also finding application, in our laboratories, in the measurement of surface charge distribution on oxidized alloy surfaces, such as those used as load-bearing implant materials, where surface charge can play an important role in interactions with proteins and, therefore, in biocompatibility.

4 SURFACE FORCES ON POLYMERS

In an effort to determine the influence of both polymer and tip surface chemistry on adhesion hysteresis, and therefore on the chemical component of friction on polymers, we have made many tens of thousands of force curve measurements on polymer surfaces (**11**). Since the forces involved are small, these experiments can neither be carried out under ambient conditions (where the situation is dominated by the capillary forces associated with adsorbed water), nor under polar solvents, where dipole-dipole interactions complicate the picture. On the other hand, perfluorinated solvents, such as perfluorodecalin, having a low dielectric constant and little interaction with the studied polymers, were found to be particularly convenient for this kind of measurement (Figure 10).

Another critical factor in these studies was obtaining well-defined surface conditions, both for polymer surface and for tip. High-purity polymer samples (polystyrene (PS), polyacrylonitrile (PAN), poly(methyl methacrylate) (PMMA), and poly(acrylic acid) (PAA)) were spin-coated on to silicon wafers from appropriate solvents. Silicon nitride tips were either cleaned in an oxygen plasma, producing a 'polar', SiO_x -coated tip, or coated with gold, yielding a 'non-polar' tip. All surfaces were characterized

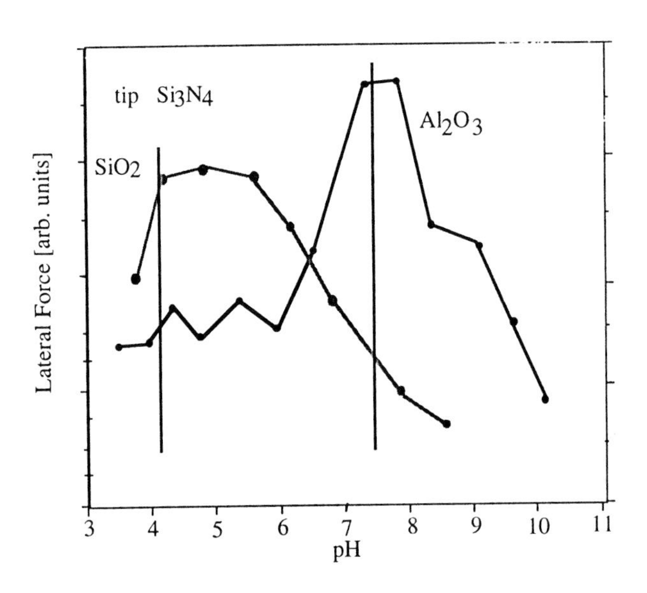

Fig. 8 Lateral force (LFM) as a function of pH (in lmM NaCl) for the systems Si_3N_4-SiO_2 and Si_3N_4-Al_2O_3.

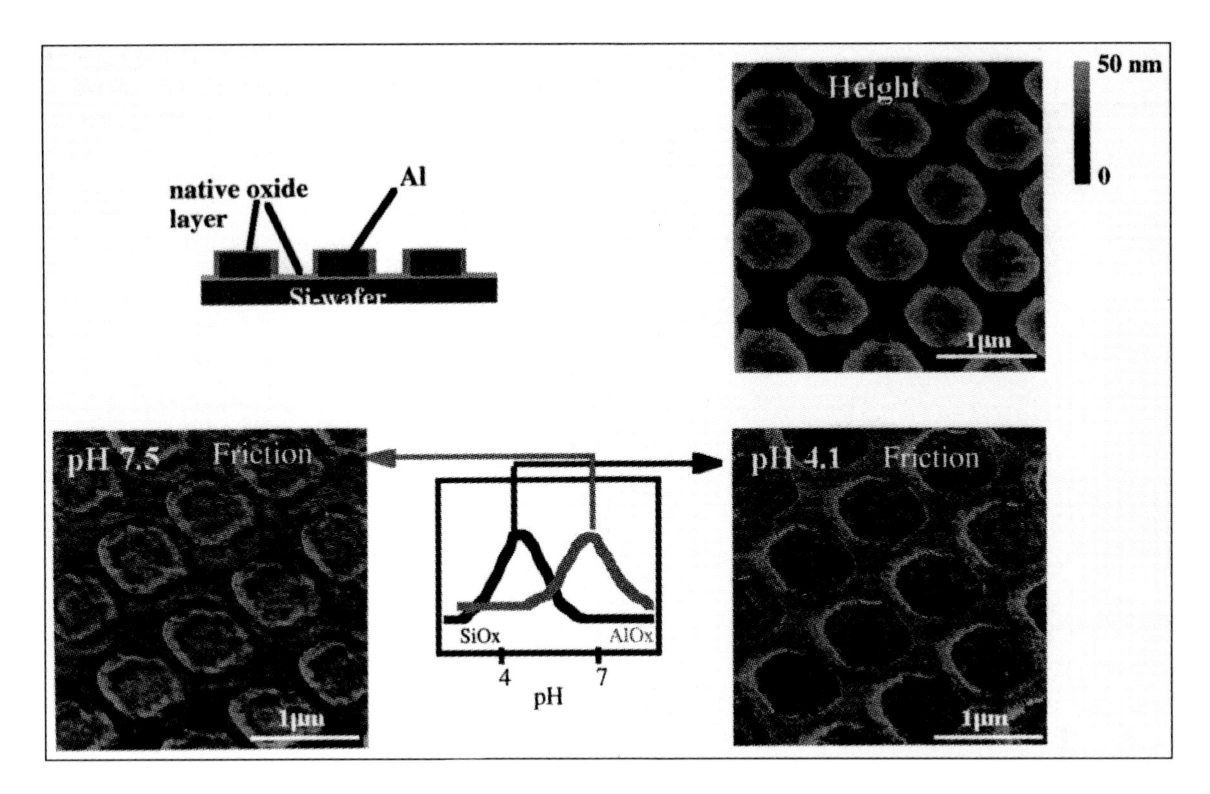

Fig. 9 AFM (top) and LFM (bottom) friction-loop-subtracted images of a microfabricated aluminum pattern on a silicon substrate, covered with native oxide, observed under lnM NaCl. Shifting pH from 4 to 7 results in an inversion of frictional contrast (**10**).

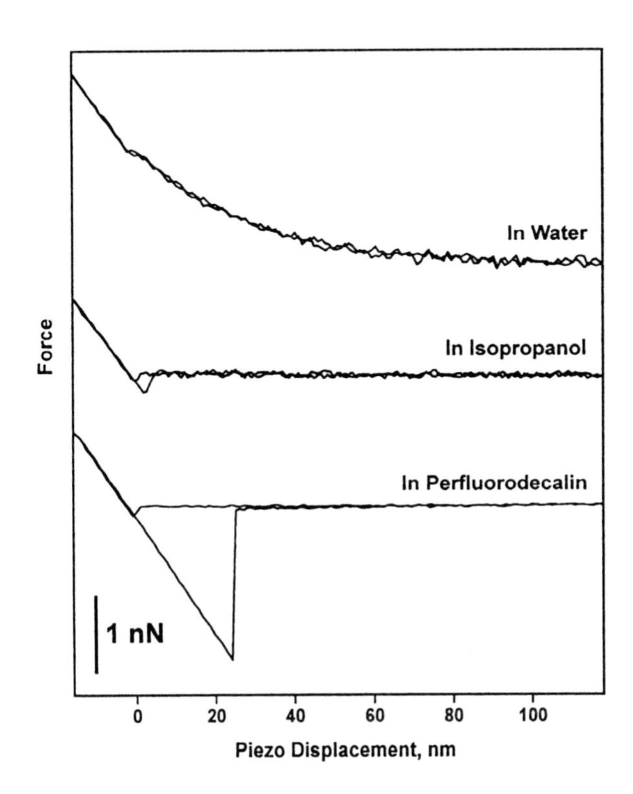

Fig. 10 Representative force curves acquired for a SiO_x probe in water, isopropanol and perfluorinated decalin (**11**).

by means of XPS. The histograms of adhesive force (which correlates with adhesion hysteresis, and, in turn, with frictional behaviour) for the various polymers and a single tip of each kind are shown in Figure 11. It is clear from these results that the more polar polymers have stronger interactions with the polar tip, while the less polar ones have stronger interactions with the non-polar one. In the first case the forces involved presumably include both dipole-dipole components of the Van der Waals force as well as possible H-bonding, whereas in the second, the interactions are limited to dispersion forces.

In Figure 11 it is interesting to note that for PS and PMMA the relative magnitudes of the adhesion force are reversed upon switching from a polar to a non-polar tip. This, in turn, has consequences for frictional (LFM) images of a surface comprising PS islands in a PMMA matrix (Figure 12), where the frictional contrast is seen to reverse upon changing from a polar to a non-polar tip. Thus, by enabling us to compare the magnitude of the Van der Waals interactions, the AFM and LFM can be used to perform chemically sensitive, high-resolution imaging of polymer blend surfaces.

5 SUMMARY AND CONCLUSIONS

AFM and LFM are valuable tools to the tribologist, providing fundamental information on the forces involved

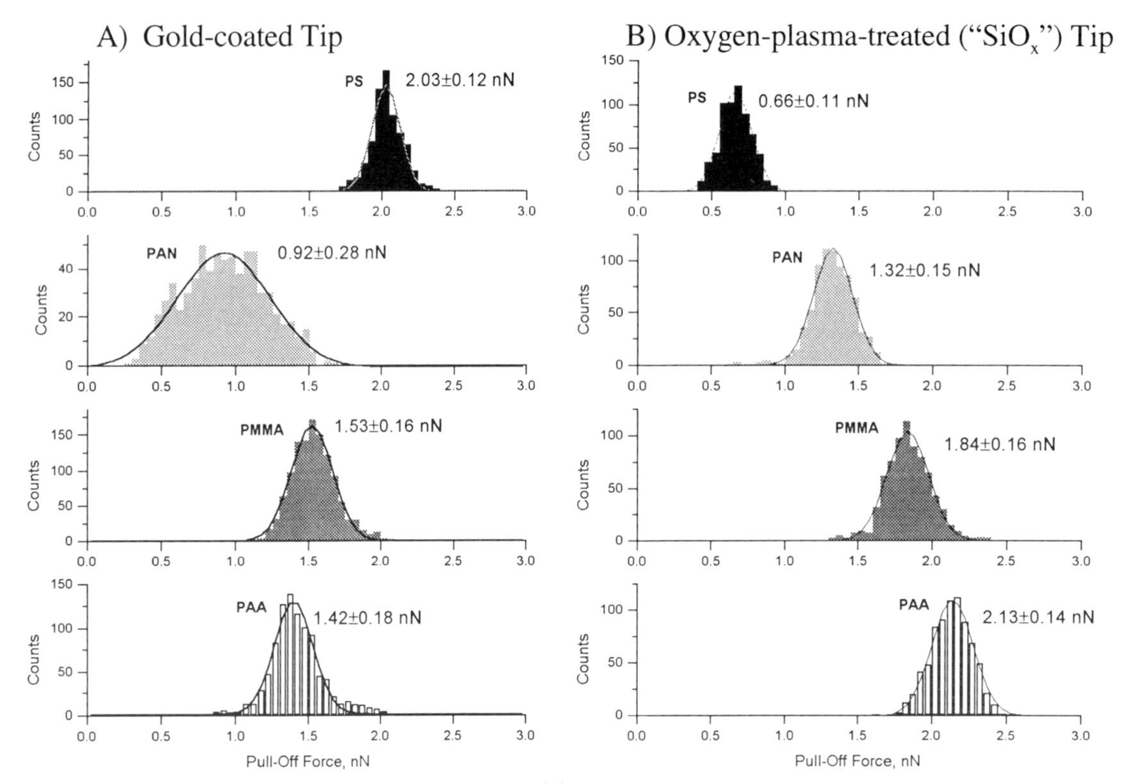

Fig. 11 Histograms of pull-off forces obtained during the acquisition of force-distance curves on polystyrene (-CH_2-CH(C_6H_5)-), polyacrylonitrile (-CH_2-CH(CN)-), poly(methyl methacrylate) (-CH_2-C(CH_3)($COOCH_3$)-), and polyacrylic acid (-CH_2-CH(COOH)-) with: a) Gold-coated, and b) Oxygen-plasma-treated ("SiO_x"), tips (**11**).

between surfaces in contact, under a variety of conditions, both lubricated and unlubricated. Thanks to these techniques, the tribological behaviour of surfaces can also be exploited for surface chemical analysis, especially in cases where the overall composition of the surface is known, and the distribution needs to be determined with high spatial resolution. This is of great potential use for the characterization of oxidic or oxide-covered surfaces, as well as surfaces of polymer blends.

6 ACKNOWLEDGEMENTS

The authors would like to thank the Council of the Swiss Federal Institutes of Technology, through the MINAST and PPM programs, and the Swiss National Science Foundation, through the NFP36 program, for their generous support of this work.

7 REFERENCES

1 **Binnig. G., Quate. C.F.,** and. **Gerber. Ch.,** *Phys. Rev. Lett.,* **56,** (1986), 930.

2 **Mate. C.M., Erlandsson. R., McClelland. G.M.,** and **Chiang. S.,** *Phys. Rev. Lett.,* **59,** (1987), 1942.

3 **Binnig. G., Rohrer. H., Gerber. Ch.,** and **Weibel. E.,** *AppA Phys. Lett.* **40,** (1982), 178.

4 **Meyer. G.,** and **Amer. N.M.,** *Appi. Phys. Lett.* **57,** (1990), p 2089; **Marti. O., Colchero. J.,** and **Mlynek. J.,** *Nanotechnology* **1,** (1990), 141.

5 **Yoshizawa. H., Chen. Y.-L.,** and **Israelachvili. J.N.,** J. *Chem. Phys.,* **97,** (1993), 4128; **Israelachvili. J.N., Chen. Y.-L., Yoshizawa. H.,** *J. Adhesion Sci. Technol.* **8,** (1994), 1234.

6 **Johnson. K.L., Kendall. K.,** and **Roberts. A.D.,** *Proc. Roy. Soc. London,* **A 324,** (1971), 301.

7 **Marti. A., Hähner. G.,** and **Spencer. N.D.,** *Langmuir* **11,** (1995), 4632.

8 **Burnham. N.A., Kulik. A.J.,** and **Gremaud. G.,** in *Procedures in Scanning Probe Microscopy,* Colton. R. J., ed., John Wiley and Sons, NY (1997).

9 **Marti. A., Hähner G.** and **Spencer N.D.,** in *Reibung und Verschleiß,* Zum Gahr K.-H., ed., DGM Verlag, Oberursel, (1996).

10 **Hähner, G., Marti. A.,** and **Spencer, N.D.** *Tribology Letters,* in press.

11 **Feldman. K., Tervoort. T., Smith. P.,** and **Spencer. N.D.,** in preparation.

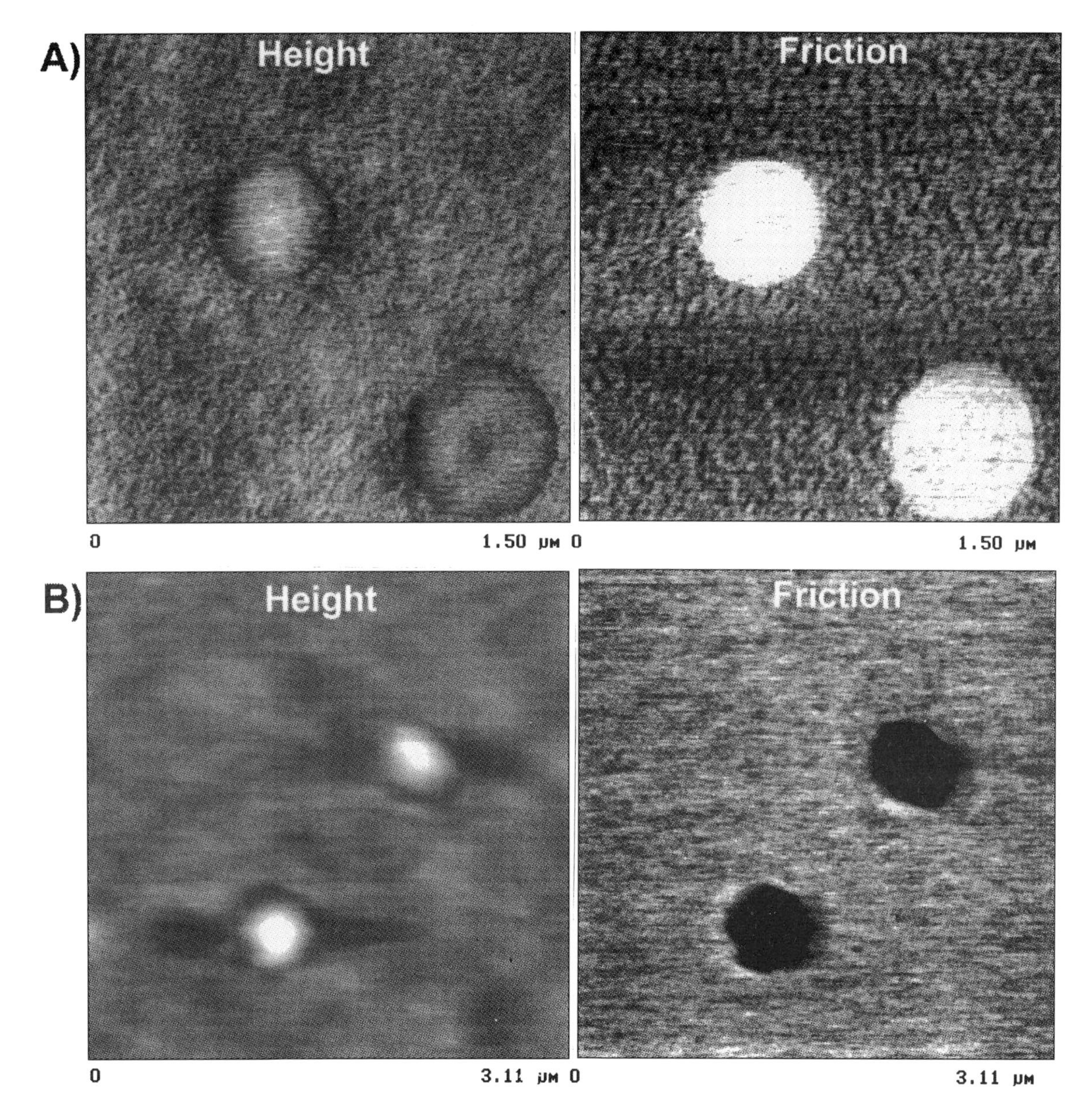

Fig. 12 Height (AFM) and friction (LFM) images of a spin-cast PS:PMMA (1:10) polymer blend, obtained with: a) Gold-coated, and b) Oxygen-plasma-treated ('SiO_x'), tips under perfluorodecalin (**11**).

Advances in the study of thin lubricant films

H SPIKES
Department of Mechanical Engineering, Imperial College of Science, Technology, and Medicine, London, UK

1 INTRODUCTION

Despite its great practical importance, boundary lubrication remains by far the least well understood lubrication regime. This is because, notwithstanding a century of research, we still know surprisingly little about the origin, nature and properties of the films which provide this type of lubrication. There are two main problems. First, it is very difficult to carry out conventional chemical analysis of boundary lubricating films since these are both extremely thin and are immured between a pair of bulk contacting solids. Second, there appear to be several different types of boundary film, ranging from quite thick inorganic layers to subtle changes in the viscosity of liquids adjacent to solid surfaces, and these different types of film may coexist. In consequence it is often difficult to disentangle the various components and this complexity has tended to hinder the emergence of a clear understanding of the nature of boundary lubricating films.

In the last few years, however, the clouds which have obscured boundary lubrication for so long have started to disperse. This is largely because of the development and application of a number of new experimental techniques which are able to study boundary films in ways not previously possible; to measure their properties and composition at oil-covered surfaces and even within rubbing contacts. These techniques are still in their infancy and much remains to be done, in particular in combining the various techniques to examine specific types of boundary lubricant film in a systematic fashion. The aim of this paper is to review these new techniques, what they have so far accomplished and to indicate the possibilities that they hold for our future understanding of boundary lubrication.

To set the scene, the paper first gives a brief history of the development of boundary lubrication, identifying the four main categories of boundary lubrication additives: oiliness, EP, antiwear and molybdenum friction modifier additives. It then outlines the main conventional experimental techniques used in the past to study the nature of boundary films.

The main body of the paper then describes in further detail four types of new, advanced techniques for studying boundary films and describes a few of the main findings each has made to date.

2 TYPES OF BOUNDARY LUBRICATION

Figure 1 shows pictorially the evolution of the field of boundary lubrication over the last hundred years. Hydrodynamic lubrication was discovered and analysed in the 1880s (**1**). It was then almost immediately recognised that lubricants, despite having the same viscosity and thus producing identical hydrodynamic films, could give quite different performance in terms of wear and friction. In particular natural oils (animal and vegetable oils) were found to produce much lower friction than the then newly-available, refined mineral oils. This difference was ascribed to a mystery property, initially known as 'oiliness' and later renamed 'boundary lubrication' when it was recognised to be a specifically surface behaviour. A key development was made in 1914, when it was found that oiliness could be imparted by surfactant additives; so that the addition of fatty acids to a mineral oil could enhance the latter's friction properties close to those of a natural oil (**2**).

From the outset, a number of different hypotheses was proposed concerning the origins of boundary lubrication (**3**). These included the possibility that the viscosity of some oils might be raised (in a manner not defined) next to solid surfaces (**4**), the possibility that oiliness might result from the reaction of components of the lubricant with metals during rubbing, to form a paste-like, protective film (**5**) and the suggestion that oiliness was related to wetting and thus depended upon the surface tension of the lubricant (**2**). Most of these proposed mechanisms were speculative. One key finding made in the 1920s was that single monolayers of surfactant are able to produce reductions in friction down to levels common for high oiliness lubricants (**6, 7**). This and key work on simple surfactant solutions (**8**) supported the concept that boundary lubricating films are adsorbed or reacted monolayers and led to this model of oiliness becoming dominant for the next sixty years (**3**). It should be noted, however, that there remained, and remains, significant minority support for the idea that oiliness films are considerably thicker than monolayers: the multilayer hypothesis (**9**).

In the late 1920s, hypoid gears were developed by the automotive industry, primarily as a means of lowering the shaft position in road vehicles. It was immediately found that, due to the high sliding speeds and loads present, such gears could not be satisfactorily protected against scuffing by existing oils, even if the latter were fortified by oiliness additives. The problem was solved by the addition of so-called 'extreme-pressure' or EP additives, based upon sulphur, chlorine and lead-containing additives (**10, 11**). For some years there was argument as to whether EP additives acted like oiliness additives but by forming stronger protective film on surfaces (**12**) or whether they conferred protection by forming a film which was easily worn away to reduce the contact pressure (**13**). Nowadays there is reasonable consensus that EP additives, at least of the sulphide type, react with hot steel surfaces under incipient scuffing conditions to form thin coatings of metal oxides and sulphides which serve to inhibit direct metal/metal contact and thus prevent catastrophic welding of rubbing components (**14**). Little is known, however, about the thickness or kinetics of formation of these coatings, or the chemical processes involved in their formation.

An important offshoot of boundary lubrication took place in the 1940s. Advancing technology had produced a progressive increase in both the steady-state running temperature of crankcase engines and other lubricated systems and also in the life expected of them. High operating temperatures were resulting in reduced oil viscosities and thus thinner hydrodynamic films, leading in turn, to unacceptable levels of wear. This problem led, in the early 1940s, to the search for additives specifically to reduce wear under high temperature conditions. Initial work by Beeck in 1941 looked at tricresylphosphate and found significant wear reduction, especially when the additive was used in combination with a conventional oiliness additive (**15**). Beeck ascribed this antiwear action to chemical polishing of the surface involving the formation of an iron phosphide eutectic but later work preferred the formation of films based on iron phosphates (**16, 17**).

In the last few years, surface infrared analysis and other research to be discussed later in this paper have led to a general acceptance of the principle that phosphorus antiwear additives form quite thick iron or zinc phosphate glass films on rubbed surfaces. On top of this material there may also be present a more fluid polyphosphate. This concept, that antiwear additives form glassy films, is supported by the practical observation that the only other class of chemical additive found to be generally effective in controlling wear at high temperatures is based on boron, another glass-producing element.

As shown in Figure 1, the fourth and most recent fork in the branch of boundary lubrication has been the development of soluble molybdenum-based additives. Oil-supply crises in the 1970s focused attention upon reducing friction in crankcase engine oils and thereby lowering fuel consumption. This led to the development of 'friction-modifying additives'. Application of these additives languished in the 1980s but in the last five years has become a critical feature of modern engine oil design.

Many friction modifiers are very similar to the oiliness additives discovered in the first quarter of the century: essentially surfactants based on fatty acids or other strongly adsorbing, long chain polar species. In the 1970s however, it was also found that some organic molybdenum compounds could produce a remarkable reduction in friction coefficient in rubbing contacts, from the 0.08 to 0.1 level typical of classical oiliness additives down to about 0.06 (**18**). The precise mechanism by which these additives produce such low friction is not yet understood. It is generally believed that MoS_2, which, like graphite, is a lamellar solid with low shear strength, forms on rubbing metal surfaces (**19**) but very little is known about the chemical pathways involved or the thickness or homogeneity of reaction film formed.

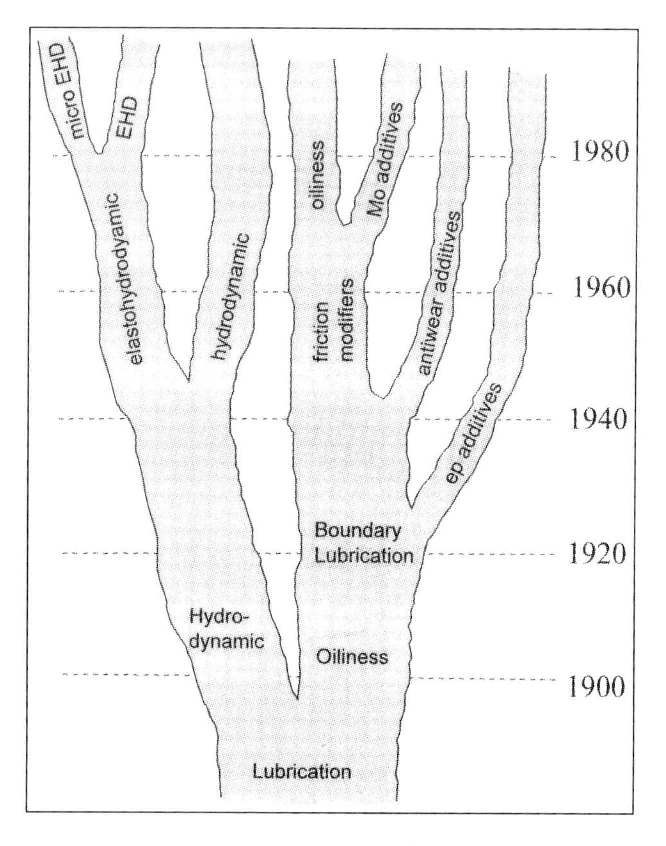

Fig. 1 Tree of lubrication.

3 TRADITIONAL APPROACHES TO STUDYING BOUNDARY FILMS

The developments in boundary lubrication described above were made prior to the availability of the modern tools to be described later in this paper. They relied upon two main types of experimental approach to examine the nature and properties of boundary films.

(a) Black box approach – performance measurement

By far the commonest means of studying boundary lubrication has always been simply to observe how changes in factors such as additive type, concentration or test temperature influence friction and wear and from this to infer the molecular basis of the film-forming process. The weakness of this, essentially black box approach, is that, because it provides no direct information about the boundary film, it is limited by the imagination or prejudice of the researcher and is also quite unsuited to examining systems where several different types of boundary film may coexist.

A typical example of this approach is research carried out in the 1960s which showed that, with many oiliness additives, friction rises sharply at a critical temperature whose value is dependent upon additive type and concentration. This was taken to represent the desorption of an adsorbed monolayer in thermodynamic equilibrium with the supernatant solution (**20–23**). Unfortunately many chemically-based film-forming and film weakening processes might be expected to show very similar temperature and concentration dependence. Similarly, the mechanism of film formation and nature of EP and antiwear films has been extensively studied by comparing the influence of variations in additive structure on wear and scuffing performance and, from this, attempting to deduce the molecular decomposition process of the additive and also the nature of the protective boundary film formed (**14, 24–27**).

(b) Out-of-contact boundary film analysis

The second approach which has traditionally been used to study boundary films has been chemical analysis of solid surfaces, out-of-contact, after rubbing. Prior to the 1960s, the techniques available to characterise such thin films were very limited, the main one being radiochemistry, and this was employed from the 1940s using the radioactive isotopes S^{35} and P^{31} to study EP and antiwear films. A limited amount of X-ray (**28**) and electron diffraction (**29, 30**) work was also carried out.

The 1960s saw the development and application to boundary film studies of analytical techniques such as infrared analysis (IR) (**31**) and X-ray fluorescence (XRF) (**32**). In the 1970s, specifically surface spectroscopic methods such as x-ray photoelectron spectroscopy (XPS), Auger and secondary ion mass spectrometry (SIMS) became available and were also used to analyse boundary films (**33**). These techniques confirmed that EP additives react on iron surfaces to form sulphides and chlorides. The findings with antiwear additives were more equivocal. Some techniques, notably IR and XRF, found quite thick phosphorus-containing films whilst others, such as SIMS, tended to show only very thin layers, a few atoms thick.

A major limitation of this out-of-contact approach is the likelihood that any boundary film present in a contact will be changed to some extent as the rubbing surfaces are cooled, separated, washed free of supernatant oil and, with some techniques, subjected to vacuum, prior to analysis. This means that the boundary films that are subsequently analysed are likely, in many cases, to be only an unrepresentative remnant of the original surface material, particularly when studying oiliness-type films which are believed to be only weakly bonded to the surface.

Over the years a number of attempts has been made to examine boundary lubricating films within contacts. In general these have made little contribution to our understanding, usually because the techniques used were operating at the very limit of applicability or because of difficulty in interpreting the measurements made. In the last decade, however, the situation has changed quite dramatically, with the development of a number of important techniques for studying boundary films *in situ*, i.e. either within rubbing contacts or on solid surfaces immersed in supernatant lubricant. This paper describes four such techniques.

4 FORCE BALANCE

One technique which is able to examine the physical nature of boundary lubricating films on solid surfaces is the force balance or surface force approach. This was first developed in the late 1960s, primarily to study the attractive forces between solid surfaces in close proximity (**34**). It involves bringing two extremely smooth surfaces very close together to produce a known and controllable gap down to fractions of a nanometer. One of the surfaces is mounted so that the normal force exerted on it across the gap by the other surface can be accurately measured. Typically the surfaces are two crossed-cylinders of mica, but a ball on flat geometry has also been used. Since the early 1980s, modern electronics and piezo-electrics has made this technique easier to apply and more sensitive and it is now used quite extensively to study both the normal and the shear forces transmitted across a thin film of lubricant between the two surfaces. By using not just the slow approach of two surfaces but also, superimposed upon this, a higher frequency, low amplitude oscillation, both viscous and elastic properties of liquid films down to just one or two molecular layers in thickness can be obtained. It is also possible to move one surface laterally against the other to impose shear on the film and measure the resulting friction force. Figure 2 shows a typical experimental arrangement where a metal-coated glass ball is moved both normally and laterally against a plane surface, (**35**).

In early work, Israelachvili showed that, for simple molecular organic liquids on very smooth mica surfaces, the normal force is an oscillatory function of distance from the surface. This indicates that the liquid molecules can form an effectively layered structure, several molecular diameters out from the surface (**36, 37**), (Figure 3).

Both Israelachvili and Granick have also studied the shear properties of simple molecular liquids between two mica surfaces. They have shown that these retain their bulk properties so long as the separation is equivalent to at least three monolayers. However when the film is thinner than this, there are gross deviations from bulk properties with, for one or two layers, effective viscosities up to several orders of magnitude higher than normal (**38, 39**), although

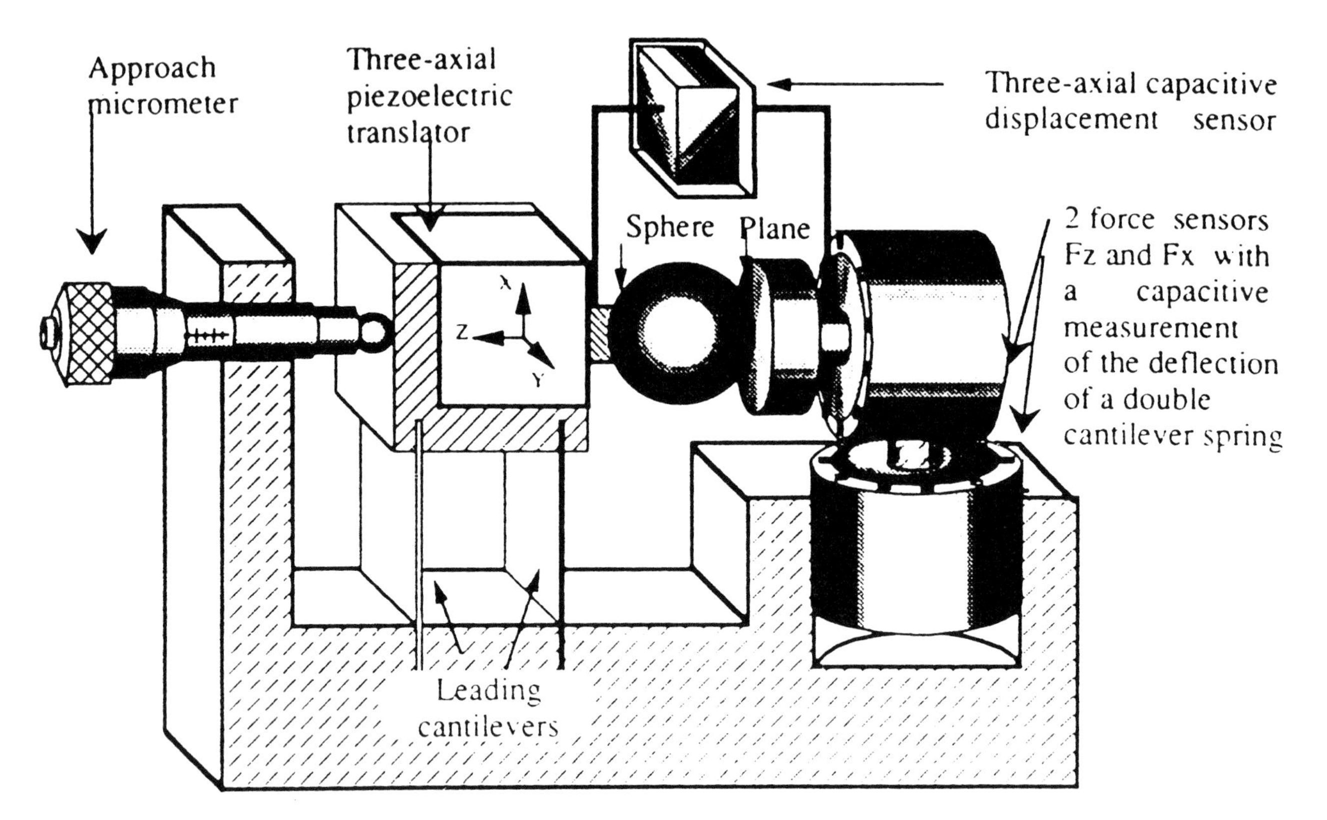

Fig. 2 Schematic view of ball on plane force balance apparatus. Three capacitance sensors detect and control the relative displacements between the sphere and plane to 0.01 nm. The plane is fixed to two sensors which measure normal F_z and tangential F_x forces with a resolution of 10^{-8} N, (from ref. (**35**)).

this viscosity falls rapidly as shear rate is increased. It also appears that the film is effectively quantised, with shear stress tending to jump in value as the film decreases from three to two or two to one molecular layer (**38**).

A limitation of much force balance work from the point of view of studying lubricant films is that, because the two surfaces generally consist of mica fastened on glass, the pressures that can be reached are quite low, being limited by the modulus of the adhesive. Georges and co-workers have applied the force balance technique to the contact of a ball on a flat with both bodies made of steel or cobalt-coated glass (**40**). This has enabled him to reach quite high pressures and also to examine the effect of roughness. Georges has shown that simple molecular liquids on surfaces form 'immobile layers' with anomalous viscometric properties typically 1 nm out from the surface. Interestingly, he did not observe the clear layering of molecules found by Israelachvili, which may indicate that this is absent with surfaces which are rough, even if only on a nanometer scale.

Georges has applied the technique to examine both polymer solutions in oil and also simple oiliness additive films. With polymers, he found 'immobile layers' of thickness between one and two times the hydrodynamic diameter of the polymer coil in solution (**41**). The elasticity of these films varied depending upon the polymer type, with polar polymers forming stiffer, more uniform layers (**42**). Quite recently, similar behaviour has been found on carbon surfaces, directly indicating the existence of steric repulsion behaviour which is believed to be the basis of engine oil dispersant action (**43**).

Georges has also applied the force balance to examine the thickness and shear behaviour of adsorbed monolayers of stearic acid formed from dodecane on cobalt ball on flat surfaces (**35**). This simple oiliness additive was found to form stiff, immobile layers 4.8 nm thick between the surfaces in a static contact, representing 2.4 nm films on each surface. When sliding was applied, this thickness increased very slightly, by less than a tenth of a nanometer, but this separation resulted in a very significant drop in friction, possibly because it resulted in the loss of overlap and thus interaction of the ends of stearic acid chains on opposing surfaces (Figure 4).

From the above it can be seen that some quite important and revealing information about boundary films is beginning to emerge from the force balance technique. The strength of the technique is that it is extremely accurate and is able to detect and measure some of the rheological

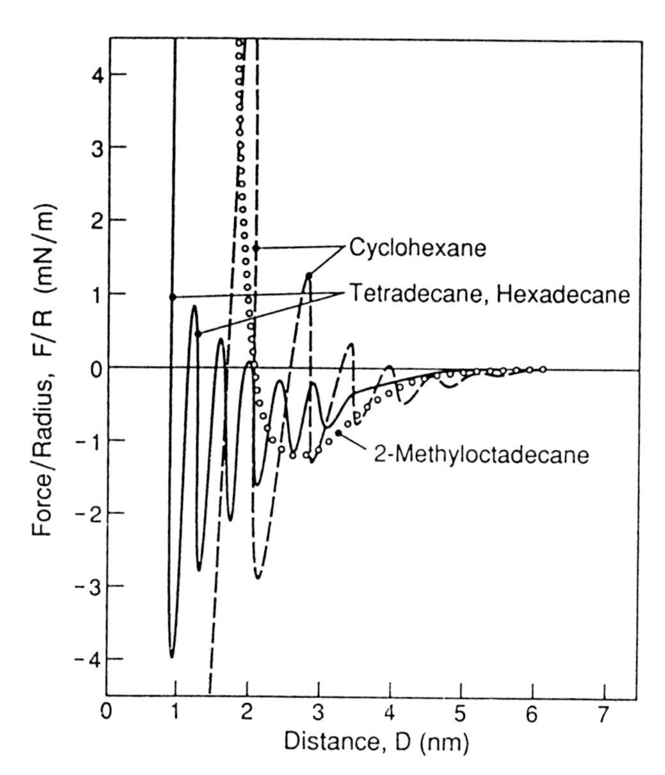

Fig. 3 Measured forces between mica surfaces. Cyclohexane of mean molecular diameter 0.55 nm (solid curve), n-alkanes of molecular width 0.4 nm (dashed curve) and 2-methyl-octadecane (dotted curve), (from ref. (**37**)).

properties of films less than 1 nanometer thick in fully-flooded conditions. It has demonstrated the mechanical strength of adsorbed monolayers of surfactant and also some of the friction properties of these films. It has also shown that quite simple liquids can achieve a considerable degree of order when confined as very thin layers between smooth surfaces; ordering which at low shear rates leads to a very large increase in viscosity but which appears to be lost as shear rate is increased.

The main limitation of the force balance technique is that it involves a very idealised contact; between very smooth surfaces and generally of very low pressure and shear rate. It is likely that many of the effects seen may disappear in more realistic contact conditions: indeed both Granick's enhanced viscosity and Israelachvili's molecular layers are not observed in the higher pressure, slightly rough metallic contact used by Georges. Finally, of course, the force balance technique does not actually involve any solid/solid rubbing which precludes the possibility of observing any boundary films which might be chemically produced in response to such rubbing contact.

5 ULTRA-THIN FILM INTERFEROMETRY

The second technique to be discussed is ultra-thin film interferometry. This measures lubricant film thickness between a steel ball rolling against a transparent flat using an extended form of optical interferometry (**44**). The principle is shown in Figure 5. White light enters the contact region where some is immediately reflected whilst the rest passes through any oil film present and also a silica spacer layer before being reflected back from the steel ball surface. Because the two beams of light thus travel different distances they interfere constructively at wavelengths, λ which obey the relationship:

$$(N-\phi)\lambda = 2n_{oi\lambda}h_{oil} + 2n_{spacer}h_{spacer} \qquad N=1,2,3... \qquad (1)$$

where N is the fringe order, h_{oil} and h_{spacer} are the thickness of oil and spacer layer film, n_{oil} and n_{spacer} the refractive indices of the two films and ϕ is a phase change that is characteristic of the surfaces and takes place upon reflection. Clearly if the wavelength of maximum constructive interference, λ can be measured and h_{spacer} is

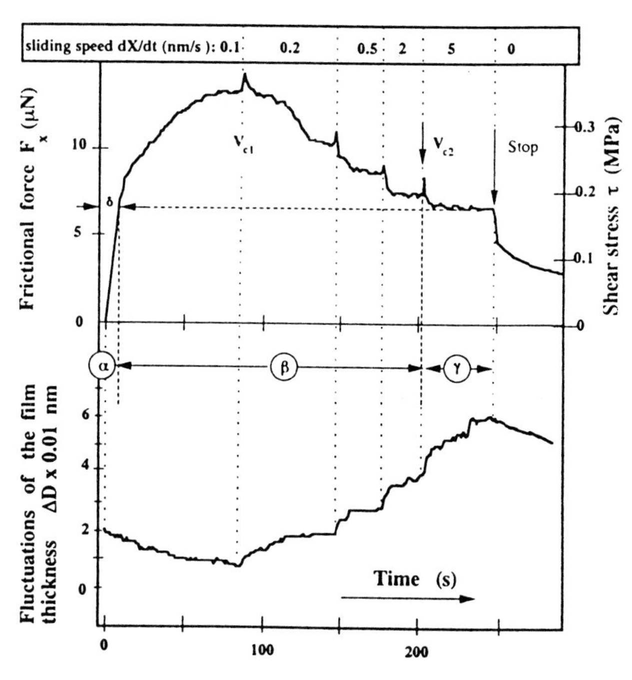

Fig. 4 Frictional force measured as a function of sliding time between a plane and a sphere coated with adsorbed monolayers of stearic acid. In the test, sliding velocity is increased in steps and ΔD is the measured variation in film thickness compared to the static case. There are three stages, (i) a short elastic period corresponding to displacement d = 0.4 nm, (ii) yield where film shear strength depends upon speed, (iii) a period above a critical speed where the yield strength of the film is constant, (from ref. (**35**)).

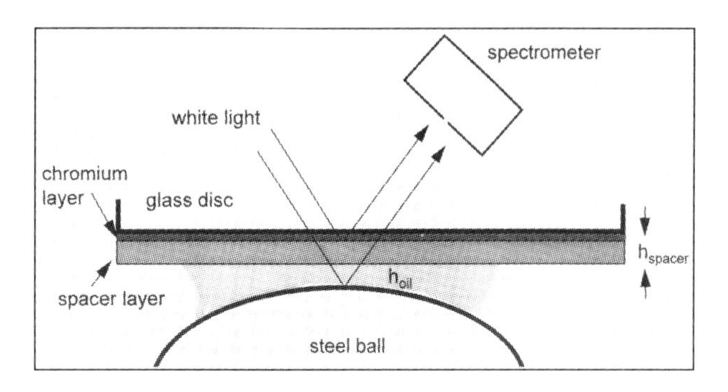

Fig. 5 Principle of ultra-thin film interferometry. A lubricated contact is formed between a smooth, reflective steel ball and a coated, glass disc. White light is shone into the contact through the glass disc. Some is immediately reflected from a thin semi-reflecting chromium layer whilst the rest passes through a silica spacer layer and any oil film present before being reflected from the steel ball surface. The extent of interference of the two beams depends upon the sum of the spacer layer and oil film thickness.

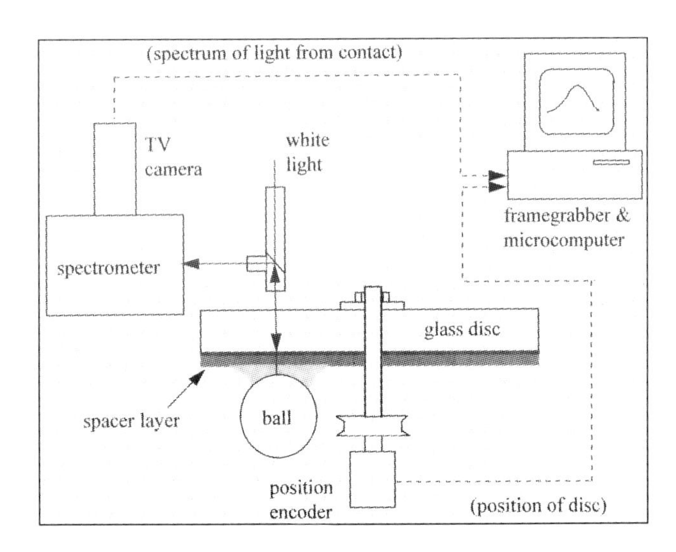

Fig. 6 Schematic diagram of ultra-thin film method. Light from the central region of the contact is dispersed by wavelength in a spectrometer and the resulting spectrum is frame-grabbed and analysed to determine the wavelength of maximum constructive interference and thus the oil film thickness. The position encoder on the disc shaft ensures that the film thickness is measured at a disc position where the spacer layer coating has been previously measured.

known, h_{oil} can be determined. To determine this wavelength, interfered light from the central region of the contact dispersed in a spectrometer and the resulting spectrum is frame-grabbed and image-analysed (Figure 6). The technique can measure film thicknesses above 10 nm to ± 5% and below 10 nm to ± 1 nm.

The technique is complementary to the force balance approach in that, like the latter, it can be used to measure the presence and rheological properties of boundary films, with considerably less accuracy than the force balance but under much more realistic contact conditions. Because the load applied to the ball is quite high, typically 20 N, (much higher that in force balance where it is of the order of one mN) a high pressure, elastohydrodynamic-type contact is formed. In such a contact, the EHD film thickness is strongly dependent on the lubricant viscosity in the immediate contact inlet, so that film thickness measurements in the very thin film region can be used to determine effective viscosities of lubricant layers of the same order of thickness.

Figure 7 shows log(film thickness) plotted against log(rolling speed) for purified hexadecane. This gives a straight line plot down to 1 nm in accord with elastohydrodynamic theory, indicating that this simple hydrocarbon maintains constant viscosity down to within 1 to 2 nm of the two rolling surfaces. Other pure hydrocarbons give similar behaviour but some base fluids form thicker than expected films at slow speeds, indicative of the presence of a viscous boundary film (**45**). Figure 8 shows results for solutions of a range of polymeric viscosity index improvers in a mineral oil at 80°C. Three of the polymers show a very pronounced increase in film thickness at slow speeds, indicative of a boundary layer of between 15 and 25 nm thickness. Further analysis suggests that this results from adsorption of these three polymers on the solid surfaces to produce layers of about one polymer coil diameter thickness on each surface having higher polymer content and consequently a viscosity about 40 times higher than the bulk solution (**46, 47**). These films have been shown to reduce both friction and wear (**48**). Both the force balance and ultra-thin film interferometry show similar boundary film behaviour with polymer solutions.

The ultra-thin film technique has also been used to study the behaviour of other types of boundary film, showing quite thick, solid-like film formation by zinc dialkyldithiophosphate antiwear additives (**49**) and also that oiliness additives can form not just monolayers of surfactant but also, under some conditions, considerably thicker films, up to 20 nm thick, (Figure 9) (**50**), lending support to the multilayer hypothesis of oiliness lubrication.

One recent application of ultra-thin film interferometry has been to demonstrate that binary blends of polar and non-polar base fluid fractionate close to polar surfaces due to differences in the van der Waals attraction of the two types of liquid molecules to solid surfaces (**51**). This was shown by studying blends where the two components had

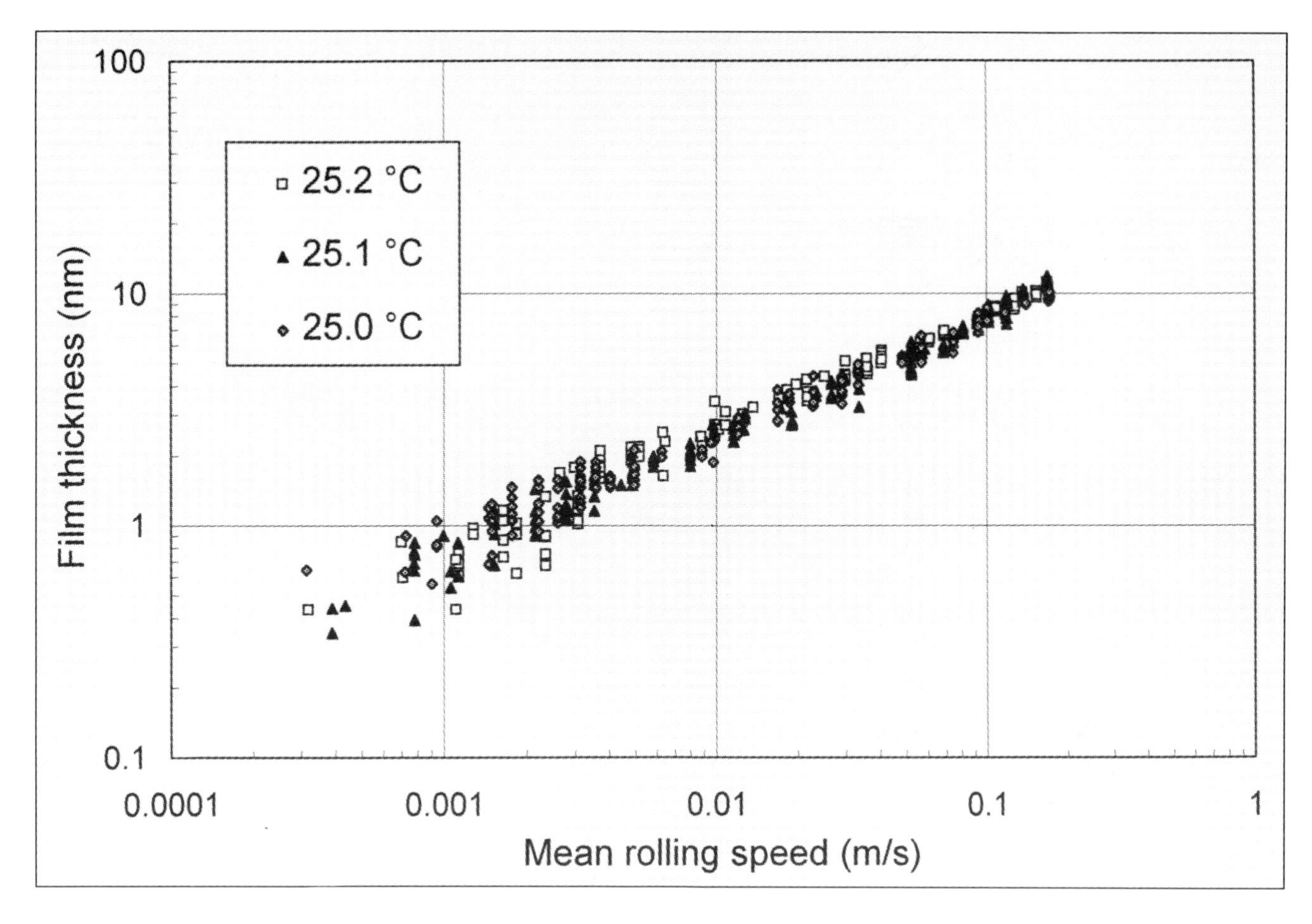

Fig. 7 Plot of film thickness versus mean rolling speed for purified hexadecane. There is no evidence of boundary film formation down to 1 nm, (**45**).

both different polarity and different viscosity. Figure 10 compares the EHD film formation of a blend of 10 wt. %of a viscous ester in low viscosity hydrocarbon with the behaviour of the two neat components. At high speeds, and thus in the thick film region, the film thickness is as predicted from the blend viscosity and close to that formed by the neat hydrocarbon. However in the thin film, boundary region, the film thickness moves towards that of the neat ester. This is believed to reflect the fact that ester molecules are preferentially attracted to the polar solid surfaces to produce surface layers about 2 to 3 nm thick comprising almost 100% ester and thus having high viscosity. The same effect was also seen with ester/mineral oil blends and at elevated temperatures (**52**). This is a new type of boundary lubrication; the production of viscous surface layers due to the formation of a gradient of base fluid molecular composition and it represents one method of manipulating and reducing the friction coefficient in thin film lubricated contacts (**52**).

6 SCANNING NANOPROBES

One group of techniques which shows promise of considerably advancing our understanding of boundary films are the various scanning nanoprobe methods. These include scanning tunnelling microscopy (STM) and atomic force microscopy (AFM). In both of these, the basic idea is to move a very sharp tip, with a radius of a few nanometers, across a solid surface. In STM, the tip is arranged to move slightly up and down as it traverses the surface, so as to maintain a constant tunnelling electric current between the tip and surface (**53**) (Figure 11). The distance the tip moves up and down is measured and this provides a very detailed map of the surface height. The resolution of the technique is such that it can detect the arrangement of individual atoms in molecules on surfaces and Figure 12 shows the pattern formed by an adsorbed layer of liquid crystal molecules on a graphite surface (**54**). The measurements obtained are not true heights, since the tunnelling current depends upon the nature of material present as well as its dimensions, but they are sufficiently detailed in many cases to indicate the alignment and even the mobility of molecular films (**55**).

In AFM, the tip is mounted on a cantilever and pressed against the surface so that the cantilever is strained. As the tip then traverses the surface, its height is adjusted to

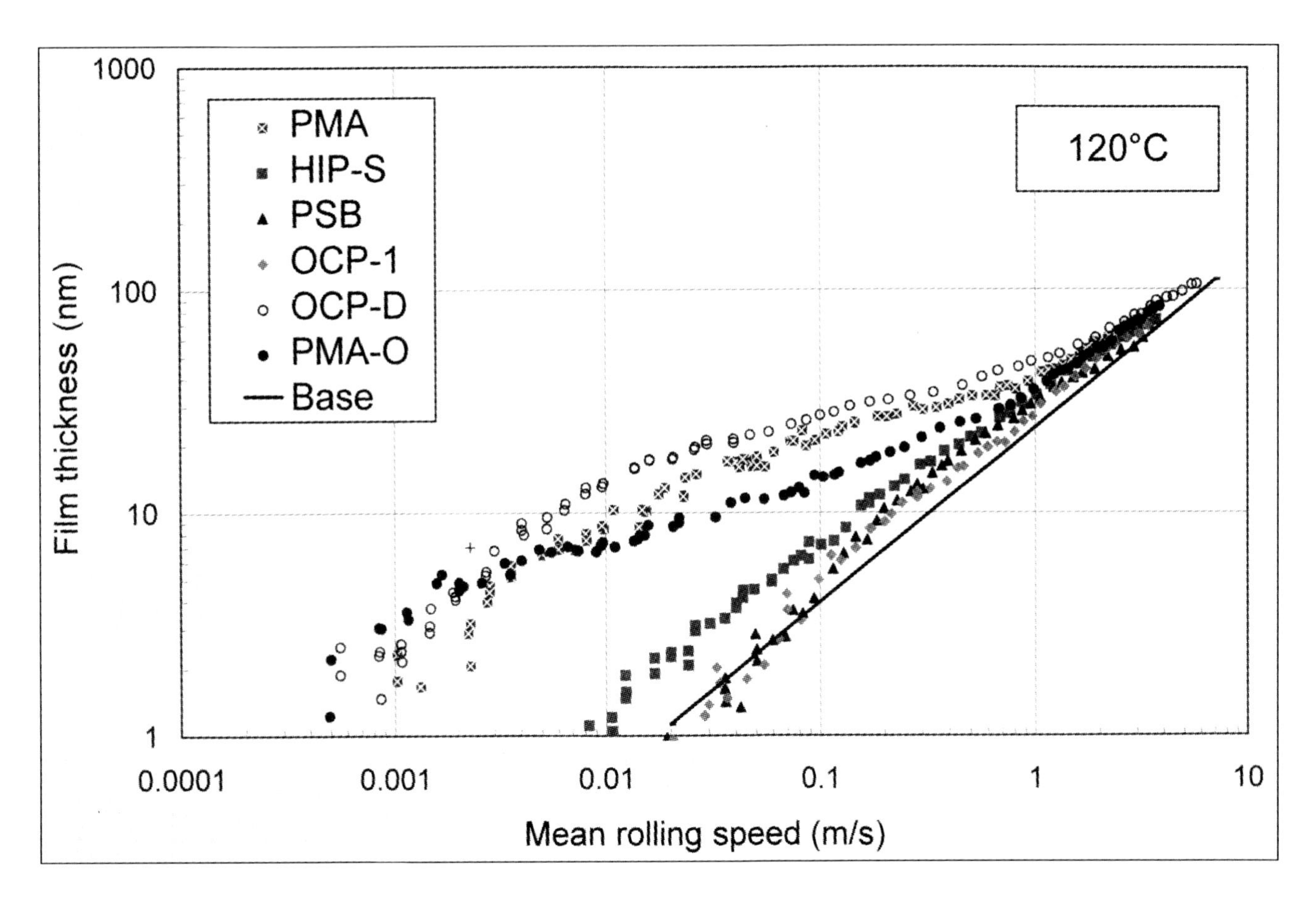

Fig. 8 EHD film formation by a range of solutions of different viscosity index improvers in a mineral oil. The solid line shows the behaviour of the polymer-free base fluid. Three of the polymers form thick boundary films, (**47**).

maintain a constant cantilever strain perpendicular to the surface. The topographic map produced is thus one of constant repulsive force. The cantilever can also bend in the sliding direction parallel to the surface and this is used in friction force microscopy (FFM) to determine how friction between surface and tip varies at the nanometer scale (**56**). AFM has been extensively used to investigate the thickness, alignment and packing of monolayers of Langmuir-Blodgett films on solid surfaces, using oiliness-type surfactants (**57, 58**). It has also been combined with FFM to investigate the friction and durability of deposited monolayers on surfaces (**59**). This work suggested that individual monolayers are very effective in reducing friction but have poor wear resistance, which supports much earlier findings on deposited multilayers, i.e. that whilst one monolayer can reduce friction very effectively in the short term, some means of repairing the film against damage, either from multilayers or from solution is needed to provide durable protection (**60**).

A particular valuable feature of both STM and AFM from the point of view of studying oiliness additives is that the tip will generally pierce thick layers of supernatant oil to image only the immediate surface film and can thus, in principle, be used to look at boundary films submerged in oil. Surprisingly little work has yet been published in this area, although STM has been employed to investigate the mobility of polymeric molecules containing benzene rings on solid surfaces (**61**) and to map the thickness of thin layers of liquid lubricant (**62**). Figure 13 shows the shift in location with time of the benzene rings in a single molecule of a perfluorinated polymer on an MoS_2 surface (**54**). AFM has also been used to study the loss of self-assembled monolayers from a surface submerged in aqueous solution under oxidising conditions (**63**).

In the last few years AFM has been modified to permit new types of tip motion, including 'tapping mode' which is able to explore the mechanical and rheological properties of surface films of lubricants and also greases (**64**).

Nanoprobe methods hold great promise for understanding boundary lubricating films. Their key strength lies in their ability to characterise surface films spatially at sub-nanometer resolution. This makes it possible to examine boundary films at the individual molecular level.

One limitation is that it is not always clear precisely what level of surface feature is being measured by methods

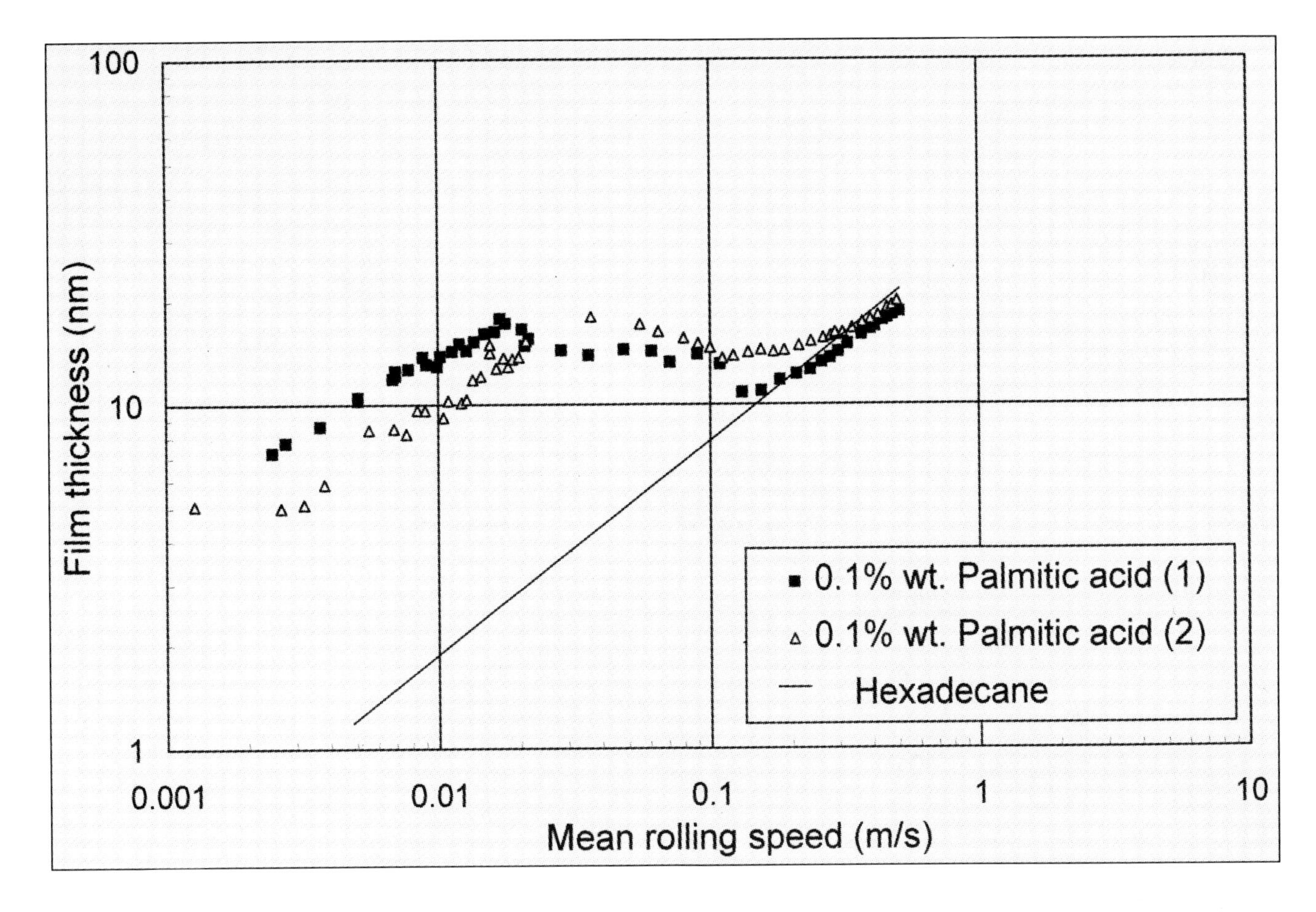

Fig. 9 Thick boundary film formation at slow speeds in two tests by a 0.1 wt. % solution of the oiliness additive, palmitic acid in hexadecane at 25°C. Also shown is the film-forming behaviour of pure hexadecane (**50**).

such as AFM. This is not a problem when the likely nature of the surface film is already well understood, such as with adsorbed or deposited monolayers of surfactants. For more complex films, which may well be heterogeneous with depth, interpretation is much more difficult. A second problem is that the contact conditions produced at a tip/substrate contact are very severe and also quite difficult to reproduce since the tip geometry itself is often not well characterised or repeatable.

7 X-RAY ABSORPTION AND DIFFRACTION

The final set of advanced techniques to be discussed in this paper are those based on x-rays. X-ray techniques including x-ray diffraction and XPS have been applied to the analysis of lubricant films for many years. A key recent advance has, however, been the availability to researchers of synchrotron radiation sources. Synchrotron sources have x-ray intensities several orders of magnitude higher than standard x-ray tubes, which makes it possible to obtain measurable absorption or diffraction effects from highly monochromatic x-ray beams interacting with small quantities of sample within a reasonably short time.

The biggest advances so far have been in the application of the two near edge x-ray absorption techniques, EXAFS (extended x-ray absorption fine structure) and XANES (x-ray absorption near edge structure). Figure 14 shows how x-ray absorption varies with energy for copper metal (**65**). The sharp increase in absorption at just over 9000 eV corresponds to the x-rays having just enough energy to fully eject an electron from the innermost shell of a copper atom. (This is the K-edge; a further, L-edge corresponds to ejection of electrons from the second shell). In Figure 13 it can be seen that the spectrum has some fine structure just above the absorption edge. The very first part of this (up to about 50 eV from the absorption edge) forms the basis of the XANES technique whilst the rest (from about 50 eV up to about 1000eV above the edge) is used for EXAFS. It should be noted that XANES is also sometimes called NEXAFS (near edge x-ray absorption fine structure).

EXAFS results from scattering of the ejected electron by surrounding atoms. Because of this it only appears in condensed matter (**65**). It provides information about the arrangement and proximity of neighbouring atoms and has been particularly widely used to identify the structure of

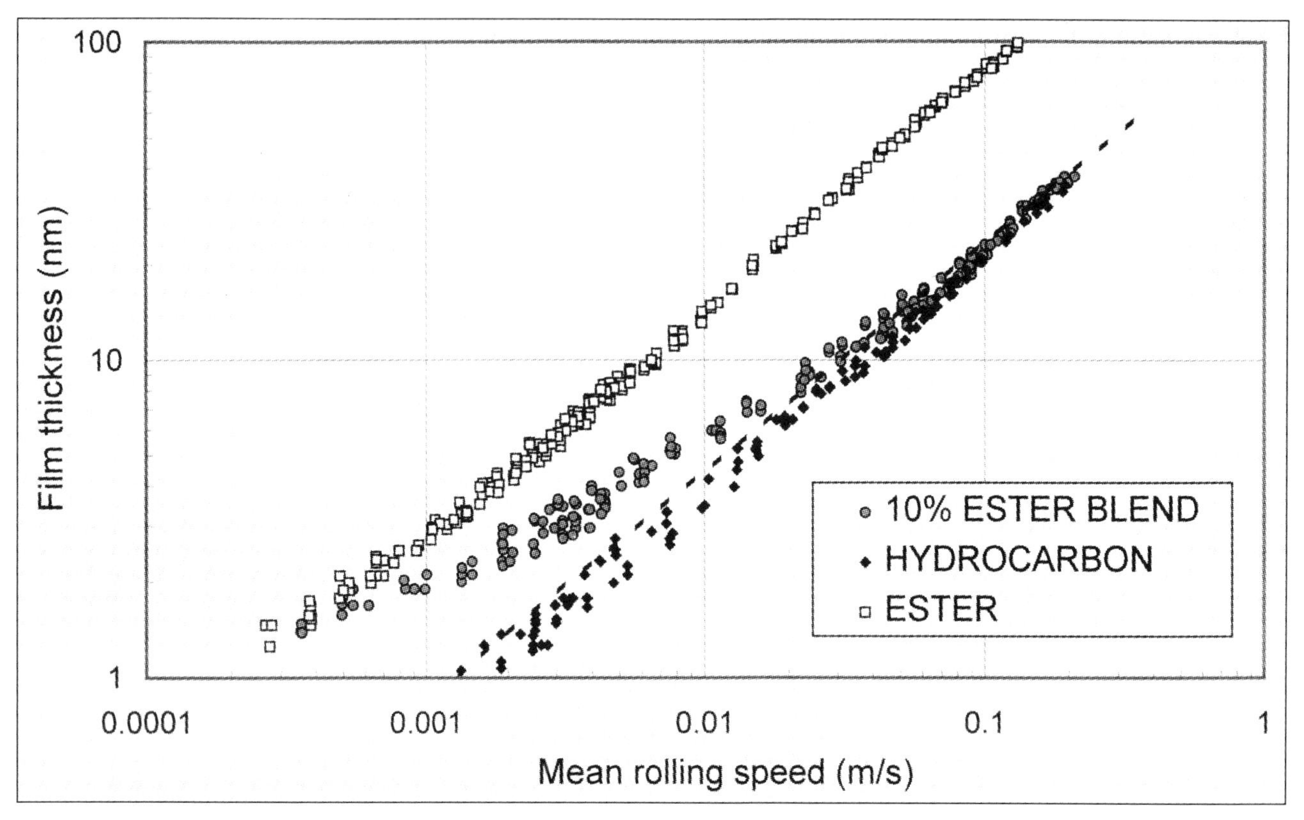

Fig. 10 Comparison of film formation by a 10 wt. % blend of a low viscosity ester in a high viscosity synthetic hydrocarbon with its neat components. The dashed line represents the theoretical EHD behaviour of the blend. In the thin film region, the blend film-forming behaviour moves progressively towards that of the neat ester (**51**).

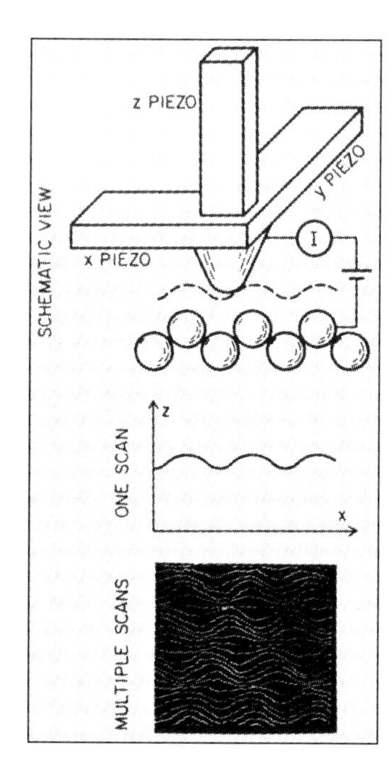

Fig. 11 Schematic view of STM method. As the tip follows the dotted constant tunnelling current path, one scan is recorded. Multiple scans are then plotted together to form images, (from ref. (**53**)).

metal complexes in biological systems. In tribology (**66, 67**) EXAFS has been applied to the study of ZDDP antiwear films on steel surfaces and has shown that the film formed consists of an amorphous iron(III)-containing phosphate structure based on poly- or pyrophosphate chains.

XANES is determined not so much by the presence of neighbouring atoms but by the bonding and oxidation state of the atom from which the electron is ejected. When the energy provided by an x-ray photon is just above the threshold required to eject an electron from an atom, the energy level to which the electron will be excited will correspond to unoccupied atomic or molecular orbital and will thus be strongly dependent upon the bonding environment of the atom itself (**65**). This makes the technique valuable for exploring the chemical nature of boundary lubricant films. In practice it is not yet possible to predict theoretically the relationship between molecular composition and XANES structure so surface films are characterised by comparison with model, synthesised compounds. Figure 15 shows XANES spectra taken from steel surfaces rubbed in solutions of zinc dialkyl and diaryl-dithiophosphate (ZDDP) antiwear additives (**68**). Also shown are spectra taken from two synthesised zinc polyphosphate glasses which indicate that the film formed

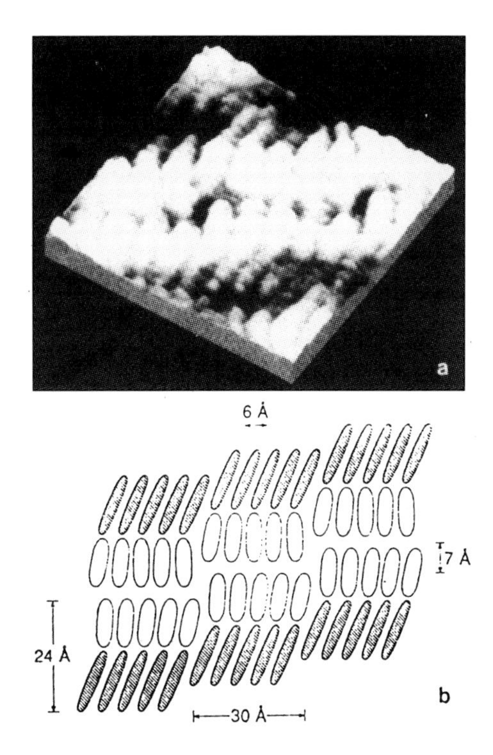

Fig. 12 (a) STM image (57Å by 57Å) of 10CB liquid crystal on graphite.
(b) Model showing packing of 10CB molecules, the shaded and unshaded segments represent the alkyl tails and cyanobiphenyl heads respectively, (from ref. (**54**)).

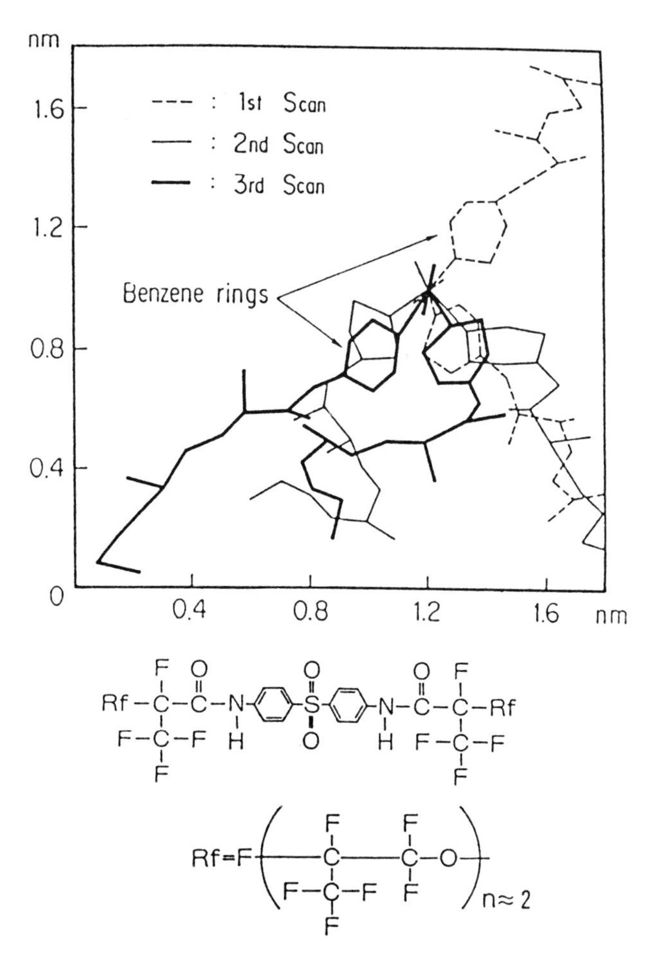

Fig. 13 Molecular configuration and shift of perfloro-based lubricant in three, sequential, 10 sec. STM scans (from ref. (**61**)).

by the dialkyl additive is more similar to the long chain polyphosphate and vice versa.

Studies have also shown that the films formed by ZDDP appear to consist of a substrate of very short chain polyphosphate with a thin coating of longer chain polyphosphate on top of this, that they generally contain very little sulphur bonded to phosphorus and that the composition of the film formed is strongly influenced by the presence of dispersant additive (**69–71**).

The XANES and to a lesser extent EXAFS approach shows considerable promise in being able to analyse in some detail very thin films. The technique is not just applicable to antiwear additives but can also examine oiliness films using electron transitions from the carbon atom. It is also possible to use polarised synchrotron radiation and thus examine molecular orientation effects, such as the molecular tilt of adsorbed or deposited monolayers of both surfactant and hydrocarbons on solid surfaces (**72**). A particular strength of both techniques is that they do not require a hard vacuum and can be used with unrinsed, rubbed surfaces, and, in principle, even look directly into rubbing contacts.

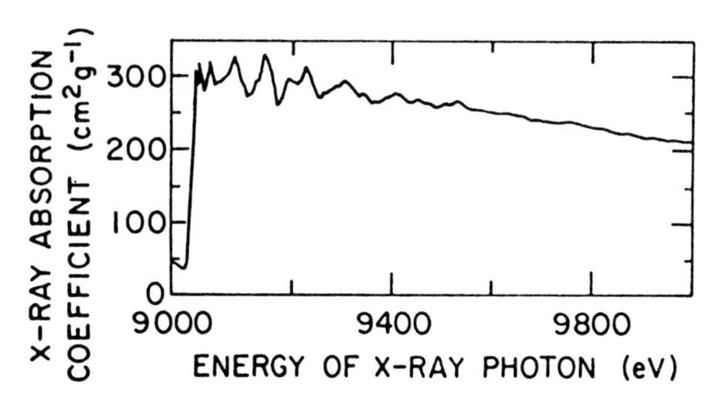

Fig. 14 The x-ray absorption coefficient for the K-edge of copper metal, (from ref. (**65**)).

One other x-ray technique of interest is x-ray diffraction. This was first used in the 1930s to study lubricant films (**28**) but has recently been applied by Israelachvili in an x-ray surface forces apparatus (XSFA). In this, an x-ray beam is able to pass through the mica/mica contact in an otherwise conventional surface force apparatus, permitting x-ray diffraction and possibly x-ray

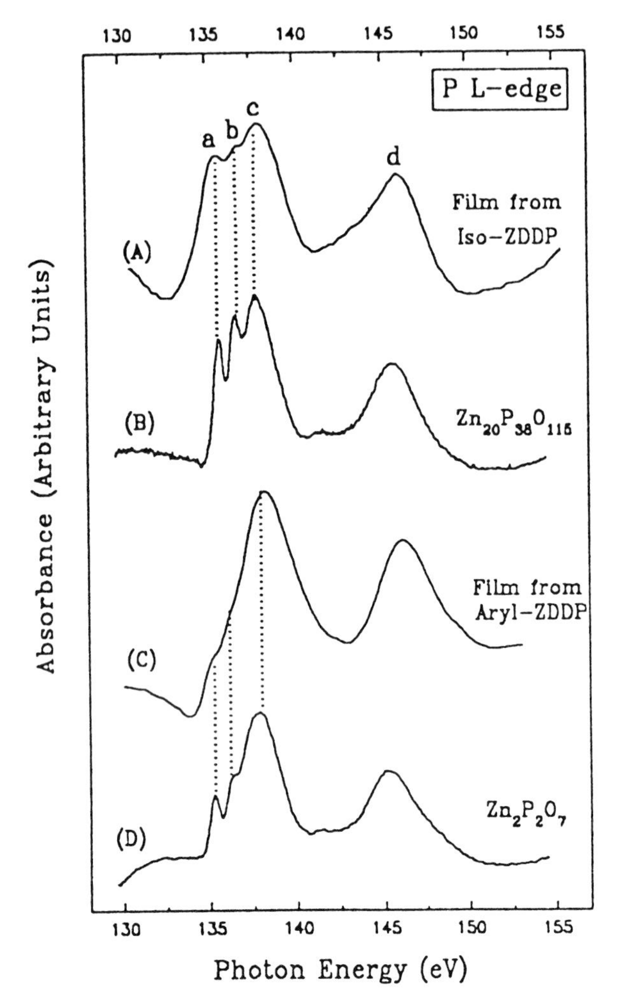

Fig. 15 P L-edge XANES spectra of tribochemical films and model compounds, (from ref. (**68**)).

absorption analysis to be applied to a thin lubricant film between the two surfaces. The method is based on the availability of high intensity, monochromatic synchrotron radiation and has so far been applied to study the orientation of quite thick smectic liquid crystal fluids under shear (**73**).

8 DISCUSSION

Four modern, experimental techniques have been described which are beginning to shed light upon the nature and properties of boundary lubricating films. It must be emphasised that this list is not exhaustive. There are a number of other, potentially important techniques, including the vibrational spectroscopies such as infrared and Raman, which have recently been employed successfully to analyse thin lubricant films within rubbing contacts (**74–76**) and also ellipsometry, which has been used to examine out-of-contact films on rubbed surfaces (**77, 78**). In addition to experimental tools there is also the whole new armoury of molecular modelling and molecular dynamics simulation methods, which are able to explore the likely behaviour of individual molecules or ensembles of molecules at solid surfaces. These too are beginning to yield valuable insights into the possible nature of boundary films (**79**).

It should be clear from the above that the four techniques described are complementary; each examining different features of surface films. Force balance and ultra-thin film interferometry are both essentially tools which indicate the presence and physical characteristics such as the rheology of surface films. Force balance operates in a very precise fashion between idealised surfaces whilst ultra-thin film interferometry studies much more realistic, high pressure lubricated contacts, at the expense of considerably less precision.

AFM and FFM examine surface films at the molecular level and so far have been most valuable in illustrating the arrangement and mobility of boundary lubricating film molecules on surfaces. In the future, using methods such as tapping mode, they may also extend the types of rheological data being produced by force balance and interferometry. They are also, of course, able to examine the topography and friction of antiwear and other boundary films. Because of the severity of the contact conditions at the very sharp tip/flat contacts, these nanoprobes do not currently mimic the bulk conditions of a lubricated contact but they may well relate to the conditions present at asperity/asperity conjunctions.

The x-ray absorption techniques, and in particular XANES, are the first techniques able to produce reliable molecular (as opposed to elemental) information about the very thin reaction films produced in real, rubbed contacts. At present the method has been confined mainly to ZDDP antiwear additives but, hopefully its scope will be broadened to examine other types of boundary film, including oiliness films and also those formed by molybdenum additives.

None of the techniques alone provides the whole solution to investigating boundary films. Force balance and ultra-thin film interferometry cannot chemically analyse films. Also, because they both rely on smooth, coated solid surfaces, neither is suited to studying films formed in contacts where extensive solid/solid rubbing has occurred. AFM and FFM are good at identifying and characterising well-defined and already partially understood boundary films but are perhaps less effective at characterising complex films formed under real rubbed contact conditions. X-ray absorption can detect and chemically analyse reaction films, but only if appropriate reference molecules can be synthesised. But although no one technique represents the philosopher's stone of boundary lubrication, a means of both physically and chemically characterising the boundary film between a pair of real, rubbing, metal surfaces, in combination they come close to this ideal.

9 CONCLUSIONS

This paper has examined four, modern experimental techniques for studying the nature and properties of boundary lubricating films: force balance, ultrathin film interferometry, the nanoprobe methods of STM, AFM and FFM and, finally, x-ray absorption methods. All have begun, in the last two or three years, to yield important information about boundary films and, in combination, hold promise that one day we may, at last, properly understand Boundary Lubrication.

10 REFERENCES

1 **Reynolds, O.**, On the Theory of Lubrication and Its Application to Mr Beauchamp Tower's Experiments, Including an Experimental Determination of the Viscosity of Olive Oil, *Phil. Trans. Roy. Soc.* **A177**, pp. 157–234, (1986).

2 **Wells, H.M.** and **Southcombe, J.E.**, The Theory and Practice of Lubrication: The 'Germ' Process, *J. Soc. of Chem. Ind.*, **39,** pp. 51T–60T, (1920).

3 **Spikes, H.A.**, Boundary Lubrication and Boundary Films, *Proc. 19th Leeds-Lyon Symposium on Tribology,* Leeds, Sept 1992, Thin Films in Tribology. Publ. Elsevier 1993.

4 **Kingsbury, A.**, A New Oil-Testing Machine and Some of Its Results, *Trans. ASME,* **24,** pp. 143–160, (1903).

5 **Stanton, T.E.**, *Friction*, publ. Longmans, Green & Co., London, 1923.

6 **Desvaux, H.**, Le Frottement des Solides: Epaisseur Minimum d'un Enduit Lubréfiant, *J. de Phys. et la Radium,* **5,** p.845, (1924).

7 **Langmuir, I.**, The Mechanism of the Surface Phenomena of Flotation, *Trans. Farad. Soc.*, **15,** pp. 62–74, (1920).

8 **Hardy, H.B.** and **Doubleday, I.**, Boundary Lubrication – The Paraffin Series, *Proc. Roy. Soc. Lond.* **A100,** pp. 550–757, (1922).

9 **Allen, C. M.** and **Drauglis, E.**, Boundary Layer Lubrication: Monolayer or Multilayer, *Wear,* **14,** pp. 363–384, (1969).

10 **Mougey, H.C.** and **Almen, J.O.**, Extreme Pressure Lubricants, *Automotive Ind.* **65,** no 26, pp. 758–761, (1931).

11 **Miller, F.L.**, Extreme Pressure Lubricants and Lubrication, *I.Mech.E. Proceedings of the General Discussion on Lubrication and Lubricants,* vol. **2,** pp .107–117, Oct 1937.

12 **Evans, A.E.**, Extreme Pressure Lubricants, *I.Mech.E. Proceedings of the General Discussion on Lubrication and Lubricants,* vol. **2,** pp. 55–59, Oct 1937.

13 **Southcombe, J.E.**, **Wells, J.H.**, and **Waters, J.H.**, An Experimental Study of Lubrication Under Conditions of Extreme Pressure, *I.Mech.E. Proceedings of the General Discussion on Lubrication and Lubricants,* vol. **2,** pp. 400–411, Oct 1937.

14 **Davey, W.** and **Edwards, E.D.**, The Extreme Pressure Lubricating Properties of Some Sulphides and Disulphides in Mineral Oil as Assessed by the Four Ball Machine, *Wear,* **1,** pp. 291–304, (1057/8).

15 **Beeck, O.**, **Givens, J.W.** and **Williams, E.C.**, On the Mechanism of Boundary Lubrication. II. Wear Prevention by Addition Agents, *Proc. Roy. Soc. Lond,* **A177,** pp. 103–118, (1940).

16 **Barcroft, F.T.** and **Daniel, S.G.**, The Action of Neutral Organic Phosphates as E.P. Additives, *Trans. ASME,* **87,** p. 761, (1965).

17 **Lacey, I.N.**, **Macpherson, P.B.**, **Kelsall, G.**, and **Spikes, H.A.**, Thick Anti-Wear Films in Elastohydrodynamic Contacts. Part II Chemical Nature of the Deposited Film. *ASLE Trans.*, **29,** pp. 306–311, (1986).

18 **Isoyama, H.** and **Sakurai, T.**, The Lubricating Mechanism of Di-thio-dithio-bis(diethyldithiocarbamate)dimolybdenum During EP Lubrication, *Trib. Intern.*, **7,** p. 151, (1974).

19 **Yamomoto, Y.** and **Gondo, S.**, On Properties of Surface Films Formed With MoTDC Under Different Conditions, *J. JSLE,* **36,** p. 235, (1991).

20 **Frewing, J.J.**, The Influence of Temperature on Boundary Lubrication, *Proc. Roy. Soc. Lond.*, **A181,** pp. 23–42, (1942).

21 **Frewing, J.J.**, The Heat of Adsorption of Long-Chain Compounds and Their Effect on Boundary Lubrication, *Proc. Roy. Soc. Lond.*, **A181,** pp. 270–285, (1944).

22 **Rowe, C.N.**, Role of Additive Adsorption in the Mitigation of Wear, *ASLE Trans.*, **13,** pp.179–188, (1970).

23 **Spikes, H.A.** and **Cameron, A.**, A Comparison of Adsorption and Boundary Lubricant Failure, *Proc. Roy. Soc. London.*, **A336**, pp. 407–419, (1974).

24 **Hiley, R.W.**, **Spikes, H.A.** and **Cameron, A.**, Polysulphides as Extreme Pressure Lubricant Additives. *Lubrication Engineering,* **37,** pp. 732–737, (1981).

25 **Allum, K.G.** and **Forbes, E.S.**, The Load-Carrying Properties of Organic Sulphur Compounds. Influence of Chemical Structure on the Antiwear Properties of Organic Disulphides, *J. Inst. Pet,* **53,** p. 173, (1967).

26 **Forbes, E.S.** and **Silver, H.B.**, The Effect of Chemical Structure on the Load-Carrying Properties of Organo-Phosphorus Compounds, *J. Inst. Pet.*, **56,** p. 90 (1970).

27 **Allum, K.G.** and **Forbes, E.S.**, The Load-Carrying Properties Metal Dialkyl Dithiophosphates. The Effect of Chemical Structure, *Proc. I. Mech. E. Tribology Convention,* Gothenburg, 1969.

28 **Clark, G.L.**, **Sterrett, R.R.** and **Lincoln, B.H.**, X-ray Diffraction Studies of Lubricants, *Ind. Eng. Chem.*, **28,** pp. 1318–1328, (1936).

29 **Andrew, L.T.**, Electron Diffraction Analysis of the Orientation of the Molecules of Lubricating Oils, *Trans. Farad. Soc.*, **32,** pp. 607–616, (1936).

30 **Brockway, L.O.** and **Karle, J.**, Electron Diffraction Studies of Oleophobic Films on Copper, Iron and Aluminium Surfaces, *J. Coll. Sci.*, **2,** pp. 277–287, (1947).

31 **Francis, S.A.** and **Ellison, A.H.**, Reflection Infrared Studies of Zinc Dialkyldithiophosphate Films Adsorbed on Metal Surfaces, *J. Chem. Eng. Data,* **16,** p 83, (1961).

32 **Rounds, F.G.**, Effect of Additives on the Friction of Steel on Steel. (I) Surface Topographic and Film Composition Studies, *ASLE Trans.*, **7,** pp. 11–23, (1964).

33 **Ferrante, J.**, Practical Applications of Surface Analytical Tools in Tribology, *Lub. Eng.*, **38,** pp .223–231, (1981).

34 **Israelachvili, J.N.** and **Tabor, D.**, The Measurement of van der Waal Dispersion Forces in the Range 1.5 to 130 nm, *Proc. Roy. Soc. Lond.*, **A331,** pp. 19–38, (1972).

35 **Georges, J-M.**, **Tonck, A.** and **Mazuyer, D.**, Interfacial Friction of Wetted Monolayers, *Wear,* **175,** pp. 59–62, (1994).

36 **Israelachvili, J.N.**, **Fisher, L.R.**, **Horn, R.G.** and **Christenson, H.K.**, Measurement of Adhesion and Short-Range Forces Between Molecularly Smooth Surfaces in

Undersaturated Vapours and in Organic Liquids, Microscopic Aspects of Adhesion and Lubrication, *Tribol Ser. 7*, J.M. Georges, Ed. Elsevier, Amsterdam, pp 55–69, (1982).

37 **Gee, M.L., McGuiggan, P.M., Israelachvili, J.N.** and **Homola, A.M.**, Liquid to Solidlike Transitions of Molecularly Thin Films in Shear, *J. Chem. Phys.*, **93**, pp. 1895–1906, (1990).

38 **Israelachvili, J.N., McGuiggan P.M.** and **Homola, A.M.**, Dynamic Properties of Molecularly Thin Liquid Films, *Science*, **240**, pp. 189–191, (1988).

39 **Granick, S.**, Motions and Relaxations of Confined Liquids, *Science*, **253**, pp. 1374–1379, (1991).

40 **Tonck, A., Georges, J.M.** and **Loubet, J.L.**, Measurements of Intermolecular Forces and the Rheology of Dodecane between Alumina Surfaces, *J. Coll. and Interface Sci.*, **126**, pp. 150–163, (1988).

41 **Georges, J.M., Millot, S., Loubet, J.L.** and **Tonck, A.**, Drainage of Thin Liquid Films Between Relatively Smooth Surfaces, *J. Chem. Phys.*, **98**, pp. 7345–7360, (1993).

42 **Georges, E., Georges, J-M.** and **Diraison, C.**, Rheology of Olefinic Copolymer Layers Adsorbed on Solid Surfaces, *Trib. Trans.*, **39**, pp. 563–570, (1996).

43 **Georges, E., Georges, J-M.** and **Diraison, C.**, Steric Repulsion of Polymethacrylate Layers, SAE 961218, (1996).

44 **Johnston, G.J., Wayte, R.** and **Spikes, H.A.**, The Measurement and Study of Very Thin Lubricant Films in Concentrated Contacts, *Trib Trans*, **34**, pp. 187–194 (1991).

45 **Guangteng, G.** and **Spikes, H.A.**, Boundary Film Formation by Lubricant Base Fluids, *Trib. Trans.*, **39**, pp. 448–454, (1996).

46 **Smeeth, M., Gunsel, S.** and **Spikes, H.A.**, The Formation of Viscous Surface Films by Polymer Solutions: Boundary or Elastohydrodynamic Lubrication. *Trib. Trans.*, **39**, pp. 720–725, (1996).

47 **Smeeth, M., Gunsel, S.** and **Spikes, H.A.**, Boundary Film Formation by Viscosity Index Improvers. *Trib. Trans.*, **39**, pp. 726–734, (1996).

48 **Smeeth, M., Gunsel, S.** and **Spikes, H.A.**, Friction and Wear Reduction by Boundary Film-Forming Viscosity Index Improvers, SAE 962037, 1996.

49 **Tripaldi, G., Vettor, A.** and **Spikes, H.A.**, Friction Behaviour of ZDDP Films in the Mixed Boundary/EHD Regime, SAE 962036, 1996.

50 **Anghel, V., Cann, P.M.** and **Spikes, H.A.**, Direct Measurement of Boundary Lubricating Films. Presented at *Leeds-Lyon Symposium*, Leeds, Sept. 1996. Accepted for publication in the Proceedings, Elsevier 1997.

51 **Guangteng, G.** and **Spikes, H.A.**, Fractionation of Lubricants at Solid Surfaces, *Wear*, **200**, pp. 336–345, (1996).

52 **Guangteng, G.** and **Spikes, H.A.**, The Control of Friction by Molecular Fractionation of Base Fluid Mixtures at Metal Surfaces, to be presented at STLE Annual Meeting, Kansas City, May 1997, accepted for publication in *Trib. Trans.*

53 **Sonnenfeld, R., Schneir, J.** and **Hansma, P.K.**, Scanning Tunneling Microscopy: A Natural for Electrochemistry, *Mod. Aspects Electrochem.*, **21**, pp 1–28, (1990).

54 **Smith, D.P.E., Hörber, H., Gerber, Ch.** and **Binnig, G.**, Smectic Liquid Crystal Monolayers on Graphite Observed by Scanning Tunneling Microscopy, *Science*, **245**, pp 43–45, (1989).

55 **Andoh, Y., Oguchi, S., Kaneko, R.** and **Miyamoto, J.**, Evaluation of Very Thin Lubricant Films. *J. Phys.*, **D 25**, pp. A71–A75, (1992).

56 **Ruan, J-A.** and **Bhushan, B.**, Atomic-Scale Friction Measurements Using Friction Force Microscopy: Part I – General Principles and New Measurement Techniques, *Trans. ASME J. of Trib.*, **116**, pp. 378–388, (1994).

57 **Schwartz, D.K., Viswanathan, R.** and **Zasadzinski, J.A.N.**, Coexisting Lattice Structures in a Langmuir-Blodgett Film Identified by Atomic Force Microscopy, *Langmuir*, **9**, pp. 1384–1391, (1993).

58 **Mori, O.** and **Imae, T.**, Atomic Force Microscope Observation of Monolayers of Arachnidic Acid, Octyldimethylamine Oxide and Their Mixtures, *Langmuir*, **11**, pp. 4779–4784, (1995).

59 **Bhushan, B., Kulkarni, A.V., Koinkar, V.N., Boehm, M., Odini, L., Martelet, C.** and **Belin M.**, Microtribological Characterization of Self-Assembled and Langmuir-Blodgett Monolayers by Atomic and Friction Force Microscopy, *Langmuir*, **11**, pp. 3189–3198, (1995).

60 **Bowden, F.P.** and **Leben, L.**, The Friction of Lubricated Metals, *Phil. Trans. Roy Soc.*, **A239**, pp. 1–27, (1940),

61 **Kaneko, R.**, Microtribological Applications of Probe Microscopy, *Trib. Intern.*, **28**, pp. 195–202, (1995).

62 **Bhushan, B.** and **Blackman, G.S.**, Atomic Force Microscopy of Magnetic Rigid Disks and Sliders and Its Applications to Tribology, *Trans. ASME*, **113**, pp. 452–457, (1991).

63 **Pan, J., Tao, N.** and **Lindsay, S.M.**, An Atomic Force Microscopy Study of a Self-Assembled Octadecyl Mercaptan Monolayer Adsorbed on Gold (III) Under Potential Control, *Langmuir*, **9**, pp. 1556–1560, (1993).

64 **Tamayo, J.** and **Garcia, R.**, Deformation, Contact Time and Phase Contrast in Tapping Mode Scanning Force Microscopy, *Langmuir*, **12**, pp. 4430–4435, (1996).

65 *X-Ray Absorption, Principles, Applications, Techniques of EXAFS, SEXAFS and XANES*, ed. D C Koningsberger and R Prins, *Chemical Analysis Series*, Vol. 92, publ. J Wiley and Sons, New York, 1988.

66 **Martin, J.M., Belin, M., Mansot, J.L., Dexpert, H.** and **Lagarde, P.**, Friction-Induced Amorphization with ZDDP – an EXAFS Study, *ASLE Trans.*, **29**, pp. 523–531, (1986).

67 **Belin, M., Martin, J.M.** and **Mansot, J.L.**, Role of Iron in the Amorphization Process in Friction-Induced Phosphate Glasses, *Trib. Trans.*, **32**, pp. 410–413, (1989).

68 **Kasrai, M., Fuller, M., Scaini, M., Yin, Z., Brunner, R.W., Bancroft, G.M., Fleet, M.E., Fyfe, K.** and **Tan, K.H.**, Study of Tribochemical Film Formation Using X-ray Absorption and Photoelectron Spectroscopies, Presented at the *21st Leeds-Lyon Symposium on Tribology*, Leeds, Sept 1994, Lubricants and Lubrication. Publ. Elsevier 1995.

69 **Yin Z., Kasrai, M., Fuller, M., Bancroft, G.M., Fyfe, K.** and **Tan K.H.**, Application of Soft X-ray Absorption Spectroscopy in Chemical Characterization of Antiwear Films Generated by ZDDP. Part I: the Effect of Physical Properties, *Wear*, **202**, pp. 172–191, (1997).

70 **Yin Z., Kasrai, M., Bancroft, G.M., Fyfe, K. Colaianni, M.L.** and **Tan K.H.**, Application of Soft X-ray Absorption Spectroscopy in Chemical Characterization of Antiwear Films Generated by ZDDP. Part II: the Effect of Detergents and Dispersants, *Wear*, **202**, pp. 192–201, (1997).

71 **Fuller, M., Yin, Z., Kasrai, M., Bancroft, G.M., Yamaguchi, E., Ryason, P.R., Willermet, P.** and **Tan, K.H.**, Chemical Characterization of Tribochemical and

Thermal Films from Neutral and Basic ZDDPs Using X-ray Absorption Spectroscopy, *Trib. Intern.*, **30**, pp. 305–315, (1997).

72 **Hastie, G.P.** and **Roberts, K.J.**, Structural Characterization of Long-Chain Hydrocarbon Thin Films Using Ultra-Soft Polarised Near Edge X-ray Absorption Spectroscopy, *Langmuir,* **11,** pp. 4170–4172, (1995).

73 **Idziak, S.H.J., Koltover, I., Davidson, P., Ruths, M., Li, Y., Israelachvili, J.N.** and **Safinya, C.R.**, Structure Under Confinement in a Smectic-A and Lyotropic Surfactant Hexagonal Phase, *Physica,* **B 221**, pp. 289–295, (1996).

74 **Cann, P.M.** and **Spikes, H.A.**, In Lubro Studies of Lubricants in EHD Contacts Using FTIR Absorption Spectroscopy, *Trib. Trans.*, **34,** pp. 248–256, (1991).

75 **Cann P.M.** and **Spikes H.A.**, Fourier-Transform Infrared Study of the Behavior of Greases in Lubricated Contacts, *Lubrication Engineering,* **48,** pp. 335–343, (1992).

76 **Westerfield, C.** and **Agnew, S.**, IR Study of the Chemistry of Boundary Lubrication with High Temperature and High Pressure Shear, *Wear,* **181–183,** pp. 805–809, (1995).

77 **Çavdar, B.** and **Ludema, K.C.**, Dynamics of Dual Film Formation in Boundary Lubrication of Steels. Part I. Functional Nature and Mechanical Properties, *Wear,* **148,** pp 305–327, (1991).

78 **Çavdar, B.** and **Ludema, K.C.**, Dynamics of Dual Film Formation in Boundary Lubrication of Steels. Part III. Real Time Monitoring and Ellipsometry, *Wear,* **148,** pp 305–327, (1991).

79 **Bhushan, B., Israelachvili, J.N.** and **Landman, U.**, Nanotribology: Friction, Wear and Lubrication at the Atomic Scale, *Nature,* **374,** pp. 607–616, (1995).

Numerical characterization of wear particle morphology and angularity of particles and surfaces

G W STACHOWIAK
Tribology Laboratory, Department of Mechanical and Materials Engineering, University of Western Australia, Perth, Australia

SYNOPSIS

The morphology of wear particles, in terms of their shape, size and surface texture, reflects the complex nature of wear processes involved in particle formation. Currently, wear particle morphology is usually assessed qualitatively based on microscopy observations. This procedure, since it relies upon the 'expert's' experience, is not always objective and it can also be expensive. Hence, recent research efforts concentrate on the development of numerical descriptors and the application of computer technology and image analysis techniques to the assessment of particle morphology. Wear particle boundaries often exhibit rugged characteristics and fractal methods appear to be suitable for their description.

Another problem is associated with the characterization of particle surface texture. Also in this case, fractal methods seem to give promising results, especially in characterizing the surface anisotropy and directionality.

An inherent feature of many tribological systems are abrasive particles. It is intuitively felt that the shape of abrasive particles is related to their abrasivity and subsequently influences the wear rates. Research efforts have been directed to the development of new parameters relating the particle shape to its ability to abrade. It is also intuitively recognized that the shape of surface profiles, i.e. surface texture, affects the abrasive wear rates. Recently, new parameters relating the angularity of both the particles and surface profiles (i.e. surface texture) to their abrasivity have been developed.

In this paper an overview of recent advances and developments in the numerical characterization of wear debris, surface profiles and abrasive particles is presented with a particular emphasis on fractal methods. The advantages and limitations of fractal techniques used in the characterization of both boundaries and surfaces of wear particles found in tribological systems are examined and discussed. The relationship between the angularity of abrasive particles, surface textures and abrasive wear is also demonstrated. Practical implications of these developments are briefly discussed.

1 INTRODUCTION

Wear particles found in tribological systems (both mechanical and biological) can be captured and their morphology examined (e.g. **1–9**). A significant amount of information about the wear processes occurring in these systems, i.e. machinery or synovial joints, can be obtained by microscopic examination of the surface and the boundary textures and structures of the particles (e.g. **5, 7–10**). This information can be used in the assessment of the current 'health' status of the tribological system and in predicting wear anomalies or state of wear. A commonly employed technique in wear particle analysis is the visual inspection of a particle microscopy image, a process which is time-consuming, expensive and, since it requires an 'expert', is not always reliable due to its subjectivity.

Recent thrust in particle analysis research is directed towards the quantitative description of particle morphology by numerical parameters. Significant advances have recently been made in the development of numerical descriptors of particle boundaries (e.g. **5, 8, 9, 11–15**). However, numerical characterization of particle surface texture, because of its complexity, still remains largely unresolved and is the subject of intense research.

The outline shape of particles is traditionally described by a set of conventional parameters, such as aspect ratio, shape factor, convexity, elongation, curl, roundness, etc. (e.g. **9, 16, 17**). The calculation methods and the influence of measurement conditions on these parameters such as the effects of accelerating voltage, filtering, noise, etc., are described in detail in (**18, 19**). Recently, fractal geometry has been introduced to describe the rugged characteristics of particle boundaries and, as a result, fractal particle boundary descriptors have been developed (**5, 7, 13, 14, 16, 20–24**). Various strategies devised for calculating fractal dimensions are reviewed in (**14, 20, 79**). It has been shown that fractals are useful in describing powders (**15, 25–27**), pigments (**25**) and wear particles (e.g. **5, 7, 22**). Fractal techniques can also be employed to characterize surface irregularities (**28–30**) and also surface anisotropy and directionality (**31**).

Characterization of particle abrasivity based on the particle shape has also been conducted (**32**). Progress in this area has been greatly accelerated with the application of fast computers resulting in new numerical parameters describing the angularity of particles (**33–35**) and surface profiles (**36**). It has been demonstrated that the angularity of the abrasive particles and surface profiles is strongly related to abrasive wear rates (**36**).

In this paper recent advances and techniques used in numerical characterization of particle boundaries and surface profiles are described. The relationship between the angularity of particles and profiles and abrasive wear rates is examined and discussed. The limitations of these techniques are also described.

2 CHARACTERIZATION OF PARTICLE BOUNDARY BY FRACTAL METHODS

A technique which is often employed to determine the complexity of a particle boundary is known as a 'structured-walk' technique and, also, as a Richardson method (e.g. **5, 12, 14, 21, 37**). In this method the perimeter or the boundary of a particle is 'walked' around at different step sizes. After completion of each round the perimeter length is estimated by summing the Euclidean length of the steps around the particle boundary. The average step length is found by dividing the perimeter estimate by the number of steps. The difference between the step size and step length is illustrated schematically in Figure 1 (**29**).

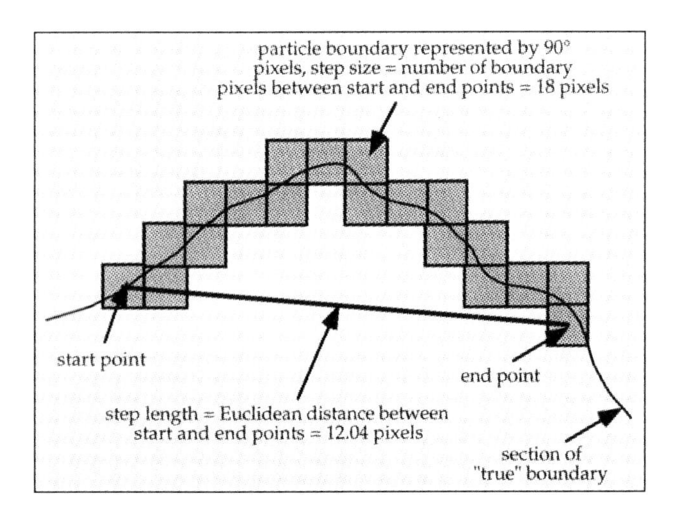

Fig. 1 Schematic illustration of a difference between step size and step length shown on a section of digitized particle boundary (adapted from (**29**)).

It can be seen from Figure 1 that as the step length decreases the perimeter estimate increases as finer details are measured. The perimeter estimates obtained at different step lengths are then plotted on a log-log scale resulting in a so-called 'Richardson plot'. From the slope of the line of best fit to the plot a fractal dimension 'D' is calculated as $D = 1 - m$, where 'm' is the slope of the line. For boundaries 'D' varies between 1 and 2, i.e. a more complex boundary exhibits a higher fractal dimension as illustrated in Figure 2.

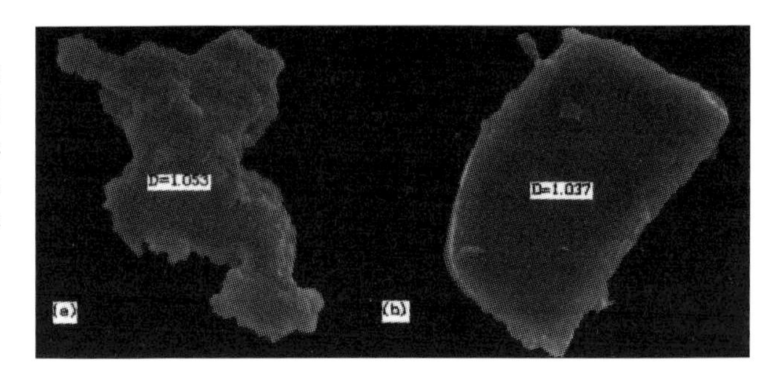

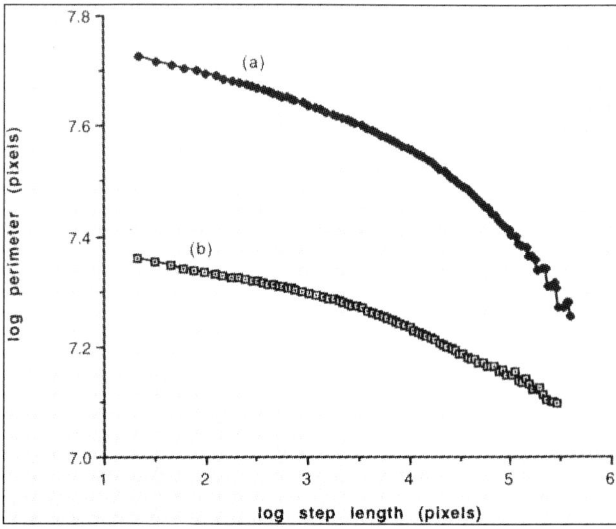

Fig. 2 SEM micrographs and Richardson plots of wear particles obtained from total hip replacement prosthesis operating in human body.

It can be assessed visually that the boundary of the particle 'a' is more complex than the boundary of the particle 'b'. This is confirmed by the slopes of the corresponding Richardson plots, i.e. the slope of curve 'a' is steeper than the slope of curve 'b', and fractal dimensions calculated for these particles.

It is often found that the Richardson plot exhibits bi-modal characteristics, as shown for example in Figure 2 (plot 'a'). The upper part of the plot at small step lengths has usually been associated with 'textural' characteristics of the boundary while the lower part of the plot at large step lengths has been associated with boundary 'structural' characteristics (e.g. **5, 14, 37**). It was found that the large step size region usually exhibits a higher fractal dimension compared to the small step size region. This is because the perimeter is increasingly underestimated due to longer line segments spanning the object. Although in some cases, e.g. soot agglomerates, particle boundaries can exhibit two distinct physical scales (**14**) and the bi-modal behaviour of Richardson plots can indeed be explained in terms of complexity of the particle boundary, in most of the cases this bi-modal behaviour can be attributed to an artefact produced by the limited data size (i.e. the object behaves as

Euclidean objects exhibiting finite data errors at large step sizes) (**13, 37**). Therefore, the interpretation of the bi-modal behaviour of a Richardson plot must be treated with caution.

2.1 Optimization of calculation techniques

One of the problems associated with the method of calculating a fractal dimension from a particle boundary is that it can give a range of fractal dimensions for a single boundary (e.g. **12, 14, 37, 38**). This is because there are many possible starting points for each 'walk' around the boundary which in turn result in perimeters of different length (**12, 37**). This problem has been rectified by developing a special computer program which calculates the perimeter estimates for all possible starting points and averages the results obtained. The process is repeated at different step sizes. The average values of perimeter estimates are then plotted against the step length on a log-log scale and from the slope of the plot the fractal dimension is calculated. This modified technique of calculating the fractal dimension from a particle boundary has a distinctive advantage of being insensitive to translation and/or rotation of the boundary studied (**12, 37**). Richardson plots calculated for one and all starting points for the boundary of the particle in Figure 2a are shown in Figure 3.

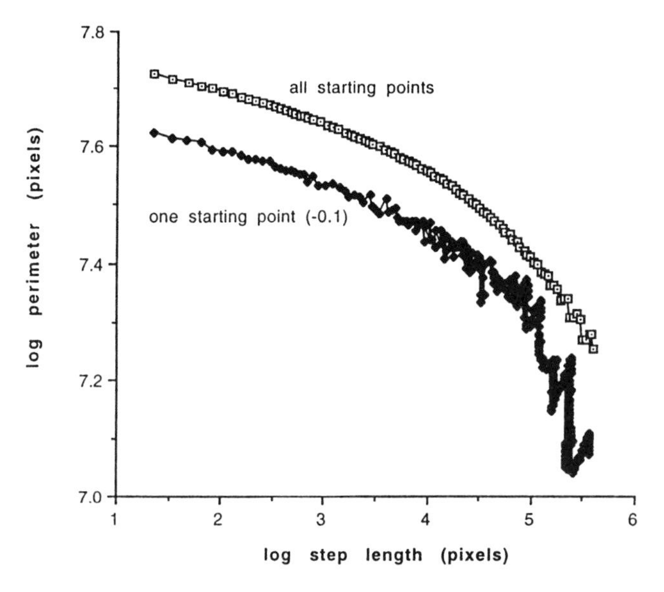

Fig. 3 Richardson plots calculated for one and all starting points for the boundary of the particle shown in Fig. 2a (plots are offset by -0.1 on a vertical axis for clarity).

It has been found that the least square fit to the data points on the Richardson plot is heavily biased towards the large step sizes if arithmetic step size increments are used, i.e. 4, 6, 8, etc., pixels (**39**). This problem has been rectified by implementing an exponential step size increment (**12, 37**). Since the choice of the initial step size is a compromise between digitization errors and loss of fine scale information, the initial step size of 5 pixels is usually selected for the analysis (**12, 37**).

Another problem is associated with curve fitting to the plot obtained. Manual curve fitting can be very subjective, leading to errors in the estimates of boundary fractal dimensions as illustrated schematically in Figure 4 where the Richardson plot for a circle is shown.

The theoretical fractal dimension for a circle is $D = 1.000$ which corresponds to a slope equal to zero. It can clearly be seen from Figure 4 that the manual curve fitting to all the data points would produce a range of slopes. Therefore the minimum curve fit length was arbitrarily chosen as half of the square root of the number of pixels in the boundary. Particles with less than 500 pixels in their boundaries are usually ignored as having too low resolution (**37**).

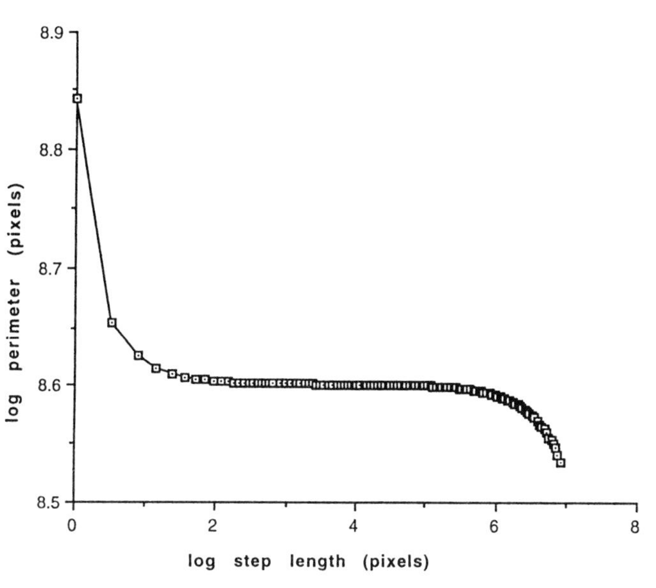

Fig. 4 Richardson plot for a circle (initial step size = 1).

2.2 Optimization of imaging techniques

Boundary fractal dimension strongly depends on the imaging technique used and image quality. When using SEM, particles can be viewed in secondary electron (SE) and back-scattered electron (BSE) modes. The effects of both modes on boundary fractal dimension were investigated (**12, 37**). It was found that the best results were achieved when the boundaries of the projected images of particles were imaged in the SE mode while the boundaries of the sectioned particles were imaged in the BSE mode (**24**). Although sectioned particles viewed in both SE and BSE modes gave a clear distinction between the particle and the background it was noticed that the particle boundary on the SE images from the sectioned particles can be affected by edge highlighting due to surface scratches. This is undesirable and results in subsequent errors in the calculated boundary fractal dimension (**37**).

Image quality is strongly affected by the level of focusing, digital smoothing, accelerating voltage and amount of image noise. De-focusing of the image has a similar effect on boundary fractal dimension to that of digital smoothing, i.e. it yields lower fractal dimension. These effects are shown in Figure 5 where the image analyzed was deliberately defocused and smoothed.

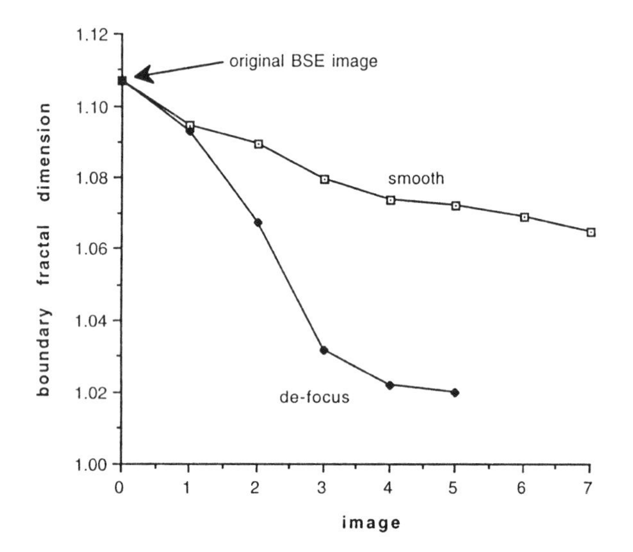

Fig. 5 Effects of image smoothing and de-focusing on the boundary fractal dimension of a sectioned particle (**12, 37**).

Another important imaging factor is the accelerating voltage. Boundary fractal dimension of a particle remains relatively constant over the range of the accelerating voltages between 15 kV and 30 kV. At lower voltages, signal to noise ratio is high and corrupted images are often obtained. It has been decided that 15 kV is the optimal accelerating voltage for viewing non-organic particles since at higher voltages the beam penetration could pose some problems, especially during the analysis of smaller particles. For synovial joint wear particles 15 kV accelerating voltage is still too high and field emission SEM can be used instead. This technique gives very sharp images even at a low accelerating voltage of 1 kV.

Boundary fractal dimension is sensitive to image noise as illustrated in Figure 6. A typical fractal shape such as the Koch curve exhibits a boundary fractal dimension of $D = \log 4/\log 3 = 1.262$. The fractal dimension of this curve, evaluated by the program developed, is slightly higher, i.e. $D = 1.2665$, due to digitization errors. The introduction of simulated noise in the background adds extra details to the particle boundary, as shown in Figure 6b, causing an increase of the boundary fractal dimension to 1.2755. The increase in boundary fractal dimension with noise is greater for shapes with low fractal dimension, e.g. Euclidean shapes (**24**).

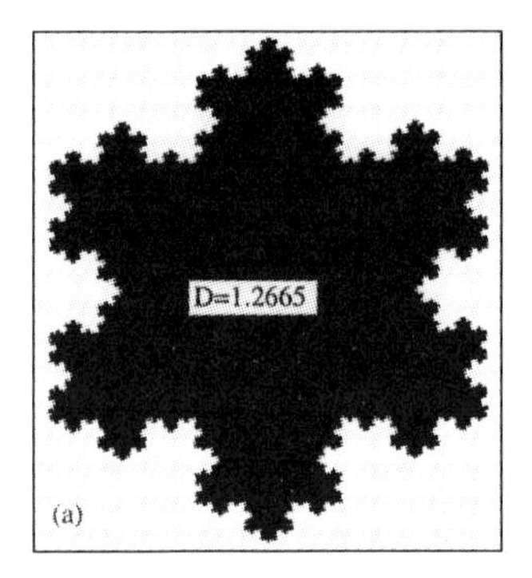

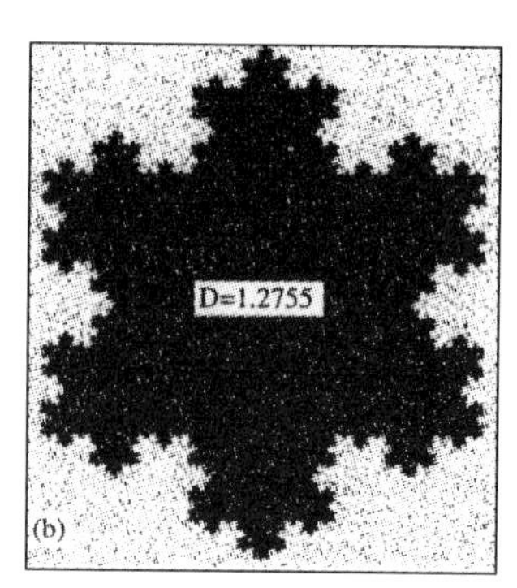

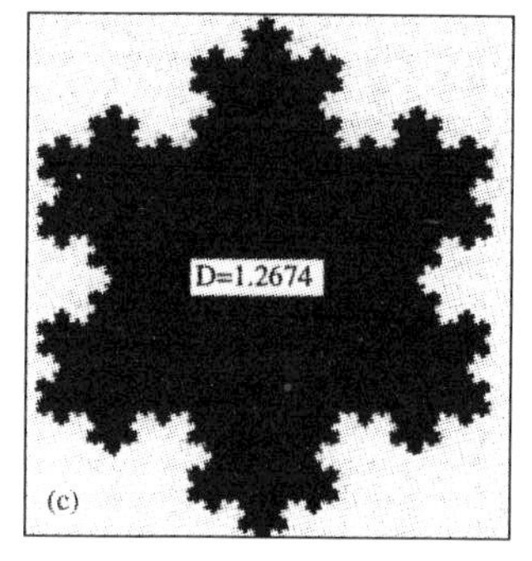

Fig. 6 Effect of image noise on boundary fractal dimension; (a) Koch island, (b) Koch island with the addition of shot and Gaussian noise, (c) image after the application of median-sigma filter.

Therefore the elimination of image noise is critical and digital filters are applied for this purpose. None of the existing filters yet works perfectly and research efforts are directed towards developing new image filters. Recently, a median-sigma filter based on the combination of two filters, median and sigma, using the modulation function has been developed (**40**). The filter is effective in reducing the image noise and is virtually free from shape distortion (**40**). The effectiveness of this filter is demonstrated in Figure 6c where this filter has been applied to the image shown in Figure 6b.

2.3 Optimization of viewing techniques

Analysis of particle shape involves extracting the particle boundary from its image. In most cases projected images of the particles are used for this purpose because of the relative ease of sample preparation for SEM examination. Apart from the projected images, images of the sectioned particles can also be used in the analysis. The question is whether the boundary from the projected image gives the same information as the boundary from the sectioned image and which of these viewing methods should be used in analysis? 'Projection' and 'section' images from a number of particles were examined and changes in boundary fractal dimension investigated. Tests conducted revealed that in most cases the boundary fractal dimensions from the projected images of particles were significantly lower than from the sectioned images as illustrated, for example, in Figure 7 (**13**). This can be explained by the fact that many surface features such as valleys are hidden by surface protrusions leading to a less irregular appearance of the projected boundary, and only become visible when the particle is sectioned as shown in Figure 8 (**13**).

Analysis showed that particles with more complex boundaries exhibit greater differences between the boundary fractal dimensions calculated from projected and sectioned images as illustrated in Figure 9 (**23**).

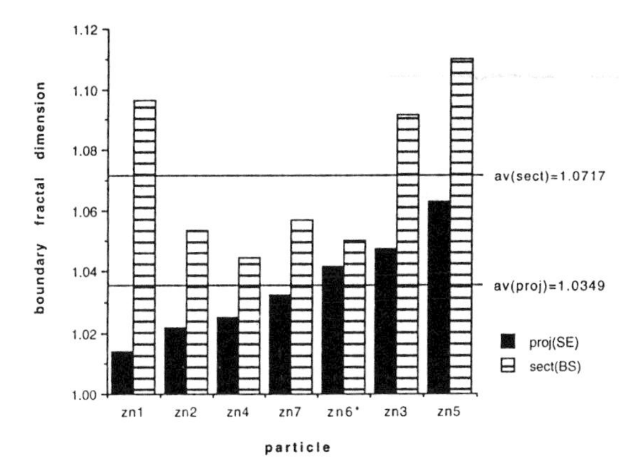

Fig. 7 Changes in boundary fractal dimension of projected and sectioned sprayed zinc particles. (* indicates multiple boundaries which develop due to sectioning, slopes of Richardson plots averaged) (**13**).

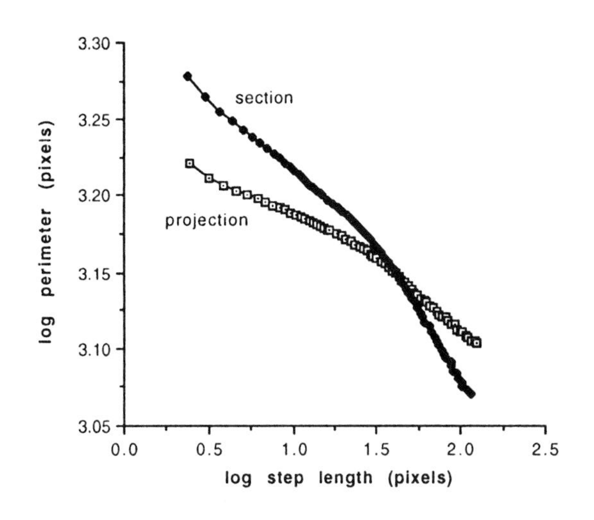

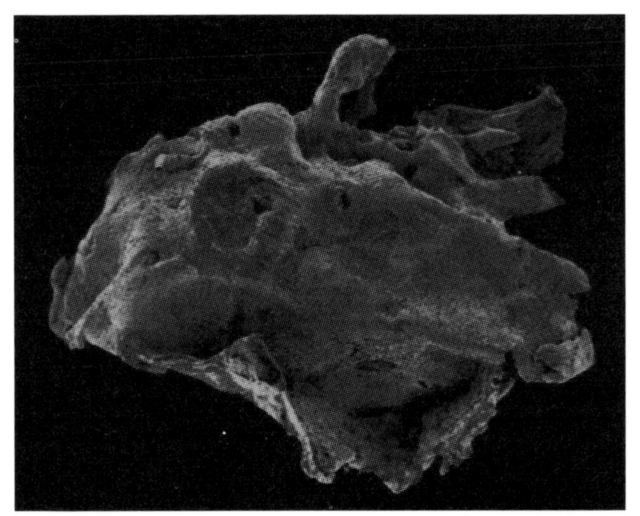

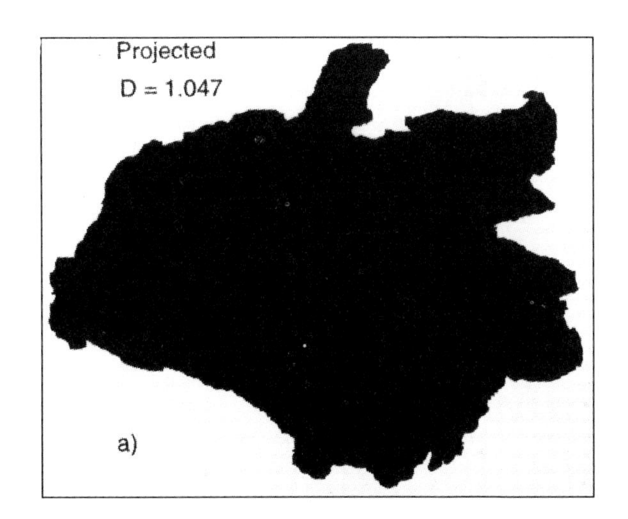

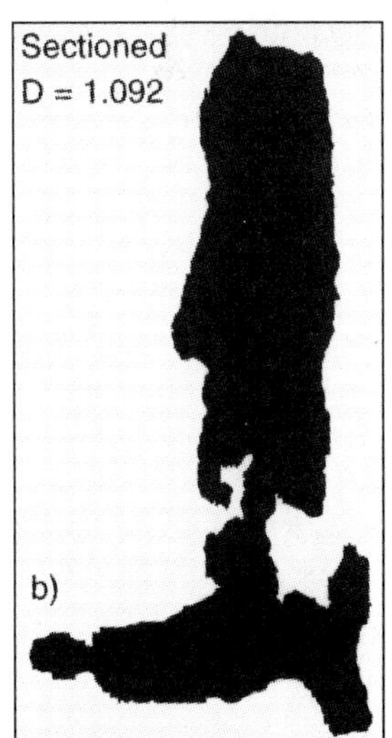

Fig. 8 Typical sprayed zinc particle with its projected (a) and sectioned images (b) and corresponding Richardson plots (c) (**13**).

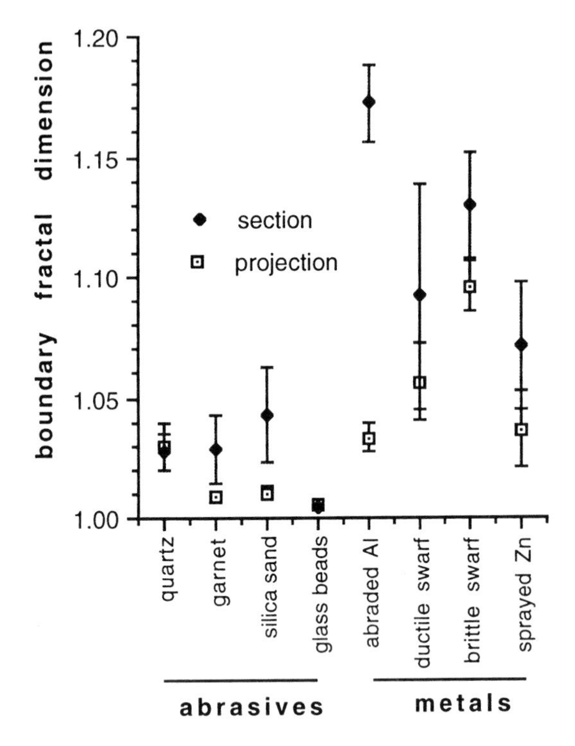

Fig. 9 Boundary fractal dimensions calculated for projected and sectioned images of typical metallic and mineral particles (error bars ±1 SD) (**23**).

The experiments demonstrated that the answer to the question, which of these viewing techniques should be used in analysis, is not simple and that there is no simple quantitative method for comparing fractal dimension from projected and sectioned boundaries (**23**). It seems that the fractal dimension from a sectioned particle is more related to surface fractal dimension. In fact, for isotropic self-similar particles surface fractal dimension can be obtained by simply adding 1 to the calculated boundary fractal dimension obtained from a sectioned particle (**21**). On the other hand in the case of abrasive particles the analysis of a projected image seems to give data more accurately related to particle abrasivity (**33**).

Boundaries of particles present in tribological systems, both mechanical and biological, have been characterized by boundary fractal dimension (e.g. **5, 7, 22, 41**). It has been shown that the boundary fractal dimension is quite effective in distinguishing among different particles and hence in detecting changes occurring in tribological systems. For example, our own work demonstrated that boundary fractal dimension is related to subtle changes occurring in arthritic synovial joints (**41**).

3 CALCULATION OF FRACTAL DIMENSION FROM SURFACE PROFILES – TECHNIQUES

Numerical characterization of 3-D topography of engineering and biological surfaces and also surfaces of wear particles still remains an unresolved and challenging problem and is a subject of intense research (e.g. **5, 42–44**). Surfaces can be represented by a series of profiles obtained, for example, by profilometry techniques, and then analyzed. Traditionally, statistical functions and parameters, such as autocorrelation function, correlation length, autocovariance function, structure function, etc., have been used to characterize the surface topography (**45–48**). Recently, however, fractal methods have also been applied to describe the complexity of surface topography (e.g. **29, 31, 49–53**).

There is a fundamental difference between a particle boundary and surface profile. The particle boundary is self-similar while the profile is self-affine. Self-affine profiles scale by different amounts in vertical (z) and horizontal (x) directions. Also, the units of the measurements may not be the same, i.e. the profile can be treated as a time series, where the elevation varies with time. It has been first indicated by Mandelbrot (**54**) that 'the compass method' used in calculation of fractal dimensions from particle boundaries cannot be used in calculation of fractal dimension from profiles.

Hence, the Richardson method used to calculate the fractal dimension from particle boundaries has been adapted (1-D Richardson) to calculate the fractal dimension from profiles. Profile data in the vertical direction is first normalized by its range, rendering all values between 0 and 1. Next, the absolute vertical difference between every point along the horizontal axis (i.e. step size = 1) is summed giving the maximum vertical variation (**29**). Naturally, when every point is used this gives the highest resolution available. The procedure is then repeated for increased step sizes, i.e. every second, third, etc., point, until some limit value is reached as illustrated schematically in Figure 10b (**29**). In our work one tenth of the number of data points was used as the maximum step size (limit value) (**29**).

It should be noted that with the increased step size value the number of possible starting points also increases. For example, when every second point is used then there are two possible starting points, when every third point is used then there are three possible starting points, etc., as is schematically illustrated in Figure 10b (**29**). Similarly to the case of calculating the fractal dimension from a particle boundary, all starting points for a particular step size are used to evaluate the vertical variation of the profile and then an average vertical sum is calculated. The results obtained are then plotted versus step size on a log-log scale. From the slope 'm' of the curve the fractal dimension of a profile 'D' is calculated, i.e. $D = 1\text{-}m$. An example of such plots obtained from electro-myography (EMG) data of a hamstring muscle, when one and all starting points were used, is shown in Figure 11.

It should be mentioned that when dealing with real profiles the number of steps is not always an integer division into the number of data point (apart from the step size = 1). In such cases the vertical variation at the incomplete steps at the start and finish of the profile is first multiplied by the incomplete step sizes (i.e. fraction of a step size), summed, then divided by the step size and added to the sum of vertical variations. The step size is also incremented exponentially. This process is schematically illustrated in Figure 10b (**29**).

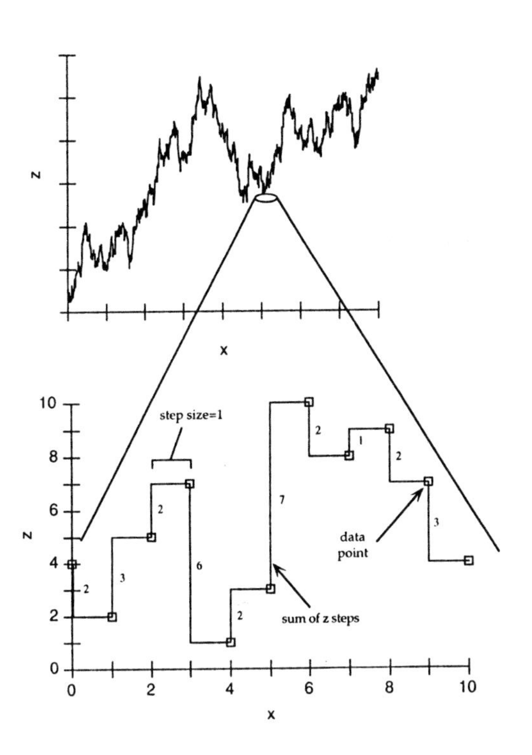

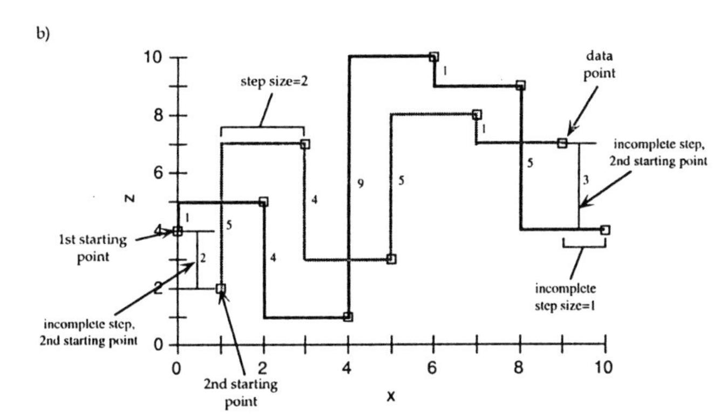

Fig. 10 Schematic illustration of evaluating the vertical variation of a profile;
(a) a profile at step size = 1 and vertical variation of $\Sigma z_{ss=1} = 30$,
(b) a profile at step size = 2, first starting point: $\Sigma z = 20$, second starting point: $\Sigma z = 15 + (2 \times 1 + 3 \times 1)/2 = 17.5$ (where 1 is the incomplete step size).
Average vertical variation:
$\Sigma z_{ss=2} = (20 + 17.5)/2 = 18.75$ as there are two possible starting points (adapted from (**29**)).

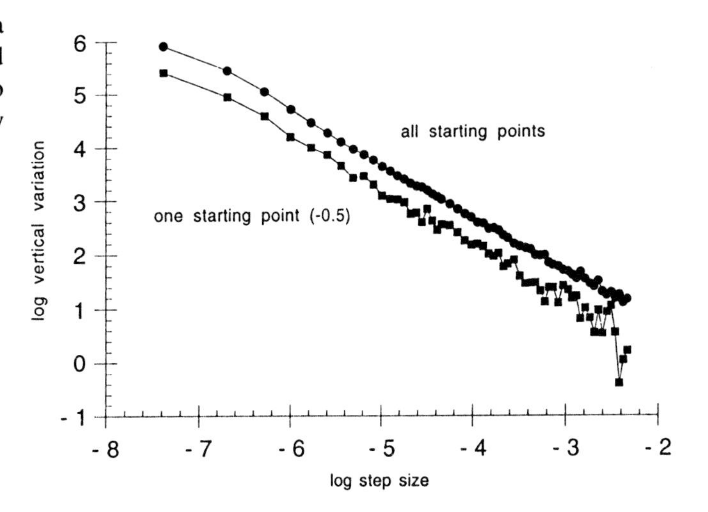

Fig. 11 Richardson plots obtained for one starting point and all starting points from EMG hamstring muscle data (EMG data courtesy of Dr M. Kuster, Kantonsspital, Baden, Switzerland).

There are also other techniques that can be used to evaluate the fractal dimension from a profile, e.g. horizontal structuring element method (HSEM) (**55**), correlation integral (**56, 57**) and fast Fourier transform (FFT) (e.g. **31**). These techniques together with the modified 1D Richardson method were tested on artificially generated profiles (**28**). The results obtained demonstrated that the problem of choosing any particular algorithm for the calculation of fractal dimension from a profile is not a simple one (**28**). The algorithm which may give the best results when applied to artificially generated profiles may not, necessarily, be the optimum algorithm for the 'real' data. Since there is no way of knowing the 'true' or even 'nominal' dimension of the surface profile under consideration, the use of any technique must be treated with caution.

3.1 Effect of noise

Similarly to the case with particle boundaries most of the techniques used to calculate the fractal dimension from profiles are very sensitive to noise (**28, 29, 31**). It has been found that with more than 10% noise added to the profile most of these techniques do not work, i.e. they provide erroneous results (**29**). Therefore it is important to understand and eliminate any noise prior to the analysis, i.e. amplifier noise, vibrations, etc.

It was found that in some rare cases manually curve-fitted log-log plots of FFT spectra may allow for separation of the signal and noise regions of the profile (**29**). For example, if the noise exhibits different fractal nature, i.e. for white noise fractal dimension $D \approx 2$, from the underlying signal, then the plot will exhibit a bi-modal nature, i.e. different slopes. In such cases the visual examination of the Richardson plot may allow to differentiate between the noise and the underlying signal. Because of the inherent inaccuracies associated with manual curve fitting to log-log plots of FFT spectra, this type of analysis is quite limited (**28, 29**).

3.2 Effect of underlying surface features

Surface profiles, except those obtained from flat surfaces, are often superimposed on large scale features, e.g. waviness, which significantly affect the fractal dimension calculated. For example, Richardson plots calculated from a profile containing 20,000 data points and the same profile superimposed on a 90° arc are shown in Figure 12 (**29**).

It can be seen from Figure 12 that the effect of underlying large-scale surface features is strong. This effect has also been reported in the literature (**58**). Thus the fractal analysis of surface profiles must be conducted carefully in order to obtain meaningful results.

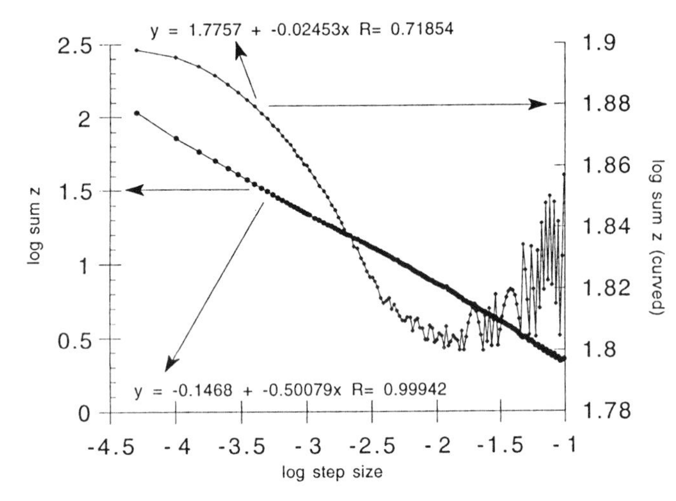

Fig. 12 Richardson plots obtained from a 20,000 data point profile (straight line) and the same profile superimposed on a 90° arc (adapted from (**29**)).

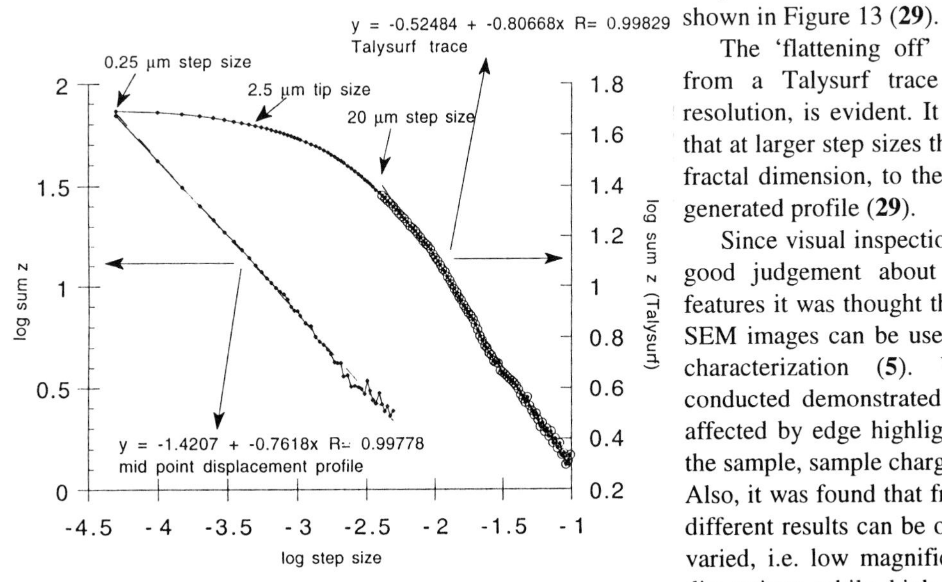

Fig. 13 Richardson plots acquired from a Talysurf trace (20,000 data points) obtained from a sand blasted stainless steel surface and from an artificial profile generated by mid-point displacement method (29).

3.3 Effects of profile acquisition techniques on fractal dimensions

Contact profile meters, e.g. Talysurf, provide the most exact surface profile data because of their direct contact with the surface. However, one of the main limitations of contact profile meters is that the finite size of the tip filters the profile data and subsequently a significant proportion of high frequency information is lost (**29, 49, 59**). Since the radius of the tip is nominally 2.5 μm, high frequency surface information, up to a wavelength of the order of five times the tip radius, is significantly smoothed. Despite these shortcomings, at resolutions below the tip filtering limit, Talysurf provides the most reliable (although slow to collect) data.

It is often assumed that a sand-blasted surface is fractal (e.g. **29, 50**). This is confirmed by comparing the Richardson plots calculated from an artificial fractal profile generated by the mid-point displacement method and a Talysurf trace obtained from a sand-blasted steel surface shown in Figure 13 (**29**).

The 'flattening off' of the Richardson plot obtained from a Talysurf trace at small step sizes, i.e. high resolution, is evident. It can also be seen from Figure 13 that at larger step sizes the plot exhibits a similar slope, i.e. fractal dimension, to the plot obtained from an artificially generated profile (**29**).

Since visual inspection of an SEM image can provide a good judgement about the relative heights of surface features it was thought that perhaps grey levels taken from SEM images can be used as a 'surface profile' in surface characterization (**5**). Unfortunately, the experiments conducted demonstrated that the grey levels are strongly affected by edge highlighting, varying atomic structure of the sample, sample charging, detector orientation, etc. (**28**). Also, it was found that from the SEM images, significantly different results can be obtained when the magnification is varied, i.e. low magnification images exhibit high fractal dimensions while high magnification images show low fractal dimensions (**28**). Thus the questions arise whether the underlying assumption that the surface is ideally fractal is valid or whether grey level surface profiles from SEM images can be applied to surface fractal characterization (**37**). The data obtained seem to indicate that this simple use of grey levels as surface elevation data for calculation of surface fractal dimension is incorrect (**29**).

Surface profiles can also be obtained from Laser Scanning Confocal Microscope (LSCM) images (**28**). LSCM has been developed to provide 3-D surface data. However, major limitations of these instruments are: limited magnification, poor resolution and sensitivity to areas of low reflectance which cause noise and 'holes' in the image (**29, 60**). Problems can also occur with surfaces which are transparent or translucent, though these can be overcome to some extent through the application of suitable coatings (**29**). Despite certain drawbacks, within its *x-y-z* resolution limits, LSCM image can provide useful information about surface characteristics. Recently developed Atomic Force Microscope (AFM) can also be used in acquisition of 3-D surface topography data at nano scale. However, due to the high magnification used, these instruments exhibit small dynamic (vertical) range and small view area (e.g. **61**).

Unfortunately, characterization of surface topography based on surface profiles, provides only limited information about the 3-D characteristics of real surfaces. The problem is that a surface is a 3-D object and its characterization, based on 2-D profiles taken from certain regions of the surface, is not adequate.

4 CHARACTERIZATION OF PARTICLE 3-D SURFACE TOPOGRAPHY

Characterization of 3-D surface topography of wear particles is a challenging problem. There are two issues associated with this problem, i.e. the acquisition of accurate 3-D data from a particle surface and the numerical description of 3-D surface features.

Although there are several techniques that could be used to acquire 3-D data from engineering surfaces, the number of techniques that can be used to acquire 3-D data from a particle surface is limited. This is due to a small particle size, usually ranging from about 1 μm to 500 μm and the presence of surface features with high relief, often greater than 4 μm. These restrictions limit the choice of techniques used for surface data acquisition to three: atomic force microscopy (AFM), optical interferometry and SEM stereoscopy.

Although the AFM has excellent vertical (~0.1 nm) and horizontal (~0.2 nm) resolutions it also has severe limitations (**62**). The main disadvantage is that its maximum vertical range is limited to about 4 μm. Typical wear particles exhibit surface features with height relief often much greater than 4 μm and average size greater than 10 μm. There may also be some difficulties associated with fixing particles so that they will not move during the measurements when examined under AFM. This problem might become particularly acute when 'biological' wear particles, i.e. from synovial joints, need to be studied.

The optical interferometry technique is limited by poor lateral resolution and the use of reflected light (**63**). For example, typical resolution is about 1 μm, unless near-field techniques are used (**63**). Since optical interferometry uses the reflected light in measurements a moderate reflectivity of a specimen surface is required and, as a result, the technique does not work well with rough surfaces. This means that the technique cannot be employed in the analysis of particles with complicated surface features.

4.1 Imaging of particle surfaces by SEM stereoscopy

The application of SEM stereoscopy imaging of the surface may provide the desired elevation map of surface features (**64**). The technique involves a sequential capturing of two images of the same particle either by tilting the specimen by a known angle or by translating the specimen by a known distance. The relative displacement (called disparity) of the images is measured by calculating the distance between the corresponding points in the two images. Based on this displacement the surface elevation is calculated. Although it has been shown that a stereoscopic pair of SEM images can be used to obtain information about 3-D topographical surface data (**64**), this technique is still not widely used because of the inadequate image matching programs and lengthy computation time required to obtain the 3-D image of the surface. Recently an attempt has been made to apply the SEM stereoscopy method to analysis of wear particle surface topography (**65**). Special software enabling the accurate acquisition of 3-D wear particle surface topography data has been developed (**65**). An example of the 3-D image of a wear particle obtained by SEM stereoscopy is shown in Figure 14.

4.2 Characterization of particle surfaces

One of the major difficulties associated with characterization of surface topographies, e.g. engineering, (machined or worn), biological (e.g. surface of the bone), surfaces of wear particles, etc., is the accurate description of their spatial properties, i.e. their anisotropy and directionality. Characterization of the spatial properties of surfaces is still, to a large extent, an unresolved problem because of the random and multi-wavelength nature of surface topography components which cannot effectively be characterized in all directions by existing techniques to the extent that something can be learned from it (**42, 66**).

Surface topography is essentially a non-stationary random process for which the variance of height distribution depends on the sampling length (**67**). This means that the same surface can exhibit different values of the statistical parameters when a different sampling length or an instrument with a different resolution is used. Naturally, this results in great inconsistencies in surface characterization (**68**). The main problem is associated with the discrepancy between the large number of length scales that a rough surface exhibits and the small number of particular length scales, i.e. sampling length and instrument resolution, that are used to define the surface parameters.

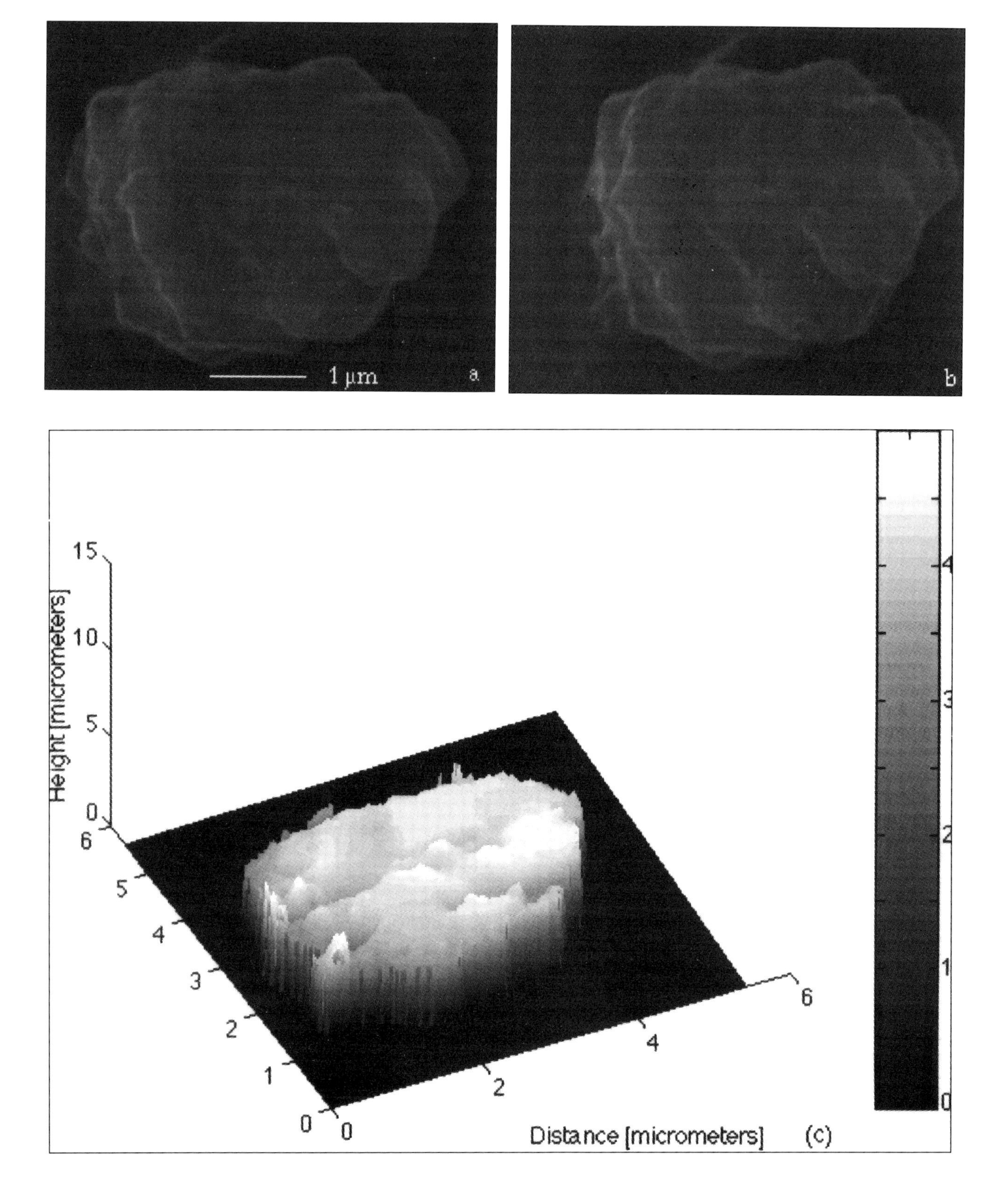

Fig. 14 Stereoscopic images of wear particle obtained from synovial sheep joint; (a) and (b) SEM stereoscopic pair of images of the particle, (c) shaded perspective view of the reconstructed elevation map (**65**).

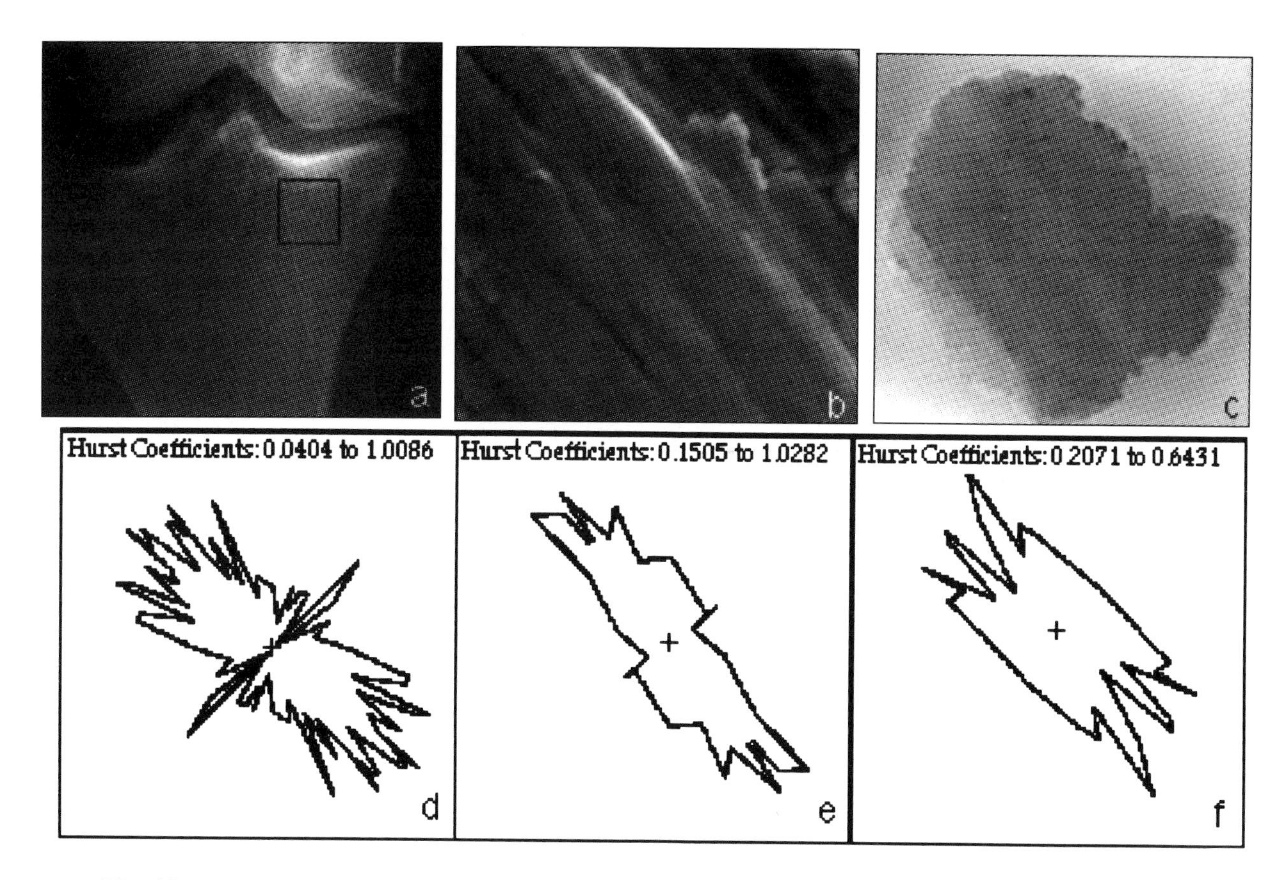

Fig. 15 Examples of typical surfaces and calculated Hurst coefficients; (a) X-ray image of osteoarthritic knee joint, (b) SEM image of the machined surface of brass, (c) interferometric image of metallic wear particle, (d) Rose plots of the Hurst coefficients as a function of orientation obtained from the Hurst Orientation Transform calculated for marked square region in (a) and entire image (b) and wear particle surface shown in (c).

Thus conventional surface parameters, such as the variance of surface heights, surface slope and curvature and the root mean square (RMS) roughness, strongly depend on the scan length and the measurement technique and hence they are not unique for a particular surface (**8**). In order to characterize a surface at all scales, scale-independent parameters must be developed and used.

The unique feature of real surfaces is that, if repeatedly magnified, increasing details are observed down to nanoscales. Attempts have therefore been made to apply fractal dimensions, since they are 'scale invariant', to the characterization of rough surfaces (e.g. **31, 52, 58**). In our work, a specially modified Hurst Orientation Transformation (HOT) has been developed and applied to reveal the surface anisotropy.

The calculation of the HOT involves searching all pairs of pixels in a large neighbourhood and building the table of the maximum differences. From this table, the Hurst coefficients are calculated from log-log plots constructed for all directions and plotted as a function of orientation to reveal surface anisotropy. The Hurst coefficient is related to fractal dimension and is a measure of surface roughness, i.e. a rougher surface is represented by lower Hurst coefficients (**51, 69**). Examples of rose plots of the Hurst coefficients obtained from the X-ray image of a bone surface of an arthritic knee joint, a SEM image of the machined surface of brass and an interferometric image of metallic wear particle are shown in Figure 15. It can be seen from Figure 15 that the surfaces analysed clearly exhibit anisotropic features, i.e. they are 'smoother' in a diagonal direction than in all other directions and this is reflected in the Hurst coefficients calculated. It seems that this technique, when fully developed, may find applications in the characterization of surfaces of engineering components in modern manufacturing industry and also in medicine.

5 CHARACTERIZATION OF PARTICLE ANGULARITY AND ITS RELATION TO TWO-BODY ABRASIVE WEAR

It is intuitively felt that, in addition to hardness and size, particle shape affects its ability to abrade, i.e. a particle boundary contains useful information on its abrasivity. Laboratory tests have confirmed that increases in particle angularity significantly increase the abrasive or erosive wear rates (**32, 70–74**). Traditionally, qualitative descriptors of particle visual appearance such as 'rounded', 'semiangular' or 'angular' have been used to classify and differentiate among various groups of abrasive particles (**71, 75–77**).

Due to the limitations of qualitative analysis, several attempts have been made to date to characterize abrasive particle shapes using various quantitative numerical descriptors and to correlate the shape with the ability to abrade or erode. For example, roundness factor (Perimeter^2/4π x Area) and the statistical parameter R_{ku} (Kurtosis) describing edge detail have been employed to differentiate among five groups of abrasives (**9**). SiC, Al_2O_3 and SiO_2 particles have been characterized using aspect ratio (Width/Length) and Perimeter^2/Area shape parameters (**73**). It was found that the erosion rate increased with increasing P^2/A and decreasing W/L for the three types of abrasive particles (**73**). Invariant Fourier Descriptors (**78**) have been used to distinguish abrasive particles of differing angularities (**32**). An empirical equation relating the angularity of particles, in terms of Fourier Descriptors, to abrasion rates has been proposed (**32**).

It seems that conventional shape parameters such as roundness or aspect ratio are good indicators of particle irregularity but do not provide a satisfactory measure of its angularity, a very important characteristic in abrasive and erosive wear. An attempt to quantify the angularity of abrasive particles is described in the next section. The descriptor developed takes into account the sharpness of particle angles and their size which reflects the probability of these angles abrading the countersurface. This descriptor is then used to correlate the particle shapes, in terms of angularity, with their ability to abrade in two-body abrasive wear tests. A similar technique is applied to characterize surface profiles and to relate the shape of the profile with the abrasive wear rates.

5.1 Characterization of particle abrasivity based on its shape

In order to characterize the shape effect on particle abrasivity, a particle angularity descriptor, based on representing the particle boundary as a set of triangles constructed at different scales, has been developed (**33**). The shape descriptor developed, called the 'spike parameter', is calculated in the following manner. At each step around the boundary the start and end points are found, in a similar way to that used in the calculations of a boundary fractal dimension. The difference is that instead of a straight line connection the boundary between the start and the end point is represented by a 'triangle'. It is assumed that the sharpness and size of these triangles is directly related to particle abrasivity, i.e. the sharper (smaller apex angle) and larger (perpendicular height) the triangles the more abrasive the particle.

First, the triangle with the greatest 'spike value', defined as:

$$sv = \cos\left(\frac{\theta}{2}\right)h$$

(where: 'h' is the perpendicular height of the triangle while 'θ' is the apex angle as shown in Figure 16), for a particular step size around the boundary is found and recorded. In the procedure all possible starting points for a particular step size around the boundary are used. The product of cos(θ/2) and the height 'h' allows selection of the triangle based on both its sharpness and its size. It can be noticed that the term SV/height = cos(θ/2) is a measure of angularity, and can vary between 0 and 1 (**34, 35**).

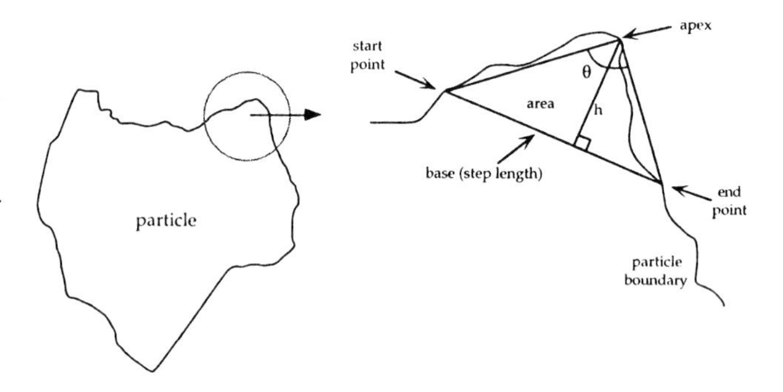

Fig. 16 Constructing triangles round the boundary (adapted from (**33**)).

Next, the process is repeated for different step sizes. For each 'walk' maximum spike values are normalized by their height, i.e. SV/height = cos(θ/2), and then averaged. The 'spike parameter' is then calculated according to the following formula (**33–35**):

$$SP = \frac{\Sigma[\Sigma(sv_{max} / h_{max}] / m}{n}$$

where:

$$SV_{max} = \left[\cos\left(\frac{\theta}{2}\right)h\right] \quad \text{for a given step size}$$

h_{max} = height at $SV_{max'}$,
m is the number of valid 'SV' for a given step size,
n is the number of different step sizes used.

A computer program developed for spike parameter calculation was tested on six artificial shapes. These shapes were constructed by adding triangles of varying angles and heights to the base circle as illustrated in Table I. Intuitively one would rank these shapes as increasing in angularity from 1 to 6 (**33**). Other typical shape parameters such as boundary fractal dimension, aspect ratio, reciprocated shape factor were also calculated. The changes in the calculated spike parameter and typical shape parameters are shown in Table I (**33**).

It can be seen from Table I that the 'spike parameter' increases with the 'spikiness' of the particles, which agrees

shape	spike parameter	boundary fractal dimension	1/shape factor	aspect ratio
	0.1332	1.0046	1.1145	1.0000
	0.1633	1.0064	1.1323	1.0078
	0.1721	1.0063	1.2933	1.0556
	0.1951	1.0115	1.1426	1.0514
	0.2119	1.0095	1.6127	1.6800
	0.7243	1.0155	3.1458	1.0000

Table I Artificial shapes with intuitively increasing angularity together with calculated spike parameters, boundary fractal dimensions, reciprocated shape factors and aspect ratios (adapted from (**33**)).

with the intuitive prediction.

Recently, another particle angularity parameter, *SPQ* (spike parameter, quadratic fit), which correlates slightly better with the wear rates than spike parameter has been developed (**35**). When calculating *SPQ* the particle boundary is approximated by a quadratic curve fit instead of triangles. Initially, a particle boundary centroid $(\bar{x}, \bar{y})$ (**80**) is located (**35**). The areas outside the circle, 'spikes', are deemed to be the areas of interest while those inside are not. For each 'spike' outside the circle the local maximum radius is found and this point is treated as the spike's apex (**35**). The sides of the 'spike', which are between the points 'sp-mp' and 'ep-mp' as illustrated in Figure 17, are then represented by fitting quadratic polynomial functions. Differentiating the polynomials at the 'mp' point yields the apex angle (θ) and the spike value '*SV*', i.e. $SV = \cos(\theta/2)$. The new spike parameter *SPQ* is then calculated according to the formula (**35**):

$$SPQ = SV_{\text{average}}$$

The calculation details of *SPQ* are described in (**35**). The advantage of *SPQ* over *SP* is that it considers only the boundary features, i.e. protrusions, which are likely to come in contact with the opposing surface. The *SP* parameter at small step sizes is sensitive to digitisation errors, and therefore a limiting apex angle of 2.9 rads had to be chosen for a 'zero' value. This leads to some 'smooth' shapes being assigned an artificially high *SP* (**33, 34**). Also for boundaries with convex curved sides, the apex angle, when calculating *SP* parameter, is underestimated (i.e. it becomes sharper than it really is) as the step size increases. The newly developed *SPQ* is almost insensitive to image focus (**35**).

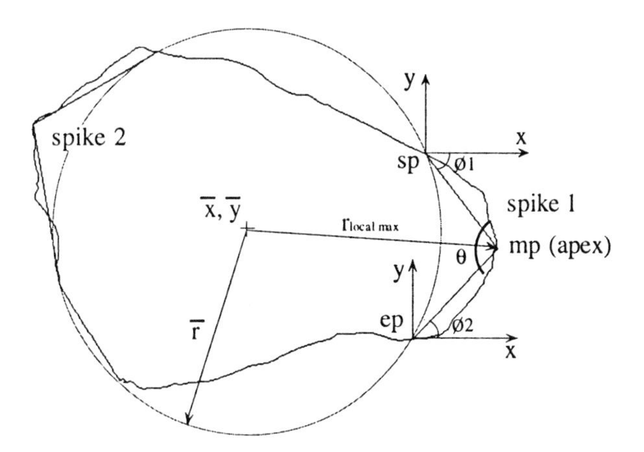

Fig. 17 Method of *SPQ* calculation by fitting quadratic segments (adapted from (**35**)).

5.2 Relationship between the particle shape and abrasive wear rates

In order to find the relationship between particle abrasivity and its shape, two-body abrasive wear tests were conducted on a pin-on-disk machine (**33, 34**). Seven types of abrasive grits were used in the experiments: glass beads, silica sand, garnet, natural industrial diamonds, silicon carbide, crushed quartz, crushed sintered alumina (**35**). For each of these abrasives 20 particles were randomly selected. Secondary electron SEM images were collected for each individual particle and spike parameters, *SP* and *SPQ*, were calculated. The results obtained are shown in Table II.

Special disks were manufactured from perspex with a 50 to 80 μm deep slot machined into the disks. The slot was filled with epoxy resin and then covered with abrasive grits of average size between 150-300 μm as schematically illustrated in Figure 18 (**33, 34**). This procedure assured that the disks were uniformly covered with particles of known size, strongly bonded to the disk surface (**33, 34**).

Table II Average spike parameter values calculated for abrasive particles used in the experiments (adapted from (**35**)).

abrasive	average wear rate (mm/min)	SD	SP	SD	SPQ	SD
glass beads	0.8048	0.0157	0.1369	0.0039	0.0231	0.0148
silica sand	2.0551	0.1171	0.2077	0.0151	0.1919	0.0871
garnet	2.2303	0.1283	0.2168	0.0271	0.2515	0.1149
diamond	2.7039	0.2317	0.2971	0.0277	0.3958	0.1236
silicon carbide	2.9499	0.1676	0.2942	0.0360	0.4247	0.1473
quartz	3.4341	0.2553	0.3239	0.0245	0.5336	0.1294
crushed alumina	3.8993	0.5192	0.3591	0.0342	0.6008	0.1271

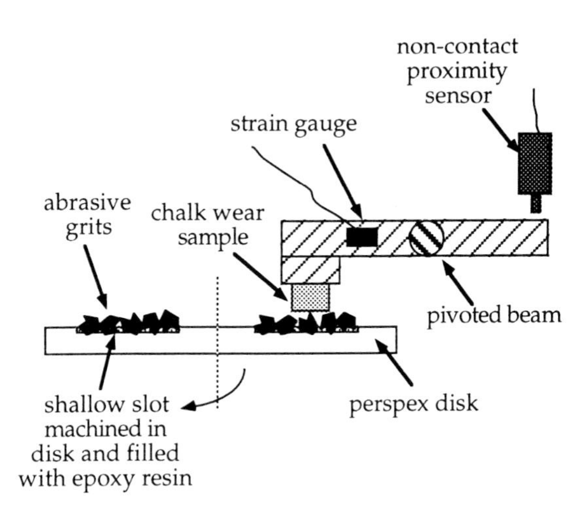

Fig. 18 Schematic diagram of the abrasive disk and the experimental apparatus (**34**).

In order to ensure that abrasion was a dominating wear process in the experiments, chalk was selected as the material for a counter-sample. The hardness of the chalk is well below the hardness of abrasives, the material is weak and its particles did not clog the disks. Tests conducted initially with aluminium and brass counter-samples failed to provide convincing data (**34**). It was found that metallic wear particles were clogging the disks (**34**). Also in the cases of quartz and crushed alumina grits, fracture of the abrasives occurred. This effect was particularly pronounced in the case of quartz particles which tended to fracture off level with the resin in which they were set (**34**). All these processes resulted in the diminished ability of grits to abrade and obscured the effect of particle shape on its ability to abrade.

Test operating conditions were maintained at the same level and the only variable was the difference between the shapes of abrasive particles. It was anticipated that these experiments would show the effects of particle shape on abrasive wear and allow to correlate the developed 'spike parameter' with the wear data.

Abrasive wear experiments demonstrated that the spike parameter calculated for the abrasives tested correlated well with the abrasive wear rates of chalk, indicating that there is a strong relationship between particle shape and its ability to abrade. The relationship between the spikes parameters developed, '*SP*' and '*SPQ*', and the abrasive wear rates is shown in Figures 19 and 20 (**35**).

It can be seen from Figures 19 and 20 that both parameters exhibit almost a linear relationship with the wear rates, i.e. spike parameter is strongly related to particle abrasivity.

6 CHARACTERIZATION OF SURFACE ABRASIVITY AND ITS RELATION TO TWO-BODY ABRASIVE WEAR

It is also intuitively felt that the shape of surface profiles is related to surface abrasivity. A method, similar to the method employed in the characterization of particle angularity has been developed and applied to the

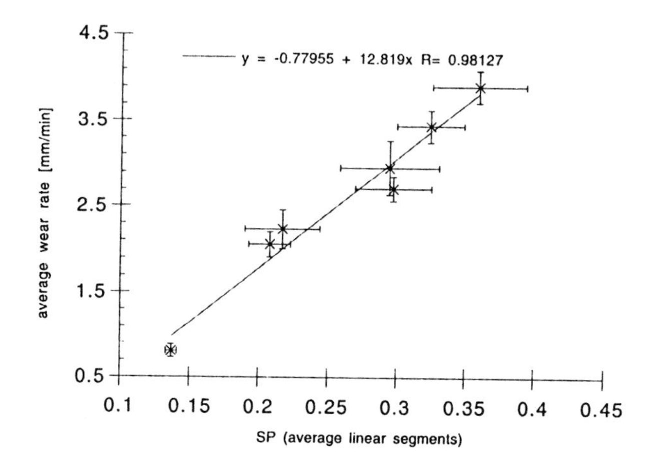

Fig. 19 Plot of average wear rate versus average *SP* (average linear segments) for different abrasive types (10 wear experiments for each abrasive type, 20 particles measured for each abrasive type, error bars are ±1 standard deviation) (**35**).

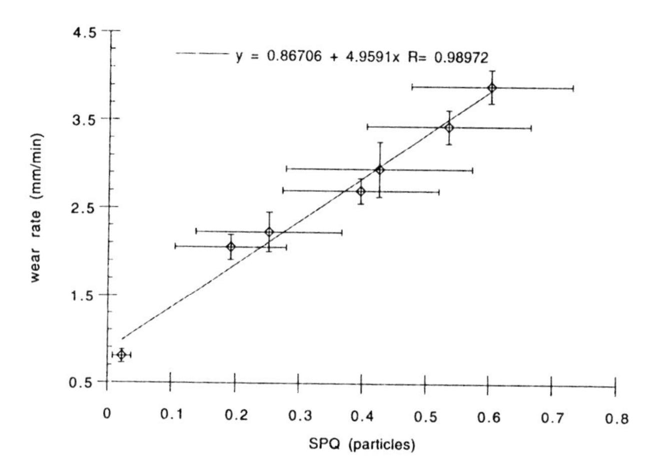

Fig. 20 Plot of *SPQ* (particles) against wear rates for seven types of mineral abrasive (10 wear experiments for each abrasive type, 20 particles measured for each abrasive type, error bars are ±1 standard deviation) (**35**).

characterization of angularity of surface profiles. A mean line running through the centre of the surface profile, measured by, e.g. Talysurf, is first fitted. This line can be visualized as a 'water mark' with asperities which may cause abrasion as 'islands' rising above it as illustrated in Figure 21 (**36**). For a particular spike the profile segments from the left point (lp) and right point (rp) on the spike to the apex are represented by quadratic functions. Differentiating these functions at the apex yields an apex angle 'θ'. Spike parameter-quadratic fit is calculated according to the formula (**36**):

$$SPQ = \frac{1}{n}\sum \cos\frac{\theta_i}{2}$$

where:
SPQ is the spike parameter-quadratic fit,
θ is the apex angle,
n is the number of spikes analysed.

From the abrasion view point apex sharpness is the most important feature. On any apex there are usually numerous 'little spikes', i.e. turning points, present. In order to obtain the true measure of the apex angularity, curve fitting must not be biased on these 'little spikes'. To eliminate the effects of 'little spikes' the computer program developed searches on the left and the right of the apex and then designates new 'lp' or 'rp' points on the condition that the vertical amplitude of the turning point pair is greater than 10% of the apex height from the mean line. This is schematically illustrated in Figure 21. Spikes whose height is less than 25% of the maximum spike height found in the profile are rejected (**36**).

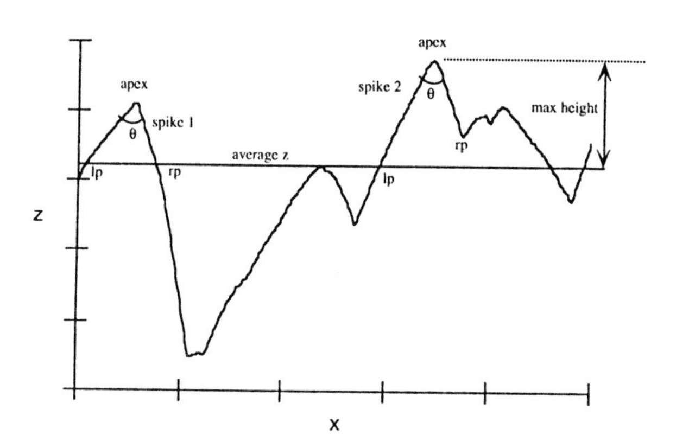

Fig. 21 Method of evaluating the spike parameter-quadratic fit from a surface profile (**36**).

6.1 The relationship between the profile shape and abrasive wear rates

In order to find the relationship between profile shape and abrasive wear rates, two-body abrasive wear tests were conducted on a pin-on-disk machine with specially manufactured disks (**36**). Chalk was selected as the material for a counter-sample to ensure that abrasion was a dominating wear process. In the experiments abrasive disks with artificially produced model surfaces and disks covered with abrasive grits were used. Disks with artificially produced model surfaces had three special well defined surface profiles machined on them. One profile was modelled by a segment of a circle with a radius of 2.5 mm, while the other two were modelled by 120° and 60° ridges respectively (**36**). Since all the ridges were of the same height the only variable in two-body experiments was the angularity or 'spikiness' of the surface profiles. The examples of Talysurf profiles obtained from model surfaces are shown in Figure 22.

Seven different types of abrasive grits, i.e. glass beads, silica sand, garnet, natural industrial diamond, silicon carbide, quartz and crushed sintered alumina, were used to manufacture abrasive disks. All the grits used in the experiments were of the same size, i.e. 250-300 μm. As an example, Talysurf profiles obtained from surfaces covered with glass beads and diamond are shown in Figure 23 (**36**). It can be seen that the differences in these profiles are clear.

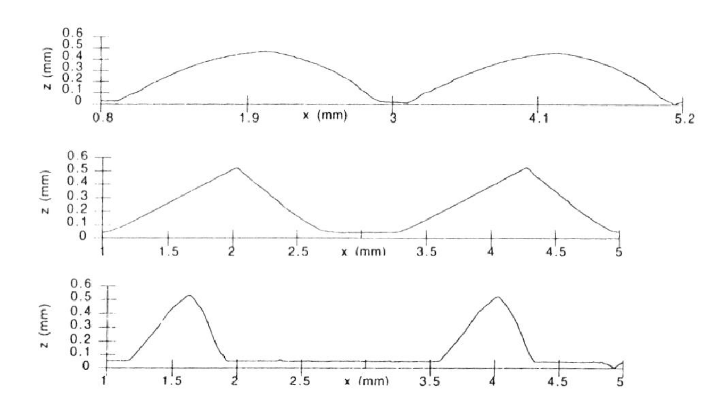

Fig. 22 Examples of Talysurf traces obtained from artificially produced model surfaces (**36**).

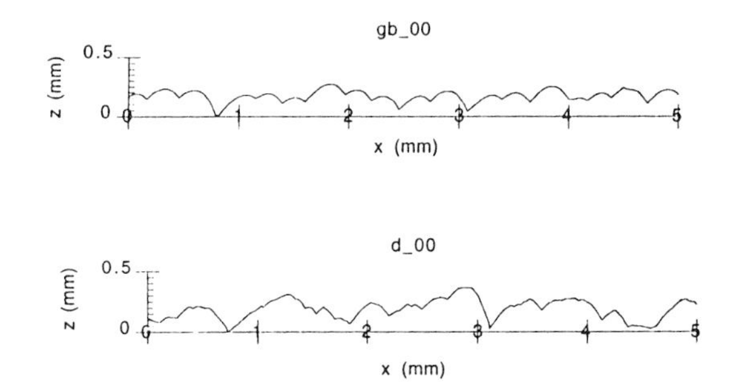

Fig. 23 Examples of Talysurf traces obtained from abrasive disks (**36**).

Low loads and velocities were used in the experiments in order to eliminate the fracture of the chalk counter-samples. careful selection of the operating conditions and the selection of chalk for counter samples ensured that 'pure' two-body abrasive wear took place.

The relationship between the shape of the profile characterized by the quadratic spike parameter (*SPQ*-profiles) and abrasive wear rates is shown in Figures 24 and 25 for artificially produced model surfaces and abrasive disks respectively.

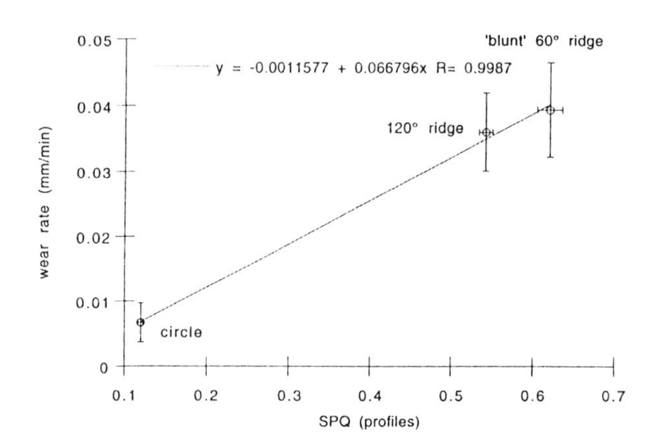

Fig. 24 Relationship between spike parameter calculated for profiles obtained from artificially produced model surfaces and abrasive wear rates (**36**).

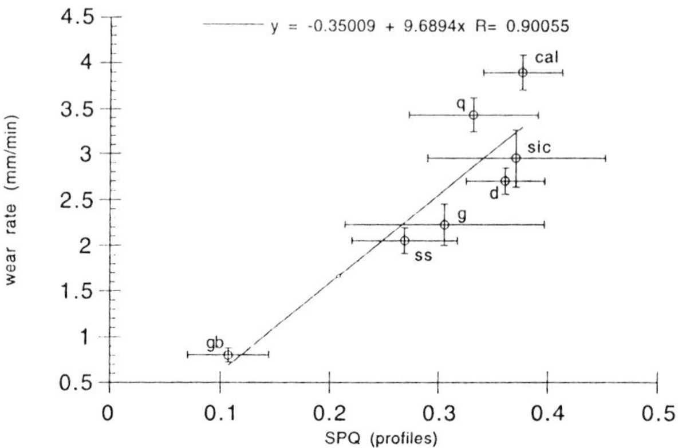

Fig. 25 Relationship between spike parameter calculated for profiles obtained from surfaces of disks manufactured with abrasive grits and abrasive wear rates (**36**).

It can be seen from Figures 24 and 25 that the quadratic spike parameter exhibits almost a linear correlation with the abrasive wear rates. However, the *SPQ* values for some of the abrasives, i.e. quartz and crushed alumina, were lower than expected (**35**). This was caused by the fixed radius of the Talysurf stylus which was unable to follow exactly the highly angular and re-entrant surfaces made of these abrasives (**35**). This problem was found in earlier work (**28**) and was also reported by other researchers (**81**). It was also found that both the surface roughness 'R_a' and the fractal dimension calculated from the profiles obtained from both artificially generated and abrasive surfaces did not correlate well with two-body abrasive wear rates.

7 CONCLUSIONS

From the work conducted the following conclusions can be drawn:

- The boundaries of particles found in tribological systems and surface profiles can be characterized by fractal dimensions. When calculating the fractal dimension from a particle boundary or surface profile all starting points on the boundary or profile should be used. In order to improve the objectivity and repeatability of results automated curve fitting should be used to assess the slope of a Richardson plot.
- Boundary fractal dimension is sensitive to focusing and background noise, and therefore sharp, clear images are required for analysis. Median-sigma filtering appears to be the most effective in reducing the noise while maintaining a virtually distortion-free image.
- The boundary fractal dimension calculated from the projected images of particles is usually lower than the boundary fractal dimension calculated from their sectioned images. In the case of estimating the surface fractal dimension from the boundary fractal dimension it appears that the image of the sectioned particle should give more accurate results.
- The structured-walk technique used to calculate boundary fractal dimension cannot be used to evaluate fractal dimension from a surface profile. This is because surface profiles are self-affine, i.e. they scale by different amounts in the vertical (z) and horizontal (x) directions. For self-affine profiles the vertical variation is measured at different scales while in the case of self-similar boundaries the Euclidean length is measured at different scales.
- The Richardson technique used in the calculation of profile fractal dimension is sensitive to non-planar underlying structure of the profile, e.g. waviness.
- 3-D maps of particle surface topography features can be obtained by SEM stereoscopy imaging.
- A modified Hurst Orientation Transformation (HOT) can be applied to reveal the surface anisotropy.
- Angularity of particles can successfully be described by a numerical parameter called the 'spike parameter', which is based on the representation of the projected particle boundary by positive, i.e. convex, triangles at many scales.
- The value of spike parameter which is characteristic for a particular abrasive grit shows some variation within a population of particles and correlates well with the wear rates.
- The spike parameter can be used in predicting the wear rates in pure abrasion. It becomes less reliable when wear processes other than abrasion, e.g. adhesion, are involved.
- Abrasivity of surfaces is related to the shape of surface profiles and can be described numerically. Quadratic spike parameter, describing the angularity of surface profiles, exhibits an almost linear correlation with the abrasive wear rates.

8 REFERENCES

1 **Seifert, W.W.** and **Westcott, V.C.,** A method for the study of wear particles in lubricating oil, *Wear,* **21,** 1972, pp. 27–42.

2 **Mears, D.C., Hanley Jr, E.N., Rutkowski, R.,** and **Westcott, V.C.,** Ferrography: its application to the study of human joint wear, *Wear,* **50,** 1978, pp. 115–125.

3 **Evans, C.H., Bowen, E.R., Bowen, J., Tew, W.P.,** and **Westcott, V.C.,** Synovial fluid analysis by ferrography, *Journal of Biochemical and Biophysical Methods,* **2,** 1980, pp. 11–18.

4 **Kirk, T.B., Stachowiak, G.W., Warton, A.,** and **Carroll, G.J.,** Ferrographic and cytological analysis of asymptomatic and arthritic human synovial fluid: a preliminary report, *Journal of Orthopaedic Rheumatology,* **2,** 1989, pp. 91–103.

5 **Kirk, T.B., Stachowiak, C.W.,** and **Batchelor, A.W.,** Fractal parameters and computer image analysis applied to wear particles isolated by ferrography, *Wear,* **145,** 1991, pp. 347–365.

6 **Yeung, K.K., McKenzie, A.J., Liew, D.,** and **Luoma, G.A.,** Development of computer-aided image analysis for filter debris analysis, *Lubrication Engineering,* **50,** 1994, pp. 293–299.

7 **Kirk, T.B.** and **Stachowiak, G.W.,** Fractal computer image analysis applied to wear particles from arthritic and asymptomatic human joints, *Journal of Orthopaedic Rheumatology,* **4,** 1991, pp. 13–30.

8 **Roylance, B.J.** and **Raadnui, S.,** The morphological attributes of wear particles – their role in identifying wear mechanisms, *Wear,* **175,** 1994, pp. 115–121.

9 **Raadnui, S.** and **Roylance, B.J.,** The classification of wear particle shape, *Lubrication Engineering,* **51,** 1995, pp. 432–437.

10 **Roylance, J., Albidewi, I.A., Laghari, M.S., Luxmore, A.R.,** and **Price, A.L.,** The development of a computer-aided systematic particle analysis procedure – CASPA, *Lubrication Engineering,* **48,** 1992, pp. 940–946.

11 **Roylance, B.J., Albidewi, I.A., Laghari, M.S., Luxmoore, A.R.,** and **Deravi, F.,** Computer-aided vision engineering (CAVE) – quantification of wear particle morphology, *Lubrication Engineering,* **150,** 1994, pp. 111–116.

12 **Hamblin, M.G.** and **Stachowiak, G.W.,** Comparison of boundary fractal dimension from projected and sectioned particle images, Part I – Technique evaluation, *Journal of Computer Assisted Microscopy,* **5,** No. 4, 1993, pp. 291–300.

13 **Hamblin, M.G.** and **Stachowiak, G.W.,** Comparison of boundary fractal dimension from projected and sectioned particle images, Part II – Dimension changes, *Journal of Computer Assisted Microscopy,* **5,** No. 4, 1993, pp.

14 **Kaye, B.H.,** *A random walk through fractal dimensions,* 1989, VCH, Verlagsgesellschaft, Weinheim, Germany.

15 **Kaye, B.H.,** *Applied fractal geometry and the fine particle specialist, Part I: Rugged boundaries and rough surfaces, Particle System Characteristics,* **10,** 1993, pp. 99–110.

16 **Russ, J.C.,** *Image Processing Handbook,* CRC Press, Inc., 1995.

17 **Hausner, H.H.,** Characterization of the powder particles shape, *Planseeberichte fur Pulvermetallurgie,* **14,** 1966, pp. 75–84.

18 **Russ, J.C.,** *Computer Assisted Microscopy,* Plenum Press, New York, 1990.

19 **Roylance, B.J., Wang, G.,** and **Bovington, C.H.,** The determination of particle morphological parameters to assist in the elucidation of the wear process, *Proc. 18th Leeds-Lyon Symposium,* 1991, pp. 75–79.
20 **Clark, N.N.,** Three techniques for implementing digital fractal analysis of particle shape, *Powder Technology,* **46,** 1986, pp. 45–52.
21 **Mandelbrot, B.B.,** *The Fractal Geometry of Nature,* W.H. Freeman, New York, 1982.
22 **Stachowiak, G.W., Kirk, T.B.,** and **Stachowiak, G.B.,** Ferrography and fractal analysis of contamination particles in unused lubricating oils, *Tribology International,* **24,** 1991, pp. 329–334.
23 **Hamblin, M.G.** and **Stachowiak, G.W.,** Variation of particle shape parameters when viewed as a projection and a section, *Journal of Computer Assisted Microscopy,* **6,** No. 2, 1994, pp. 51–59.
24 **Hamblin, M.G.** and **Stachowiak, G.W.,** Fractal dimensions from images, *Image Analysis '94,* editors: A. van Riessen and J.R. Hatch, WASEM, 1994, p. 4 and pp. 43–61.
25 **Kaye, B.H.,** Fractal description of fine particle systems, *Modern Methods in Fine Particle Characterization,* J.K. Bedow (editor), CRC Press, Boca Raton, FL, 1983.
26 **Chan, L.C.** and **Page, N.W.,** Fractal measures of particle shape and their relation to interparticle friction in granular materials, *Proc. of the First Australasian Congress on Applied Mechanics,* 21–23 February 1996, Melbourne, Australia, pp. 475–480.
27 **Vallejo, L.E.,** Fractal analysis of granual materials, *Geotechnique,* **45,** pp. 159–163.
28 **Hamblin, M.G.** and **Stachowiak, G.W.,** Measurement of fractal surface profiles obtained from scanning electron and laser scanning microscope images and contact profile meter, *Journal of Computer Assisted Microscopy,* **6,** No. 4, 1994, pp.181–194.
29 **Hamblin, M.G.** and **Stachowiak, G.W.,** Application of the Richardson technique to the analysis of surface profiles and particle boundaries, *Tribology Letters,* **1,** 1995, pp. 95–108.
30 **Pfeifer, P.** and **Avnir, D.,** Chemistry in non-integer dimensions between 2 and 3, Part 1: Fractal theory of heterogeneous Surfaces, Part 2: Fractal surfaces of Absorbants, *I. Chem. Phys.,* **79,** 1983, pp. 3558–3571.
31 **Russ, J.C.,** *Fractal Surfaces,* Plenum Press, New York, 1994.
32 **Swanson, P.A.** and **Vetter, A.F.,** The measurement of abrasive particle shape and its effect on wear, *ASLE Transactions,* **28,** 1984, pp. 225–230.
33 **Hamblin, M.G.** and **Stachowiak, G.W.,** A Multi-scale measure of particle abrasivity, *Wear,* **185,** 1995, pp. 225–233.
34 **Hamblin, M.G.** and **Stachowiak, G.W.,** A multi-scale measure of particle abrasivity and its relation to two body abrasive wear, *Wear,* **190,** 1995, pp. 190–196.
35 **Hamblin, M.G.** and **Stachowiak, G.W.,** Description of abrasive particle shape and its relation to two-body abrasive wear, *Tribology Transactions,* **39,** No. 4, 1996, pp. 803–810.
36 **Hamblin, M.G.** and **Stachowiak, G.W.,** Description of surface abrasivity and its relation to two-body abrasive wear, accepted for publication in *Wear,* 1996.
37 **Stachowiak, G.W.** and **Hamblin, M.G.,** Application of fractals to the description of shape of the particles found in tribological systems, Mechanical Engineering Transactions, *Journal of the Inst. of Engineers,* Australia, **ME 20,** No. 2, 1995, pp. 127–136.
38 **Korvin, G.,** *Fractal Models in the Earth Sciences,* 1992, Elsevier.
39 **Judd, K.,** An improved estimator of dimension and comments on providing confidence intervals, *Physica D.,* **56,** 1992, pp. 216–228.
40 **Podsiadlo, P.** and **Stachowiak, G.W.,** Median-sigma filter for SEM wear particle images, *Journal of Computer Assisted Microscopy,* **7,** 1995, pp. 67–82.
41 **Podsiadlo, P.** and **Stachowiak, G.W.,** Analysis of shape of wear particles found in synovial joints, *Journal of Orthopaedic Rheumatology,* **8,** 1995, pp. 155–160.
42 **Dong, W.P., Sullivan, P.J.,** and **Stout, K.J.,** Comprehensive study of parameters for characterising three-dimensional surface topography, III: Parameters for characterising amplitude and some functional properties, *Wear,* **178,** 1994, pp. 29–43.
43 **Dong, W.P., Sullivan, P.J.,** and **Stout, K.J.,** Comprehensive study of parameters for characterising three-dimensional surface topography, IV: Parameters for characterising spatial and hybrid properties, *Wear,* **178,** 1994, pp. 45–60.
44 **Stout, K.J., Sullivan, P.J., Dong, W.P., Mainsah, E., Luo, N., Mathia, T.,** and **Zahouani, H.,** The development of methods for the characterization of roughness in three dimensions, *Publication No. EUR 15178 EN of the Commission of the European Communities,* Luxembourg, 1993.
45 **Thomas, T.R.,** (editor), *Rough Surfaces,* Longman Group Limited, 1982.
46 **Whitehouse, D.J.,** *Handbook of Surface Metrology,* Institute of Physics Publishing, Bristol and Philadelphia, 1994.
47 **Stachowiak, G.W.** and **Batchelor, A.W.,** *Engineering Tribology,* Elsevier, 1993.
48 **Sayles, R.S.** and **Thomas, T.R.,** The spatial representation of surface roughness by means of the structure function: a practical alternative to correlation, *Wear,* **42,** 1977, pp. 263–276.
49 **Vandenberg, S.** and **Osborne, C.F.,** Digital image processing techniques, fractal dimensionality and scale-space applied to surface roughness, *Wear,* **159,** 1992, pp. 17–30.
50 **Dubuc, B., Zucker, S.W., Tricot, C., Quiniou, J.F.,** and **Wehbi, D.,** Evaluating the fractal dimension of surfaces, *Proc. Royal Society, London,* Series A, **425,** 1989, pp. 113–127.
51 **Russ, J.C.,** Characterizing and modelling fractal surfaces, *Journal of Computer Assisted Microscopy,* **4,** 1992, pp.73–126.
52 **Majumdar, A.,** and **Bhushan, B.,** Role of fractal geometry in roughness characterization and contact mechanics of surfaces, *Transactions ASME, Journal of Tribology,* **112,** 1990, pp. 205–216.
53 **Majumdar, A.** and **Tien, C.L.,** Fractal characterization and simulation of rough surfaces, *Wear,* **136,** 1990, pp. 313–327.
54 **Mandelbrot, B.B.,** Self-affine fractals and fractal dimensions, *Physica Scripta* **32,** 1985, pp. 257–260.
55 **Borgerfors, G.,** Distance transforms in arbitrary dimensions, *Comp. Vision,* Graphics Image Proc., **27,** 1984, pp. 321–345.
56 **Grassberger, P.** and **Procaccia, I.,** Characterisation of strange attractors, *Phys. Rev. Letters,* **50,** 1983, pp. 346–349.
57 **Judd, K.,** An improved estimator of dimension and

comments on providing confidence intervals, *Phys. D.,* **56,** 1992, pp. 216–228.

58 **Ganti, S.** and **Bhushan, B.**, Generalized fractal analysis and its applications to engineering surfaces, *Wear,* **180,** 1995, pp. 17–34.

59 **Dong, W.P., Sullivan, P.J.**, and **Stout, K.J.**, Comprehensive study of parameters for characterizing three-dimensional surface topography II: Statistical properties of parameter variation, *Wear,* **167,** 1993, pp. 9–21.

60 **Tricot, C., Ferland, P.**, and **Baran, G.**, Fractal analysis of worn surfaces, *Wear,* **172,** 1994, pp. 127–133.

61 **McWald, T.H., Vorburger, T.V., Fu, J., Song, J.F.**, and **Whitenton, E.**, Methods divergence between measurements of micrometer and sub-micrometer surface features, *Nanotechnology,* **5,** 1994, pp. 33–43.

62 **Wickramasinghe, H.K.**, Scanned-probe microscopes, *Scientific American,* **261,** 1989, pp. 98–105.

63 **Caber, P.J.**, Interferometric profiler for rough surfaces, *Applied Optics,* **32,** 1993, pp. 3438–3441.

64 **Boyde, A.**, Quantitative photogrammetric analysis and quantitative stereoscopic analysis of SEM images, *Journal of Microscopy,* **98A,** 1973, pp. 452–471.

65 **Podsiadlo, P.** and **Stachowiak, G.W.**, Characterization of surface topography of wear particles by SEM stereoscopy, accepted for publication in *Wear,* 1996.

66 **Lynch, J.A., Hawkes D.J.**, and **Buckland-Wright, J.C.**, Analysis of texture in macroradiographs of osteoarthritic knees using the fractal signature, *Phys. Med. Biol.,* **36,** No. 6, 1991, pp. 709–722.

67 **Sayles, R.S.** and **Thomas, T.R.**, Surface topography as a non-stationary random process, *Nature,* **271,** 1978, pp. 431–434.

68 **McCool, J.I.**, Relating profile instrument measurements to the functional performances of rough surfaces, *Transactions ASME, Journal of Tribology,* **109,** 1987, pp. 264–270.

69 **Russ, J.**, Surface characterization: fractal dimensions, Hurst coefficients, and frequency transforms., *Journal of Computer-Assisted Microscopy,* **2,** No. 3, 1990, pp. 161–183.

70 **Swanson, P.A.** and **Klann, R.W.**, Abrasive wear studies using the wet sand and dry sand rubber wheel tests, *Proc. Int. Conf. Wear of Materials,* ASME, San Francisco, CA, pp. 379–389, 1981.

71 **Liebhard, M.** and **Levy, A.**, The effect of erodent particle characteristics on the erosion of metals, *Wear,* **151,** 1991, pp. 381–390.

72 **Raask, E.**, Tube erosion by ash impaction, *Wear,* **13,** 1969, pp.301.

73 **Bahadur, S.** and **Badruddin, R.**, Erodent particle characterization and the effect of particle size and shape on erosion, *Wear,* **138,** 1990, pp. 189–208.

74 **Moore, M.A.** and **Swanson, P.A.**, The effect of particle shape on abrasive wear: a comparison of theory and experiment, *Proc. Int. Conf. Wear of Materials,* ASME, NY, pp.1–11, 1983.

75 Glossary of terms relating to powders, British Standard No 2955, London, 1965.

76 *Metals Handbook,* 9th Ed., ASM, **7,** pp. 233–236, 1984.

77 **Prasad, S.V.** and **Kosel, T.H.**, A study of carbide removal mechanisms during quartz abrasion II: Effect of abrasive particle shape, *Wear,* **95,** 1984, pp. 87–102.

78 **Luerkens, D.W,. Beddow, J.K.**, and **Vetter, A.F.**, Invariant Fourier descriptors, *Powder Technology,* **31,** 1982, pp. 209–215.

79 **Allen, M., Brown, B.J.**, and **Miles, N.I.**, Measurement of boundary fractal dimensions: review of current techniques, *Powder Technology,* **84,** 1995, pp. 1–14.

80 **Gupta, L.** and **Srinath, M.D.**, Contour sequence moments for the classification of closed planar shapes, *Pattern Recognition,* **20,** No. 3, 1987, pp. 267–272.

81 **Poon, C.Y.** and **Bhushan, B.**, Comparison of surface roughness measurements by stylus profiler, AEM and non-contact optical profiler, *Wear,* **190,** 1995, pp. 76–88.

The influence of oxidation on the wear of metals and alloys

F H STOTT
Corrosion and Protection Centre, UMIST, Manchester, UK

SYNOPSIS

Sliding wear can be influenced significantly by heat, either frictional or externally applied, since it can facilitate oxidation of the contacting metal or alloy surfaces. This can result in a decrease in wear rate, usually associated with a change from metallic debris to oxide debris. The present paper reviews some of the models developed to account for the generation of oxide during sliding and the effects of such oxides on the rates of wear. Emphasis is placed on high speed unidirectional sliding, where frictional heat can lead to surface temperatures that are sufficiently high to result in relatively thick oxides on the contacting surfaces, and low speed reciprocating sliding, where frictional heat is low but externally applied heat can lead to oxidation of the surfaces. Under the former conditions, wear is caused by spallation of oxide from the contacting asperities; this occurs when the oxide attains a critical thickness, leading to mild-oxidational wear. At very high speeds, surface temperatures may be high enough for the oxide to melt, leading to severe-oxidational wear. Under low speed, reciprocating sliding conditions, oxide and oxidized metal debris can be retained and compacted onto the contacting surfaces, giving wear protection. Here, increased temperature facilitates the generation of oxide debris and assists in compaction of the debris to give the wear-protective layers. At low temperatures, such layers consist mainly of loosely-compacted particles; at higher temperatures, typically >250°C, smooth 'glaze' surfaces develop on top of solidly compacted particle layers, giving even more effective protection against wear damage.

1 INTRODUCTION

Most metals are thermodynamically unstable in air and react with oxygen to form an oxide. As this oxide usually develops as a layer or scale on the surface of the metal or alloy, it can give protection by acting as a barrier that separates the metal from the gas. Further oxidation requires the metal ions and/or oxygen to penetrate across the barrier layer and combine chemically to form new oxide at the scale/gas or scale/metal interface. This process is usually controlled by the rate of diffusion of the reactants across the barrier layer and, hence, is very dependent on temperature, reflecting the exponential relationship between rates of diffusion and temperature. It is also very dependent on the defect nature of the barrier layer. Thus, iron and mild steel form layers of Fe_3O_4 on exposure to air at temperatures up to 570°C. These are effective barriers to diffusion and steel components can be used successfully at such temperatures for long periods. However, at higher temperatures, the phase FeO becomes stable and develops as a layer at the metal/Fe_3O_4 scale interface. This is a very defective oxide and is a poor barrier to diffusion of reactants. Hence, the rate of oxidation of mild steel increases very rapidly at temperatures above 570°C. Thus, for application at higher temperatures, alloys (usually based on iron, iron-nickel or nickel) are designed to develop a slow-growing barrier layer at the base of the oxide scale. For instance, a typical gas turbine alloy, based on Ni-20wt%Cr, contains sufficient chromium to develop a barrier layer of Cr_2O_3 at the base of the NiO-rich scale during oxidation at temperatures up to 900°C. This oxide layer gives effective protection against further oxidation and metal losses are usually acceptably low.

As well as protecting against loss of metal due to chemical oxidation, oxides can also protect against loss of metal due to mechanical damage caused by sliding wear. Indeed, the beneficial effects of oxygen in helping to reduce wear have been recognized for many years. Even at room temperature, metal-on-metal sliding is often accompanied by an initial period of severe wear followed by a change to a lower wear rate, known as mild wear (**1**). This change is favoured by various parameters, such as low sliding speed (**2**), low load (**3**) and high rates of oxidation (**4, 5**). Also, for many metals and alloys, there is a transition temperature (either ambient temperature or surface temperature due to frictional heat) above which the wear rate in the mild wear regime becomes very low compared with the corresponding rate at lower ambient or surface temperatures (**5–7**). This is due to the establishment of a continuous oxide layer (sometimes known as a 'glaze') which gives reduced resistance to sliding and good protection against wear damage (**8**). As discussed in this paper, the mechanisms of establishment of this layer, its nature and composition can be very different from those of the barrier layer which gives oxidation protection. Moreover, although a smooth 'glaze' layer may not develop under all mild-wear conditions, particularly at ambient or surface temperatures below the transition temperature, the relatively low wear rate in the mild wear regime can be accounted for by the formation of a compact layer of oxide or oxide debris particles (**9**).

This paper presents some of the principles of oxidation

and discusses the effects of oxidation on sliding wear. The section on sliding wear is divided into two parts: high speed unidirectional sliding, where friction provides much of the heat for oxidation, and low speed reciprocating sliding, where the heat for oxidation is mainly external heat. Most research has been concerned with the former, particularly at room temperature, while the present author has studied extensively the effects of oxides under the latter conditions. Although there are some common features in the role of oxidation on wear under the two sets of conditions, it is convenient to review them separately here.

2 OXIDATION OF ALLOYS AT ELEVATED TEMPERATURES

The establishment of an oxide scale on an alloy occurs by a nucleation and growth process. When the clean component is exposed to an oxygen-rich gas, small, impinging nuclei of all the thermodynamically stable oxides develop on the surface. In a conventional alloy, such as one based on Ni-20%Cr, in 1 atm air or oxygen, this involves single component oxides, such as NiO and Cr_2O_3, or more complex oxides, such as $NiCr_2O_4$, of all the alloying elements. These initial nuclei of oxide coalesce rapidly to give a complete layer. During this initial or transient stage, the rate of oxidation is rapid; all the elements in the alloy oxidize and the amounts of the various oxides in the layer are approximately proportional to the concentration of the elements in the alloy.

Once the transient oxide layer has been established, it continues to grow following diffusion of metal ions to the scale/gas interface or oxygen to the scale/alloy interface. The rate of thickening of the layer is determined by the temperature, the oxygen pressure and the spatial distribution, the amount, the composition and the structure of the initial oxide phases (**10**). Thus, for Ni-20%Cr, where approximately 80% of the surface is covered by the NiO phase and 20% by the Cr_2O_3 phase in the initial stages, (neglecting any $NiCr_2O_4$ that may develop), the faster-growing NiO phase overgrows the Cr_2O_3 phase and a transient oxide scale layer of essentially NiO develops, incorporating the other transient oxide phases.

At the same time, the thermodynamically favoured oxide (Cr_2O_3 on Ni-20%Cr) attempts to establish as a complete layer at the base of the transient oxide scale layer. This is achieved after a short period for this alloy at high temperatures (>600°C). Sufficient Cr_2O_3 nuclei develop in the alloy, at or close to the alloy/scale interface, to coalesce and form a complete layer, the oxygen activity at the transient oxide scale layer/alloy interface being sufficiently high to oxidize selectively chromium in the alloy. This is shown schematically in Figure 1 (**11**). The rate of oxidation is then controlled by transport of reactants across this Cr_2O_3-rich layer, a much slower process than across the initially formed NiO-rich layer.

However, in alloys containing less than 10 to 15%Cr, there may be insufficient chromium to establish a complete

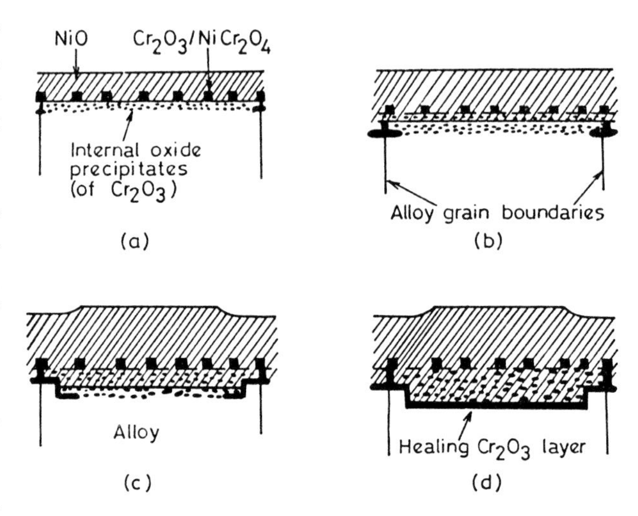

Fig. 1 Schematic representation showing transient oxidation and subsequent development of a healing Cr_2O_3-rich layer on Ni-20%Cr at high temperature (**11**).

Cr_2O_3-rich layer; instead, Cr_2O_3 forms as discrete internal oxide particles in the alloy matrix and the steady-state scale remains essentially NiO. Although the Cr_2O_3 particles are incorporated eventually into the scale, as shown schematically in Figure 2 (**12**), they are insufficient in density to act as effective barriers to transport of reactants and the oxidation rate is controlled by transport through the less protective NiO matrix. Hence, the steady-state scale developed on Ni-20%Cr is much more protective than that developed on alloys containing less than 10 to 15%Cr, leading to the design of high-temperature alloys based on the former composition. In general, the oxide that forms the

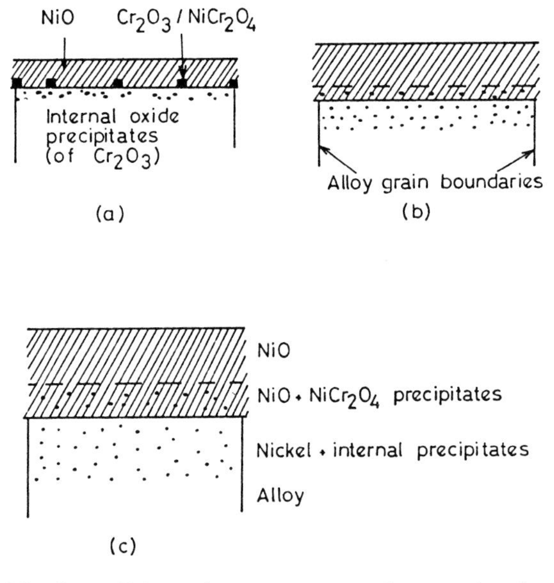

Fig. 2 Schematic representation showing transient oxidaiton and subsequent development of the steady-state scale on Ni-5%Cr at high temperature (**12**).

steady-state scale depends on various factors, such as alloy composition, oxygen solubility and diffusivity in the alloy, alloy interdiffusion coefficients, growth rates of the various oxides, microstructure and surface condition of the alloy, mechanical properties of the scale and the oxidation conditions, particularly temperature and oxygen partial pressure (**13**).

3 GENERATION OF OXIDES DURING SLIDING WEAR

3.1 High-speed unidirectional sliding

Although there is good understanding of oxidation of metals and alloys, and mechanisms of development of oxides, this is less true in situations where relative motion occurs and the contacting metal or alloy components are subjected to sliding stresses as well as chemical degradation. However, even at room temperature, the development of oxide on the load-bearing contacts can reduce to some extent metal-to-metal interactions and, hence, wear rate, giving mild wear conditions.

There have been numerous studies of such oxidational wear in the past 35 years, since it is relevant to many practical sliding situations. Most have been concerned with the oxidation induced by frictional heat during room temperature unidirectional sliding, although oxidation induced by external heat is important in high-temperature sliding, such as occurs in jet engines. In their review of oxidational wear, with particular reference to sliding of steels at room temperature, Lim and Ashby (**14**) distinguished two regimes: mild-oxidational wear and severe-oxidational wear, where the prefixes, 'mild' and 'severe' refer to the extent of oxidation rather than wear; indeed the wear rate is often lower in the severe-oxidational wear regime and both regimes can give mild wear. They indicated that sliding speeds of at least 1 m s^{-1} were required to give sufficient frictional heat for oxidation to become important, with a flash temperature approaching 700°C being possible at 1 m s^{-1}. At lower speeds, the wear debris is mainly metallic; a drop in wear rate is associated with a change to oxide debris (**3, 15**). At the higher velocities, an oxide film forms on the contacting surfaces (**16**) and, above a critical velocity, wear is caused by removal of the oxide (**17**). Good evidence has been provided to show that, in the mild-oxidational wear regime, oxidation at the points of contact is caused by frictional heat and the oxide formed grows until, at a critical oxide film thickness, about 10 μm for steel, it spalls off as wear debris (**18**). A schematic representation of this mechanism of oxide generation is shown in Figure 3 (**14**).

Lim and Ashby (**14**) also proposed that, at velocities above about 10 m s^{-1}, oxidation becomes more extensive, resulting in a transition to severe-oxidational wear. This is associated with general oxidation and, even, with the oxide becoming plastic and melting locally to a viscous liquid that can flow under the sliding action, as shown schematically in Figure 4.

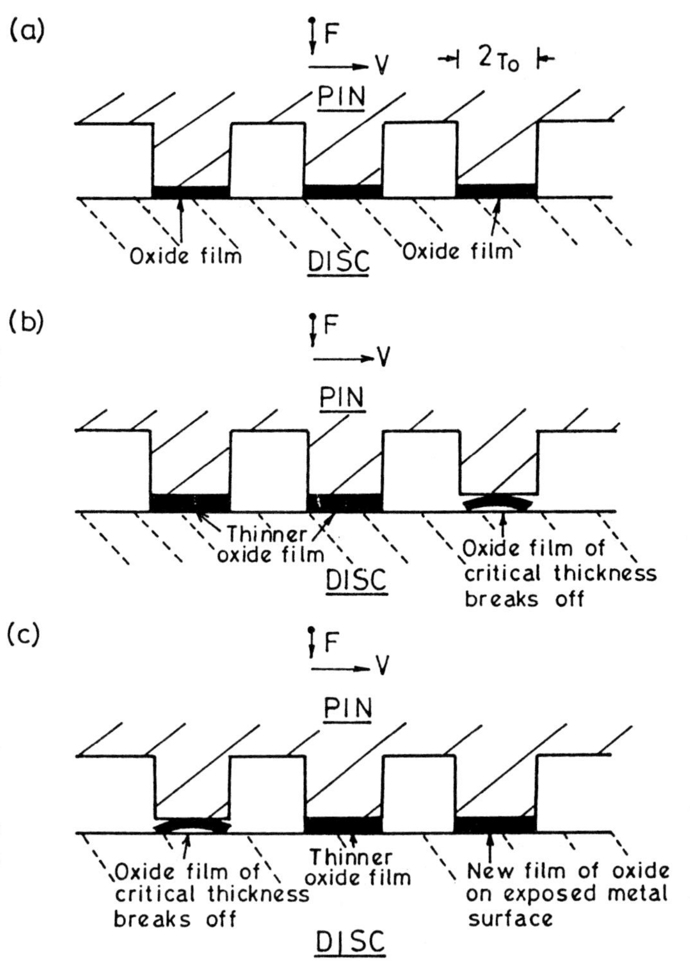

Fig. 3 Schematic representation of idealized mild-oxidational wear model for pin-on-disc specimen configuration (**14**).
(a) Oxide films grow on the asperity contacts.
(b) The critical oxide film thickness is attained on one of the contacts and the oxide layer breaks off as wear debris.
(c) New oxide grows on the metal exposed following loss of the oxide film while the oxide film on another asperity reaches the critical thickness and breaks off.

3.2 Low speed reciprocating sliding at elevated ambient temperatures

The present author has studied extensively the effects of oxides on wear during reciprocating sliding at relatively low speeds (75 to 100 mm s^{-1} on average) at ambient temperatures from 20° to 800°C (eg **5, 8, 19–21**). The mechanisms of oxide generation showed some similarities

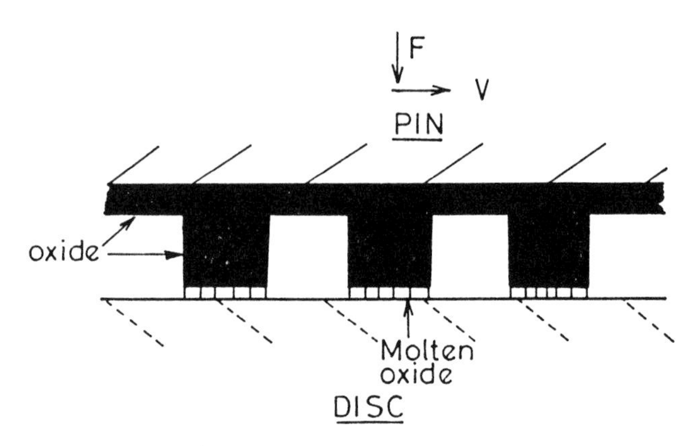

Fig. 4 Schematic representation of idealized severe-oxidational wear model for pin-on-disc specimen configuration (**14**).

to those during high speed sliding but there was much less influence of frictional heat under these low speed conditions and no indication of oxide melting in the wear process, even at ambient temperatures of 800°C.

In particular, three limiting cases for the generation of oxide were identified (**20**):

(i) Oxide may be formed by oxidation of the metal asperities while in contact, the extent of such oxidation in one traversal being dependent on the temperatures developed at the asperity contacts, the duration of the contact and the oxidation characteristics of the metal. At the same time, more general oxidation of all the apparent area of contact occurs, with the amount depending on the surface temperatures. Oxide from both sources may be removed completely during subsequent traversals, exposing clean metal to the environment which is re-oxidized in the subsequent traversal. This is similar to the mild-oxidational wear model and had been proposed previously (**22**). The resulting wear debris may be swept aside or cause abrasion of the metal substrate, resulting in wear, or may be compacted between the sliding surfaces to give wear-protection (**8**). This was termed an 'oxidation-scrape-reoxidation' mechanism.

(ii) In some conditions, particularly at high ambient temperatures, the oxide developed during sliding (or present on the surface prior to sliding) may be removed only partially or not at all by the sliding action and continues to thicken with time (termed a 'total oxide' mechanism). This can give protection against further wear, as discussed for the severe-oxidational model earlier. However, here, melting of the oxide is not observed and the mechanism of protection is associated with deformation of the oxide.

(iii) A third source of oxide during sliding is oxidation of metallic debris. Thus, such wear debris particles, produced in the early stages, may be broken up and reduced in size by the sliding action, thereby exposing fresh areas of clean metal for further oxidation (termed a 'metal debris' mechanism). The extent of such oxidation can be considerable since the rate of initial or transient oxidation can be relatively rapid while the surface area of exposed metal may be large. Thus, for iron or steel, a limiting oxide thickness of 2 nm is achieved in less than 0.1 s at 20°C (**23**). Applying such data to a collection of iron particles of 0.1 μm diameter, about 6% of the debris is oxidized in a fraction of a second at 20°C (**21**). During sliding, the oxidation process is promoted by the heat of deformation and the increased energy of the particles due to the increased defect density and surface energy. Also, if the heat of oxidation released is taken into account, fine metallic particles may be oxidized spontaneously and completely under some conditions (**24**). Hence, a considerable amount of oxide can be generated during low speed sliding, even at low temperatures (**21**). The resulting oxide debris can develop into a wear-protective layer, as considered later.

4 THE INFLUENCE OF OXIDES ON SLIDING WEAR

4.1 High-speed unidirectional sliding

The model for mild-oxidational wear (Fig. 3) was developed by Quinn, *et al.* (**17, 18, 25–28**) and an iterative technique was used to ascertain the values of the parameters from experimental observations (**29**). Lim and Ashby (**14**) presented a slightly modified version for wear of mild steel, as follows:

Assuming that oxidation of iron follows parabolic kinetics, then the mass of oxygen taken up by the oxide film per unit area in time, t, is:

$$m^2 = K_p t \tag{1}$$

where:
K_p is the parabolic oxidation rate constant,

$$K_p = B\exp-\left[\frac{Q}{RT}\right], \tag{2}$$

B is the Arrhenius constant,
Q is the activation energy for oxidation,
R is the gas constant,
T is the absolute temperature.

It was assumed that the oxide film forms on the asperities and thickens with time until it reaches a critical thickness, x_c, at which it spalls off as wear debris. It was also assumed that the average composition of the oxide was Fe_3O_4; hence, if a volume, V_{Fe} of iron is oxidized per unit area, the mass gained per unit area as a result of formation of oxide is:

$$m = \frac{2}{3} V_{Fe\rho} \frac{M_o}{M_{Fe}} \qquad 3$$

where:

ρ is the density of iron,
M_o is the molecular weight of oxygen,
M_{Fe} is the molecular weight of iron.

Assuming that the thickness of the oxide, x, is equal to the thickness of iron from which it originates (i.e. is equal to V_{Fe}, and neglecting the volume expansion that normally occurs on oxidation:

$$x^2 = C^2 K_p t \qquad (4)$$

where:

$$C = \frac{3 M_{Fe}}{2 M_o \rho}$$

In the mild-oxidational wear regime, it is assumed that oxidation is caused by frictional heat and is confined to the asperity tips. Then, the time to reach the critical oxide thickness, x_c, is given by:

$$t_c = \frac{x_c^2}{C^2 B \exp\left[-Q / RT_f\right]} \qquad (5)$$

where T_f is the flash temperature.

Wear is caused by this oxide spalling, so a volume, $A_r x_c$, (where A_r is the real area of contact) is lost in every time period, t_c. During this period, the sliding distance is vt_c (where v is the velocity), so the wear rate, W, is:

$$W = \frac{A_r x_c}{vt_c} = \frac{A_r C^2 B}{v x_c} . \exp\left[-\frac{Q}{RT_f}\right] \qquad (6)$$

or, in dimensionless variables:

$$\tilde{W} = \left(\frac{C^2 B r}{x_c a}\right) . \exp\left[-\frac{Q}{RT_f}\right] \frac{\tilde{F}}{\tilde{v}} \qquad (7)$$

where:

$\tilde{W}$ is the normalized wear rate = $\dfrac{W}{A_n}$,

$\tilde{F}$ is the normalized pressure on the sliding surface = $\dfrac{A_r}{A_n} = \dfrac{F}{A_n H}$,

$\tilde{v}$ is the normalized velocity = $\dfrac{v r}{a}$,

r is the radius of the circular nominal contact area,
A_n is the nominal area of contact,
a is the thermal diffusivity of the metal,
H is the room temperature hardness of the metal.

Although this equation is relatively simple, it is very difficult to fit to measured wear data. It has been proposed that B and Q should be treated as adjustable parameters, chosen to fit measurements of wear rate (**14**). By assuming a critical oxide thickness of 10 μm, using the value of Q obtained from static oxidation measurements and selecting a high value of B (since literature values of B for static oxidation range from 10^{-2} to 10^6 kg^2 m^{-4}s^{-1}), reasonable agreement between experimental and predicted wear rates was obtained, as shown in Figure 5. Here, contours of normalized wear rate, calculated from equation (7), are plotted, using data specified in reference (**14**). Superimposed on the contours are experimental datum points for steels tested at velocities below 5 m s^{-1}, taken from several sources.

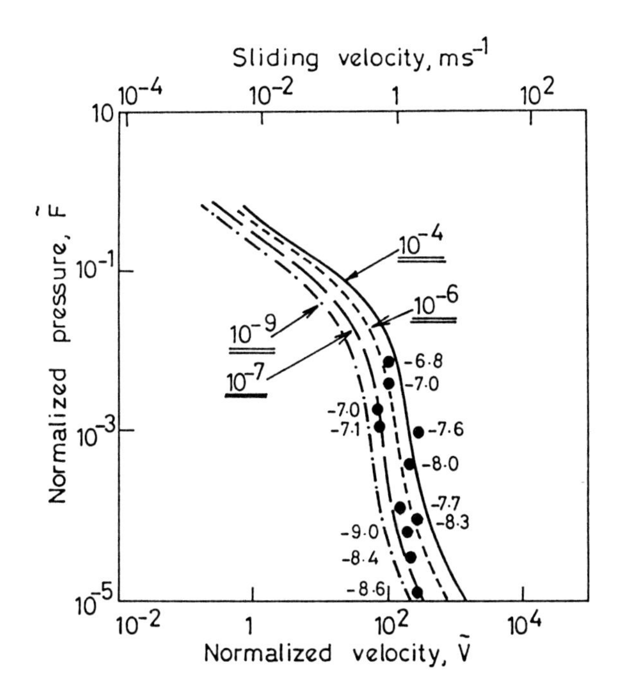

Fig. 5 Calibration of the mild-oxidational wear model. The contours of normalized wear rate, $\tilde{W}$, for values from 10^{-4} to 10^{-9} are indicated, calculated from equation (7). The numbers given against the experimental points are $\log_{10} \tilde{W}$ (**14**).

However, in the present author's view, the uncertainties regarding the magnitude of the flash temperature, the temperature gradient from the tip into the bulk substrate, the rate of temperature decay when the asperities come out of contact, and the number and mean size of the load-bearing junctions, make it difficult to use data from static oxidation studies. In any case, there is a change in mechanism for oxidation of iron under static conditions on increasing the temperature through 570°C. At lower temperatures, the oxidation rate is relatively low, with the Fe_3O_4 scale giving good protection. However, at higher temperatures, FeO becomes thermodynamically stable and develops as a layer between the steel and the Fe_3O_4 layer. Diffusion across the FeO layer then controls the rate of oxidation; this oxide is very defective and grows much faster than the Fe_3O_4 layer. Hence, the relationship:

$$K_\rho = B\exp-\left[\frac{Q}{RT}\right]$$

is only valid over the temperature range up to 570°C. A similar relationship would hold at 570° to 1000°C, but the values of B and Q would be very different. Presumably, this accounts for the wide range of B values referred to from the literature (**14**).

A further difficulty in comparing experimental data with predicted values relates to the assumed critical oxide film thickness for spallation under the stresses generated during sliding (10 µm). There have been many studies of the criteria for spallation of scales under static conditions, particularly during cooling where differential thermal contraction between the metal and scale can result in large compressive stresses and strains in the scale, often leading to scale spallation. The critical compressive strains for spallation of oxide layers decrease with increasing layer thickness (**30**), while numerous other factors, such as impurities in the alloy, curvature of the specimen surface, rate of application of strain, and the extent of voidage and number of defects in the layer also influence the value of the critical strain. The complex issue of mechanical properties of protective oxide scales has been reviewed at a recent conference, published in reference (**31**).

Lim and Ashby (**14**) have also modelled the severe-oxidational wear regime (Fig. 4); here, continuous oxides develop over the surface and the frictional heat is reported to be sufficient to melt locally the oxide at the asperity tips. The melting asperity absorbs latent heat of melting and the molten material flows and spreads across the adjacent cooler surface where it resolidifies, releasing its latent heat. In the model, it is assumed that a fraction, f_m, of all the material that melts is lost as wear fragments and that, since smearing of the liquid redistributes the heat, the immediate subsurface temperature is at the bulk surface temperature. Hence, the heat lost by conduction can be calculated, allowing the amount retained in the pin and disc surfaces to be determined. The total volume of oxide that melts and, thus, the wear rate can then be estimated as follows:

$$\tilde{W} = f(m)\left[\frac{K_o(T_o - T_b)}{L_o\,a}\frac{(\tilde{F}N)^{1/2}}{\beta\tilde{v}}\right]\left[\alpha\mu\frac{aH\beta}{K_o(T_o - T_b)}\left(\frac{\tilde{F}}{N}\right)^{1/2}\tilde{v} - 1\right] \quad (8)$$

where:

K_o is the thermal conductivity of the oxide,
T_o is the melting point of the oxide,
T_b is the bulk temperature,
L_o is the latent heat of fusion per unit volume for oxide,
N is the total number of contacting asperities,
β is a dimensionless parameter for bulk heating,
α is the heat distribution coefficient,
μ is the coefficient of friction.

Reasonable fit was found between the model and experimental wear data (**32–34**), for a value of $f_m = 0.01$, as shown in Figure 6 (**14**). Here, contours of normalized wear rate, calculated from the model equations, are plotted while experimental datum points, taken from several sources, are superimposed on the contours.

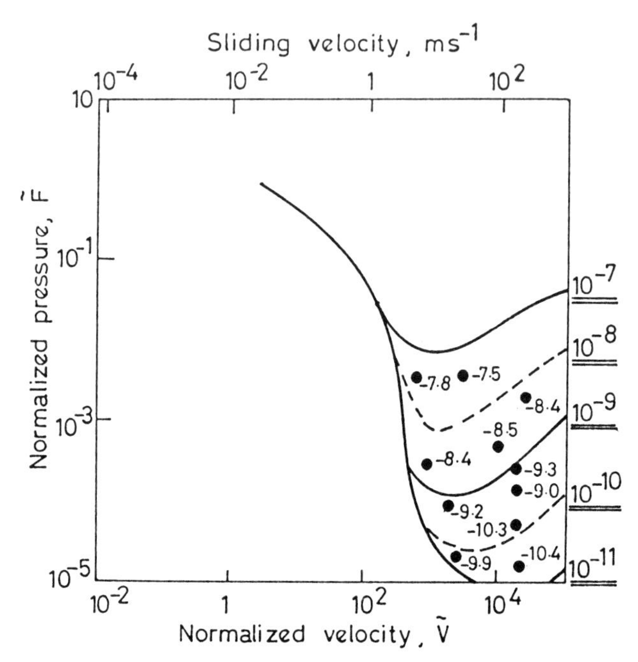

Fig. 6 Calibration of the severe-oxidational wear model. The contours of normalised wear rate, $\tilde{W}$, for values of 10^{-7} to 10^{-11} are indicated, calculated from the model equations and assuming f_m = 0.01. The numbers given against the experimental points are $\log_{10} \tilde{W}$ (**14**).

4.2 Low speed reciprocating sliding at elevated ambient temperatures

4.2.1 Establishment of wear-protective layers

There is significant evidence that the wear of alloys is affected by the presence of debris particles, particularly under reciprocating conditions where such debris can be retained on the contacting surfaces, as shown schematically in Figure 7 (**21**). Thus, as described earlier, during sliding, wear debris particles are generated by the relative motion under load. Although some are lost, resulting in wear, others are retained within the tracks where they are comminuted by repeated plastic deformation and fracture while moving freely between the sliding surfaces. However, once they have been reduced to a sufficiently small size, they are agglomerated at certain locations, particularly in grooves, due to adhesion forces between solid surfaces arising from surface energy (**35**), and develop compact layers which eventually become load-bearing (Fig. 7(c)). During these processes, the fine particles are sintered together to some extent to form more solid layers; such sintering of fine particles can occur at temperatures only slightly above 20°C (**36**), but is more significant at higher ambient or surface temperatures. This reduces material loss since newly formed debris particles are re-cycled into the layers while, as the particles are heavily deformed and oxidized, the layers are hard and give protection against wear damage. As sliding continues, two competitive processes occur: break-down of the layers, resulting in wear (Fig. 7(e)), and sintering/cold welding between the particles, which consolidates the layer (Fig. 7(d)). The latter process, and oxidation of the particle layers, occur more rapidly as the temperature is increased. If the surfaces become solid before breakdown occurs, wear decreases to a very low value due to the creation of a 'glaze' layer on top of the debris particles. On the other hand, if the particle layers are not very well compacted and sintered, the loose hard particles are removed more easily and the amount of wear damage is higher. This accounts for the concept of a critical temperature, above which the wear rate in the mild wear regime is very low and below which it is relatively higher. The critical temperature is that temperature required to establish a 'glaze' surface under a given set of conditions.

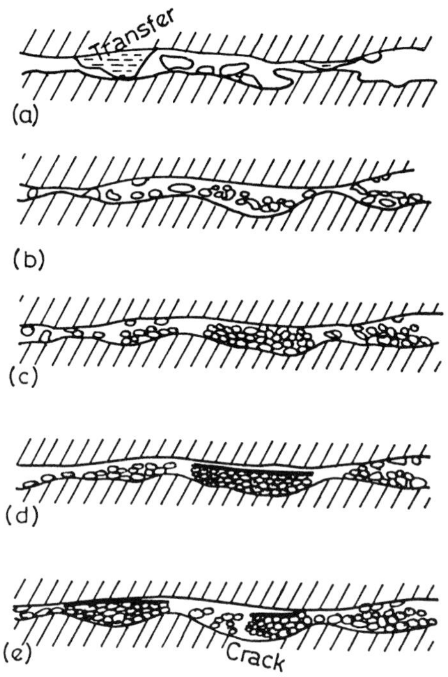

Fig. 7 Schematic representation of a model for the development of wear-protective oxide layers from wear debris particles (**21**).
(a) Generation of metallic wear debris.
(b) Comminution, oxidation and agglomeration of debris particles.
(c) Compaction of agglomerated debris particles.
(d) Development of 'glaze' layer over compacted particle layer under some conditions.
(e) Breakdown of wear-protective layers and development of new protective layers in other areas.

A model to account for the establishment of such wear-protective layers has been proposed (**9**). It was assumed that the wear rate of the wear-protective layers is negligible compared to that of other areas and wear debris particles are only generated from the other areas. Thus, the wear volume after sliding time, t, is:

$$V_t = \frac{\pi}{6}\int_0^t \{A_t\, N_t\, [1 - C_e(t)] \int_0^\infty [D^3 f(D)\, P_r(D)]\, dD\}\, dt \qquad (9)$$

where:

N_t is the number of wear debris particles formed per unit time at time, t.

A_t is the apparent area of the wear scar at time, t.

$C_e t$ is the effective coverage by wear-protective layers at time, t.

$f(D)\,dD$ is the percentage of newly generated particles that fall in the diameter range of D to $(D+dD)$.

$P_r(D)$ is the probability that a wear particle of diameter, D, is removed from the wear track.

The volume of wear debris particles retained within the sliding surfaces at time, t, is the same expression except that $[1-P_r(D)]$ is substituted for $P_r(D)$ in equation (9). If it is assumed that all retained particles are eventually compacted into wear-protective layers, of average thickness, δ, then the coverage of the wear surface by such layers is:

$$C_{comp} = \frac{V_r(t)}{2A\delta} \qquad (10)$$

These layers are particularly effective if a 'glaze' oxide of critical thickness, δ_c can be established on their surfaces. The oxidation kinetics of the wear debris can be expressed in terms of a scale thickness, x, as:

$$x = g\,(t)$$

where $g(t)$ is the relevant function of time.

Hence, the critical time, t_c, for a 'glaze' of critical thickness, δ_c, to develop on the compacted debris is:

$$t_c = g^{-1}(\delta_c) \quad (11)$$

where $g^{-1}(\delta_c)$ is the inverse function of $g(t_c)$.

If an area of compacted particle layer $dA_c(\tau)$ is established at time, τ, then an area of 'glaze' equal to $dA_c(\tau)$ should have developed after a further period, t_c, ie at a time $(\tau + t_c)$. The total area of 'glaze' surface at time, t, is:

$$A_g(t) = \int_o^{t-t(c)} dA_c(\tau) \quad (12)$$

From the above, it is apparent that both a compacted particle layer and a 'glaze' layer can co-exist; each is wear-protective, although the latter is more effective. From experimental observations (**21**), the fractional coverage of the wear surface by 'glaze' layer needed to cause the severe to mild wear transition is much less than that by the non-'glaze' compact particle layer.

4.2.2 Criteria for removal of a wear debris particle from the sliding surfaces

Wear of the surface under sliding conditions involves removal of wear debris particles. A model has been developed to account for the wear observed during reciprocating sliding (**9**), based on the concept that, when a protruding asperity from a sliding surface comes into contact with a loose debris particle, the latter may roll forward, skid at the point of contact or become entrapped in the contacting surfaces, as shown in Figure 8 (**21**); the first two result in removal of the particle from the surfaces. A fourth mechanism, rotation of the particle, does not cause wear as the particle does not move in the direction of the sliding counterface.

It is assumed that the particle is a sphere of diameter, D (Fig. 9 (a)); the static equilibrium conditions for the particle are given in Figure 9 (b), where T_1, T_2, N_1 and N_2 are the local contact forces. It has been shown (**9**) that the critical criterion for the particle to be removed by skidding or rolling occurs when:

$$\sin \alpha < \frac{1-f^2}{1+f^2} \quad (13)$$

where:
$f = \min(f_1, f_2)$,
$T_1 \max = f_1 N_1$,
$T_2 \max = f_2 N_2$,
T_1 max and T_2 max are the maximum frictional forces that can be attained.

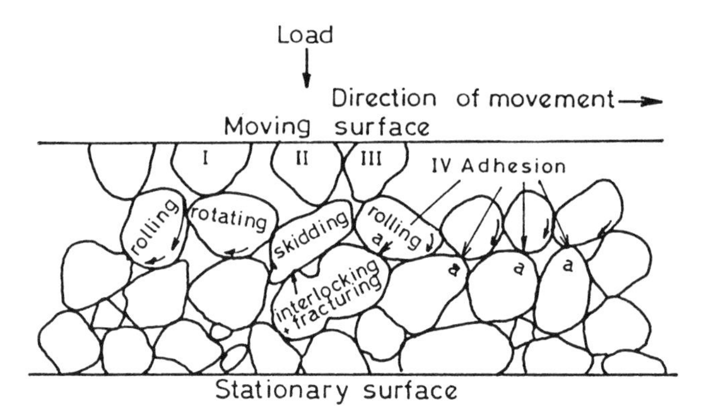

Fig. 8 Schematic diagram showing the four possible mechanisms of relative movement of particles during sliding; mechanism I (rotation), II (skidding), III (rolling), and IV (adhesion or entrapment).

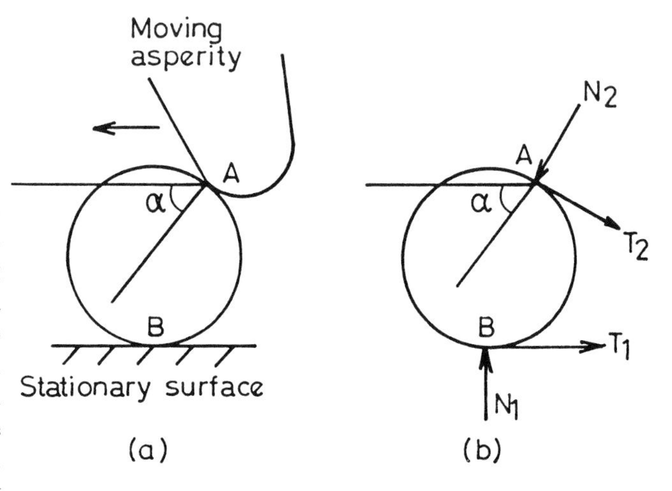

Fig. 9 Schematic diagram showing the contact conditions for a spherical debris particle between a plane surface and a protruding asperity from the countersurface.

Hence, particles that satisfy the criterion at formation will move with the countersurface and may be removed from the wear track or may become entrapped at some location where the contact conditions no longer favour movement. Also, particles that are trapped initially may be removed later if they are unable to agglomerate into a layer. Overall, there are two main conditions leading to removal of the particle: if it is unable to become incorporated into a compacted particle layer (compaction criterion) and if it satisfies the removal criterion of equation (13), even when it remains at the lowest point on the wear surface (contact-geometry criterion). These two conditions have been analyzed separately (**9**), giving the following:

The compaction criterion for removal of a particle, P_1, is:

$$P_1 = \begin{cases} 1, \text{ when } D \geq D_{c1} \\ 0, \text{ when } D < D_{c2} \end{cases}$$

where: $$D_{c1} = \frac{3\pi}{8} \frac{\gamma\phi A}{L(P_a / N_p)_{crit}}$$

$$(P_a / N_p)_{crit} = \frac{1-k}{1-k^n}\left(\frac{1}{c} \cdot \frac{c+fs}{s-fc} - \frac{1}{s}\right)$$

$$k = \frac{s+f(1-c)}{s-f(1+c)}$$

$(P_a/N_p)_{crit}$ is the critical value of the relative adhesion force for a stable particle layer to form,
L is total normal load,
γ is the energy of adhesion between particles originating from surface energy of the solid particles,
ϕ is the ratio of real area of contact to apparent area of contact,
$s = \sin \alpha$,
$c = \cos \alpha$.

The compact-geometry criterion, P_2, follows from equation (13), using the Greenwood-Williamson surface contact model (**37**) and assuming that a rough surface with hemispherical asperities is in contact with a rigid smooth countersurface. The height distribution of asperities is assumed to be Gaussian, with 95% being higher than $Z = -2\sigma$; this is taken as the lowest height on the surface. From the analysis:

$$P_2 = \begin{cases} 1 \text{ when } D \geq D_{c2} \\ 0 \text{ when } D < D_{c2} \end{cases}$$

where:
$D_{c2} = (1 + f^2)(d + 2\sigma)$
d is the separation between the two sliding surfaces
σ is the standard deviation of the asperity height distribution.

Overall, the probability of particle removal, P_r, is obtained from the above two probabilities, P_1 and P_2

$$P_r = P_1 \vee P_2 = \begin{cases} 1, \text{ when } D \geq D_c \\ 0, \text{ when } D < D_c \end{cases}$$

where D_c is min (D_{c1}, D_{c2}).

Using the model, the wear rates for like-on-like reciprocating sliding of a nickel-base alloy have been calculated and compared with those observed for Nimonic 80A (an alloy based on nickel-20% chromium) during experiments at ambient temperatures of 150° to 600°C. It was assumed that most of the oxide produced in the tests was nickel oxide, since growth of this oxide predominates in the early stages, especially at low temperatures, that such oxide thickens in a logarithmic relationship with time at temperatures below 350°C and in a parabolic relationship with time at higher temperatures, and that the size generation of wear debris particles when they have just been generated is Gaussian (**9**).

Figure 10 gives plots of the calculated variations of wear volume and surface coverage by compacted debris particles and by 'glaze' layers as functions of time at several ambient temperatures. In the early stages, metal-metal contact and large contact stresses, due to the hemispherical pin-on-disc specimen geometry, produce large particles. This and high contact pressures favour a high probability of removal of particles from between the sliding surfaces. As sliding continues, the wear scar size increases and the contract pressures decrease while the debris particles entrapped between the sliding surfaces are comminuted, until they eventually become smaller than the critical size for compaction. Thereafter, the surface coverage by compacted particle layers increases rapidly, resulting in a sharp decrease in wear rate. At higher temperatures, 'glaze' layers develop on top of the compacted debris as a result of increased sintering and oxidation of the particles, leading to more solid and wear-protective surfaces.

Figure 11(a) compares the calculated and experimentally measured wear volumes as a function of temperature. The wear rate decreases rapidly with increase in temperature at about 250°C. This cannot be accounted for entirely by the increased rates of oxidation on increasing temperature, as apparent from the Arrhenius plot shown in Figure 11(b). Here the same wear data have been plotted against reciprocal temperature. From the gradient of the slope, the activation energy for the wear process is approximately 3000 cal mol^{-1}. This value is more than one order of magnitude less than that for oxidation of a metal but is within the range of activation energy for physical adsorption interactions between solid surfaces (**38**). It is also within the range of activation energy for adhesion between contacting asperities in the stick-slip static friction process (**39**).

Although tribo-oxidation and static oxidation are different processes, the activation energies, in both cases, are essentially the energy barrier for kinetic processes within the oxide layers and should be similar (**29**). Hence, the low activation energy for the wear process indicates that the development of wear-protective layers is not controlled by the oxidation process but is related to adhesion between debris particles in the wear surfaces, as stipulated in the model. Thus, the main effects of

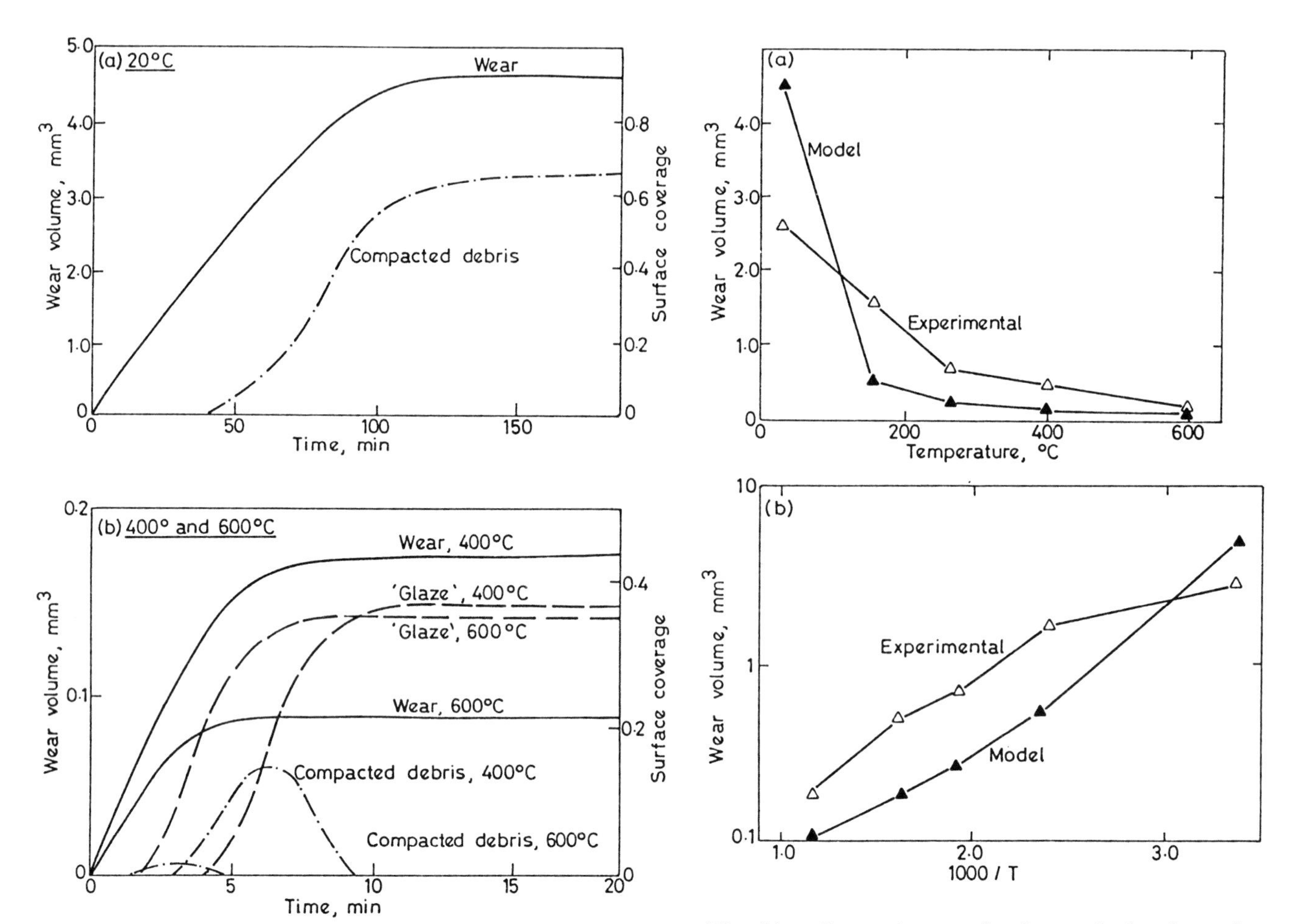

Fig. 10 Calculated plots of wear against sliding time under a load of 15 N at (a) 20°C and (b) 400° and 600°C, showing surface coverage by compacted debris and by 'glaze' layers.

Fig. 11 Comparison of the calculated and experimentally determined variations of wear volume (6 h sliding) under a load of 15 N.
(a) with temperature,
(b) with reciprocal temperature.

increasing temperature are that the rate of generation of wear debris particles, particularly oxide debris, is increased, more debris particles are involved in establishment of the compacted wear-protective layers and the compaction of such particles into a surface layer is facilitated.

5 CONCLUSIONS

1. Oxidation can have a beneficial effect in reducing wear during sliding of metals and alloys, particularly by preventing metal-metal contact. The heat required for oxidation can be induced by friction or can be applied externally.
2. During high speed, unidirectional sliding, a model based on spallation of oxide of thickness greater than a critical value from the contacting asperities has been proposed to account for the wear observed in mild-oxidational wear conditions. At very high speeds, the surface temperatures may cause melting of the oxide, leading to severe-oxidational wear.
3. During low speed, reciprocating sliding, under conditions where debris can be retained between the contacting surfaces, such debris can be compacted into the surfaces, giving effective wear protection. A model to account for the observed wear rates has shown that increased temperatures facilitate the generation of oxide debris and assist in compaction of the debris to develop the wear-protective layers.

6 ACKNOWLEDGMENTS

The author is grateful to his colleagues at UMIST who have collaborated on the low speed, reciprocating sliding research, particularly G.C. Wood, J. Jiang, J. Glascott and M.M. Stack, and the various groups who have studied high speed, unidirectional sliding.

7 REFERENCES

1 **Archard, J.F.** and **Hirst, W.,** *Proc. Roy. Soc. London,* 1956, **A236,** 397.

2 **Lancaster, J.F.**, *Proc. Roy. Soc. London,* 1963, **A273,** 466.
3 **Welsh, N.C.**, *Phil. Trans. Roy. Soc. London,* 1965, **A257,** 31.
4 **Eyre, T.S.** and **Maynard, D.**, *Wear,* 1971, 18, 301.
5 **Stott, F.H., Glascott, J.**, and **Wood, G.C.**, *Wear,* 1985, **101,** 311.
6 **Smith, A.F.**, *Tribology Int.,* 1986, **19,** 65.
7 **Newman, P.T.** and **Skinner, J.**, *Wear,* 1986, **112,** 291.
8 **Stott, F.H.** and **Wood, G.C.**, *Tribology Int.,* 1978, **11,** 211.
9 **Jiang, J., Stott, F.H.** and **Stack, M.M.**, *Wear,* 1995, **181–183,** 20.
10 **Chattopadhyay, B.** and **Wood, G.C.**, *Oxidation Metals,* 1970, **2,** 373.
11 **Stott, F.H., Bartlett, P.K.N.**, and **Wood, G.C.**, *J. Mater. Sci. Eng.,* 1987, **88,** 163.
12 **Wood, G.C.** and **Hodgkiess, T.**, *Nature,* 1966, **211,** No.5056, 1358.
13 **Stott, F.H.**, *Reports on Progress in Physics,* 1987, **50,** 783.
14 **Lim, S.C.** and **Ashby, M.F.**, *Acta Metall.,* 1987, **35,** 1.
15 **Uetz, H.** and **Sommer, K.**, *Wear,* 1977, **43,** 375.
16 **Nakajima, K.** and **Mizutani, Y.**, *Wear,* 1969, **13,** 283.
17 **Quinn, T.F.J.**, *Wear,* 1971, **18,** 413.
18 **Quinn, T.F.J.**, *Br. J. Appl. Phys.,* 1962, **13,** 33.
19 **Stott, F.H., Glascott, J.**, and **Wood, G.C.**, *Corrosion-Erosion-Wear of Materials at Elevated Temperatures,* ed. A.V. Levy, 1987, 263, (NACE, Houston).
20 **Stott, F.H., Glascott, J.**, and **Wood, G.C.**, *Proc. Roy. Soc. London,* 1985, **A402,** 167.
21 **Jiang, J., Stott, F.H.**, and **Stack, M.M.**, *Wear,* 1994, **176,** 185.
22 **Tao, F.F.**, *ASLE Trans.,* 1969, **12,** 97.
23 **Kruger, J.** and **Yolken, H.T.**, *Corrosion,* 1994, **20,** 29t.
24 **Davies, D.E., Evans, U.R.**, and **Agar, J.N.**, *Proc. Roy. Soc. London,* 1954, **A225,** 443.
25 **Quinn, T.F.J.**, *ASLE Trans.,*1978, **21,** 78.
26 **Quinn, T.F.J., Dowson, D.M.**, and **Sullivan, J.**, *Wear,* 1980, **65,** 1.
27 **Rowson, D.M.** and **Quinn, T.F.J.**, *J. Phys. D, Appl. Phys.,* 1980, **13,** 209.
28 **Quinn, T.F.J.**, *Tribology Int.,* 1983, **15,** 257.
29 **Quinn, T.F.J., Sullivan, J.L.**, and **Dowson, D.M.**, *Wear,* 1984, **94,** 175.
30 **Nagal, M.M., Saunders, S.R.J.**, and **Guttman, V.**, *Materials at High Temperature,* 1994, **12,** 163.
31 *Materials at High Temperature,* 1994, **12,** 83–256.
32 **Cocks, M.**, *J. Appl. Phys.,* 1957, **28,** 835.
33 **Powell, D.G.** and **Earles, S.W.E.**, *ASLE Trans.,* 1968, **11,** 101.
34 **Earles, S.W.E.** and **Kadhim, M.J.**, *Proc. Inst. Mech. Eng.,* 1965, **180,** 531.
35 **Johnson, K.L., Kendall, K.**, and **Roberts, A.D.**, *Proc. Roy. Soc. London,* 1971, **A324,** 301.
36 **Zhou, Y.H., Harmelin, M.** and **Bigot, J.**, *Script Metall.,* 1989, **23,** 1391.
37 **Greenwood, J.A.** and **Williamson, J.B.P.**, *Proc. Roy. Soc. London,* 1966, **A295,** 300.
38 **Yamamoto, N.** and **Nakajima, K.**, *Wear,* 1981, **70,** 321.
39 **Brockley, C.A.** and **Davis, H.R.**, *Trans. ASME,* 1968, **F90,** 35.

A survey of tribology problems in the process industry and a suggested methodology for problem solving

J D SUMMERS-SMITH
Tribology Consultant, Guisborough, Cleveland, UK

SYNOPSIS

Examples are given of the types of problems most frequently encountered in tribological mechanisms in machines, with particular reference to those machines operating in the process chemical industry. A methodology for failure investigation is presented.

1 INTRODUCTION

My experience has largely been in the chemical process industry and thus my remarks will be addressed particularly to tribology problems in that industry. These remarks should, however, have relevance to other industries, particularly the petroleum and petrochemical industries, where much of the equipment is essentially similar: turbines, compressors, electric motors, pumps, gears, etc. Although inevitably a few design problems occur with new machines, by far the greatest number of problems occur through interactions with the system of which the machine is an integral part. Purchasing specifications include the operating conditions, but it is rare for the machine manufacturer to be able to test under realistic conditions using the process fluid, particularly the effect of lubricant contamination or process constraints that are likely to be outside the experience of the machine designer, the precise mounting arrangements and the transient conditions that are likely to occur in practice.

While problems of this nature provide new and challenging technical analysis, it remains the case that many of the problems encountered are self-inflicted. Numerous surveys have been published showing patterns of failure for different tribological components (for example, rolling bearings) and the actions that have to be taken to reduce the incidence of failure. It is a sad reflection of human nature that corporate memories are short and the need for remedial action, even within the same organisation, quickly forgotten. Failures of simple tribological components, such as bearings and seals, through fitting or assembly errors, poor lubrication, maloperation, remain a major cause of machine breakdown.

I give typical examples to highlight different tribological problem areas in the process industries. Any selection must invariably reflect the experience of the author. The examples are based on my own experience, but I hope they provide a fair sample of the current scene.

Recognising that problems will continue to be a significant factor in plant reliability and output, some thoughts will be given to failure investigation aimed at giving a fundamental understanding that will provide a cure rather than treating the symptoms, all too common in the stressful situation of getting a plant back on line.

PART I: THE FAILURE SCENE

2 PROBLEM AREAS

2.1 System effects

Many large machine trains involve a driver, gearbox and driven machine with a common system supplying lubricant to bearings, seals, couplings and gears. A turbine-type lubricating oil containing an anti-oxidant and corrosion inhibitor is normally used. Such oils have an FZG Rating of 6/7 and are suitable for the lubrication of the turbine-quality gears normally used. However, should scuffing of the gears occur, either through an incorrect run-in procedure or misalignment, and, even when incorrectly diagnosed, the immediate reaction of the gear manufacturer is to specify an anti-wear (AW) or extreme pressure (EP) gear oil. However, such oils with AW and EP additives are more reactive than normal turbine oils and may cause problems in other parts of the system. I give three examples.

2.1.1 Attack of white metal by zinc dialkyldithiophosphate (ZDDP)

This problem involved a hydrocarbon process gas compressor with white metal-lined floating-bush shaft seals. At the high temperature in the seal (probably > 120°C), caused by the high rate of shear in the oil film (seal bore 0.15 m, shaft speed 83 rev/s) and the low flow rate through the seal, the ZDDP antiwear additive reacted with the copper in the white metal lining of the seal rings forming a deposit of copper sulphide that took up the clearance, resulting in seal failure.

2.1.2 Thermal breakdown of EP additive

On another hydrocarbon processing plant, there were two compressor trains, each compressor driven by an electric motor through speed-increasing helical gears. A scuffing incident was suspected and the gear manufacturer stated that an EP oil must be used, although he had originally agreed to the use of a normal double-inhibited turbine oil. The compressor shafts were fitted with radial face seals using the system lubricant as a barrier liquid. In this case the high temperature in the seal caused breakdown of the EP additive that led to lock-up of the seal and excessive leakage of lubricant into the process. Seals on similar compressors in the company operated satisfactorily. Moreover, the compressors' manufacturer had no experience elsewhere with his machines using EP oils.

2.1.3 Seal and bearing failures on natural gas compressor

The third case involves a natural gas plant with compressors driven by gas turbines through speed-increasing gears. Scuffing was experienced at start-up (possibly because of an incorrect run-in procedure) and the gear manufacturer recommended the use of an EP oil. No further gear problems were reported, but seal and thrust bearing failures occurred through attack by the EP additive.

2.2 Contamination by process gas

Where the system lubricant is used as a barrier liquid for the shaft seals of a centrifugal compressor, it is inevitable that some contamination of the lubricant by the process gas takes place.

2.2.1 Ammonia

With increasing scale of production, ammonia manufacture moved from reciprocating to centrifugal process gas compressors. When centrifugal compressors were adopted, ammonia leaked into the lubricating oil through the shaft seals and then reacted with the normal succinic acid corrosion inhibitor used in turbine-type oils, forming ammonium succinate. The ammonium succinate, which is insoluble in hydrocarbon oil, formed deposits in sensitive locations, such as seals, bearings and turbine control systems, causing failures (**1**). This caused most problems in the control system and in one incident led to a major fire when there was a failure to trip on a process upset and the machine overspeeded until the seals failed and the leaking high-pressure hydrogen ignited spontaneously. The formation of ammonium succinate was a problem on ammonia synthesis plants world-wide that was solved by developing a lubricating oil that contained a non-reactive corrosion inhibitor, but it would have been a prescient lubricant supplier, machine manufacturer or plant designer to have anticipated it. This type of problem could well crop up as new processes are developed.

2.2.2 Hydrochloric acid vapour

An investigation of thrust bearing failures on a screw compressor involving contamination of the lubricant by hydrochloric acid vapour has already been described (**2**). It is not necessary to repeat the details here, but it is worth mention as it brings out an important point. The problem concerned not only contamination, but a change of the lubricant used from a straight oil without additives to an additive-treated one. The compressor shafts were sealed by carbon-ring seals using nitrogen as a blanketing gas. The compressor manufacturer correctly identified the problem as one of contamination, though this had not been a problem when the straight oil was being used, and recommended an increase in the pressure of the nitrogen to the seals. This was not an acceptable solution as it would have meant increased leakage of nitrogen that would have interfered with the process. *All relevant parties should be involved in problem solving, but only the operator can decide on an acceptable solution.* When the operator did not accept the solution proposed, the machine manufacturer lost interest.

2.2.3 'Sour' hydrocarbon gases

Natural gas contains varying amounts of sulphur compounds depending on the field from which it is extracted. As already mentioned, some leakage of process gas occurs through oil barrier floating-bush seals into the lubricating oil. The seal rings are lined with white metal to prevent damage to the shaft at starting or transient rubs during operation. At the high temperature in the oil film in the seal, the sulphur in the gas can react with the copper in tin-rich white metal forming a deposit of copper sulphide that takes up the clearance and leads to seal failure. The use of copper-free, lead-rich white metal can give some protection, but at the most severe conditions (high sulphur contents, higher temperature in machines rotating at higher speeds) the sulphur reacts with lead forming a deposit of lead sulphide with the same end result. Fig. 1 is a guidance chart produced from user experience giving the limits of satisfactory performance of tin-rich and lead-rich white metals as a function of gas sulphur content and the peripheral speed of the shaft in the seal, which is assumed to give an indication of the temperature in the seal. This has given useful guidance to machine operators, but as oil fields are worked out there is a tendency for the sulphur content of the natural gas to rise and sulphur attack of seal rings is becoming an increasing problem. Cast iron rings have been proposed for seal rings operating in Zone C of Fig. 1, but it is too early to say if this will provide a satisfactory solution.

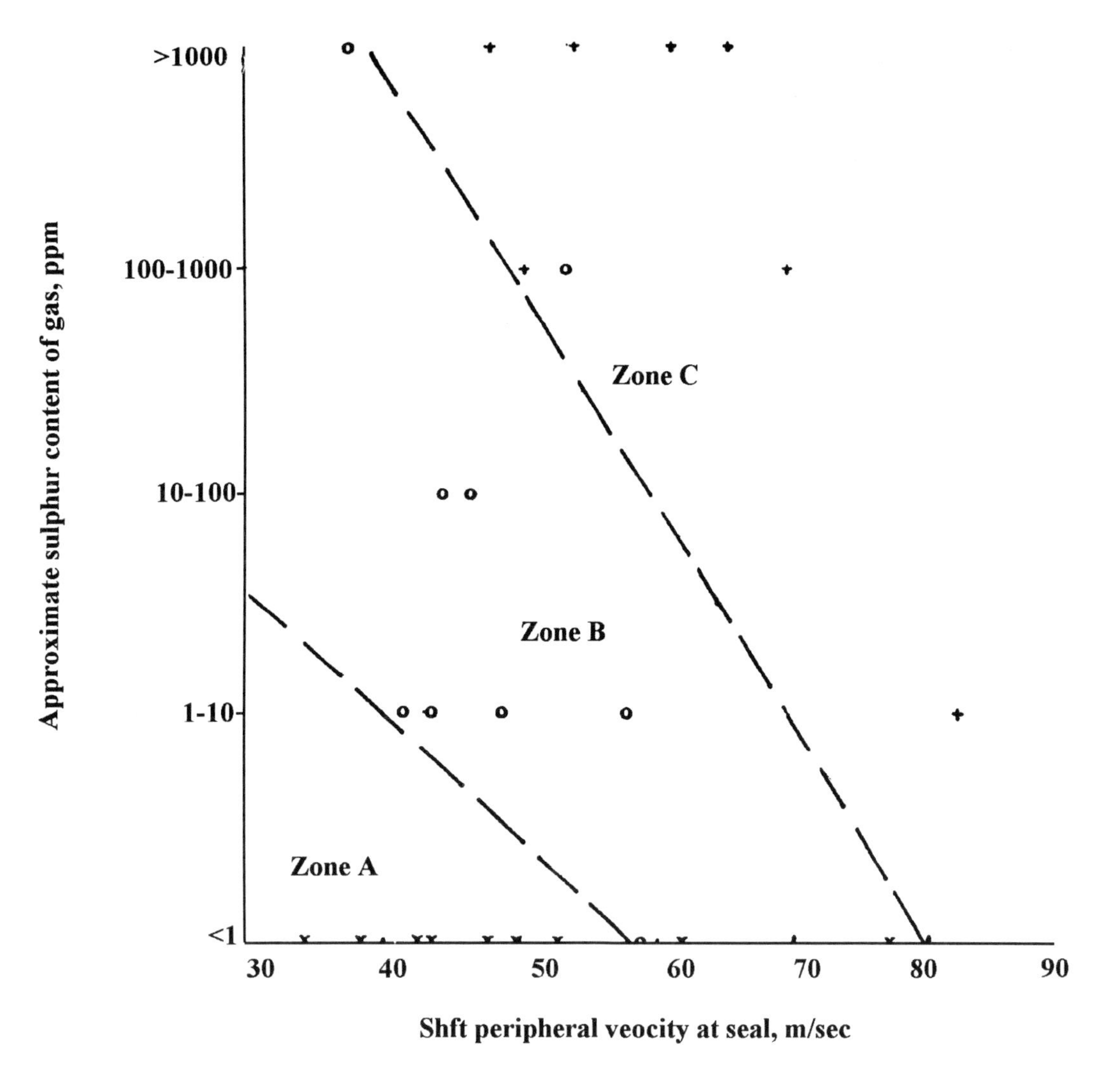

Fig. 1 Sulphur attack of white metal lined seal rings in compressors handling 'sour' hydrocarbon gases:
x no problems;
o problems with tin-rich white metal, lead-rich white metal satisfactory;
+ severe problem with both tin-rich and lead-rich white metal.

3 ROTOR INSTABILITIES

Increasingly rotodynamic machines tend to be operated above their first, or even higher, lateral critical frequency. This has led to the almost universal specification of tilting-pad journal bearings for such machines. Tilting-pad bearings are immune to oil-film induced instabilities; however, they can give rise to other problems – there is no one design of bearing that gives the best performance for all operating conditions. A considerable amount of heat is generated in a high-speed journal bearings. Part of this is removed from the pads by the oil flow, part by conduction through the pad pivot. The latter is a poor heat path and with direct-on-line machines this can result in thermal growth at the centres of the pads and complete loss of clearance. Fig. 2 shows a start-up failure of a set of five pads; the characteristic is that all pads are wiped in the centres, whereas in the case of an overload or lubrication failure wiping would be confined to the pads in the direction of the load line. The problem can be overcome by bringing the machine up to operating speed more slowly or by using increased clearance; the latter, however, reduces the damping capacity of the bearing and its capacity to carry out-of-balance loads.

4 ROLLING BEARINGS

Rolling bearings are by far the commonest type of bearing in industrial machines. The bearing manufacturers' catalogues are unrivalled in the guidance they give in bearing selection, installation and lubrication. Despite this, failures arising through ignorance of the advice given in the catalogues are still a common cause of machine breakdown. It is not proposed to discuss this aspect further

Fig. 2 Loss of clearance in tilting-pad journal bearing in high-speed machine with direct on-line start.

as the remedy is clear, but to concentrate on other less obvious effects.

4.1 Bearing design changes

Design changes are introduced from time to time by the rolling bearing manufacturers, sometimes to give enhanced performance, other times for commercial reasons. Such modified designs are subject to extensive bench testing before the bearings are introduced to the market place, but as already mentioned the manufacturer is at one remove from the user and does not always foresee the possible effect these changes may have on machines operating in service.

4.1.1 Increased load-carrying capacity

In 1986, a client purchased three 130 kW electric motors operating at 60 rev/s. These motors were fitted with oil-lubricated cylindrical roller bearings at the drive end. The motor manufacturer recommended that the bearings be changed every two years, suggesting that he, at least, suspected that they were operating near their design limit. The motors ran satisfactorily under this regime until 1991; since then three failures have occurred. The original bearings were NU3127, which according to the manufacturer's catalogue had a maximum limit for oil-lubrication of 66 rev/s. Some time in the 1980s the manufacturer introduced a new design with increased roller sizes giving enhanced load-carrying capacity, but at the expense of a reduction in the maximum speed for oil-lubrication to 60 rev/s, the actual operating speed. In 1989, the original design of bearing was withdrawn and replaced entirely by the modified design. There had been no problems on the plant as far as load-carrying capacity was concerned, but problems in lubrication dated from the time of the change. The user was given no choice.

4.1.2 Plastic cages

Reinforced nylon cages were introduced in the 1980s for a range of bearing types. Extensive bench testing showed that they gave a number of advantages: reduced weight, reduced cost, better lubrication. In the case of angular-contact bearings this improved performance was not realised in service, the cages breaking up and melting. Many bearings suffer misalignment in operation. Misalignment subjects the cage in angular-contact bearings to alternating stresses and eventually with metallic cages the cage fails by fatigue. While cage fatigue is a common cause of failure in angular-contact bearings, in many cases the life obtained is considered acceptable by the user; the change to nylon cages, however, gave a much reduced life. As a consequence a number of users specifically banned the use of angular-contact bearings with nylon cages. The reaction of the bearing manufacturer has been slow, but user experience has now been recognised and there has been a recent move to reintroduce angular-contact bearings with machined brass cages for heavy duty applications.

4.2 Machine design change

It is common practice for the manufacturer of a successful design of machine to produce a range of increased frame sizes. While the bearing arrangement on the original design may have been validated by the bearing manufacturer, the practice with the new designs is frequently to use larger bearings of the same type without reference to the bearing manufacturer, merely checking from the catalogue that the load-carrying capacity and speed ratings are satisfactory. A point is reached, however, when the lubrication arrangements are no longer adequate; for example, the grease-lubrication limit or the limit for oil-bath lubrication is exceeded and the new design suffers lubrication failures.

5 LUBRICANT FAILURE

Failures of lubricated components are invariably blamed on the lubricant. Failures of industrial lubricants – as distinct from failure of lubricant supply – are extremely uncommon and where the lubricant is involved it can almost always be traced back to the use of the wrong lubricant or changes in lubricant formulation as already discussed in Section 2.

My experience has been that such failures are predominantly caused by additives acting in a way not envisaged by the lubricant formulator. The additives used in standard turbine oils seldom cause problems (the case discussed in 2.1.2 is an outstanding exception), but the more reactive load-carrying additives and foam suppressants can be double-edged swords and should only be used when they are absolutely essential.

PART II: FAILURE INVESTIGATION AND PROBLEM SOLVING

6 FAILURE INVESTIGATION

A systematic approach is essential in any failure investigation. This is particularly so in the stressful circumstances associated with a major problem when the pressure to get a plant back on line can be overwhelming. In such circumstances half truths and prejudices abound and all concerned support different hypotheses.

I use simple A4 forms as the basis for collecting information in all failure investigations. Fig. 3 gives an example of such a form for use with journal bearings. There are similar forms for other tribological components, e.g. plain thrust bearings, rolling bearings, mechanical seals, gears, floating-bush seals, glands for high-pressure reciprocating pumps. Side 1 of the form lists the essential data that has to be acquired if a subsequent analysis of the component is to be carried out. This is factual information that has to be obtained from manuals and plant records. Side 2 provides space in which the history of the machine can be recorded. This is preferably obtained from plant records, though frequently one has to rely on verbal statements. The single most important piece of information is whether we are dealing with a new machine, in which case there exists the possibility of a design weakness or operator error through unfamiliarity, or whether the machine has been in service for some time, when such factors are less likely and one should concentrate on finding any change in operation that has taken place (uprating, change in process conditions, lubricant, etc.).

A key slogan is "*Trust is good; Check is better*". At this stage assume nothing. Visit the site. Handle the hardware. Speak directly to those involved, whatever their status. Develop a feel for the real situation.

It is worth noting that the form is headed "Design Check". Design Guides, such as those available in the *ESDU* (Engineering Sciences Data Unit) *Data Items*, the *Tribology Handbook* (**3**), and *Mechanical Seal Practice for Improved Performance* (**4**) are extremely useful in this connection. Many operators of machines tend to ignore such documents as they are not primarily concerned with design. The value of design guides in failure investigation is the facility they provide for checking an existing design to determine whether it is operating well within the design envelope or at the margins where failure is more possible as the result of minor changes (e.g. lubricant viscosity, temperature, bearing clearance, wear).

It is particularly useful for the specialist liable to be involved in failure investigation to develop a contact list. One's own expertise and experience are inevitably limited. There is no shame in accepting this and consulting others.

7 A METHODOLOGY FOR PROBLEM SOLVING

The forms, like the example given in Fig. 3, may be all that is required when investigating a straightforward bearing failure. In a major investigation much more detail is required. Following the investigation described in 2.2.2, my colleague and friend Dr. G Taylor and I summarised the rationale we had employed in reaching a successful outcome in a particularly complex failure investigation.

A more generalised version of this is given in an Appendix. This is intended as an aide-memoir to ensure that nothing of significance is overlooked in the heat of the investigation. It also looks at the role of specialists in a failure investigation and how they interface with the owner of the problem.

Job No.
Date

JOURNAL BEARING DESIGN CHECK

Company/Plant/Contact

Machine
Ref.

Details

Speed

Critical Frequency

Bearing
Type

Rotor Weight

Bearing Load (W)

Diameter (d)

Width (b)

b/d

Clearance (c_d)

c_d /d

Specific Loading (W/bd)

Oil Grooves: type
width
circumferential length

Lubrication
Oil

Inlet Temperature
Drain Temperature
Effective Temperature
Feed Pressure
Effective Viscosity

Filter

History

2

Fig. 3 Form for recording data on journal bearing failure.

8 ACKNOWLEDGEMENT

I am grateful to Dr G Taylor for allowing me to reproduce the methodology given in the Appendix.

9 REFERENCES

1 **Summers-Smith, J.D.** and **Livingston, J.,** Compressor oil system problems. *Am. Instn. Chem. Engrs. Ammonia Plant Safety* 1976, **75**, 70–71.

2 **Summers-Smith, J. D.** and **Taylor, G.,** Screw compressor bearing failures on process plant: a case study and some general lessons. *Proc. Instn. Mech. Engrs.*, Part J, 1995, **209**, 77–83.

3 **Neale, M.J.,** (ed.). *A tribology handbook*, 2nd edition, 1996 (Butterworth-Heinemann, Oxford).

4 **Summers-Smith, J.D.,** (ed.). *Mechanical Seal Practice for improved performance,* 2nd Edition, 1992 (Mechanical Engineering Publications, London).

APPENDIX

A methodology for failure analysis

A checklist of questions.

1 History

Is the equipment new to the plant?
Is it a new or changed design?
How long has it been in service?
Are failures repetitive or is this a one-off?
What myths have become established over the years?
Is there a competent maintenance record?
Has a competent record been kept of the failure(s)?

2 Failure evidence

Are broken parts, plant reports, plant records of operating experience (e.g. running time to failure) available?
Can evidence be obtained at an early stage of failure, i.e. by deliberately shutting down before complete destruction or complete failure?

3 Experience

3.1 General
Are several items of equipment subject to same behaviour?
What experience is there available elsewhere on similar equipment?
What experience is available from supplier?

3.2 Operating conditions
Do the failures occur under specific operating conditions?
Is the equipment operated in the correct manner?
Are the operating conditions clearly understood?
Have any modifications been made to the equipment or operating conditions?

3.3 Condition monitoring
What operating parameters are measured/monitored during normal operation?
Is there a gradual deterioration in the machine or does the failure occur without warning?
Can the good operating conditions be identified?
Is there a baseline or standard against which the performance can be judged?

4 Methodology

Have attempts been made to solve the problem using a logical procedure?
Have all the attempted fixes been fully documented?

5 Investigators

Have the right people been involved in the investigation?
Are you using a consultant (either internal or external)?
If so: (1) Have you a clear understanding of what you want from the consultant?
(2) Does the consultant have a clear understanding of your requirements?
(3) Do you have a competent interface with your consultant?

Recent analyses and designs of thick-film bearings

M TANAKA
Department of Mechanical Engineering, University of Tokyo, Japan

SYNOPSIS

This paper is a state-of-the-art report on recent analyses and designs of hydrodynamically lubricated journal and thrust bearings. Recent papers, mostly published after 1991, are referred to; these are taken from journals and proceedings on tribology. Not only bearings for turbomachinery but also automotive engine bearings are covered.

1 INTRODUCTION

Thick film bearings, mostly hydrodynamic bearings, are widely used in various machinery. Since the principles of hydrodynamic bearings were discovered, and the Reynolds equation was presented late last century, these bearings have made great progress in performance and operational life, which should be attributed to a large amount of research work and engineering experience. It is often said that hydrodynamic lubrication or hydrodynamic bearings are no longer the target for further research work, but the contents of recent journals and proceedings on tribology show the contrary, and engineers in many industries are now eagerly seeking new, reliable design methods for bearings which have to survive increasingly more severe operational conditions.

The papers are categorised into six topics, that is: turbulent lubrication, THD lubrication of journal bearings, thrust bearings, automotive engine bearings, bearing design, and cavitation. Understandably all the papers listed in the references cannot be summarised separately for want of space. Furthermore it would be tedious to read a report which simply consists of neutral summaries of published papers, so the author of this review will give a collective summary on each topic of thick film bearings. Consequently this report is inevitably biased more or less by the author's view.

2 TURBULENT LUBRICATION

Steam turbines and generators for electric utilities increased tremendously in power capacity in the 60s and 70s, resulting in an increase in shaft diameter. Consequently, journal surface speeds and mean bearing clearances also increased, which led to a laminar-to-turbulent transition in the lubricating films. These bearings should be designed not by means of conventional laminar lubrication models but by turbulent lubrication models. References (**1**) to (**5**) listed are related to this field.

Ng and Pan (**1**) and Aoki and Harada (**2**) each presented a turbulent lubrication theory based on a different turbulent flow model. These theories are now widely used to design partial arc bearings, two-lobe bearings, and tilting pad journal or thrust bearings operated in the turbulent regime for steam turbines, gas turbines and generators. The bearing performance predicted by the theories has been confirmed to be in good agreement with measurements. Consequently it seems bearing design engineers are satisfied with the turbulent lubrication theories available as far as the static performance of bearing is concerned.

However the laminar-to-turbulent transition has not been well defined theoretically or identified experimentally, except for Couette flow in the annular clearance between two cylinders, which corresponds to full circular bearings. Journal or thrust pad bearings are widely used in actual machines, so not only practical bearing engineers but also academics are interested in how the transition in those types of bearing takes place. Innes and Leutheusser (**3**) studied experimentally the transition of the film flow between a moving flat plate and an inclined stationary plate. They found that the transition takes place around a Reynolds number of 1200, defined by the inlet film thickness as the representative length, and also that the transition is independent of the pad-length-to-inlet-gap ratio. In their test rig, the inlet gap was 25 mm which is too thick compared to the ordinary lubricating film thickness, that is, some hundred micrometres even in the case of large journal bearings used for steam turbines. Design engineers are greatly interested in whether or not the structure of the turbulent flow in the 'very thick' film tested by them is the same as that of the flow in the ordinary, hydrodynamic lubricating film, which has not yet been fully explored. Another problem yet to be solved is whether or not the turbulent flow at thick film positions changes into laminar flow with the decrease of film thickness in the convergent film, and how the bearing performance is obtained with, if any, partially laminar and partially turbulent lubrication. Measurements across the lubricating film thickness of even several hundreds of a micrometre are still difficult and challenging.

Turbulent lubrication is found when the kinematic viscosity of the lubricant is very low. Liquid oxygen (LOX) is one such fluid, and San Andres (**4**) analysed flexure-pivot hybrid journal bearings lubricated by LOX intended to support cryogenic turbopumps, presumably for the next generation of space shuttle main engines. San Andres applied his own model to the lubricating film and conducted thermo-hydrodynamic analysis in the turbulent region. The main target of the analysis was the rotordynamic stability of the bearing, not the tribological problems. However, because cryogenic fluids are known to have poor boundary lubrication performance, possible surface damage of bearing and journal during startups and shutdowns will require some new design of the mating surface materials.

High speed operation of a sliding bearing brings not only turbulent flow of the lubricating film but also high operating temperatures of the film and bearing surfaces. Furthermore the linear rotordynamic coefficients of the lubricating film in a bearing depend on the film shape at equilibrium for the operating condition. The three-dimensional viscosity variation in the film strongly affects the film shape itself, so large-size, high-speed bearings should be designed by means of turbulent, thermo-hydrodynamic lubrication models which will be discussed below.

3 THD LUBRICATION OF JOURNAL BEARINGS

High speed operation of sliding bearings results in high operating temperatures of lubricating films and bearing pad surfaces, which may dangerously decrease the safety margin against seizure, because the former lowers the viscosity of the lubricating film, decreasing the minimum film thickness, and the latter decreases the strength of soft bearing metals.

On the other hand, conventional isoviscous hydrodynamic lubrication models cannot predict the temperatures of lubricating films or bearing surfaces inherently. Consequently thermo-hydrodynamic (THD) lubrication models are required to design high speed bearings because the models incorporate the generation of frictional heat in the films and the removal of heat by convection in the oil film flow and also by conduction through the solid walls of the mating surfaces. Thus, THD models can give the three-dimensional temperature variation in the films and predict the bearing performance, and also the temperature variation on the bearing surfaces. References (**6**) to (**58**) are related to THD lubrication.

Since Dowson (**6**) gave the generalised Reynolds equation for lubricating films in which viscosity varies three-dimensionally, numerous original or review papers have been published. Recent activities are concentrated mostly on identifying the boundary conditions for temperatures, including the temperature in the oil mixing region, developing Elasto-THD models which take into consideration the dilation and deformation of solid walls due to thermal effects or pressure effects, finding the best oil feeding method to decrease the operating temperature of the bearing, and extending or identifying the dynamic characteristics of the lubricating film under THD conditions. Design engineers have gradually started applying THD models to the design of actual sliding bearings.

3.1 Turbulent THD model

Taniguchi, *et al.* (**7**) presented a turbulent THD model based on Ng and Pan's turbulent theory (**1**) to design large tilting pad journal bearings used for steam or gas turbines for electric utilities and confirmed that the load-dependent variations of minimum film thickness and journal eccentricity ratio predicted by their theory were in good agreement with measurements in their test rig of four-pad tilting pad journal bearing 479 mm in diameter (Fig. 1). The predictions of the maximum pad surface temperature and frictional loss of the bearing were also confirmed to agree well with measurements (Fig. 2). Bearing engineers usually recognise the transition from laminar to turbulent flow of the oil film when the bearing temperature rise with shaft speed slows down, and their predictions and measurements of pad surface temperature show the transition clearly.

However, the film shapes measured near the leading and trailing edges show elastic (pressure and thermal) deformation of the pad (Fig. 3) and the measured film pressure which remains zero (ambient pressure) near the trailing edge corresponds to the measured diverging film shape. The figure shows that the film thickness in a large bearing should be estimated with elastic deformation of bearing pad being considered. Nevertheless, their THD model predicted bearing performance precisely without elastic deformation being considered. Consequently there remains a question of what bearings or what operating conditions require elasto-THD models in designing sliding bearings.

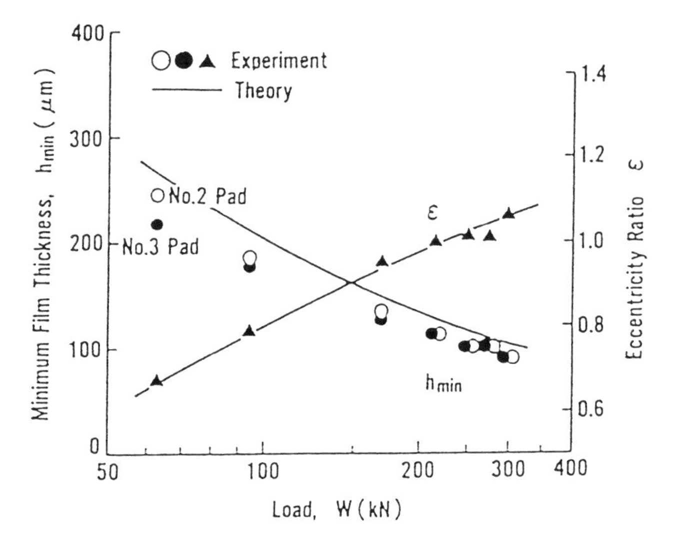

Fig. 1 Eccentricity ratio and minimum film thickness ($N = 3000$ r/min) (**7**).

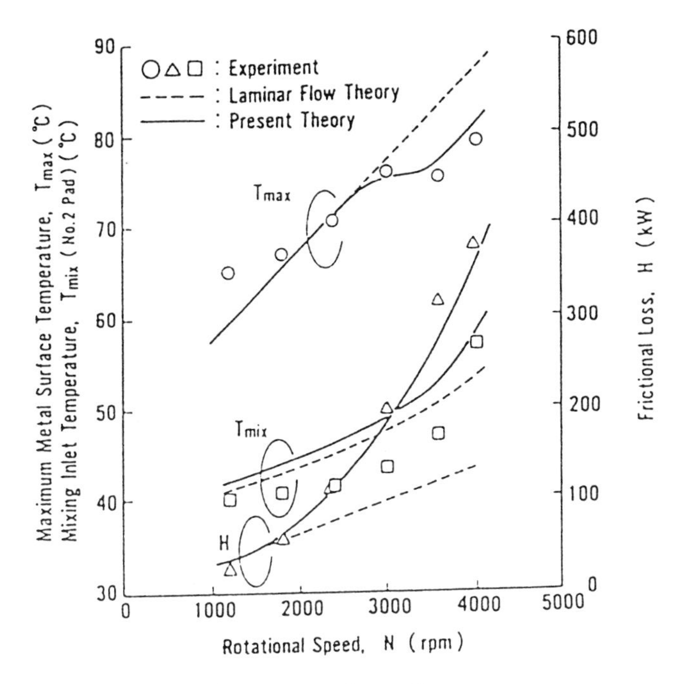

Fig. 2 Maximum metal surface temperature, mixing inlet temperature of no. 2 pad and bearing frictional loss ($W = 180$ kN) (**7**).

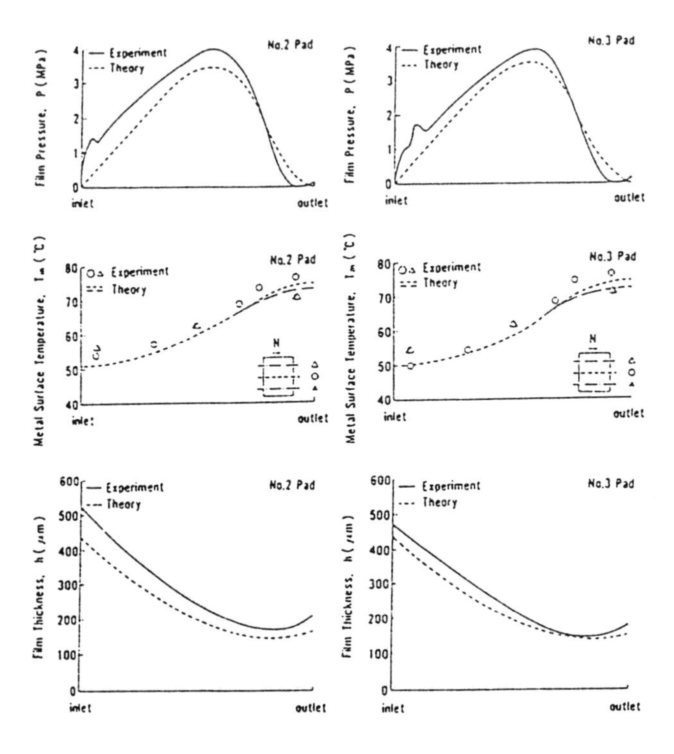

Fig. 3 Comparison between theory and experiment of film pressure, metal surface temperature and film thickness (N = 3000 r/min, W = 180 kN) (**7**).

Bouard, *et al.* (**8**) theoretically studied the THD performance of tilting pad journal bearings operating in the turbulent regime, and invoked a better heat exchange between the film and the bearing surface in explaining the fact that the bearing surface temperature stagnates at the transition from the laminar to turbulent flow regime, though the frictional loss increases monotonously.

Bouard, *et al.* (**9**) compared three turbulent models (Ng and Pan, Elrod and Ng, and Constantinescu) in analysing the THD performance of tilting pad journal bearings, and found the predictions of the three models were similar, though the Constantinescu model, which is the simplest of the three, could save computing time.

3.2 THD performance behaviour

Fitzgerald and Neal (**10**) presented measurements of the performance of circular bearings over a wide range of load and speed. The variations of maximum bush surface temperature and journal surface temperature with load at various constant speeds are quite different for two different values of bearing length-to-diameter ratio (Figs. 4 and 5). In the case of the short bearing, the maximum bush surface temperature monotonically increases with load at constant speeds, while it does not vary much and has the lowest value in the case of the long bearing. In general, the journal surface temperature for the short bearing increases with load, while that for the long bearing decreases with load.

Loading direction for bearings often varies in some rotating machinery because of variation in the transmitted torque or valve forces. Consequently bearings used for such machinery should be designed with this factor being considered. El-Deihi and Gethin (**11**) studied theoretically and experimentally the THD performance of a circular bearing with a twin axial groove at the horizontal position, and specifically the effect of loading direction. They found that the maximum bush surface temperature was strongly affected by the loading direction.

Fillon, *et al.* (**12**) measured the pad surface temperatures of four-pad tilting pad journal bearings of two different preload factors, and found that the higher the preload factor, the higher the temperature. However this may mislead some engineers because they increased the bearing preload by decreasing the bearing clearance with the pad clearance being nearly constant. Bearing preload is often increased by increasing pad clearance with bearing clearance being constant. In this case the effect of increasing bearing preload would be opposite.

Basri and Gethin (**13**) studied experimentally the THD performance of three-lobe journal bearings of three different types. Three-lobe bearings are known to have better oil whip stability than two-lobe bearings and also to be suitable to support vertical rotors because of the high radial rigidity of the oil film, and three-lobe bearings are, in general, used for high speed operation. Consequently the THD performance of the bearing should be identified

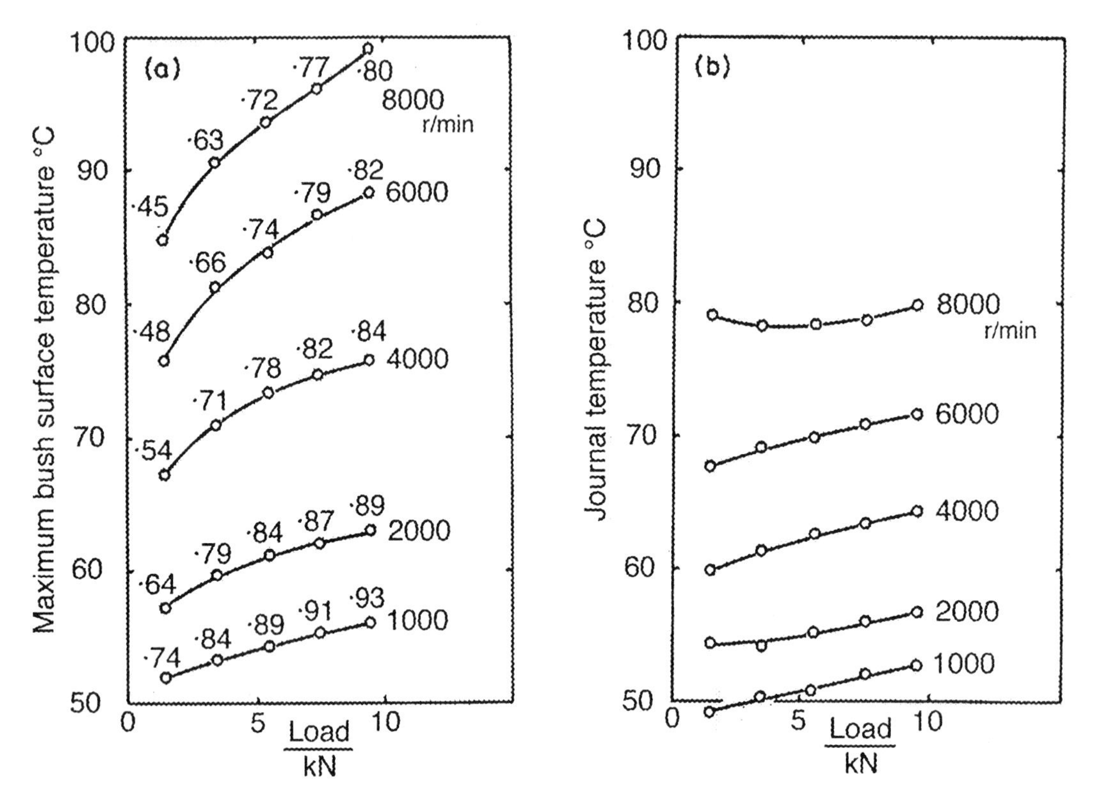

Fig. 4 Bearing operational characteristics. $b/d = 0.5$, $c/d = 0.002$ (**10**):
(a) Bush surface maximum temperature (estimated ε values are shown),
(b) Journal temperature.

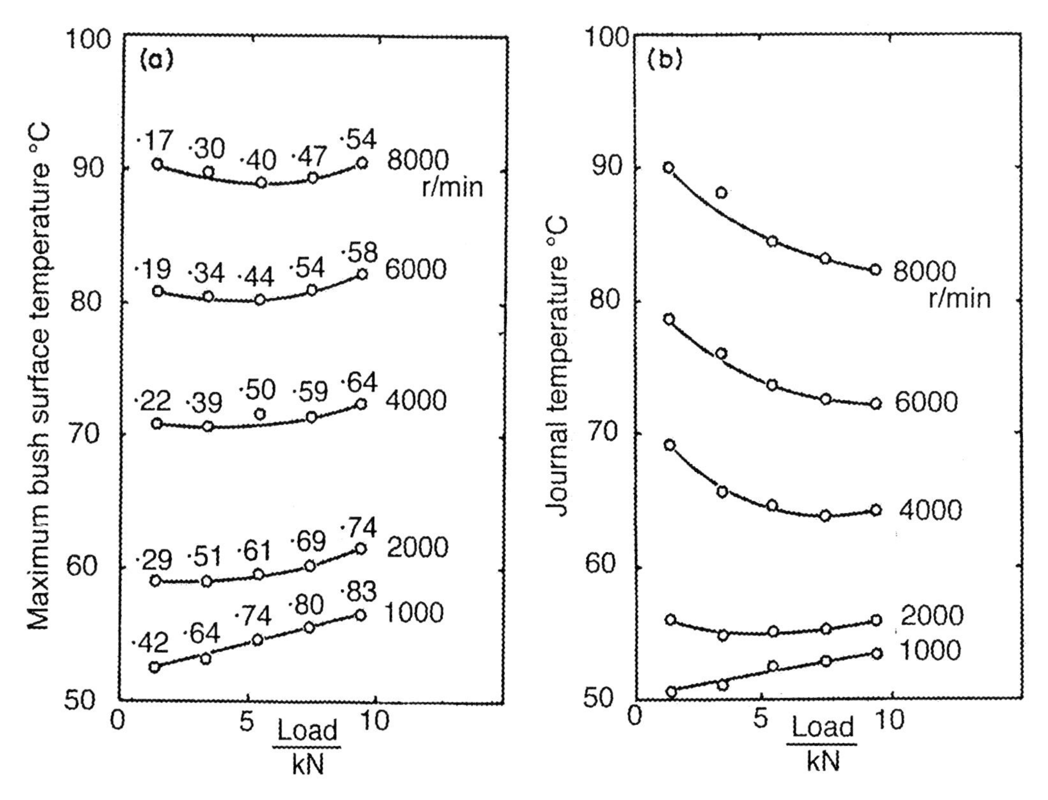

Fig. 5 Bearing operational characteristics. $b/d = 1.0$, $c/d = 0.002$ (**10**):
(a) Bush surface maximum temperature (estimated ε values are shown),
(b) Journal temperature.

to choose a reasonable design. They tested a symmetric three-lobe bearing, a tilted one, and a tilted one with circular side rail. Preload factor was 0.8 for all three bearings. They found that pad surface temperatures for the tilted one was lower than those for the other two bearings, that the maximum pad surface temperatures increased monotonically with load at constant speeds, and also that the maximum temperature was affected by the load direction.

Simmons and Dixon (**14**) and Simmons and Lawrence (**15**) conducted experiments on the THD performance of five pad tilting pad journal bearings 200 mm in diameter with sealed (flooded) lubrication under practical operating conditions for high-speed gearbox bearings, and presented a large amount of performance data over a wide range of operating conditions. In the first paper, the measured maximum pad surface temperatures clearly show the inflection of the temperature variation with speed (Fig. 6), and they surmised that this phenomenon was due to the laminar-to-turbulent transition in the surrounding oil flow, rather than that in the lubricating film flow, because the measured temperatures of the pad leading edges started decreasing with speed before the temperature inflections take place (Fig. 7). However this phenomenon may be alternatively explained as follows: the lubricating film under each pad is thickest at the leading edge. Consequently the laminar-to-turbulent transition in the film takes place at the leading edge first. The turbulent heat transfer is known to be greater than the laminar one, and the pad surface temperatures decrease.

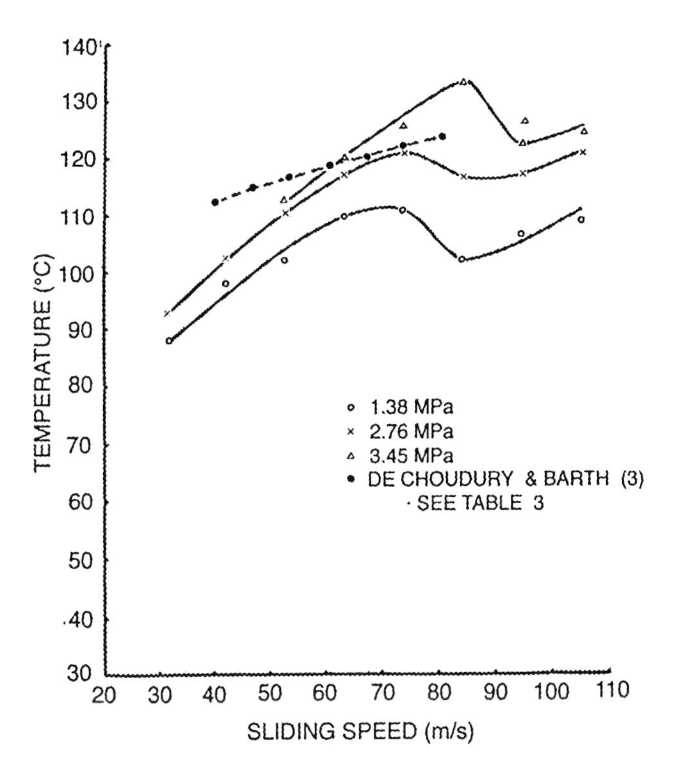

Fig. 6 Maximum pad temperature vs. speed for Configuration 1, diametric clearance 0.23 mm, preload ratio 0.52, load on pad (**14**).

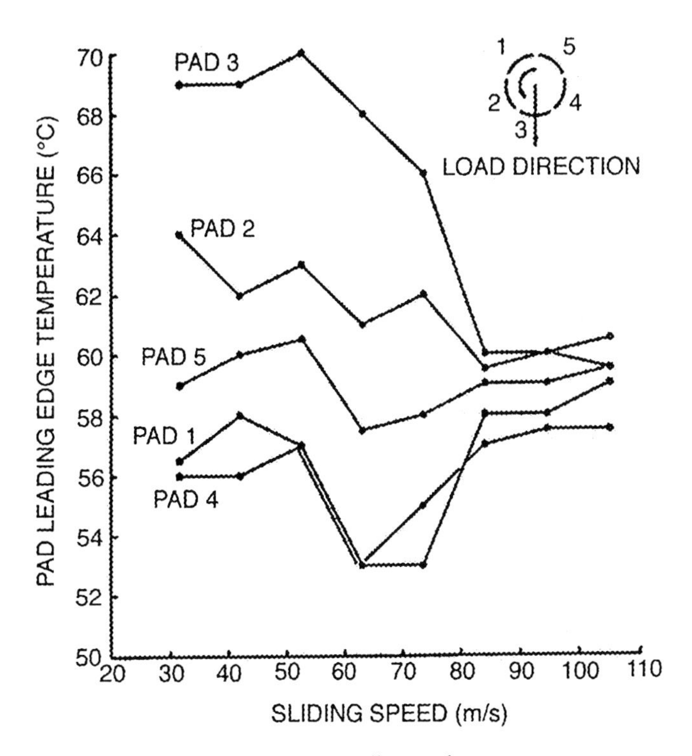

Fig. 7 Journal pad leading edge temperature against speed for Configuration 1 (**14**).

In the second paper (**15**), the authors presented similar experimental results in the case of offset pivot pads, while the pads in the first paper (**14**) were centre-pivoted. They found that the maximum pad surface temperatures were lower by 10 to 20°C than those for centre-pivoted pad (Fig. 8). This can be attributed to the increase in the lubricant flow under pads because the pads' attitude angles increase due to the offset pivot position.

Bouchoule, *et al.* (**16**) conducted experiments on the THD performance of five-pad tilting pad journal bearings with LBP load direction, and compared the results with predictions of their laminar elasto-THD model. The measured maximum pad surface temperatures increased at first but decreased beyond a certain shaft speed, which means the laminar-to-turbulent transition took place (Fig. 9). Predictions by the elasto-THD model are shown not to be in good agreement with the measurements, though the pad surface temperature profiles in the circumferential direction were well predicted. The three bearings tested have different pivot positions (A: 0.5, B: 0.55, C: 0.6), and Fig. 9 shows that the closer to the trailing edge the pivot position, the lower the maximum pad surface temperature, which is the same as the results in reference (**15**).

Makino, *et al.* (**17**) presented experimental results of the THD performance of a six-pad tilting pad journal bearing and also an offset-halves bearing in a bearing test

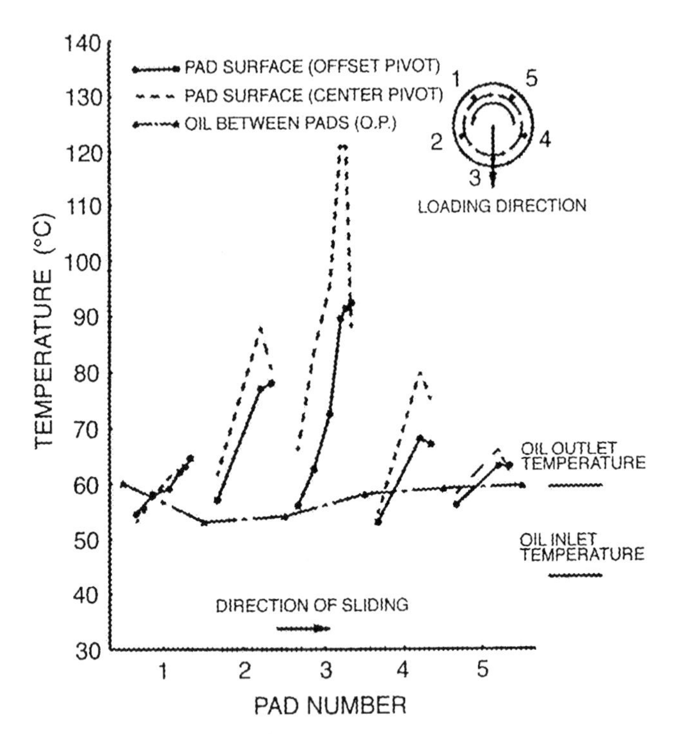

Fig. 8 Thermocouple readings obtained at 73 m/s and 2.76 MPa together with equivalent center pivot pad results from Ref. (**1**). Load-on-pad (**15**).

rig operated up to the high speed of 70000 r/min. They found that the pad surface temperature stagnated and the discharge oil temperature increased markedly around a shaft speed of 30000 r/min, and concluded that the transition from laminar to turbulent flow had taken place at this speed. The experimental results were compared with predictions given by the turbulent THD model (**7**), and in general, satisfactory agreement was obtained. However, with regard to the temperature profile on the pad surface, predictions were noticeably lower than measurements (Fig. 10), though they assumed 100% hot oil carry-over in the model because of the small diameter of the bearing (40 mm). They surmised that the mixing process in the space between two adjacent pads needed to be improved with the lubricant inertia effect being included.

3.3 Inlet pressure rise

Taniguchi, *et al.* (**7**) measured a clear pressure hump slightly inside the leading edge under the operating condition for actual bearings (Fig. 3). This hump is often called the ram pressure or inlet pressure rise. Their THD model did not cover the phenomenon.

Kim and Rodkiewicz (**18**) theoretically studied the effect of inlet pressure rise on the THD performance of tilting pad journal bearings, and predicted that the pressure hump near the pad inlet would increase the pad tilt angle significantly and affect the bearing performance.

Ha, *et al.* (**19**) theoretically and experimentally studied the effects of inlet pressure rise on the THD performance of four-pad tilting pad journal bearings. Predicted inlet pressure rise is confirmed to agree well with measurements (Fig. 11). They conclude that the inlet pressure rise decreases the pad surface temperature because it increases the tilt angle of the pad, decreasing hot oil carry-over and increasing the incoming oil flow rate. Measurements of the THD performance agreed well with their turbulent THD model with the inlet pressure rise effect being incorporated.

Mori, *et al.* (**20**) and Zhang (**21**) theoretically studied the inlet pressure rise under isoviscous, laminar flow conditions, and showed the effects of various bearing variables and operating variables on the pressure rise.

3.4 Oil feeding method

Tanaka (**22**) theoretically and experimentally studied the effect of a new oil feeding method on the THD performance of high-speed tilting pad journal bearings. He removed two floating ring seals at both ends of the bearing assembly in order to expedite the outflow of hot oil from the pad chamber, and installed nozzles, placed in the spaces between two adjacent pads, which eject pressurised lubricant oil towards the journal surface. The maximum pad surface temperature was found to reduce by 10°C, compared to conventional sealed lubrication, and the effect of this spot feeding becomes pronounced with the increase in shaft speed (Fig. 12), which agreed well with his theoretical predictions. With this new method, high-speed journal bearings could recover a considerable safety margin against seizure.

Dmochowski, *et al.* (**23**) studied the effect of leading

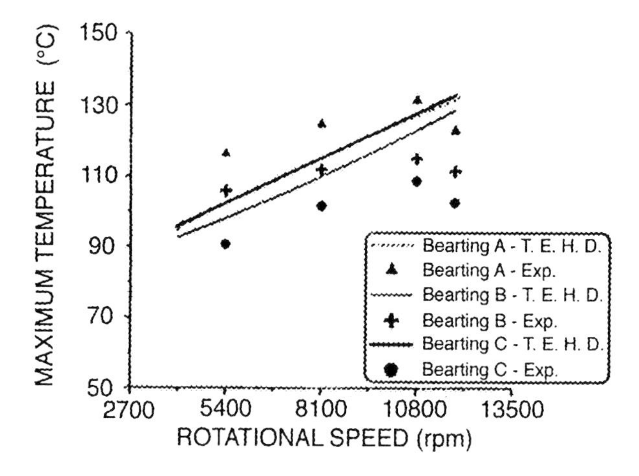

Fig. 9 Experimental and theoretical maximum babbitt temperatures versus rotational speed, for bearings A, B, and C (W = 88 kN) (**16**).

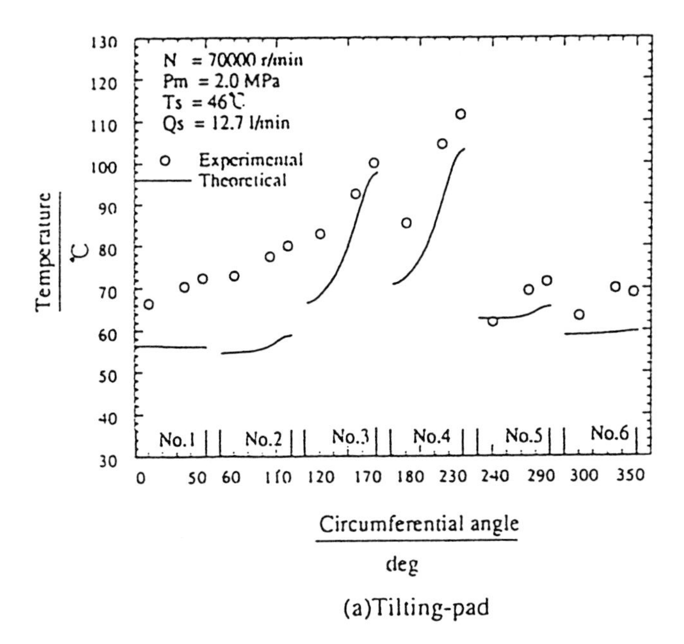

Fig. 10 Surface temperature distributions (**17**).

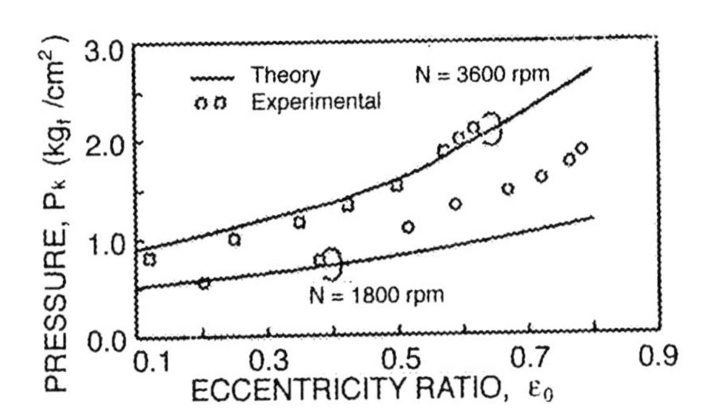

Fig. 11 Comparison between theory and experiment of inlet pressure of the No. 2 pad in the mid-plane (**19**).

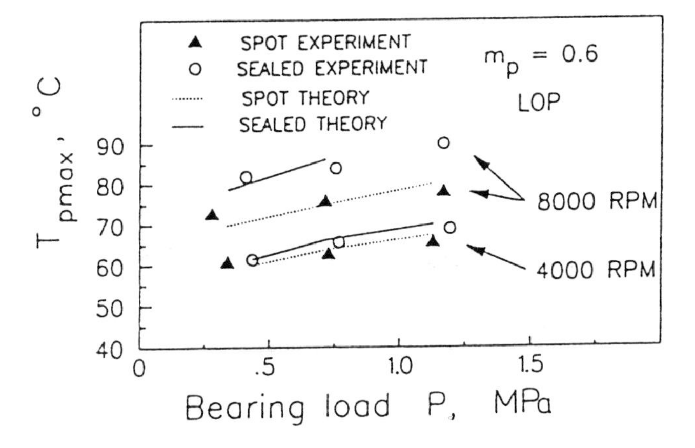

Fig. 12 Measured and predicted variation of T_{pmax} with bearing load (seald and spot: $m_p = 0.6$, LOP) (**22**).

edge groove feeding method on the THD performance of five-pad tilting pad journal bearings, and found that this method reduced pad surface temperatures significantly, compared to conventional bearings of the sealed lubrication type, particularly at high speeds (Fig. 13). They assumed that this effect was caused by the efficient feeding of cold oil from the leading edge groove into the pad clearance. They also carried out Elasto-THD analysis and obtained good agreement between predictions and measurements.

Harangozo, *et al.* (**24**) compared experimentally the effects of three different lubrication methods on bearing power loss and pad surface temperature in a four-pad tilting pad journal bearing 127 mm in diameter. The three methods are conventional sealed lubrication (flooded lubrication), directed lubrication (spot lubrication) and leading-edge lubrication. In the case of low load (3.4 kN), power losses and pad surface temperatures for the three methods showed no significant difference up to a shaft speed of 7000 r/min. On the other hand, in the case of high load (22 kN, 1.3 MPa), the directed lubrication was found to give the lowest maximum temperature on the loaded pad surfaces, while the flooded lubrication showed the highest temperature. With regard to generating lower bearing power losses, the three methods are arranged in the same order as in the case of lower pad surface temperature. However the authors recommend application of the 'improved' leading-edge lubrication, though the details of the 'improvement' are not explained here. It must be noted that the tested bearing pads had negative preloads. Different results may have been obtained if the pads had had ordinary positive preloads.

3.5 Boundary conditions for temperature

3.5.1 Inlet mixing temperature

Taniguchi, *et al.* (**7**) assumed a uniform oil temperature across the film thickness resulting from simple mixing of hot oil discharged from a pad outlet and cool oil supplied in the space between two adjacent pads. The temperature of oil at the pad inlet predicted by their model agreed well with measurements.

However Tanaka (**22**) could not obtain good predictions with the similar boundary condition of uniform temperature across the inlet film thickness. Instead, he assumed a linearly varying temperature across the inlet oil film with the mixed oil temperature and the journal surface temperature at each of the two boundaries, and obtained predictions which agreed well with measurements. On the other hand Frene and his team use a parabolically varying temperature across the film thickness. Tanaka assumed that 100% of cool oil was mixed uniformly with 50% of the discharged oil from the previous pad in the case of spot lubrication, and 90% in the case of conventional sealed lubrication.

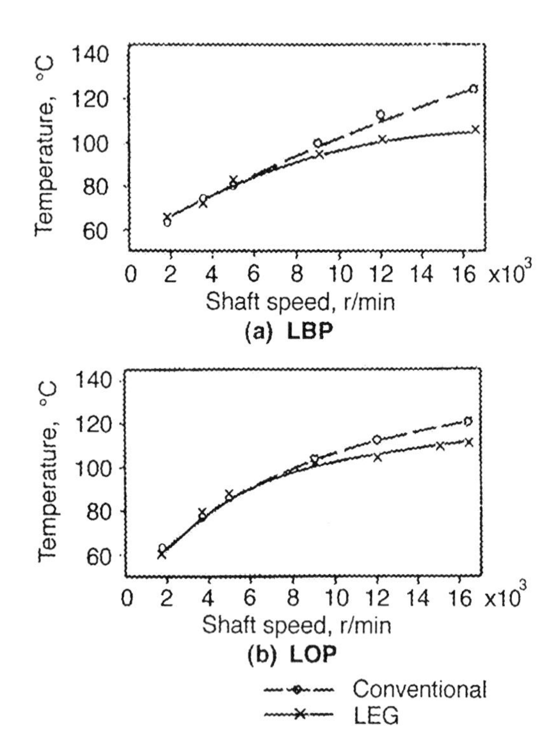

Fig. 13 Maximum pad temperature versus shaft speed. Load 11.2 kN (**23**).

El-Deihi and Gethin (**11**) studied the THD performance of a circular bearing with twin axial grooves at the horizontal position. They obtained fairly good agreement between predictions and measurements of bush surface temperature excursion in the circumferential direction, but some of the measured temperatures at the pad inlet were in poor agreement with predictions. It shows that even simple mixing in the ordinary groove has not been completely identified.

Fillon, *et al.* (**12**) measured the oil temperatures in the spaces between two adjacent pads of four-pad tilting pad journal bearings, and found that, in some cases, the temperatures were higher than those measured at the trailing edge of the upstream pad, though cool oil was supplied into the space (Fig. 14). They assumed this phenomenon was caused by the reverse flow of hot oil from the downstream pad. Partial reverse flow is believed to take place in the lubricating film when pad preload factor and/or journal eccentricity ratio is high. Their measurements show that oil mixing models should be established with this factor being considered in some cases.

Dmochowski, *et al.* (**23**) studied the effect of the oil feeding method from a groove placed at each leading edge of the pad, and assumed that 100% of cold oil fed to the groove was pulled into the film by the action of the rotating journal, resulting in a decrease in hot oil carry-over from the previous pad. However their measurements of pad surface temperature show that the pad leading edge temperatures of the conventional bearing and the bearing with leading edge grooves are nearly the same, though the two temperatures tend to depart from each other towards the trailing edge (Fig. 15). On the other hand, Tanaka (**22**) obtained significant difference in pad leading edge temperatures between conventional sealed lubrication and spot lubrication (Fig. 16).

Bouard, *et al.* (**8**) studied the effect of inlet temperature profile across the film thickness on the THD performance, and compared the constant profile and the parabolically varying one. They found that the pad surface temperature was strongly affected by the profile, and also that the parabolic variation gave better agreement with measurements at high speeds.

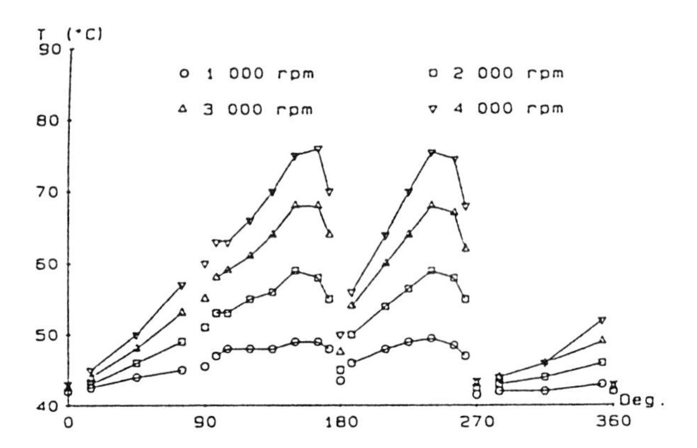

Fig. 14 Internal pad surface temperatures of bearing A versus angular coordinate, for different speeds (W = 10,000 N) (**12**).

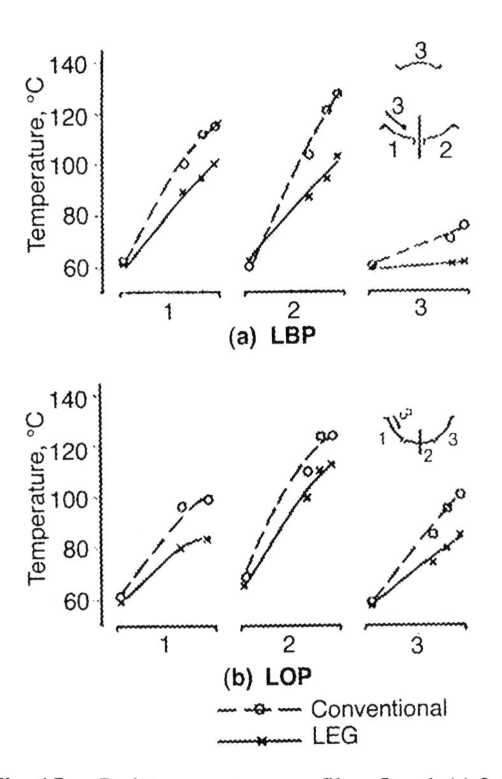

Fig. 15 Pad temperature profiles. Load 11.2 kN, shaft speed 16,500 rpm (**23**).

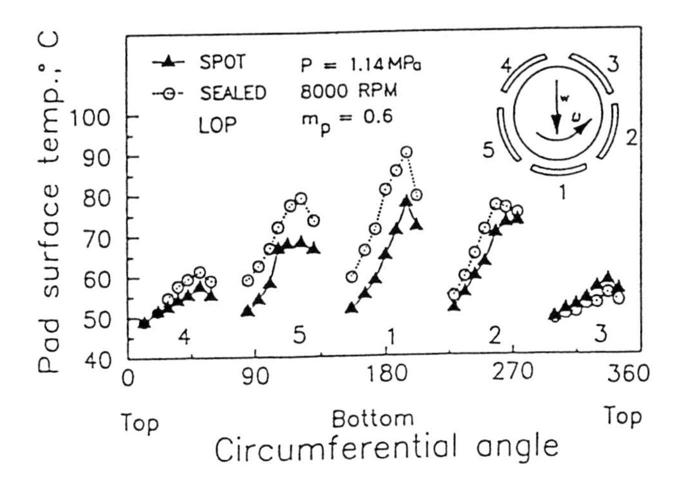

Fig. 16 Measured pad surface temperature along pad centerline (8000 rpm, 1.14 MPa, sealed and spot: $m_p = 0.6$, LOP) (**22**).

3.5.2 Journal surface temperature

Taniguchi, *et al.* (**7**) calculated a uniform journal surface temperature, based on the assumption of zero net heat flow between journal and surrounding oil films, but the model was not verified because the journal surface temperature was not measured.

Tanaka (**22**) found experimentally that the surface temperature of the journal in a five-pad tilting pad bearing increased with shaft speed and also that the temperature rise in the case of spot feeding was significantly lower than in the case of conventional sealed lubrication (Table 1).

Table 1 Measured journal surface temperature (P = 1.13 MPa) (**22**).

m_p	Bearing Type	RPM 4000	6000	8000
0.2	LOP, SPOT	56.0°C	63.2	70.7
0.2	LOP, SEALED	58.5	65.5	86.6
0.2	LBP, SPOT	54.8	67.3	67.8
0.6	LOP, SPOT	55.8	59.2	64.6
0.6	LOP, SEALED	58.9	72.0	81.6
0.6	LBP, SPOT	57.1	59.1	58.4
0.6	LBP, SEALED	59.8	58.8	80.0

In the elasto-THD analysis of Fillon, *et al.* (**12**), they calculated the journal surface temperature on the zero net heat flow between journal and oil films. Their measurements were slightly higher than predictions with the maximum discrepancy of 7°C which they assumed was caused by the churning loss in the spaces between the pads.

Journal surface temperature is one of the dominant factors which determines the temperature of oil at the pad inlet, so more effort should be made to establish theoretical models which predict precisely the temperature on the moving surface and also to verify the models experimentally.

3.5.3 Double layer treatment

Practical journal or thrust bearings for rotating machinery are not made of a single material but usually have a double-layered structure, that is, soft babbitt on the surface and hard steel for the backing. Babbitts are known to have a thermal conductivity lower than that of steel, which increases the temperature for the same amount of incoming heat flux, so the bearing surface temperatures can be predicted more precisely with this factor being considered. Khonsari and Wang (**25**) showed theoretically that the maximum pad surface temperature for a double-layered bearing was 4°C higher than that of a single material bearing. This effect is more pronounced for a thicker babbitt layer on the bearing surface consequently bearings of large diameter for large machinery should be designed with this effect being considered, but the effect of babbitt layers of 2 mm or less in thickness for bearings of small or medium size could be, by and large, assumed to be the same as a single layer of steel.

3.5.4 Thermal model for cavitation region

Fillon, *et al.* (**12**) measured sharp drops of pad surface temperature near the trailing edge of loaded pads (Fig. 14), which may be due to possible oil film rupture. When bearing load and/or geometric preload is high, the film may become divergent near the trailing edge, and the oil film may rupture, which requires solving the energy equation with this phenomenon being considered. Otherwise, this sharp drop may be difficult to predict theoretically.

Knight and Ghadimi (**26**) presented modified effective length models to analyse the temperature profiles of a bearing surface in the cavitated region. They assumed that the cavitated lubricant film consists of two different parts, that is, the oil film occupying the full space between the journal and the bearing and another oil film, thinner than the local clearance. The former has the Couette flow velocity distribution across the film thickness, while the latter separates from the bearing surface and adheres to the journal surface, moving at the journal surface speed. The ratio of the axial lengths of the two films and the ratio of the latter film to the local clearance are free parameters. With these parameters selected properly, they obtained good agreement with measurements given by other researchers.

Ma and Taylor (**27**) studied theoretically and experimentally the surface temperature drop in the cavitation region of a twin-axial-groove circular bearing and a two-lobe bearing. Their measurement of bush surface temperature in the cavitation region was not in satisfactory agreement with the prediction of the conventional THD model with the effective length cavitation assumption. Consequently they proposed a new approach to calculate the bush surface temperature in the region, assuming the separation model for cavitation, no energy dissipation there, and reverse flow of cold oil from the downstream groove into the region. The separation model assumes that the oil film in the cavitation region is separated from the bearing surface and moves with the rotating journal. This model could predict the temperature drop in the cavitation region more precisely.

3.6 Elasto-THD models

Elasto-THD models incorporate not only elastic deformation of solid walls due to thermal and pressure effects but also dilation. A journal dilates with increase in its uniform temperature, and a bearing bush or pad dilates with increase in bush temperature or gets deformed with the spatial variation of temperature or due to the hydrodynamic pressure developed in the film and the reaction force at the pivot point. These factors vary the shape of the lubricating film and in turn the bearing performance.

Khonsari and Wang (**28**) studied the effects of elastic deformation or dilation of shaft and bearing bush on bearing performance. They found the predictions of their elasto-THD model to be in good agreement with measurements, though they also found various simulation results were significantly different from each other when the dilation or the deformation was considered individually (Table 2).

Fillon, *et al.* (**12**) theoretically and experimentally studied the THD performance of four-pad tilting pad journal bearings, and showed that their Elasto-THD model could give better agreement with measurements of pad surface temperature than their THD model (Fig. 17). They also found theoretically that the mean bearing clearance decreases with shaft speed at constant load.

3.7 Numerical calculation technique

When Fillon, *et al.* (**12**) conducted elasto-THD analysis of tilting pad journal bearings, they saved computation time by solving the thermal and elastic problems only in the mid-plane of the bearing on the assumption that the axial variations of both temperatures and solid displacements are negligibly small. This assumption is sometimes well justified by measurement, but sometimes should be carefully studied for particular cases. For example, there may be some heat source near the bearing, and some heat flux may come into the bearing oil film through the shaft or casing.

Tanaka and Hatakenaka (**29**) applied the power law method, widely used in computational fluid dynamics, to THD analysis of titling pad journal bearings. When journal eccentricity ratio and/or pad preload is high, conventional methods often have some difficulties in obtaining sufficient convergence of solutions, mainly because the temperatures to be solved oscillate and do not converge. This new method can give solutions without difficulty even for pad preload higher than 0.99 and eccentricity ratio higher than 0.99, which is particularly useful in analysing the THD performance of guide bearings used for vertical rotors of hydraulic pumps/turbines in electric power stations.

Bouard, *et al.* (**8**) theoretically studied the THD performance of tilting-pad journal bearing operating in the turbulent flow regime. They found that very severe convergence criteria and very thin slicing (l/l00 or less) of the calculation domain were required to obtain the convergence of solutions. They also found that the solutions were greatly dependent on the thermal properties of both wall materials and lubricant. However Taniguchi, *et al.* (**7**) easily obtained convergence of solutions for turbulent operation of tilting pad journal bearings with the lubricating film being sliced thick (l/8) and also with the convergence criteria milder than Bouard's case. Though both models are based on the same Ng and Pan turbulence theory, the differences obtained, have not yet been fully explained.

Vijayaraghavan (**30**) presented an efficient numerical calculation method for THD analyses of cavitated bearings. The original method was developed by Elrod (**31**), and was extended to incorporate cavitation. The properties across the film thickness are determined at

Table 2 Comparison of simulation results (**28**).

	T_{max} (°C)	P_{max} (N/m^2)	ϕ (deg)	t'	Q (gm/min)
Elastothermohydrodynamic (ETHD)	51.4	2.4E6	35.4	0.76	2.591E3
Shaft thermal expansion only	53.6	11.6E6	25.8	0.87	2.256E3
Bush thermal expansion only	49.0	1.36E6	43.97	0.66	2.925E3
Bush elastic deformation only	51.3	3.55E6	29.3	0.8	2.585E3
Shaft and bush thermal expansion only	51.57	2.86E6	37.6	0.76	2.479E3
Thermohydrodynamic (THD)	51.69	4.9E6	32.46	0.8	2.399E3

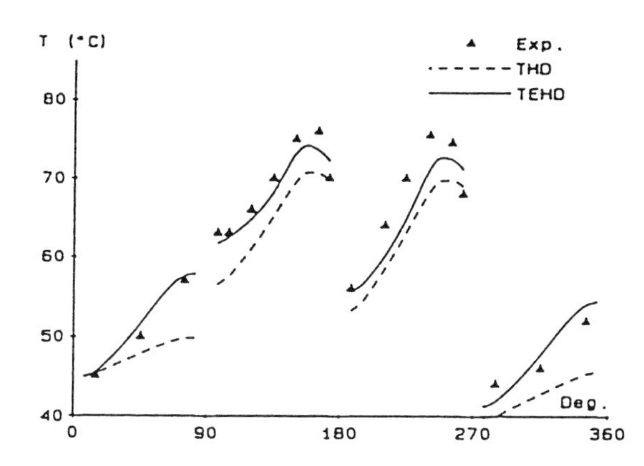

Fig. 17 Experimental and theoretical temperatures of the internal surface of pads (bearing A; $N = 4000$ rpm; $W = 10{,}000$ N) (**12**).

Lobatto points and their distributions are expressed by collocated polynomials. This method was found to be very fast, stable and accurate. Predictions agreed well with experimental data presented by other researchers.

Kim, *et al.* (**32**) presented an ETHD model. They employed the upwind finite element technique to avoid the oscillation of temperature in solving the energy equation. The variations of temperature and elastic deformation in the axial direction were assumed to be negligibly small, and the calculation domain was reduced to the oil-film cross section of bearing. The prediction was found to be in good agreement with experimental data in the laminar region.

Tucker and Keogh (**33**) applied the computational fluid dynamics (CFD) technique to the static THD analysis of a journal bearing, and obtained results which were in good agreement with measurements. The full three-dimensional transport equations were solved instead of the Reynolds equation. This approach allows the calculation domain to be extended out of the lubricating length of the bearing, and also the journal surface temperature need not be assumed to be isothermal in the circumferential direction. Understandably, the disadvantage of this method is the computing time needed to obtain converged solutions.

3.8 THD bearing design

THD or ETHD models can give more accurate prediction than conventional isoviscous hydrodynamic lubrication models at the cost of much computing time needed to repeat numerical calculations over and over again until convergence of solutions is obtained. Furthermore it would take much time for engineers to prepare running computer programs of the models. On the other hand engineers in industry want handy, fast and accurate methods or tools to predict bearing performance or design bearings. Consequently it is essential to provide engineers with some shortcuts to the final solutions or simplify models to reduce the computing time required, with the least accuracy being sacrificed.

Khonsari, *et al.* (**34**) tried to generalise THD analyses for journal bearings. In spite of, or because of, numerous publications of research work on THD analyses, engineers are at a loss to know which results should be applicable to the design of specific bearings. This paper provides the first answer to the question. They proposed a simple and fast method of using THD design charts to predict the maximum operating temperatures of bothpad surface and journal surface by using two temperature-rise parameters. The values predicted by this method were compared with rigorous THD solutions and experimental data given in previous papers published by other authors, and good agreement was obtained. All the bearings referred to here are plain journal bearings, and this method will be a strong and efficient design tool for bearings when it is extended to cover bearings of other types used for high speed operations.

Pilakas and Parkins (**35**) gave algebraic functions which predict temperature profiles on the bearing bush, the maximum bush surface temperature and its circumferential position. The functions include bearing length-to-diameter ratio, shaft speed and load as variables, and the coefficients were determined from experimental data in the case of a single-axial groove journal bearing by means of regression analysis techniques. It is very easy to use the functions and it takes little time to predict the THD performance of a journal bearing. The versatility of this approach needs investigation.

Fillon and Khonsari (**36**) extended their work on plain journal bearings (**34**) and presented design charts to predict easily and rapidly the maximum pad surface temperature and the effective temperature of five-pad tilting pad journal bearings. Comparisons between the predictions of this method and experimental data were shown and discussed. The discussion of this paper, which is found separately (**37**), would provide readers with some more information on this new approach to quick and reliable design methods anticipated by design engineers.

3.9 Dynamic problems

Bearings for rotating machinery should be designed not only to maintain their tribological functions during operation but also to enable rotors to rotate smoothly up to the maximum operating speed without a large amplitude rotor vibration or unstable vibration. To this end, bearings should be carefully designed to obtain reasonable values of rotordynamic coefficients of the lubricating film. Linear dynamic characteristics of the lubricating film depends on the film shape at equilibrium and the viscosity of the lubricant. Consequently, first of all, predictions of the film shape at equilibrium and the viscosity are essential. In this respect, THD or ETHD models can give more accurate predictions than conventional isoviscous models. If the static THD characteristics of a bearing can be identified precisely, then the next step would be easy.

Tucker and Keogh (**38**) applied the CFD technique

developed in their previous work (**33**) to analyse the THD performance of a journal bearing with the journal being rotating and also whirling in the bearing clearance. They found that the journal surface is not isothermal, though conventional THD analyses assume the contrary.

Some more papers (**39**) to (**43**) related to this topic are listed in the references. These papers concentrate on transient or steady-state response of a rotor supported in journal bearings under THD or ETHD conditions.

3.10 Concluding remarks for Section 3

In spite of plenty of results obtained in the papers mentioned in this section, there still remain some questions yet to be fully answered in THD or ETHD lubrication, and also new questions have emerged.

Many of the design variables and properties of the materials concerned have been revealed to influence the THD performance of bearings. However there may be some more overlooked factors which need to be incorporated. One of them would be heat sources outside the lubricating region. Conventional THD models regard viscous dissipation in the lubricating film as the sole heat source, but in actual machinery like steam turbines or gas turbines, some amount of heat from the hot working fluid is conducted through shafts and casings to the lubricating films. Tor and Tanaka (**57, 58**) calculated the THD performance of a circular journal bearing with this factor considered, and presented some analytical and experimental results. An external heat source was found to have a significant effect on the temperature profile on the bush surface in the circumferential direction, and consequently, on the journal eccentricity ratio. Future THD models need to incorporate this factor in order to obtain more precise predictions which agree well with measurements obtained in actual machinery.

On the other hand, similar theoretical or experimental results from different sources could lead to different conclusions which are often mutually contradictory. Thus not only design engineers but researchers themselves cannot find the path through the jungle of so much information on the THD performance of bearings.

It seems that the exciting first stage for the first generation pioneers is ending and that the boring but inevitable second stage is starting for tenacious followers of the second generation with sharp eyes to sort out, tag and rearrange various findings steadily and patiently and make a bird's-eye but precise map of the complicated world of THD. To this end, interrelated, co-operative research programmes would be needed to be organised and started.

4 THRUST BEARINGS

Thrust bearings for horizontal rotors are usually larger in size than journal bearings, but bearing loads are not so large compared with vertical rotors. Exceptions are thrust bearings for recent gas turbines of large output capacity where thrust loads are increasing in magnitude because the cancellation of thrust unbalance is very difficult. On the other hand, thrust bearings for vertical rotors usually take very large loads because they support all the weight of the rotor and often downward fluid forces too. A thrust bearing is designed to minimise the power loss, keep the minimum film thickness above a lower limit and keep the maximum pad surface temperature lower than an upper limit for the operating condition. In tilting pad thrust bearings the maximum pad surface temperature and the minimum film thickness are strongly linked by the thermal- and pressure-induced deformation of pad. Thermal deformation always makes a pad surface convex (crowning), but pressure deformation makes a pad surface convex or concave, depending on how the pad is supported. Consequently, not only THD or turbulent analysis of the lubricating films but also thermal and pressure deformation analysis of pad-support device assembly are required to design thrust bearings. Papers (**59**) to (**68**) related to this topic are listed in the references.

Ettles and Anderson (**59**) carried out a three-dimensional elasto-THD analysis of thrust pad bearings for vertical rotors with pads being supported by a disc. Figs. 18 and 19 show that the smaller the relative size of the pad support disc, the thinner the minimum film thickness at a specific load and also the higher the maximum pad surface temperatures. Partial cutaway of the support disc and water cooling of the back metal of the bearing assembly is shown to reduce the maximum pad surface temperature and increase the minimum film thickness significantly.

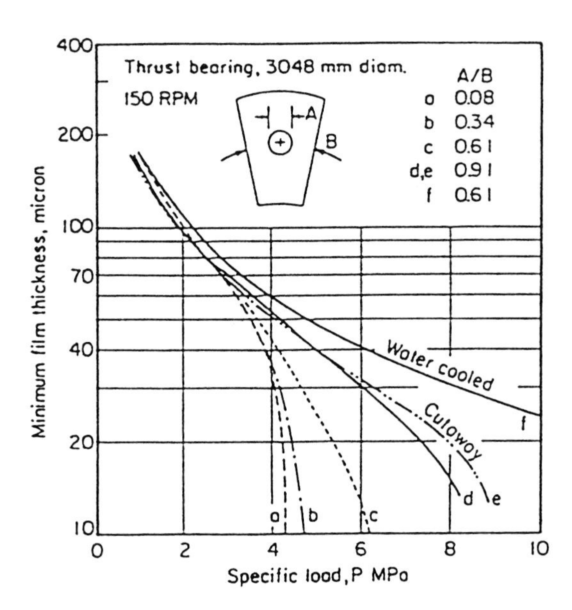

Fig. 18 The effect of load and disk support size on the minimum film thickness of the example bearing (**59**).

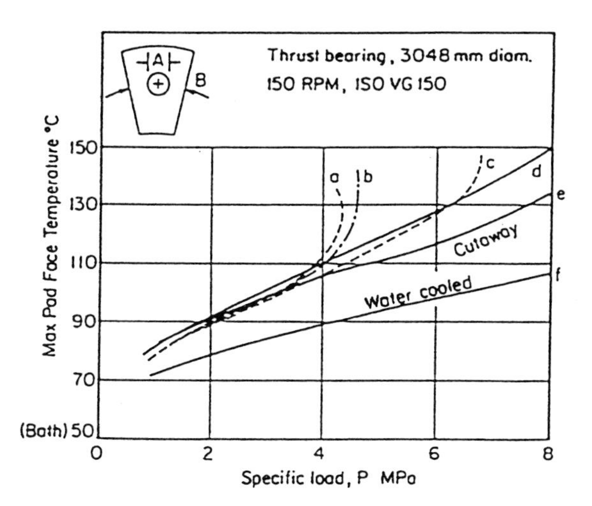

Fig. 19 The effect of load and disk support size on the maximum face temperature of the example bearing (**59**).

Ettles (**60**) studied the effects of pad dimensions and support spring arrangements on the THD performance of spring-supported thrust bearings by means of a three-dimensional elasto-THD model. He showed that the larger the bearing size, the more sensitive to pad deformation the bearing performance. The pad crowning ratio due to the pressure effect increases with the square root of the pad arc length, while the thermally-induced crowning ratio increases with a power of 3/2, which means that thermal deformation is more sensitive to the variation of pad arc length. The thermal and pressure deformation of the pad must be well controlled by selecting pad dimensions and/or support spring arrangements properly.

Brockett, *et al.* (**61**) conducted a three-dimensional ETHD analysis of parallel tapered-land thrust bearing performance with dilation and deformation of not only pad but runner being included. The minimum film thickness with runner deformation included is shown to be lower than that with runner deformation neglected. The difference becomes larger with an increase in applied load, and with a decrease in pad thickness. On the other hand, the frictional loss showed no significant effect of runner deformation being included. In this paper, the runner is assumed to be isothermal, and consequently to be deformed only by pressure action. However runner surface temperature is known to vary in the radial direction, as shown in reference (**62**). Future bearing design needs to study incorporating the three-dimensional temperature variation in the runner and also the thermal dilation and deformation of the runner due to the temperature variation.

Kim, *et al.* (**63**) carried out an experimental study of the THD performance of tilting pad thrust bearings with spot (directed) lubrication. They obtained good agreement with their THD model (**64**), and found that the minimum film thickness decreases with the shaft speed, particularly at higher loads (Fig. 20).

Uno, *et al.* (**65**) experimentally studied the effects of using poly-tetrafluoroethylene (PTFE) for pad surface material on the performance of sector-shape tilting pad thrust bearings used for vertical rotors of hydro-generators. Conventional pads use whitemetals for lubricated surfaces, and the pad surfaces often get more or less damaged during startups and shutdowns of the machinery. To avoid this problem, externally pressurised oil lift systems are needed to form a sufficient thickness of hydrostatic lubricating film at low shaft speeds. However this solution raises another technical problem of increased maintenance work. Consequently they replaced the conventional babbitt on the pad surface with PTFE, and found that the maximum pad surface temperature decreased by 16°C over a wide range of load and also that the maximum film pressure decreased by about 30%, as shown in Fig. 21. Surface wear was also found to reduce drastically to 1/50 of that for conventional metal pads, and the friction at start-up was found to be less than 1/3 of a conventional bearing. Consequently, PTFE-coating of thrust pad can lead to a simpler design of bearing without an oil lift system, size reduction with lower manufacturing cost, longer life of operation, and more reliable operation with less maintenance cost.

Papers on thrust bearings are tremendously outnumbered by those on journal bearings. The impact of demanding needs from industry would change this situation. Simpler design, lower manufacturing and maintenance costs and high operation reliability will be the keywords of practical needs for future small or medium size thrust bearings, and the realization of better bearing performance corresponding to higher shaft speeds and higher thrust loads are those for future large-size rotating

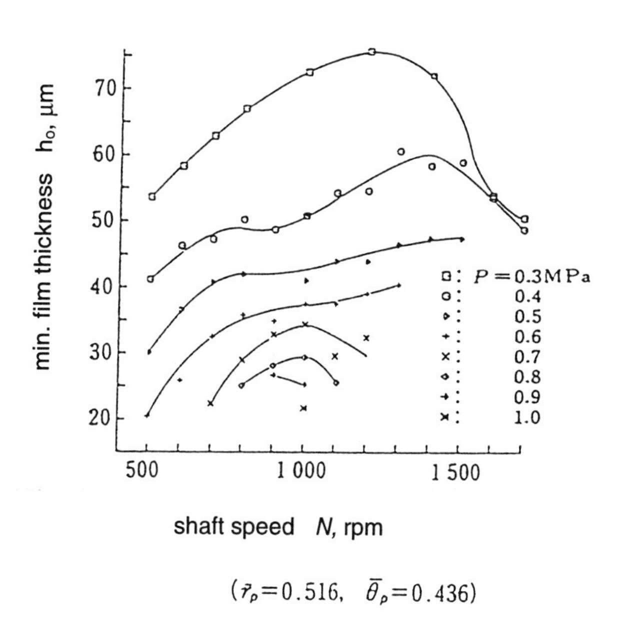

Fig. 20 Shaft speed and minimum film thickness $(\bar{r}_p = 0.516, \bar{\theta}_p = 0.436)$ (**63**).

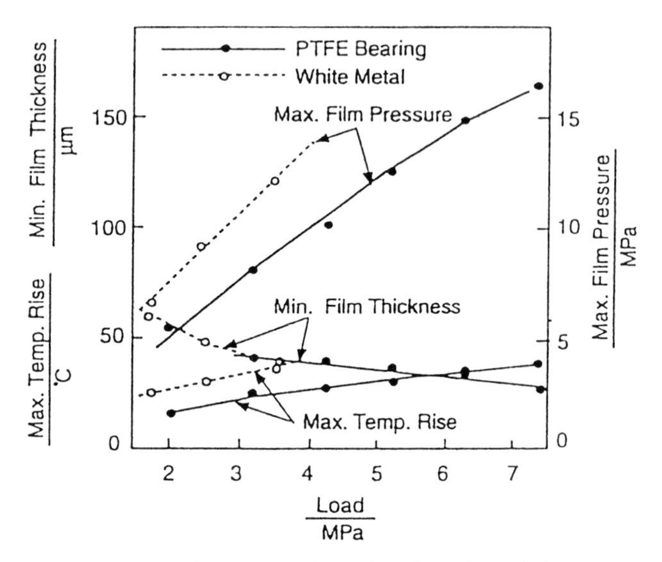

Fig. 21 Performance of test bearing (800 r/min or 27.3 m/s) (**65**).

machinery like pumped-up hydraulic generator sets or gas turbines for electric utilities.

5 AUTOMOTIVE ENGINE BEARINGS

Recent bearings for automotive engines are running out of performance margins under increasingly more severe operating conditions (higher speeds, higher loads), which enhances the danger of seizure or fatigue. Consequently, it is urgently required to develop new bearing designs which can survive increasingly harsher operating conditions and also to establish lubrication models which can predict bearing performance more precisely.

5.1 New bearing designs

Matsuhisa, *et al.* (**69**) studied the effects of the composition of the bearing alloy on the thermal and mechanical characteristics of engine bearing metals. Cu-Pb-Sn bearing metals with Pb-Sn-In overlay are widely used for engine bearings. They found that the melting temperature of the overlay monotonously increases with a decrease of the amounts of Sn and In but the tensile strength reaches a maximum at 5% Sn. They could raise the melting temperature of the overlay considerably by choosing the optimum amounts of Sn and In, compared with the conventional overlay. Furthermore they found the diffusion layer rich in Sn and In to be thinner than conventional overlays. This layer, formed between the overlay and the diffusion barrier layer of Ni, is known to have a low melting temperature; consequently the thicker the layer, the weaker the bearing at high operating temperature. Cu-Pb-Sn bearing metals are known to increase in thermal conductivity with the decrease in Sn content but decrease in strength. Under higher operating conditions, the more quickly the conducted heat from the bearing surface is dissipated, the lower the temperatures of the bearing metal and the overlay. To this end, they decreased the weight content of Sn, and minimised the decrease of strength by modifying the rolling process in manufacturing the bearing metal. They confirmed that the combination of the new overlay and the new bearing metal have a higher anti-seizure performance than conventional bearings (Fig. 22). This paper received the 1996 Best Paper Award of JAST.

Kumada, *et al.* (**70**) designed a microgrooved engine bearing which has a continuous spiral groove on the lubricating surface. The grooved surface was found to improve the running-in performance and also anti-seizure performance when the pitch and the depth of the groove were optimised. Fig. 23 shows that the bearing maintains a lower temperature compared to a smooth surface bearing over a wide range of shaft speed and that the temperature difference increases with shaft speed. The grooved bearings did not seize even 50 minutes after the lubricant supply was stopped. The excellent performance is considered to be attributable to the reduced metal contact area because of the groove ridges and also to the sustained lubrication by the trapped oil in the groove. This bearing design received the 1995 Tribo-Technology Award of JAST.

5.2 Transient THD analysis

THD analyses are widely studied for statically loaded bearings, but very few papers on transient THD analyses of dynamically loaded engine bearing performance have been published so far. It is obvious that transient THD analyses take much more computing time and need various, new calculation techniques.

Paranjpe and Han (**71**) studied the transient THD analysis of sinusoidally loaded engine bearings. They calculated transient variations of oil flow rate, power loss and temperatures of oil film, bearing bush and journal, using computation grids moving in the oil film and

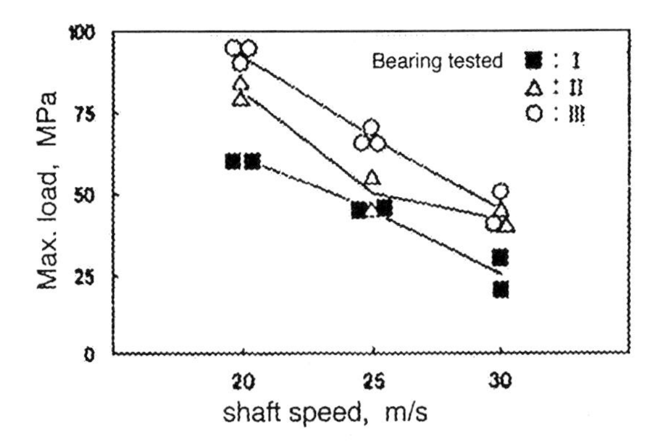

Fig. 22 Experimental results of seizure test (**69**).

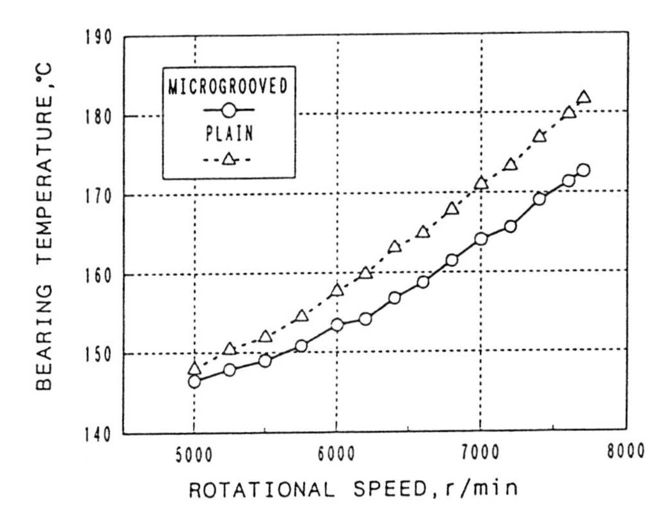

Fig. 23 Bearing temperature with increasing rotational speed (**70**).

different time scales in calculating temperatures in the oil film and in the solids. They found that the temperature in the oil film shows considerable variation over time and space (Fig. 24), while the time scales for thermal variation in the bearing and the journal are several orders of magnitude greater than those for the oil film. Consequently the journal and the bearing bush act as heat reservoirs, absorbing or releasing a large fraction of the total heat generated, and they can be treated as being in quasi-steady state during one loading cycle.

Paranjpe (**72**) extended the transient THD analysis of engine bearing under sinusoidal loading (**71**) to the case of abearing subjected to firing engine loading. Fig. 25 shows the temperature variation in the oil film. He compared the results of the full THD analysis, the adiabatic THD analysis and the simplified thermal analysis (Table 3, Fig. 26). The adiabatic THD analysis agrees well with the full THD analysis as a whole, but the simplified analysis also agrees partially.

Table 3 Results for different analysis methods (**72**).

	h_{min} µm	P_{max} Mpa	Power Loss Watts	Oil Flow cc/s
Main bearing				
Full THD	1.83	97	611	15.6
Adiabatic THD	1.53	100	585	16.1
Simplified Thermal	2.02	91	600	16.4
Isothermal	2.86	82	782	11.8
Connecting Rod Bearing				
Full THD	0.78	57	351	2.85
Adiabatic THD	0.64	55	347	2.92
Simplified Thermal	0.95	57	272	3.50
Isothermal	2.43	52	432	2.19

5.3 Bearing performance

Moteki, *et al.* (**73**) found experimentally that the frictional torque of a bearing with a superfinished journal is less than that in the case of a lapped journal. They found also that the bearing surface profile changed greatly during the running-in period, so that the initial roughness of the bearing surface has little effect on the bearing performance.

Suzuki, et al (**74**) measured the surface temperatures of connecting rod big end bearings of a diesel engine under

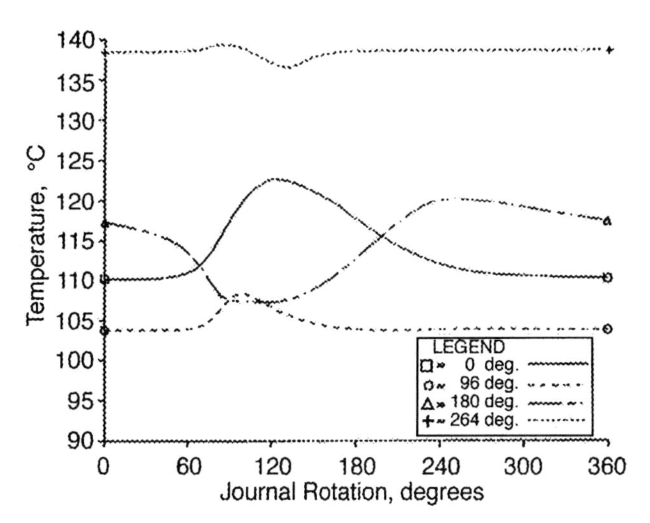

Fig. 24 Temperature in the oil film at various circumferential locations shown as degrees from the X axis (see Fig. 1) in the last load cycle for a bearing with sinusoidal loading running at 5000 r/min (**71**).

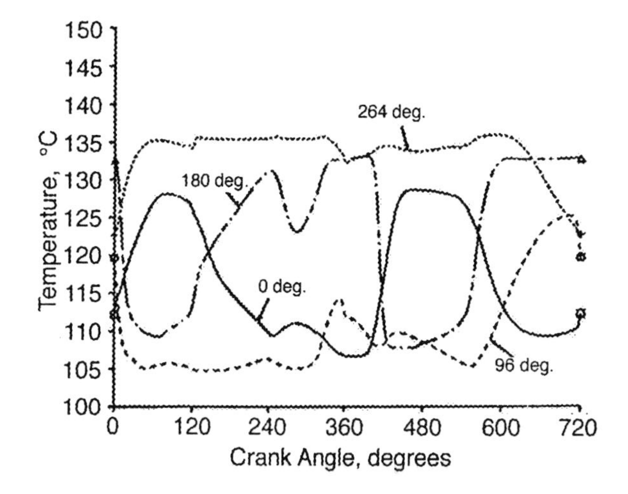

Fig. 25 Temperature in the oil film at various circumferential locations shown as degrees from the x-axis (see Fig.1) in the last load cycle for a dynamically loaded automotive main bearing running at 5000 r/min (**72**).

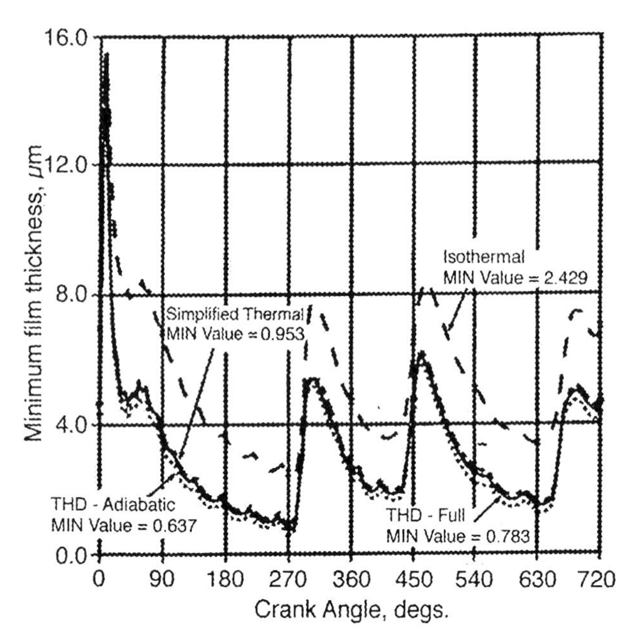

Fig. 26 Comparison between the predictions of the full THD analysis, an adiabatic THD analysis, a simplified thermal analysis and an Isothermal analysis for a dynamically loaded automotive connecting rod bearing running at 4800 r/min (**72**).
(a) minimum film thickness.

motoring or firing operations. The temperatures were found to be greatly affected by the rotating speed of the journal and to vary in the circumferential and axial directions. The temperatures on the rod side were higher than those on the cap side, which can be explained by the fact that the journal centre orbits near the rod side surface longer than on the cap side surface during a loading cycle. With regard to the axial variation, the high temperature regions are found near the bearing edge on the rod side (Fig. 27), which can be explained by the predictions by means of the authors' EHD analysis which show that the minimum film thickness near the bearing edges are smaller than that at the midcentre of bearing (Fig. 28).

Suzuki, *et al.* (**75**) studied experimentally the effect of oil supply rate on the temperature and the minimum oil film thickness of a connecting rod bearing for a petrol engine. The bearing temperature increased with the decrease in supply oil flow rate because the bearings were less cooled by the decreased oil flow rate. They found that the contact of journal and bearing increased only at one third or less of the standard oil flow rate. This phenomenon was explained by the experimental observation that some part of the axially discharged hot oil was sucked into the cavitated film region that is, axial recirculation of oil. The supply oil flow rate to the engine bearings can be reduced more than previously estimated, which could contribute to a greater reduction in bearing friction.

5.4 EHD analysis

Classical hydrodynamic lubrication theory assumes rigid surfaces of the mating walls. However engine bearing structures are becoming more and more flexible, and the bearing loads are getting higher. Consequently bearing performance should be analysed with the assumption of elastic walls.

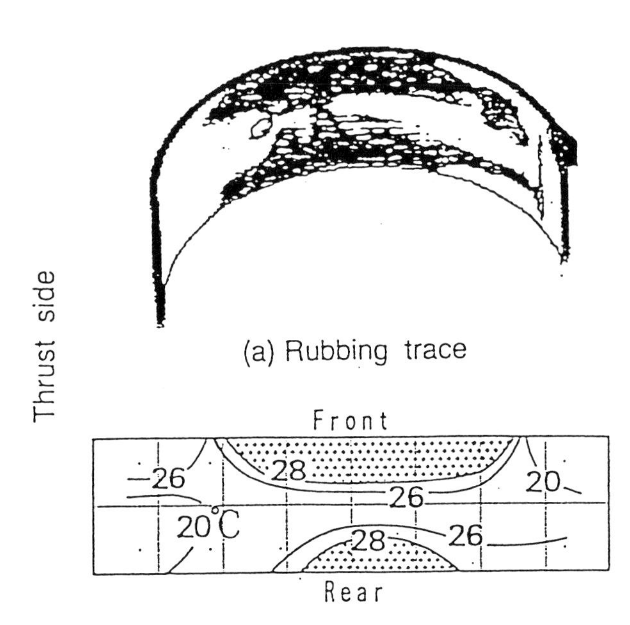

Fig. 27 Relationship between photograph of bearing surface and distribution of bearing surface temperature (full load, 5000 rpm) (**74**).

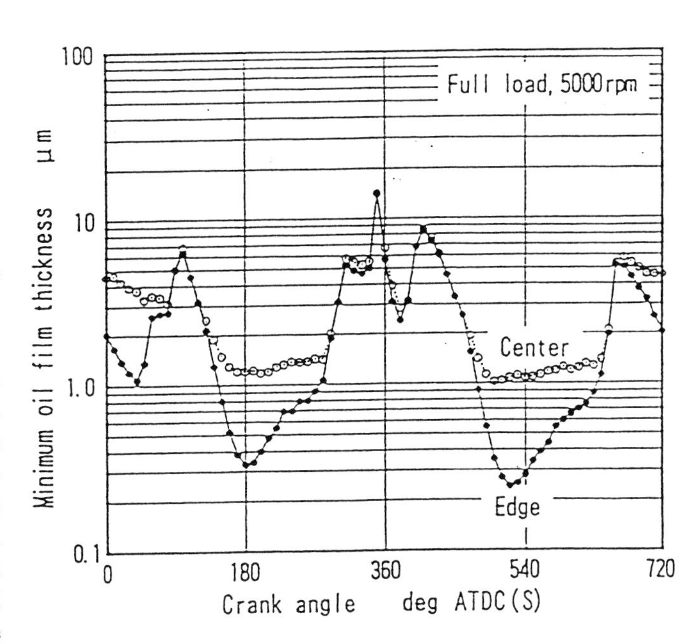

Fig. 28 Comparison of minimum oil film thickness on the center and edge in bearing width (**74**).

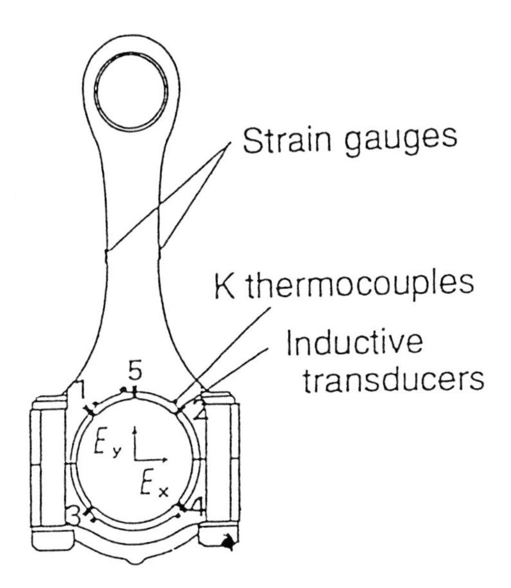

Fig. 29 Test connecting rod (**77**).

Masuda, *et al.* (**76**) measured oil film pressure distribution in a connecting rod big-end bearing with a test rig. The pressure sensor was embedded in the surface of the crankpin journal. They found that the maximum pressure measured on the cap side was lower than that on the rod side, and also that the measured pressure angle was wider than that on the rod side. This implies that under dynamic loading the cap side deformation is larger than that on the rod side, which is reasonably understood from the shape of the connecting rod.

Ozasa, *et al.* (**77**) ran an EHD analysis of connecting rod big-end bearings theoretically and experimentally under motoring conditions. Fig. 29 shows the test bearing with various sensors installed. Fig. 30 shows that measurements of the journal centre orbit agree well with predictions of the EHD model. The broken lines show the initial clearance circle of the bearing, and the journal centre orbit exceeds the limit on the cap side, which shows that the bearing is deformed when the engine is in operation. This paper shows that EHD models are essential to predict bearing performance.

Xu (**78**) studied the EHD performance of engine bearings theoretically and experimentally. He showed that the elastic deflection of the bearing and housing have a significant effect on the distribution and maximum value of the film pressure developed. The minimum oil film thickness is found at the bearing edges in the high pressure region because the bearing surface profile in the axial direction becomes a concave shape due to the pressure. These findings are in good agreement with Figs. 27 and 28 in reference (**74**). The position of the maximum shear strain predicted by the EHD model agreed well with experimental measurements.

Ushijima, *et al.* (**79**) calculated the EHD performance of an engine bearing with both elastic deformation of the

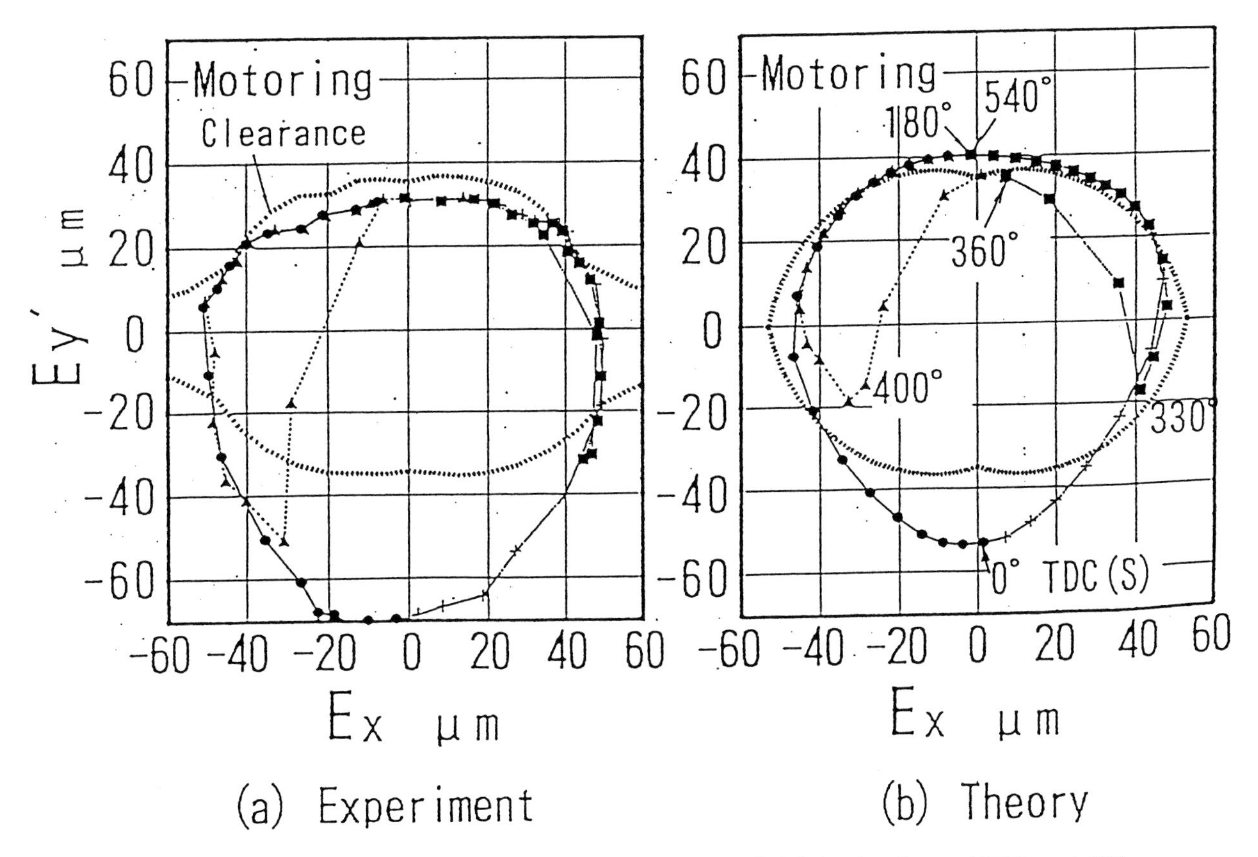

Fig. 30 Journal center orbits, experiment versus EHD lubrication theory (4000 rpm) (**77**).

bearing housing and the pressure-dependent viscosity increase of lubricant being considered. The maximum pressure is related to the bearing fatigue life, and the maximum torque is related to seizure. Figs. 31 and 32 show that the maximum values of film pressure and the frictional torque during one cycle decrease, compared to the predictions of the conventional rigid film model. They also studied the individual effect of each deformation and the viscosity change on the performance, and found that the viscosity increase leads to an increase in the temperature and also in frictional torque, though the elastic deformation causes a decrease in these.

6 BEARING DESIGN

References (**80**) to (**93**) are related to the practical design

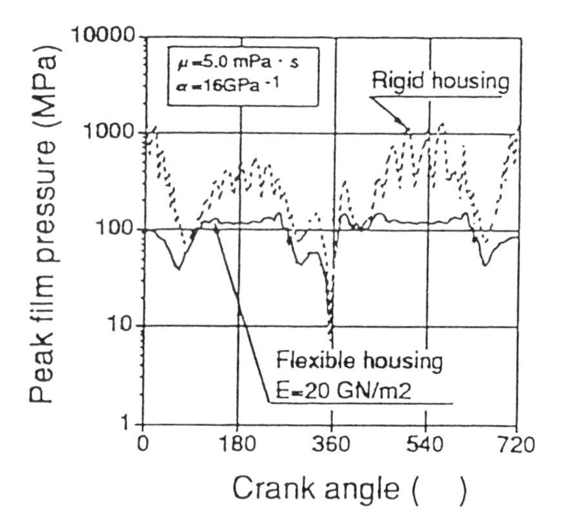

Fig. 31 Influence of elastic deformation on peak oil film pressure (**79**).

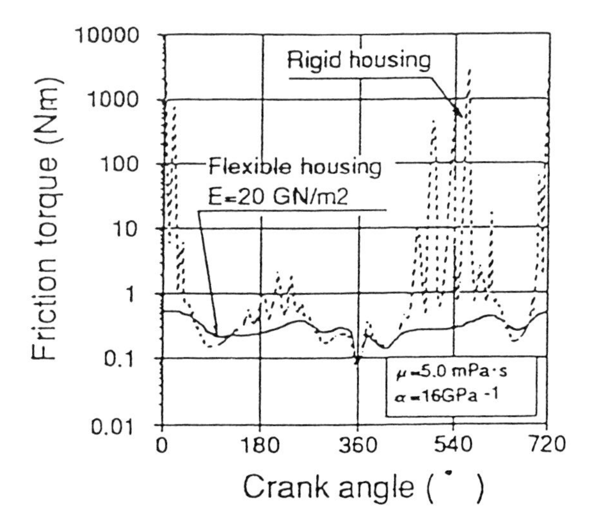

Fig. 32 Influence of elastic deformation on friction torque (**79**).

of bearings, bearing performance data for design and bearing design methods.

Anderson and Ettles (**80**) calculated journal loci in contra-rotating journal bearings for naval propulsion systems. Twin propellers rotating in directions contrary to each other, on the same axis, at the same speed can increase propulsion efficiency and save fuel cost. In that system of double spool shafting, an intershaft bearing is needed. A full circular, hydrodynamic bearing is well known to have no load capacity under such operating condition. Consequently rolling element bearings or hydrostatic bearings are normally used for these systems. They studied the dynamic responses of the journal under static and dynamic loading in various multi-lobe journal bearings for the system, and found that satisfactory load capacity could be obtained through the use of offset or preloaded arc of bearing pads.

Taylor, *et al.* (**81**) presented measurements of the static performance of a three-lobe journal bearing with a high preload factor of 0.75. Bearing performance depends strongly on the journal eccentricity from the bearing centre which is always very difficult to determine. They successfully established the centre of the test bearing in operation with a technique based on the assumed symmetry of the three lobes.

Hargreaves (**82**) calculated the static performance of a tri-taper journal bearing used for a high-speed (max. 13,000 r/min) step-up gear box. From the figure of the bearing in the paper, it appears to be a tilted three-lobe bearing. The data presented cover wide ranges of operating conditions, which is useful for bearing design engineers. He also mentions the important effect of manufacturing error of bearing dimension on the performance.

Kang, *et al.* (**83**) calculated the static and dynamic characteristics of an oil lubricated herringbone-grooved journal bearing (HGJB) which is replacing rolling element bearings in some high speed machinery where lower noise level is preferable and also high radial bearing stiffness is needed. They proposed a new design of HGJB with eight circular-profile grooves, while many analyses were published for rectangular-grooved HGJBs. They obtained the optimal values of design variables to maximise the radial stiffness of the bearing, and also found the stiffness to be slightly less than that of the rectangular-grooved ones but significantly larger than that of a plain journal bearing operating at low eccentricity ratio. They also found that the stability of HGJB is better than that of a plain journal bearing.

Nii, *et al.* (**84**) designed ferro-fluid lubricated journal bearings for vertical rotors of a polygonal mirror motor operated at up to 30,000 r/min. High speed polygon mirror motors are used for laser printers of high quality, and the motor rotors need to rotate at high speeds with low acoustic noise level and also with a small size of rotor centre orbit. It is increasingly difficult for rolling element bearings to satisfy the operational performance. They designed a three-lobe journal bearing combined with viscous seal and magnetic seal to contain lubricant inside the bearing unit, and confirmed that the bearing unit satisfies the performance needed. This design received the JSME Award.

Hashimoto (**85, 86**) presented the concept of optimum design of journal bearings and showed a general method to solve the optimum problem of bearing design. He applied his method to designing a plain journal bearing operating at high speeds and showed the solution obtained by means of the method.

Kurita and Tanaka (**87**) presented a computer-aided optimum design tool for tilting pad journal bearings widely used for high speed rotating machinery. These bearings have many more design variables than plain journal bearings, so it takes design engineers much time to obtain even a non-optimum solution which only removes various restrictions on bearing dimensions and performance; it takes much more time to obtain an optimum solution. Their optimum design template gives an easy and fast optimum design method for the bearing and a solution obtained for an example problem was confirmed to be an optimum solution.

Summaries of references (**88**) to (**93**) must be omitted due to lack of space.

7 CAVITATION

Lubricating oil films are well known to rupture in the divergent clearances between journal and bearing. Cavitated films make no contribution to the load-carrying capacity of the bearing because there is no positive hydrodynamic pressure. However there remains some oil film, dragged by the rotating journal or runner. Consequently the oil film produces some frictional torque, which should be taken into consideration to establish more precise THD models. On the other hand, starved lubrication may occur in some high speed bearings when lubricant oil is supplied insufficiently to bearings, and the oil film shape and the cavitation boundary are understandably different from those in the case of flooded lubrication. Finally, dynamic characteristics of the lubricating film is strongly dependent on the behaviour of the cavitated part of oil film.

Heshmat (**94**) studied a wide range of cavitation problems of the lubricating oil film, and compared predictions with available measurements, discussing the discrepancies. He presented three basic modes of cavitation patterns possible in hydrodynamic lubrication to explain the discrepancy (Fig. 33).

Hashimoto (**95**) calculated the static and dynamic performance of a journal bearing operating in the turbulent flow region with 0-type cavitation (Fig. 34) being assumed in the divergent film. The performance of a hydrodynamic bearing is often calculated with the half Sommerfeld condition which assumes the axially-linear cavitation boundary to be at the minimum film thickness position.

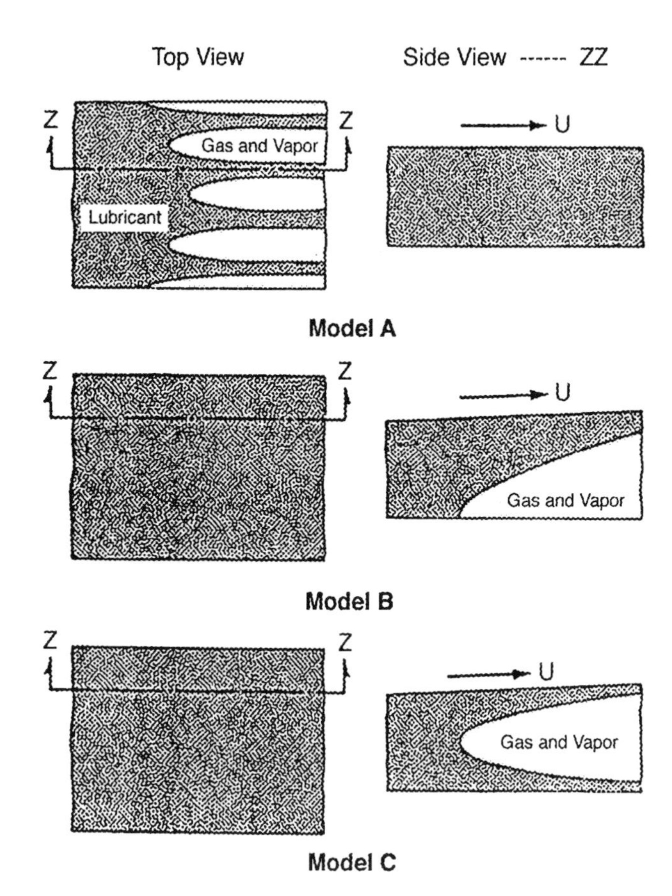

Fig. 33 Models of cavitation zone (**94**).

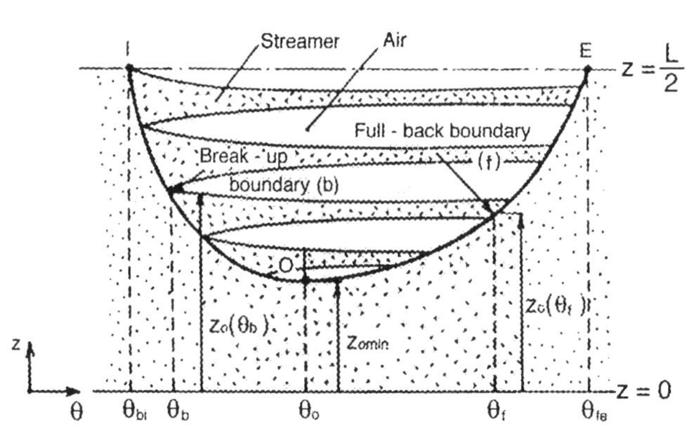

Fig. 34 'O-type' cavitation pattern (**95**).

Hashimoto found that the calculated performance with the O-type cavitation was quite different from that in the case of the half-Sommerfeld condition (Fig. 35).

Lubricant running through a lubricating system inevitably mixes with the surrounding air. Chamniprasart, *et al.* (**96**) presented a mathematical model of the mixture of oil and dissolved air, and calculated the performance of the bearing lubricated with the mixture over a wide range of variables constituting the model. They found that the

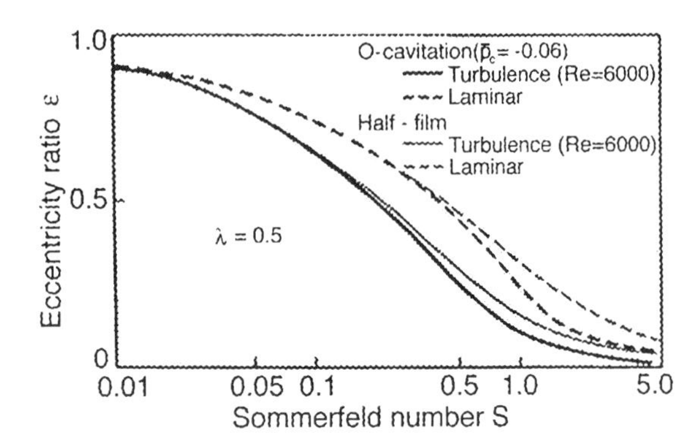

Fig. 35 Variation of eccentricity ratio with Sommerfeld number (**95**).

amount of dissolved air significantly affects the hydrodynamic pressure in the film.

Koeneke, *et al.* (**97**) proposed axial oil film rupture for lightly loaded journal bearings operating at very high speeds, like floating bush bearings used for turbocharger rotors. The hydrodynamic pressure is assumed to vary across the film thickness because of the centrifugal force acting on the film due to the very high rotational speed of the journal. Some of the dissolved air emerges and coalesces in the film where the pressure decreases below a critical point, which partially cavitates the film. In the case of journal bearings with a full circumferential oil groove at midplane, lubricant flows axially under the pressure difference, with the hydrostatic pressure being gradually lost in the axial direction of bearing. Consequently the film rupture grows near the journal surface and toward the bearing end. This axial film rupture was found to affect the velocity profile across the film thickness, and consequently to explain the decreasing speed ratio of floating bush to journal observed in actual floating bush bearings.

Mori and Mori (**98**) and Ota, *et al.* (**99**) studied the effect of lubricant inertia forces on the cavitation boundary, and Yu and Keith (**100**) applied the boundary element method to predict cavitation in journal bearings with some variations of bearing clearance in the axial direction.

8 ACKNOWLEDGEMENT

The author is very grateful to the Organising Committee for kindly giving him the chance of presenting this invited paper at the First World Tribology Congress held in London in 1997.

9 REFERENCES

1 **Ng, C.W.** and **Pan, C.H.T.**, A linearized turbulent lubrication theory, *Trans. ASME, J. Lubrication Technology,* 1965, **87,** p. 675.

2 **Aoki, H** and **Harada, M.**, Turbulent lubrication theory for full journal bearings, *J. JSLE,* 1971, **16,** pp. 348–356.

3 **Innes, G.E.** and **Leutheusser, H.L.**, An investigation into laminar-to-turbulent transition in tilting-pad bearings, *Trans. ASME J. Tribology,* 1991, **113,** pp. 303–307.

4 **Andres, L.S.**, Turbulent flow, flexure-pivot hybrid bearings for cryogenic applications, *Trans. ASME, J. Tribology,* 1996 **118,** pp. 190–200.

5 **Vaidyanathan, K.** and **Keith Jr., T.**, Performance characteristics of cavitated noncircular journal bearing in the turbulent flow regime, *Tribology Trans.,* 1991, **34,** pp. 35–44.

6 **Dowson, D.**, A generalised Reynolds equation for fluid film lubrication, *Int. Mechanical Science,* 1962, **4,** pp. 159.

7 **Taniguchi, S.**, **Makino, T.**, **Takeshita, K.**, and **Ichimura, T.**, A thermohydrodynamic analysis of large tilting-pad journal bearing in laminar and turbulent flow regimes with mixing, *Trans. ASME, J. Tribology,* 1990, **112,** pp. 542–550.

8 **Bouard, L.**, **Fillon, M.**, and **Frene, J.**, Thermohydrodynamic analysis of tilting-pad journal bearings operating in turbulent flow regime, *Trans. ASME, J. Tribology,* 1996, **118,** pp. 225–231.

9 **Bouard, L.**, **Fillon, M.**, and **Frene, J.**, Comparison between three turbulent models – application to thermohydrodynamic performances of tilting pad journal bearings, *Tribology International,* 1996, **29,** pp. 11–18.

10 **Fitzgerald, M.K.** and **Neal, P.B.**, Temperature distributions and heat transfer in journal bearings, *Trans. ASME J. Tribology,* 1992, **114,** pp. 122–130.

11 **El-Deihi, M.K.I.** and **Gethin, D.T.**, A thermohydrodynamic analysis of a twin axial groove bearing under difference loading directions and comparison with experiment, *Trans. ASME J. Tribology,* 1992, **114,** pp. 304–310.

12 **Fillon, M.**, **Bligoud, J.C.**, and **Frene, J.**, Experimental study of tilting-pad journal bearings – comparison with theoretical thermoelastohydrodynamic results, *Trans. ASME J. Tribology,* 1992, **114,** pp. 579–588.

13 **Basri, S.B.** and **Gethin, D.T.**, An experimental investigation into the thermal behaviour of a three-lobe profile bore bearing, *Trans. ASME J. Tribology,* 1993, **115,** pp. 152–159.

14 **Simmons, J.** and **Dixon, S.**, Effect of load direction, preload, clearance ratio and oil flow on the performance of a 200 mm journal pad bearing, *Tribology Trans.,* 1994, **37,** pp. 227–236.

15 **Simmons, J.E.L.** and **Lawrence, C.D.**, Performance experiments with a 200 mm, offset pivot journal pad bearing, *Tribology Trans.,* 1996, **39,** pp. 969–973.

16 **Bouchoule, C.**, **Fillon, M.**, **Nicolas, D.**, and **Barret, F.**, Experimental study of thermal effects in tilting-pad journal bearings at high operating speeds, *Trans. ASME, J. Tribology,* 1996, **118,** pp. 532–537.

17 **Makino, T.**, **Morohoshi, S.**, and **Taniguchi, S.**, Thermohydrodynamic performance of high-speed journal bearings, *PIME, J. Engineering Tribology,* 1996, **210,** pp. 179–188.

18 **Kim, K.W.** and **Rodkiewicz, C.M.**, On the thermal effects

in the design of tilting-pad bearings subjected to inlet pressure build-up, *Trans. ASME J. Tribology,* 1991, **113,** pp. 526–532.

19 **Ha, H.C., Kim, H.J.,** and **Kim, K.W.,** Inlet pressure effects on the thermohydrodynamic performance of a large tilting pad journal bearing, *Trans. ASME, J. Tribology,* 1995, **17,** pp. 160–165.

20 **Mori, A., Makino, T.,** and **Mori, H.,** Entry flow and pressure jump in submerged multi-pad bearings and grooved bearing *Trans. ASME J. Tribology,* 1992, **114,** pp. 370–378.

21 **Zhang, J.X.,** Mechanisms of pressure variation at the leading edge of a bearing pad, *PIME, J. Engineering Tribology,* 1996, **210,** pp. 125–134.

22 **Tanaka, M.,** Thermohydrodynamic performance of a tilting pad journal bearing with spot lubrication, *Trans. ASME J. Tribology,* 1991, **113,** pp. 615–619.

23 **Dmochowski, W., Brockwell, K., DeCamillo, S.,** and **Mikula, A.,** A study of the thermal characteristics of the leading edge groove and conventional tilting pad journal bearings, *Trans. ASME, J. Tribology,* 1993, **115,** pp. 219–226.

24 **Harangozo, A., Stolarski, T.,** and **Godzawa, R.,** The effect of different lubrication methods on the performance of a tilting-pad journal bearing, *Tribology Trans.,* 1991, **34,** pp. 529–537.

25 **Khonsari, M.M.** and **Wang, S.H.,** On the maximum temperature in double-layered journal bearings, *Trans. ASME J. Tribology,* 1991, **113,** pp. 464–469.

26 **Knight, J.** and **Ghadimi, P.,** Effects of modified effective length models of the rupture zone on the analysis of a fluid journal bearing, *Tribology Trans.,* 1992, **35,** pp. 29–38.

27 **Ma, M.T.** and **Taylor, C.,** Prediction of temperature fade in the cavitation region of two-lobe journal bearings, *PIME, J. Engineering Tribology,* 1994, **208,** pp. 133–140.

28 **Khonsari, M.M.** and **Wang, S.H.,** On the fluid-solid interaction in reference to thermoelastohydrodynamic analysis of journal bearings, *Trans. ASME J. Tribology,* 1991, **113,** pp. 398–404.

29 **Tanaka, M.** and **Hatakenaka, K.,** Thermohydrodynamic analysis of journal bearings with partial reverse flow in the film, *Proceedings of JAST Tribology Conference,* Kanazawa, 1994, pp. 82–84.

30 **Vijayaraghavan, D.,** An efficient numerical procedure for thermohydrodynamic analysis of cavitating bearings, *Trans. ASME, J. Tribology,* 1996, **118,** pp. 555–563.

31 **Elrod, H.G.,** Efficient numerical method for computation of the thermohydrodynamics of laminar lubricating films, *Trans. ASME, J. Tribology,* 1991, **113,** pp. 506–511.

32 **Kim, J, Palazzolo, A.,** and **Gandagi, R.,** TEHD analysis for tilting-pad journal bearings using upwind finite element method, *Tribology Trans.,* 1994, **37,** pp. 771–783.

33 **Tucker, P.G.** and **Keough, P.S.,** A generalized computational fluid dynamics approach for journal bearing performance prediction, *PIME, J. Engineering Tribology,* 1995, **209,** pp. 99–108.

34 **Khonsari, M.M., Jang, J.Y.,** and **Fillon, M.,** On the generalization of thermohydrodynamic analysis for journal bearings, *Trans. ASME, J. Tribology,* 1996, **118,** pp. 571–589.

35 **Pliakas, P.** and **Parkins, D.W.,** Single-axial groove journal bearings: static characteristics and a temperature prediction technique, *PIME, J. Engineering Tribology,* 1996, **210,** pp. 45–54.

36 **Fillon, M.** and **Khonsari, M.,** Thermohydrodynamic design charts for tilting-pad journal bearings, *Trans. ASME, J. Tribology,* 1996, **118,** pp. 232–238.

37 **Tanaka, M.,** Discussion on Reference (36), *Trans. ASME, J. Tribology,* 1996, **118,** pp. 702–703.

38 **Tucker, P.G.** and **Keogh, P.S.,** On the dynamic thermal state in a hydrodynamic bearing with a whirling journal using CFD techniques, *Trans. ASME, J. Tribology,* 1996, **118,** pp. 356–363.

39 **Desbordes, H., Fillon, M., Chan Hew Wai, C.,** and **Frene, J.,** Dynamic analysis of tilting-pad journal bearing-influence of pad deformations, *Trans. ASME, J. Tribology,* 1994, **116,** pp. 621–628.

40 **Desbordes, H., Fillon, M., Frene, J.,** and **Chan Hew Wai, C.,** The effects of three-dimensional pad deformations on tilting-pad journal bearings under dynamic loading, *Trans. ASME, J. Tribology,* 1995, **117,** pp. 379–384.

41 **Fillon, M., Desbordes, H., Frene, J.,** and **Chan Hew Wai, C.,** A global approach of thermal effects including pad deformations in tilting-pad journal bearings submitted to unbalance load, *Trans. ASME, J. Tribology,* 1996, **118,** pp. 169–174.

42 **Gandagi, R.K., Palazzolo, A.B.,** and **Kim, J.,** Transient analysis of plain and tilt pad journal bearings including fluid film temperature effects, *Trans. ASME, J. Tribology,* 1996, **118,** pp. 423–430.

43 **Ramesh, J., Majumdar, B.C.,** and **Rao, N.S.,** Nonlinear transient analysis of submerged oil journal bearings considering surface roughness and thermal effects, *PIME, J. Engineering Tribology,* 1995, **209,** pp. 53–62.

44 **Clarke, D. M., Fall, C., Hayden, G.N.,** and **Wilkinson, T.S.,** A steady-state model of a floating ring bearing including thermal effects, *Trans. ASME J. Tribology,* 1992, **114,** pp. 141–149.

45 **Freund, N.O.** and **Tieu, A.K.,** A thermo-elasto-hydrodynamic study of journal bearing with controlled deflection, *Trans. ASME, J. Tribology,* 1993, **115,** pp. 550–556.

46 **Wang, N.Z.** and **Seireg, A.A.,** Thermohydrodynamic lubrication analysis incorporating thermal expansion across the film, *Trans. ASME, J. Tribology,* 1994, **116,** p. 681–689.

47 **Rodkiewicz, C.M.** and **Yang, P., A** non-Newtonian TEHL analysis of tilting-pad bearings subjected to inlet pressure build-up, *Trans. ASME, K. Tribology,* 1995, pp. 461–467.

48 **Hussain, A., Mistry, K., Biswas, S.,** and **Athre, K.,** Thermal analysis of noncircular bearings, *Trans. ASME, J. Tribology,* 1996, **118,** pp. 246–254.

49 **Ramesh, J., Majumdar, B.C.,** and **Rao, N.S.,** Thermohydrodynamic analysis of submerged oil journal bearings considering surface roughness effects, *Trans. ASME, J. Tribology,* 1997, **119,** pp. 100–106.

50 **Sheeja, D.** and **Prabhu, B.**, Thermal and non-Newtonian effects on the steady state and dynamic characteristics of hydrodynamic journal bearings – theory and experiments, *Tribology Trans.*, 1992, **35**, pp. 441–446.

51 **Paranjpe, R.** and **Han, T.**, A study of the thermohydrodynamic performance of steadily loaded journal bearings, *Tribology Trans.*, 1994, **37**, pp. 679–690.

52 **Lin, J.** and **Chen, Y.**, Thermohydrodynamic analysis of a journal bearing in a turbulent flow regime – Part I: theory, *Tribology Trans.*, 1994, **37**, pp. 727–734.

53 **Lin, J.** and **Chen, Y.**, Thermohydrodynamic analysis of a journal bearing in a turbulent flow regime – Part II: application to an axial groove bearing, *Tribology, Trans.*, 1994, **37**, pp. 820–828.

54 **Swanson, E.** and **Kirk, G.**, An experimental comparison of two steadily loaded plain journal bearings, *Tribology Trans.*, 1994, **37**, pp. 843–849.

55 **Swanson, E.** and **Kirk, G.**, Experimental temperature and pressure profiles for two steadily loaded journal bearings, *Tribology Trans.*, 1995, **38**, pp. 601–606.

56 **Sheeja, D.** and **Prahbu, B.S.**, Thermohydrodynamic analysis of journal bearings lubricated by non-Newtonian fluids – theory and experiments, *PIME, J. Engineering Tribology*, 1994, **208**, pp. 173–182.

57 **Tor, S.** and **Tanaka, M.**, Experimental investigation of the effect of shaft heating and cooling on single bore journal bearing, *Proceedings Nordtrib* 1996, Vol. II.

58 **Tor, S.** and **Tanaka, M.**, Thermohydrodynamic performance of journal bearings with external heat source being considered (2nd report), *Proceedings of JAST Tribology Conference*, Kita-Kyushu, 1996, pp. 81–83.

59 **Ettles, C.M.** and **Anderson, H.G.**, Three-dimensional thermoelastic solutions of thrust bearings using Code Marmacl, *Trans. ASME J. Tribology*, 1991, **113**, pp. 405–412.

60 **Ettles, C. M.**, Some factors affecting the design of spring supported thrust bearings in hydroelectric generator, *Trans. ASME, J. Tribology*, 1991, **113**, pp. 626–632.

61 **Brockett, T.S.**, **Barrett, L.**, and **Allaire, PE.**, Thermo-elasto-hydrodynamic analysis of fixed geometry thrust bearings including runner deformation, *Tribology Trans.*, 1996, **39**, pp. 555–562.

62 **Tanaka, M.**, **Hori, Y.**, and **Ebinuma, R.**, Measurement of the film thickness and temperature profiles in a tilting pad thrust bearings, *Proceedings of the International Tribology Conference Tokyo*, 1985, pp. 553–558.

63 **Kim, K.W.**, **Tanaka, M.**, and **Hori, Y.**, An experimental study on the thermohydrodynamic lubrication of tilting pad thrust bearings, *J. JAST*, 1995, **40**, pp. 70–77.

64 **Kim, K.W.**, **Tanaka, M.**, and **Hori, Y.**, A three-dimensional thermohydrodynamic performance of sector-shaped, tilting-pad thrust bearings, *Trans. ASME, J. Lubrication Technology*, 1983, **105**, pp. 406–413.

65 **Uno, S.**, **Andoh, M.**, **Namba, S.**, and **Mukai, K.**, Overview of recent tendencies in thrust bearings for hydrogenerators, *J. JAST*, 1997, **42**, pp. 129–135.

66 **Prashad, H.**, An approach to evaluate capacitance, capacitive reactance and resistance of pivoted pads of a thrust bearing, *Tribology Trans.*, 1992, **35**, pp. 435–440.

67 **Dabrowski, L.** and **Wasilczuk, M.**, A method of friction torque measurement for a hydrodynamic thrust bearings, *Trans. ASME, J. Tribology*, 1995, **117**, pp. 674–678.

68 **Rodkiewicz, C.** and **Yang, P.**, Proposed TEHL solution system for the thrust bearings inclusive of surface deformations, *Tribology Trans.*, 1995, **38**, pp. 75–85.

69 **Matsuhisa, H.**, **Yamamoto, K.**, **Sakamoto, M.**, and **Tanaka, T.**, Improvement of anti-seizure property on three layer bearings for high speed and high load automotive engines, *J. JAST*, 1994, **39**, pp. 792–799.

70 **Kumada, Y.**, **Hashizume, K.**, and **Kimura, Y.**, Performance of plain bearings with circumferential microgrooves, *Tribology Trans.*, 1996, **39**, pp. 81–86.

71 **Paranjpe, R.S.** and **Han, T.**, A transient thermohydrodynamic analysis including mass conserving cavitation for dynamically loaded journal bearing, *Trans. ASME, J. Tribology*, 1995, **117**, pp. 369–378.

72 **Paranjpe, R.S.**, A study of dynamically loaded engine bearings using a transient thermohydrodynamic analysis, *Tribology Trans.*, 1996, **39**, pp. 636–644.

73 **Moteki, K.**, **Ushijima, K.**, **Tasaki, Y.**, and **Aoyama, S.**, A method of analyzing lubrication conditions in engine bearings and influence of journal and bearing surface roughness on friction loss, *Proceedings of the International Tribology Conference, Yokohama* 1995, pp. 1011–1015.

74 **Suzuki, S.**, **Ozasa, T.**, **Yamamoto, M.**, **Nozawa, Y.**, **Noda T.**, and **Ohori, M.**, Temperature distribution and lubrication characteristics of connecting rod big end bearings, *SAE Trans.* 1995, 952550, pp. 2025–2034.

75 **Suzuki, S.**, **Ozasa, T.**, **Noda, T.**, and **Konomi, T.**, Effect of oil supply rate on temperature and contact of a con-rod big end bearing, *Proc. J. SAE*, 1996, 9638789, pp. 133–136.

76 **Masuda, T.**, **Ushijima, K.**, and **Hamai, K.**, A measurement of oil film pressure distribution in connecting rod bearing with test rig, *Tribology Trans.*, 1992, **35**, pp. 71–76.

77 **Ozasa, T.**, **Yamamoto, M.**, **Suzuki, S.**, **Nozawa, Y.**, and **Konomi, T.**, Elastohydrodynamic lubrication model of connecting rod big end bearings: comparison with experiments by diesel engine, *SAE Trans.*, 1995, 952549, pp. 2011–2024.

78 **Xu, H.**, Effects of EHD contacts upon the bearing and housing behaviour, *SAE Paper #960987*, 1996.

79 **Ushijima, K.**, **Moteki, K.**, **Goto, T.**, and **Aoyama, S.**, A study on engine bearing performance focusing on the viscosity-pressure characteristic of the lubricant and housing stiffness, *SAE Paper #961144*, 1996.

80 **Anderson, H.** and **Ettles, C.**, Contrarotating journal bearings for naval propulsion systems, *Tribology Trans.*, 1992, **35**, pp. 509–515.

81 **Taylor, D.**, **Kostrzewsky, G.**, **Flack, R.**, and **Barret, L.**, Measured performance of a highly pre-loaded three-lobe journal bearing – Part I: static characteristics, *Tribology Trans.*, 1995, **38**, pp. 507–516.

82 **Hargreaves, D.J.**, Predicted performance of a tri-taper journal bearing including turbulence and misalignment

effects, *PIME, J. Engineering Tribology,* 1995, **209,** pp. 85–98.

83 **Kang, K., Rhim, Y.,** and **Sung, K.,** A study of the oil-lubricated herringbone-grooved journal bearing – Part 1: Numerical analysis, *Trans. ASME, J. Tribology,* 1996, **118,** pp. 906–911.

84 **Nii, K., Kawaike, K., Uno, S., Matsubayashi, J.,** and **Nakajima, T.,** Fero-fluid lubricated bearings for a polugon mirror motor, *PIME, J. Engineering Tribology,* 1996, pp. 199–204.

85 **Hashimoto, H.,** Optimum design for sliding bearings (lst report), *Proceedings of JAST Tribology Conference, Kanazawa,* 1994, pp. 785–788.

86 **Hashimoto, H.,** Optimum design for sliding bearings (2nd report), *Proceedings of JAST Tribology Conference, Kanazawa,* 1994, pp. 789–792.

87 **Kurita, M.** and **Tanaka, M.,** An optimum design method for tilting pad journal bearings, *Proceedings of JAST Tribology Conference, Kanazawa,* 1994, pp. 781–784.

88 **Tripp, H.** and **Melodick, T.,** Steady load eccentricity measurements in a two axial groove journal bearing, *Tribology Trans.,* 1991, **34,** pp. 292–300.

89 **Metha, N.,** Static and dynamic characteristics of orthogonally-displaced pressure dam bearings, *Tribology Trans.,* 1993, **36,** pp. 201– 206.

90 **Ashman, D.,** Investigation of the failure of heavily loaded journal bearings, *PIME, J. Engineering Tribology,* 1994, **208,** pp. 167–172.

91 **Nataraj, C., Ashrafiuon, H.,** and **Arakere, N.,** Effect of fluid 5-inertia on journal bearing parameters, *Tribology Trans.,* 1994, **37,** pp. 784–792.

92 **Pascovici, M.D., Khonsari, M.M.,** and **Jang, J.Y.,** On the modelling of a thermomechanical seizure, *Trans. ASME, J. Tribology,* 1995, **117,** pp. 744–747.

93 **Summers-Smith, J.D.** and **Taylor, D.,** Screw compressor bearing failure on process plant: a case study and some general lessons, *PIME, J. Engineering Tribology,* 1995, **209,** pp. 77–84.

94 **Heshmat, H.,** The mechanism of cavitation in hydrodynamic lubrication, *Tribology Trans.,* 1991, **34,** pp. 177–186.

95 **Hashimoto, H.,** The effects of 'O-type' cavitation on the performance characteristics of turbulent journal bearings, *Tribology Trans.,* 1991, **34,** pp. 100–106.

96 **Chamniprasat, K., AL-Sharif, A., Rajagopal, K.R.,** and **Szeri, A.Z.,** Lubrication with binary mixtures: bubbly oil, *Trans. ASME, J. Tribology,* 1993, **115,** pp. 253–260.

97 **Koeneke, C.E., Tanaka, M.,** and **Motoi, H.,** Axial oil film rupture in high speed bearings due to the effect of the centrifugal force, *Trans. ASME, J. Tribology,* 1995, **117,** pp. 394–398.

98 **Mori, A.** and **Mori, H.,** Re-examination of film rupture boundary condition in hydrodynamic lubrication under inertia effect, *Trans. ASME J. Tribology,* 1991, **113,** pp. 604–608.

99 **Ota, T., Yoshikawa, H., Hamasuna, M., Motohashi, T.,** and **Oi, S.,** Inertia effects on film rupture in hydrodynamic lubrication, *Trans. ASME, J. Tribology,* 1995, **117,** pp. 685–690.

100 **Yu, Q.** and **Keith Jr., T.G.,** Prediction of cavitation in journal bearing using a boundary element method, *Trans. ASME, J. Tribology,* 1995, **117,** pp. 411–421.

Friction and wear of PTFE-filled PPS and the role of PTFE filler

Y UCHIYAMA, T IWAI, and **Y UENO**
Kanazawa University, Faculty of Engineering, Kanazawa, Japan
Y UEZI
Starlite Company Limited, Tribosystems Laboratory, Osaka, Japan

SYNOPSIS

The friction and wear properties of polyphenylene sulfide (PPS), polytetrafluoroethylene (PTFE), PTFE-filled PPS and PPS composite filled with PTFE and zinc oxide (ZnO) fine particles were examined when rubbed against austenitic stainless steel at various temperatures from 24°C to 250°C.

The coefficient of friction for unfilled PPS was relatively high and around 0.6 in the temperature range from 24°C to 150°C. The specific wear rate for PPS was around 10^{-5} mm^3/Nm in the temperature range from 24°C to 200°C and steeply increased to 10^{-3} mm^3/Nm at 250°C. PTFE showed a relatively low friction coefficient from 0.2 to 0.3 and a higher specific wear rate of around 10^{-4} mm^3/Nm. When PTFE 20 vol.% is filled into PPS, the PTFE filler has the effect of reducing the coefficient of friction and the specific wear rate to around 10^{-6} mm^3/Nm. PPS composites filled with both PTFE and ZnO showed a similar friction coefficient to the PTFE, and had a very low specific wear rate of 10^{-7} mm^3/Nm at 24°C. At around 100°C the lowest value of 5×10^{-8} mm^3/Nm was experienced. Even at temperatures from 130°C to 200°C the specific wear rates were of the order of 10^{-6} mm^3/Nm. When the low wear rates for the PPS composites filled with PTFE and ZnO were experienced, thin and uniformly transferred films were seen and metal fluoride was detected on the mating stainless steel. The transferred films are thought to adhere strongly to the mating surface by chemical bonding. It is therefore concluded that the PTFE filler is effective in reducing the friction and wear not only by its molecular smoothness and banded structure, but also by strongly adhered films on mating surfaces.

1 INTRODUCTION

PTFE exhibits very low friction at low sliding speeds. On the other hand, among plastics PTFE shows a relatively high wear rate. On the mating surface, PTFE films are easily formed by the destruction of the banded structure and the molecular smoothness of PTFE (**1–4**). It is known that the friction and wear properties of PTFE vary with sliding speed, temperature and contact pressure. The wear curves obtained for different temperatures and contact pressures as a function of sliding speed are fitted to a reference curve to make a master curve (**5**). The basic wear processes involved in the formation of wear particles of PTFE have been reported previously (**6**). The wear rates are explained by the wear processes and the detachment rate of the wear fragments. Transferred films also affect the wear rate of polymers. When PTFE is added to other plastics as a filler, beneficial friction and wear properties are recognized (**7–11**).

In this paper, the effects of temperature on the adhesive wear of PPS, PTFE, PTFE-filled PPS, and PPS filled with PTFE in combination with ZnO are investigated. The role of PTFE filler in the friction and wear is also discussed.

The transferred films on the mating disk were also examined by XPS (X-ray photoelectron spectroscopy) analysis.

2 EXPERIMENTAL

In this experiment, a pin-on-disk type wear apparatus was used as shown in Fig.1. A polymer pin specimen (**7**) 3 mm in diameter is rubbed against an austenitic stainless steel disk (**8**) in a circular motion with a mean diameter of 40 mm. The stainless steel was 7 mm thick and 64 mm in diameter. It was finished by buffing using abrasives 1 μm in size and the center-line average roughness R_a of the surface was 0.02 μm. The surface temperature was measured by a non-contact type thermosensor (**6**), and was automatically controlled by the heater (**9**) under the disk. The friction force was measured using strain gauges (**4**) attached to the leaf springs (**3**), and the wear depth of the polymer pin was measured with a differential transformer (**9**). Experiments were carried out at a load W of 9.8 N (this load gives a contact pressure p of 1.39 MPa) and various temperatures from room temperature to 250°C. The sliding speed was 30 cm/s.

The specimen codes and compositions of the polymer pins are given in Table 1. The PPS and PTFE specimens were made by cutting from plate-shaped specimens. The specimen PPS/PTFE contains PTFE powder (ca. 15 μm in mean diameter) in the PPS matrix, and the PPS/PTFE/ZnO contains ZnO (ca. 2 μm in diameter) one vol. part per hundred parts of the PPS/PTFE. They were made by

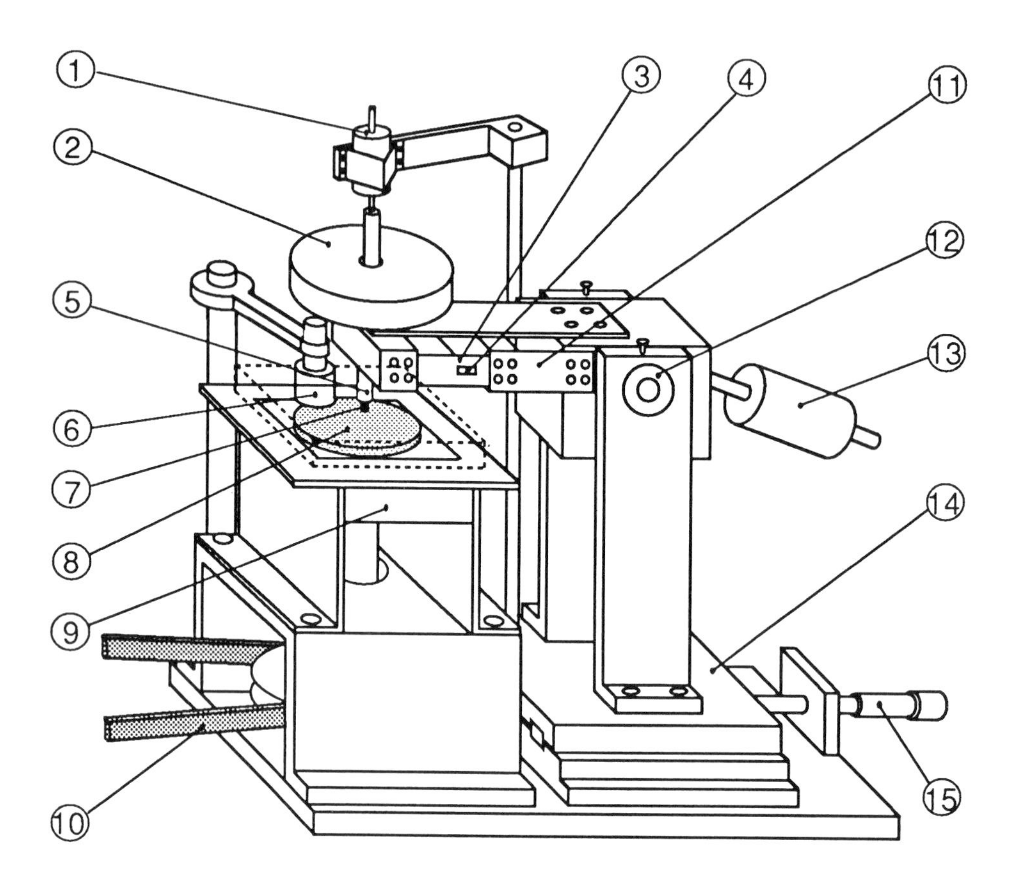

Fig. 1 Pin on disk type wear apparatus: 1. differential transfomer; 2. dead weight; 3. leaf spring; 4. strain gauge; 5. pin specimen holder; 6. non-contact type thermosensor; 7. pin specimen; 8. disk specimen; 9. heater; 10. belt; 11. arm; 12. bearing; 13. counter balance; 14. moving stage; 15. micrometer.

cutting from pin-shaped specimens which were made by injection moulding. The mechanical properties of the specimens are shown in Table 2. In Fig. 2(a) and (b), fracture surfaces of the PPS/PTFE and PPS/PTFE/ZnO were shown respectively. In Fig. 2(a) PTFE fillers from 10 to 20 μm in diameter are observed in the PPS matrix. In this figure there are many holes. The holes seem to be made by detachment of the PTFE fillers which accompanied the counter-fracture surface. In Fig. 2(b) small particles of ZnO about 2 μm in diameter are seen in the PPS matrix together with large particles of PTFE.

Table 1 Specimen codes and compositions.

Specimen code	Polymer and filler		
	PPS	PTFE	ZnO
PPS	100%		
PIFE		100%	
PPS/PTFE	80 vol.%	20 vol.%	
PPS/PTFE/ZnO	80 vol.%	20 vol.%	1 vol. part [1]

1. 1 vol. part per hundred parts of the PPS/PTFE

The transferred films on the disk were observed by an optical microscope, and the thickness of the film was measured using a non-contact type optical profilemeter. The transferred films on the disk were also analysed by XPS to examine the chemical composition, and the effects of transferred PTFE films on the friction and wear were discussed.

Table 2 Mechanical properties of the specimens.

Specimen code	Elastic modulus	Tensile strength	Elongation
	(GPa)	(MPa)	(%)
PPS	4.20	140.0	1.6
PTFE	0.50	20.7	220.0
PPS/PTFE	1.86	42.3	2.3
PPS/PTFE/ZnO	1.65	56.4	3.0

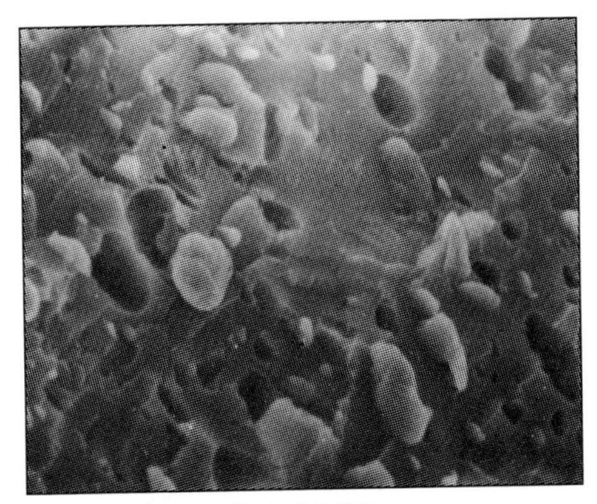

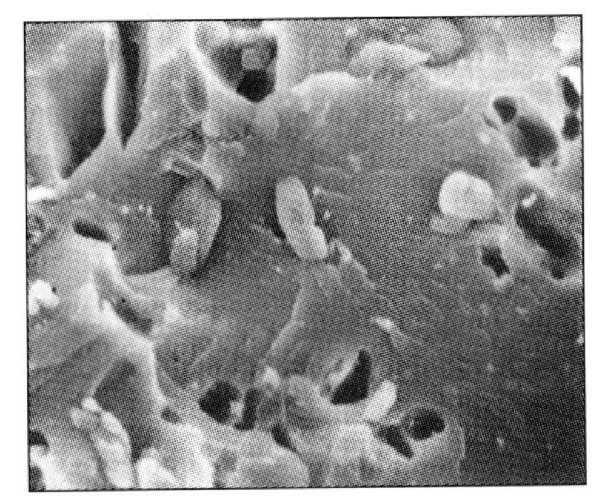

Fig. 2 Fracture surfaces of the filled PPS specimens:
(a) PPS/PTFE; (b) PPS/PTFE/ZnO.

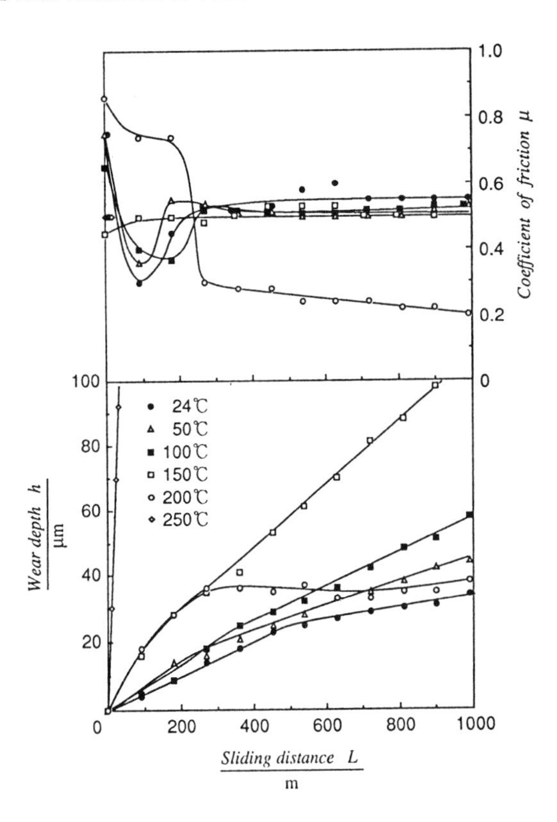

Fig. 3 Variations in the coefficient of friction and wear depth for PPS with sliding distance under various temperatures: $W = 9.8$N ($p = 1.39$ MPa); $v = 30$ cm/s.

3 RESULTS AND DISCUSSION

3.1 Effect of temperature on friction and wear

Figure 3 shows the variation of the coefficient of friction and wear depth for PPS with sliding distance at various temperatures. After the initial stage, steady-state rubbing was attained. In steady-state rubbing at 24°C, the coefficient of friction of PPS was 0.55 and the wear rate was relatively low compared with those at higher temperatures. At 250°C the highest wear rate was observed. In the initial stage at 200°C relatively high friction coefficient and wear rate were experienced. However a very low friction coefficient of 0.2 and the lowest wear rate were attained at steady state rubbing. For the range of temperature examined, the friction coefficient of PPS was around 0.5 except for that at 250°C.

The coefficient of friction and wear depth for the PTFE and the PTFE-filled PPS were also examined. The wear depth for the PTFE increases almost linearly with sliding distance. The coefficient of friction is relatively low ranging from 0.18 to 0.28 overthe temperature range examined. The wear rate for the PTFE-filled PPS is initially large. Following this a steady state exists during which the wear depth increases linearly with sliding distance. At 24°C and 250°C, the wear rate is high. However at other temperatures examined the wear rates were low. The friction coefficient is low compared with the unfilled PPS. At 24°C and 50°C, the friction coefficient is around 0.3. At high temperatures from 100°C to 150°C the friction coefficient showed a very low value of 0.12.

Figure 4 shows the coefficient of friction and the wear depth for the specimen PPS/PTFE/ZnO *versus* sliding distance. From 24°C to 125°C, the wear rates were very low for steady-state rubbing. At 135°C and 150°C, it showed relatively high wear rates. Above 200°C the wear rates increased drastically. At temperatures ranging from 24°C to 125°C, the friction coefficient varied from 0.2 to 0.3. Above 135°C, low friction coefficients from 0.07 to 0.1 were observed.

Figure 5 shows the variations with temperature of the coefficient of friction and specific wear rates of the four polymer specimens for steady state rubbing. The friction coefficient for the PPS showed values from 0.4 to 0.6 at temperatures from 24°C to 150°C, and the specific wear rates were of the order of 10^{-5} mm^3/Nm. At 200°C, the friction coefficient drastically decreased to 0.2 and the specific wear rate decreased to $6\text{x}10^{-6}$ mm^3/Nm. At 250°C the friction coefficient increased to 0.5 and the specific wear rate also increased steeply to 10^{-3} mm^3/Nm.

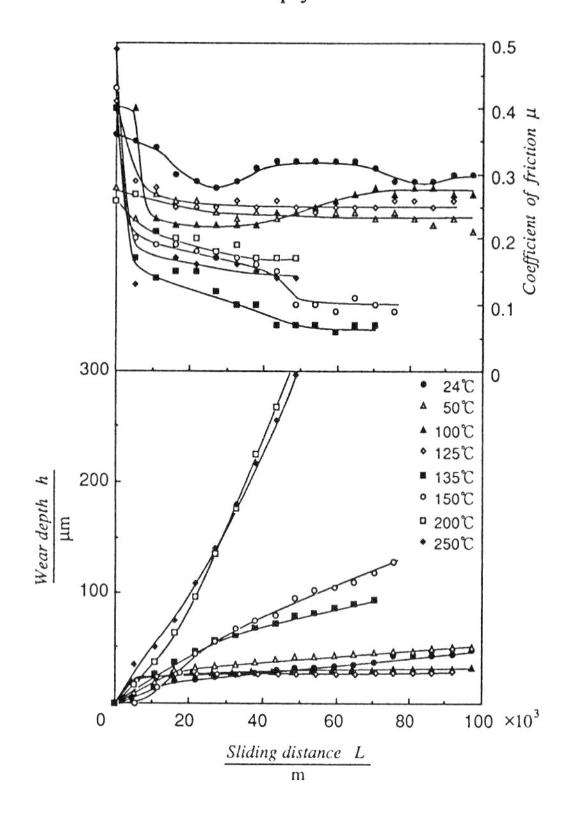

Fig. 4 Variations in the coefficient of friction and wear depth for PPS/PTFE/ZnO with sliding distance under various temperatures: W = 9.8N (p = 1.39 MPa); v = 30cm/s.

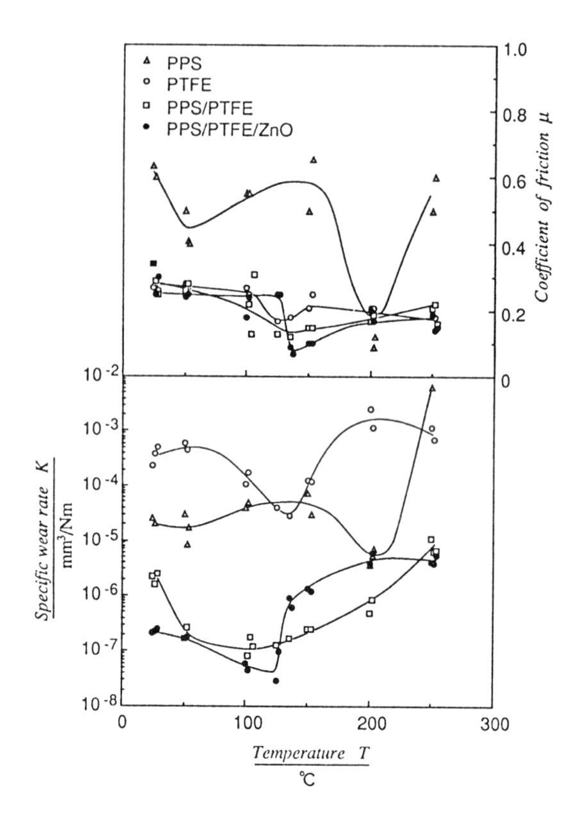

Fig. 5 Variations in the coefficient of friction and specific wear rate for specimens with temperature: W = 9.8N (p = 1.39 MPa); v = 30cm/s.

In this way the PPS shows relatively higher friction coefficients and wear rates. PTFE shows low friction coefficient of around 0.2, decreasing slightly with increasing temperature. From 24°C to 100°C the specific wear rates for the PTFE were of the order of 10^{-4} mm^3/Nm. Around the glass transition temperature T_g of 139°C, there is a minimum friction coefficient of 0.15 and a specific wear rate of 3×10^{-5} mm^3/Nm. At 200°C and 250°C, high specific wear rates of 10^{-3} mm^3/Nm were observed. When PPS was filled with PTFE, the friction coefficient decreased to around 0.2. Around 120°C to 130°C, a very low friction coefficient of 0.13 was experienced. The specific wear rate was 2×10^{-6} mm^3/Nm at 24°C, and then it decreased to 2 or 3×10^{-7} mm^3/Nm at temperatures from 50°C to 150°C. Above 200°C, the specific wear rate increased monotonically from 10^{-6} mm^3/Nm to 10^{-5} mm^3/Nm with increasing temperature.

The addition of PTFE filler has an effect of minimizinf the friction coefficient. The frictional behaviour was similar to that of PTFE. At 24°C the specific wear rate was reduced by about one order of magnitude compared with that of PPS and two orders compared with that of PTFE.

Very low wear rates were experienced for PPS composites which contained PTFE and ZnO at temperatures from 24°C to 125°C. The specific wear rate was 2×10^{-7} mm^3/Nm at 24°C. The lowest wear rate of 5×10^{-8} mm^3/Nm was observed at 100°C to 125°C. Above 135°C the specific wear rate increased steeply to the order of 10^{-6} mm/Nm. The friction coefficient varied from 0.08 to 0.22 over the temperature range examined. Minimum friction coefficient was observed at 135°C.

The PTFE and ZnO fillers reduced the specific wear rate at temperatures from 24°C to 125°C. However the wear rate was raised in the range from 135°C to 200°C. At 250°C, the wear rate was comparable to that of the PPS/PTFE. When the wear properties of the PPS/PTFE/ZnO were compared with the PPS/PTFE, it is seen that ZnO filler reduced the wear rate of the PPS/PTFE by about one order at 24°C. At temperatures from 135°C to 200°C, the wear rates were raised about five times over those of the PPS/PTFE. However the friction coefficient was similar to that of the PPS/PTFE.

Therefore the PTFE filler is effective in reducing friction and wear. When ZnO filler was added to the PPS/PTFE, a remarkable reduction of wear rate was observed at temperatures from 24°C to 125°C.

3.2 Observation of the transferred films

Transferred films on the mating stainless steel disk were observed using an optical microscope. Figure 6(a)–(h) show the mating disks which were rubbed against PPS, PTFE, PPS/PTFE and PPS/PTFE/ZnO at 24°C and 100°C. As shown in Fig. 6(a), thin PPS films were observed on the mating surface at 24°C. However at 100°C only slight PPS films were observed on the mating disk. When PTFE was rubbed against the mating disk, transferred fragments were observed on the disk at 24°C and 100°C (Fig. 6(c) and (d)). Many fragments were observed in the rubbing track at 24°C compared with those at 100°C. When the PPS/PTFE pin was rubbed against the mating disk at 24°C, the mating surface was almost covered with thick transferred films. At 100°C the specific wear rate of the PPS/PTFE was $2x10^{-6}$ mm^3/Nm and the value was about two orders less than that of the PPS. At 100°C the transferred films were relatively thin compared with those at 24°C. At this time the lowest specific wear rate of 10^{-7} mm^3/Nm was experienced. The addition of the PTFE filler into the PPS matrix has an effect of forming thick transferred films on the mating surface. Figure 6(g) shows the transferred films from the PPS/PTFE/ZnO pin at 24°C. Many transferred fragments were seen on the mating disk. Then a very low wear rate of $2x10^{-7}$ mm^3/Nm was observed. At 100°C relatively thin transferred films uniformly covered the mating disk. The wear rate then was very low and was around $5xl0^{-8}$ mm^{-3}/Nm. When the thin transferred films were observed at 100°C for these two PPS composites, lower wear rates were experienced compared with those at 24°C.

The surface roughness of the transferred film on the disk was measured using a non-contact type profilemeter and maximum height of the profile, including transferred films, R_{max}, was obtained at various temperatures. As shown in Fig. 7, slight PPS transferred films were observed over the temperature range examined except at 200°C. PTFE transferred fragments about 4–5 μm thick were observed below 100°C. However at 125°C uniformly thin transferred PTFE films 0.5 μm thick were seen. At 150°C flake-like transferred films also appeared on the mating surface. At 200°C and 250°C transferred films or fragments were not observed by optical microscopy. Very thin PTFE films seemed to adhere to the mating disk.

At 24°C uniform PPS/PTFE films were observed. At 125°C the films were extended largely towards the sliding direction and flakes also appeared. At 135°C there were films and flakes on the disk. At 150°C thick films and many flake-like fragments were seen. At 250°C, large, thick transferred films and thin films were simultaneously observed.

The features of the PPS/PTFE/ZnO were very similar to those of the PPS/PTFE. However at 24°C relatively larger fragments and films were observed on the mating disk. On the rubbing surface of the PPS/PTFE/ZnO pin, many fine ZnO particles were observed and distinct scratches were not found at 24°C. On the other hand there were fine scratching scars on the PPS/PTFE pin, and plastic flow was observed.

Addition of the fine ZnO particles, therefore prevented the PPS/PTFE/ZnO surface from easy ploughing. The lower wear rate then seems to arise from mechanical reinforcement of the pin surface. From 135°C to 200°C relatively thick transferred fragments of the PPS/PTFE/ZnO were observed on the mating surface compared with those of the PPS/PTFE. The wear rate for the PPS/PTFE/ZnO was then higher than that for the PPS/PTFE.

3.3 Surface chemical analysis of the mating surface

There have been many studies of the chemical reactions of fillers and polymer matrices with mating surfaces **(12–15)**.

The mating stainless steel disks were examined by XPS after steady-state rubbing against the pin specimens. Spectra of PPS and PTFE were detected on the mating disks when rubbed at 100°C and 150°C against the PPS/PTFE and PPS/PTFE/ZnO. The intensity of the PTFE for these two transferred disks at 100°C showed higher levels compared with those at 150°C. Therefore the transferred films of PTFE were dominant at 100°C, and PPS seems to be enriched at 150°C. On the film transferred disks Fe_2O_3 and $FeSO_4$ were simultaneously detected. The intensity of $FeSO_4$ at 150°C was higher than that at 100°C both on the PPS/PTFE and PPS/PTFE/ZnO transferred disks. Moreover the intensity of $FeSO_4$ on the PPS/PTFE/ZnO transferred disk was higher than that on the PPS/PTFE transferred disk. Therefore the ZnO filler was effective in placing $FeSO_4$ onto the mating disk. The sulfur is thought to come from the PPS. When the PPS pin was rubbed against stainless steel at room temperature, $FeSO_4$ was not detected. The sulfur in the PPS seems to react only at higher temperatures and ZnO accelerated the formation of $FeSO_4$.

Additional experiments were made to examine the transferred films of the PTFE, PPS/PTFE and PPS/PTFE/ZnO on the disk at room temperature. On the disk rubbed against PTFE, PTFE, decomposed PTFE and metal fluoride (CrF_3, FeF_2) were identified in the XPS spectra. For the PPS/PTFE transferred disks, decomposed PTFE and metal fluoride were observed. However the intensity of the two species was relatively low compared with that on the PTFE transferred disk. For the PPS/PTFE/ZnO transferred disk, the intensity of metal fluoride was higher than that on the PPS/PTFE transferred disk. Compared with CrF_3, a high intensity of FeF_2 was detected. From the XPS analysis ZnO seems to have an effect of removing surface contamination and metal oxides to form more metal fluorides such as FeF_2.

Even at the high temperature of 100°C, static contact between PTFE and copper did not form metal fluoride. However the rubbing of PTFE against copper formed metal fluoride even at room temperature. When the PPS

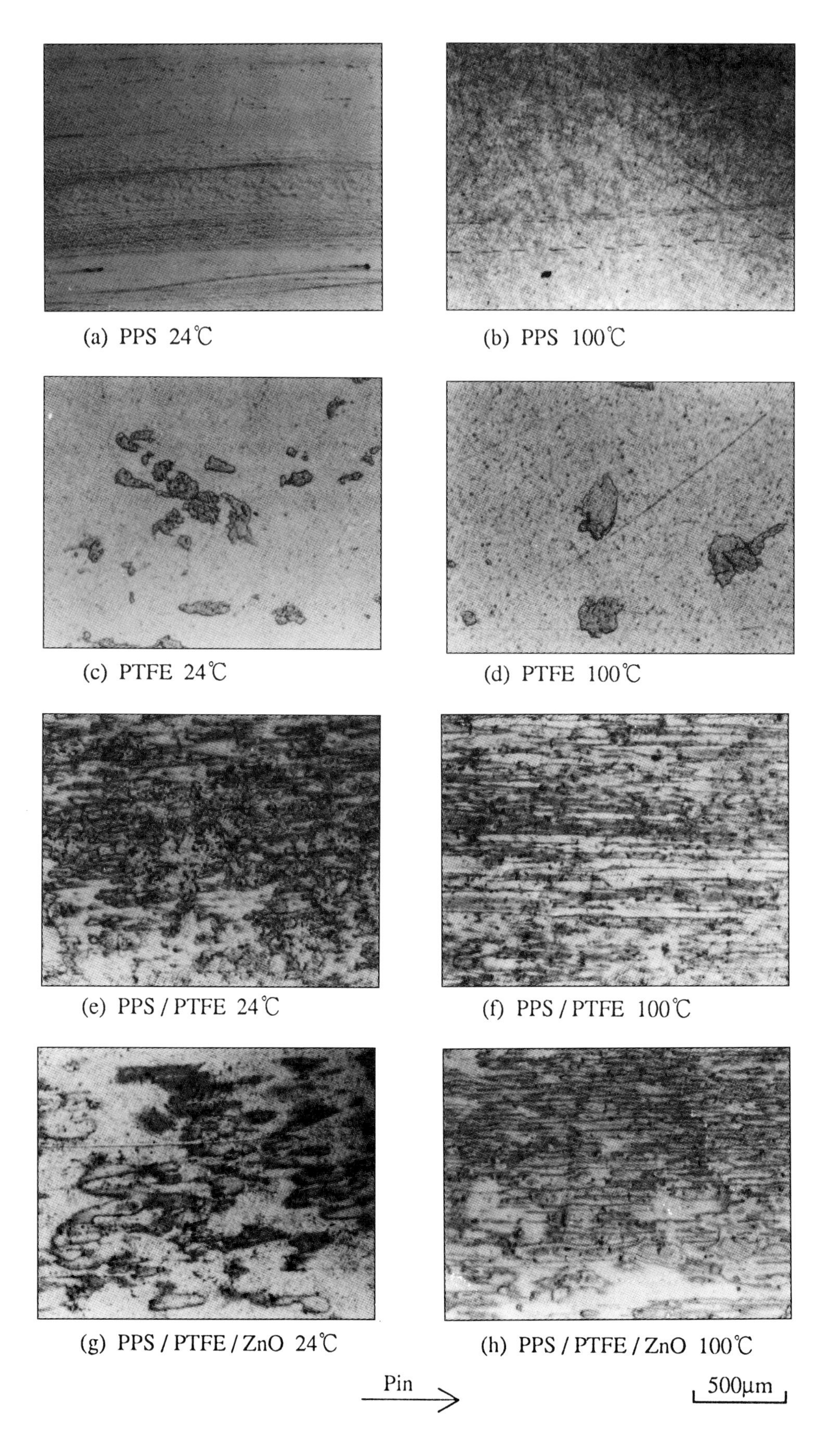

Fig. 6 Rubbing surfaces of a stainless steel disk when rubbed against the four pin specimens PPS, PTFE, PPS/PTFE, and PPS/PTFE/ZnO at 24°C and 100°C.

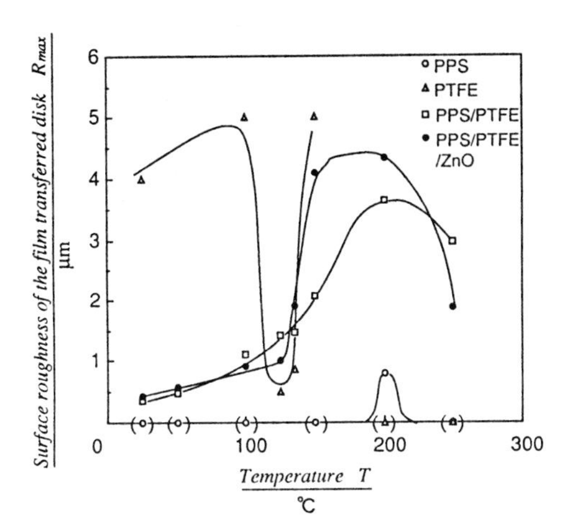

Fig. 7 Variations in the surface roughness of the film transferred disks which were rubbed against four pin specimens with temperature.

composites containing PTFE were rubbed against a mating stainless steel disk, transferred films or fragments were observed together with metal fluoride. The metal fluoride is thought to favour the formation of uniform and stable transferred films on the mating disk.

4 CONCLUSIONS

The friction and wear of PPS, PTFE, PTFE-filled PPS and PPS filled with both PTFE and fine ZnO particles were examined when rubbed against an austenitic stainless steel disk at various temperatures from 24°C to 250°C. The transferred films were also examined. From the results using XPS analysis, the following conclusions were drawn:

1. Unfilled PPS showed relatively high friction coefficients of around 0.6 at a temperature range from 24°C to 150°C. The specific wear rate was around 10^{-6} mm^3/Nm over the temperature range from 24°C to 200°C and increased steeply to 10^{-3} mm^3/Nm at 250°C.
2. PTFE showed relatively low friction coefficients from 0.2 to 0.3 and higher specific wear rate around 10^{-4} mm^3/Nm. When PTFE 20 vol.% was filled into PPS, the PTFE filler had the effect of reducing the friction coefficient to the level of PTFE and the specific wear rate to around 10^{-6} mm^3/Nm. At around 100°C, a specific wear rate of 10^{-7} mm^3/Nm was observed. PPS filled with both PTFE and ZnO showed a similar effect in minimizing the friction coefficient to the level of PTFE, and had a very low specific wear rate of 10^{-7} mm^3/Nm at 24°C. At around 100°C the lowest value of $5x10^{-8}$ mm^3/Nm was observed. Even at temperatures from 130°C to 200°C the specific wear rates were in the order of 10^{-6} mm^3/Nm.
3. When very low wear rates were experienced for the PTFE-filled PPS and PPS filled with both PTFE and ZnO, thin and uniformly transferred films were seen and metal fluoride was detected together with PTFE on the mating stainless steel. The transferred films are thought to adhere strongly to the mating surface by chemical bonding. At 24°C decomposition of PPS was not observed. However at 100°C and 150°C, $FeSO_4$ was detected together with Fe_2O_3, indicating that sulfur from the PPS reacted with iron in the stainless steel.

Thin transferred films which contain PTFE and PPS are useful in reducing friction and wear. ZnO filler has an effect of reinforcing the PTFE-filled PPS and protecting the surface from easy ploughing. Moreover ZnO is effective in removing surface contamination and metal oxides to form metal fluorides.

5 ACKNOWLEDGEMENTS

The authors are indebted to Messrs. S. Ueda, S. Araki and S. Nakabayashi for their assistance in the experiments. Thanks are extended to Starlite Co., Ltd. for manufacturing the filled PPS specimens.

6 REFERENCES

1 **Bunn, C.W., Cobbold, A.J.,** and **Palmer, R.P.,** The fine structure of polytetrafluoroethylene, *J. Polymer Sci.,* 1958, **38,** 365–376.
2 **Rachel Makinson, K.** and **Tabor, D.,** The friction and transfer of polytetrafluoroethylene, *Proc. Roy. Soc.(London),* 1964, **A281,** 49–61.
3 **Steijn, R.P.,** The sliding surface of polytetrafluoroethylene: An investigation with electron microscope, *Wear,* 1968, **12,** 193–212.
4 **Tanaka, K., Uchiyama, Y.,** and **Toyooka, S.,** The mechanism of wear of polytetrafluoroethylene, *Wear,* 1973, **23,** 153–172.
5 **Uchiyama, Y.** and **Tanaka, K.,** Wear law for polytetrafluoroethylene, *Wear,* 1980, **58,** 223–235.
6 **Uchiyama, Y.,** The mechanism of formation of wear particles of polytetrafluoroethylene, *Wear,* 1981–1982, **74,** 247–262.
7 **Uchiyama, Y., Yamada, Y.,** and **Miura, H.,** Fundamental studies on the wear of the plastic materials for sliding bearings against aluminum shafts (Part 1), Friction and wear properties of various plastic materials and surface damage of mating aluminum, *J. JSLE International Edition,* 1989, **10,** 5–10.
8 **Uchiyama, Y., Yamada, Y.,** and **Miura, H.,** *ibid.* (Part 2), *Frictional properties and wear lives of the transferred films, ibid.,* 1989, **10,** 11–16.
9 **Uchiyama, Y., Ogawa, T., Uezi, Y., Kudo, A.,** and **Kimura, T.,** Friction and wear of liquid crystal polymers, *Proc. Japan International Tribology Conference Nagoya,* 1990, 1359–1364.
10 **Uchiyama, Y., Uezi, Y., Kudo, A.,** and **Kimura, T.,** Effect of temperature on the wear of unfilled and filled liquid crystal polymers, *Wear,* 1993, **162–164,** 656–661.

11 **Uchiyama, Y., Iwai, T.,** and **Morimoto, A.,** Friction and wear properties of thermoplastic polyimides, *Proc. International Tribology Conference Yokohama,* 1995, 367–372.
12 **Wheeler, D.R.,** The transfer of polytetrafluoroethylene studied by X-ray photoelectron spectroscopy, *Wear,* 1981, **66,** 355–365.
13 **Blanchet, T.A., Kennedy, F.E.,** and **Jayne, D.T.,** XPS analysis of the effect of fillers on PTFE transfer film development in sliding contacts, *Tribology Transactions,* 1993, **36,** 535–544.
14 **Gong, D., Zhang, B., Xue, Q.,** and **Wang, H.,** Investigation of adhesion wear of filled polytetrafluoroethylene by ESCA, AES and XRD, *Wear,* 1990, **57,** 25–39.
15 **Bahadur, S., Gong, D.,** and **Anderegg, J.,** Investigation of the influence of CaS, CaO and CaF_2 fillers on the transfer and wear of nylon by microscopy and XPS analysis, *Wear,* 1996, **197,** 271–279.

Lubrication and wear of ultra-high molecular weight polyethylene in total joint replacements

A WANG, A ESSNER, V K POLINENI, D C SUN, C STARK, and **J H DUMBLETON**
Howmedica, Inc., Pfizer Medical Technology Group, Rutherford, New Jersey, USA

SYNOPSIS

This paper reviews recent advances in tribology of artificial joints. Emphases are given to the latest developments in understanding wear mechanisms and in wear testing of ultra-high molecular weight polyethylene (UHMWPE) materials and components. Two major advances have been made. One is the discovery and recognition of the importance of multi-directional motion in wear mechanisms and wear testing of UHMWPE. The other is the development of an orientation-softening wear concept. The significance of joint kinematics in wear of acetabular and tibial components is discussed. New ways of improving the wear resistance of UHMWPE are proposed based upon the theory of orientation softening. Crosslinking of UHMWPE by ionizing radiation has been shown to significantly improve the wear resistance. The degree of improvement is greater with higher doses of irradiation and is much more effective for the hip than for the knee. A starved lubrication mechanism is proposed for the conforming contact between a UHMWPE acetabular cup and a CoCr or alumina ceramic head. Soluble proteins are found not to be an effective boundary lubricant for UHMWPE. Criteria for validating wear and joint simulator testing are discussed. Focuses are given on the choice of lubricant, the degree of motion and the positioning of components in joint simulator testing.

1 INTRODUCTION

The success of total joint arthroplasty during the past thirty years is largely due to the use of ultra-high molecular weight polyethylene as a bearing surface. A property of UHMWPE that distinguishes it from other polymers is that the molecular chains of UHMWPE are extremely long and highly entangled which makes it resistant to wear. However, wear does occur in UHMWPE and sometimes can be excessive in younger and more active patients. Factors affecting wear are complex including design, material, surgical and patient factors. This paper focuses on the material aspects of wear and emphasizes in particular recent developments in the elucidation of wear mechanisms and wear testing.

A UHMWPE molecule consists of numerous repeating units of ethylene monomers. The backbone structure of the molecule is the C-C covalent bond. In the solid state, UHMWPE molecules are arranged in both ordered and disordered regions, i.e., crystalline and amorphous regions (**1**). In the crystalline region, the chains are folded with the chain axis, i.e., the C-C direction, oriented perpendicularly to the chain fold-interface. The area of the fold-interface is much larger than the chain fold length which gives rise to a characteristic lamellar shape for each crystallite. In the amorphous region, there is no regular chain folding but adjacent chains are interconnected through random mechanical entanglements and occasionally chemical crosslinks (for unirradiated UHMWPE, chemical crosslinking is almost non-existing). The connections between crystalline and amorphous regions are provided by tie-molecules. The volumetric percentage of the crystallites is the volume crystallinity of the material.

The mechanical properties of UHMWPE are determined by a) the balance between crystalline and amorphous regions, i.e., crystallinity, b) the number of connections between the crystalline regions, i.e., number of tie molecules, which is inversely proportional to the thickness of the crystallite, c) the number and nature of connections within the amorphous regions, i.e., degree of mechanical entanglements and crosslinks, and d) the presence or absence of orientation of the crystallites (**1, 2**). The elastic modulus and yield strength of UHMWPE are much more sensitive to changes in crystallinity and crystalline orientation than any other properties. Elastic modulus is particularly sensitive to crystallinity and crystallite orientation, increasing almost linearly with increasing crystallinity (**2**). Increasing elastic modulus without proportionally increasing yield strength increases the probability of plastic contact in the UHMWPE which may increase the wear rate.

The relationship between mechanical properties and wear resistance for UHMWPE is extremely complex. For polymers in general, the values of tensile strain energy at break have been considered as a measure of the abrasion resistance. Ratner, *et al.* (**3**) and Lancaster (**4**) have found independently that the abrasive wear rates of various polymers can be correlated with the reciprocal of the product of ultimate tensile strength and elongation to failure. A similar correlation was found for UHMWPE using a reciprocating wear tester under serum lubricated conditions where the wear mechanism was predominantly surface fatigue (**5**). However, recent studies using multi-axial hip joint simulators have indicated otherwise (**6–8**). For example, both the ultimate tensile strength and elongation to failure decrease with increasing the degree of crosslinking of the UHMWPE (**7, 8**). The wear resistance, on the other hand, increases significantly as the degree of crosslinking increases (**7, 8**).

Recent advances in the understanding of UHMWPE structure and its relationship to wear behavior have provided some explanation for the missing link between wear resistance and conventional mechanical properties, i.e., the discovery of crystalline anisotropy induced by molecular orientation during multi-directional joint articulation (**9–11**). As will be discussed later in detail, this finding has changed our ways of thinking about mechanical properties and wear behavior of UHMWPE. It has also changed the ways we conduct our wear testing and has highlighted the key areas that we will be focusing on in the future in order to develop a more wear-resistant UHMWPE. In the remainder of this paper, we will be focusing on the following specific areas:

- The molecular origin of wear.
- Wear and lubrication mechanisms.
- Effect of joint kinematics on wear.
- Methods of wear testing.

2 THE MOLECULAR ORIGIN OF WEAR

The phenomenon of contact sliding-induced molecular orientation was reported by Pooley and Tabor in 1972 **(12).** They found that linear semi-crystalline polymers such as polytetrafluoroethene (PTFE) and high density polyethylene (HDPE) possess frictional characteristics that differ from those of polymers that have branched molecular chains. While the coefficient of friction between a linear polymer and a smooth hard surface is sensitive to the way in which the two surfaces move, this motion-dependent behavior is absent for branched polymers. This finding occurs because linear polymers can orient at the sliding interface while branched polymers, because of steric hinderance, cannot.

Subsequent to Pooley and Tabor's study on polymer friction, Briscoe and Stolarski (**13, 14**) reported that the wear behavior of PTFE and HDPE is also motion-dependent. Higher wear rates were associated with linear motion while non-linear motion resulted in lower wear rates in PTFE and HDPE under dry sliding conditions against smooth steel counterfaces. They observed that during dry sliding the transfer film formed on the steel counterface was lumpy and thicker for non-linear motion than for linear motion. These authors believed that the thicker transfer film adhered more strongly to the

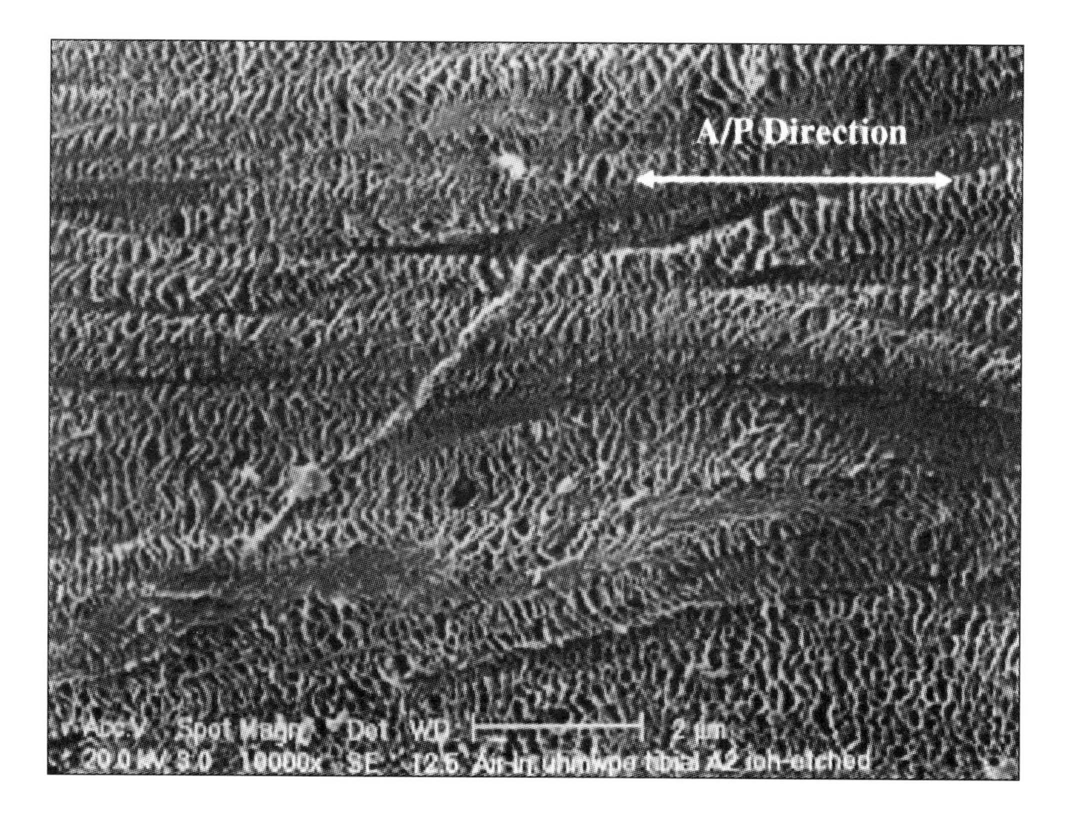

Fig. 1

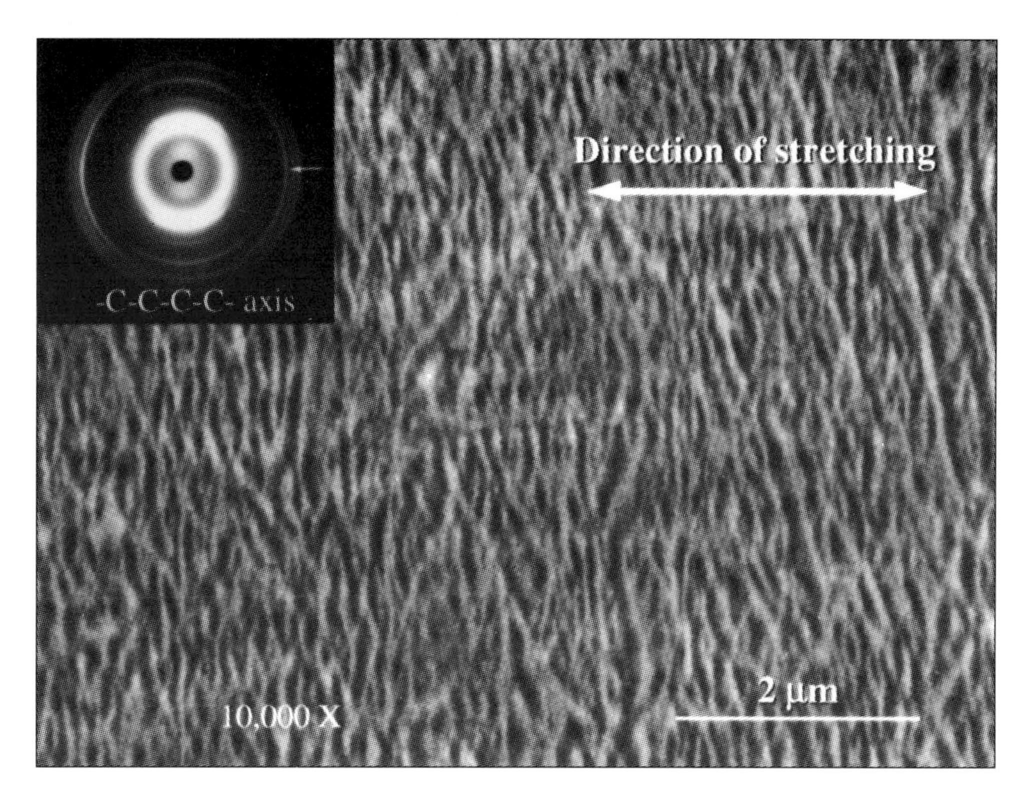

Fig. 2

counterface and was less likely to become loose wear debris. Recent studies on UHMWPE under lubricated conditions have indicated that the wear behavior of UHMWPE is very sensitive to the mode of relative motion (**15–17**). Linear motion, whether unidirectional or reciprocating, results in very low wear rates while multi-directional motion results in very high wear rates in non-crosslinked linear UHMWPE. This motion-dependent wear behavior for UHMWPE under lubricated conditions is quite opposite to that observed for the dry sliding wear for PTFE and HDPE.

Direct evidence of molecular orientation on worn surfaces of UHMWPE was identified by Wang, *et al.* (**10, 11**) by means of plasma etching and scanning electron microscopy. These authors examined the surfaces of acetabular and tibial components that had been tested on hip and knee simulators and observed extensive plastic deformation on the worn surfaces. By etching the surfaces with ion plasma, they were able to reveal a preferential orientation of crystalline lamellae in a direction perpendicular to the direction of principal motion, i.e., the direction of flexion/extension, an example of which is shown in Fig. 1. Wang, *et al.* (**11**) also applied this plasma etching method to both tensile stretched and shear fractured specimens and observed a similar crystalline orientation phenomenon to that found on the worn surfaces of tibial components, an example of which is shown in Fig. 2. In addition, a flat-plate X-ray diffraction method was used to verify the orientation morphologies as revealed by plasma etching. Bellare, *et al.* (**18**), in a separate study, analyzed the crystalline morphologies of retrieved UHMWPE tibial components using small angle X-ray scattering (SAXS). They also observed various degrees of molecular orientation at different locations of the worn tibial component.

Other forms of evidence of molecular orientation were provided by McKellop, *et al.* (**19**), Jasty, *et al.* (**20**), and Wang, *et al.* (**9**), who examined retrieved or simulator tested components of UHMWPE with SEM. A common observation by all these investigators was that all cups examined exhibited a characteristic appearance of surface stretching in the form of fibril formation or fibrillation. McKellop, *et al.* (**19**) believed that this fibrillar morphology was a result of micro-adhesive wear while Jasty, *et al.* (**20**) associated it with third-body abrasive wear. Wang, *et al.* (**9**) proposed that this fibrillar appearance simply reflects the occurrence of molecular orientation on the worn surfaces which can be a result of either adhesive wear or abrasive wear.

The significance of molecular orientation in UHMWPE wear was acknowledged by both Jasty, *et al.* (**20**) and Wang, *et al.* (**9**). Jasty, *et al.* believed that scratching by third-body particles causes a breakup of 'spherulites' to become a chain-extended fibrillar structure. This chain-extended structure is brittle and therefore can be easily broken off to become loose wear particles by subsequent encounters with third-body particles. Wang, *et al.*, on the other hand, proposed that the fibrillar structure is readily formed on worn surfaces by surface traction forces even in the absence of three-body abrasion. This oriented structure is stronger or harder (not brittle) in the orientation direction but weaker or softer in the transverse direction which is perpendicular to the orientation direction. Under the conditions of multi-directional motion, which as will be seen later apply to both the hip and the knee joint, it is the orientation softening phenomenon that is predominantly responsible for the detachment of fibrous wear debris from the worn surfaces.

Experimental support for the concept of 'orientation softening' was later provided by Wang, *et al.* **(10)** in a separate study. In their experimental setup, they first stretched a thin sheet of UHMWPE in uniaxial tension at various stresses and then machined smaller tensile specimens from the prestretched sheet in two orthogonal directions: one parallel to the pre-stretching direction and the other perpendicular to it, Fig. 3. Uniaxial tensile tests were then performed on these smaller specimens. The results are shown in Fig. 4. For the parallel-specimens, the ultimate tensile strength increased significantly with increasing pre-stretching stress. For the perpendicular-specimens, the ultimate strengths decreased substantially as the pre-stretching stress increased. The yield strength showed a similar trend. Therefore, the phenomena of both orientation hardening and orientation softening were observed in this experiment. Based on these findings, Wang, *et al.* proposed that the most important material property for wear resistance is the resistance to orientation

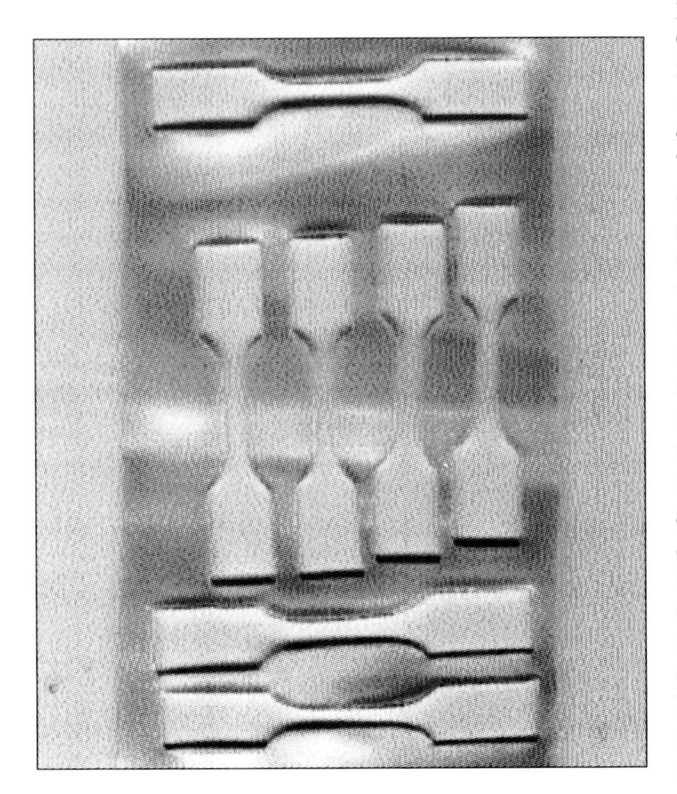

Fig. 3

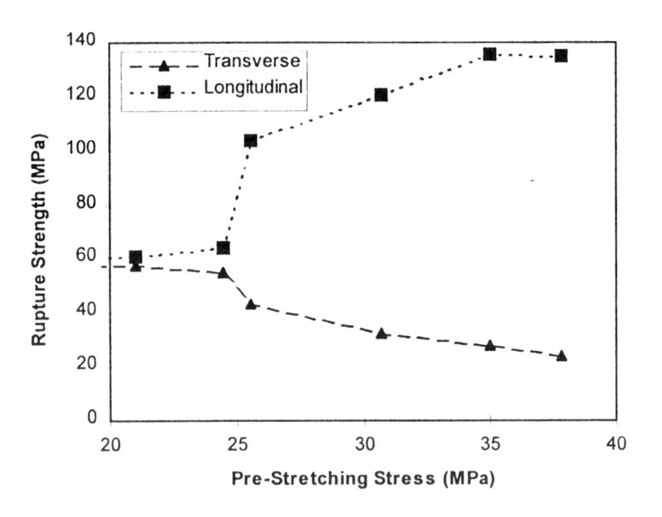

Fig. 4

softening. Since molecular orientation is associated with linear polymers rather than branched or crosslinked polymers, one way to improve the wear resistance of UHMWPE is to convert its linear molecular structure to a crosslinked structure. Experimental data will be presented later in this paper to substantiate this hypothesis.

3 WEAR AND LUBRICATION MECHANISMS

The pre-requisite for the occurrence of molecular orientation is the accumulation of plastic strain on the worn surfaces of UHMWPE acetabular or tibial components. For plastic strain to occur and accumulate, the nature of sliding contact must be plastic. For a conforming hip joint, the maximum nominal contact stresses are approximately within the range of 3 to 6 MPa which are well below the compressive yield strength of the UHMWPE (~ 12 MPa). Therefore, the nature of contact in the hip would be entirely elastic without a significant contribution from frictional stresses (surface traction). Even under a very high level of sliding friction, the bulk of the acetabular component would still deform elastically although the surface layer may experience extensive plastic flow. Three different contact conditions that have been seen clinically favor the occurrence of surface plastic flow. One very obvious condition is that one or both of the surfaces in contact are sufficiently rough so that contact occurs predominantly at asperity levels which often results in contact stresses exceeding the yield strength of the asperities. Such a condition may be satisfied by the contact of a smooth femoral head on a rough acetabular surface, a rough femoral head on a smooth acetabular surface or a rough femoral head on a rough acetabular surface. The second contact condition which favors plastic flow is the situation in which third-body abrasive particles are trapped between the two solid surfaces. The third contact condition, which is the least understood or the most misunderstood, is the situation in which both surfaces are nominally smooth and no third-body particles are involved.

Experience with clinically retrieved total hip implants has shown that high clinical wear rates were seen not only on components that showed gross damage or extensive scratching but also on components that exhibited excellent surface finish, i.e., polishing. Laboratory hip simulator testing has also shown that high wear rates occur readily on components that are nominally smooth throughout the entire period of testing **(9, 19, 21)**. SEM examinations of retrieved or simulator tested acetabular cups often reveal the following four types of surface features: a) regular and irregular arrays of surface ripples and bumps; b) oriented and non-oriented loose fibrils; (c) oriented fibrils with no loose ends; d) multi-directional scratches within which loose fibrils are sometimes seen. It is our observation that features b), c), and d) are associated with high wear rates while feature a) with low wear rate. McKellop, *et al.* **(19)** have attributed feature b) to micro-adhesive wear while Jasty, *et al.* **(20)** associated features c) and d) with three-body abrasive wear. Wang, *et al.* **(5, 22)** proposed a micro-fatigue wear mechanism for feature a) and agreed with the classifications of both McKellop, *et al.* and Jasty, *et al.* on features b) to d). Examples of the occurrence of these three wear mechanisms in THR are shown in Fig. 5. The relative importance of these mechanisms in acetabular cup wear is shown schematically in Fig. 6. From the wear mechanism point of view, we believe that both the adhesive and abrasive wear mechanisms are predominantly responsible for higher clinical wear rates. In other words, high wear rates can result from the contact of both smooth and rough surfaces.

The classic description of adhesive wear states that the interfacial shear stress must exceed the shear strength of the softer surface so that material can be either transferred to the harder counterface or detached from the softer surface to become loose wear particles **(23)**. In other words, interfacial friction plays a key role in the adhesive wear process. If the surfaces are effectively lubricated with joint fluid, adhesive wear would be very unlikely to occur. Then why does adhesive wear occur frequently in artificial

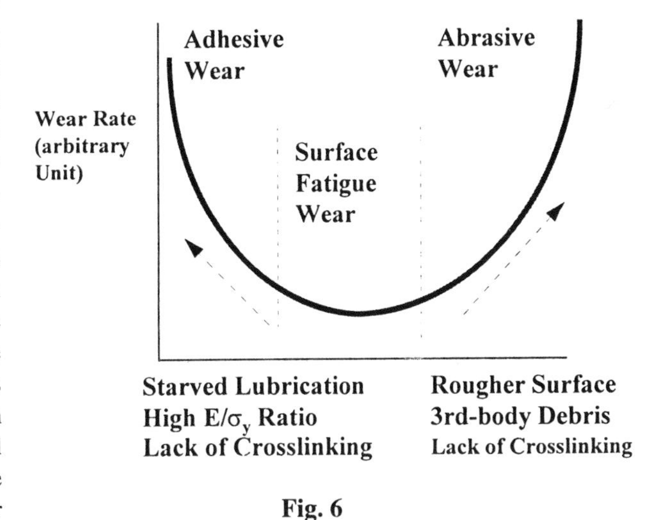

Fig. 6

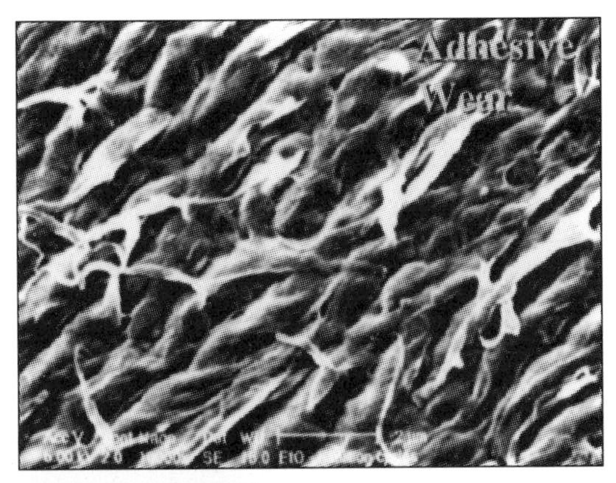

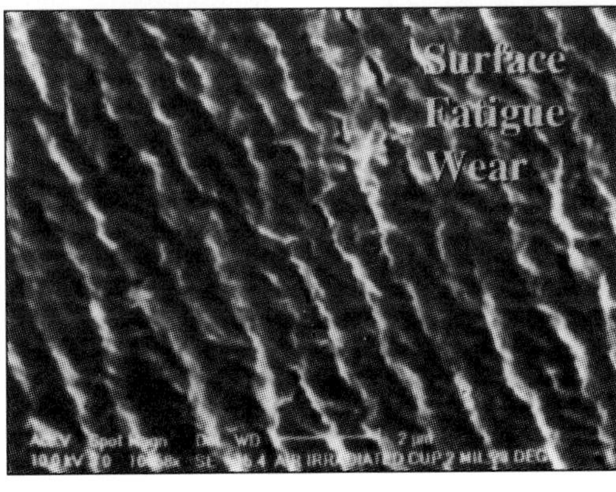

Fig. 5

joints? To answer this question from a fundamental point of view, we must assume that artificial joints of the type of metal/polyethylene or ceramic/polyethylene are not so effectively lubricated as we normally believe. We will present the following experimental evidence to support our hypothesis.

In May 1995 at the ASTM Workshop on the characterization and performance of articular surfaces, Wang, *et al.* presented an experiment to investigate the effect of contact area on the wear of UHMWPE in conforming contact situations **(24)**. This experiment involved the contact of a convex CoCr cylinder (75 mm in diameter) oscillating (±30°) on a conforming concave UHMWPE surface, Fig. 7. A constant contact load of 1100 N was applied to all specimens. Bovine serum was used as lubricant. The test results were surprising, Fig. 8. We found that for conforming contact, the wear rate increased significantly as the contact area or contact width increased beyond a critical value, although the contact stresses decreased with increasing the contact area. SEM examinations of the worn surfaces revealed that when the contact width was below this critical value the surfaces were extremely smooth with no tearing or rupture of the surfaces; however, when the contact width was above this critical value, the worn surfaces showed extensive tearing and rupture which is a typical characteristic of adhesive wear. We then proposed that a starved lubrication regime exists for conforming contact with large contact areas. Artificial hip joints satisfy this condition for starved lubrication. Therefore, the larger the femoral head, the worse the lubrication condition. This finding led us to

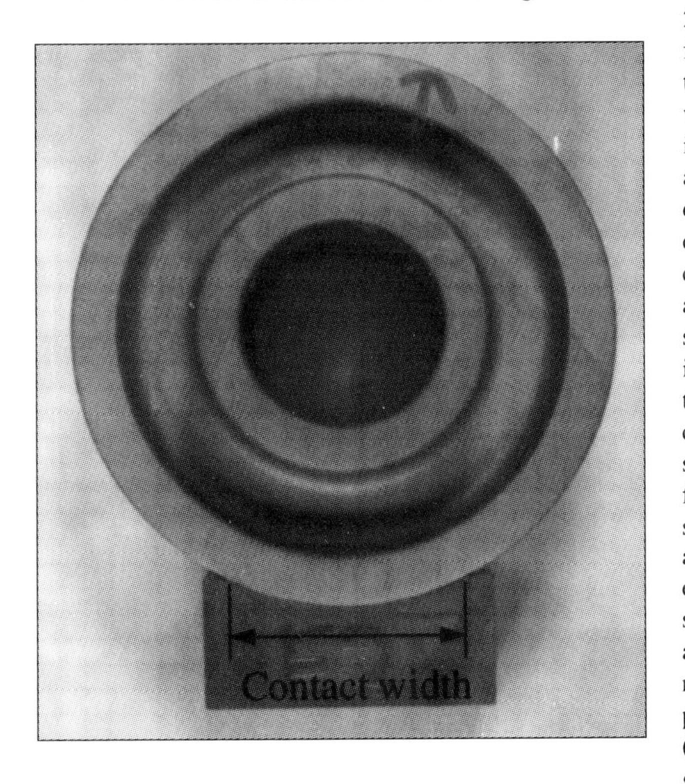

Fig. 7

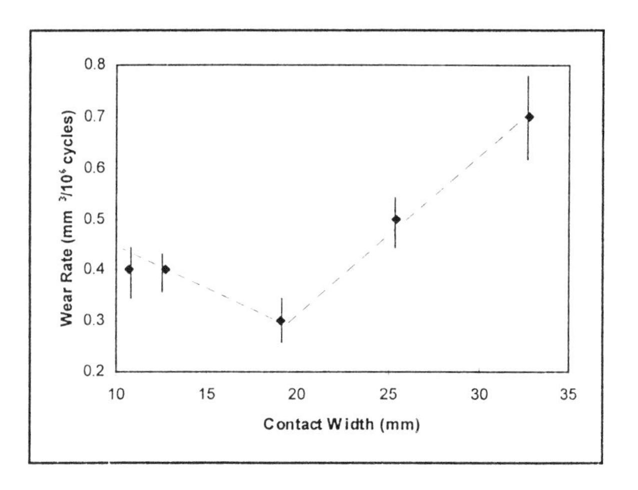

Fig. 8

question the conventional ways of thinking about artificial joint lubrication.

In a very recent study on the effect of protein concentration in bovine serum on the wear of UHMWPE acetabular cups, our hypothesis of the existence of starved lubrication in artificial hip joints was further strengthened **(25)**. In this experiment, a hip joint simulator was used. Bovine sera with three different protein concentrations were used as lubricants. Both EtO sterilized and gamma-irradiated/stabilized UHMWPE cups were tested against both CoCr and alumina ceramic heads with each of the three serum solutions as lubricant. The gamma-irradiated/stabilized cups were prepared by sterilizing the components in nitrogen with gamma-irradiation at 2.5 Mrads followed by a free-radical reduction process to further crosslink the UHMWPE **(26)**. Fig. 9 summarizes the test results. Among the three solutions used, the highest wear rate was found with the lubricant that had an intermediate concentration of proteins. Either below or above this concentration, the wear rate went down. The effect was different for alumina and CoCr heads and also different for EtO-sterilized and gamma-irradiated/stabilized cups although the overall trend was similar. Against both alumina and CoCr heads, the wear rates of the EtO sterilized cups were consistently higher than those of the irradiated/stabilized cups regardless of the concentration of the proteins in the serum. This is in agreement with the clinical results reported by Oonishi, *et al.* **(27)** who found significantly higher wear rates for EtO-sterilized cups than for gamma-irradiated (100 Mrads) cups against both stainless steel and alumina ceramic heads. Oonishi, *et al.* also observed that the use of alumina heads was much more effective in reducing the clinical wear rates for the EtO-sterilized cups than for the gamma-irradiated cups. This is again in agreement with the data shown in Fig. 9. In a recent review comparing laboratory and clinical wear performance of metal and ceramic femoral heads, Dowson **(28)** showed that the use of alumina ceramic heads in general results in approximately 50% reduction in UHMWPE wear compared to the use of metal femoral heads. The data shown in Fig. 9 on the wear of the EtO-sterilized cups seem to support Dowson's conclusions. However, the data on the irradiated/stabilized do not. It appears that the use of alumina ceramic heads may effectively reduce the wear rates for non-crosslinked linear UHMWPE. This may not be the case for crosslinked UHMWPE.

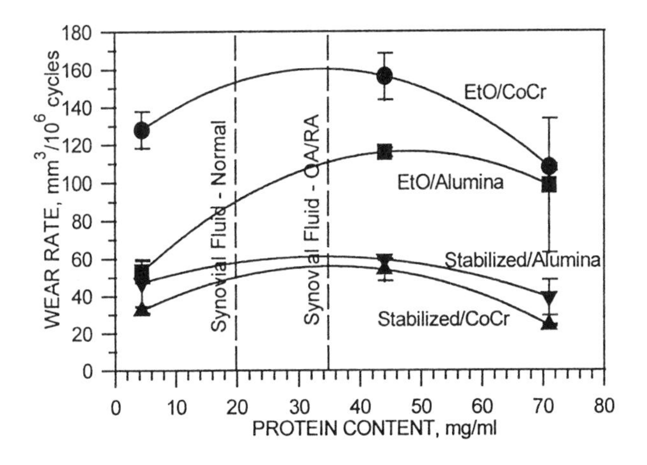

Fig. 9

We were surprised to find that the protein concentrations in synovial fluids of both normal **(29)** and diseased **(30)** joints, including joints after total arthroplasty, fall within the range that was associated with the highest wear rates reported in Fig. 9. Therefore, we conclude that soluble proteins are not an effective boundary lubricant for polyethylene joints. During our tests, we noticed that the serum degraded more quickly for the lubricants with higher initial protein concentrations. The degradation products were in the form of insoluble gel-like precipitates that tended to settle at the bottom of test chambers as shown in Fig. 10. We postulated that this by-product of degradation may be a very effective solid lubricant which could shield the polyethylene surface from direct contact with the femoral head surface. This hypothesis explained successfully why the wear rates dropped as the protein concentration in the serum exceeded a critical value.

We also examined the worn surfaces of both the acetabular and femoral components after testing. We found no transfer film formation on either the femoral head or the acetabular cup surface with the lubricants having higher protein concentrations. However, we observed various degrees of discrete patches of transfer films (reprocessed wear debris) on the worn surfaces of all acetabular cups with the lubricant having the least amount of proteins. The degree or frequency by which this transfer film occurred was much higher for the cups articulating against alumina heads than that against CoCr heads. We believe that the presence of the back-transferred film on the acetabular cup surface protected the underlying substrate from direct contact with the femoral head and thus provided a solid lubrication effect similar to that provided by the degradation by-products of the insoluble proteins. In conclusion, this study indicates that soluble proteins are not an effective boundary lubricant for UHMWPE and therefore artificial hip joints of the polyethylene/metal or polyethylene/ceramic type are not protected by boundary lubrication *in vivo*. This study also indicates that the current laboratory practice of using bovine serum as lubricant is far from ideal although it is still the best possible solution so far. With the limitation of the current understanding of artificial joint lubrication, it is important

Fig. 10

that we choose our lubricant as close as possible to synovial fluid. One very important variable to control is the protein concentration in the serum and the rate of its degradation during wear testing. It is advised that a sufficient amount of serum be used for each test chamber and the serum be changed frequently. It is also advised that an anatomical test position be adopted since the degradation by-products of the insoluble proteins tend to settle down rather than float to the top; therefore, the chances of entrapping these insoluble proteins between the articulating surfaces are much more reduced with an anatomical test position than with an inverted, i.e., cup-at-bottom, position. More detailed discussions on wear test methods will be given in a separate section.

The three surface wear mechanisms discussed above apply to the knee joint also. For the knee joint, however, the most serious clinical wear mechanism is subsurface cracking and delamination. This subsurface wear mechanism is perhaps more design, dependent than material specific. Thin polyethylene, incongruent geometry, or component malalignment may increase the chances for delamination wear (**31–33**). The subject of delamination wear itself deserves a separate study which is beyond the scope of this paper.

4 EFFECT OF JOINT KINEMATICS ON WEAR MECHANISMS

In the previous sections, we discussed the three major wear mechanisms, the lubrication conditions in the joints and the phenomenon of molecular orientation and orientation softening. In this section, we will discuss the impact of joint kinematics on the wear mechanisms and wear rates of UHMWPE bearing surfaces. The importance of joint kinematics on UHMWPE wear has long been recognized in total joint replacements. Until very recently, the knowledge was limited to the recognition of the effect of different joint geometries on hip and knee component wear. Before 1995, studies had been concentrated in computing and analyzing contact stresses in acetabular and tibial components. The most significant contribution was made by Bartel, *et al.* (**34, 35**) using finite element models to depict the effect of polyethylene thickness, congruency and the presence or absence of metal backing on UHMWPE wear. This approach is still very valuable from the design and engineering point of view. Its limitations are becoming more apparent if one tries to use it to understand the fundamental mechanisms of wear. As discussed previously, UHMWPE wear is predominantly a plastic strain-driven process involving very high levels of localized plastic deformation. Most of the wear particles produced are extremely small, on the order of micrometre or even submicrometre levels (**36, 37**). Without looking into the molecular origins of wear, it would be impossible to fully understand the mechanisms of wear let alone to develop a better UHMWPE that would offer some real significant clinical improvement.

The discovery of molecular orientation and the recognition of the importance of orientation softening marked a breakthrough towards a better understanding of UHMWPE wear mechanisms. This current understanding began with the pioneering work of Ramamurti, *et al.* (**38**)

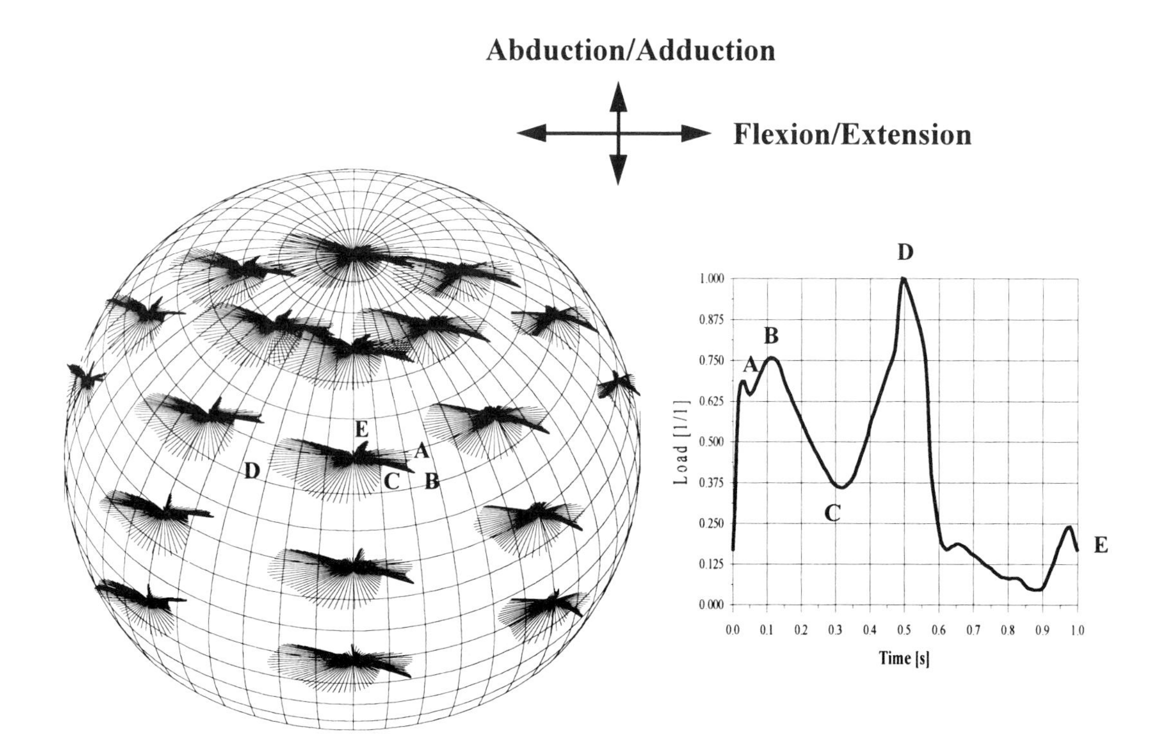

Fig. 11

who discovered the phenomenon of so-called 'crossed-shear' on the articulating surface of the human hip joint by a computer simulation. Shortly after the publication of Ramamurti's work at the 1995 Society For Biomaterials Annual Meeting, Wang, *et al.* proposed the orientation softening concept in May 1995 at the ASTM workshop on UHMWPE (**24**). Meanwhile, more computer simulation work was initiated in our laboratory to verify the findings of Ramamurti, *et al.*. In our simulation model, we used a different set of kinematic data from that used by Ramamurti, *et al.*. However, the results of our simulation were surprisingly similar to that of Ramamurti *et al.*, although some minor differences exist between these two simulations. We found that the trajectories of motion at various locations on the articulating surface were more like irregular ellipses rather than rectangles as found by Ramamurti, *et al.* Overlapping and crossing of the motion trajectories occur everywhere on the articular surface. Again hip joint motion can be characterized as 'crossed-shear'. In addition to simulating hip joint motion, we also studied the distribution of surface shear stresses on various locations on the articular surface as a function of normal gait (**39**). We obtained a so-called 'butterfly' shear stress pattern for each point of contact, see Fig. 11. We found that the magnitude of the maximum shear stress occurs in the butterfly wing direction while the shear stresses in the tail direction are approximately 80% of the maximum shear stress found in the wing direction. The frequency by which the shear stress occurs in the wing direction is at least five times higher than that in the tail direction. These findings indicated that molecular orientation would occur preferentially in the butterfly wing direction while orientation softening would most likely occur in the tail direction. In other words, the butterfly shear stress patterns associated with the hip joint satisfy the conditions for the occurrence of orientation softening in UHMWPE acetabular cups.

The motion in the knee joint is, however, much more linear. Although crossed-shear motion is still a possibility because of the combined motions of flexion/extension and internal/external rotation, the maximum possible angles at which these shear stresses cross are deemed much smaller than the maximum angle of 90° found for the hip joint. For a functional knee joint replacement, this angle is probably less than 10°. This results in a maximum transverse shear stress component of less than 20% of the maximum longitudinal shear stress component. Therefore, orientation softening would not be as severe for the knee joint as for the hip joint although it is still a significant wear mechanism for the knee joint (**9, 15**).

If we treat an oriented UHMWPE surface as a unidirectionally reinforced composite, we find that the failure strength of the 'composite' is highly anisotropic, Fig. 12. The strength depends to a great extent on the angle of loading which is determined by the angle of the crossed-shear. The extreme condition, i.e., the 90° off-axis loading situation, may correspond to that associated with the hip joint while the region within 0 and 10° may be related to the conditions associated with the knee joint. It is obvious that the hip joint experiences a much higher degree of orientation softening than the knee joint. From a practical point of view, this finding implies that the impact of crosslinking on UHMWPE wear would be much greater for the hip than for the knee.

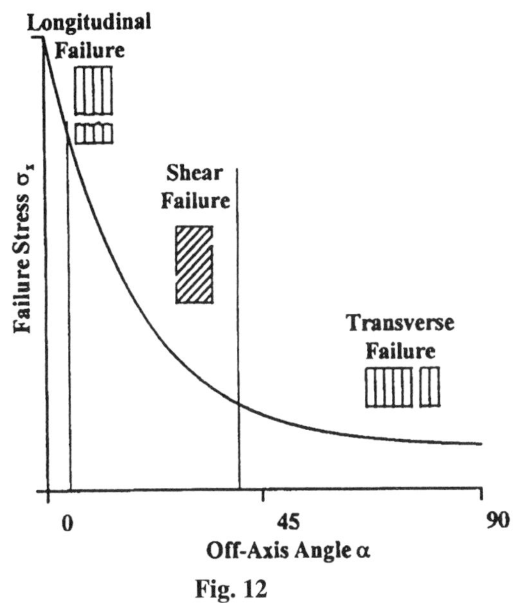

Fig. 12

To verify this hypothesis, a series of experiments was carried out using both hip and knee joint simulators with UHMWPE components that were crosslinked to various degrees by gamma irradiation in nitrogen at different doses followed by stabilization (**8**). The hip simulator was an 8-station servo-hydraulic system (MTS Systems Corporation, Eden Prairie, MN, USA). Computer simulation of the kinematics of the hip simulator indicated that it produced a butterfly shear stress pattern similar to the natural hip joint during gait (**16**). Fig. 13 shows a photograph of one of the 8 stations. The knee simulator used in this study was a 6-station MTS servo-hydraulic system. The kinematic conditions of the simulator consist of flexion/extension, A-P translation, internal/external rotation and axial loading (**40**). A photograph of the MTS knee simulator is shown in Fig. 14. Test conditions for both the hip and knee simulator tests are summarized in Table 1. Fig. 15 shows the volumetric wear rate as a function of radiation dosage for both the acetabular and tibial components. For the acetabular components, the wear rate decreased almost exponentially as the radiation dose increased. For the tibial components, the wear rate was independent of the dose of radiation between 0 and 5 Mrads; beyond 5 Mrads, the wear rate decreased again but to a much lesser extent compared to the acetabular cups. Furthermore, at low levels of radiation the acetabular components showed higher wear rates than the tibial components. The reverse was found at the highest levels of radiation (>5 Mrads). These results indicate very strongly that radiation-induced crosslinking is much more effective in reducing the wear rate for the hip than for the knee.

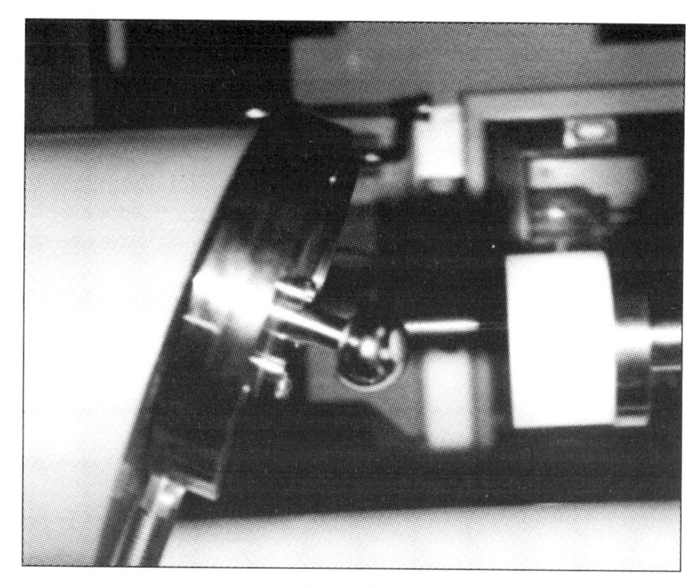

Fig. 13

Fig. 14

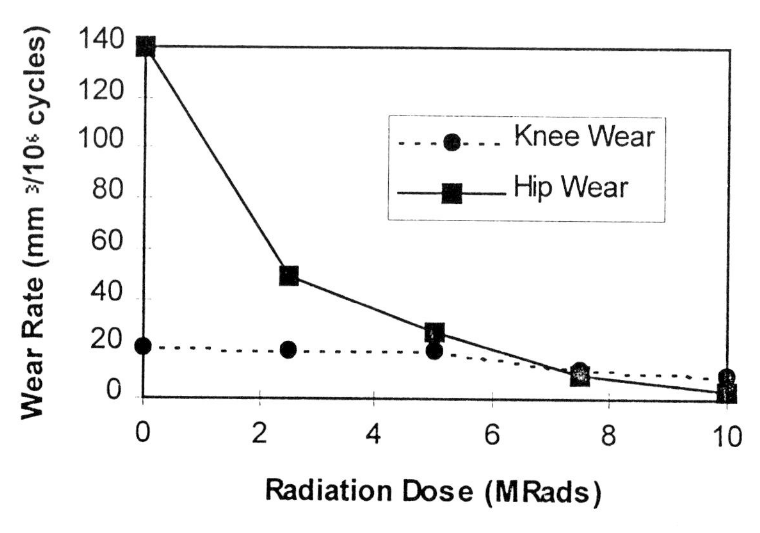

Fig. 15

Table 1 Test conditions for hip and knee simulator testing.

	Hip Simulator	Knee Simulator
Motion:	± 23° Biaxial Rocking (Crossed-Shear)	Flex./Ext.: 23°; A/P Translation: 12 mm; Int./Ext. Rotation: +6/-7.5°
Loading:	Physiological (2450 N Max.)	Physiological (2700 N Max.)
Position:	Anatomical	Anatomical
Lubricant:	Serum	Serum
Frequency:	1 Hz	1 Hz
UHMWPE:	Irradiated (N_2)/Stabilized Cups	Irradiated (N_2)/Stabilized Tibial Inserts
Counterface:	32 mm CoCr Heads (forged Vitallium® alloy)	Duracon® Large Femoral Component (CoCr) (cast Vitallium® alloy)

5 METHODS OF WEAR TESTING

5.1 Methods of screening wear testing

The ultimate objective of laboratory wear testing is to predict the clinical wear performance of existing or new bearing materials and designs. This has been an extremely difficult task to perform because of the very complex and multi-factorial nature of the tribological conditions experienced *in vivo*. Historically, the role of wear testing has been much narrowed down to screen 'bad' materials from ever being used clinically. Even this limited objective has rarely been achieved. During the past three years or so, many questions have been raised regarding the validity of various types of wear test machines and test methods (**24, 38, 39**). Thanks to the unprecedented interest and heavy research activities in UHMWPE wear from orthopaedic surgeons, implant manufacturers, and academic institutions around the world since the early 1990s, much progress has been made in both the understanding of wear mechanisms and the improvement of wear testing methodology. As far as wear testing is concerned, two major breakthroughs have been made. One is the realization of a fundamental design flaw in all types of screening wear machines that are based on reciprocating or linear motion. The other is the resurgence of interest in the use of joint simulators that are designed upon the kinematic principles of natural human joints.

The fundamental design flaws associated with pin-on-plate or pin-on-disc type of wear testers were first acknowledged by McKellop, *et al.* (**41**), Wang, *et al.* (**24**) and Bragdon, *et al.* (**42**). At the 1995 ASTM Workshop on UHMWPE in Denver, Colorado, McKellop (**41**) pointed out that screening wear testers underestimated wear rates by one to three orders of magnitude compared to average clinical wear rates for acetabular cups. He suspected that the low wear rates may be associated with the linear motion of common screening wear machines. At the same meeting, Wang, *et al.* (**24**) presented a set of experimental data comparing linear wear machines with multi-axial joint simulators. They found that not only were the wear rates generated by screening wear machines orders of magnitude different than joint simulators but also the wear rankings were different. For instance, a non-crosslinked UHMWPE wore less than a crosslinked UHMWPE on a linear-motion wear tester; the exact opposite was found on a multi-axial hip joint simulator. To explain these fundamental differences, Wang, *et al.* proposed the concept of 'orientation softening' at the ASTM Workshop. At other meetings, Bragdon, *et al.* (**42, 43**) presented their findings comparing a hip simulator based on curvilinear motion with a hip simulator based on crossed-shear motion. Up to 7 million cycles of testing, no wear could be measured from cups that were run with the linear-motion simulator while a clinically significant amount of wear was produced by the simulator with the crossed-shear motion. These independent discoveries achieved at three non-related laboratories all pointed out that wear testers based on linear motion should not be used as a screening tool for UHMWPE. Since then, the use of linear wear testers has been abandoned by all the three research groups. This, however, does not mean that all wear testers other than simulators are useless. In fact, both Bragdon, *et al.* of the Massachusetts General Hospital (MGH) and Wang, *et al.* of Howmedica have designed and built new screening wear machines based on the principles of crossed-shear motion. The new MGH machine was a biaxial pin-on-disk system adopting the motion of a rectangular loop which was derived from their computer model of human hip joint kinematics (**42**). This may be a very valuable and economical system for screening materials that would be intended for the hip joint. The new Howmedica screening machine was a 12-station biaxial non-conforming line-contact system. Each of the 12 stations consists of a convex CoCr cylinder oscillating on a flat polymer specimen which can also oscillate to various degrees about its vertical axis, Fig. 16. This system simulates the geometry and motions of the knee joint. Preliminary data have indicated that at a constant flexion/extension angle the wear rates of UHMWPE specimens can be increased significantly by increasing the angle of internal/external rotation.

First, a valid simulator should reproduce the correct wear resistance rankings for materials that have a well-documented clinical record. This is the most important criterion for validating a hip simulator. In the past, such a validation was done by comparing UHMWPE with PTFE **(41, 44)**. We have been questioning this approach since we began conducting our first hip simulator test five years ago. We believed that the structural differences between UHMWPE and PTFE were so huge that any test using any wear machine would easily give a correct ranking. Since PTFE was abandoned decades ago and UHMWPE has been the only polymer used for orthopaedic bearing applications ever since, a rational way to validate a simulator is to compare different types of UHMWPE that have been used and their clinical wear performances have been well documented. One obvious choice is to compare so-called conventional UHMWPE with so-called enhanced UHMWPE since both types of UHMWPE have been used extensively during the past five or six years and clinical results are now available.

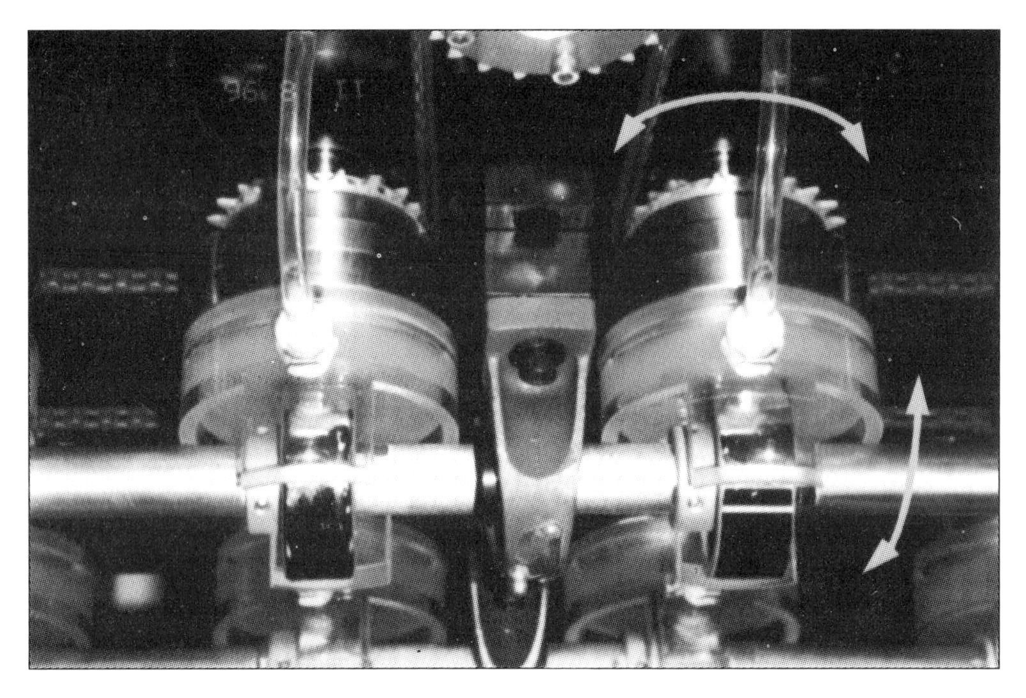

Fig. 16

5.2 Methods of joint simulator testing

A wear test machine that is capable of testing actual prostheses has historically been called a joint simulator. In definition, a joint simulator should be able to test not only actual prosthetic components but also should simulate the correct kinematic, physiological and anatomical conditions as seen in the human joints. In reality, not a single simulator in existence meets such a precise definition. Compromises have to be made in the laboratory if progress is to be made. The difficult question is which compromises can be made and which cannot. Both ASTM and ISO have been working for the past five years to introduce some standards for hip and knee joint simulator testing. So far, neither one of them has been able to formally introduce a standard. The following specific areas have caused the most controversy:

a) choice of lubricant;
b) degree of motion;
c) position of components in the simulator;
d) measurement of wear: gravimetric or volumetric?

Before we discuss each of the areas in detail, let us take a look at some of the important criteria for validating a joint simulator and its test protocol.

In the authors' laboratory, Hylamer® enhanced UHMWPE acetabular cups (Depuy-DuPont Orthopaedics, Warsaw, Indiana, USA) and conventional UHMWPE (extruded GUR4150, PolyHi Solidur, Fort Wayne, Indiana, USA) acetabular inserts (Howmedica, Inc., Rutherford, New Jersey, USA) were gathered. All components were opened at the same time and then soaked in deionized water at room temperature. According to the types of packaging, the Hylamer® cups were divided into two groups: one group was from packages filled with air (Hylamer®-Air) and the other group from packages that appeared to be vacuum packed (Hylamer®-Vacuum). All the Howmedica cups were packaged in air (Conventional-Air). According to the markings on the packages, both the Hylamer® cups and the Howmedica cups were sterilized by gamma-irradiation. All cups were of 32 mm ID and 52 mm OD and all had a 10° hood. An 8-station hip joint simulator as shown in Fig. 13 was used to compare the wear performance of the three groups of cups, i.e., Hylamer®-air, Hylamer®-vacuum, and Conventional-air. The test conditions are summarized as follows: anatomical positioning, i.e., cup on top of the head; crossed-shear motion, i.e., biaxial rocking of the head inside the cup; physiological loading, i.e., Paul-curve with 2450 N maximum and 50 N minimum load; motion and loading synchronized at 1 Hz; double-filtered bovine serum used as lubricant (Hyclone Laboratories, Logan, UT, USA). 32mm CoCr femoral heads (forged Vitallium® alloy, Howmedica, Rutherford, NJ, USA) were used as the counterface against all cups. Two tests were conducted. For each of the tests, at least two Conventional-Air cups were used as controls. Both tests lasted two million cycles. A total of four Hylamer®-Air, four Hylamer®-Vacuum, and four Conventional-Air cups were evaluated. Soak-control cups for each type of specimens were used for the purpose of correcting fluid absorption. Wear was measured by weighing each cup at 500,000 cycle intervals using an electronic balance (resolution: 0.01 mg). Volume losses were obtained by dividing the weight loss values by the density of each material (0.955 g/cm^3 for Hylamer®; 0.938 g/cm^3 for Conventional UHMWPE).

The results are shown in Fig. 17. At 2 million cycles, the Hylamer®-Air cups lost 300–616 mm^3, the Hylamer®-Vacuum cups lost 239–324 mm^3, and the Conventional-Air cups lost 176–190 mm^3. The differences in total wear among all three groups are statistically significant. Volumetric wear loss values were converted to linear wear rates using the equation $V = \pi r^2 P$, where V is the volume loss, r is the cup radius and P is the linear penetration. Assuming one million cycles is approximately equivalent to one year of *in vivo* use, the linear wear rates with standard deviations and ranges were as follows: Hylamer®-Air, 0.27 ± 0.08 mm/yr (0.18-0.36); Hylamer®-Vacuum, 0.17 ± 0.02 mm/yr (0.15-0.20); Conventional-Air, 0.12 ± 0.004 mm/yr (0.11-0.12). These data indicate that the Hylamer®-Air cups had a 132% increase in wear rate and the Hylamer®-Vacuum cups had a 40% increase in wear rate compared to the Conventional-Air cups. Differences in the linear wear rates were statistically significant at the 95% confidence level.

Chandler and Smith **(45)** reported an average clinical wear rate of 0.27 mm/yr for Hylamer® cups (range: 0-0.83 mm/yr) and an average clinical wear rate of 0.11 mm/yr for conventional UHMWPE cups. Livingston, *et al.* **(46)** reported an average clinical wear rate of 0.25 mm/yr for Hylamer® cups against non Depuy heads and 0.19 mm/yr with Depuy heads while the conventional UHMWPE control cups had a clinical wear rate of 0.12 mm/yr. Chmell, *et al.* **(47)** reported a high clinical incidence of Hylamer® cup failure due to excessive wear (4.2% failure at 37 months, estimated to be 14% at 48 months). The current simulator test results demonstrate higher Hylamer® wear rates and are in excellent agreement with clinical findings. Not only were the rankings correct but also the amount of wear for both the Hylamer® and the Conventional UHMWPE cups.

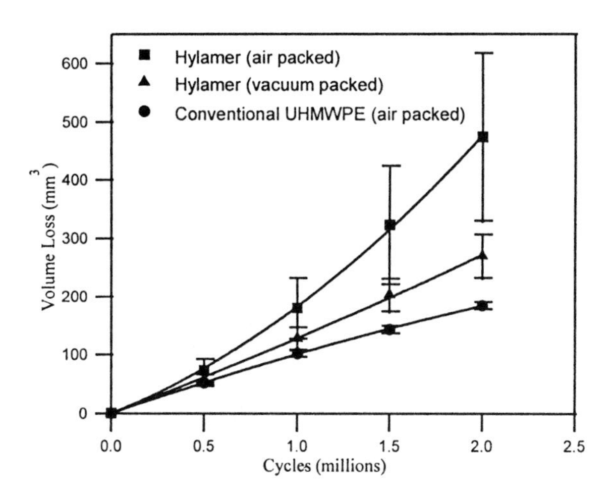

Fig. 17

Other criteria for simulator testing are much more qualitative and subjective to the interpretations of the person who conducts the test. These include the ability of the simulator to reproduce similar to clinical wear patterns, wear mechanisms and wear debris. Experience has shown that these criteria can be met by most hip simulators that employed crossed-shear motion, physiological loading and serum lubrication. Fig. 18 shows a SEM micrograph of wear debris produced by the Howmedica/MTS hip simulator. Both the size and morphology of the wear particles are similar to those observed clinically **(36, 37)**. Because of the qualitative nature and subjectiveness of these criteria, meeting these criteria alone does not necessarily mean that a simulator will produce correct clinical rankings. But it is unlikely that correct clinical rankings can be reproduced without meeting these qualitative criteria first.

One of the most important parameters that affect all the criteria discussed above is the choice of lubricant. Pure water or saline solutions without the presence of proteins have been shown to be unsuitable as a lubricant **(44, 48)**. Even the qualitative criteria cannot be met by these non-protein containing lubricants. Bovine serum has been shown to satisfy all the qualitative criteria discussed above. However, as discussed earlier, the protein concentration in bovine serum has a significant effect on wear rates and wear rankings. Higher protein-containing serum degrades faster and the degradation by-products, insoluble protein precipitates, provide a solid lubricant for the articular surfaces. Unfortunately, this is a laboratory artifact. If the protein concentration is too low, transfer films form on the acetabular cup surface, which again help the lubrication of

Fig. 18

the surfaces, introducing another laboratory artifact. Therefore, a good serum lubricant should have a protein concentration within the range of human synovial fluids, which is approximately from 20 mg/ml to 35 mg/ml (**29, 30**).

The nature of motion in articulation is now recognized as a fundamental design parameter for any hip simulator. The crossed-shear motion or more specifically the butterfly shear stress pattern should be reproduced by any hip simulator. The latter requires that the simulator loading be physiological. Linear motion should be avoided because it fundamentally changes the wear mechanism, e.g. causing orientation hardening rather than orientation softening.

The positioning of the components in the simulator is another parameter that may have serious effect on the test results (**49**). Anatomical positioning is recommended. The arguments for this are as follows. First, it reduces the chances of entrapping the degradation by-products of serum proteins which can dramatically change the lubrication mechanisms of the joints. Second, an anatomical position with a stationary cup offers the correct loading pattern on the component during articulation. This leads to the so-called tunneling wear pattern often observed on retrieved acetabular cups. The inverted position with a moving cup on the bottom spreads the load over a much greater area of the cup. This not only makes the overall wear pattern more uniform but also distorts the butterfly shear stress pattern because the two peak forces associated with heel-strike and toe-off no longer hit the same spot at the cup during each loading cycle. Our experience (**49**) has shown that the inverted test position significantly underestimates the wear rates, especially for non-crosslinked UHMWPE (e.g. EtO-sterilized UHMWPE cups).

The method of measuring wear can also affect the accuracy of the results. Our experience with gravimetric techniques has been very satisfactory. For the amount of wear produced with our simulators, the effect of fluid absorption is negligible (less than 2% of the total weight change) (**14**). Volumetric techniques such as coordinate measuring machines (CMM) measure volume changes due to both wear and creep. The accuracy of these techniques is usually not as good as the gravimetric method for UHMWPE. However, the CMM technique may be a better method for materials that do not creep or tend to absorb an excessive amount of fluid, making gravimetric method unreliable.

6 CONCLUSIONS

Significant progress in the understanding of wear mechanisms and wear testing has been achieved during the past five years. In two specific areas, fundamental advances have been made. One is the discovery and recognition of the importance of multi-directional motion in wear and wear testing. The other is the development of the orientation softening concept in UHMWPE wear. These developments have provided solid theoretical explanations for the laboratory proven beneficial effect of crosslinking in wear resistance. In various laboratories across the US, efforts have been made to develop advanced versions of UHMWPE that would offer some real improvement in clinical wear performance. A recently developed post-radiation stabilization process based on crosslinking principles has shown great promises in reducing UHMWPE wear. A better understanding of the art and science of wear testing is emerging. This has led to the development of better screening wear machines in at least two independent laboratories. One of the areas in artificial joint tribology that is still the least understood is the mechanism of lubrication. The debate on the choice of lubricant for wear testing is likely to continue for a long time to come. This area deserves more studies than it has received so far. The relationship between mechanical properties and wear resistance is still not fully explained. Such a relationship is difficult to establish in view of the anisotropic nature of UHMWPE and the very complex stress patterns experienced by artificial joints. Wear testing especially joint simulator testing should remain the ultimate test for material screening. Mechanical testing, oxidation resistance testing, or any other testing should be considered important, but secondary to wear testing.

7 ACKNOWLEDGEMENTS

The authors would like to thank their colleauges B. Edwards, S.-S. Yau and M. Sokol for performing the plasma etching experiment, the biaxial tensile test, and the computer simulation, respectively. The X-ray diffraction work was conducted by Miss M. Choudhury and Dr I. M. Hutchings at the University of Cambridge.

8 REFERENCES

1 **Bassett, D.C.,** Principles of polymer morphology, *Cambridge University Press,*1981.

2 **Bueche, F,** Physical properties of polymers, *Interscience Publishers,* New York, 1962.

3 **Ratner, S.B., Farberova, I.I., Radyukevich, O.V.,** and **Lur'e, E.G.,** Connection between wear resistance of plastics and other mechanical properties, *Adrasion of Rubber*, D.I. James (ed.), (MacLaren & Sons, London, 1967) 145.

4 **Lancaster, J.K.,** Abrasive wear of polymers, *Wear,* 1969, **14,** 223.

5 **Wang, A., Sun, D.C., Stark, C.,** and **Dumbleton, J.H.,** Wear mechanisms of UHMWPE in total joint replacements, *Wear* 1995, **181-183,** 241-249.

6 **Wroblewski, B.M., Siney, P.D., Dowson, D.,** and **Collins, S.N.,** A prospective clinical and joint simulator study of the performance of 22 mm alumina ceramic heads and corsslinked polyethylene acetabular cups in total hip arthroplasty, *JBJS* , 1996, **2,** 280-285.

7 **Muratoglu, O.K., Bragdon, C.R., Jasty, M.,** and **Harris, W.H.,** The effect of radiation damage on the wear rate of UHMWPE components, *ASTM Symposium on Characterization and Properties of UHMWPE*, 1996, New Orleans,USA.

8 **Wang, A, Sun, D.C., Stark, C.,** and **Dumbleton, J.H.,** Effects of sterilization methods on the wear of UHMWPE acetabular cups, *5th World Biomaterials Congress*, 1996, Toronto, Canada.

9 **Wang, A., Stark C.,** and **Dumbleton, J.H.,** Mechanistic and morphological origins of ultra-high molecular weight polyethylene wear debris in total joint replacement prostheses, *Proc. Instn.Mech. Engrs., Part H: Eng. in Med.,* 1996, **210,** 141-155.

10 **Wang, A, Sun, D.C., Yau, S.-S., Edwards, B., Sokol, M., Essner, A., Polineni, V.K., Stark, C.,** and **Dumbleton, J.H.,** Orientation softening in the deformation and wear of ultra-high molecular weight polyethylene, *Wear*, 1997, **203–204,** 230–241.

11 **Wang, A., Essner, A., Polineni, V.K., Edwards, B., Yau, S.-S., Sun, D.C., Stark, C.,** and **Dumbleton, J.H.,** Orientation softening as a mechanism of UHMWPE wear, *ASTM Symposium on Characterization and Properties of UHMWPE*, 1996, New Orleans, USA.

12 **Pooley, C.M.** and **Tabor, D.,** Friction and molecular structure: the behaviour of some thermoplastics, *Proc. Royal Soc. London*, 1972, **329A,** 251-258.

13 **Briscoe B.J.** and **Stolarski T.A.,** Combined rotating and linear motion effects on the wear of polymers, *Nature*, 1979, **281,** 206-208.

14 **Briscoe, B.J.** and **Stolarski, T.A.,** Transfer wear of polymers during combined linear motion and load axis spin, *Wear,* 1985, **104,** 121-137.

15 **Wang, A., Sun, D.C., Stark, C.** and **Dumbleton, J.H.,** Wear testing based on unidirectional motion: fact or artifact?, *5th World Biomaterials Congress*, 1996, Toronto, Canada, 583.

16 **Wang, A., Polineni, V.K., Essner, A., Sokol, M., Sun, D.C., Stark, C.,** and **Dumbleton, J.H.,** The significance of non-linear motion in the wear screening of orthopaedic implant materials, *J. Testing & Evaluation,* Vol. **25,** No. 2, 1997, 239–245.

17 **Bragdon, C.R., O'Connor, D.O., Lowenstein, J.D., Jasty, M.,** and **Syniuta, W.D.,** The importance of multidirectional motion on the wear of polyethylene, *Proc. Instn. Mech. Engrs., Part H: Eng. in Med.*, 1996, 157-166.

18 **Bellare, A., Spector, M.,** and **Cohen, R.E.,** Morphological alterations in retrieved tibial inserts of UHMWPE: a small angle X-ray scattering study, *42nd Annual Meeting, Orthopaedic Research Society*, 1996, Atlanta, Georgia, USA, 495.

19 **McKellop, H.A., Campbell, P., Park, S-H., Schmalzried, T.P., Grigoris, P., Amstutz, H.C.,** and **Sarmiento, A,** The origin of submicron polyethylene wear debris in total hip arthroplasty, *Clinical Orthop. Rel. Res.*,1995, **311,** 3-20.

20 **Jasty, M., James, S., Bragdon, C.R., Goetz, D., Lee, K.R., Hanson, A.E.** and **Harris, W.H.,** Patterns and mechanisms of wear in polyethylene acetabular components retrieved at revision surgery, *20th Annual Meeting of the Society for Biomaterials*, Boston, Massachusetts, 1994, 103.

21 **McKellop, H.A., Schmalzried, T.A., Park, S-H.,** and **Campbell, P.,** Evidence for the generation of submicron polyethylene wear particles by micro-adhesive wear, *Transaction of the 19th Annual Meeting of the Society For Biomaterials*, The Society For Biomaterials, Minneapolis, USA, 1993, 184.

22 **Wang, A.**, **Stark, C.**, and **Dumbleton, J.H.**, Role of cyclic plastic deformation in the wear of UHMWPE acetabular cups, *J. Biomed. Mater. Res.*,1995, **29**, 619-626.

23 **Lancaster, J.K.**, Basic mechanisms of friction and wear of polymers, *Plastics & Polymers*, 1973, 297-306.

24 **Wang, A.**, **Sun, D.C.**, **Stark, C.** and **Dumbleton, J.H.**, Factors affecting wear screening, *ASTM Workshop on Characterization and Performance of Articular Surfaces*, 1995, Denver, Colorado,USA.

25 **Polineni, V.K.**, **Wang, A.**, **Essner, A.**, **Stark, C.**, and **Dumbleton, J.H.**, Effect of lubricant protein concentration on the wear of UHMWPE acetabular cups against Co-Cr and alunina femoral heads, submitted to *23rd Annual Meeting of the Society For Biomaterials*, 1997, New Orleans, USA.

26 **Sun, D.C.** and **Stark, C.**, Non-oxidizing polymeric medical implant, US Patent No. 5414049.

27 **Oonishi, H.**, **Takayama, Y.**, and **Tsuji, E.**, Improvement of polyethylene by irradiation in artificial joints, *Radiat. Phys. Chem.*,1992, **39,** 495-504.

28 **Dowson, D.**, A comparative study of the performance of metallic and ceramic femoral head components in total replacement hip joints, *Wear,* 1995,**190,** 171-183.

29 **Dumbleton, J.H.**, *Tribology of Natural and Artificial Joints,* Elsevier, London, 1981.

30 **Saari, H.**, **Santavirta, S.**, **Nordstrom, D.**, **Paavolainen, P.**, and **Konttinen, Y.T.**, Hyaluronate in total hip replacement, *J. Rheumatol.*, 1993, **20,** 87-90.

31 **Hood, R.W.**, **Wright, T.W.**, and **Burstein, A.H.**, Retrieval analysis of total knee prostheses, *J. Biomed. Mater. Res.*, 1983, **17,** 829-842.

32 **Bartel, D.L.**, **Burstein, A.H.**, **Toda, M.D.**, and **Edwards D.L.**, The effect of conformity and plastic thickness on contact stresses in metal-backed plastic implants, *J. Biomech. Engng.*,1985, **107,** 113.

33 **Landy, M.M.** and **Walker, P.A.**, Wear of ultra-high molecular weight polyethylene components of 90 retrieved knee prostheses, *J. Arthroplasty,*1988,**3,** S73-85.

34 **Bartel, D.L.**, **Bicknell, V.L.**, **Ithaca, M.S.**, and **Wright, T.M.**, The effect of conformity, thickness, and material on stresses in ultra-high molecular weight omponents for total joint replacements, *J. Bone & Joint Surg.*, 1986, **68-A,** 1041-1051.

35 **Bartel, D.L.**, **Wright, T.M.**, and **Edwards, D.L.**, The effect of metal-backing on stresses in polyethylene acetabular components, *The Hip: Proc. of 11th Open Scientific Meeting of the Hip Society*, C. V. Mosby (ed.), St. Louis,USA, 1983, 229-239.

36 **Campbell, P.**, **Ma, S.**, **Yeom, B.**, *et al.*, Isolation of predominantly submicron sized UHMWPE particles from periprosthetic tissues, *J. Biomed. Mater. Res.*,1995, **29,** 27-131.

37 **Schmalzried, T.P.**, **Campbell, P.**, **Brown, I.C.**, **Schmitt, A. K.**, and **Amstutz, H.C.**, Polyethylene wear particles generated *in vivo* by total knee replacements compared to total hip replacements, *21st Annual Meeting of the Society For Biomaterials*, San Francisco, USA, 1995, 310.

38 **Ramamurti, B.S.**, **Bragdon, C.R.**, and **Harris, W.H.**, Loci of selected points on the femoral head during normal gait, *21st Annual Meeting of the Society For Biomaterials*, San Francisco, USA, 1995, 347.

39 **Wang, A.**, **Sokol, M.**, **Stark, C.**, and **Dumbleton, J.H.**, Distribution of frictional stresses on the articular surfaces of human hip joint and its effect on wear of UHMWPE acetabular cups, *5th World Biomaterials Congress*, 1996, Toronto, Canada, I-874.

40 **Essner, A.**, **Wang, A.**, **Stark, C.**, and **Dumbleton J.H.**, A simulator for the evaluation of total knee replacement wear, *5th World Biomaterials Congress*, 1996, Toronto, Canada, I-580.

41 **McKellop, H.A.**, Comparison between laboratory wear tests and clinical performance of past bearing materials, *ASTM Workshop on Characterization and Performance of Articular Surfaces*, 1995, Denver, USA.

42 **Bragdon, C.R.**, **O'Connor, D.O.**, **Lowenstein, J.D.**, **Jasty, M.**, and **Harris, W.H.**, Development of a new pin on disk testing machine for evaluating polyethylene wear, *5th World Biomaterials Congress*, 1996, Toronto, Canada, II-788.

43 **Bragdon, C.R.**, **Jasty, M.**, **Lowenstein, J.D.**, **Elder, J.**, and **Lee, K.R.**, Mechanism of wear of retrieved polyethylene acetabular components, *63rd Annual Meeting of the American Academy of Orthopaedic Surgeons*, 1996, Atlanta, Georgia, 376.

44 **Good, V.D.**, **Clarke, I.C.**, and **Anissian, L.**, Water and bovine serum lubrication compared in simulator PTFE/CoCr wear model, *J. Biomed. Mater. Res.: Applied Biomaterials*, 1996, **33,** 275-283.

45 **Chandler, H.P.** and **Smith, S.**, Comparison of *in vivo* rates of wear of Hylamer and conventional polyethylene acetabular liners, *26th Annual Harris Hip Course, Total Hip Replacement - Polyethylene: where are we now?,* Cambridge, USA, 1996.

46 **Livingston, B.J.**, **Chmell, M.J.**, **Reilly, D.T.**, **Spector, M.**, and **Poss, R.**, The wear rate of Hylamer cups is higher than conventional PE and differs with heads from different manufacturers, *26th Annual Harris Hip Course, Total Hip Replacement -Polyethylene: where are we now?,* Cambridge, USA, 1996.

47 **Chmell, M.J.**, **Poss, R.**, **Thomas, W.H.**, and **Sledge, C.B.**, Early failure of Hylamer acetabular inserts due to eccentric wear, *J. Arthroplasty*, 1996, **3,** 351-353.

48 **Wang, A.**, **Essner, A.**, **Stark, C.**, and **Dumbleton, J.H.**, Comparison of the size and morphology of UHMWPE wear debris produced by a hip simulator under serum and water lubricated conditions, *Biomaterials,* 1996, **17,** 865-871.

49 **Polineni, V.K.**, **Essner A.**, **Wang, A.**, **Sun, D.C.**, **Stark, C.** and **Dumbleton, J.H.**, Effects of radiation induced oxidation and crosslinking on the wear of UHMWPE acetabular cups, *ASTM Symposium on Characterization and Properties of UHMWPE*, 1996, New Orleans, USA.

Materials-based concepts for an oil-free engine

M WOYDT
BAM, Berlin, Germany

SYNOPSIS

From the tribological point of view, unlubricated engines can be produced if wear coefficients of sliding couples lower than $5x10^{-8}$ mm³/(Nm) can be demonstrated, which are independent of ambient temperature and sliding speed. A review of published tribological data and a search in a database yielded no couples with such a low wear coefficient at 22°C and 400°C. Unlubricated sliding couples are however known with wear coefficients of about 10^{-7} mm³/(Nm), which opens up the possibility of developing such engines for basic studies or to formulate life-time stable lubricants with enhanced lubricity. For an engine lubricated for life the materials choice as well as the design concepts will determine the properties of the lubricating fluid.

1 INTRODUCTION

About 20 years ago, all over the world, programmes were initiated to develop unlubricated engines. The main difficulty involved in concepts such as adiabatic engines (**1, 2**) low heat rejection engines (**3, 4**), two-stroke engines (**5**) and hydrogen engines (**6**) is the tribology of multiple tribosystems, especially that of the piston ring/cylinder liner system (**7–9**).

The driving forces to develop unlubricated engines are nowadays related to strict exhaust emission regulations (**10**), environmental legislation (**11**), waste management restrictions for used oils (**12**), and the new German product recycling law (**13**), as well as demand from customers to reduce operating costs. Awareness of the environmental impact of lubricants has now been codified. Progressive release usually causes the lubricant to enter the environment in small amounts spread out over time or space, or both, from millions of vehicles, rather than in a large amount which enters the environment at one time and in one place, as would result from an accident.

Mineral-based or synthetic engine oils with additive packages form an indispensable and reliable adjunct to all materials used nowadays in internal combustion engines, but are classified as ecotoxic and water-polluting. The question thus arises whether it is possible, from the tribological point of view, to convert from conventional, fully oil-lubricated engines to unlubricated ones. An unlubricated combustion engine is of interest because:

a) causal relationships exist between the exhaust emission of hydrocarbons and the piston ring/cylinder and valve shaft/valve guide tribosystems (**14**),
b) such relationships also exist between some engine oil additives and the conversion efficiency and life-time of the catalyst, and
c) the oil-free engine will be more compact in its basic construction, with fewer parts, and avoids environmental problems due to oil leakage, oil burning and the disposal of used oil.

From these considerations it is clear that tribology can make important future contributions to a low- or zero-emission engine through new wear-resistant materials and new lubricants. The unlubricated engine is only one tribology-related topic. The expression 'oil-free' can imply several possible definitions:

- oil-free for the customer (i.e. life-time filling),
- oil-free operation (i.e. dry or unlubricated), or
- use of a liquid lubricant other than a petroleum oil.

2 TRIBOLOGY

2.1 System analysis of tribosystems

Before the feasibility of unlubricated tribosystems in combustion engines can be discussed, the necessary volumetric wear coefficient must be defined. Data describing the operating conditions were collected by polling experts from automotive companies. This analysis provided the average volumetric wear coefficients which occur nowadays in conventional oil-lubricated four-stroke gasoline engines. In order to select new materials, based on published data from the literature or from any kind of non-engine screening test (e.g. DIN 50324 or ASTM G-99), these candidate materials must exhibit a lower minimum wear level in these unlubricated screening tests than lubricated, real tribosystems in engines.

Table 1 summarises the calculated volumetric wear coefficient (or specific wear rate) for the main relevant tribosystems of a conventionally oil-lubricated combustion engine, for steady-state running for an overall period of 2000 hours. For this evaluation it is assumed that the engine runs at 3000 rpm, and the normal force is either the maximum force (calculated from the combustion pressure) or the average force during one crankshaft cycle. During this operation time, each tribosystem experiences around 400 million sliding or rolling cycles. The calculated specific wear rates (volumetric wear coefficient as defined by DIN 50323) indicate that the wear limit for the selection of unlubricated candidate materials is less than $5x10^{-8}$ mm³/(Nm) and should be independent of sliding velocity and ambient temperature.

2.2 Unlubricated wear coefficient

The quickest way to obtain an overview of candidate materials for an unlubricated engine is to search the quantitative tribological database Tribocollect V1.1 (**15**) which contains more than 11800 data sets.

For the present study, all data sets for unlubricated non-polymeric materials tested at both 22°C and 400°C were selected. 400°C is the minimum level of ambient temperature which must be considered for piston rings in unlubricated engines. A component temperature of 600°C or 800°C would be more realistic (**2, 8**).

Figs. 1 and 2 show the 722 resulting data points, plotted as volumetric wear coefficient versus friction coefficient and sliding velocity, respectively. It can readily be seen that, with respect to the requirements of Table 1, no appropriate sliding couple is found in the tribological database with a volumetric wear coefficient below $5x10^{-8}$ mm³/(Nm) at 400°C and at a sliding velocity above 0.5 m/s

Note, however, that at sliding speeds below 1 m/s and 22°C, some couples exhibit volumetric wear coefficients around $5x10^{-8}$ mm³/(Nm), but that the wear increases when the temperature reaches 400°C.

2.3 Unlubricated coefficient of friction

It is well known that engine oils lubricate under hydrodynamic or EHD conditions, leading to a coefficient of friction between 10^{-2} and 10^{-3}. The question of whether unlubricated tribocouples can exhibit similarly low coefficients of friction to those obtained under hydrodynamic or EHD conditions can be readily answered by a search in the Tribocollect database.

Figure 3 shows a plot of coefficient of friction versus sliding speed. Above 22°C, no data with coefficient of friction lower than 10^{-2} could be found. In order to minimize the frictional losses, it can be concluded that an unlubricated engine will not run efficiently. Considerable work has been done to apply solid lubricants in engines. However, solid lubricants exhibit low friction only in vacuum, whereas the friction and wear rates under the operating conditions in engines are higher.

In the last few years, some results showing coefficients of friction in the range 10^{-2} to 10^{-3} have been published for dry lubricated couples (**16**). Such 'superlubricity' requires the following conditions:

- atomistically smooth surfaces (roughness < 2 nm),
- frictional power losses 1 mW/mm²,
- elastic microcontact,
- no surface contamination,
- defined and constant ambient conditions,
- small surface interactions,

Table 1 Estimation of the volumetric wear coefficients of tribosystems in an oil lubricated internal combustion engine.

n = 3000 rpm; life = 2000 h; 4 cylinder, 2000 cm³

Tribosystem	Normal force [N]		Sliding speed [m/s]	Wear volume [mm³]	Sliding distance in 2000 h [km]	Necessary volumetric wear coefficient [mm³/(N*m)]	
	Cycle average	Maximum				Maximum	Minimum
1. Piston ring[2]	283	3956	8	113	57600	$7 * 10^{-9}$	$5 * 10^{-10}$
2. Piston ring[3]	42	593	8	113	57600	$4,7 * 10^{-8}$	$3,3 * 10^{-9}$
Valve guide [1]	700	1000	0,5	67	3600	$2,6 * 10^{-8}$	$1,8 * 10^{-9}$
Cam / follower	1000	3000	2,3	6	16500	$3 * 10^{-10}$	$1,2 * 10^{-10}$

1 Gas pressure max. 15 % of 1st piston ring.

2 Bore = 90 mm; stroke = 80 mm; thickness piston ring = 2 mm; wear length radial = 0.200 mm; gas pressure max. = 70 bar.

3 Wear of valve shaft set = 0; shaft diameter = 7 mm; guide length = 30 mm; stroke = 10 mm; wear length radial = 0.100 m.

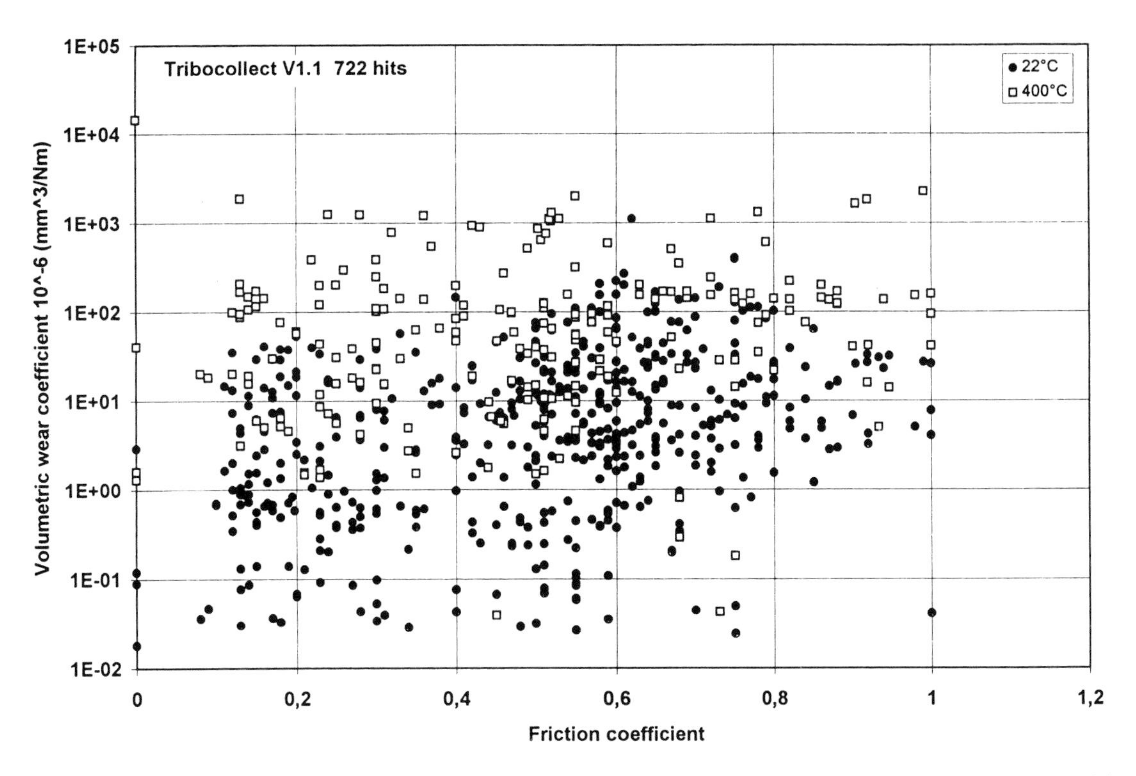

Fig. 1 Plot of the specific wear rate versus solid state friction coefficient of sliding couples found in the BAM tribological database.

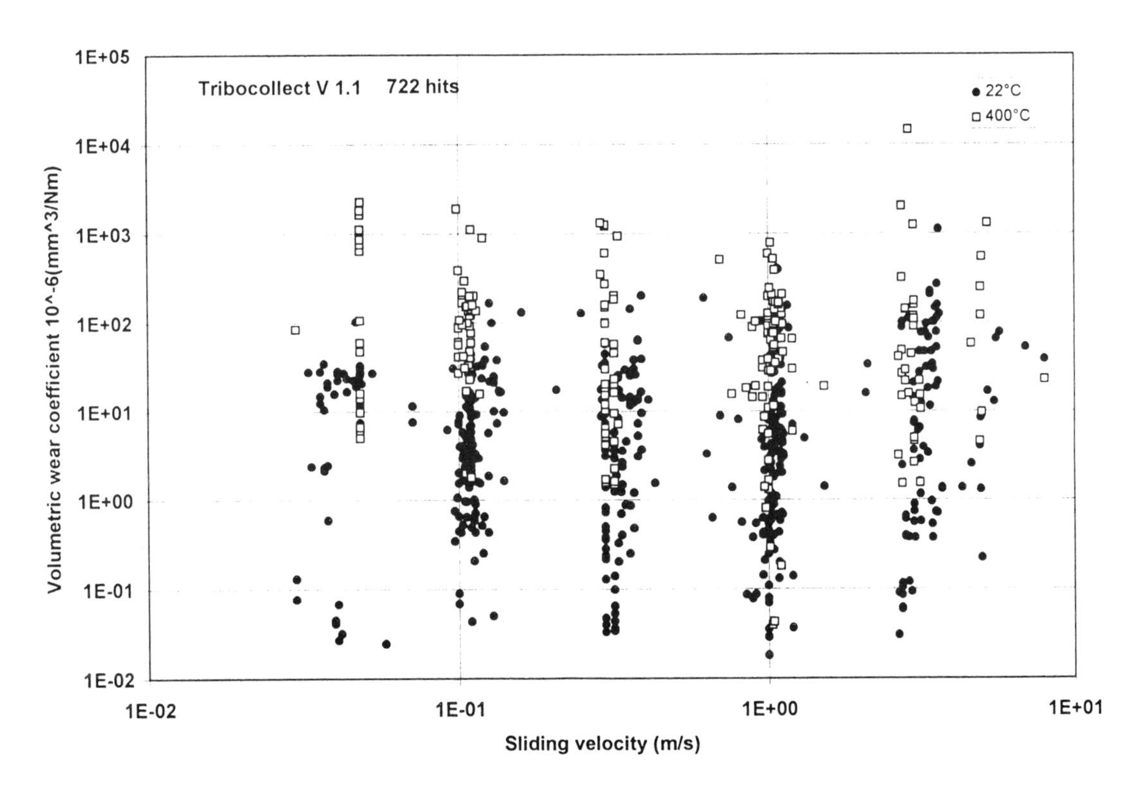

Fig. 2 Plot of the specific wear rate versus sliding speed of couples found in the tribological BAM database of solid state friction.

- shear layers at the surface which are crystallographically oriented parallel to each other.

These conditions can be found only in micromechanics or space applications, not in automotive applications or machinery where the tribosystem runs over a wide range of normal force, temperature and sliding speed.

Another promising technology to reduce dry friction, theoretically possible in oxidizing atmospheres, is the use of so-called Magnèli-phases: oxides which exhibit crystallographic layers with a crystalline slip ability (**17**). TiO_2 represents such a substance and is stable in air. The shear strength, in the range from 8 to 81 MPa, depends on the stoichiometry and is comparable to that of MoS_2.

It is unclear, though, whether these phases can be synthesised and exhibit stable tribological properties under the operating conditions of an automotive engine.

3 TRIBOLOGICAL HIGH-TEMPERATURE APPLICATIONS

The main operational problems of tribosystems operating at temperatures above 400°C are caused by the fact that liquid lubricants are not stable at continuous bulk temperatures above 200°C or at short peak temperatures above 400°C, so that the triboelements must operate oil-free.

Liquid lubricants are essential in tribosystems:

- to separate the mating surfaces for low wear and friction,
- to remove wear particles,
- to cool the surfaces and components, and
- to avoid adhesive wear (mainly for metals).

If liquid lubricants cannot be used, their functions must be taken over by other mechanisms or materials which are insensitive to temperature. The shearing of the upper surface area of micro-contacts is the only possible mechanism under dry friction to accommodate the velocity between two sliding surfaces.

Fig. 4 lists high-temperature tribosystems with some materials used in each case for unlubricated sliding.

Only ceramic materials were used in the tribosystems listed in Fig. 4, because they operate between 600°C and

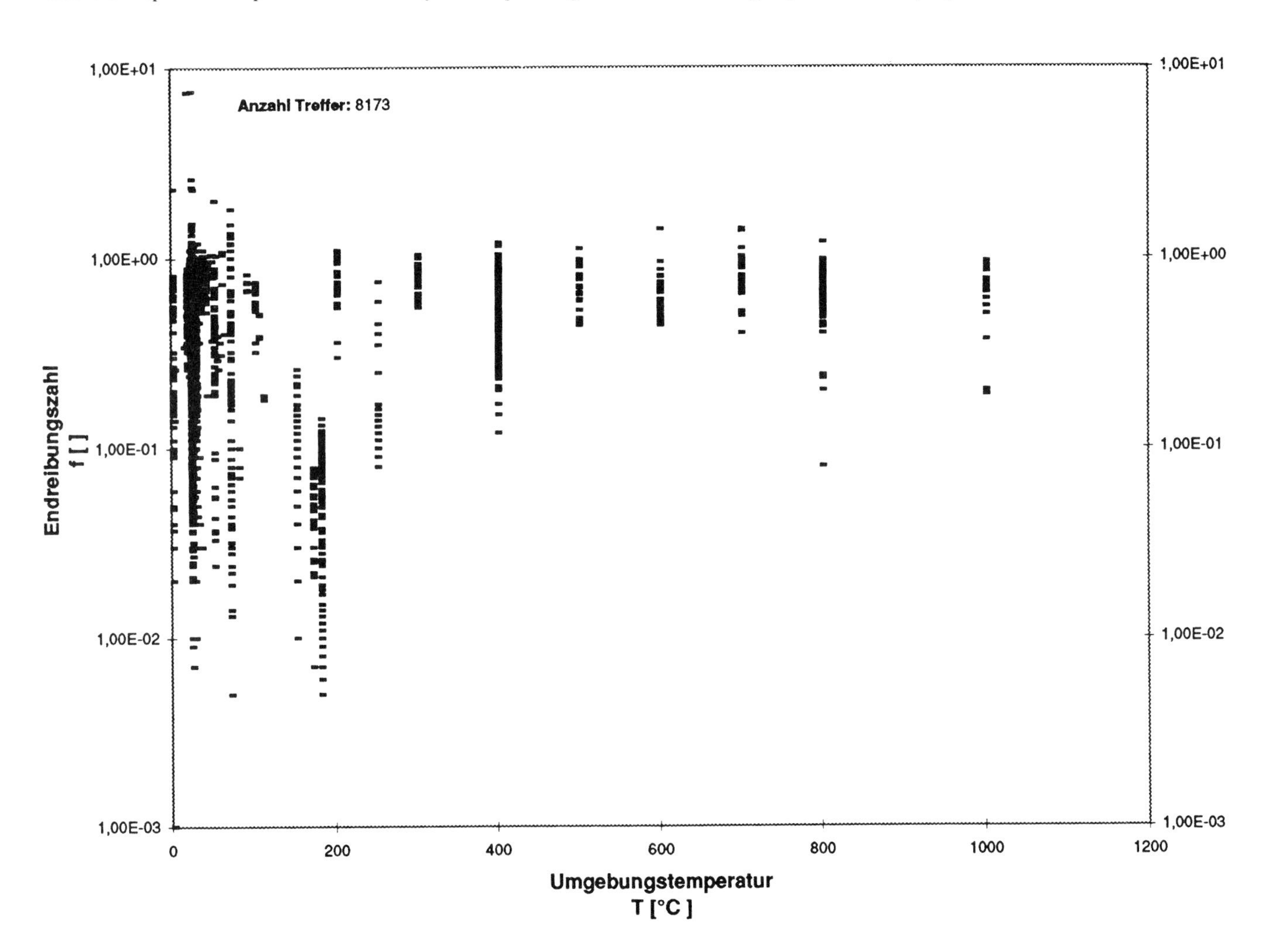

Fig. 3 Overview plot of final coefficient of friction versus ambient temperature found in 'TRIBOCOLLECT V1.1'.

Product	Tribosystem	Component temperature °C	Gaseous environment	Materials
Adiabatic diesel engine or oil-free engine	Piston ring / cylinder	700	CO, HC_x, NO_x, CO_2, N_2, C	ZrO_2, SiC,Al_2O_3-SiC(whisker), Si_3N_4, PS200 Si_3N_4-TiN, Si_3N_4-TiC, vapour phase lubrication, carbon
Gas turbine heat exchanger	Shroud / Cordierite wheel	600–900	CO_2, CO, N_2	Cordierite, NiO-CaF_2, Graphite, PbO-SiO_2
Stirling engine	Piston ring / cylinder	600–1000	H_2, He	PS200 (Cr_3C_2-Ag-BaF_2/CaF_2-NiAl), SiC, SiSiC
Hypersonic engine	Wall / seal ring	1000	H_2, O_2, H_2O	Al_2O_3-SiC-Whisker/Fiber, SiC
Gas sliding seal	Ring / ring	900	CH_x	TiC, PS 200, Si_3N_4, SiC
Oil-free turbine rolling bearing	Ball / rings / cages	600	air	Ga/In/WSe_2, Si_3N_4, M50, vapour phase lubrication

Fig. 4 High-temperature tribosystems in applications and prototypes.

1000°C. Experience in these applications has shown that metallic materials fail by adhesive wear and are not suitable. Fully ceramic materials were applied not only for tribological reasons, but additionally for structural, thermodynamic and design considerations.

It is noticeable that all components were built from monolithic materials or from sintered solid lubricant composites. These technical solutions give a high wear reserve and provide the potential for the loss of large amounts of material by wear, which must be taken into account through high clearances, because at high temperatures the wear coefficients are high.

The wear behaviour of couples can be improved by intrinsic solid lubricants only in those applications where reducing or oxygen-free atmospheres exist (see Fig. 4). In the compilation in Fig. 4 the gas film-lubricated seal ring represents an exception, because the mating surfaces contact only under starting and stopping conditions.

If the wear-reducing velocity accommodation between two sliding surfaces by liquid lubrication is not possible or must be avoided, dry lubrication concepts must be developed to achieve specific wear coefficients below 5.10^{-8} mm³/Nm.

4 DRY LUBRICATION CONCEPTS FOR HIGH TEMPERATURES

Table 2 summarizes four main dry lubrication mechanisms

Table 2 Dry lubrication mechanisms for tribosystems operating at high temperatures.

Tribo-oxidative formation of lubricious oxides	**Incorporation of intrinsic solid lubricants in a matrix**
Incorporation of Ti-compounds, which can form a thin TiO_{2-x} layer in oxidative environments by tribo-oxidation. The mechanical properties are slightly affected. The wear reserve is high. Hard substrates achieves the maximum effect.	The incorporated, soft solid lubricant particles weaken the mechanical properties of matrix and disturb sintering or thermal spraying. The non-oxide phases lose their properties by oxidation. The wear reserve is high.
TiC, TiN, (Ti, Mo) (C, N)	hex-BN, MoS_2, $(CF_x)_n$, graphite, WSe_2
Transfer film formation	**Vapour phase Lubrication**
Depending on surface roughness, sliding velocity and normal force, the triboelement forms an adhesive, thin film on the counterpart surface. Any liquid interrupts this mechanism.	Surface active substances can be delivered through a vapour or as vapour and form a film by decomposition on the surface or by reaction with the surface. The mechanism operates effectively up to very high temperatures. This technique requires an additional supply system. Many of the known substances are toxic or neurotoxic.
Polymers, glasses, graphite, C- and SiC-fibers	Tricresylphosphate, ethylene, CO-H_2, phthalocyanin, O-phatolonitrile, PFPE, arylphosphate, tricyclophosphazene.

applicable to engines, which have already been applied to prototypes or test rigs or, rarely, to actual products. The following dry lubrication concepts will be discussed with reference to the operating conditions in an engine.

4.1 Lubricious oxides

Publicly funded European and German research projects have investigated, by modern surface analysis, the surface composition from low wear regions of dry sliding couples of both commercial ceramics and newly-developed ceramic composites. From these results, the formation of 'lubricious oxides' as wear-reducing dry lubricating mechanisms was detected (**18, 19**). In this context, tribo-oxidation will be treated here as a mechanism of lubrication and not of wear.

A tribochemically formed lubricating soft oxide layer on a hard substrate, with a low shear strength, results under unlubricated conditions in considerable wear reduction and sometimes in decreased friction (**19**). The self-formation of lubricious oxides thus provides a mechanism of dry lubrication to enhance wear resistance when oils must be avoided.

The friction force F_1 of such tribosystems is defined by equation (1):

$$F_1 = \frac{F_N \tau_s}{H(T)} \tag{1}$$

where F_N is the normal force, τ_s is the shear strength of the thin oxide layer and $H(T)$ is the hardness of the substrate as a function of temperature.

Figure 5 summarizes pin-on-disk results with new self-mated ceramic-ceramic and ceramic fibre-reinforced glasses against ceramics, compared with commercial ceramics, and illustrates the progress made in reducing the

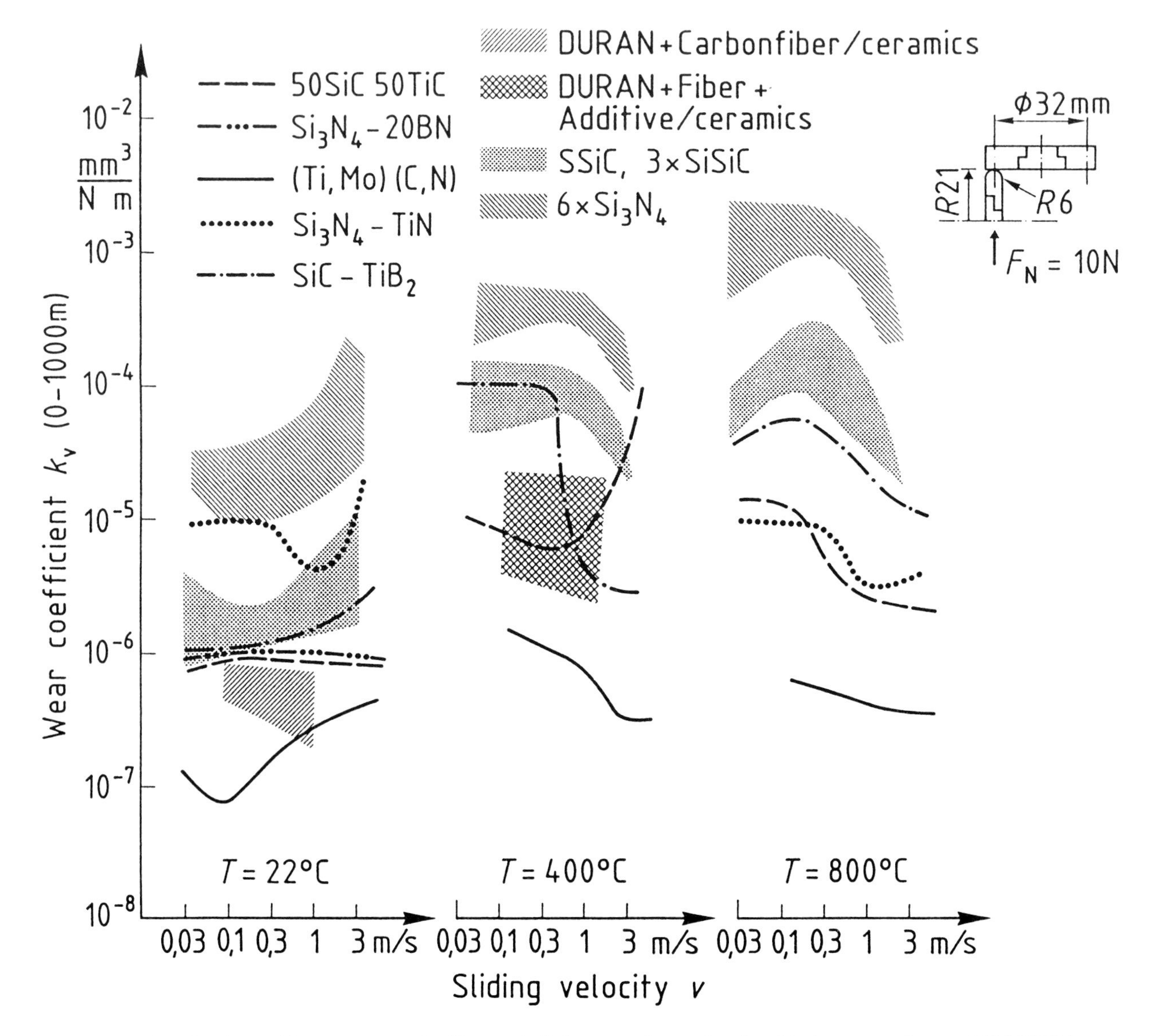

Fig. 5 Total specific wear rate of self-mated sliding couples under dry friction according to DIN 50324 or ASTM G-99

total volumetric wear coefficients of ceramics and ceramic composites. At room temperature, the total volumetric wear coefficient (TVWC = the sum of pin and disc wear coefficient) has been significantly reduced below the value of 10^{-8} mm³/(Nm) previously widely regarded as the application limit for dry sliding couples. The wear behaviour often depends only marginally on the sliding velocity. Materials are of interest for sliding applications, when the volumetric wear coefficient is lower than 10^{-8} mm³/(Nm). At temperatures between 400°C and 800°C the TVWC approaches or falls below the 10^{-8} mm³/(Nm) threshold value. The incorporation of Ti-compounds, which can form a self-lubricating TiO_{2-x} surface layer by tribo-oxidation, has considerably improved the wear resistance. The materials characterized were: SiC-TiC, Si_3N_4-TiN, (Ti, Mo) (C, N) and SiC-TiB_2, which gave the same bend strength and higher toughness than the Ti-free ceramics and are electrical conductive (**18, 19**).

4.2 Incorporation of solid lubricants in a matrix

For some decades, the incorporation of specific solid lubricants in materials and matrices for high temperature applications has been pursued in the United States. These solid lubricants achieved no success in long-term operation (except in space applications), but are still considered as candidate materials for e.g. dry turbine main shaft rolling bearings, unlubricated engines etc.

This situation is not surprising, because on the one hand any solid lubricant weakens the material's structure and therefore its strength or fatigue life. On the other hand, under normal atmospheres the typical effective non-oxide solid lubricants oxidize and lose their beneficial properties. As an example, the tribological results achieved with Si_3N_4-hex. BN can be cited (**19**) (see Fig. 5).

Overall, many intrinsic or extrinsic solid lubricants start to lose their tribological effectiveness with increasing temperature, through oxidation, melting or decomposition.

4.3 Transfer film lubrication

The favourable tribological behaviour of many polymer materials under dry sliding conditions is determined by the fact that they form a thin transfer film on the countersurface, which depends on the surface roughness of the countersurface and on the sliding velocity. High PV-values, as well as temperatures above 300°C, hinder these mechanisms.

Glasses (**21**), carbon materials (**22**) and ceramic fibre-reinforced glass or cordierite matrices (**21**) can operate as alternatives at up to 500°C. With these materials at 1 m/s and room temperature, friction coefficients as low as 0.1 and sliding wear coefficients around 10^{-7} mm³/Nm have been reported (**21**).

Any type of liquid interrupts this transfer film mechanism. As a consequence of the lubricious transfer film formation, a certain running-in wear in the range of 1 µm to 5 µm of clearance, sometimes more, must be taken into account.

Fine-grained turbostratic carbon prepared from mesophase offers promising potential for design, because it exhibits both appropriate tribological and mechanical properties for the requirements of a piston, and also plays a key role in the way to an oil-free engine (**23, 24**). The thermal expansion of carbon is only one quarter of that of aluminium and the bending strength is independent of temperature. The main lubrication mechanism of carbon is the formation of a transfer film. Recent two-stroke engine results (**24**) have shown a guaranteed 1000 hour endurance life with a 1:200 oil-to-fuel ratio using a combination of wear-resistant and self-lubricating sliding couples (e.g. a carbon piston with coated or ceramic liners). Carbon pistons operate successfully in four-stroke gasoline and diesel engines and offer reduced fuel consumption, noise and exhaust emission (see Fig. 6) (**24**), mainly caused by a lower piston weight and clearance.

4.4 Vapour phase lubrication

A gaseous transport medium or a gas mixture can be supplied to the tribocontact and form a lubricious substance on the rubbing surfaces by decomposition or reaction. This mechanism can be applied over a wide range of temperatures, up to the highest temperatures, and is effective only under dry friction (**26–28**).

Many substances which have been mixed into the gas have disadvantages in terms of toxicity and ecological aspects. In addition, this mechanism requires a supply and control system which adds weight and needs to be refilled. Such a system can only be used in military and space applications.

The volumetric wear coefficient published by Klaus *et al.* (**28**) for plasma-sprayed ceramic coatings against cast iron with vapour-delivered lubricants was greater than 10^{-7} mm³/(Nm). Another principle proposed by J. Lauer, *et al.* (**27**) is the catalytic decomposition of gaseous hydrocarbons on hot ceramic surfaces to form graphitic lubricants. No wear data from these tests have however been found in the literature.

4.5 Gas lubrication

In a non-liquid lubricated engine, the separation of the surface could be achieved by a gaseous film. Such a gas film can support engine components only if the relative velocity is very high, because the viscosity of a gas is typically 1/10000 that of a liquid. This inconvenience can be minimized with a very low gap. Tests performed with a ringless piston made in Si_3N_4 and running with a clearance of 10 µm offered some benefits, but the system failed within some hours due to severe scuffing (**29**). At least in small-scale fully-ceramic engines, ringless pistons seem to operate. UTI have described a 5 kW oil-free two-stroke

Fig. 6 Piston made in fine-grained carbon for a four cylinder 1.9 litre TDI (Photo: Motoren GmbH Greiner, D-88267 Vogt i.A., August 1996).

engine made in SiSiC successfully running with a clearance of 2 μm (**32**).

4.6 Rolling bearings

The high frictional losses involved in dry sliding contacts could be reduced by the use of rolling element tribocontacts, as proposed in the Ford unlubricated engine (**3**). However, a crankshaft with rolling element bearings is difficult to manufacture and the dynamic load carrying capability of a dry-running rolling bearing is significantly reduced (**33**). An oil-lubricated ceramic ball bearing can support a load 30 times greater than an unlubricated one.

5 OVERVIEW OF ENGINE CONCEPTS BASED ON NEW MATERIALS

The previous sections have illustrated that unlubricated tribosystems, even with high performance materials and coatings, cannot attain the low level of friction coefficient and wear rate of oil-lubricated tribosystems in conventional engines.

This conclusion is consistent with the current state of the art. So far, no low heat rejection engine has found widespread practical use. The leading companies in the development of unlubricated engines have abandoned this research direction, and their attention is now focussed more on long-term lubricated engines.

5.1 ISUZU

The ISUZU low heat-rejection engine combines traditional lubricated tribosystems and a thermo-structure in the form of a ceramic engine (**4**). The divided structure, with the cylinder liner in an upper section operating at high temperatures and a lower section at more normal temperatures, reduces the tribological problems. For this, a high-strength and low thermal conductivity silicon nitride, as well as a Si_3N_4 with better lubricant wettability, have been developed. The piston rings and the lower section of the cylinder liner were manufactured from a gas-pressure sintered Si_3N_4 containing Fe_3O_4 (**7**). According to ISUZU, this ceramic engine concept, lubricated with a low additive polyolester oil, exhibits promising results (see Fig. 7).

5.2 Adiabatics

The main part of the engine consists of a hybrid piston with solid lubricated top compression rings operating at 540°C and a hydrodynamically lubricated oil ring with a synthetic polyolester high-temperature lubricant (**29**). Adiabatics expect that such high-temperature tribological system for engines will be operational within the next ten years.

5.3 Daimler Benz

The development of a life-time lubricated and fully sealed engine characterizes the target of Daimler Benz (**30**) and represents one approach to the development of a fully environmentally friendly engine. This engine concept already involves new materials combined with highly wear-resistant coatings and materials (**31**). The liquid lubricant will be rapidly biodegradable and yet stable over the life-time of the engine, with a low additive content.

Fig. 7 Cylinder liner, piston rings and bearings made in Fe_3O_4 – alloyed Si_3N_4 for long-life lubrication (Photo: ISUZU Ceramic Research Institute, Ltd. (Japan)), May 1996).

6 CONCLUSIONS

From the tribological point of view, unlubricated engines can be produced if wear coefficients of sliding couples lower than $5x10^{-8}$ mm³/Nm can be demonstrated, which are independent of ambient temperature and sliding speed. Materials are known with unlubricated wear coefficients around 10^{-7} mm³/Nm, which opens the possibility of building such engines for basic studies. To improve the wear resistance further, different dry lubrication mechanisms must be evaluated with respect to the required life-time, the allowable costs and operating conditions as well as customer requirements.

The outstanding wear resistance of some materials can be used to increase the operating times of oils, up to 'life-time' operation. An energy-efficient and reliable engine, with lower frictional power losses, will only be able to be realized, from the point of view of current state-of-the-art, if the tribosurfaces are separated by a liquid film. The materials choice and the design concept determine the nature of the fluid, which will thus influence the whole engine construction.

7 REFERENCES

1 **N.N.,** *A review of the state of the art and projected technology of low heat rejection engines, National Academy Press*, Washington, D. C. 1987.

2 **Kamo, R.,** Ceramic engines and their cost effectiveness, *Ceramica Acta*, **Vol. 3,** No. 1, 1991, pp. 49–65.

3 **Havstad, H., Garwin, I.J.,** and Wade, W.R., A ceramic insert uncooled diesel engine. *SAE paper* **860447.**

4 **Kawamura, H.,** and **Matsuoka, H.,** Low heat rejection engine with thermos structure, *SAE paper* **950978.**

5 **Greiner, P.,** Development of engines and apparatuses with ceramic components International Congress, *High-Tec-Materials and Finishing*, March 1989, Praxis-Forum, Berlin, Germany.

6 **Tanaka, A.,** New types engines and their tribological assignments, *Proc. of the Int. Tribology Conference, Nagoya, Japan,* 1990, p. 1339.

7 **Kita, H., Kawamura, H., Unno, Y.,** and **Sekiyama, S.,** Low frictional ceramic materials, *SAE paper* **950981.**

8 **Kamo, R.** and **Bryzik, W.,** High temperature lubrication of adiabatic engine, *Synopses 3S1-7, International Tribology Conference*, November 1995, Japan.

9 **Woydt, M.** and **Habig, K.-H.,** Tribological criteria and assessments for the life of unlubricated engines, *Lubrication Engineering*, 1994, **50,** 519–522.

10 *Schadstoffemission leichte Kraftwagen*, PkW, Stufe 2.70/220/EWG, EG-Amtsblatt **Nr. L. 100** vom 1994.

11 **N.N.,** *Committee of experts for environmental questions Waste Management. Special Expertise,* 1990, Metzler-Poeschel-Verlag Stuttgart, Germany.

12 *Verordnung über das Einsammeln und Befördern sowie über die Überwachung von Abfällen und Reststoffen,* AbfResÜberV vom 1990.

13 *Rückstand-und Abfallwirtschaftsgesetz* 1994, Bundesgesetzblatt (BGBl. I. S. 2705).

14 **Platzer, W.,** Auswirkungen zukünftiger Emissionsvorschriften für NFZ-Dieselmotoren auf die Tribologie im Motor, *Congress Tribotechnische Werkstoffe im Kraftfahrzeug*, 1990, Praxis-Forum, Berlin, Germany.

15 *Tribocollect – A numerical, tribological data base for materials, coatings and lubricants,* BAM, Berlin, Germany.

16 **Martin, J-M., Donnet, C., Le Mogne, T.,** and **Belin, M.,** How to reduce friction coefficients in the millirange by solid lubrication, *Proc. of the Int. Tribology Conference*, 1995, Japan.

17 **Gardos, M.N.,** The effect of anion vacancies on the tribological properties of rutile. *Tribol. Trans.* **32,** 427-436.

18 **Woydt, M., Kadoori, J., Hausner, H.,** and **Habig, K.-H.,** Werkstoffentwicklung an Ingenieurkeramik nach tribologischen Gesichtspunkten, *Cfi/Ber.* Dt. keram. Gesellschaft, **67,** 4, 123–130.

19 **Woydt, M.,** and **Skopp, A.,** Ceramic and ceramic composite materials with improved friction and wear properties, *Tribol. Trans,* **38,** 2, 233–242.

20 **Wäsche, R.** and **Habig, K.-H.,** Physikalisch-chemische Grundlagen der Feststoffschmierung, *BAM-Research Report,* **158,** 1989.

21 **Brückner, R., Klug, T., Skopp, A.,** and **Woydt, M.,** Friction and wear behaviour of C- and SiC-fibre reinforced glass composites against ceramic materials; *Wear,* **169,** 243–250.

22 **Thiele, W.,** Tribologisches Verhalten von Kohlegraphit (Tribological behaviour of graphite) SGL Carbon, GmbH, Germany, 1993.

23 **Minford E.,** and **Prewo, K.,** Friction and wear of graphite fiber-reinforced glass matrix composites, *Wear,* **102,** 253–264.

24 **Greiner, P.,** *Use of ceramic parts of SiSiC and carbon for engines and condensers and securing of function in long-time tests*, Annual Report, BMFT, 1991.

25 **Koch, M.,** Pistons made of carbon, *Auto, Motor, Sport*, **3,** 36–38, 1993.

26 **Klaus, E.E., Jeng, G.S.,** and **Duda, J.L.,** A study of tricresyl phosphate as a vapour delivered lubricant. *Lubrication Engineering*, **1989,** 717–723.

27 **Lauer, J.L.,** and **Dwyer, S.R.,** Tribochemical lubrication of ceramics by carbonaceous vapors, *Tribo Trans*, 1989, **45, 11,** 717–723.

28 **Groeneweg, M., Barber, G.C.,** and **Klaus, E.,** Vapor delivered lubrication of diesel engines cylinder kit rig simulation. *J. of the Soc. Tribologists and Lubrication Engineers,* 1991, 1035–1039.

29 **Bryzik, W.** and **Kamo, R.,** High temperature tribology for future diesel engines. *Proc. 82nd NATO-AGARD-Meeting,* Portugal, **CP 589**, 1996, 589.

30 **Müller, B.,** Battle of materials under the bonnet, *Bild der Wissenschaft*, 1996, p. 104–106.

31 **Mörgenthaler, K.D.,** *Ceramic components in engines, requirements and properties. Keramik in Wissenschaft und Praxis;* ed: G. Grathwohl; DGM Informationsgesellschaft mbH Verlag, Oberursel, Germany, 1993, pp 17–28.

32 **Poeschl, G.,** Piston-machine without lubrication, *Company brochure* 1995, UTI AG, Frankfurt, Germany.

33 **Rombach, M.,** Ceramic ball bearings made in silicon nitride, *Maschinenmarkt,* 1996, **102,** 40, 40–43.

State of the art of polymer tribology

S W ZHANG
University of Petroleum, Beijing, China

1 INTRODUCTION

With the application of polymers for tribological purposes increasing and extending into more new areas, much research on the tribology of polymers and its application in various industries, has been made. The intention of this review is to present comprehensive information on entirely new phenomena which have been discovered and investigated in the past few years on the tribological properties of polymers and their applications. The progress in understanding of polymer tribology during the past decade is reviewed here under three headings: rubber tribology, plastics tribology, and tribology of polymer composites.

2 RUBBER TRIBOLOGY

2.1 Rubber Friction

Recently, a number of tribologists have turned their attention to the friction of rubber.

Some measurements of friction were reported for long thin blocks of natural rubber (NR), both unchlorinated and with surface chlorination, sliding against smooth glass (**1**). It is found that the frictional coefficient is independent of applied load over a wide range. Moreover, it is found to increase with sliding speed and decrease with temperature. However, chlorination of a thin surface region could reduce the coefficient of friction and its dependence on speed and temperature, similar to the effect found with thermoplastics.

Because the mechanics of a tyre in contact with a frictional flat or curved surface is of great significance in understanding the road/vehicle load transfer characteristics, the finite element analysis of a rubber block in friction contact has been carried out (**2**). The stick-slip phenomenon was predicated as a function of frictional coefficient. Moreover, a viscoelastic finite element contact analysis to predict the hysteretic friction between a sliding rubber block and equally spaced triangular asperities was also conducted, and its application to studies of pavement skid resistance was considered (**3**). The analyses conclude that the pressure dependence of friction is affected by the amount of contact of rubber with the asperity, and the coefficient of hysteresis friction increases with speed, increasing to a maximum value and then decreases with a further increase in speed. The frictional coefficient is heavily dependant on the geometry of the asperity. This discovery has been successfully applied to studying the friction mechanism of grooved road pavements (**4–6**).

Since the low thermal conductivity of rubbers can result in a very high temperature at the interface, it is important to investigate the effects of frictional heating on the sliding friction of rubbers. Ettles and Shen (**7**) have presented a paper concerning the effects of heat generation on the level of friction at the interface.

In order to explain the wet frictional coefficient of rubber in terms of some viscoelastic parameters, the relation between temperature and wet friction due to hysteresis was studied (**8**) with a British Pendulum tester (BPT). This relation was deduced from the theory of rolling resistance between a viscoelastic plane and a rigid spherical body.

An axially symmetric rotational test, using an annular specimen, could eliminate the frictional lift phenomenon responsible for the high pressure gradients in conventional tests. For the purpose of putting the idea into practice, a rotational friction tester was developed (**9**), which effectively eliminated the severe normal contact pressure gradients inherent in many conventional test apparatus. The uniform pressure distribution of the annular test specimen provided a pointwise value of the coefficient of friction essential to full-field traction and wear prediction.

Theories have been developed and refined to provide reasonable predictions of rubber friction for a wide variety of operating conditions, but some questions remain to be answered, such as: what is the source of friction within the real contact area? How, and to what extent, does viscoelasticity of the rubber affect the friction? With the support of some examples illustrating how this multitude of factors come together to give a certain level of friction in a rubber article, Roberts (**10**) has proposed a guide to estimate the friction of a rubber component.

Barquins has studied the adhesion and friction of elastomers extensively (**11–15**). The underlying mechanisms of adhesion and of sliding and rolling friction of rubber-like materials were studied on a microscopic level, by scrutinizing the phenomena occurring inside both static, and rolling, or sliding contact areas, between a single, hard, transparent spherical or cylindrical asperity, and the smooth and flat surface of a elastomer sample. It has been shown that the visualization of the contact area between a blunt, rigid, asperity and a smooth, flat, surface of a rubber-like material, can be applied to the investigation of various frictional mechanisms and accompanying phenomena.

Run-in friction is nascent friction at the initial stage of a friction process. The basic characteristics of run-in friction have been examined between a hemispherical, silicone rubber sample and a smooth glass disc (**16**). Observation shows that the run-in frictional coefficient increases with sliding time until a steady maximum value is reached. This characteristic does not change with the operating conditions and test procedures (Fig. 1).

Slip generated at the boundary of the Hertzian contact during tangential displacement is called Mindlin slip. Sasada, Hiratsuka, and Tomita (**17**) give a brief explanation of this phenomenon, claiming that the slight reciprocating movement of a rubber eraser over a sheet of paper enables us to observe the appearance of the Mindlin slip phenomenon easily. The possibility of observing Mindlin slip with the naked eye by such a simple method as these slight hand movements with a rubber eraser provides a valuable demonstration.

2.2 Rubber wear

2.2.1 Dry abrasion

A comprehensive survey of the research on this subject was proposed by Zhang in 1989 (**18**). However, much more has been achieved in this field since then.

Because of the lack of an acceptable explanation for some aspects of rubber abrasion, a hypothetical mechanism for creating subsurface cracks during frictional sliding has been put forward (**19**), as a part of the process of rubber abrasion which is supplementary to the crack growth mechanism (**20**). It consists of the unbounded elastic expansion of microscopic precursor voids until they burst open as cracks, under the action of internal pressure or of triaxial tension in the surrounding rubber. The most probable mechanism responsible for the generation of a sufficiently large inflation pressure or triaxial tension seems to be thermal decomposition of rubber.

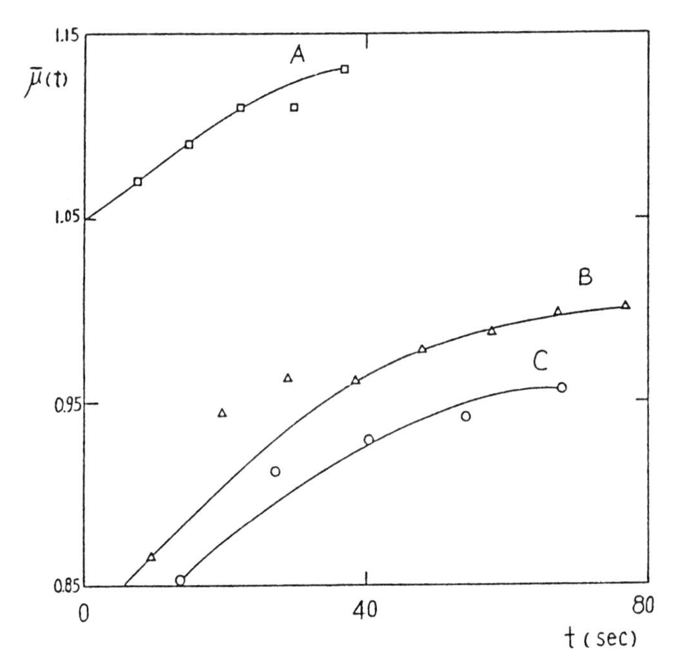

Fig. 1 Friction coefficient plotted against run-in time: (W = 1.18 N; A – V = 8.1 cm/s; B – V = 6.2 cm/s; C – V = 4.4 cm/s); (after Ref. (**16**)).

The mechanisms of the pattern abrasion of rubber have been investigated (**21**) with an unfilled isoprene rubber wheel rubbing against a steel cylinder. It has been concluded that the wear rate of rubber is affected not only by the crack propagation rate along the low slope, but also by the crack propagation rate when the crack moves upwards.

Recently, Fukahori and Yamazaki (**22–24**) studied the process of rubber abrasion by a line contact using a steel slider of the razor blade type against a rubber block specimen. The latter is fixed on a high damping steel plate which moves backwards and forwards along a horizontal linear path. It has been discovered that the periodic formation of abrasion patterns can be ascribed to the following two kinds of periodic motions generated in the frictional sliding of rubber: stick-slip oscillation of low frequency, of the order of 10–20 Hz, and microvibration of high frequency of the order of 500–1000 Hz (Fig. 2).

Figure 2(a) is a typical spectrum, representing stick-slip motion. Figure 2(b) is an acceleration spectrum of the normal vibration measured simultaneously with the frictional force. As Figure 2(b) shows, the most prominent phenomenon is the violent microvibrations generated at the slip stage of stick-slip motions. The microvibration generates the initial abrasion patterns and the stick-slip motion propagates them to the final abrasion patterns.

Their experimental observations have reconfirmed that the formation pattern of abrasion is propagated from primary ridges, with small spacing, to secondary ridges, with large spacing, as described previously by Zhang (**25**).

Recently, several studies have shown that the fractal concept can be applied to the study of wear surfaces and the debris of rubber abrasion by a line contact (**26–28**). A wear test was performed with a modified blade abrader (**20**, **29**). It is found that the wear surface profiles of three rubber compounds, NR, polybutadiene (PBD) and NR/PBD/SBR, are fractal over a limited range and the fractal parameters (fractal dimension and scale factor) quantitatively link the wear topography with the wear rate. The fractal dimension depends on the wear mechanism. Computer-aided fractal analysis of profilometer traces of wear surfaces provides a quantitative description of the worn surface and improves our understanding of wear processes.

The fractal dimension of the debris increases as the wear load grows and is taken to be the result of the effects

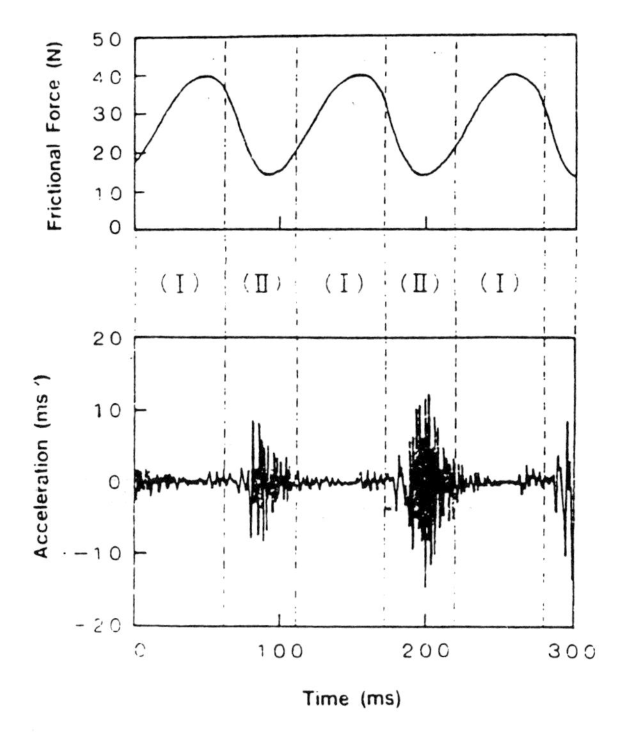

Fig. 2 Periodic motions genenated during frictional sliding of natural rubber:
(a) Spectrum of frictional force against sliding time,
(b) Spectrum of acceleration against sliding time (after Ref. (**21**)).

of the wear load on the agglomeration mechanism. However, the essential correlation between the fractal dimension and the characteristic of wear processes still remains obscure.

Grosch (**30**) used the theory of the wear of slipping wheels to discuss both tyre-wear data obtained under controlled slip and laboratory abrasion of slipping wheels. He has discovered that the discrepancies between the predictions of the theory and the experimental results are similar in respect of tyre wear and laboratory abrasion. The theory did not bring the temperature effects on abrasion into consideration. Hence, it is proposed to link abrasion to the energy density at break, measured at very high temperatures.

More recently, a theoretical wear equation for rubber abrasion by a line contact has been derived by Zhang and Yang (**31**) from the point of view of energy on the basis of the experimental results (**25**, **29**, **32**):

$$R_s - R_u = r\exp[-(P f K_i)/(e^* S)] \sin\theta \qquad (1)$$

where R_s, R_u are the wear rates in steady and unsteady states respectively, r is the crack growth per stress cycle, P is normal load, f is coefficient of friction, K is coefficient of energy accumulated, e^* is wear energy density, S is the cross section of the tongue tip of the ridge, θ is crack-growth angle, and i is the number of the stress cycle.

This wear equation shows a marked improvement on the wear mechanism of rubber abrasion by a line contact obtained previously. It can be applied to calculate the wear rate and to predict the working life of rubber products.

The wear mechanisms of NR, SBR (styrene-butadiene-rubber) and HNBR (hydrogenated nitrile rubber) vulcanizates when abraded against: hard rock, a knurled aluminium disc, and a silicon carbide abrader were studied (**33**). It was interesting to find that the wear of NR and SBR vulcanizates against hard rock at low normal load (6 KPa), took place by a fatigue wear mechanism and switched over to a frictional wear mechanism at high normal load (above 18 KPa). However, in HNBR vulcanizates the wear took place by an abrasive wear mechanism.

2.2.2 Oily abrasion and wet abrasion

The effect of lubrication on rubber abrasion was investigated with a blade abrader (**34**). It has been observed that when a lubricant is applied, a much finer pattern develops and the wear rate is much lower. Moreover, the horizontal force on the blade does not decrease so much, indicating a shortcoming in the theory deduced earlier for dry abrasion (**20**). The horizontal force, as measured by a blunt slider, is substantially lowered in the presence of the lubricant (**35**). Blade abrasion experiments (**36**) with unfilled rubbers incorporating an internal lubricant, have also corroborated the conclusion of ref. (**20**), the pattern spacing getting smaller, and the abrasion decreasing.

As for the oily abrasion, some evidence of the possible importance of ambient temperature on smear production was obtained (**37**). A study was made of the smearing of a peroxide-cured NR wheel (Akron abrasion type) when run on a smooth glass plate with interfacial slip. It has been found within the temperature range from -30–60°C, the higher the ambient temperature, the greater the production of smear, and the lower the temperature, the greater the amount of debris left on the glass with less smear.

The mechanisms of wet abrasion for nitrile rubber (NBR) were investigated with a pin-on-disk-type test machine (**38**, **39**). The specimens were immersed in drilling mud. Based on the experimental observations, it seemed that wet abrasion of NBR involves two different mechanisms: a local micro-tearing process and a general micro-layering or micro-polishing process.

2.2.3 Erosive wear

When the wear process of a solid surface is initiated by a fluid medium containing particles flowing in a direction approximately parallel to the surface of solid at a certain speed, this type of erosive wear is termed abrasion erosion. It is also considered elsewhere as a type of sliding erosion (**40**).

The mechanism of abrasive erosion for polyurethane (PU) and SBR has been investigated using a special abrasive erosion test machine (**41**). Based upon the analysis of the experimental results, the mechanism of abrasive erosion of PU materials is considered to be microdeformation and/or microtearing of the surface layer under the microcutting action of the abrasives. The wear mechanism for SBR is both chemomechanical degradation and thermal decomposition of the material resulting in a sticky layer formed in the surface regions, rather than fatigue delamination of material. This is also ascribed to the fact that the strength and hardness of SBR material are much lower than those of PU materials.

Arnold and Hutchings (**42**) studied the erosive wear of unfilled elastomers. They found that there are two separable mechanisms; one dominating at glancing angles of impact, the other taking over under conditions of normal impact. Moreover, the material is removed from the surface by fatigue crack propagation in both mechanisms. At high impact angles, tensile stresses in the surface, arising from the frictional forces due to particle impact, cause the growth and intersection of fine cracks on the surface, and as a result, material loss occurs. At glancing angles of impact, the mechanism of material removal was found to be very similar in nature to that of abrasion by a sharp blade or by a smooth indentor (**42**).

A model (**43**) to predict the erosion behaviour of rubber under conditions of glancing impact by small hard particles was developed from that of Southern and Thomas (**20**) for sliding abrasion, by combining their treatment of the growth of surface cracks with a model for particle impact. An expression for the erosion rate was developed. Good qualitative agreement was obtained between the predictions of the model and the experimental measurements for NR, SBR materials and a highly cross-linked polybutadiene rubber. However, quantitative agreement was less good.

The physical process of abrasive erosion in annular pipes has been studied (**44, 45**). It is found that the abrasive erosion in annular pipe is developed in two stages: impact erosion and wet abrasion, with the later as the main wear process. Moreover, the mechanism of wet abrasion for PU materials is mainly ascribed to surface fatigue wear. Based on these results, a model of abrasive erosion in annular pipes was developed to predict the erosion behaviour of rubbers.

Mostly, the researches on mechanisms of rubber erosion were confined to the study of the physical process of wear. Investigating the process of erosion, as viewed from the chemical aspect, involves new research thinking and methods, the results of which are a great help in achieving a deep overall understanding of the wear mechanism of rubbers.

In recent years, surface effects during abrasive erosion of rubbers have been studied by Zhang and colleagues (**46–48**).

From the analysis of eroded surfaces of PU and NR materials by scanning electron microscopy (SEM), X-ray photoelectron spectroscopy (XPS), and Fourier transform infrared spectrometer (FT–IR), it has been found that the surface chemical effects in abrasive erosion for PU materials, exposed to water containing quartz particles are: fracture of macromolecular chains, thermal decomposition of groups hydrolysis of carbonate groups, and surface oxydation of methane and ether groups (**46, 47**).

The surface chemical effects for abrasive erosion of NR materials in three different fluid mediums (H_2O, polyacrylamide and NaOH) containing quartz particles were also investigated (**48**). It has been concluded that the surface chemical effects in NR during abrasive erosion are mainly: mechanical fracture of macromolecular chains and oxidative degradation of weight groups. The main products are low molecular weight polymers containing not quite the same groups in a different medium during oxidative decomposition.

2.3 Rubber lubrication

The thickness of boundary liquid films in lubricated rubber contacts is an important parameter for designing rubber components used in lubricated conditions. A new apparatus has been developed to measure simultaneously the frictional force and thickness of the liquid film in static and sliding contact (**49**), which uses a laser to measure the film thickness and to resolve the thickness of film down to 1 nm. Stable films in static contact have been recorded at thicknesses up to 25 nm at an average pressure of 10 KPa. It has been observed that these films collapse at higher average pressure, up to 50 KPa. This fact brings into question the role of electric double layers in supporting the normal load.

A study on the friction of hydrogel and PU elastic layers, when sliding against each other under a mixed lubrication regime was carried out (**50**). It has been shown that start-up friction in PU-PU contacts is extremely high and significantly greater than that for metal on PU or PU on hydrogel. It was concluded that the most appropriate material combination for cushion-form joints was a hard surface sliding on a compliant layer which should preferably be porous.

3 TRIBOLOGY OF PLASTICS

3.1 Friction of plastics

The friction behaviour of different molecular weight virgin poly(ether ether) ketones (PEEK) and of PEEK blends with polytetrafluoroethylene (PTFE) was observed under dry sliding conditions against hard steel (**51**). It has been found that the spherulite size and hardness of PEEK play an important role in friction performance. Concerning the friction property of PEEK-PTFE blends, the lowest friction coefficient was measured for the blend with a PTFE volume fraction of 15%.

Experimental studies were conducted on PTFE rubbing

against metallic surfaces (**52**). It has been shown that PTFE friction at the initial stages of sliding is, to a great extent, dependent on molecular-kinetic processes taking place in the contact zone. It has a viscoelastic origin under certain conditions; the friction parameters at the initial stage, and after the transfer-film formation, are associated with the rheological properties of the polymer. Their change depends on the test conditions.

The surface layer structure of ultrahigh molecular weight polyethylene (UHMWPE) was investigated in contact with a steel counterface (**53**). A surface layer was found to be generated, varying with the intensity of tribochemical processes through the thickness. The thinnest superficial layer (up to 2 nm) of UHMWPE was found to contain fewer oxidized fragments after sliding. It was concluded that the transfer of iron which forms cluster structures in superficial layers seemed to play an active role in the tribochemical processes.

The effect of sliding velocity on the friction of UHMWPE against a steel counterface was investigated using bovine serum as the lubricant (**54**). The coefficient of friction was found to be in the range 0.07–0.2 and not dependent on the sliding velocity.

Experimental studies on the influence of the roughness of polyacetal, with no filler, on tribological properties and the influence of unfilled mating plastics, such as polyethylene (PE), polyamide (PA), polycarbonate (PC), and epoxy resin (EP), were conducted (**55**). It was found that the average coefficient of friction between polyacetal and S45C steel decreases, accompanied by the increase in the surface roughness of S45C, showing a minimum value when the roughness is approximate 0.11 μm. However, roughness is not so influential on the average coefficient of friction when polyacetal or PE having different roughnesses and S45C are rubbed against each other.

The effect of surface roughness on the friction of polyimide (PI) when slid against mild steel in vacuum was investigated (**56**). It has been concluded that for the occurrence of frictional coefficient reduction in vacuum, relative surface roughness of the friction pair is identified to be the essential factor. Undesirable friction reduction in vacuum might be inhibited when the mild steel surface roughness is set to be slightly greater than that of the PI.

Cross-linked epoxy polymers were tested to investigate the effect of structure and physical state on their tribological behaviour (**57**). The structural details were found to dominate the tribological properties of cross-linked polymers, depending on their physical state. It has been found that the coefficient of friction in the glassy state depends on the hydroxyl content in the intermodular segments and on the network density in the highly-elastic state.

The friction of polymers generally depends to a large extent upon interfacial adhesion. With modified PE, a linear relationship exists between the coefficient of friction and the potential adhesion on a model aluminium substrate (**58, 59**).

3.2 Wear of plastics

3.2.1 Sliding contact

The effect of roughness and sliding speed on the wear and friction of UHMWPE against stainless steel was studied under sliding condition (**60**). It was observed that the effect of sliding speed on wear and friction was mostly affected by its contribution to interfacial temperature. Moreover, the effect of increased surface roughness at high sliding speed was to raise the friction coefficient and consequently the interfacial temperature. The wear behaviour of UHMWPE during water lubrication was investigated (**61**) under reciprocating sliding conditions. The effects of molecular weight, filler, lubrication, counterface roughness and sliding configuration on the polymer's transfer characteristics were studied. It has been found that the transfer of the UHMWPE material to the metal counterface during sliding wear involves interlamellar shear of the polymer and results in the development of a highly orientated transfer film. Specifically, significant differences were observed in the degree of crystallinity, crystallite size and orientation in the deformed surface layers of the polymer and debris, compared with those of the bulk polymer. An increase in molecular weight did not significantly improve the wear resistance of the UHMWPE but the addition of fillers did. The sliding wear behaviour of UHMWPE has been studied on a stainless steel counterface under water lubrication conditions (**62**) using a specially modified vertical drill stand. The UHMWPE showed a significant drop in wear rate, under low pressures, at high velocity.

The effect of different types of lubricants and varied counterface roughness on the wear of UHMWPE against AISI 431 was examined using a reciprocating sliding wear rig (**63**). The test shows that the logarithm of the specific wear rate in water is proportional to the surface roughness. However, in 5% oil-in-water, there is a significant transition in specific wear rate at a surface roughness of approximately 0.35 μm. At surface roughness of less than 0.35 μm, the specific wear rate in oil-in-water is much lower than that in water; otherwise, the specific wear rate in oil-in-water approximates that in water. The wear mechanisms are considered to be a function of type of lubricant and topography of the counterface.

Since high density polyethylene (HDPE) and PTFE are widely used as bearing materials, a study was undertaken to examine the effects of atmospheric humidity and external electrical field on the wear behaviour of these materials in polymer-metal sliding contact (**64**). It was found that humidity influenced the wear of both HDPE and PTFE. An electrical field of strength 3 V/cm did not measurably affect the wear of PTFE but in the case of HDPE changes were observed in the wear factor, the magnitude and sign of the alterations depending upon the humidity.

A study on PTFE transfer and wear mechanism was carried out (**65**). PTFE transferred films were found to decrease or increase the friction force depending on variations in adhesional interaction on the contact area during the layer formation. After a PTFE transferred layer has formed on the contacting surface, PTFE begins to wear by a thermoactivation mechanism with an activation energy of 40–60 kJ/mol. The low values of activation energy indicate that the wear of PTFE is associated predominantly with the breaking of weak intermolecular bonds and with interplanar shear and slipping of crystalline aggregates formed in the structure of the materials.

The mild-severe transition for unfilled PTFE was investigated as a function of sliding speed and temperature (**66**). It has been found that mild sliding wear gives way to severe upon an increase in sliding speed, or a decrease in temperature. This transition is related to kinetic friction reaching a threshold.

The wear mechanism of PTFE and fluoropolymer alloy (F50-I) under lubricated sliding contact conditions was also investigated (**39**).

Based on experimental results obtained in a series of studies, Stolarski presented a comprehensive discussion on the tribological performance of polyetheretherketone (PEEK) (**67**, **68**). An important observation emerging from the dry sliding contact studies is that, regardless of the contact configuration, the wear rate is very small. Another important finding is that lubrication of the PEEK sliding contact is not very effective and can produce effects detrimental to the polymer performance.

Concerning the wear property of PEEK-PTFE blends (**51**), the wear rates of the blends with PTFE volume fraction varying from 5–85% were lower than that of virgin PEEK. A minimum value in wear rate existed at a PTFE volume fraction of 5%.

As effective techniques to monitor the mechanophysical and mechanochemical variations, thermal analyses including differential scanning calorimetry (DSC) and thermogravimetry analysis (TGA) were used to study the thermal behaviour of the wear debris, and the worn surface of PEEK after unlubricated sliding wear tests (**69**). It has been found that the frictional temperature at the sliding interface might ranged from 300 to 345°C, favouring plastic flow to occur and providing a lubricating effect. Moreover, different wear testing conditions affect the microstructure of the wear surface and wear debris, as well as the tribological behaviour of the bulk PEEK, the wear mechanism, and the morphological features of the wear debris. It may be that the fractal dimensions of the wear debris are a function of the loading or the wear process involved (**70**).

In order to find an effective way to diagnose and predict the wear performance of polymers, the following investigations have been carried out to study the relationship between the wear debris morphology and wear behaviour of bulk materials. Fractal geometry has been applied to the quantitative analysis of the boundary texture of the wear debris (**71**). It is concluded that a combined mechanism consisting of microcutting, plastic flow, and fatigue-delamination effects, is responsible for the wear behaviour of PEEK. The wear debris are fractals and can be quantitatively characterized with fractal dimensions which are determined by the slit island method. As the loading dependence of the fractal dimension D_F is similar to that of the specific wear W_S (Table 1), the fractal dimension of the wear debris might be regarded as a measure of wear rate. Thus, the wear behaviour of polymers could be predicted on the basis of the morphological investigation of wear debris.

Table 1 Specific wear rate of PEEK and fractal dimension of PEEK debris as a function of the apparent contact pressure (sliding velocity $v = 1\ ms^{-1}$) (**71**).

Pressure (MPa)	1	3	7	8
Ws ($10^{-6}\ mm^3N^{-1}\ m^{-1}$)	7.20	7.00	12.92	18.13
D_F	1.22	1.23	1.34	1.38

A fractal analysis of worn surfaces has been undertaken for two polymers used as matrix materials in restorative dental composites (**72**). The resins are: 60/40 bisphenol, a glycidyl methacrylate/triethylene glycol dimethacrylate (BIS), and ethoxylated bisphenol, and a dimethacrylate (EBP), which contains 3% vol fumed silia with a mean particle size of 40 nm. It has been concluded that confocal microscopy provides data that require smoothing in regions where no reflection from the rough surface is obtained, as fractal characterization of a rough surface is much more affected by the data quality than by the information loss due to profile analysis.

A dimensional analysis was carried out to develop an empirical equation for the wear of polymers (**73**). In this wear equation, the volume loss of polymer material, while sliding in a pin-on-disc machine, is expressed in terms of the operating conditions, material properties, and counterface roughness. Both linear, and non-linear, relationships of volume loss with other variables were considered in evaluating a dimensionless wear coefficient. It was concluded that the non-linear relationship is much better.

The performance of some 24 different polymers subjected to cavitation erosion, solid particle erosion, abrasion and sliding wear has been identified with the help of laboratory facilities which simulate the conditions encountered in mining and industrial machinery (**74**). The modes and mechanisms of the wear of different polymers under these various conditions were discussed.

The influence of certain factors on PE oxidative wear was analysed (**75**). The preliminary oxidation which arose from technological processing was considered along with the direct oxidation resulting from the friction caused by atmospheric oxygen and metal oxides, which were found on the metallic counterface during sliding.

3.2.2 Rolling contact

The rolling contact behaviour of polymers has received much less attention. A preliminary study on the rolling contact performance of polymers was published several years ago (**76**). More recently Stolarski (**67**) went further and pointed out that the main mode of failure of PEEK in rolling contact was plastic flow, not fatigue. The limit of rolling contact performance of polymers seems to be defined by its ability to dissipate the energy generated at a particular combination of contact pressure and rolling velocity in order to avoid excessive softening. Lubrication of PEEK rolling contacts results in an even further improvement of its already satisfactory performance.

3.2.3 Impacting contact

The impacting contact performances of PTFE and nylon-6 have been investigated with a special abrasive erosion test machine (**41**). Ploughing, scratches, and the accumulation of plastic deformation were found on the eroded surface of PTFE, and scratches, as well as milder plastic deformation were found on the abraded surface of nylon-6. It has been concluded that the wear mechanism of PTFE is primarily plastic fracture. As for the erosion of nylon-6, it is considered to be the result of microcutting by flowing abrasives. The influential factors on the wear rates of polymers are mainly the flow velocity and the size and concentration of particles.

3.2.4 Fretting contact

The sliding behaviour in a fretting contact between poly-(methylmethacrylate) (PMMA) and a glass counterface has been investigated by means of both experiments and finite element (FE) computations (**77**), in which studies were focused on the initial loading conditions, i.e. at low number of fretting cycles.

It has been proved that FE computation is an effective way to predict the initial fretting conditions in the tribopair. The conditions for the transition between partial slip and gross slip have been determined as a function of the loading parameters of the tribopair. The FE model was also proved to fit well with the experimental data and to be used to simulate fretting loops, with the assumption that the polymer behaved elastically and that the friction obeyed Coulomb's law.

The behaviour of polycarbonate (PC) and PMMA exposed to oscillating conditions has been investigated (**78**). Results were interpreted in terms of velocity accommodation mechanisms. It was found that cracks propagated until the fretting amplitude was taken up by the elastic deformation of the cracked substrate, and particle detachment led to third-body formation which protected the substrate. Moreover, lubricating polymers with solvents of equal or close solubility lead to rapid bulk deterioration under fretting contact.

The characteristics of the top surface layer of PC at the nm order were examined using an atomic force microscope (AFM) (**79**). This examination has clarified the mechanism by which the surface layer of PC rises when the whole surface is scratched (scan-scratching), whilst the surface is fed in a perpendicular direction to the scratch. It has been proposed that a groove is formed with upheavals (ridges) formed in the forward direction of the tip and at both sides of the groove. When the feeding pitch of scan-scratching is reduced, the adjoining upheavals of scratch marks interfere with each other. At the same time as the gradual reduction of the feeding pitch, the first apparent feature is that, upheavals are layered over each other and the scratched surface looks as if it is expanded evenly. Secondly, great upheavals occur on the scratched surface as if the same line had been scratched repeatedly. This is the mechanism of upheavals on the scratched surface caused by scan-scratching.

4 TRIBOLOGY OF POLYMER COMPOSITES

Ten years ago, Briscoe and Tweedale (**80**) provided a general review of composite tribology which was oriented towards polymeric matrices. Particularly interesting was an overview given of the tribological behaviour of highly resistant polymers, in particular PEEK, against smooth steel counterparts (**81**), in which the effects of internal lubricants, especially PTFE, and short fibre reinforcements (glass vs carbon) were outlined. A full account will not be repeated here, but suitable references will be given.

4.1 PEEK composites

Friedrich and his group have studied the tribological behaviour of PEEK composites extensively (**51, 81–87**). Different types and amounts of fibre reinforcement and/or lubricants were investigated as a function of steel counterface roughness and testing temperature under dry sliding conditions. It is found that the effect of steel counterface roughness on the wear rate for short fibre-reinforced PEEK is not as pronounced as for the virgin matrix. An increase in testing temperature results in high specific wear rates and lower coefficients of friction. Moreover, the wear rates are highly affected by the type and amount of lubricants incorporated. It has been found that the incorporation of carbon fibres proves more beneficial than glass fibres with respect to both friction and wear performance. Ovaert and Cheny also investigated the counterface roughness effect on the wear of a carbon fibre reinforced PEEK composite (**88**).

Recently, PEEK composites have been processed with plasma to improve their tribological behaviour (**89**). Argon was used as the treatment gas. The treatment time and voltage of the plasma were in the range of 1.0–5.0 min and 0.8–1.3 kV respectively. It was shown that the friction coefficient and the specific wear rate of the composite against a steel ring were decreased significantly under sliding conditions. It is concluded that the improvement of tribological behaviour of PEEK composites with plasma-treated surface comes from the improvement of the interface strength of carbon fibre-reinforced PEEK composites. Therefore, this surface treatment technique for polymers and their composites has great potentialities.

Experimental study of PEEK composites filled with CuS and PTFE powders while rubbing against tool steel was undertaken under ambient sliding conditions (**90–92**). The addition of PTFE to these composites was found to reduce both the wear rate and friction coeffficient, while the coefficient of friction increased with the addition of CuS. Obviously, the wear rate of the composites depends on their ability to form transfer films on the steel counterfaces. It has been observed that the growth of transfer film is a mechanical process in which the fragments of the removal material are locked into the crevices of counterface asperities. Moreover, it has been found that the compounds of FeF_2, $FeSO_4$ and FeS which are formed near the interface of the transfer film with its steel substrate, enhance the bonding between the transfer film and the counterface. The cracking and delamination of the transfer film will be affected by the bonding of the transfer film to the counterface.

Tribological properities of nanometre Si_3N_4-filled PEEK were investigated through sliding the PEEK composite block against a carbon steel ring (**93, 94**). It was found that the nanometre Si_3N_4-filled PEEK composite exhibited much lower frictional coefficient and wear rate than pure PEEK. Moreover, a thin, uniform, and tenacious transfer film was observed to be formed on the ring surface, which improved the tribological behaviour of this filied PEEK composite.

4.2 PTFE composites

The adhesion behaviour of Zn-PTFE composite on carbon steel was studied in an ultrahigh vacuum system (**95**). It was found that the adhesion coefficient increased with the load and temperature. The transfer film was accumulated into multilayers on the counterface.

The correlation between the wear behaviour and microstructure of graphite-PTFE composites was studied over a wide graphite volume content range from 0 to 50% (**96**). The properties of microscopic imperfections were analysed by positron annihilation lifetime spectroscopy. It has been found that the wear behaviour of graphite-PTFE composites is in correspondence with the physical properties of imperfections in a macro- and microscopic sense and is reflected fairly well by the interfacial properties between crystalline and amorphous regions in PTFE.

The wear debris produced during severe friction processes were examined (**97**). It was observed that the wear debris were in the form of complete, unbroken, wave-like ribbons for good ductility composites, and in the form of flake, for poor ductility composites. Based on the observations of wear debris, it was concluded that the mechanism of the formation of wear debris mainly resulted from friction extrusion of the polymer-based composites in the contacting region.

A material which can reduce the adhesion and friction of soil-engaging components of machines for land locomotion, Al_2O_3+PTFE+PPS (polyphenylene sulfide) composite has been developed. The water wettability, soil adhesion and abrasion of this composite were investigated (**98**). The advancing contact angles of distilled water on the composites were found to increase as a result of the dispersion of Al_2O_3 particles with a certain content and size in the PTFE and PTFE+PPS matrix, and the behaviour of anti-adhesion of soil to the surface of the composites remains much better.

The friction and wear mechanisms of PTFE composites, sliding against stainless steel in an aqueous environment, were examined (**99**). It was found that the wear of PTFE composites filled with only glass fibres was much greater than of that other composites in water. This was probably due to the increased mating surface roughness resulting from the adhered glass powders. Microscopic observation testified that the glass fibres were abraded easily, and finely broken glass fibres adhered to the mating metal surface in water.

A type of PTFE-based multilayer, self-lubricating composite, made with a steel backing, a sintered, porous, bronze middle layer, and a surface layer consisting of a mixture of PTFE and Pb powders, has been developed. The tribological behaviours of this composite sliding against stainless steel as studied under oil lubrication conditions (**100, 101**). It was discovered that the friction and wear properties can be greatly improved with lubrication by gasoline engine lubricant oil, glycerol or triethanolamine.

4.3 Other composites

In order to answer some questions regarding the effectiveness of particulate fillers, solid lubricants, metal powders and oxides, in modifying the tribological behaviour, the functions of these three kinds of fillers have been discussed, on the basis of a overall survey of the data published in the literature, as well as a general view of the mechanisms and hypotheses presented by a number of researchers for wear reductions (**102**). This discussion is limited to sliding situations with the filled polymer composite sliding against a smooth counterface under dry, ambient conditions. Unfortunately, a conclusive statement concerning the tribological behaviour of filled polymers has not been made.

Bahadur and Gong (**103**) seem to claim that there is a critical packing state in polymer-based composites incorporating inorganic powder fillers, in which the maximum mechanical strength, and the highest abrasion resistance of the composite, are obtained. Based on this concept of critical packing, a theoretical model was presented to calculate the optimal filler proportion.

The effect of reinforcement and synergism between

CuS and carbon fibre on the wear of nylon-11 was studied under sliding and ambient conditions (**104**). In that study wear was found to be reduced considerably when nylon was reinforced with carbon fibre and wear reduction was even more significant when carbon fibre was used for reinforcement along with CuS as the filler. These differences in wear behaviour have been investigated in terms of the synergism between carbon fibre and CuS.

In a study on the tribological behaviour of MoS_2-filled nylon-6 (**105**), it was found that MoS_2 was not very effective for friction reduction but could cause an increase in wear.

In order to develop practically applicable novel plastic resin materials with a high strength and high heat resistance, effects of the addition of solid lubricants (PTFE, MoS_2 and graphite), on the tribological properties of the new thermosetting resins, condensed polycyclic aromatic (COPNA) resin (hereafter referred to as SK resin), were examined (**106**). It was shown that graphite is the most effective filler for wear reduction among these three solid lubricants. However, the frictional coefficient is 0.35–0.8 times that of SK resin itself.

Tribological properties of plain weave fabric-reinforced poly(vinyl butyral), modified phenolic resin matrix composites. with three different reinforcing fabric materials as components, (E-glass, high-strength carbon and Kevlar-49) were investigated, against a cast iron disc under dry sliding conditions (**107**). Of the three reinforcing fabric materials, the composite made with glass fabric reinforcement showed poor wear resistance. The best wear resistance was discovered in the Kevlar-49 fabric reinforced composite. In an experimental study of tribological behaviour of a unidirectional oriented E-glass fibre-reinforced epoxy composite against a cylindrical counterface under sliding conditions, friction and wear were found to be improved when sliding took place against either a clean or a wet counterface (**108**).

Wear behaviour of polyamide (PA) 66 and PC containing glass fibres, UHMWPE and PTFE/2% silicone oil has been studied against a stainless steel counterface under reciprocating dry sliding conditions (**109**). It was shown that the incorporation of UHMWPE and PTFE fillers and glass reinforcement into PA 66 and PC results in greatly improved sliding wear behaviour. The PA 66/PTFE/2% silicone oil composite presented the best overall wear behaviour, whilst the high glass-filled variants of PA66 and PC were good for both wear resistance and mechanical strength.

The tribological performance of other polymer composites under dry sliding conditions was also studied, such as polyoxymethylene (POM) composite (**110**), polyethernitrile (PEN) composite (**111**, **112**), and polyethersulphone (PES) (**113**), as well as the polymer-based model composites reinforced with ceramic particles (**114**).

In addition, the solid particle erosion behaviour of different types of polymer composites reinforced with glass fibres has been examined at two impact angles (90° and 30°) (**115**).

5 CONCLUDING REMARKS

With so many advantages, polymers are now utilized for numerous purposes, such as seals (**116–118**), gears (**119–120**), bearings (**121–133**), brakes and clutches (**134–135**), transmission belts (**136**), rollers (**137**), tank track pads (**138–141**), artificial joints (**142–144**), grinding mills (**145**), engines (**146–147**), space instruments (**148**), office automation machinery (**149**), and audio-visual machinery (**150**). A summary of the tribological applications of plastics and rubbers is given (**151**).

Obviously, along with the extensive application of polymers for tribological purposes, the understanding of polymer tribology is becoming increasingly important. The reverse is also true in that an increased understanding of polymer tribology will further promote the tribological application of polymers.

This paper has traced the progress of achievements in understanding of polymer tribology over the past decade. Although the intention here has been to provide as inclusive an overview as possible of this important field, some issues are clearly neglected, such as polymeric films and coatings, polymer alloys and polymeric additives, owing to the limited space.

6 ACKNOWLEDGEMENTS

The author is grateful to Dr Yang Zhaochun and Sun Xudong for their assistance in the preparation of this article.

7 REFERENCES

1 **Extrand, C.W., Gent, A.N.**, and **Kaany, S.Y.**, Friction of a rubber wedge sliding on glass, *Rubber Chem.Technol,* 1990, **64,** 108–117.

2 **Tabaddor, F.**, Finite element analysis of rubber block in friction contact, *Computers and Structures*, 1989, **32** (3/4), 549–562.

3 **Purushothaman, N.** and **Moore, I.D.**, Hysteresis sliding friction on rubber – finite element analysis, *J. Engineering Mechanics,* 1990, **116** (1), 217–232.

4 **Purushothaman, N.**, *Numerical analysis of sliding of rubber over triangular and rectangular grooved asperities – tyre pavement interaction.* Thesis presented to the University of Newcastle, at Newcastle, Australia, in partial fulfilment of the requirements for the degree of Doctor of Philosophy, 1987.

5 **Purushothaman, N.** and **Moore, I.D.**, Finite element analysis of a viscoelastic solid sliding over triangular asperities, *Int. Conf. on Numerical Methods in Engineering. Theory and Application,* No 2, 1987, Swansea, UK.

6 **Moore, I.D., Purushothaman, N.**, and **Heaton, B.S.**, Three dimensional elastic finite element study of the skid resistance of grooved pavements, *Int. J. Numerical Methods in Energy,* 1988, **26,** 437–452.

7 **Ettles, C.M.McC.** and **Shen, J.H.**, The influence of frictional heating on the sliding friction of elastomers and polymers, *Rubber Chem. Technol.*, 1987, **61,** 119–136.

8 **Kawakami, S., Misawa, M.**, and **Hirakawa, H.**, The relation between temperature dependence of vircoelasticity and friction due to hysteresis, *Rubber World,* 1988, **198** (5), 30–36.

9 **Lazeration, J.J.**, Determination of the coefficient of friction of rubber at realistic tyre contact pressures, *Rubber Chem. Technol.*, 1987, **60,** 966–974.

10 **Roberts, A.D.**, A guide to estimating the friction of rubber, *Rubber Chem. Technol.,* 1989, **65,** 673–686.

11 **Barquins, M.** and **Roberts, A.D.**, Adhesion and friction of expoxidized natural rubber vulcanizates, *J. Chem. Phys.*, 1987, **84,** 225–230.

12 **Barquins, M.**, Adherence and rolling kinetics of a rigid cylinder in contact with a natural rubber surface, *J. Adhes.*, 1988, **26,** 1–12.

13 **Barquins, M.** and **Felder, E.**, Adherence and friction of rubber-like materials: some new observations, in Holmberg K. and Nieminen, I. (eds.), *Proc. 5th Int. Congr. on Tribology,* Espoo, 1989, Finish Soc. Tribology, Espoo, 1989, Vol. 3, pp. 295–301.

14 **Barquins, M.**, Adherence, friction and contact geometry of a rigid cylinder rolling on the flat and smooth surface of an elastic body (NR and SBR), *J. Nat. Rubber Res.*, 1990, **5,** 199–210.

15 **Barquins, M.**, Adherence, friction and wear of a rubber-like materials, *Wear,* 1992, **158,** 87–117.

16 **Zhang, S.W.**, Examination of the effect of run-in on rubber friction, in Holmberg, K. and Nieminen, I. (eds.), *Proc. 5th Int. Congr. on Tribology,* Espoo, 1989, Finish Soc. Tribology, Espoo, Vol. 3, 1989, pp. 341–346.

17 **Sasada, T., Hiratsuka, K.**, and **Tomita, H.**, Mindlin slip on a rubber eraser, *Japanese J. Tribol.*, 1995, **40** (1), 35–40.

18 **Zhang, S.W.**, Advances in studies on rubber abrasion, *Tribol. Intern.*, 1989, **22** (2), 143–148.

19 **Gent, A.N.**, A hypothetical mechanism for rubber abrasion, 1988, **62,** 750–756.

20 **Southern, E.** and **Thomas, A.G.**, Studies of rubber abrasion, *Rubber Chem. Technol.*, 1979, **52,** 1008–1018.

21 **Uchiyama, Y.** and **Ishino, Y.**, Pattern abrasion mechanism of rubber, *Wear,* 1992, **158,** 141–155.

22 **Fukahori, Y** and **Yamazuki, H.**, Mechanism of rubber abrasion. Part I: Abrasion pattern formation in natural rubber vulcanizate, *Wear,* 1994, **171,** 195–202.

23 **Fukahori, Y** and **Yamazuki, H.**, Mechanism of rubber abrasion. Part II: General rule in abrasion pattern formation in rubber-like materials, *Wear,* 1994, **178,** 109–116.

24 **Fukahori, Y** and **Yamazuki, H.**, Mechanism of rubber abrasion. Part III: How is friction linked to fracture in rubber abrasion?, *Wear,* 1995, **188,** 19–26.

25 **Zhang, S.W.**, Investigation of abrasion of nitrile rubber, *Wear,* 1984, **57,** 769–778.

26 **Stupak, P.R.** and **Donovan, J.A.**, Fractal analysis of rubber wear surfaces and debris, *J. Mater. Sci.*, 1988, **23,** 2230–2242.

27 **Stupak, P.R., Kang, J.H.**, and **Donovan, J.A.**, Fractal characteristics of rubber wear surfaces as a function of load and velocity, *Wear,* 1990, **141,** 73–84.

28 **Stupak, P.R., Kang, J.H.**, and **Donovan, J.A.**, Computer-Aided fractal analysis of rubber wear surfaces, *Materials characterization,* 1991, **27,** 231–240.

29 **Zhang, S.W.**, Mechanisms of rubber abrasion in unsteady state, *Rubber Chem. Technol.*, 1984, **57,** 755–768.

30 **Grosch, K.A.**, Abrasion of rubber and its relation to tyre wear, *Rubber Chem. Technol.*, 1991, **65,** 78–106.

31 **Zhang, S.W.** and **Zhaochun Yang**, Energy theory of rubber abrasion by a line contact, in *Proc 7th Conference on Friction, Lubrication and Wear,* Bucharest, Sept. 10–12, 1996, Vol. **1**, pp. 347–352.

32 **Zhang, S.W.**, Theory of rubber abrasion by a line contact, in 'Polymer Wear and its Control', Lee, L.H. (ed.), ACS, Washington, DC, 1985, pp. 189–196.

33 **Thavamani, P., Khastgir, D.**, and **Bhowmick, A.K.**, Microscopic studies on the mechanisms of wear of NR, SBR and HNBR vulcanizates under different conditions, *J. Mater. Sic.* 1993, **28,** 6318–6322.

34 **Muhr, A.H., Pond, T.J.**, and **Thomas, A.G.**, Abrasion of rubber and the effect of lubricants, *J. Chem. Phys.*, 1987, **84,** 331.

35 **Muhr, A.H.**, Lubrication of model asperities on rubber, in Dowson, D., Godet, M. and Taylor, C.M. (eds.), Vehicle Tribology, *Proc. 17th Leeds-Lyon Symp. on Tribology,* Elsevier, Amsterdam, 1991, pp. 195–204.

36 **Muhr, A.H.** and **Richards, S.C.**, Abrasion of rubber by model asperities, *Kautschuk and Gummi Kunststoffe,* 1992, **45,** 376–379.

37 **Muhr, A.H.** and **Roberts, A.D.**, Rubber abrasion and wear, *Wear,* 1992, **158,** 213–228.

38 **Zhang, S.W.** and **Xibin Chang,** Wet abrasion of nitrile rubber, *Proc. Japan Int. Tribology Conf.,* Nagoya, Oct. 29–Nov. 1, 1990, JST, Osaki Printing Co. Ltd., 1990, pp. 249–254.

39 **Zhang, S.W.**, Wet abrasion of polymers, *Wear,* 1992, **158,** 1–13.

40 **Scieszka, S.F.**, Sliding erosion mechanism and measurement, *STLE Tribology Trans.*, 1992, **35,** 59–64.

41 **Zhang, S.W., Wang Dego,** and **Yin Weihua,** Investigation of abrasive erosion of polymers, *J. Mater. Sci.*, 1995, **30,** 4561–4566.

42 **Arnold, J.C.** and **Hutchings, I.M.**, The mechanisms of erosion of unfilled elastomers by solid particle impact, *Wear,* 1990, **138,** 33–46.

43 **Arnold, J.C.** and **Hutchings, I.M.**, A model for the erosive wear of rubber at oblique impact angles, *J. Phys. D: Appl. Phys.* 1992, **25,** A222–A229.

44 **Zhang, S.W.** and **Yang, Z.**, Theoretical wear model of abrasive erosion in annular pipes, *Proc. Int. Tribology Conf. Yokohama,* 1995, pp. 217–221.

45 **Yang, Z.** and **Zhang, S.W.**, Theoretical study in physical process of abrasive erosion on annular pipe wall, *J. Tribol.*, 1995, **15,** 306–309, (in Chinese).

46 **He Renyang, Zhang, S.W., Wang Deguo,** and **Fan Qiyun,** Surfacial chemical effects of polyurethane abrasive erosion, *J. Tribol.*, 1995, **15,** 45–51 (in Chinese).

47 **Zhang, S. W.**, Studies of rubber wear, Proc. IMechE., Part J., *J. Engineering Tribology* (to be published).

48 **He Renyang, Zhang, S.W., Fan Qiyun,** and **Wang Deguo,** Surfacial chemical effects of abrasive erosion of natural rubber, *J. Tribol.*, (to be published) (in Chinese).

49 **Richards, S.C.** and **Roberts, A.D.**, Boundary lubrication of rubber by aqueous surfactant, *J. Phys. D: Appl. Phys.*, 1992, **25,** A76–A80.

50 **Caravia, L., Dowson, D., Fisher, J., Corkhill, P.H.,** and **Tighe, B.J.,** Friction of hydrogel and polyurethane elastic layers when sliding against each other under a mixed lubrication regime, *Wear,* 1995, **181–183,** 236–240.

51 **Lu, Z.P.** and **Friedrich, K.,** On sliding friction and wear of PEEK and its composites, *Wear,* 1995, **181–183,** 624–631.

52 **Senatrev, A.N., Smurugov, V.A., Biran, V.V.,** and **Savkin, V.G.,** On the initial stages of PTFE friction and transfer, *Soviet J. Friction and Wear,* 1993, **14,** 49–52.

53 **Krasnov, A.P.,** *et al.*, The surface layer structure of ultra high molecular weight polyethylene (UHMWPE) at friction against steel counterface, *Soviet J. Friction and Wear,* 1993, **14,** 53–57.

54 **Fisher, J., Dowson, D., Hamdzah, H.,** and **Lee, H.L.,** The effect of sliding velocity on the friction and wear of UHMWPE for use in total artificial joints, *Wear,* 1994, **175,** 219–225.

55 **Sekiguchi, I., Yamaguchi, Y., Tamura, K.,** and **Deguchi, A.,** Effect of surface roughness and mating plastics on the tribological properties of polyacetal, *Japanese J. Tribol.,* 1993, **38,** 1097–1108.

56 **Tamura, H.** and **Hiro, S.,** Effect of surface roughness of friction and wear of polyimide slid against mild steel in vacuum: analysis based on adhesive/abrasive wear model, *Japanese J. Tribol.,* 1994, **39,** 113–121.

57 **Kryzhanovskii, V.K.** and **Konova, O.V.,** Effect of structure and physical state of epoxypolymers on their tribological behaviour, *Soviet J. Friction and wear,* 1993, **14,** 74–78.

58 **Lavielle, L.,** Friction of polyethylene-terpolymer mixtures on polymer substrates: adhesion dependence, *Wear,* 1992, **157,** 181–187.

59 **Lavielle, L.,** Polymer-polymer friction: relation with adhesion, *Wear,* 1991, **151,** 63–75.

60 **Barrett, T.S., Stachowiak G.W.,** and **Batchelor, A.W.,** Effect of roughness and sliding speed on the wear and friction of ultra-high molecular weight polyethylene, *Wear,* 1992, **153,** 331–350.

61 **Marcus, K.** and **Allen, C.,** The sliding wear of ultrahigh molecular weight polyethylene in an aquaeous enviroment, *Wear,* 1994, **178,** 17–28.

62 **Clarke, C.G.** and **Allen, C.,** The water lubricated, sliding wear behaviour of polymeric materials against steel, *Tribol. Intern.,* 1991, **24,** 109–118.

63 **Lloyd, A.I.G.** and **Noel, R.E.J.,** The effect of counterface surface roughness on the wear of UHMWPE in water and oil-in-water emulsion, *Tribol. Intern.,* 1988, **21,** 83–88.

64 **McNicol, A., Dowson, D.,** and **Davies, M.,** The effect of humidity and electrical fields upon the wear of high density polyethelene and polytetrafluoroethylene, *Wear,* 1995, **181–183,** 603–612.

65 **Smurugov, V.A., Senatrev, A.I., Savkin, V.G., Biran, V.V.** and **Sviridyonok, A.I.,** On PTFE transfer and thermoactivation mechanism of wear, *Wear,* 1992, **158,** 61–69.

66 **Blanchet, T.A.** and **Kennedy F.E.,** Sliding wear mechanism of polytetrafluorothylene (PTFE) and PTFE composites, *Wear,* 1992, **153,** 229–243.

67 **Stolarski, T.A.,** Tribology of polyetheretherketone, *Wear,* 1992, **158,** 71–78.

68 **Hasseini, S.M.** and **Stolarski, T.A.,** Contact configuration effects on the dry and lubricated wear of polymers, *Suf. Eng.,* 1988, **4,** 322–326.

69 **Zhang, M.Q., Ping, Z.,** and **Friedrich,** Thermal analysis of the wear debris of polyetheretherketone, Tribol. Intern., 1997, **30,** 103–111.

70 **Zhang, M.Q.** and **Friedrich, K.,** *On the wear debris of polyetheretherketone,* IVW-Research Report 95–23, Institute for Composite Materials, University of Kaiserslautern, Germany, 1995.

71 **Zhang, M.Q., Ping, Z.,** and **Friedrich,K.,** On the wear debris of polyetheretherketone: fractal dimensions in relation to wear mechanisms, *Tribol. Intern,* 1997, **30,** 87–102.

72 **Tricot, C., Ferland, P.,** and **Baran, G.,** Fractal analysis of worn surfaces, *Wear,* 1994, 172, 127–133.

73 **Viswanath, N.** and **Bellow, D.G.,** Development of an equation for the wear of polymers, *Wear,* 1995, **181–183,** 42–49.

74 **Bohm, H., Betz, S.,** and **Ball, A.,** The wear resistance of polymers, *Tribol. Intern.,* 1990, **23,** 399–406.

75 **Pleskachersky, Y.M., Zaitsev, A.L.,** and **Smirnov, V.V.,** Oxidation and its influence on low pressure polyethylene wear, *Wear,* 1995, **181–183,** 222–226.

76 **Lawrence, K.C.** and **Stolarski, T.A.,** Rolling contact wear of polymers: a preliminary study, *Wear,* 1989, **132,** 183–191.

77 **Krichen, A., Kharrat, M.,** and **Chateauminois, A.,** Experimental and numerical investigation of the sliding behaviour in a fretting contact between poly (methylmethacrylate) and a glass counterface, *Tribol. Intern.,* 1996, **29,** 615–624.

78 **Dahmani, N.,** *et al.*, Velocity accommodation in polymer fretting, *Wear,* 1992, **158,** 15–28.

79 **Kaneko, R., Hamada, E.,** and **Andon, Y.,** Abrasive wear process of polycarbonate (part 1): projection forming by scratching, *Japanese J. Tribol.,* 1993, 38, 63–73.

80 **Briscoe, B.J.** and **Tweedale, P.J.,** A critical review of the tribology of polymers composites, *Proc. 6th Int. Conf. Composite Materials,* Malthews, *et al* (eds.) Elsevier Applied Science, London, 1987.

81 **Friedrich, K., Lu, Z.,** and **Hager, A. M.,** Recent advances in polymer composites, tribology, *Wear,* 1996, **190,** 139–144.

82 **Friedrich, K., Karger-Kocsis, J.,** and **Lu, Z.,** Effects of steel counterface roughness and temperature on the friction and wear of PE(E)K composites under dry sliding conditions, *Wear,* 1991, **148,** 235–247.

83 **Cirino, M., Friedrich, K.,** and **Pipes, R.B.,** The abrasive wear behaviour of continuous fibre polymer composites, *J. Materials. Sci.,* 1987, **22,** 2481–2492.

84 **Voss, H.** and **Friedrich, K.,** On the wear behaviour of short-fibre-reinforced PEEK composites, *Wear,* 1987, **116,** 1–18.

85 **Brockmiiller, K., Friedrich, K.,** and **Maisner, M.,** Effect of counter-component roughness on the wear of PEEK and short-fibre reinforced PEEK composites, *Kunststoffe/German Plastics,* 1990, **80,** 701.

86 **Cirino, M., Friedrich, K.,** and **Pipes, R.B.,** Evaluation of polymer composites for sliding and abrasive wear applications, *Composites,* 1988, **19,** 383.

87 **Mody, P. B., Chou, T-W.** and **Friedrich,** Effect of testing conditions and microstructure on the sliding wear of graphite fibre/PEEK matrix composites, *J. Mater. Sci.,* 1988, **23,** 4319.

88 **Ovaert, T.C.** and **Cheng H.S.,** Counterface topographical effect on the wear of polyetheretherketone and a polyetheretherketone-carbon fibre composite, *Wear,* 1991, **150,** 275–287.

89 **Zhang, R., Hager, A.M., Friedrich, K., Song, Q.,** and **Dong, Q.,** Study on tribological behaviour of plasma-treated PEEK and its composites, *Wear,* 1995, **181–183,** 613–623.

90 **Voort, J.V.** and **Bahadur, S.,** The growth and bonding of transfer film and the role of CuS and PTFE in the tribological behaviour of PEEK, *Wear,* 1995, **181–183,** 212–221.

91 **Bahadur, S.** and **Gong, D.,** The role of copper compounds as fillers in the transfer and wear behaviour of polyetheretherketone, *Wear,* 1992, **154,** 151–165.

92 **Bahadur, S., Gong, D.,** and **Anderegy, J.W.,** The investigation of the action of fillers by XPS studies of the transfer films of PEEK and its composites containing CuS and CuF_2, *Wear,* 1993, **160,** 131–138.

93 **Wang, Q., Xu, J., Shen, W.,** and **Liu, W.,** An investigation of the friction and wear properties of nanometer Si_3N_4 filled PEEK, *Wear,* 1996, **196,** 82–86.

94 **Wang, Q., Xue, Q.,** and **Shen, W.,** The friction and wear properities of nanometre SiO_2 filled polyetheretherketone, *Tribol Intern.,* 1997, **30,** 193–197.

95 **Gong, D., Xue, Q.,** and **Wang, H.,** Physical models of adhesive wear of polytetrafluoroethylene, *Wear,* 1991, **147,** 9–24.

96 **Yan, F., Wang, W., Xue, Q.,** and **Wei L.,** The correlation of wear behaviours and microstructures of graphite-PTFE composites studied by positron annihilation, *J. Appl. Polym. Sci.,* 1996, **61,** 1231–1236.

97 **Yan, F., Xue, Q.,** and **Yang, S.,** Debris formation process of PTFE and its Composites, *J. Appl. Polym. Sci.,* 1996, **61,** 1223–1229.

98 **Lu, X.C., Wen, S.Z., Tong, J., Chen, Y.T.** and **Ren, L.Q.,** Wettability, soil adhesion, abrasion and friction wear of PTFE (+PPS) + $A1_20_3$ composites, *Wear,* 1996, **193,** 48–55.

99 **Watanabe, M.,** Wear mechanism of PTFE composites in aqueous environments, *Wear,* 1992, **158,** 79–86.

100 **Zhang, Z.Z., Shen, W.C., Liu, W.M, Li, TS.,** and **Zhao J.Z.,** The tribological characteristics of JS material under lubrication of oil, *Wear,* 1996, **193,** 163–168.

101 **Zhang, Z.Z., Shen, W.C., Liu, W.M., Xue, Q.J.,** and **Li, T.S.,** Tribological properties of polytetrafluorothylene-based composite in different lubricant media, *Wear,* 1996, **196,** 164–170.

102 **Bahadur,S.** and **Gong, D.,** The action of fillers in the modification of the tribological behaviour of polymers, *Wear,* 1992, **158,** 41–59.

103 **Bahadur, S.** and **Gong, D.,** Formulation of the model for optimal proportion of filler in polymer for abrasive wear resistance, *Wear,* 1992, **157,** 229–243.

104 **Bahadur, S.** and **Gong, D.,** The effect of reinforcement and the synergism between CuS and carbon fibre on the wear of nylon, *Wear,* 1994, **178,** 123–130.

105 **Liu, W., Huang, C., Gao, L., Wang, J.,** and **Dang, H.,** Study of the friction and wear properties of MoS_2-filled nylon-6, *Wear,* 1991, **151,** 111–118.

106 **Sekiguchi, I., Kubota, K., Oyanagi, Y., Kasaka, M.,** and **Sone, Y.,** Tribological properties of COPNA resin, *Wear,* 1992, 158, 171–183.

107 **Vishwanath, B., Verma, A.P.,** and **Kameswara Rao, C.V.S.,** Effect of reinforcement on friction and wear of fabric reinforced polymer composites, *Wear,* 1993, **167,** 93–99.

108 **El-Tayeb, N.S.** and **Gadebrab, B.M.,** Friction and wear properties of E-glass fibre reinforced epoxy composites under different sliding contact conditions, *Wear,* 1996, **192,** 112–117.

109 **Byett, J.H.** and **Allen, C.,** Dry sliding wear behaviour of polyamide 66 and polycarbonate composites, *Tribol. Intern.,* 1992, **25,** 237–246.

110 **Odi-Owei, S.** and **Schipper, D.J.,** Tribological behaviour of unfilled and composite poloxymethylene, *Wear,* 1991, **148,** 363–376.

111 **Itoi, M.** and **Pipes, R.B.,** PAN and pitch-based carbon fibre-reinforced polyethernitrile composites, *J. Thermoplast. Compos. Master.,* 1990, **3,** 172.

112 **Fridrich, K.** and **Karger-Kocsis, J.,** On the sliding wear performance of polyethernitrile composites, *Wear,* 1992, **158,** 157–170.

113 **Friedrich, K.** and **Wu, J.S.,** Polymer composites with high wear resistance, in Lee, S.M. (eds.), *International Encyclopaedia of Composite Materials,* Vol.4. VCH, New York, 1990, p. 255.

114 **Durand, J.M., Vardavoulias, M.** and **Jeandin, M.,** Role of reinforcing ceramic particles in the wear behaviour of polymer-based model composites, *Wear,* 1995, 181–183, 826–832.

115 **Roy, M., Vishwanathan, B.,** and **Sundararajan, G.,** The solid particle erosion of polymer matrix composites, *Wear,* 1994, **171,** 149–161.

116 **Kanzaki, Y.,** Application of polymers to seals, *Japanese J. Tribol.,* 1992, **37,** 735–742.

117 **Karaszkiewicz, A.,** Hydrodynamics of rubber seals for reciprocating motion, *Power Intern.,* 1987, **33,** 169–172.

118 **Rivkin, M.** and **Kholodenko, A.,** Mechanical seal with elastomeric rotating element, Part 2: Experimental study, *Rubber Chem. Technol.,* 1994, **67,** 62–75.

119 **Tsukamoto, N.,** Application of polymers to gears, *Japanese Tribol.,* 1992, **37,** 743–750.

120 **Tsukamoto, N., Maruyama, H.,** and **Mimura, H.,** Water lubrication characteristics of polyacetal gears filled with carbon fibres, *JSME Int. J., Series C: Dynamics, Control, Robotics, Design and Manufacturing,* 1993, **36,** 499–506.

121 **Horiuchi, K.,** Application of polymers to sliding bearings and heat-resistant gears, *Japanese J. Tribol.,* 1992, **37,** 751–758.

122 **Guseva, M.I., Lysenkov, P.M., Sokov, E.V.** and **Vladimirov, B.G.,** Surface modification of rubber inserts in stern bearings by ion implantation technique, *Soviet J. Friction and Wear,* 1993, **14,** 742–747.

123 **Vorontsov, P.A., Muratov, Kh.I.,** and **Petrenko, V.A.,** Testing of metal-polyfluoroethylene journal bearings in turbocompressors for gas processing, *Chemical and Petroleum Engineering,* 1991, **26,** 591–595.

124 **Tochilnikov, D.G., Krasnyj, V.A., Priemskij, L.D., Ginzburg, B.M.** and **Bulatov, V.P.,** Application of polymeric coatings to increase of wear resistance bearing linear operating surface at insufficient lubrication, *Soviet J. Friction and Wear,* 1992, **13,** 689–694.

125 **Rumuza, Z., Kusznierewicz, Z.,** and **Manturzyk, G.,** Testing miniature in particular polymer-polymer journal bearings, *Wear,* 1994, **174,** 39–46.

126 **Akhramenko, N.A.** and **Melnichenko, I.M.,** On working capacity of polymeric films in fixed joints of bearings, *Soviet J. Friction and Wear,* 1993, **14,** 1111–1114.

127 **Potekha, V. L., Nevzorov, V.V.** and **Shcherek, M., A** study on thermal wear characteristics of metalpolymeric precision bearings, *Soviet J. Friction and Wear,* 1994, **15,** 78–83.

128 **D'Agostino, V., Niola, V.,** and **Caporiccio, G.,** Tribological behaviour of sintered iron bearings self-lubricated with PTFE under severe operating conditions, *Tribol. Intern.*, 1988, **21,** 105–108.

129 **Oliver, D. R.,** Load enhancement effects due to polymer thickness in a short model journal bearing, *J. Non-Newtonian Fluid Mechanics,* 1988, **30,** 185–196.

130 **Rymuza, Z.,** Designing miniature plastic bearings, *J. Tribol., Trans. of ASME,* 1990, **112,** 135–140.

131 **Rymuza, Z.,** Predicting wear in miniature steel-polymer journal bearings, *Wear,* 1990, **137,** 211–249.

132 **Marx, S.** and **Janghans, R.,** Friction and wear of highly stressed thermoplastic bearings under dry sliding conditions, *Wear,* 1996, **193,** 253–260.

133 **El-Sayed, A.A., El-Sherbiny, M.G., Abo-El-Ezz, A.S.** and **Aggag, G.A.,** Friction and wear properties of polymeric composite materials for bearing applications, *Wear,* 1995, **184/1,** 45–53.

134 **Lnoue, M.,** Application of polymers to brakes and clutches, *Japanese J. Tribol.,* 1992, **37,** 759–766.

135 **Gopal, P., Dharani, L.R.,** and **Blum, F.D.,** Load, speed and temperature sensitivities of a carbon-fibre-reinforced phenolic friction material, *Wear,* 1995, 181–183, 913–921.

136 **Hoshiro, T.,** Application of polymers to transmission belts, *Japanese J. Tribol.,* 1992, **37,** 767–774.

137 **Kon, S.,** Application of polymers to rollers, *Japanese J. Tribol.,* 1992, **37,** 775–782.

138 **Thavamani, P.** and **Bhowmick, A.K.,** Wear of tank track pad rubber vulcanizates by various rocks, *Rubber, Chem. Technol.,* 1991, **65,** 31–45.

139 **Medalia, A.I., Alesi, A.L.** and **Mead, J.L.,** Pattern abrasion and other mechanisms of wear of tank track pads, *Rubber Chem. Technol.,* 1991, **65,** 154–175.

140 **Thavamani, P., Khastgir, D.K.,** and **Bhowmick, A.K.,** Development and field performance of tank track pad compounds, *Plastics, Rubbers and Composites Processing and Applications,* 1993, **19,** 245–254.

141 **Dwyer, M.J., Okello, J.A.,** and **Scarlett, A.J.,** Theoretical and experimental investigation of rubber tracks for agriculture, *J. Terramechanics,* 1993, **30,** 285–298.

142 **Tateishi, T.,** Friction and wear of polymers for artificial joints, *Japanese, J. Tribol.,* 1992, **37,** 783–790.

143 **Wang, A., Sun, D.C., Stark, C.** and **Dumbleton, J. H.,** Wear mechanisms of UHMWPE in total joint replacements, *Wear,* 1995, **181–183,** 241–249.

144 **Barbour, P.S.M., Barton, D.C.,** and **Ficher, J.,** The influence of contact stress on the wear of UHMWPE for total replacement hip prostheses, *Wear,* 1995, **181–183,** 250–257.

145 **Marklund, G.** and **Eriksson, Klas-Goran,** Abrasion: rubber beats metal, *European Rubber J.,* 1989, 171–24–25.

146 **Ecklund, R.,** Polymer and composite use in gas turbine engines, Flight-Vehicle Materials, Structures and Dynamics Assessment and Future Directions, 1994, **2,** 385–394.

147 **Rabe, J.,** Plastic elements in and around the engine, *Int. J. Vehicle Design,* 1990, **11,** 246–271.

148 **Nishimura, M.,** Application of polymers to space instruments, *Japanese J. Tribol.,* 1992, **37,** 791–796.

149 **Suzuki, K.** and **Morita, M.,** Application of polymers to office automation machinery, *Japanese J. Tribol.,* 1992, **37,** 797–804.

150 **Okada, K.,** Application of polymers to audio-visual machinery, *Japanese J. Tribol.,* 1992, **37,** 797–804.

151 **Uchiyama, Y.,** Survey of polymer tribology, *Japanese J. Tribol.,* **37,** 657–665.

Wear by hard particles

K-H ZUM GAHR
University of Karlsruhe, Institute of Materials Science II and Karlsruhe Research Center, Institute of Materials Research I, Karlsruhe, Germany

1 INTRODUCTION

Wear by hard particles occurs in many different situations such as in earth-moving equipment, slurry pumps or pipelines, rock drilling, rock or ore crushers, pneumatic transport of powders, dies in powder metallurgy, extruders or chutes. According to Fig. 1, the wear processes may be classified by different modes depending on the kinematics and by mechanisms depending on the physical and chemical interactions between the elements of the tribosystem which result in detachment of material from the solid surfaces. Compared with unlubricated sliding wear, the value of the wear coefficient k, i.e. the dimensionless quotient of the amount of volumetric wear W_V times the hardness of the wearing material H divided by the normal load F_N and the sliding distance s, estimated from practical experience can be substantially greater in abrasive or erosive wear (**1–5**). Fig. 1 can only represent a very rough estimation of the wear coefficient because of wide variation of the wear mechanisms occurring in an actual tribosystem as a function of the operating conditions and properties of the tribo-elements involved, which can result in changes of the k value by some orders of magnitude.

In abrasive wear, material is displaced or detached from a solid surface by hard particles, or hard particles between or embedded in one or both of the two solid surfaces in relative motion, or by the presence of hard protuberances on a counterface sliding with a velocity v relative to the surface. Two-body abrasion is caused by hard protuberances or embedded hard particles, while in three-body abrasion the hard particles can move freely (roll or slide) between the contacting surfaces. According to (**6–8**), the rate of material removal in three-body abrasion can be one order of magnitude lower than that for two-body abrasion, because the loose abrasive particles abrade the solid surfaces between which they are situated for only about 10% of the time while they spend about 90% of the time rolling. Hard particles striking a solid surface either carried by a gas or a liquid stream can cause erosive wear whereby the wear mechanism depends strongly on the angle of incidence of the impacting particles. The interaction between hard particles and a solid surface can generally be accompanied by adhesion, abrasion, deformation, heating, surface fatigue and fracture.

Fig. 2 shows schematically some general trends of wear loss of materials depending on properties of the abrasive particles and the wearing materials as well as the operating conditions (**9**). With increasing hardness of the abrasive particles, the wear loss can increase by about one to two orders of magnitude from a low to a high level (Fig. 2a). This transition depends on the ratio of the hardness of the abrasive particles to the hardness of the material being worn (Fig. 2b). The increase from the lower to the higher wear level occurs for a single-phase material when the hardness of the abrasive particles is equal to the hardness of the material worn. At equal bulk hardness of a multiphase material, the matrix of the material containing hard reinforcing phases is softer than the matrix of the single-phase material. Hence, the transition from low to high wear level of the multiphase material starts and ends when the hardness of the matrix and the hardness of the reinforcing phase are respectively exceeded by the hardness of the abrasive particles (Fig. 2b). At the high wear level, the wear loss of the multiphase material can be greater than that of the single-phase material if the reinforcing phase can be detached from the matrix and in addition acts abrasively on the matrix.

Important operating parameters are the average size of the abrasive particles and the applied normal load, or during erosive wear the velocity of impinging hard particles. Wear loss of ductile and brittle materials depends on these parameters according to power laws exhibiting different values of the exponent n (Fig. 2c). For ductile materials the velocity exponent can be expected to be between 2 and 3, while on brittle materials 3 to 4 is more likely. A transition from low to high wear level can occur on brittle materials such as alumina with increasing particle size and/or normal load or applied contact pressure, respectively.

In general, spherical particles cause lower wear than angular particles at a given average particle size. However, the effect of particle shape should be smaller in three-body abrasion because the loose particles can reorient themselves during sliding and rolling contact compared with two-body abrasion.

During erosive wear, the dependence of wear loss on impact angle is influenced by the size of the impinging particles, the impact velocity and the target material. A trend has been observed that the wear loss increases, at a given impact angle, with increasing size and velocity of the impinging particles. Ductile metallic materials such as plain carbon steels exhibit maximum wear loss at impact

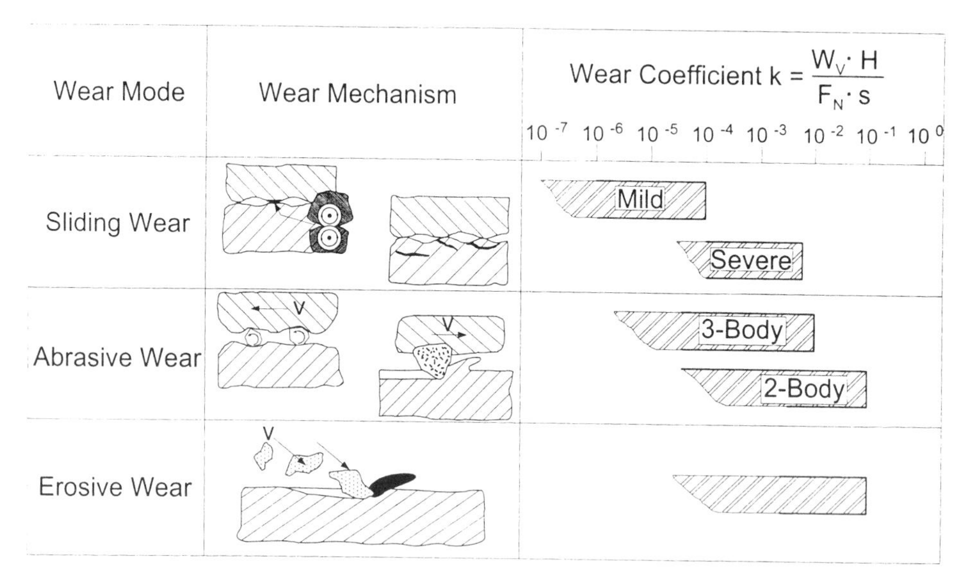

Fig. 1 Values of wear coefficient k as a function of wear mode and wear mechanism without lubricative media.

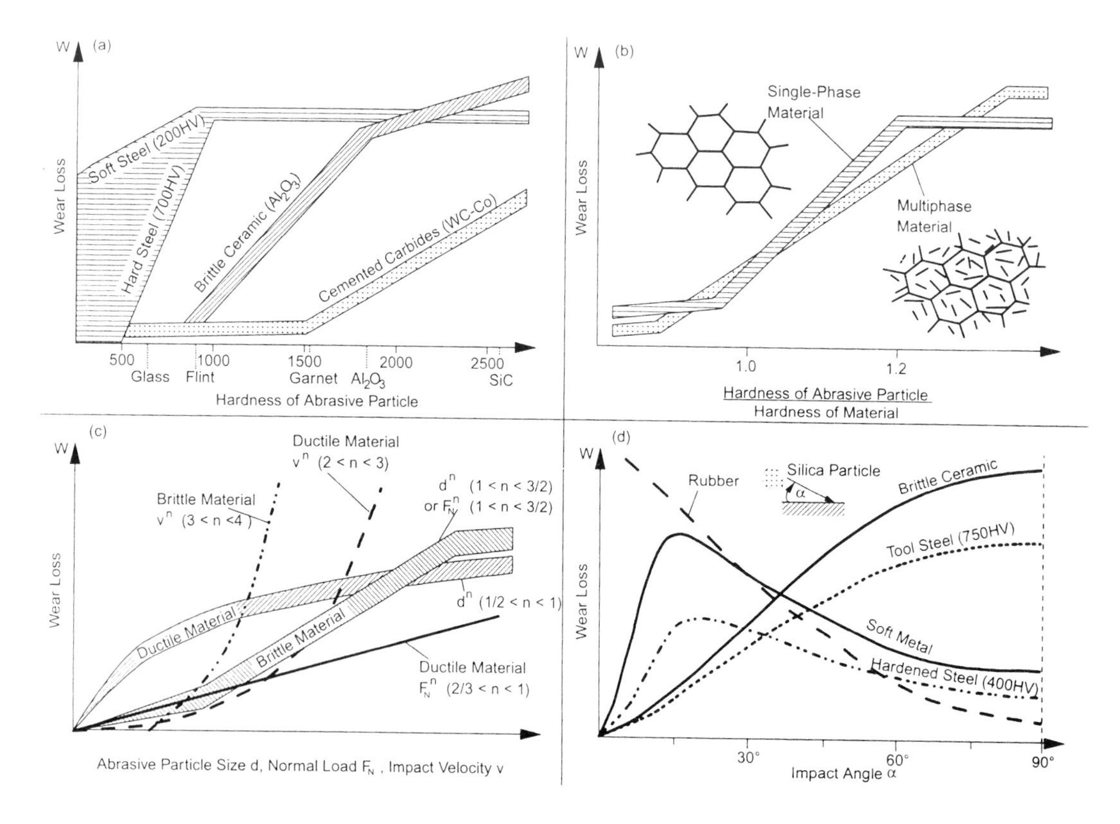

Fig. 2 Schematic representation of wear loss by hard particles as a function of material properties and operating parameters such as (a) hardness of abrasive particle, (b) ratio of hardness of abrasive particle and hardness of the wearing material, (c) abrasive particle size, normal load and impact velocity, and (d) impact angle without considering lubricating media.

angles less than about 30 degrees (Fig. 2d). Brittle materials such as ceramics or hardened tool steels show increasing wear with increasing impact angle and frequently show maximum values for normal incidence. Wear loss decreases continuously with increasing impact angle for materials of very high elasticity such as rubber. It must be considered that Fig. 2 can only represent general trends because in a given tribosystem many factors act simultaneously in a complex manner.

2 MODELLING OF TWO-BODY ABRASIVE WEAR

Models for two-body abrasion have been developed to a substantially greater depth than for three-body abrasion. Penetration of a sliding abrasive particle into a metallic surface results in microploughing or microcutting depending on the attack angle (**10**). Below a critical attack angle, the metallic material is mainly elastic-plastically deformed and flows around and beneath the sliding particle, but no material is removed from the surface. Increasing the attack angle leads to a transition from microploughing to microcutting, i.e. material flows up the front face of the abrasive particle and it is detached from the wearing surface in the form of a chip.

This model of a critical attack angle can be replaced by the criterion of a critical penetration depth of the abrasive particle or a rigid indenter. Using a spherical indenter, the degree of penetration, the ratio of penetration depth divided by the radius of contact, is directly connected with the attack angle which increases with penetration depth. According to (**5, 11**), the degree of penetration D_p as a function of the dimensionless shear strength f, i.e. the ratio between the shear strength at the contact interface and the shear yield stress of the wearing material, allows different abrasive mechanisms to be distinguished (Fig. 3a). Wedge formation means that a wedge prow forms at the front of the indenter and material is removed by propagation of a crack. Fig. 3b shows the increase of the wear coefficient k with increasing degree of penetration for metals under unlubricated sliding.

Hardness of a material can be defined as the resistance to penetration by a hard indenter. Hence, hardness of the wearing material affects the penetration depth of abrasive particles, but Fig. 4 shows that it fails to predict abrasive wear resistance of different materials fully. For example, cold working of metals can increase hardness substantially but not abrasive wear resistance, and it may even reduce it. This general relation between abrasive wear resistance and hardness was measured in many studies (**12, 13**). The increment in abrasion resistance with increasing hardness of materials is substantially larger in pure metals than in heat-treated steels or ceramics. Hardness fails to predict wear resistance because it cannot characterize the interactions between abrasive particles and the wearing materials sufficiently which, however, determine the formation of wear debris.

According to an early model (**3**), which predicts wear volume W_V only depending on the attack angle α, the normal load F_N, the hardness H of the material, the geometry of the indenter (in this case a conical indenter) and the sliding distance s, the wear volume:

$$\frac{W_V}{s} = \frac{2 \cdot \tan\alpha}{\pi} \cdot \frac{F_N}{H} \tag{1}$$

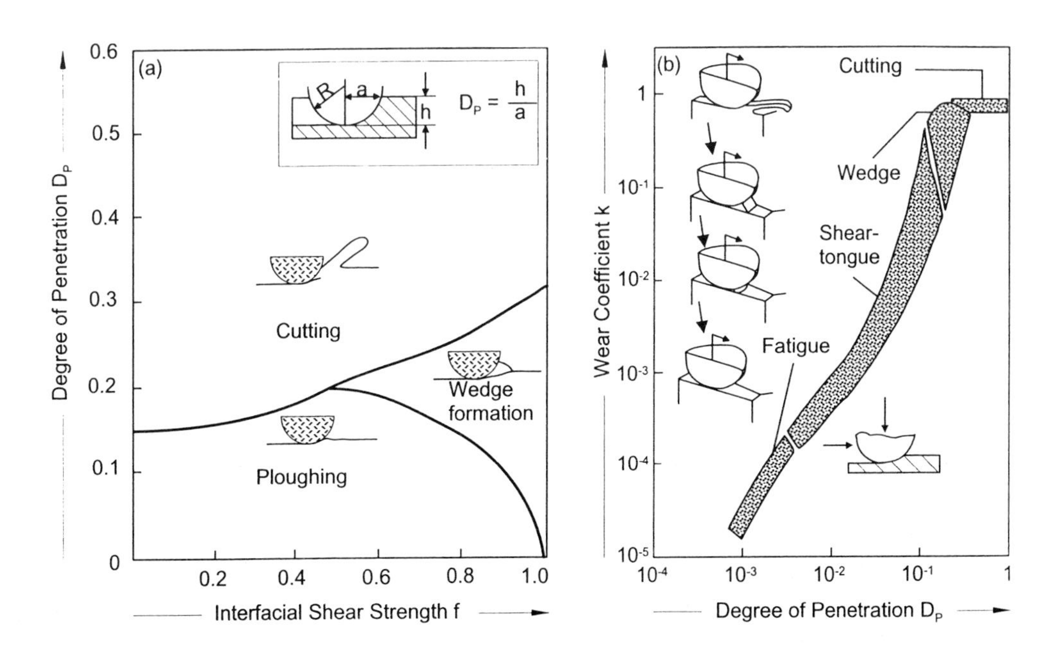

Fig. 3 Modes of interaction and wear coefficient k owing to a hard spherical indenter sliding on metallic materials as a function of (a) interfacial shear strength and (b) degree of penetration (from Hokkirigawa and. Kato (**5, 11**)).

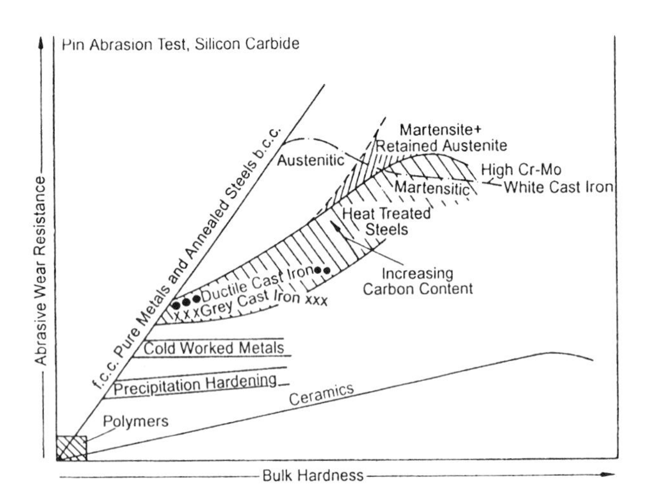

Fig. 4 Schematic drawing of abrasive wear resistance (2-body abrasion, W_v^{-1}) of different materials measured in the pin abrasion test as a function of their bulk hardness.

can be calculated from the volume of the wear groove produced.

However, it must be considered that only a portion of the volume of a wear groove formed on a metallic surface is removed as wear debris and the remainder of the groove volume is plastically displaced to the sides of the groove (**14–16**).

Hence, a more general model (**9, 17**) was developed which describes abrasive wear by distinguishing four types of interactions between abrasive particles and a wearing material (Fig. 5), namely microploughing, microcutting, microfatigue and microcracking. In the ideal case, microploughing due to a single pass of one abrasive particle does not result in any detachment of material from the wearing surface. A prow is formed ahead of the abrading particle and material is continuously displaced sideways to form ridges adjacent to the groove produced. Volume loss can, however, occur owing to the action of many abrasive particles or the repeated action of a single particle. Material may be ploughed aside repeatedly by passing particles and may break off by low cycle fatigue, i.e. microfatigue. Pure microcutting results in volume loss by chips equal to the volume of the wear grooves. Microcracking occurs when highly concentrated stresses are imposed by abrasive particles, particularly on the surface of brittle materials. In this case, large wear debris are detached from the wearing surface by crack formation and propagation. Micro-ploughing and microcutting are the dominant processes in ductile materials while microcracking becomes important on brittle materials.

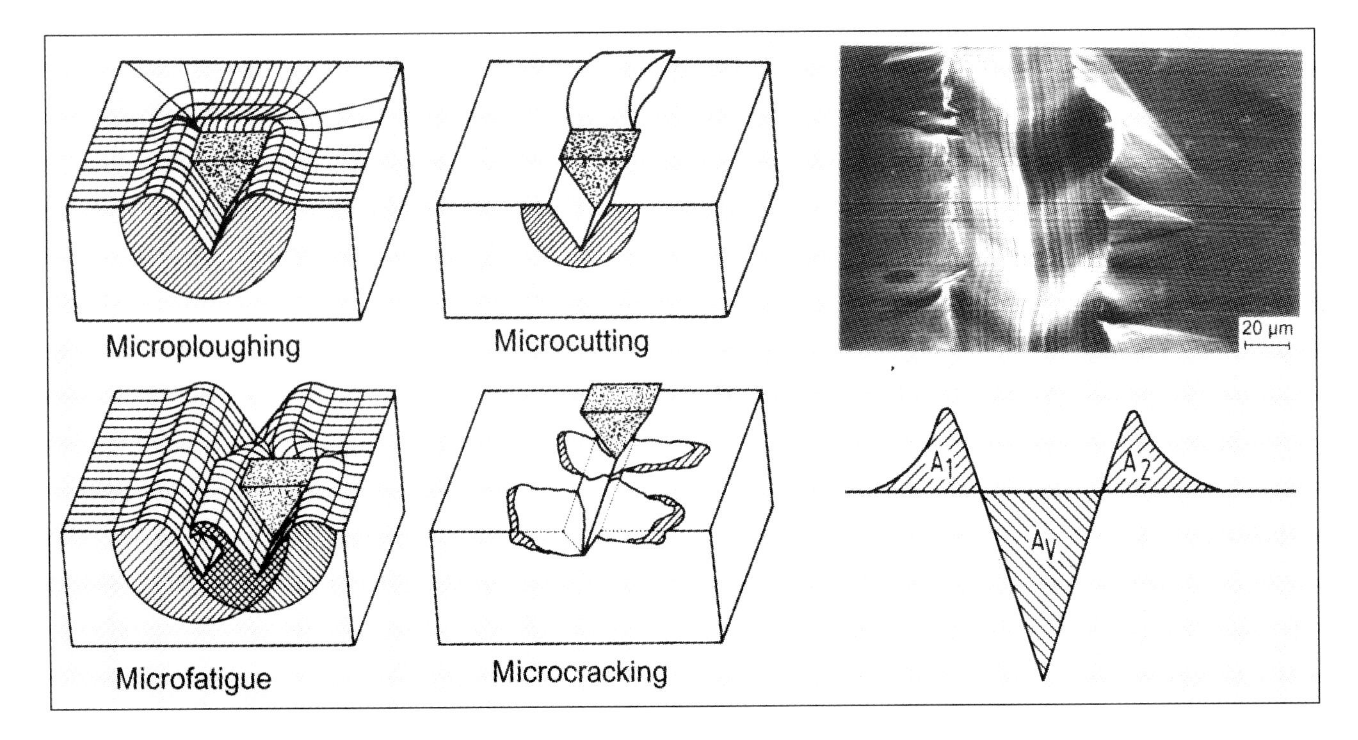

Fig. 5 Schematic representation of different interactions between sliding abrasive particles and the surface of materials, scanning electron micrograph of a wear groove on an austenitic steel and a draft of a taper section through a wear groove.

In a single scratch experiment, the ratio of volume of material removed as wear debris to the volume of the wear groove produced can be described by the f_{ab} value (Fig. 5) which is defined (**9, 17**) as

$$f_{ab} = \frac{A_V - (A_1 + A_2)}{A_V} \qquad (2)$$

where A_V is the cross-sectional area of the wear groove and $(A_1 + A_2)$ represents the amount of material which is pushed to the groove sides by plastic deformation. In the case of microcracking, spalling occurs at the edges of the wear grooves (Fig. 5) which leads to negative values of A_1 and A_2 and f_{ab} becomes greater than 1. The f_{ab} value becomes equal to unity for ideal microcutting and equal to zero for ideal microploughing. Values of f_{ab} ranging from 0.15 to 1.0 have been measured experimentally on about 30 different materials by using metallographic taper sections (**9, 18**).

A theoretical model (**9**) for calculating the f_{ab} value due to a combination of microploughing and microcutting results in:

$$f_{ab} = 1 - \left(\frac{\varphi_{lim}}{\varphi_s} \right)^{2/\beta} \qquad (3)$$

where φ_s is the effective deformation on the wearing surface and φ_{lim} is the capability of deformation of the wearing material before microfracture occurs locally during abrasion in a given tribosystem. β is a factor which describes the decay of deformation with increasing depth below the wearing surface and depends mainly on the work-hardening behaviour of the wearing material. The equation (3) is defined for $\varphi_s \geq \varphi_{lim}$ only while for $\varphi_s < \varphi_{lim}$ microploughing or elastic/plastic deformation occurs without any wear loss. Hence, the f_{ab} value is not only a material property but it is also a function of the operating conditions during abrasion which affect both φ_s and φ_{lim}.

According to this model (**17**) the linear wear intensity $W_{l/s}$ defined as the change in the length of a pin specimen or the thickness of a plate per sliding distance is given by:

$$W_{l/s} = \Phi_1 \cdot f_{ab} \cdot \frac{p}{H_{def}} \qquad (4)$$

or inserting the Eq. (3):

$$W_{l/s} = \Phi_1 \left[1 - \left(\frac{\varphi_{lim}}{\varphi_s} \right)^{2/\beta} \right] \cdot \frac{p}{H_{def}} \qquad (5)$$

where H_{def} is the hardness of the highly deformed material, e.g. of wear debris, p is the applied surface pressure and Φ_1 is a geometrical factor which depends on the shape of the abrasive particles. From Eq. (5) it follows that abrasive wear loss is strongly influenced by the capability of deformation (φ_{lim}) and the work-hardening capacity ($\beta \approx (H_{def}/H)^{1/3}$ as the first approximation) and H_{def} of the wearing material in addition to the operating conditions (p, Φ_1, φ_s). This theoretical model was strongly supported by many experimental results (**9, 17**).

Fig. 6 shows good agreement between experimental results measured on many different metallic materials and

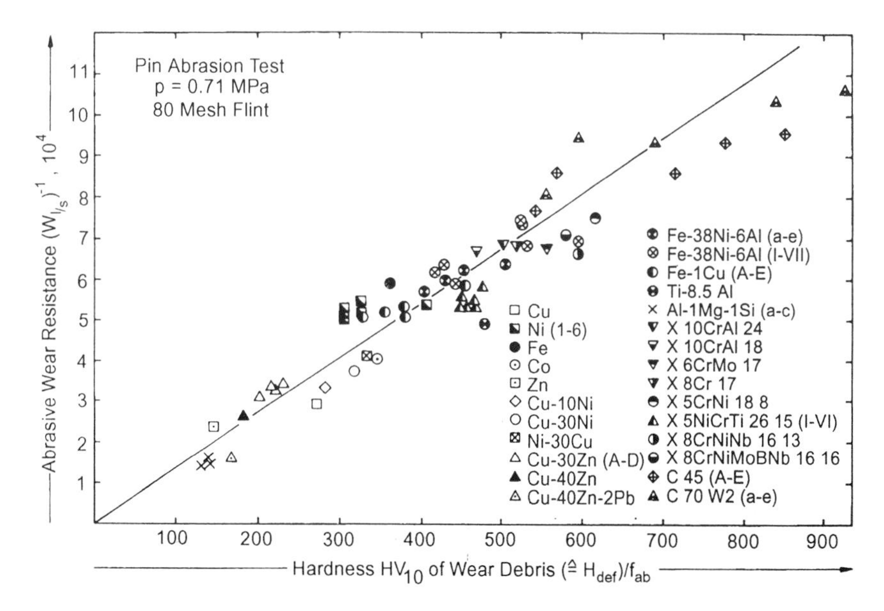

Fig. 6 Abrasive wear resistance on pure metals, single-phase alloys, precipitation-hardened ferritic and austenitic alloys, austenitic and ferritic steels and plain carbon steels against 80 mesh flint measured in a pin abrasion test with $p = 0.71$ MPa contact pressure versus the ratio of hardness of wear debris to f_{ab} values measured in a scratch test by using a sliding diamond loaded by 2 N.

the prediction of Eq. (4). Abrasive wear resistance $(W_{l/s})^{-1}$ of these materials against flint abrasive grits (80 mesh) increased linearly with the ratio $(f_{ab}/H_{def})^{-1}$ in the pin abrasion test used. The deviation of the test data from straight line at high ratios was caused by a change in wear mechanism from microploughing/microcutting to some microcracking on the steels of high hardness.

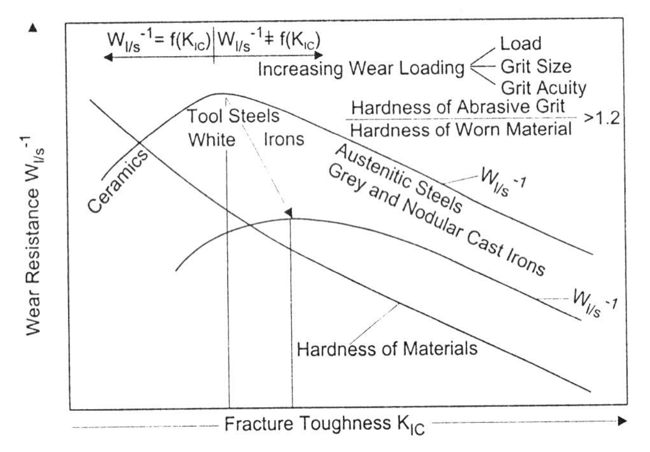

Fig. 7 Schematic drawing of the relation between fracture toughness and abrasive wear resistance $(W_{l/s})^{-1}$ of different metallic and ceramic materials (2-body abrasion, pin abrasion test).

Varying degrees of microcracking can occur during abrasion of inherently brittle materials such as highly hardened steels or ceramics or materials containing internal notches such as cracks, pores, graphite lamellae, embrittled grain boundaries, inclusions or large carbides, which promote crack formation and/or crack propagation. The extent of cracking depends on the fracture toughness (K_{Ic}, K_{IIc}) of the stressed material, loading conditions, the size and shape of the penetrating abrasive particles, and the type of internal notches. Using models of fracture mechanics, a critical applied load or surface pressure can be estimated above which flaking of material occurs (**9, 17**). For a material containing cracks or thin graphite lamellae the critical surface pressure p_{crit} is given by:

$$p_{crit} = \Phi_2 \cdot \frac{\lambda \cdot K_{IIc}^2}{D_{ab}^2 \cdot H \cdot \mu^2} \tag{6}$$

and for the linear wear intensity due to microcutting ($f_{ab} = 1$) and microcracking we obtain:

$$W_{l/s} = \Phi_1 \cdot \frac{P}{H_{def}} + \Phi_3 \cdot A_f \cdot D_{ab} \frac{p^{3/2} \cdot H^{1/2}}{K_{Ic}^2} \mu^2 \cdot \Omega \tag{7}$$

where:

$$\Omega = 1 - \exp\left\{ -\left(\frac{p}{p_{crit}} \right)^{1/2} \right\}$$

Φ_2 and Φ_3 are geometrical factors which depend on the shape of the abrasive particles and the shape of the cracking during abrasive wear. These factors can be calculated for special cases (**9**). λ is the mean free path between and A_f the area fraction or density of the inherent defects, e.g. cracks or graphite lamellae. μ is the coefficient of friction at the leading face of the abrasive particles, D_{ab} is the effective size of the abrasive particles and K_{Ic}, K_{IIc} the fracture toughness of the wearing material under mode I (tension) or II (shear) loading.

Other models (**19, 20**) using fracture mechanics to describe removal of material by lateral cracking result in similar equations for linear wear intensity such as:

$$W_{l/s} = \Phi \cdot \frac{D_{ab}^{1/2} \cdot p^{5/4}}{K_c^{3/4} \cdot H^{1/2}} \tag{8}$$

where Φ is a constant. As many studies have shown, fracture toughness K_{Ic} decreases with increasing hardness or yield stress of the materials tested, i.e. fracture toughness and hardness of a given material can be connected to each other. These models suggest that wear loss of brittle materials increases more rapidly than linearly with applied surface pressure or normal load ($F_N = p\ A$, where A = area of contact) in accordance with Fig. 2c.

Fig. 7 summarises the general effect of fracture toughness and hardness of materials on their abrasive wear resistance in a given tribosystem. Brittle materials such as ceramics under severe wear conditions show an increasing wear resistance with increasing fracture toughness despite simultaneously decreasing hardness. The surfaces of these materials worn by hard particles exhibit more or less microcracking that can result in spalling of individual grains of the wearing materials and in addition microcutting in accordance with the Eq. (7). Materials with values of fracture toughness greater than a critical value, which depends on the operating conditions (load, size and acuity of the abrasive grits, sliding speed etc.), are worn by a combination of microploughing and microcutting. Hence, the wear loss of these materials is determined with others by the hardness (see Eq. (5)) but is independent of fracture toughness. In this regime, wear resistance decreases with decreasing hardness due to enhanced penetration depth; experimentally a large variety of values of wear resistance can be measured on different materials but equal hardness owing to differences in the capability of deformation φ_{lim}.

The aim of the foregoing modelling of two-body abrasion is to show the most important parameters influencing wear resistance of homogeneous materials but not to calculate exact values of wear. The theoretical models discussed above do not consider the interaction between parallel scratches or grooves which can result in enhanced material removal (**22, 23**), or temperature (**24**) or liquid media (**25**). It is evident, and supported by many experimental results, that hardness of a wearing material alone is insufficient to describe resistance to abrasive wear. Microstructural parameters such as grain size (**26–28**) or reinforcing phases can substantially affect the amount of wear. Abrasive wear tests on different oxide ceramics showed to a first approximation an increase in wear resistance with an increase in the reciprocal of the square root of the grain size.

3 EFFECT OF SECOND PHASES ON ABRASIVE WEAR

Composites can offer the answer for achieving high hardness and sufficient fracture toughness to avoid brittle fracture. Second phases, e.g. hard ceramic particles or fibres, can be incorporated into a softer and more ductile matrix. The abrasive wear resistance of such composites depends on different microstructural parameters such as the hardness, shape, size, volume fraction and distribution of the embedded phases, the properties of the matrix and the interfacial bonding between the second phase and the matrix.

Fig. 8 shows different interactions between abrasive particles and a reinforcing phase. It distinguishes between hard and soft abrasive particles, i.e. harder or softer than the reinforcing phase, and also between small and large sizes of the reinforcing phase. Hard abrasive particles can easily dig out small phases and cut or crack larger ones. Soft abrasive particles are able to dig out small phases or produce large pits. The indentation depth of soft abrasive particles is substantially reduced by hard reinforcing phases if the mean free path between them is smaller than the size of the abrasive particles, in particular. Large phases deficiently bonded to the matrix can be pulled out. However, large phases strongly bonded to the matrix can blunt or fracture soft abrasive particles.

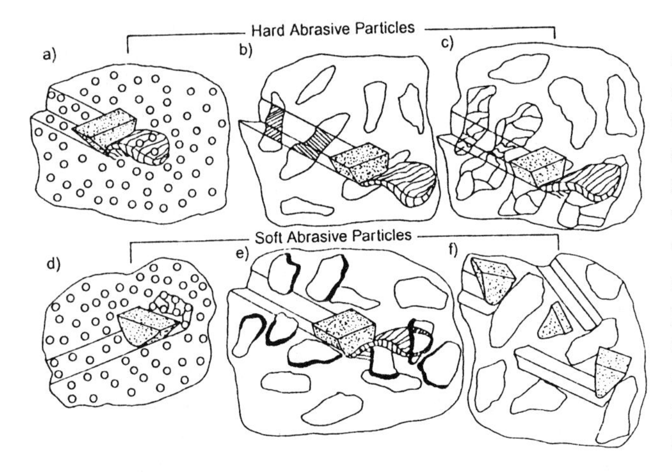

Fig. 8 Interactions between sliding hard or soft abrasive particles and reinforcing phases.

Fig. 9 shows wear grooves caused by a sliding SiC tip (attack angle and wedge angle of 90 degrees) on the surfaces of TiC and TaC steel composite layers which were produced on the die steel 90MnCrV8 (0.9 % C) by laser cladding (**29**). The reinforcing TiC (≈ 2600 HV) and TaC (≈ 1800 HV) particles were embedded in the martensitic steel matrix (810 HV30). The soft TaC phase was grooved (microcutting primarily) by the hard SiC tip (Fig. 9a) and was smashed under high normal loads (Fig. 9c). Resulting debris of the TaC were impressed into the matrix under the sliding SiC tip (Fig. 9b). The harder TiC particles, with an average size of about 30 μm, were not or only moderately grooved by the SiC (Fig. 9e, f). With increasing normal loads, the large TiC particles were broken. TiC particles of a size smaller than that of the wear groove were dug out of the matrix (Fig. 9d).

The effect of size and volume fraction of a hard reinforcing phase was studied on SiC dispersoids of varying size (5 to 200 μm) which were galvanically embedded in a nickel matrix. Fig. 10 shows the wear intensity (amount of linear wear divided by length of the wear path) of the composites related to that of the pure nickel matrix as a function of the volume fraction of the SiC reinforcing dispersoids. Wear intensities were measured by using 80 mesh (grit size of about 200 μm) SiC abrasive paper. Small dispersoids were easily dug out and resulted in enhanced wear loss with increasing volume fraction owing to their own abrasive action as loose particles. At a given volume fraction, wear intensity decreased with increasing size of the dispersoids. A minimum of wear intensity was measured for 15 and 120 μm SiC dispersoids as a function of volume fraction. Microcracking connected with spalling of dispersoids occurred at high volume fraction. Composites containing the large 200 μm SiC dispersoids exhibited continuously decreasing wear intensity with increasing volume fraction. In this case, the ratio of size of wear grooves to size of the dispersoids was substantially smaller than unity.

In agreement with the results of Fig. 10, studies on SiC or Al_2O_3 particle-reinforced aluminium alloys showed decreasing abrasive wear loss with increasing volume fraction and size of the reinforcing particles (**30, 31**). In studies on a polyester resin which was reinforced by continuous steel fibres perpendicularly oriented to the wearing surface, wear intensity also decreased continuously with increasing fraction of the reinforcing fibres up to about 45 vol.% (**32**). However, in other studies on polymeric composites containing dispersed Al_2O_3 particles (**33**) or short fibres (**34**) of glass or carbon, wear loss decreased only below a critical volume fraction or even increased continuously with increasing amount of reinforcing phase.

From the foregoing results and also studies on white cast irons (**9, 35**) it follows that under mild operating conditions abrasive wear loss decreases with increasing

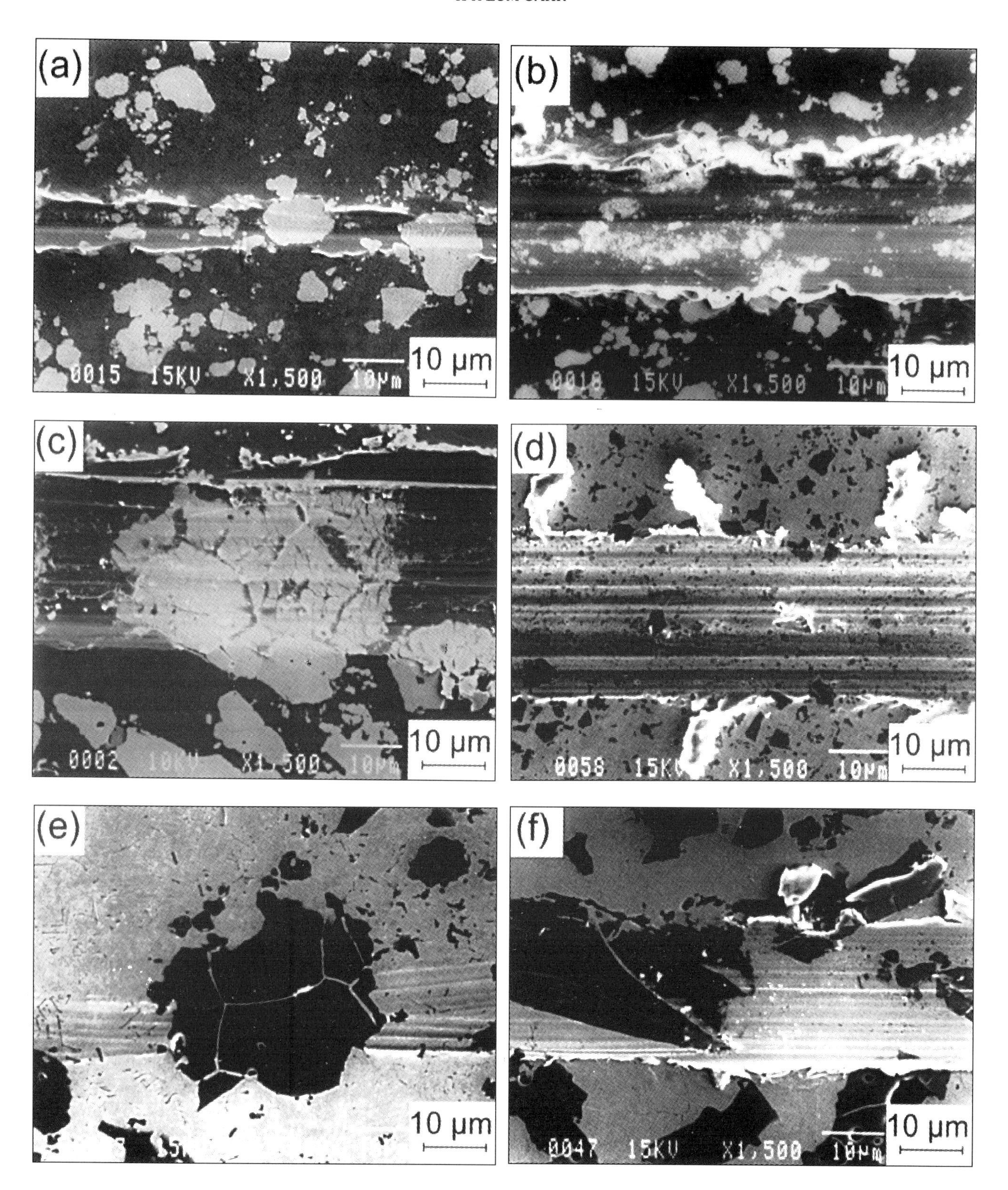

Fig. 9 Scanning electron micrographs of polished surfaces of heat treated (A) composite layers scratched by a sliding silicon carbide tip: (a) 3 μm TaC, $F_N = 1$ N; (b) 3 μm TaC, $F_N = 2$ N; (c) 30 μm TaC, $F_N = 5$ N; (d) 3 μm TiC, $F_N = 5$ N; (e) 30 μm TiC, $F_N = 1$ N; and (f) 30 μm TiC, $F_N = 5$ N, respectively.

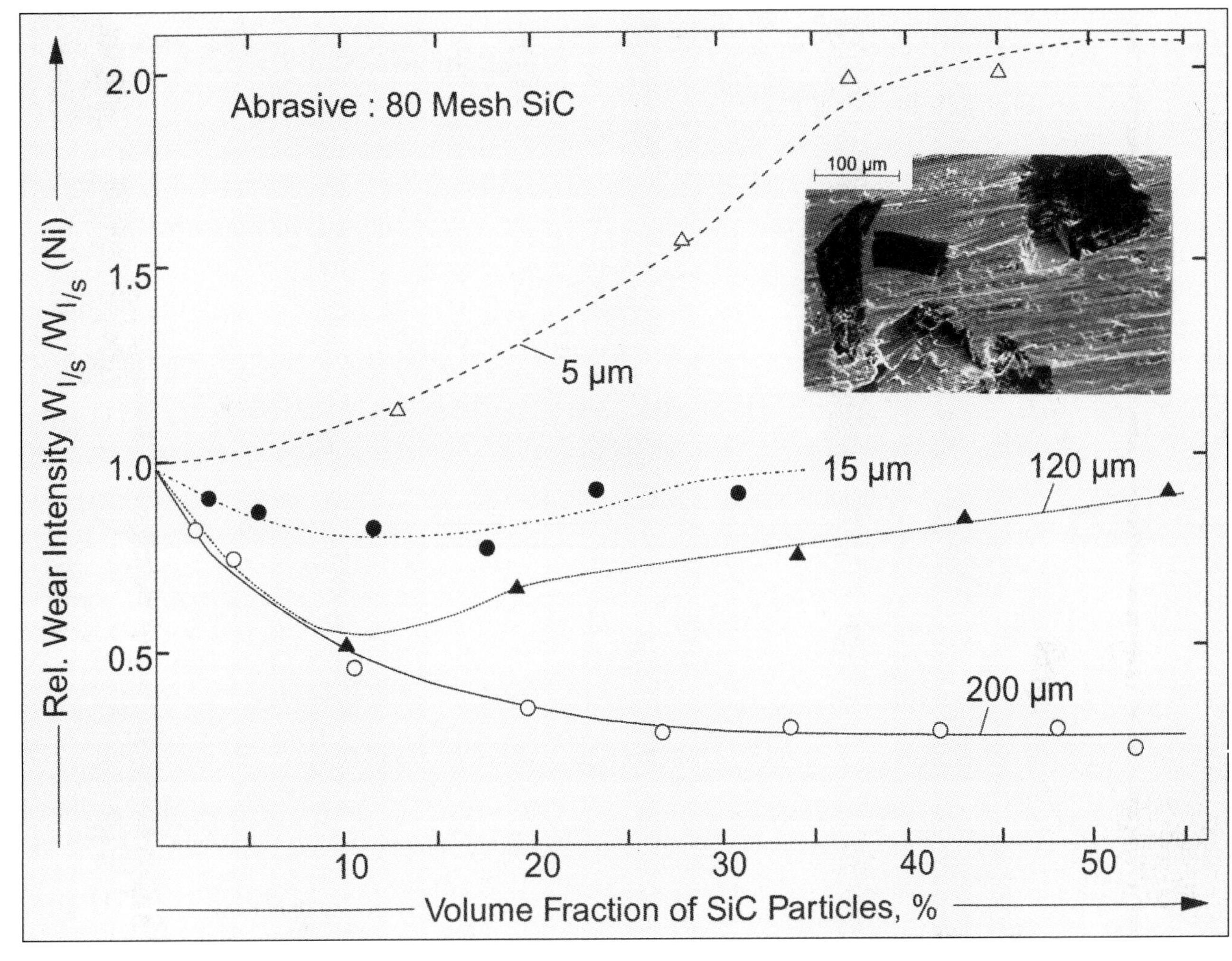

Fig. 10 Relative linear wear intensity (related to that of pure nickel) of SiC dispersion-hardened galvanic nickel versus SiC volume fraction. Average size of SiC dispersoids 5, 15, 120, and 200 μm.

volume or area fraction of the reinforcing phase. This means that the reinforcing phase is strongly embedded in the matrix and spalling does not occur. However, under severe conditions with microcracking and spalling of the reinforcing phase, wear loss can increase with increasing volume fraction of reinforcing particles or short fibres. These conclusions are also in agreement with orientation effects in fibre-reinforced materials. Fibres perpendicularly oriented to the wearing surface of polymeric composites lead to the lowest wear loss (**34**). Fibres parallel to the surface may be dug out more easily than those perpendicular to it, if the indentation depth of the abrasive particles is larger than about half the fibre diameter.

For brittle materials such as ceramics, resistance to severe abrasion is influenced among others by fracture toughness of the wearing material according to Eqs. (6) to (8), Fig. 7 and experimental results (**9, 20, 21, 36**). Fracture toughness of brittle materials can generally be enhanced by mechanisms of energy dissipation. Such toughening mechanisms are stress-induced phase transformation, stress-induced microcracking, crack pinning with crack deflection, crack bridging or pullout of reinforcing phases. However, all tribologically induced interactions between abrasive particles and ceramics are concentrated on a relatively thin surface zone. Hence, some of these toughening mechanisms can fail under tribological contact because they act in larger volumes only. Also, an appropriate fracture toughness value for the relevant surface zone of the wearing has to be inserted in the theoretical models, material which can differ from the bulk value.

Laser surface alloying of ceramics can be very effective in increasing fracture toughness and wear resistance of brittle ceramics (**37, 38**). Fig. 11 shows the volumetric wear of a HfO_2 alloyed alumina measured by using an abrasive wheel covered with 120 mesh SiC abrasive paper and sliding back and forth across the ceramic surface. The alloyed alumina surface contained a cellular eutectic HfO_2-Al_2O_3 phase with a hardness only about half of that of the alumina crystallites. The volume fraction of this soft phase was varied up to an amount of about 50 %. Despite the decreasing hardness of the composite, the volumetric wear loss decreased substantially with increasing amount of the soft phase. This behaviour can be explained by enhanced fracture toughness compared with that of the monolithic alumina owing to grain refinement, eliminating of porosity and introduction of the soft phase between the hard and brittle Al_2O_3 crystallites.

experimental results on polyester resin reinforced by continuous steel fibres and the rule of mixtures of type A by results on ferritic-pearlitic steels (**9**). Experimental values of wear resistance of multiphase structures are frequently measured between the limits given by both rules of mixtures.

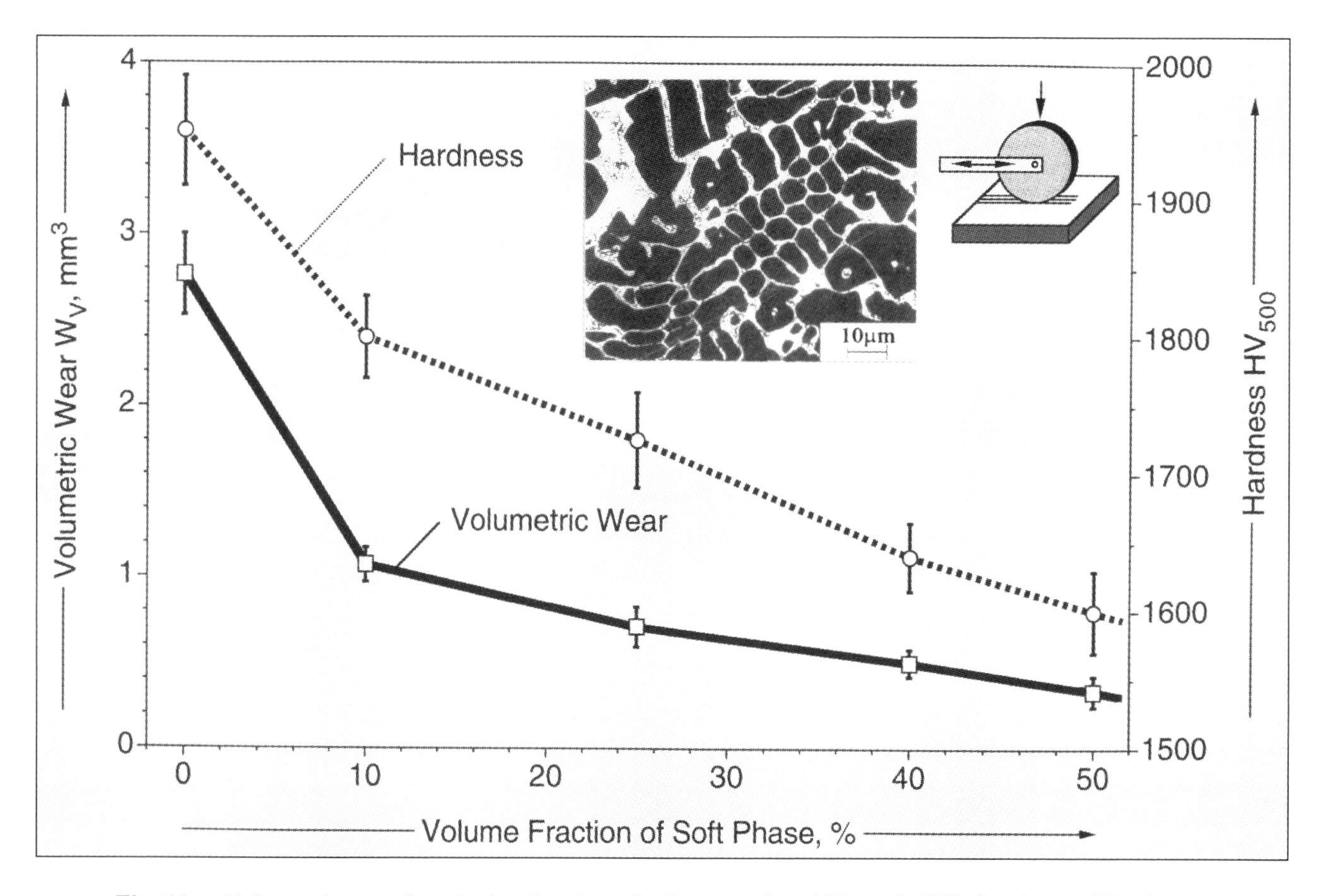

Fig. 11 Volumetric wear loss in the abrasive wheel test against 120 mesh SiC abrasive and hardness versus the volume fraction of a soft eutectic phase of au Al_2O_3 ceramic surface alloyed by HfO_2.

Rules of mixtures may offer a simple method for a first estimation of wear resistance of multiphase structures or composites if the reinforcing constituents are strongly fixed in the matrix and are not pulled or dug out by abrasively acting particles. Fig. 12 shows the wear resistance of two-phase structures as a function of the volume fraction of the reinforcing constituent. The wear resistance of structures of type B increases strongly with volume fraction of the reinforcing constituent even at small volume fractions. In contrast, wear resistance of structures of type A is substantially increased only by large volume fractions of the reinforcing phase. In the case of type B, the reinforcing constituent predominantly determines the wear behaviour and the phases are worn consecutively. The effect of the reinforcing phase depends on the identity, size and the distribution of this phase and also on size and identity of abrasive particles used, or more generally on the wear system. The rule of mixtures of type B is supported by

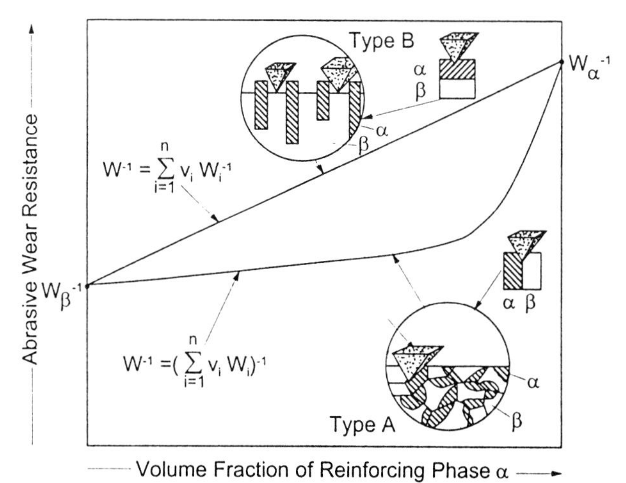

Fig. 12 Schematic representation of abrasive wear resistance of two-phase structures as a function of volume fraction of a reinforcing phase.

4 REFERENCES

1 **Burwell, J.T.** and **Strang, C.D.**, Metallic wear. *Proc. Roy. Soc.*, **A 212,** (1952), 470–477.

2 **Archard, J.F.**, Contact and rubbing of flat surfaces. *J. Appl. Phys.*, **24**, (1953), 981–988.

3 **Rabinowicz, E.**, *Friction and wear of materials*, John Wiley, New York 1965.

4 **Hutchings, I.M.**, *Tribology: Friction and wear of engineering materials,* Edward Arnold, London, 1992.

5 **Hokkirigawa, K.**, Advanced techniques for in-situ observations of microscopic wear processes, in *Proc. Surface Modification Technologies VIII,* T.S. Sudarshan and M. Jeandin (eds.), The Institute of Materials, London, 1995, pp. 93–105.

6 **Rabinowicz, E., Dunn, L.A.**, and **Russell, P.G.**, A study of abrasive wear under three-body conditions. *Wear,* **4,** (1961), 345–355.

7 **Fang, L., Kong, X.L., Su, J.Y.**, and **Zhou, Q.D.**, Movement patterns of abrasive particles in three-body abrasion. *Wear,* **162–164,** (1993), 782–789.

8 **Xie, Y.** and **Bushan, B.**, Effects of particle size, polishing pad and contact pressure in free abrasive polishing. *Wear,* **200,** (1996), 281–295.

9 **Zum Gahr, K.-H.**, Microstructure and Wear of Materials. *Tribology Series* 10, Elsevier, Amsterdam 1987.

10 **Mulhearn, T.O.** and **Samuels, L.E.**, The abrasion of metals: a model of the process. *Wear,* **5,** (1962), 478–498.

11 **Hokkirigawa, K.** and **Kato, K.**, An experimental and theoretical investigation of ploughing, cutting and wedge formation during abrasive wear. *Tribology International,* **21,** (1988), 51–57.

12 **Richardson, R.C.D.** The maximum hardness of strained surfaces and the abrasive wear of metals and alloys. *Wear,* **10,** (1967), 353–382.

13 **Moore, M.A.**, *Abrasive wear, in Fundamentals of Friction and Wear,* D.A. Rigney (ed.),. ASM, Metals Park, Ohio, 1980, pp. 73–118.

14 **Stroud M.F.** and **Wilman, H.**, The proportion of the groove volume removed as wear in abrasion of metals. *Br. J. Appl. Phys.*, **13,** (1962), 173–178.

15 **Zum Gahr, K.-H.**, Formation of wear debris by the abrasion of ductile metals. Wear, **74,** (1981/82), 353–373.

16 **Moore, M.A.** and **Swanson, P.A.**, The effect of particle shape on abrasive wear: a comparison of theory and experiment, in *Wear of Materials* 1983, K.C. Ludema (ed.), ASME, New York 1983, pp 1–11.

17 **Zum Gahr, K.-H.**, Modelling of Two-Body Abrasive Wear. *Wear,* **124,** (1988), 87–103.

18 **Zum Gahr, K.-H.** and **Mewes, D.**, Severity of material removal in abrasive wear of ductile metals, *in Wear of Materials* 1983, K.C. Ludema (ed.), ASME, New York 1983, pp. 130–139.

19 **Evans, A.G.** and **Wilshaw, T.R.**, Quasistatic particle damage in brittle solids. – I. Observations, analysis and implications. *Acta Metall.*, **24,** (1976), 939–956.

20 **Moore, M.A.** and **King, F.S.**, Abrasive wear of brittle solids. Wear, **60,** (1980), 123–140.

21 **Evans, A.G.** and **Marshall, D.B.**, Wear mechanisms in ceramics, in *Fundamentals of Friction and Wear of Materials,* D.A. Rigney (ed.), Amer. Soc. Met., Columbus, Ohio 1981, pp. 439–452.

22 **Jacobson, S., Wallen P.**, and **Hogmark, S.**, Fundamental aspects of abrasive wear studied by a new numerical simulation model, in *Wear of Materials*, K.C. Ludema (ed.), ASME, New York 1987, pp. 595–606.

23 **Xu, H.H.K., Jahanmir, S.**, and **Wang, Y.**, Effect of grain size on scratch interactions and material removal in alumina. *J. Am. Ceram. Soc.*, **78,** (1995), 881–891.

24 **Berns, H., Fischer, A.**, and **Kleff, J.**, Scratch tests on iron-, nickel- and cobalt-based alloys at elevated temperatures. Wear, **162–164,** (1993), 585–589.

25 **Backmark, U.**, Wear testing in water slurries. *Wear,* **162–164,** (1993), 1029–1032.

26 **Rice, R.W.**, Micromechanics of microstructural aspects of ceramic wear. Ceramic engineering and science. *Proc. 9th Annual Conf. On Composites and Advanced Ceramic Materials.* Am. Ceram. Soc., Westerville, Ohio 1985, pp. 940–945.

27 **Mukhopadhyay, A.K.** and **Mai, Y.Q.**, Grain size effect on abrasive wear mechanisms in alumina ceramics. *Wear,* **162–164,** (1993), 258–268.

28 **Davidge, R.W.** and **Riley, F.L.**, Grain-size dependence of the wear of alumina. *Wear,* **186–187,** (1995), 45–49.

29 **Axén, N.** and **Zum Gahr, K.-H.**, Verschleiß von TaC- und TiC-Laserdispersionsschichten durch weiche und harte Abrasivstoffe. *Mat.-wiss. u. Werkstofftech.*, **23,** (1992), 360–367.

30 **Zamzam, M.A.**, Abrasive wear of aluminium-matrix composites. *Metall,* **45,** (1991) 250–254.

31 **Garcia-Cordovilla, G., Narciso, J.**, and **Louis, E.**, Abrasive wear resistance of aluminium alloy/ceramic particulate composites. *Wear,* **192,** (1996), 170–177.

32 **Zum Gahr, K.-H.**, Einfluß des Makroaufbaus von Stahl/Polymer-Faserverbundwerkstoffen auf den Abrasivverschleiß. *Z. Werkstofftech.*, **16,** (1985), 297–305.

33 **Lu, X.-Ch., Wen, S.Z., Tong, J., Chen, Y.-T.**, and **Ren, L.-Q.**, Wettability, soil adhesion, abrasion and friction wear of PTFE (+PPS)+Al_2O_3 composites. *Wear,* **193,** (1996), 48–55.

34 **Friedrich, K.**, Wear of reinforced polymers by different abrasive counterparts, in *Friction and Wear of Polymer Composites,* K. Friedrich (ed.), Elsevier, Amsterdam 1986, pp. 233–287.

35 **Zum Gahr, K.-H.** and **Eldis, G.T.**, Abrasive wear of white cast iron. Wear, **64,** (1980), 175–194.

36 **Yamamoto, T., Olsson, M.,** and **Hogmark, S.,** Three-body abrasive wear of ceramic materials. *Wear,* **174,** (1994), 21–31.

37 **Zum Gahr, K.-H.** and **Schneider, J.,** Multiphase Al_2O_3 ceramic with high resistance to unlubricated sliding wear, in Proc. Int. Tribology Conf. Yokohama 1995, *Jap. Soc. of Tribologists,* Tokyo 1996, Vol. I, pp. 397–402.

38 **Zum Gahr, K.-H. Bogdanow, Ch.** and **Schneider, J.,** Friction and wear reduction of Al_2O_3 ceramics by laser-induced surface alloying, *Wear,* **181–183,** (1995), 118–128.

Subject Index

S

T

U

V

W

X

Z

Authors' Index

W

Z

New Directions in Tribology

UK Organizing Committee

Dr Bill Roberts (Chairman)
Dr Alex Alliston-Greiner, Plint & Partners Limited
Professor Tom Bell, University of Birmingham
Professor Brian Briscoe, Imperial College of Science, Technology & Medicine
Mr David Carnell, NSK-RHP European Technology Centre
Dr Les Hampson, National Centre of Tribology
Dr Ian Hutchings, University of Cambridge
Dr H Peter Jost, President, International Tribology Council (Ex-Officio)
Mr Stephen Maw, Neale Consulting Engineers Limited
Dr David Parkins, Cranfield Institute of Technology
Mr Julian Reed, Neale Consulting Engineers Limited
Dr Rob Rowntree, European Space Tribology Laboratory, Chairman, IMechE Tribology Group,
Dr Brian Roylance, University of Wales
Professor Hugh Spikes, Imperial College of Science, Technology & Medicine
Professor Chris Taylor, University of Leeds
Mr Bob Wood, Consultant to the Tribology Group, IMechE
Dr David Yardley, David Yardley & Associates

Patron of the Congress

HRH The Prince Philip, Duke of Edinburgh KG KT

Plenary and Invited Papers from
the First World Tribology Congress

150th Anniversary
1847 - 1997

New Directions in Tribology

8–12 September 1997

Organized by the Tribology Group of
the Institution of Mechanical Engineers (IMechE)

Sponsored by

NSK-RHP
AEA Technology plc
GKN
Climax Molybdenum Company
Pall Industrial Hydraulics

Published by Mechanical Engineering Publications Limited for
the Institution of Mechanical Engineers, Bury St Edmunds and London.

First Published 1997

ISBN 1 86058 099 8

A CIP catalogue record for this book is available from the British Library.

Printed by Bookcraft Limited, Bath, UK

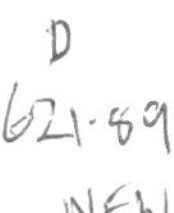